W9-ARW-328

Genetics of Populations

Jones and Bartlett Titles in Biological Science

AIDS: Science and Society, Fourth Edition, Fan/Conner/ Villarreal

AIDS: The Biological Basis, Third Edition, Alcamo

Alcamo's Fundamentals of Microbiology, Seventh Edition, Pommerville

Alcamo's Laboratory Fundamentals of Microbiology, Seventh Edition, Pommerville

Aquatic Entomology, McCafferty/Provonsha

Biology: Investigating Life On Earth, Second Edition, Avila

Botany: An Introduction to Plant Biology, Third Edition, Mauseth

The Cancer Book, Cooper

Cell Biology: Organelle Structure and Function, Sadava

Creative Evolution?!, Campbell/Schopf

Defending Evolution: A Guide to the Evolution/ Creation Controversy, Alters/Alters

Diversity of Life: The Illustrated Guide to the Five Kingdoms, Margulis/Schwartz/Dolan

DNA Sequencing: Optimizing the Process and Analysis, Kieleczawa

Early Life: Evolution on the Precambrian Earth, Second Edition, Margulis/Dolan

Electron Microscopy, Second Edition, Bozzola/Russell

Elements of Human Cancer, Cooper

Encounters in Microbiology, Alcamo

Essential Genetics: A Genomic Perspective, Third Edition, Hartl/Jones

Essentials of Molecular Biology, Fourth Edition, Malacinski

Evolution, Third Edition, Strickberger

Experimental Techniques in Bacterial Genetics, Maloy

Exploring the Way Life Works: The Science of Biology, Hoagland/Dodson/Hauck

Genetics: Analysis of Genes and Genomes, Sixth Edition, Hartl/Jones

Genetics of Populations, Third Edition, Hedrick

Genomic and Molecular Neuro-Oncology, Zhang/Fuller

Grant Application Writer's Handbook, Fourth Edition, Reif-Lehrer

How Cancer Works, Sompayrac

How Pathogenic Viruses Work, Sompayrac

How the Human Genome Works, McConkey

Human Anatomy and Physiology Coloring Workbook and Study Guide, Anderson

Human Biology: Health and Homeostasis, Fifth Edition, Chiras

Human Embryonic Stem Cells: An Introduction to the Science and Therapeutic Potential, Kiessling/Anderson

Human Genetics: The Molecular Revolution, McConkey

The Illustrated Glossary of Protoctista, Margulis/ McKhann/Olendzenski

Introduction to the Biology of Marine Life, Eighth Edition, Sumich/Morrissey

Laboratory Research Notebooks, Jones and Bartlett Publishers

A Laboratory Textbook of Anatomy and Physiology, Eighth Edition, Donnersberger/Lesak

Major Events in the History of Life, Schopf

Medical Genetics: Pearls of Wisdom, Sanger

Microbes and Society, Alcamo

Microbial Genetics, Second Edition, Maloy/Cronan/ Freifelder

Microbiology: Pearls of Wisdom, Booth

Missing Links: Evolutionary Concepts and Transitions Through Time, Martin

Oncogenes, Second Edition, Cooper

100 Years Exploring Life, 1888—1988, The Marine Biological Laboratory at Woods Hole, Maienschein

Origin and Evolution Of Intelligence, Schopf

Plant Cell Biology: Structure and Function, Gunning/Steer

Plants, Genes, and Crop Biotechnology, Second Edition, Chrispeels/Sadava

Population Biology, Hedrick

Protein Microarrays, Schena

Statistics: Concepts and Applications for Science, LeBlanc

Vertebrates: A Laboratory Text, Wessels

Genetics of Populations

Third Edition

Philip W. Hedrick

ARIZONA STATE UNIVERSITY

JONES AND BARTLETT PUBLISHERS

Sudbury, Massachusetts

BOSTON TORONTO LONDON SINGAPORE

World Headquarters
Jones and Bartlett Publishers
40 Tall Pine Drive
Sudbury, MA 01776
978-443-5000
info@jbpub.com
www.jbpub.com

Jones and Bartlett Publishers Canada
2406 Nikanna Road
Mississauga, ON
L5C 2W6
CANADA

Jones and Bartlett Publishers International
Barb House, Barb Mews
London W6 7PA
UK

Production Credits
Chief Executive Officer: Clayton Jones
Chief Operating Officer: Don W. Jones, Jr.
President, Higher Education and Professional Publishing: Robert W. Holland, Jr.
V.P., Design and Production: Anne Spencer
V.P., Sales and Marketing: William Kane
V.P., Manufacturing and Inventory Control: Therese Bräuer
Executive Editor, Science: Stephen L. Weaver
Managing Editor, Science: Dean W. DeChambeau
Associate Editor, Science: Rebecca Seastrong
Senior Production Editor: Louis C. Bruno, Jr.
Marketing Manager: Matthew Bennett
Marketing Associate: Laura M. Kavigian
Cover Design: Anne Spencer
Photo Researcher: Kimberly Potvin
Composition: Northeast Compositors, Inc.
Illustrations: Elizabeth Morales
Printing and Binding: Malloy
Cover Printing: Malloy
Cover Photo: © Photodisc

Library of Congress Cataloging-in-Publication Data

Hedrick, Philip W., 1942–
Genetics of populations / Philip W. Hedrick. — 3rd ed.
p. cm.
Includes bibliographical references and index.
ISBN 0-7637-4772-6 (alk. paper)
1. Population genetics. I. Title.
QH455.H43 2005
576.5′8–dc22 2004056666

Printed in the United States of America

08 07 06 05 04 10 9 8 7 6 5 4 3 2 1

Genetics of Populations

Third Edition

Philip W. Hedrick
ARIZONA STATE UNIVERSITY

JONES AND BARTLETT PUBLISHERS
Sudbury, Massachusetts
BOSTON TORONTO LONDON SINGAPORE

World Headquarters
Jones and Bartlett Publishers
40 Tall Pine Drive
Sudbury, MA 01776
978-443-5000
info@jbpub.com
www.jbpub.com

Jones and Bartlett Publishers Canada
2406 Nikanna Road
Mississauga, ON
L5C 2W6
CANADA

Jones and Bartlett Publishers International
Barb House, Barb Mews
London W6 7PA
UK

Production Credits
Chief Executive Officer: Clayton Jones
Chief Operating Officer: Don W. Jones, Jr.
President, Higher Education and Professional Publishing: Robert W. Holland, Jr.
V.P., Design and Production: Anne Spencer
V.P., Sales and Marketing: William Kane
V.P., Manufacturing and Inventory Control: Therese Bräuer
Executive Editor, Science: Stephen L. Weaver
Managing Editor, Science: Dean W. DeChambeau
Associate Editor, Science: Rebecca Seastrong
Senior Production Editor: Louis C. Bruno, Jr.
Marketing Manager: Matthew Bennett
Marketing Associate: Laura M. Kavigian
Cover Design: Anne Spencer
Photo Researcher: Kimberly Potvin
Composition: Northeast Compositors, Inc.
Illustrations: Elizabeth Morales
Printing and Binding: Malloy
Cover Printing: Malloy
Cover Photo: © Photodisc

Library of Congress Cataloging-in-Publication Data

Hedrick, Philip W., 1942–
Genetics of populations / Philip W. Hedrick. — 3rd ed.
p. cm.
Includes bibliographical references and index.
ISBN 0-7637-4772-6 (alk. paper)
1. Population genetics. I. Title.
QH455.H43 2005
576.5′8–dc22 2004056666

Printed in the United States of America

08 07 06 05 04 10 9 8 7 6 5 4 3 2 1

To Cathy, Leigh, and Scott

Brief Contents

Contents

Preface

To comprehend the evolutionary process, one must have a thorough understanding of population genetics. In addition, one should be familiar with many aspects of related disciplines, such as population ecology and behavior, and more applied disciplines such as conservation biology, medical genetics, and plant and animal breeding, which have implicit assumptions about the genetic nature of populations.

Since the publication of the second edition of *Genetics of Populations*, one area of population genetics that has greatly expanded is molecular population genetics and evolution. Throughout the book, I have presented current molecular genetics approaches and examples to illustrate these exciting developments. There are now two chapters explicitly devoted to molecular topics, Chapter 8, Neutral Theory and Coalesence, and Chapter 11, Molecular Population Genetics and Evolution. The latter chapter introduces three major topics: molecular phylogenetics and phylogeography, paternity analysis and individual identity, and identifying genes influencing quantitative traits.

This book is designed for graduate students and advanced undergraduates who have had a course in genetics or evolution and have an aptitude for quantitative thinking. In general, I have endeavored to integrate empirical and experimental population genetics with theory. In particular, I have given methods for estimating population genetic parameters as well as other statistical tools useful for population genetics. The level of mathematics generally required is college algebra, although introductory calculus and statistics are sometimes employed. I have also provided examples and theory relevant to studies of both animals and plants. Although many studies of population genetic phenomena have focused on *Drosophila* and human beings, I have included examples in as many other organisms as possible and given the classic examples illustrating various phenomena.

The structure of this edition is similar to that of the second edition. In Chapter 1, I begin with an introduction to both basic genetic information and quantitative approaches used in population genetics. In the first edition, this information was either included later in the text or in an appendix, but its new position focuses on the fundamental nature of this information. After introductory material in Chapter 1, I review the extent of genetic variation with examples ranging from visible polymorphisms to nucleotide variation. Chapter 2 introduces techniques to describe the amount of genetic variation in a population. Factors that influence the extent and pattern of genetic variation are first examined singly, from both an experimental and a theoretical viewpoint, and then the joint effects of two or more factors operating simultaneously are considered. Selection is discussed in Chapters 3 and 4, inbreeding in Chapter 5, genetic drift and effective population size in Chapter 6, mutation in Chapter 7, and gene flow in Chapter 9. Chapter 10 discusses linkage disequilibrium and recombination and how this information can be used in analysis of multilocus and DNA sequence data.

I made several novel changes in the presentation of material at the end of the book in the second edition that I retain here. New to these editions are the page numbers after the references in the Bibliography where the references are cited in the text. This addition should aid in finding citations of specific references in the text and replaces the author index. Also new to these editions are the boldface entries in the subject index that indicate the page in the text where a term is first defined or discussed. This addition replaces the glossary and provides a definition for a given term in the appropriate context. As in the first edition, I have given recent references in the current literature as well as classic references. A new feature of this edition, at the beginning of the bibliography, is a listing of software packages and their websites used for genetic data analysis, population genetic models, and estimation of population genetic parameters.

A special feature of the book is the examples using actual data to illustrate the concepts developed in the text. Generally these examples are either classic ones or specific, recent ones that are good illustrations of particular phenomena. Many of the statistical techniques used in population genetics are now available in software packages (see the list in Table 2.1, the Bibliography, and the text) so that more complicated calculations can be carried out. Also available are a number of software packages that illustrate the impact of particular evolutionary factors on genetic variation and can be used in teaching (see the list in Table 3.1 as well as the Bibliography). At the end of each chapter, there are problems that illustrate particular aspects of the material in the chapter. The numerical answers to these problems appear at the end of the book.

This book has developed over the years as I have taught and carried out research in population genetics. My teaching and research in population genetics are intertwined, with many research ideas coming from students' questions and with my understanding of particular areas in population genetics increasing as I have investigated various topics. In particular, I have written a number of reviews on various topics in population genetics and learned much from these intensive literature surveys. In recent years, I have used population genetic techniques and approaches in investigation of small populations, particularly those of threatened and endangered species.

My introduction to the concepts of population genetics was through the enthusiastic and excellent teaching of Ralph Comstock, Richard Lewontin, and David Merrell. I appreciate the interaction on population genetics with various colleagues over the years, particularly those with Dennis Hedgecock, William Klitz, Outi Savolainen, and Glenys Thomson. In recent years, I have profited from the collaborative research in conservation and population genetics at Arizona State University with Rich Fredrickson, Dan Garrigan, Alan Giese, Gustavo Gutierriz-Espleta, Carla Hurt, Steven Kalinowski, Sudhir Kumar, Rhonda Lee, Phil Miller, W. L. Minckley, Joel Parker, Karen Parker, Ruby Sheffer, Anne Stone, and Brian Verrelli.

Finally, thanks to the following individuals who provided comments on portions of either the second or third edition:

Fred Allendorf, University of Montana
Michael Blouin, Oregon State University
Ed Bryant, University of Houston
David Houle, Florida State University
Sudhir Kumar, Arizona State University
Michael Nachman, University of Arizona
Len Nunney, University of California at Riverside
Bret Payseur, Cornell University
Cynthia Riginos, Duke University
Kermit Ritland, University of British Columbia
Matthew Saunders, University of Arizona
Glenys Thomson, University of California at Berkeley
Brian Verrelli, Arizona State University
Mike Whitlock, University of British Columbia

Phil Hedrick
School of Life Sciences
Arizona State University
August, 2004

1

General Background and the Diversity of Genetic Variation

Webster's Third International Dictionary defines evolution as "the process by which through a series of changes or steps any living organism or group of organisms has acquired the morphological and physiological characters which distinguish it." This definition was entirely appropriate for Darwin's time and for the first half of the twentieth century. But, with all the changes that molecular biology had revealed—none of which is visible to the naked eye—a much broader definition is needed. In this book, as in earlier discussions of the neutral theory, I include in the word evolution all changes, large and small, visible and invisible, adaptive and nonadaptive. In some cases, evolution may occur even by random fixation of very slightly deleterious mutants, whose selection coefficients are comparable to or only slightly larger than the mutation rates.

Motoo Kimura (1983)

. . . the classic stages of a theory's career. First, a new theory is attacked as absurd. Then, it is admitted to be true, but obvious and insignificant. Finally, it is seen to be so important that its adversaries claim that they themselves discovered it.

William James

The Lord in His wisdom made the fly
And then forgot to tell us why.

Ogden Nash

Population genetics is a field that has had periods of great interest and growth and other periods in which there was less innovation and fewer new contributions. This is not unexpected for disciplines that develop a new paradigm and then refine and expand research within this paradigm (Kuhn, 1962). In its early history, population genetics attracted such scientific geniuses as Ronald A. Fisher, J. B. S. Haldane, and Sewall Wright, who provided the theoretical underpinnings to population genetics in the 1920s and 1930s and formed much of the paradigm still used today (Provine, 1971). Their impact was so thorough that Lewontin (1963) lamented that they "had said everything of truly fundamental importance about the theory of genetic change in populations and it is due mainly to man's infinite capacity to make more and more out of less and less, that the rest of us are not currently among the unemployed."

However, the advent of the first population molecular data from allozymes and amino acid sequence data in the late 1960s and then DNA sequence data in the 1980s produced an entire new set of questions and greatly re-energized population genetics. The need for a context in which to interpret these molecular data became fundamental, and population genetics and its evolutionary interpretations provided one. Today we are obtaining large amounts of DNA sequence data from many genes in many different species. Understanding the evolutionary significance of this DNA variation, both within and between species, is providing a great new challenge for population genetics. Furthermore, the role of this molecular variation in adaptive differences in morphology, behavior, and physiology and nonadaptive variation in complex genetic diseases is a topic to which population genetic approaches can make fundamental contributions.

Population genetics includes facets of several different disciplines and various approaches to scientific knowledge. In the first part of this chapter, we provide an introduction to the different approaches used in population genetics, general molecular genetics, and relevant quantitative tools and topics. The advanced reader may omit familiar topics and proceed on to others. Depending on their interests, readers may wish to consult more advanced treatments of specific topics by referring to the references cited.

In the last half of the chapter, we provide an introduction to the extent of genetic variation found in natural populations, including molecular variation, visible polymorphisms, lethals, and quantitative variation. Here we emphasize the extensive and diverse genetic variation found in natural populations and leave discussion of the factors influencing the amount and pattern of genetic variation to later chapters. Much of the interest in current population genetics is in the fast-expanding area of molecular population genetics, which we introduce here, discuss throughout the book, and give a more detailed current overview in Chapter 8.

I. APPROACHES TO POPULATION GENETICS

Approaches used to investigate phenomena in population genetics, and many other biological disciples, can be generally separated into three basic types: empirical, experimental, and theoretical. The traditional **empirical** (or descriptive) approach in population genetics comprises extensive observation of the genetic variation of a particular gene or genes in a population or populations, perhaps over time, and the measurement of related factors, such as environmental patterns, that may influence this genetic variation. These data may provide associations between the patterns or levels of genetic variation and other factors, thereby suggesting potential problems for further study. The genetic variants used initially in these empirical investigations initially included morphological variants, blood group polymorphisms, and chromosomal inversions, and then starting in the 1960s, **allozyme variation**—that is, genetic variation in enzymes and proteins. Some of the classic examples of genetic polymorphism, such as color or blood group variation, still have many aspects of their evolutionary genetics that need to be clarified for a comprehensive understanding of the factors influencing their frequencies. In recent years, similar empirical examinations have focused on DNA sequence variation, mainly between different species but also between individuals within the same species.

Generally only **experimental** tests can provide support for hypotheses developed from empirical data about the effect of particular factors on levels and patterns of genetic variation. Traditional experiments are exemplified by the moving or transplanting of a population to a new environment and comparing them with the nontransplanted population to examine the significance of an environment on genetic variants. However, in recent years, the definition of an experiment in evolutionary genetics has become broader and includes, for example, comparisons of DNA sequences between organisms or genes that have different histories, functions, or other characteristics.

Using information obtained either from empirical or experimental studies, one can construct a general **theoretical** model to account for the empirical and experimental observations. Theoretical models may also provide a general framework in which to generalize to other similar situations and provide an overall conceptual basis for understanding the impact of various factors on the levels and patterns of genetic variation. In addition, theoretical models can be used to make predictions about future genetic change and to provide past scenarios consistent with present-day genetic variation. However, caution is necessary because some theoretical research is based on very specific model assumptions, and it may have only tenuous connections to biological reality.

Figure 1.1 shows the interconnections among these approaches to population genetics. Generally, empirical information is obtained first, and then a **hypothesis**, ***a genetic or evolutionary explanation of the observations***, is developed. Then experiments are carried out in an effort to disprove a particular hypothesis or to provide support for its validity. In other instances, a theoretical model is developed to explain empirical observations, and then further data are collected to falsify or substantiate the biological appropriateness of the model. The feedback among the three approaches can then allow the refinement and development of particular hypotheses. Of course, new findings in population genetics, as in other areas of biology, often do not occur through such an ideal framework and come about either as serendipitous events or new insights based on comprehensive understanding of the issue being examined.

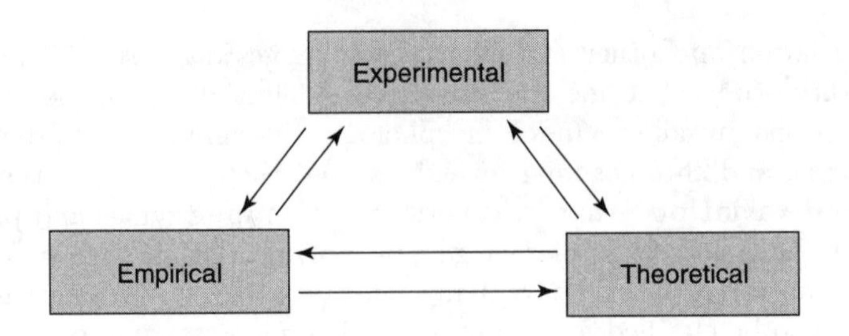

Figure 1.1. The relationships and feedback among the empirical, experimental, and theoretical approaches to population genetics.

Interestingly, Veuille and Slatkin (2002) suggested that population genetics is unique among the biological sciences because it was first elaborately developed as a theoretical discipline before experimental research had a significant impact. Furthermore, they suggested that this early emphasis on theory has ensured a continuous interchange between theoretical and experimental approaches. As a contrast, in the physical sciences, theoretical approaches often dominate research developments.

For scientific knowledge to increase, alternative hypotheses to explain given observations must be developed, and then **critical experiments**—***experiments with alternative possible outcomes that will exclude one or more of these hypotheses***—must be devised and properly executed. For an introduction to this approach, called the **method of strong inference**, see Platt (1964). The method of strong inference can result in the elimination or falsification of some hypotheses so that only one or several remain. The remaining hypotheses are then said to be consistent with the experimental and/or empirical results. Because of the general difficulty in designing critical experiments, it may not be possible to falsify all hypotheses but one. In fact, in many aspects of population genetics, the causation of particular phenomena appears to be multifaceted, making designing and executing critical experiments difficult.

In designing and carrying out experiments, it is useful to keep in mind the factors that compose a sound experimental design. First, traditional experiments should have an adequate number of independent **replicates**. Second, there should be simultaneous and appropriate **controls**. It is necessary to ensure that the replicates are truly independent and that the controls are appropriate. Third, the **sample size**, or the number of observations, should be large enough so that the likelihood of statistical anomalies is low. In determining both the necessary number of replicates and the necessary sample sizes to reveal a given evolutionary genetic effect, it is useful to determine the level of **statistical power**, the probability of rejecting a null hypothesis when it is false (p. 24; Sokal and Rohlf, 1995; Zar, 1999). Even in the "experiments" that compare DNA sequences of different types, these attributes should be kept in mind. For example, are there enough independent sequences? Do the compared groups differ only by the characteristic of interest? What is the statistical power provided by the data?

Although this philosophical approach to understanding phenomena in science sounds appealing and straightforward and although it has proven productive in population genetics, one should be cautious when generalizing to other situations. For example, many basic genetic phenomena have been first documented in **model organisms** in controlled experimental settings; the "laws" of segregation and independent assortment were demonstrated by Gregor Mendel in garden peas, and spontaneous mutation was first documented in the laboratory in *Drosophila melanogaster*, the fruit fly, by H. J. Muller. On the other hand, experimental data from model organisms, such as *Drosophila*, may not reflect the evolutionary factors important in many mammals because the population size in most mammals is generally several orders of magnitude less than that of *Drosophila*, or experimental data generated gathered in controlled laboratory or greenhouse settings may not adequately reflect the complexity of nature and the important factors operating in natural systems.

Laboratory experiments with a segregating genetic variant in *Drosophila* have provided useful demonstrations of selection and genetic drift for heuristic purposes, as illustrated in Chapters 3 and 6. However, if there are too many constraints, then such experiments can only reflect the theoretical model that they are supposedly examining and provide little information about the system in nature. On the other hand, examination of phenomena in uncontrolled natural situations may not provide clear enough results to differentiate between alternative hypotheses.

The documentation of allozyme variation using protein electrophoresis became the new empirical tool in the 1970s. In fact, Lewontin (1974) characterized this approach as scientific heaven for some researchers: "finding an experiment that works and doing it over and over and over." Determining the variation in DNA sequences of particular genes in particular organisms and their relationship to other sequences may become the new heaven for population genetics empiricists. In fact, the amount of information in DNA sequences has resulted in a new field, called **genomics**, that endeavors to compare and understand the significance of this variation. On the other hand, the selective effects on many molecular variants may be too subtle for traditional experimental examination, and only by using extensive data sets that examine the cumulative selective effects over many generations can these effects be uncovered. The borderline between empirical and experimental approaches has become hazy, and a contemporary experiment may be to obtain sequence data to falsify or support a particular hypothesis. An exciting opportunity is the availability of large DNA data sets that are now being generated using the techniques developed for rapid nucleotide sequencing. Although these data may provide new occasion to answer population genetics questions, the data may not be ideal to investigate particular evolutionary hypotheses because they are often generated for other purposes.

II. INTRODUCTION TO SOME GENETIC TOPICS

Like other disciplines, population genetics has a number of terms that are its lingua franca. Most of these should be familiar, but it is useful to briefly mention the important ones to avoid later confusion. After introducing these terms, we briefly review some molecular genetic aspects that are useful in our discussion of population genetics. For the reader interested in more details of basic and molecular genetics, most contemporary general genetics texts provide good coverage.

a. Some Genetic Terms

Our definition of terms here are designed for their application to population genetics and may not appear exactly correct to, for example, a molecular geneticist. First, we can define a **gene** as a unit of inheritance that is transmitted from parents to offspring. For a current dicussion of the criteria used in identifying genes based on molecular genetic information, see Snyder and Gerstein (2003). The place that a particular gene resides on a chromosome is called a **locus** (the plural form is **loci**). Often we use the terms gene and locus more or less interchangeably. Generally, genes are coding units that produce either polypeptides, which in turn become proteins and enzymes, or RNA. However, there are other "genes" that are in noncoding regions of the genome that are useful in population genetics. For example, most of the Y chromosome in humans and other mammals is made up of noncoding regions, but the variation among different Y chromosomes has been useful in determining paternity and relationships between different human groups. The different forms of a gene are termed **alleles**. Some genes may have only a few, quite similar, allele variants in a population, whereas other genes may have many alleles, some of which are widely divergent.

For most genes in higher organisms, each individual has two copies of a given gene, generally one from its mother and one from its father. As a result, they are **diploid,** and the number of copies of a gene in a population is therefore twice the number of individuals. Generally, a single copy of all of the genes in an organism is called a **genome**, and thus, a diploid individual would have two complete genomes. At a single gene, a diploid individual has two alleles, and these together are called the **genotype**. An individual may have two alleles that are identical, in which case they are a **homozygote**, or two copies that are different, so that they are a **heterozygote**. Many more primitive organisms, such as bacteria, have only one copy of each gene; they are **haploid** (as are egg or sperm cells, the gametes in higher organisms). Many plants and some animals are **polyploid**, meaning that they have more than two copies of a single gene or more than two genomes. For example, wheat is a hexaploid, having six genomes, and salmon are tetraploid, having four genomes. However, because the doubling event in the ancestry of salmon took place many millions of years ago, many salmon genes have since become diploidized—that is, the two diploid sets of genes segregate independently of each other.

To indicate different alleles at a particular gene, we generally use subscripts. This is particularly appropriate for DNA-based genetic variation for which there may be many DNA sequences (alleles) at a given gene. In this case, the alleles at gene A would be indicated by A_1, A_2, $A_3 \ldots A_i$, etc. Similarly, at another gene B, the alleles would be indicated by B_1, B_2, $B_3 \ldots B_j$, etc. For example, the genotype at gene A in a diploid organism could be a homozygote, for example, A_3A_3, or a heterozygote, for example, A_1A_2.

Most genes in higher organisms are on non–sex-determining chromosomes, or **autosomes**, whereas a minority are on sex chromosomes, such as the X and Y. Autosomal genes are diploid, whereas genes on the **X chromosome** in mammals and other organisms are haploid in males and diploid in females. In Hymenoptera, such as ants, bees, and wasps, just as for X chromosomes, the males are haploid and the females are diploid; hence, these organisms are called **haplo-diploid**.

The genes on the **Y chromosome** are in the haploid state in males and are not present in females. Y chromosomes show strictly **paternal inheritance**, that is, transmission from male parent to male offspring. Both **mitochondrial DNA (mtDNA)** and **chloroplast DNA (cpDNA)** are also in the haploid state and generally show **maternal inheritance**—transmission from female parent to female offspring. However, unlike Y chromosomes, mtDNA and cpDNA are transmitted to all offspring, including males, but the male offspring do not transmit them further.

A **chromosome** is a long piece of DNA that generally segregates independently of other chromosomes. The piece of DNA that makes up a chromosome can be very long; for example, the longest human chromosome consists of a DNA molecule that is over 200-million nucleotides (see below) or bases long. Genes may be on the same chromosome—that is, they show physical **linkage** to each other—or they may be on different chromosomes. If they are on different chromosomes or are far apart on the same chromosome, they will show **independent assortment**—that is, there will be equal numbers of parental and nonparental like gametes in their progeny. On the other hand, if the genes are closely linked, then nearly all of the progeny will contain parental gametes. For linked loci, nonparental gametes are produced by **recombination**, generally the result of breaking and rejoining to homologous chromosomes during meiosis. The amount of recombination between two genes is a function of their physical proximity, or their genetic map distance (**map units**) apart. For closely linked loci, the number of map units is approximately equal to the percentage of recombination expected between the two loci.

The genes in different organisms are spread over different numbers of chromosomes, and different organisms have different total numbers of chromosomes (O'Brien, 1993). For example, *D. melanogaster* has only three major chromosomes with 66, 103, and 108 map units for a total of 277 map units. On the other hand, humans have 23 chromosomes, with a total of approximately 4,000 map units, individual chromosomes having from 69 to 249 map units. If the genes are very close on a chromosome, then there will be a low amount of recombination between them, and gametic arrays will be retained over generations. It is sometimes useful to specify the alleles at linked genes on a given copy of a chromosome, an array called a **haplotype**. For example, a given haplotype may be $A_2B_1C_4D_1 \ldots$. A diploid individual has two haplotypes of a particular genetic region. The term *haplotype* is also used for the DNA sequence that may include part or all of a gene or closely linked DNA sequences on a given copy of a chromosome.

When originally describing genetic variants based on DNA sequence, researchers often use terminology based on geographic origin, name of the gene, size of the allele, or other criteria. When such sequences are deposited

in **GenBank**, the database of the National Institutes of Health of publicly available DNA sequences, these alleles are given an identifier, such as AF030867, so that other researchers can retrieve the exact sequence. Currently, there are over 28-billion nucleotides from over 250,000 species deposited in GenBank, and the number increases greatly each year. Although this data bank has become invaluable in many aspects of molecular evolution research, one should be careful when using it because the level of errors may not always be negligible (Forster, 2003; Harris, 2003).

In some cases, we use the symbol "+" to indicate the **wild-type** allele at a given gene. Generally the wild-type allele has been thought to result in the wild or normal **phenotype** (appearance) of a given organism. However, unless there is contrast between a wild-type allele and a detrimental mutant allele that alters the phenotype or reduces fitness, it is often difficult to determine which allele is the wild type because nearly all alleles found using molecular techniques appear to have the same phenotype. In a few cases, we follow a general convention that may be different from the above symbolism. For example, we symbolize the alleles at the ABO red blood cell locus as A, B, and O, the dominant M (melanic) and recessive m (typical) at the melanic locus in the peppered moth, and F (fast) and S (slow) to indicate alleles at an allozyme locus that migrate at different speeds in a electrophoretic protein gel.

In biochemical terms, a gene is a region of DNA (deoxyribonucleic acid) that is in turn made up of a sequence of nucleotides. There are four different types of **nucleotides** or **bases** in DNA: **adenine (A)**, **thymine (T)**, **guanine (G)**, and **cytosine (C)**. DNA is a double-stranded molecule; whenever an A is on one strand, it is paired with a T on the other strand, and whenever a G is on one strand, it is paired with a C on the other strand. A and G are **purines**, whereas C and T are **pyrimidines**. The nucleotide at a particular position can change by mutation. If the change is between the same type of nucleotides—that is, purine to purine (A to G or G to A) or pyrimidine to pyrimidine (C to T or T to C)—it is called a **transition**. If the change is between the two types of nucleotides—that is, purine to pyrimidine (e.g., A to T) or pyrimidine to purine (e.g., T to A)—it is called a **transversion**.

In general, there is an association between the physical distance between two genes, the number of nucleotides, and the amount of recombination observed between them. For example, in the human genome, two genes that have 0.01 recombination between them (are 1 map unit apart) are generally approximately 10^6 nucleotides (1 **megabase**, or **Mb**) apart. However, there are regions of lower-than-average recombination, such as around the centromeres or at the end of some chromosomes and other regions of higher recombination, sometimes referred to hotspots of recombination. Many genes are between 1000 base pairs (bp) and 3000 bp long (1000 bp is often called a **kilobase**, or **kb**).

b. The Genetic Code

Because there are only 4 types of nucleotides and 20 types of amino acids commonly found in proteins, there needs to be a genetic code to specify the relationship between these two types of information. In the 1960s, it was discovered that this genetic code was composed of nucleotide triplets that specify the different amino acids, and all 64 combinations of the triplet genetic code were determined (Table 1.1). Note that the pyrimidine **uracil,** U, replaces T in RNA and the genetic code, but that later, when we examine DNA sequence evolution, T is used in place of U. Of these triplet **codons**, 61 specify particular amino acids, and 3 are stop codons that result in termination of synthesis (Table 1.1). The genetic code is nearly **universal** over all organisms. The major exceptions are in mtDNA (Nei and Kumar, 2000), which has a few differences from the universal code for several closely related codons. For example, there are three differences in the vertebrate mtDNA code from the standard genetic code.

TABLE 1.1 The 64 nucleotide combinations of the genetic code (codons) specified by their position in the codon and the resulting amino acids, indicated by both the three-letter abbreviation and the one-letter symbol.

First position	Second position: U	C	A	G
U	UUU, UUC: Phe (F) UUA, UUG: Leu (L)	UCU, UCC, UCA, UCG: Ser (S)	UAU, UAC: Tyr (Y) UAA, UAG: Stop	UGU, UGC: Cys (C) UGA: Stop UGG: Trp (W)
C	CUU, CUC, CUA, CUG: Leu (L)	CCU, CCC, CCA, CCG: Pro (P)	CAU, CAC: His (H) CAA, CAG: Gln (Q)	CGU, CGC, CGA, CGG: Arg (R)
A	AUU, AUC, AUA: Ile (I) AUG: Met (M)	ACU, ACC, ACA, ACG: Thr (T)	AAU, AAC: Asn (N) AAA, AAG: Lys (K)	AGU, AGC: Ser (S) AGA, AGG: Arg (R)
G	GUU, GUC, GUA, GUG: Val (V)	GCU, GCC, GCA, GCG: Ala (A)	GAU, GAC: Asp (D) GAA, GAG: Glu (E)	GGU, GGC, GGA, GGG: Gly (G)

Because there are only 20 amino acids and 61 codons that code for amino acids, there is a high **degeneracy**, or redundancy, in the genetic code. In fact, the amino acids leucine (L, this is the single letter code for this amino acid), arginine (R), and serine (S) are each coded by six different nucleotide triplets. Five other amino acids—valine (V), proline (P), threonine (T), alanine (A), and glycine (G)—are specified by four codons each. In these cases, the third position of the codon can be any of the four different nucleotides, termed **fourfold degenerancy**, and still code for the same amino acid. At the other extreme, there are two amino acids—methionine (M) and tryptophane (T)—that are specified by one codon (AUG can be the initiation codon as well as code for M). Finally, one amino acid, isoleucine (I), is specified by three codons, and nine amino acids are specified by two codons: termed **twofold degenerancy**, phenylalanine (P), tyrosine (Y), histidine (H), glutamine (Q), asparagine (N), lysine (K), aspartic acid (D), glutamic acid (E), and cysteine (C). In these cases, either of the two pyrimidines or either of the two purines in the third position codes for the same amino acid.

When a single-base change in the DNA sequence does not result in a change in an amino acid, it is referred to as a **silent** or **synonymous** mutation or substitution. All of the cases that have multiple codons specifying the same amino acid differ in only the third nucleotide position of the codon and have silent single-nucleotide changes, except for the three amino acids that are specified by six codons. In 8 of the 16 combinations for the first two nucleotides, the nucleotide in third position makes no difference in the amino acid. For example, when the first two nucleotides are UC, then the resultant amino acid is serine for all four possible nucleotides in the third position.

On the other hand, single nucleotide changes that result in a new amino acid are called **replacement** or **nonsynonymous** mutations or substitutions. Nearly all changes in the first-codon position and all changes in the second-codon position result in new amino acids, the only exception being the two pairs of codons that specify arginine. In Table 1.1, the two pyrimidines (U and C) are given first and then the two purines (A and G). For the third position, transitions (pyrimidine to pyrimidine or purine to purine) generally result in synonymous substitutions, whereas tranversions in many cases result in nonsynonmous substitutions.

c. Structure of the Genome and a Typical Eukaryotic Gene

The genome consists of coding regions and noncoding regions. There are two primary types of coding regions: genes that code for polypeptides and genes that code for RNAs, such as transfer RNA (tRNA) and ribosomal RNA (rRNA), as their final product. Different organisms have various proportions of their genome in coding sequences, ranging from very high values in prokaryotes, to much lower proportions in vertebrates and higher plants. Some noncoding regions play important roles in chromosomal structure and gene regulation, although many noncoding parts of the genome have no known function.

A large proportion of the noncoding regions of the genome is made up of repeated sequences, some of which are very short, such as the dinucleotide repeat GCGCGCGC . . . , whereas others may be composed of very long repeats. For example, the *Alu* sequence is about 300 bp long and is repeated approximately 300,000 times in the human genome, accounting for approximately 5% of the genome. These repeated sequences may be present in different numbers of copies and have formed the basis of detecting differences among individuals in a population. For example, **microsatellite loci**, sometimes called short, tandem repeats (STR) because the repeat length is often from two to six nucleotides, have been used in mapping genomes, in forensic work, and in conservation genetics (Ellegren, 2004). The positions of 5264 highly variable, dinucleotide microsatellite loci spread throughout the human genome were documented by Dib *et al.* (1996), and many more are available now.

In coding regions, there also may be repeated sequences of related genes. These **multigene families** are generally thought to be the result of duplication of specific regions, followed by divergence through mutation and perhaps selection. Some of the members of a multigene family may have acquired a stop codon by mutation or by a deletion or addition that changes the codon-reading frame and results in a stop codon. These genes that now are not translated are termed **pseudogenes** because their sequences are still close to the ancestral active gene but they can no longer function as a gene. For example, the major histocompatibility complex (MHC) in humans, which we discuss below, is composed of nearly 300 closely linked genes, many of which appear to have descended from an ancestral gene or genes and have since evolved different functions or have become pseudogenes.

Surveys have been carried out to determine the general level of variation in the genome for humans. Generally, the sequence is the same in different individuals at a given nucleotide site, but approximately 1 in 1.9 kb is variable. Such variable positions are called **single nucleotide polymorphisms (SNPs**, pronounced snips). In the total human genome, 1.42-million SNPs have been mapped (The International SNP Working Group, 2001).

One of the surprising findings from molecular genetics is that many eukaryotic genes have internal regions of the gene that are transcribed into RNA but that are not translated into polypeptides. In other words, the functional parts of a gene are in pieces; the coding sections of a gene are interrupted by noncoding regions. Perhaps the best way to visualize the structure of an eukaryotic gene is to consider the type of diagram in Figure 1.2. Here the nucleotide sequence of the gene includes the **promoter** region, the coding regions or **exons**, the noncoding regions or **introns,** and any other flanking sequence that is important in the function of the gene. Some eukaryotic genes may have many exons and introns, but prokaryotic organisms do not have introns. For example, the 1272 genes on human chromosome 6 have an average of 7.9 exons each, and 1 gene has 101 exons (Mungall *et al.*, 2003).

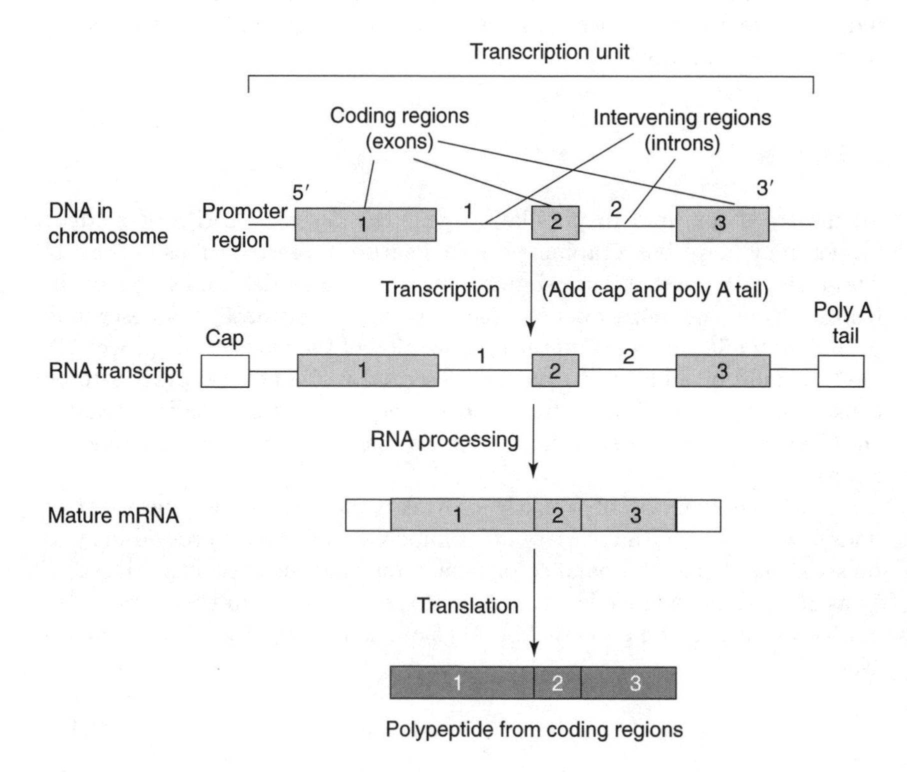

Figure 1.2. A schematic model of a gene with three exons (lightly shaded areas) and two introns, its RNA transcript, the mature mRNA, and the translated polypeptide (darkly shaded).

The DNA sequence is first transcribed into the RNA transcript, and this transcript is modified at both ends. DNA sequences are usually written in the order that they are transcribed from the upstream, or 5′, end to the downstream, or 3′, end. This transcript is then processed and becomes messenger RNA mRNA when the intervening sequences or introns are excised. During this processing, the protein-coding regions or exons are spliced together, and the intron sequences are removed to form the complete mRNA sequence to be used for translation into the polypeptide sequence. It has been suggested that the exons are functional units, such as domains in proteins, and in fact, a correlation is found in some cases, but not in general.

III. INTRODUCTION TO SOME QUANTITATIVE TOPICS

In examining and measuring the effect of different evolutionary factors in population genetics, we use symbols to indicate different effects. Generally, we use the traditional symbols, some of which are the first letter of a word that describe the effect, such as s for the selection coefficient against a homozygote or m for the rate of gene flow (from the more general term migration). The sources of other symbols are more obscure, such as w, which indicates the relative fitness of a genotype, and f, which is the inbreeding coefficient of an individual.

a. Models

To illustrate concepts in population genetics we use a series of models. Models may be verbal, graphic, or mathematical representations of the real world. Here we generally use simple mathematical models and graphical illustrations of their consequences. A model has the advantage of describing both a particular process and the organization of the parts of the process. A model should be an accurate enough description of a natural process to be consistent with actual observations. However, assumptions made for mathematical purposes can sometimes give an inaccurate picture of biological reality.

A simple example of a mathematical model might be a personal or household budget with a variety of its important aspects expressed in symbols. Let us define the balance of money on hand next month (time $t+1$) as B_{t+1}. This balance is a function of the balance from this month, B_t, and two values, income for the month I minus the expenses for the month E, or

$$B_{t+1} = B_t + I - E \tag{1.1a}$$

The amounts I and E are called **parameters** and are the true values of these two quantities. Although this expression may describe the overall situation, it does not provide the detail necessary to understand the process. For example, the parameters I and E may both be composed of several quantities. For example, E may be composed primarily of food (F), housing (H), and medical (M) expenses. The model can then be expanded to

$$B_{t+1} = B_t + I - (F + H + M) \tag{1.1b}$$

A more complete description of all of the factors that contribute to the balance in an extension of the model could allow a nearly exact understanding of the factors that determine the amount of money on hand. The degree of detail in a model in some respects depends on the researcher's interest and whether he or she is interested in general principles or an exact description of a particular situation. Often there is a tradeoff between having a general model, with a few important parameters that can be generalized and used in analogous situations, and a detailed model developed for a specific application.

The model as given above is a **deterministic** one: given a particular starting balance, then with particular values for the parameters, the balance some time in the future is precisely determined. In many situations, however, there may be **stochastic**, or chance, events that influence the level or presence of particular parameters. These factors can be introduced into the model by assuming that there is some probability that they occur or that they occur at some value. As a result of this introduced variation, it is no longer possible to predict the future balance exactly as we could when all of the parameters are deterministic. In addition, there may be unusual or even unique factors beyond the description in the basic model. For example, an individual may lose her or his job or incur unforeseen housing or medical expenses.

As a simple population example of a model, let us consider a model to predict the number of individuals in a population. First, let N_t and N_{t+1} be the numbers in generations t and $t + 1$, respectively. Let R be the ratio of the number in generation $t + 1$ over the number in generation t or

$$R = \frac{N_{t+1}}{N_t} \tag{1.2a}$$

If this ratio, which in population ecology is usually called the net replacement rate or the average number of offspring per parent is 2, then the population number would double each generation.

This expression can be rearranged to predict the number of individuals in a future generation. That is, the number in generation $t + 1$ is

$$N_{t+1} = RN_t$$

In other words, the number in the next generation is a product of the numbers in the previous generation and the net replacement rate.

If we assume that the same **recursive relationship** is present for other generations, for example, that

$$N_{t+2} = RN_{t+1}$$

then by substitution from the expression above

$$N_{t+2} = R^2 N_t$$

In general, we can show that the initial number, N_0, is related to that in generation t by

$$N_t = R^t N_0 \tag{1.2b}$$

This very useful type of relationship, which we use in examining nearly all of the different population genetic factors, allows us to make a prediction t generations in the future. For example, if a population began with 100 individuals ($N_0 = 100$) and the replacement rate is 2, the expected number of individuals after five generations is $N_5 = 2^5(100) = 3200$.

It is often useful to examine the change in numbers (or change in some other value) in a given period of time. For example, if we let

$$\Delta N = N_{t+1} - N_t$$

where the Greek capital letter Δ (delta) indicates the difference, then by substitution

$$\begin{aligned} \Delta N &= RN_t - N_t \\ &= N_t(R-1) \end{aligned} \tag{1.2c}$$

This is called a **difference equation**, and with it we can calculate the expected change in numbers at different times. We use difference equations to describe the change in allele frequencies caused by various factors such as selection and gene flow. If $R = 1$ (each individual is replaced by one individual), the term in parentheses becomes 0, and there is no change in the numbers in the population.

On the other hand, if population growth is continuous (the time intervals become very small), the change in population numbers is best described by a **differential equation** such as

$$\frac{dN}{dt} = rN \tag{1.3}$$

where r is another measure of population growth called the intrinsic rate of increase and dN/dt is the differential that gives the instantaneous change in numbers in a very small time unit t. For example, if $r = 0.001$ and $N = 100$, then $dN/dt = 0.1$, and the expected increase for this small time interval is 0.1 individuals. We do not use differential equations here very much, but they can often be solved to give general expressions of change when difference equations cannot.

Models have several advantages that are important in population genetics. First, they convey a basic understanding and enable us to focus on the factors important in a particular process. This is especially helpful in designing experiments to examine the importance of various factors and in teaching the general concepts of a new topic. Second, a model can provide an exact description of a situation or process. Finally, models can be used for the prediction of future events. For example, given a model and some basic information, one might predict the population numbers at some future time, as we did above. It should be noted that, in general, **verbal models**, and even some mathematical models, are not precise enough to make predictions that would allow the evaluation of their appropriateness for the situation. Furthermore, verbal models cannot generally predict counterintuitive results, whereas mathematical models may.

The models that we discussed above are deterministic ones, but stochastic events are often of great significance in population genetics. Sometimes a particular probability distribution can be used to determine these probabilities. If the model is complicated, however, computer simulation based on **random numbers**, often called a **Monte Carlo simulation**, can be used to determine the outcome. Monte Carlo simulation is particularly useful to mimic changes in allele frequency due to chance in a finite population—that is, genetic drift. To simulate genetic drift, random numbers that have a uniform distribution between zero and one are used. For example, for a locus with two alleles, if a uniform random number is between zero and the allele frequency in the parents, an A_1 gamete is produced. If the random number is between the allele frequency in the parents and one, an A_2 gamete is produced. This process is continued until $2N$ gametes are produced, the number in a diploid population of size N. The allele frequency in the progeny is then calculated by counting the number of A_1 and A_2 alleles generated. For the next generation, this frequency becomes the parental allele frequency, and the process is started over, with a new set of random numbers, to generate the next progeny generation.

b. Means, Variances, and Confidence Intervals

In evaluating various genetic attributes of populations, such as the amount and pattern of genetic variation, it is necessary to use some descriptive statistics (for more discussion of general statistical topics in a biological context, see Sokal and Rohlf, 1995; Zar, 1999). Because populations are often large, generally a **sample**, a smaller group thought to be representative of the entire population, is obtained from the population, and its characteristics are evaluated. For example, such sampling is widely used in political and other polls where not every individual can be interviewed. The **estimates** from the sample should reflect the parametric values in the population if the sample is randomly drawn from the population. For example, we may estimate the frequency of allele A_1 from a sample of 100 individuals. If the sample was randomly drawn from the total population that may consist of many thousands of individuals—that is, if it is a **random sample**—then the estimate should be relatively close to the allele frequency in the population if we were to determine the genotype of every individual in the population.

Let us assume that we are measuring the phenotypic value for a given trait, for example, height or weight, in a sample of n different individuals. If we symbolize the value of the trait in individual i as x_i, then we may calculate the **mean**, a measure of the average of a group of values, in a sample of n individuals for this trait as

$$\bar{x} = \frac{1}{n}\sum_{i=1}^{n} x_i \tag{1.4a}$$

where the overbar indicates the mean and the Greek capital letter Σ (sigma) indicates the summation over all of the individuals.

In a given sample, there may be different amounts of dispersion around the mean. For example, there may be no dispersion if all individuals are exactly the same height or weight, or there may be an extreme amount of dispersion if there are equal numbers of small and large individuals. The **variance**, a measure of the dispersion around the mean, is calculated as

$$V_x = \frac{1}{n-1}\sum_{i=1}^{n} (x_i - \bar{x})^2 \tag{1.4b}$$

In other words, the variance of x is the average of the sum of the squared deviations of individual values from the mean ($n - 1$ is used instead of n in the denominator because sample variance estimated in this manner is **unbiased** and thus will provide an estimate of the true parametric value when averaged over samples of any size).

Another related measure of the dispersion around the mean, the **standard deviation**, is also useful. The standard deviation is the square root of the variance

$$sd = (V_x)^{1/2} \tag{1.4c}$$

and is therefore a measurement on the same scale as the mean. As a result, the standard deviation may be used for distributions that are normal (bell shaped) to state the proportion of the values within a given part of the distribution. Normal distributions are mathematically defined, symmetrical distributions, with a given mean and variation as defined by their standard deviation. Figure 1.3 gives a theoretical normal distribution for a trait, with the mean indicated by the tickmark in the center. The total area under the curve is unity; the darkly shaded area within one standard deviation on either side of the mean encompasses 68% of the total area, and the total shaded area within two standard deviations on either side of the mean encompasses 95% of the area.

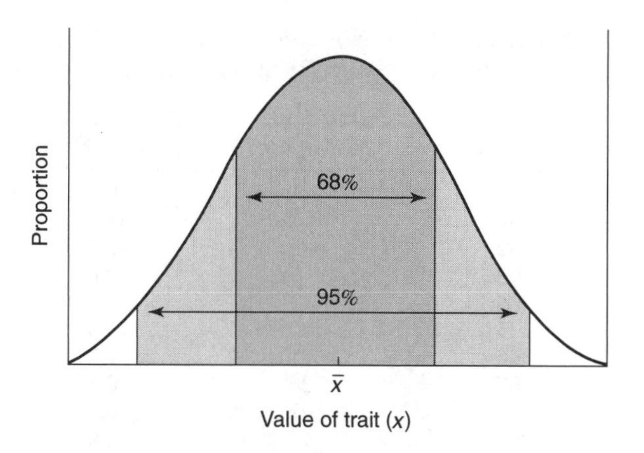

Figure 1.3. The normal distribution of a trait indicating the mean $\bar{x}$ and areas encompassing 68% (one standard deviation) and 95% (two standard deviations) of the distribution.

For statistical purposes, it is often assumed that the distribution of values around a mean is close to a normal distribution, and in fact, samples from natural populations are often close to a normal distribution. To illustrate this, the distribution of height of male students at Connecticut Agricultural College (now the University of Connecticut) from 1914 is given in Figure 1.4. Note that of the 175 students, only a few individuals are either very short or very tall and that most are in intermediate height range (Table 1.2). The mean height for this sample is 67.3 inches, which can be obtained by summing the product of values in the first two columns of Table 1.2 and dividing by 175:

$$\begin{aligned}\bar{x} &= \frac{1}{175}[58(1) + 61(1) + 62(5) + ... + 72(4) + 73(4) + 74(1)] \\ &= 67.3\end{aligned}$$

Figure 1.4. The distribution of heights, in inch categories, of male students at Connecticut Agricultural College (from Crow, 1997). (Courtesy of Albert Blakeslee, 1914.)

TABLE 1.2 The distribution of height for 175 male students at Connecticut Agricultural College.

Height (in.)	*Number*	$(x_i - \overline{x})^2$
58	1	86.7
61	1	39.8
62	5	28.2
63	7	18.6
64	7	11.0
65	22	5.3
66	25	1.7
67	26	0.1
68	27	0.5
69	17	2.9
70	11	7.2
71	17	13.6
72	4	22.0
73	4	32.4
74	1	44.8

The variance may be obtained in an analogous manner by using the second and third columns in Table 1.2. Thus

$$\begin{aligned} V_x &= \frac{1}{174}[86.7(1) + 39.8(1) + ... + 32.4(4) + 44.8(1)] \\ &= 7.3 \end{aligned}$$

and the standard deviation is 2.7. If we use the assumptions of the normal distribution, then approximately 68% of the individuals in the company should be between 64.6 and 70.0 inches in height, and 95% should be between 61.9 and 72.7 inches in height. In fact, seven individuals are less than 62 or greater than 72 inches in height. This is 4% of the group, very close to the 5% expected.

There are two other related measures that are often useful. The **coefficient of variation** ($\boldsymbol{CV}$) is the standard deviation divided by the mean in percentage or

$$CV = \frac{100(V_x)^{1/2}}{\overline{x}} \tag{1.5a}$$

Because the variance can often increase as the square of the mean, this measure allows a meaningful comparison of the amount of variation in different groups that have different mean values. The standard deviation of a statistic, such as the mean, is known as the **standard error**. The standard error of the mean, *se*, is

$$se = \left(\frac{V_x}{n}\right)^{1/2} \tag{1.5b}$$

and can be used in determining the reliability of the estimate of the mean. For example, for a normally distributed variable, approximately 95% of the means from samples of size n would be expected to fall within the interval $\pm$ 1.96se around the true mean. Therefore, the 95% **confidence interval**, $\bar{x} \pm 1.96se$, should then contain the true mean 95% of the time.

Often the shape of the distribution is not normal or cannot be assumed to be normal. When this occurs, it is useful to estimate the confidence interval around the mean using other approaches because, for example, significance levels on the two sides of the mean may be different distances from the mean. In this case, a randomization test in which the probability of all outcomes more extreme than that observed can be either calculated exactly or estimated by computer simulation. In addition, resampling techniques, such as the **bootstrap** and **jackknife**, can provide confidence intervals or significance probabilities for virtually any estimate (Efron and Tibshirani, 1993).

In this introduction, we have been concerned with only one variable, but in various aspects of population genetics, we may be interested in the simultaneous variation in two variables. To measure the relationship between two variables, we can use **regression** if we can express one variable as a function of the other, such as height of progeny as a function of the height of their parents. Estimates of **covariance** also assume a functional relationship similar to regression analysis. On the other hand, an estimate of **correlation** measures only the extent to which two variables vary together, and no functional relationship is assumed. If no association exists between the two variables, the slope of the linear regression, the covariance, and the correlation coefficient are all zero. Positive and negative associations between the two variables result in positive and negative values of the measures, respectively.

There are three types of means that we use in later chapters. The mean as we calculated it above is called the **arithmetic mean**. For example, if we assume that two relative fitnesses are 1.0 and 0.25, then their arithmetic mean is 0.625. In some cases, it is appropriate to calculate the **geometric mean** (e.g., when determining the conditions necessary for maintenance of polymorphism with fitness variation over time; see p. 215). The geometric mean is the Nth root of the product of N values or

$$\begin{aligned}\bar{x}_g &= (x_1 x_2 x_3 ... x_i ... x_N)^{1/N} \\ &= \left(\prod_{i=1}^{N} x_i\right)^{1/N} \qquad (1.6a)\end{aligned}$$

where the Greek capital letter Π (pi) indicates multiplication of all of the individual values. For our two fitness values of 1.0 and 0.25, the geometric mean is 0.5, lower than the arithmetic mean of these values.

In some cases, it is appropriate to calculate the **harmonic mean** (e.g., to determine the effective population size if the population size varies in different generations; see p. 330). The harmonic mean is based on the reciprocals of the individual values:

$$\bar{x}_h = \frac{N}{\frac{1}{x_1} + \frac{1}{x_2} \cdots \frac{1}{x_i} \cdots \frac{1}{x_N}} = \frac{N}{\sum_{i=1}^{N} \frac{1}{x_i}} \tag{1.6b}$$

Again, assuming that two fitnesses are 1.0 and 0.25, their harmonic mean is 0.4, lower than both arithmetic and geometric means. In general, $\bar{x} \geq \bar{x}_g \geq \bar{x}_h$.

c. Probability

In designing experiments to test various aspects of population genetics, it is important to understand the elements of probability (symbolized here primarily by Pr but also by P below). As a simple example, assume that there are only two possible mutually exclusive outcomes of an event, such as the flipping of a coin, that may result in either a head of a politician (or royal) or a tail (not head). Therefore, the sum of the probabilities of the first outcome, $\Pr(1) = 0.5$ or a head, and that of the second outcome, $\Pr(2) = 0.5$, or a tail, is unity, or $\Pr(1) + \Pr(2) = 1$. Of course, in another situation, such as the probability of getting or not getting an infection that one is exposed to, the probability of the two outcomes may not be equal; for example, Pr(1) could be 0.14, and Pr(2) would then be 0.86. If there are more than two outcomes, then the added probabilities of all of them should be equal to 1. The fact that the ***sum of the probabilities of mutually exclusive events is the probability that any of the events will occur*** is termed the **addition rule** of probability.

When a number of trials occur—for example, a number of coins are tossed—then the expected number of outcomes of a particular type is equal to the probability of the outcome times the number of trials N. For outcome 1, it is $E(1) = \Pr(1)N$, where E(1) is the **expected** number of outcomes of type 1. For example, if a coin is tossed 20 times, then the expected number of heads, outcome 1, is $E(1) = 0.5(20) = 10$. In a real situation, the **observed** number of a particular outcome often does not equal the expected number. For example, the observed number of heads out of 20 in the above coin toss might be 8 instead of the expected 10.

Another important aspect of probability is the probability of particular outcomes when there are two or more independent events. For example, we may want to know the probability of outcome 1 occurring in two different events. If we assume that the two events are independent of each other—that is, the outcome of the first event is unrelated to the outcome of the second event and vice versa—then the joint probability of outcome 1 occurring both times is ***the product of the probability of the event occurring each time***, or Pr(1,1) = Pr(1)Pr(1). This **multiplication rule** of probability can be extended to any number of independent events. For example, the joint probability of obtaining a head from a fair coin four times in a row (or from a simultaneous toss of four fair coins) is

$$\Pr(1,1,1,1) = \Pr(1)\Pr(1)\Pr(1)\Pr(1) = (0.5)(0.5)(0.5)(0.5) = 0.0625$$

If there are two possible outcomes of an event, such as heads or tails of a coin flip or allele A_1 or A_2 in a gamete produce by a heterozygote, then in a sample of size N there are i outcomes of the first type and j outcomes of the second type ($i + j = N$). If the probability of the first outcome is p and of the second outcome is q ($p + q = 1$), the **binomial probability** that the first type of outcome will occur i times in N independent trials is

$$\Pr(i) = \frac{N!}{i!j!}p^i q^j \qquad (1.7a)$$

where "!" means to multiply by all integers from the given value to unity, thus $4! = (4)(3)(2)(1)$, and by definition $0! = 1$. The binomial coefficient, $N!/(i!\,j!)$, gives the number of different sequences (permutations) in which a given combination of outcomes can occur. Note that we need only to specify the number of one type of outcome because all of the remaining outcomes are of the other type. For example, the probability of obtaining two heads in five flips of a fair coin ($p = q = 0.5$) is

$$\Pr(2) = \frac{5!}{2!3!}0.5^2 0.5^3 = 10(0.03125) = 0.3125$$

In this case, there are 10 different orders (permutations) in which two heads and three tails can be obtained in five coin tosses: HHTTT, HTHTT, HTTHT, etc. The expected number of outcomes of type 1 again is the product of the number of trials, N, and the probability of the event, p, or Np. The variance of the number of occurrences in a sample of size N is Npq.

A binomial distribution in which the value of p becomes small while N becomes large (the product Np remains moderate in size) approaches a **Poisson probability** distribution. In this case, the probability of i occurrences is

$$\Pr(i) = \frac{e^{-\mu}\mu^i}{i!} \qquad (1.7b)$$

where the mean number of occurrences, Np, is equal to μ (the Greek lower case letter mu). For example, assume that we are surveying N different areas to determine how many individuals are in each of them. If p, the probability of an individual being in an area, is 0.01 and $N = 100$ ($\mu = 1$), the probability of 0 individuals in a given area becomes

$$\Pr(0) = \frac{e^{-1}\mu^0}{0!} = 0.368$$

Similarly, $\Pr(1) = 0.368$, $\Pr(2) = 0.184$, and so on. An important attribute of the Poisson distribution is that the variance is the same as the mean, μ.

When there are more than two possible outcomes, the probability of a combination of outcomes can be calculated from a **multinomial probability** distribution. For example, if there are three types of outcomes, such as genotypes A_1A_1, A_1A_2, and A_2A_2, then the probability of obtaining i of genotype A_1A_1, j of genotype A_1A_2, and k of genotype A_2A_2 ($i + j + k = N$) is

$$\Pr(i, j) = \frac{N!}{i!j!k!} P^i H^j Q^k \tag{1.7c}$$

where P, H, and Q are the probabilities of the three genotypes in the population from which the sample was taken ($P + H + Q = 1$).

An important application of probability is in the evaluation of experimental results (see Sokal and Rohlf, 1995; Zar, 1999). In an experiment, we have a **null hypothesis**, that is, the hypothesis being tested. We would like to minimize both the probability of rejecting a true null hypothesis, called a **type I error**, and the probability of accepting a false null hypothesis, a **type II error**. First, before we carry out the test, we should decide what magnitude of type I error that we will allow. Generally, the level of type I error is given as a probability and is symbolized by α or P. For example, often an $\alpha = 0.05$ is used and can be given as a 5% significance level or $P \leq 0.05$.

In addition, particularly when the null hypothesis is not rejected, it is important to determine the **statistical power** of the test. If the probability of a type II error is β, then the statistical power of the test is 1 - β, that is, the probability of rejecting a null hypothesis when it is false. On the other hand, if very large sample sizes are available, then statistical significance may be observed even when there are relatively small effects. In this case, it is very important that the **effect size**, say the amount of selection that would cause such a difference, be estimated so that the magnitude of the statistically significant factor may be placed in an appropriate population genetics context.

Finally, when **multiple tests** or **multiple comparisons** are carried out, then it is necessary to adjust the α level. For example, if 20 tests are made, then it is expected by chance that one will be significant at the

$P \leq 0.05$ level. As a result, it would be inappropriate to reject the null hypothesis even though one or more tests are statistically significant. This effect can be illustrated by calculating the probability that one of more of N tests are significant at the α level, or the experimentwise error rate, α' as

$$\alpha' = 1 - (1 - \alpha)^N \tag{1.8a}$$

where $(1 - \alpha)^N$ is the probability that no individual tests are significant (Weir, 1996). For example, if $\alpha = 0.05$ and $N = 10$, then $\alpha' = 0.40$; that is, the actual experimentwise significance level is much larger than that intended.

To avoid this problem, the α values for individual tests need to be adjusted. For an overall $\alpha' = 0.05$, then equation 1.8a can be solved as

$$\begin{aligned} \alpha &= 1 - (1 - \alpha')^N \qquad (1.8b) \\ &\approx \alpha'/N \end{aligned}$$

In other words, to achieve an overall level of $\alpha' = 0.05$ with $N = 10$, then the α values for individual tests need to be 0.005 (this approach is known as the **standard Bonferroni** or **Dunn-Sidak** correction).

Another approach called the **sequential Bonferroni** that provides increased statistical power is often used in evolutionary and ecological studies (Rice, 1989). In this approach, the experiments are ranked on their observed P values low to high. The experimentwise α (usually set at 0.05) is divided by the number of statistical tests N and compared with the lowest observed P value. If the lowest $P \leq \alpha/N$, then it is significant, and if it is $> \alpha/N$, then this and all other higher P values are nonsignificant. Given significance for the lowest P value, then the next lowest P is compared with $\alpha/(N-1)$, and so on until a P value is nonsignificant. As an example, Rice (1989) gave five P values of 0.01, 0.012, 0.015, 0.02, and 0.4. In this case, only the first value would be significant with the standard Bonferroni correction, that is, $0.01 \leq \alpha/N = 0.05/5$. On the other hand, the first four values would be significant with the sequential Bonferroni: $0.01 \leq 0.01$, $0.012 \leq 0.0125$, $0.015 \leq 0.167$, and $0.02 \leq 0.02$.

The problems with multiple tests are becoming much more common with genomewide studies of linkage, gene expression, and so forth, in which often thousands of comparisons are made. One approach is simply to examine further any individual comparison that has a significance value of, for example, $\alpha = 0.01$ or 0.001, depending on the number of comparisons. Because the major problem with carrying out so many tests is the number of false positive values (truly null differences that will be called significant), Storey and Tibshirani (2003) have recommended an alternative approach based on the "false discovery rate," which they suggest provides a better measure of significance.

d. Matrices

Matrices have a number of applications in population genetics, and we use them when considering both genetic drift in Chapter 6 and gene flow in Chapter 9. A matrix, **X**, is a set of values called **elements** arranged in a rectangular form (we use boldface for the symbol for a matrix and a vector as is mathematical convention). For example, a square matrix of three rows and three columns is

$$\mathbf{X} = \begin{vmatrix} x_{11} & x_{12} & x_{13} \\ x_{21} & x_{22} & x_{23} \\ x_{31} & x_{32} & x_{33} \end{vmatrix}$$

where the first subscript indicates the row number and the second subscript the column number. A matrix with only one column is called a column vector or just a **vector**. An example is

$$\mathbf{Y} = \begin{vmatrix} y_1 \\ y_2 \\ y_3 \end{vmatrix}$$

A matrix can be multiplied by such a vector to produce a new vector as follows

$$\mathbf{Y}' = \mathbf{XY} \tag{1.9}$$

This multiplication is carried out by summing the products of the corresponding elements in the rows of the matrix and the vector so that

$$\mathbf{Y}' = \begin{vmatrix} \sum_{i=1}^{3} x_{1i}y_i \\ \sum_{i=1}^{3} x_{2i}y_i \\ \sum_{i=1}^{3} x_{3i}y_i \end{vmatrix}$$

For example, if a matrix and a vector have the following elements

$$\mathbf{X} = \begin{vmatrix} 1.0 & 0.25 & 0.0 \\ 0.0 & 0.5 & 0.0 \\ 0.0 & 0.25 & 1.0 \end{vmatrix} \qquad \mathbf{Y} = \begin{vmatrix} 0.2 \\ 0.3 \\ 0.5 \end{vmatrix}$$

then

$$\mathbf{Y}' = \begin{vmatrix} (1.0)(0.2) + (0.25)(0.3) + (0.0)(0.5) \\ (0.0)(0.2) + (0.5)(0.3) \ \ + (0.0)(0.5) \\ (0.0)(0.2) + (0.25)(0.3) + (1.0)(0.5) \end{vmatrix} = \begin{vmatrix} 0.275 \\ 0.15 \\ 0.575 \end{vmatrix}$$

In some cases, it is useful to multiply a square matrix by itself. The result is a new matrix of the same size. The elements in the new matrix are the sum of the products of the elements in a given row and a particular corresponding column. For example, when a 3 × 3 matrix is squared, the resulting matrix is

$$\mathbf{X}^2 = \begin{vmatrix} \sum_{i=1}^{3} x_{1i}x_{i1} & \sum_{i=1}^{3} x_{1i}x_{i2} & \sum_{i=1}^{3} x_{1i}x_{i3} \\ \sum_{i=1}^{3} x_{2i}x_{i1} & \sum_{i=1}^{3} x_{2i}x_{i2} & \sum_{i=1}^{3} x_{2i}x_{i3} \\ \sum_{i=1}^{3} x_{3i}x_{i1} & \sum_{i=1}^{3} x_{3i}x_{i2} & \sum_{i=1}^{3} x_{3i}x_{i3} \end{vmatrix}$$

Using the same matrix as above, then

$$\mathbf{X}^2 = \begin{vmatrix} 1.0 & 0.375 & 0.0 \\ 0.0 & 0.25 & 0.0 \\ 0.0 & 0.375 & 1.0 \end{vmatrix}$$

This process can be carried out to any power t to obtain the powered matrix $\mathbf{X}^t$.

IV. GENETIC VARIATION

Before we examine in later chapters how to quantify the amount and pattern of genetic variation and to examine the factors that influence it, we need an overview of genetic variation in natural populations. For centuries, humans have been aware that variations existed among themselves and that such variants tended to run in families. Differences among varieties or breeds of domesticated plants and animals were also assumed to be inherited. However, only when Mendelian genetics and Darwin's theory of natural selection were synthesized into the neoDarwinian theory of evolution in the 1930s was a substantial effort initiated to document the amount of genetic variation within and between natural populations.

THEODOSIUS DOBZHANSKY (1900–1975)

Dobzhansky's contributions to biology were wide-ranging (Ayala, 1977; Glass, 1980; Lewontin *et al.*, 1981; Levine, 1995), but his synthesis of modern evolutionary ideas in *Genetics and Origin of Species* (1941) was one of his most important. Born in Russia, he came to the United States in 1927. His research on *Drosophila* began in 1933, and his long-term studies on *D. pseudoobscura* exquisitely documented genetic (chromosomal) variation in natural populations of this species. These studies have been continued by his students and colleagues (Anderson *et al.*, 1991; Schaeffer *et al.*, 2003). His interests encompassed human evolution, behavior, and the philosophy of science, all areas on which he had extensive influence. Another of his major contributions was his impact on young scientists, a number of whom, including Bruce Wallace, Richard Lewontin, Timothy Prout, and Francisco Ayala, were his students. A fitting quote illustrates his dedication to the evolutionary viewpoint when considering biological problems: "Nothing in biology makes sense except in the light of evolution" (Dobzhansky, 1973). (Photo ©Science Photo Library/Photo Researchers, Inc.)

The early studies of genetic variation in natural populations concentrated on easily detected and/or quantifiable variation, such as color or morphological variants (Ford, 1940), chromosomal inversions (Dobzhansky, 1941), or blood groups (Landsteiner and Weiner, 1940). Although these are important variants and good case studies, they did not allow an estimate of the total amount of genetic variation in the genome of the populations studied. In fact, all of these variants are atypical of the majority of loci in the genome in one way or another—at most, only several color or morphological polymorphic loci exist in a population, inversions generally contain a large number of genes, and blood group variants were initially found as the result of transfusion problems. The lack of definitive knowledge concerning the extent of genetic variation led to widely differing views in the 1960s, ranging from the estimate that 50% or more of the loci in a given individual are heterozygous to the calculation that only 12 to 20 variable loci in the whole genome place a severe strain on a population (Mayr, 1963).

These disparate viewpoints concerning the amount of genetic variation, many polymorphic loci versus a few polymorphic loci in a population, were termed by Dobzhansky (1955) the **balanced** and **classical** views of the organization of the genome, respectively. More specifically, in a balanced view of the genome, the linear array of genes on the chromosomes of a typical diploid individual would appear as

$$\frac{A_1 \quad B_2 \quad C_3 \quad \cdots \quad X_1 \quad Y_1 \quad Z_3}{\overline{A_4 \quad B_2 \quad C_7 \quad \cdots \quad X_1 \quad Y_1 \quad Z_2}}$$

where a large proportion of the loci are heterozygous (A, C, and Z are heterozygous genes here). According to the balanced view, the different alleles at each locus are maintained by balancing selection, selection that generally gives an overall advantage to heterozygotes (see Chapters 3 and 4).

In contrast, the classical view of the same chromosome would be

$$\frac{+ \quad + \quad m \quad + \quad \cdots \quad + \quad + \quad +}{\overline{+ \quad + \quad + \quad + \quad \cdots \quad + \quad + \quad +}}$$

where + and m indicate wild-type and mutant alleles, respectively. Here only a small proportion of the loci are heterozygous, and they are heterozygous for recessive, deleterious mutants. According to the classical view, the mutants are maintained at a low frequency by the joint effect of recurrent mutation constantly generating new variants and selection acting to eliminate them; this is sometimes termed mutation–selection balance (see Chapter 7). In these two different viewpoints, only variation that affected fitness was generally considered.

We must have two types of measures in order to understand the level and pattern of genetic variation and determine whether one of the viewpoints is more appropriate. First, there needs to be some way to measure genetic variation in an unbiased manner, and second, there needs to be some way to measure selective effects of different genotypes, such as homozygotes and heterozygotes. In theory, if extensive homologous sequences of DNA in a population were compared, then the degree of genetic variation would be known. Only in recent years has it become feasible to sequence large amounts of DNA, and we briefly introduce some of these data below and discuss them in more detail in later chapters, particularly Chapter 8. Ultimately, one would also like to know the extent of genetic variation that is selectively important—that is, related to fitness—but it is unlikely that a significant amount of DNA variation is related to fitness differences (Chapter 8). After all, a fundamental goal in determining the extent of genetic variation is to document the variation that results in selective differences among individuals, the so-called **stuff of evolution** (Lewontin, 1974).

As a first approach to obtaining an unbiased estimate of genetic variation, Lewontin and Hubby (1966) and Harris (1966) studied the variation at a series of enzyme-coding genes, generally called **allozymes**, by examining the electrophoretic mobility of their products with the assumption that this assessment of variation would be independent of the function of these genes. Their surveys and, subsequently, those of many other researchers indicated that an extensive amount of genetic variation exists for a number of allozyme loci in virtually every species except those in which it appears that genetic drift has greatly reduced genetic variation. From these studies, the level of variation makes it appear that the balanced view is a more accurate representation of the variation at allozyme loci. This view of the situation is oversimplified, however, because the connection between the description of genetic variation and the factors responsible for its maintenance is likely to be more complex than these two viewpoints imply.

The amount and kind of genetic variation in a population are potentially affected by a number of factors, such as selection, inbreeding, genetic drift, gene flow, and mutation (see the discussions in subsequent chapters). These factors may have general particular effects; for example, genetic drift and inbreeding can be considered to always reduce the amount of variation and mutation to always increase the amount of variation. Other factors, such as selection and gene flow, may either increase or reduce genetic variation, depending on the particular situation. Combinations of two or more of these factors can generate virtually any amount or pattern of genetic variation. In other words, the balanced-view explanation that the amount of genetic variation present is due to balancing selection appears to be a great oversimplification of the actual forces determining the amount of genetic variation in a particular situation. In fact, a quite different explanation called the **neutral theory**, or **neutrality** (see Chapter 8), in which allele differences are neutral with respect to selection, combined with mutation and genetic drift is consistent with many observations of genetic variation. At this point, we recognize that the factors mentioned can affect genetic variation but reserve discussion of their relative importance in increasing, maintaining, or reducing genetic variation.

We discuss below the amount of genetic variation without giving any detailed background because the primary intent of this introduction is to illustrate the diversity of genetic variation. In Chapter 2, we discuss how genetic variation within populations can be quantified. Briefly, the measures used below are the **proportion of polymorphic loci**, P, and the average **heterozygosity** over all loci in the particular population $\bar{H}$, where the overbar indicates the mean over all examined loci and individuals. Polymorphism refers to the occurrence of different genetic forms of a locus in the same population, such as the A, B, and O alleles for the ABO red blood cell locus in humans. If 50 such loci are examined and 10 have more than one allele, then the proportion of polymorphic loci is $P = 10/50 = 0.2$.

Likewise, the heterozygosity is the proportion of diploid genotypes composed of two different alleles. To illustrate this concept, let us use another polymorphic human blood group locus, the *MN* locus. At this locus there are three possible genotypes: two homozygotes, *MM* and *NN,* and one heterozygote, *MN.* Let us assume that in a population of 200 individuals, 90 are *MN* heterozygotes. Therefore, the observed proportion of heterozygotes for this locus in this population is $H = 90/200 = 0.45$. If the heterozygosity is known for a number of loci, then these values can be averaged over loci to give a mean value of heterozygosity, $\bar{H}$.

The two primary measures to quantify the amount of DNA variation are similar to those for single genes (in Chapter 2 we discuss them further). Let us assume that we have a number of homologous DNA sequences that are a given number of nucleotides sites long. First, we can calculate the **proportion of variable nucleotide sites,** p_S, over all of the sequences. Second, we can calculate the **nucleotide diversity,** π (Greek lower case letter pi), or the proportion of nucleotide sites that are different when any two of the sequences are randomly compared. This measure is equivalent to the average heterozygosity, or diversity, for the nucleotides examined.

As a simple example of how to calculate these measures, let us use two of the alcohol dehydrogenase (*Adh*) sequences from Kreitman (1983) that we discuss below. Sequences Ja-S and F1-F differ at 5 of the 2379 nucleotide sites compared so that $p_S = 5/2379 = 0.0021$. Let us assume that we have a sample of three sequences, two of which are Ja-S, called Ja-S(1) and Ja-S(2), and one is F1-F; thus, there are three comparisons of two sequences at a time. The heterozygosity for the Ja-S(1) – Ja-S(2) comparison is 0, and that for the Ja-S(1) – F1-F comparison is $5/2379 = 0.0021$. The Ja-S(2) – F1-F comparison is also 0.0021 so that the average heterozygosity per site for this sample of three sequences becomes $\pi = 0.0042/3 = 0.0014$.

a. Allozyme Variation

Many population genetic surveys focused on the variation in proteins as an estimate of the variation in the DNA sequence that determines the amino acid sequence of these proteins. The application of protein electrophoresis to population genetic problems by Lewontin and Hubby (1966) and Harris (1966) was a landmark development in evolutionary genetics. The basic technique of protein electrophoresis, as applied to evolution, has been described in detail elsewhere (Hillis *et al.*, 1996; Hoelzel, 1998; Avise, 2004). In general, electrophoresis makes possible the separation of different proteins extracted from blood, tissues, or small, whole organisms. The process is carried out by running an electric charge through a supporting medium (usually either a starch or acrylamide gel) into which the protein has been placed. The proteins are allowed to migrate for a specific amount of time and then are stained with various protein-specific chemicals so that the relative mobility of a specific protein can be determined. Relative mobility is generally a function of the size, charge, and shape of the molecule. If two proteins have different amino acid sequences, then they often have different mobilities because the differences in sequence result in a change in size and/or charge of the molecule. Lewontin and Hubby (1966) pointed out, however, that this approach does not detect all of the variation because some amino acid differences do not result in charge or size differences detectable by electrophoresis (see discussion in Barbadilla *et al.*, 1995; Veuille and King, 1995).

In surveys of electrophoretic variation, proteins from a group of individuals are run usually simultaneously on a single gel. Figure 1.5 is an example of such a gel showing variation in two allozyme loci that code for the enzyme leucine amino peptidase (LAP) for nine individual brown snails (Selander, 1976). Because LAP is a monomeric enzyme, homozygotes show a single stained region or band, and heterozygotes are double banded. The upper enzyme locus (*Lap-1*) is polymorphic for two electrophoretic bands, and the lower enzyme locus (*Lap-2*) is polymorphic for three. Because migration of the proteins goes from the bottom to the top of the gel pictured, the phenotypes (presumed genotypes) are as given in Figure 1.5, where *F*, *M*, and S indicate fast, intermediate, and slow migrating alleles, respectively. For example, the individual on the left is a heterozygote *FS* at the *Lap-1* locus and homozygote *SS* at the *Lap-2* locus. It is generally assumed that each band represents a different allele at a locus, but as discussed above, there may be substantial amino acid variation that does not result in different bands.

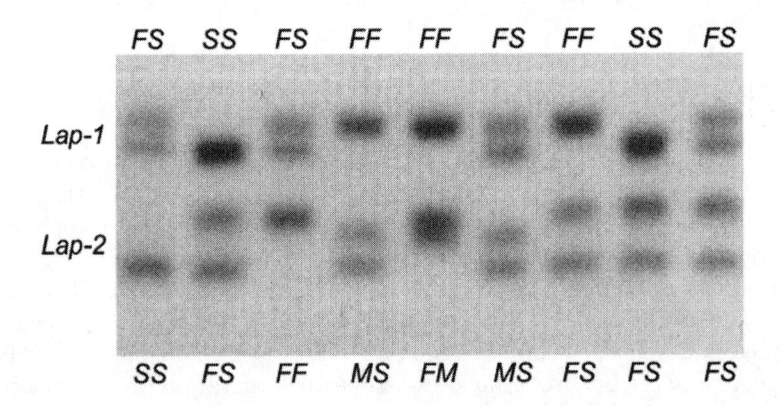

Figure 1.5. Variation in two leucine amino peptidase enzymes in the brown snail, *Helix aspersa* (from Selander, 1976). The upper system (*Lap-1*) is polymorphic for two alleles (*F* and *S*) and the lower system (*Lap-2*) is polymorphic for three alleles (*S*, *M*, and *F*). The genotypes are indicated above and below the gel for the nine individuals pictured.

The use of electrophoretic techniques to investigate population genetic problems reinvigorated the field of population genetics in the 1970s because this approach allowed studies of genetic variation in virtually any species with relatively little equipment and expertise. A number of allozyme surveys examined the amount of variation in different populations of the same or related species. It appears that some allele variants are often widespread in a species and in closely related species. On the other hand, species or populations may differ in the presence or frequency of various variants. Additionally, populations may differ in the amount of heterozygosity averaged over several loci, or they may be very similar. A particular emphasis in allozyme studies has been the documentation of both spatial (on both microgeographic and larger spatial scales) and temporal genetic variation; we give examples of types of studies below.

An example of spatial variation is the study of Selander and Kaufman (1975), who systematically collected the brown snail, *Helix aspersa,* in two city blocks of Bryan, Texas, and found 2218 snails in 43 colonies. The brown snail was introduced to Bryan in the early 1930s and is now found widely in gardens and yards. An example of their data is given in Figure 1.6, where the frequencies of various alleles for the *Mdh*-1 (malic dehydrogenase) locus are indicated in the pie diagrams. Even over such a small area, they found extensive variation. For example, the frequency of *Mdh*-1^{120} varied from approximately 0.30 to greater than 0.75 in different colonies. The heterozygosities for the two blocks were similar for *Mdh*-1 and some other loci, but for other loci, they varied considerably. For example, the heterozygosity of locus *Got*-1 in block A was 0.306 and in block B was 0.002.

Gaines *et al.* (1978) examined five polymorphic loci over a 30-month period in the prairie vole, *Microtus ochrogaster.* This small rodent undergoes large fluctuations in population density in approximately 3-year cycles. The average generation length is estimated to be 6 to 8 weeks so that during the

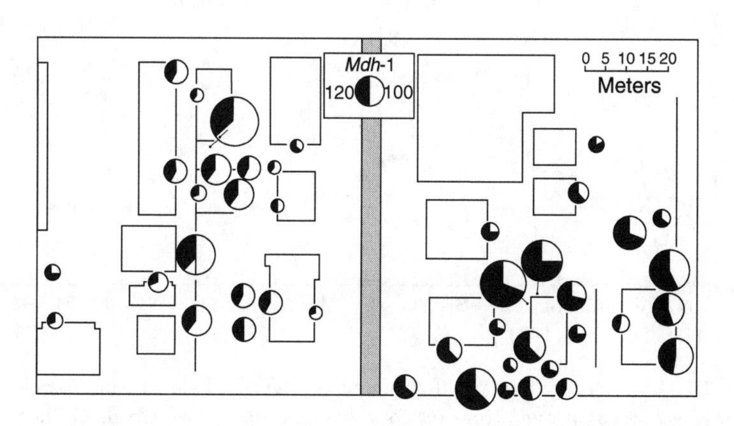

Figure 1.6. The allele frequencies at the *Mdh*-1 locus in brown snail colonies in two city blocks separated by an alley (shaded). Circle size is proportional to colony size, and proportions within the circles indicate allele frequency (from Selander and Kaufman, 1975).

study period there were approximately 20 generations. Data for an allele at the *Lap* locus for four sites near Lawrence, Kansas, are given in Figure 1.7. Sites A, B, and D were within 1.2 km of each other, and site C is 3.6 km from the main study area. There was considerable variation in the allele frequency over time and among sites, varying from zero early in the sampling period to nearly 0.5 at other times. Although some of the variation appears to be due to sampling effects, other loci that were monitored, such as *Tf* and *G6pd,* did not show nearly as much variation in frequency. There was considerable variation in average heterozygosity at one site (varying from approximately 0.10 to 0.35 at different times), whereas at the other three sites, there was much less variation in heterozygosity.

One of the major discoveries from the extensive electrophoretic studies was the documentation of extensive genetic variation at the molecular level in many different organisms. Overall, the greatest amount of variation appeared to be in invertebrates and plants and the smallest amount in vertebrates. Most species were generally examined for approximately 30 loci, but in humans and a few other organisms, many more loci were surveyed. The 71 enzyme loci surveyed in humans are given in Table 1.3 (Harris and Hopkinson, 1972). Of these loci, 51 were monomorphic, whereas 20 were polymorphic with heterozygosity values ranging from 0.02 to 0.53. From these data, the proportion of polymorphic loci, P, is 0.282, and the average heterozygosity over all loci, $\bar{H}$, is 0.067.

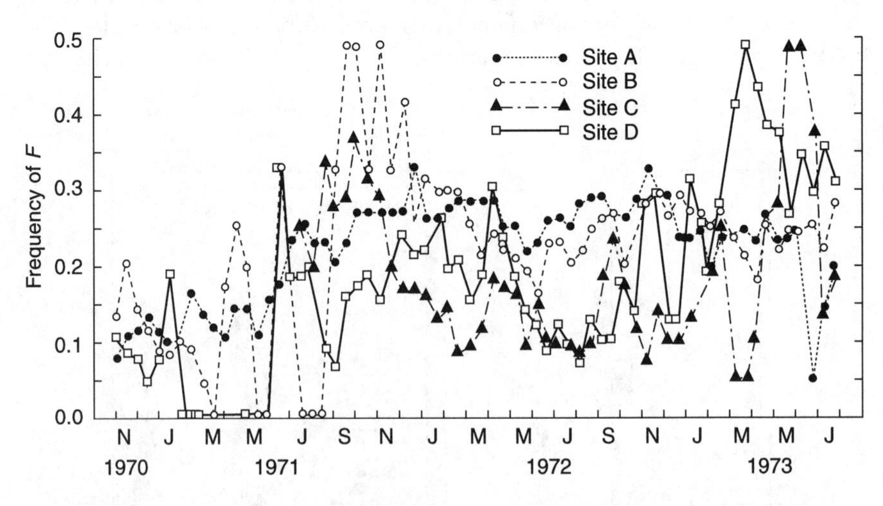

Figure 1.7. The changes in the frequency of the F allele at the *Lap* locus in *Microtus ochrogaster* over time on four live-trapped grids (from Gaines *et al.*, 1978).

TABLE 1.3 The heterozygosity for 71 allozyme loci in humans (Harris and Hopkinson, 1972).

Locus	*Heterozygosity (H)*
51 monomorphic loci	0.000
Peptidase C	0.002
Peptidase D	0.020
Glutamate-oxaloacetate transaminase	0.030
Leucocyte hexokinase	0.050
6-Phosphogluconate dehydrogenase	0.050
Alcohol dehydrogenase-2	0.070
Adenylate kinase	0.090
Pancreatic amylase	0.090
Adenosine deaminase	0.110
Galatase-1-phosphate uridyl transferase	0.110
Acetyl cholinesterase	0.230
Mitochondrial malic enzyme	0.300
Phosphoglucomutase-1	0.360
Peptidase A	0.370
Phosphoglucomutase-3	0.380
Pepsinogen	0.470
Alcohol dehydrogenase-3	0.480
Glutamate-pyruvate transaminase	0.500
RBC acid phosphatase	0.520
Placental alkaline phosphatase	0.530
$\overline{H}$	0.067

b. Nucleotide and Amino Acid Sequence Variation

In the past decade, the techniques to sequence large amounts of DNA have been greatly improved and automated. As a result, the fundamental amount of genetic variation can be obtained for many genes in many organisms, and this DNA sequence data can then be used to document DNA variation and infer the amino acid variation present in different individuals, populations, and species. Many of these technical advances can be attributed to the extraordinary effort to sequence the total human genome. The complete genomic sequence of about 100 prokaryotes is now completed (Lynch and Conery, 2003), including the important genetic model bacterium *Escherichia coli.* The eukaryotes sequenced include the worm (*Caenorhabditis elegans*), sea squirt (*Ciona intestinalis*), fruit fly (*D. melanogaster*), mosquito (*Anopheles gambiae*), wild mustard (*Arabidopsis thaliana*), rice, puffer fish, mouse, rat, human, and most recently, the sequence of the dog has been finished (Kirkness *et al.*, 2003). Other genomes now being sequenced include the honeybee, chicken, cow, pig, macaque, and chimpanzee (Couzin, 2003).

Examination of the similarities and differences between genomes of related organisms, **comparative genomics**, should give detailed history of evolution changes and insight into disease, development, and other genetically based phenomena. For example, comparison of the genomes of the nematodes *Caenorhabitis elegans* and *C. briggsae*, which are morphologically very similar, show that they have undergone more than 4,000 chromosomal breakages since they diverged, have high levels of synonymous substitutions, and have very different numbers of various categories of genes (Blaxter, 2003; Stein *et al.*, 2003).

We do not give the techniques used to determine DNA variation here but suggest Hillis *et al.* (1996), Hoelzel (1998), and Avise (2004) as introductions to the fundamentals and McPherson *et al.* (2000) and Dieffenback and Dveksler (2003) for details about recent DNA technology. One of the first widely used approaches to measure variation in DNA used enzymes present in bacteria that recognize foreign DNA, such as that from bacteriophages, and cleave it. These enzymes, called **restriction endonucleases**, recognize specific sequences in the DNA and cleave the DNA in a particular manner. Restriction enzymes have been widely used to characterize mtDNA, a small circular DNA molecule (16,569 nucleotides in humans) with around 16 genes, that is generally maternally inherited. mtDNA is large enough that the likelihood of different restriction sites in animals of the same species is high. For example, the enzyme EcoR1 recognizes three sites in human mtDNA and five sites in honey bee mtDNA (Hillis *et al.*, 1996). Individuals or populations may be variable for the presence of different sites resulting in fragments of different lengths called **restriction fragment length polymorphisms (RFLPs)**.

One of the first applications of this approach was by Avise *et al.* (1979) in the pocket gopher, *Geomys pinetis*, from the southeastern United States. They examined 87 individuals with six different restriction enzymes and found 23 different mtDNA haplotypes (Figure 1.8). The number of restric-

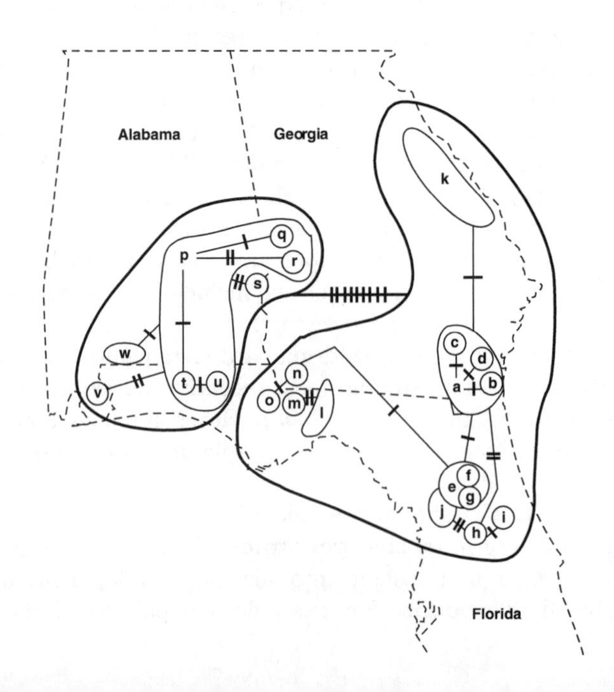

Figure 1.8. The relationship of 23 different mtDNA haplotypes for 87 pocket gophers (from Avise *et al.*, 1979). A network connecting the most related haplotypes is superimposed over the geographic sources of the animals, where the slashes reflect the numbers of inferred differences between haplotypes.

tion site differences between haplotypes is given as a slash so that, for example, haplotypes p and q differ by one site and p and r differ by two. Obviously, there is a strong concordance between the relationship of haplotypes and their geographic location, similar haplotypes being present in adjacent populations. The major exception is the estimated nine-restriction-site division between the eastern and western populations.

In 1977, techniques to sequence DNA were introduced. When DNA was sequenced manually (this is very infrequent now with the widespread use of automated sequencers), it was put into four separate tubes, each of which identifies the presence of one of the four nucleotides. For example, Figure 1.9 gives a 59-base sequence on an autoradiograph using the Sanger *et al.* (1977) method (from a MHC allele in the Gila topminnow, Hedrick and Parker, 1998a). In this case, the sequence read from the bottom is TACGCCCGGT . . . . This approach has been replaced by automated DNA sequencers in recent years. An example of the graphical output of the same 59-base sequence of DNA from an automated DNA sequencer is presented in Figure 1.10. Again, there are four different sequencing reactions, but they are placed on a single lane of a gel. The plots are given in four colors indicating the different nucleotides (only the dark G can be easily differentiated in the figure) so that the highest peak at each nucleotide position indicates which nucleotide is present.

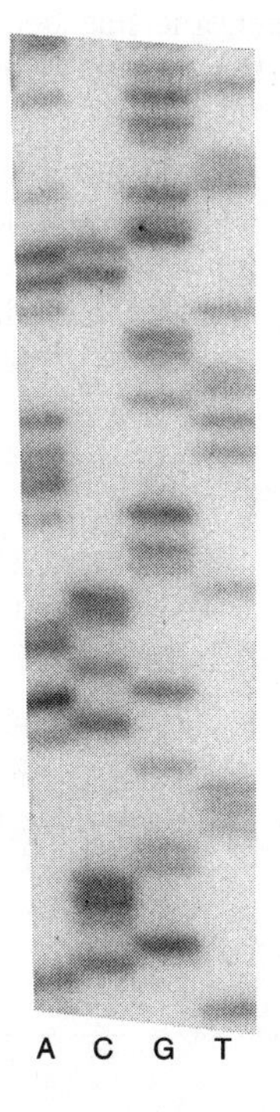

Figure 1.9. An example of a sequencing gel radiograph, where the different columns indicate the presence of the four nucleotides. The 59-base sequence is from MHC allele *Pooc-6* from the Gila topminnow (Hedrick and Parker, 1998a).

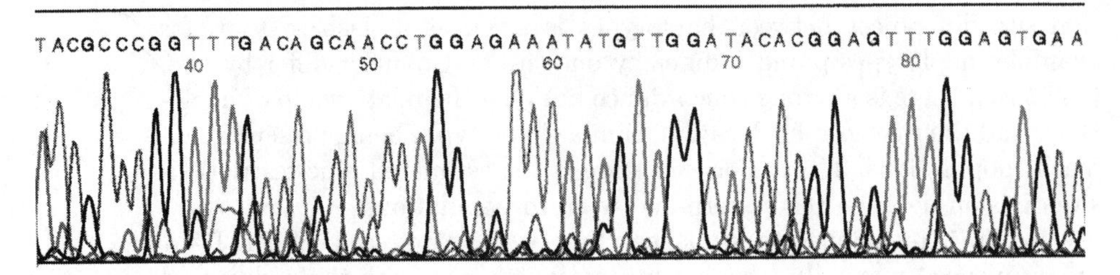

Figure 1.10. An example of the graphical output from an automated sequencer. different positions indicate the presence of different nucleotides (the four different nucleotides from the actual printout are given in different colors). This is the same sequence as given in Figure 1.9 read from bottom to top.

One of the first studies to document the amount of DNA variation was that for the *Adh* gene in *D. melanogaster* (Kreitman, 1983). Table 1.4 gives the 49 variable sites for the 11 different sequences over the complete region he sequenced, 43 of which had variation in the nucleotides present and 6 had either insertions or deletions. If we do not include the insertions and deletions, then the region sequenced was 2379 nucleotides. Therefore, the proportion of variable sites, $p_S = 43/2379 = 0.018$. Because there are 11 different sequences, then there are $(11)(10)/2 = 55$ comparisons between sequences. Later (in Table 2.19), we give the values for all these comparisons, but let us just give the average here over all comparisons, which is the nucleotide diversity, $\pi = 0.0065$.

Variation in the human genome has been calculated in large samples representing individuals from European, African American, and Chinese

TABLE 1.4 Variable nucleotide sites in the 11 sequences of the alcohol dehydrogenase (*Adh*) locus in *D. melanogaster* (after Kreitman, 1983). Dashes indicate nucleotides identical with the consensus sequence, triangles indicate sites of insertions (downward) and deletions (upward), and the asterisk in exon 4 indicates the amino acid difference between the *F* (Fast) and *S* (Slow) alleles.

Sequence	*5′*	*Intron 1*	*Larval leader*	*Exon 2*	*Intron 2*	*Exon 3*	*Intron 3*	*Exon 4*	*3′*
								*	
Consensus	CCG	CAATATGGG▼C▼G	C	T	AC	CCCC	GGAAT	CTCCACTAG	A ▼ C AGC▼C ▼ T▲
Wa-S	- - -	- - - - - -AT- - - - - -	-	-	- -	TT-A	CA-TA	AC- - - - - - -	- - - - - - - - - - - -▲
Fl-1S	- - C	- - - - - - - - - - - - -	-	-	- -	TT-A	CA-TA	AC- - - - - - -	- - - - - - - - - - - -▲
Slow Af-S	- - -	- - - - - - - - - - - - -	-	-	- -	- - - -	- - - - -	- - - - - - - - -	A - - - - - -T▼- 1 A-
Fr-S	- - -	- - - - - - - - - - - - -	-	-	GT	- - - -	- - - - -	- - - - - - - - -	A - -1- TA- - - - - -
Fl-2S	- - -	AG- - -A-TC- - - -	-	G	GT	- - - -	- - - - -	- - - - - - - - -	- C 3 - - - - - - - - -
Ja-S	- - C	- - - - - - - - - - - - -	-	G	- -	- - - -	- - - - -	- - -T-T-CA	C 4 - - - - - T - - -
Fl-F	- - C	- - - - - - - - - - - - -	-	G	- -	- - - -	- - - - -	- -GTCTCC-	C 4 - - - - - - - - -
Fr-F	TGC	AG- - -A-TC▼G▼-	-	G	- -	- - - -	- - - - -	- -GTCTCC-	C 4 G - - - - - - - -
Fast Wa-F	TGC	AG- - -A-TC▼G▼-	-	G	- -	- - - -	- - - - -	- -GTCTCC-	C 4 G - - - - - - - -
Af-F	TGC	AG- - -A-TC▼G▼-	-	G	- -	- - - -	- - - - -	- -GTCTCC-	C 5 G - - - - - - - -
Ja-F	TGC	AGGGGA- - -▼- -T	-	G	- -	- - - -	- -G- -	- -GTCTCC-	C 4 - - - - - - -1- -

descent (The International SNP Map Working Group, 2001). Table 1.5 gives the average amount of variation from this survey for the 22 different autosomes and the two sex chromosomes, X and Y. Overall, there is a SNP every 1.91 kb, and the average nucleotide diversity π per site is 0.000751. The average diversity for 20 of the 22 autosomes is within 10% of this mean, with the lowest diversity for chromosome 21 (0.000519) and highest diversity for chromosome 15 (0.000879). Both of the sex chromosomes had significantly lower variation than the autosomes, as expected because of their lower effective population sizes and consequently greater genetic drift (see p. 322 and p. 329).

TABLE 1.5 The length and the amount of variation for the different human chromosomes as measured from a survey of 1.42 million SNPs.

Chromosome	*Length* ($bp/10^6$)	*kb per SNP*	$\pi(\times 10^4)$
1	214	1.65	7.72
2	223	2.15	7.37
3	187	2.01	7.52
4	169	2.00	8.08
5	171	1.45	7.23
6	165	1.71	7.44
7	149	2.08	7.59
8	125	2.16	7.74
9	107	1.73	8.13
10	128	2.09	8.25
11	129	1.53	8.38
12	125	2.11	7.55
13	94	1.77	8.03
14	89	2.03	7.40
15	73	1.94	8.79
16	74	1.91	8.29
17	73	2.12	7.83
18	73	1.62	8.14
19	56	2.18	7.64
20	63	2.15	7.15
21	34	1.62	5.19
22	34	1.19	8.53
X	131	3.77	4.69
Y	22	5.19	1.51
Total or mean	2,710	1.91	7.51

To examine the diversity over each chromosome, they were divided into contiguous 200,000-bp bins. Throughout the genome, 95% of these bins had nucleotide diversities between 2.0×10^{-4} and 15.8×10^{-4}. For example, Figure 1.11 gives the diversity for the 825 bins on chromosome 6. Note that only a few bins have diversity less than the 95% region but that a larger number on this chromosome have a diversity greater than the 95% region. The region with the most variation is centered around 34 Mb and

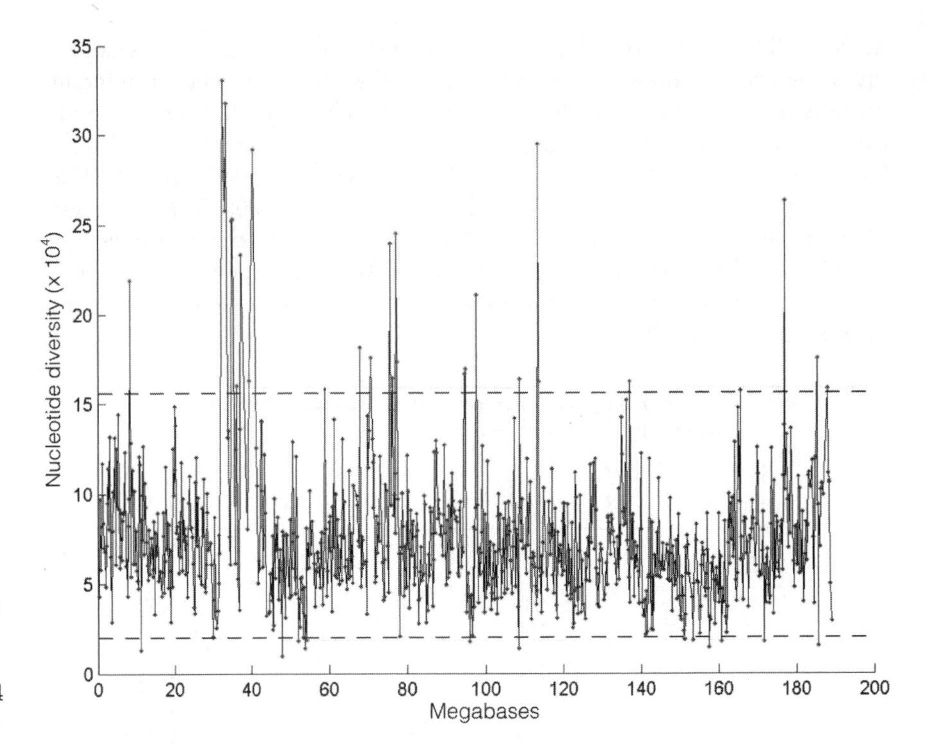

Figure 1.11. The distribution of the nucleotide diversity ($\times 10^4$) across human chromosome 6 as calculated in 200,000-bp bins. The two broken horizontal lines indicate the values within which 95% of the bins fall. The highly variable region around 34 Mb encompasses the 4 Mb of the *HLA* genes.

is the site of the highly variable human MHC (in humans these genes are known as the human leukocyte antigen, *HLA*, genes, see above).

For most protein-coding genes, there is little variation in the amino acid sequence. For example, in the *Adh* sequences given in Table 1.4, all of the amino acids in the coding regions are identical except for the one position that distinguishes the *F* from the *S* allele (the top six sequences are *S*, and the bottom five are *F*). If we calculate the proportion of amino acids that are polymorphic in the same manner as the proportion of variable nucleotide sites, then only 1 of 255 sites are polymorphic and $p_{S.aa} = 1/255 = 0.0039$. We can also calculate the average amino acid heterozygosity (or diversity) by determining the number of comparisons that give different amino acids at the variable site and dividing by the total number of amino acid sites. Because there are 30 different combinations that are heterozygous, the amino acid heterozygosity is $\pi_{aa} = 30/(255)(55) = 0.0021$, where there are 30 comparisons that give different amino acids at the variable site.

Genes of the MHC are the most variable loci known in humans (Beck and Trowsdale, 2000) and are also highly variable in many other vertebrates (Edwards and Hedrick, 1998). Variation in the genes of the human MHC have been the subject of intensive study for many years because of their importance in determining acceptance or rejection of transplanted organs, their role in many autoimmune diseases, and their general importance in recognition of pathogens. With sequence-determined alleles in worldwide human surveys, genes *HLA-A*, *HLA-B*, and *HLA-DR1* have 243, 499, and 321 alleles, respectively (Garrigan and Hedrick, 2003), with *HLA-B* being the most variable gene in the human genome (Mungall *et al.*, 2003). Also, unlike that in *Adh* in *D. melanogaster* and most other genes, most of the nucleotide variation is in functionally important parts of the MHC genes and results in amino acid variation.

Figure 1.12 gives the amino acid heterozygosity for the 366 (*HLA-A*) or 363 (*HLA-B*) amino acid sites in humans (Hedrick *et al.*, 1991). A large proportion of these sites are polymorphic and for *HLA-A*, $p_{S.aa} = 69/366 = 0.189$ and for *HLA-B*, $p_{S.aa} = 58/363 = 0.160$. Overall, the amino acid heterozygosity (π_{aa}) for *HLA-A* and *HLA-B* are 0.064 and 0.058, respectively. However, there are a number of individual amino acid sites that have heterozygosities that are greater than 0.5 for both genes. Most of these amino acid sites appear, judging by the three-dimensional structure of the HLA molecule, to have important functions involved with initiating the immune response. In fact, the 29 amino acid sites that have such important functions have average heterozygosities per site of 0.264 and 0.337 for *HLA-A* and *HLA-B*. On the other hand, the sites that do not have these functions have heterozygosities about an order of magnitude lower: 0.036 and 0.031 for *HLA-A* and *HLA–B*, respectively.

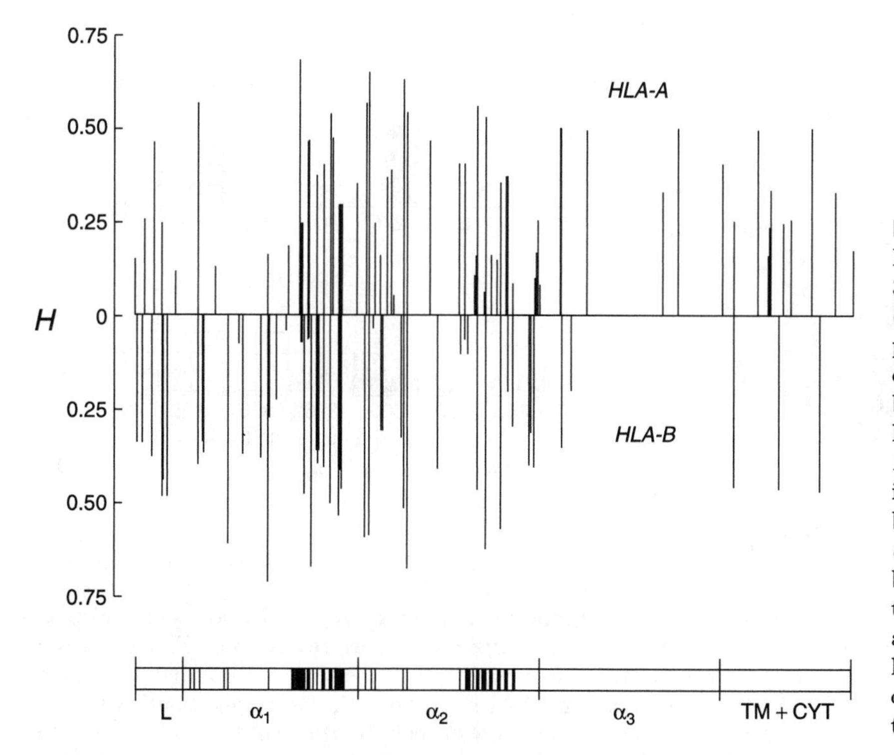

Figure 1.12. Average heterozygosity for the 366 (*HLA-A*) or 363 (*HLA-B*) amino acid positions. *HLA-A* heterozygosity is indicated by the bars above the horizontal axis, and *HLA-B* heterozygosity is indicated by the bars below (from Hedrick *et al.*, 1991). Indicated by vertical bars along the bottom are the 54 amino acid sites postulated to interact with other molecules and the HLA domains.

One of the recent breakthroughs in evolutionary genetics is the ability to obtain DNA sequence information from **ancient DNA**, that is, DNA from preserved teeth, bones, or other tissue. Under good preservation conditions, sequence from DNA up to 100,000 years old may be retrieved, but beyond that age, and much more quickly in poor preservation conditions, DNA would be broken up into such small pieces that it is virtually impossible to reliably sequence (Hofreiter *et al.*, 2001). Because the amounts of ancient DNA may be very small, laboratory protocols have been developed to ensure that the DNA being examined is not contaminated with DNA from other research or, in the case of ancient human DNA, DNA from the researchers. For a list of authenticity criteria used in ancient DNA research, see Hofreiter *et al.* (2001).

An example of the use of ancient DNA where other information was shown to be misleading is in moas, a large, flightless, extinct bird from New Zealand (Bunce *et al.*, 2003; Huynen *et al.*, 2003). Until this research, three species of *Dinornis* moas were recognized that differed markedly in size from 1 to 2 m in height at the back and in weight from 34 to 242 kg (Figure 1.13a). However, by developing DNA probes specific for the female sex chromosome W (female birds are ZW, and males are ZZ), it was determined that the differences in size were the result of sexual dimorphism in size, with female *Dinornis* moas on average 1.4 times the size of males from the same area (Figure 1.13b). Furthermore, using mtDNA, they found

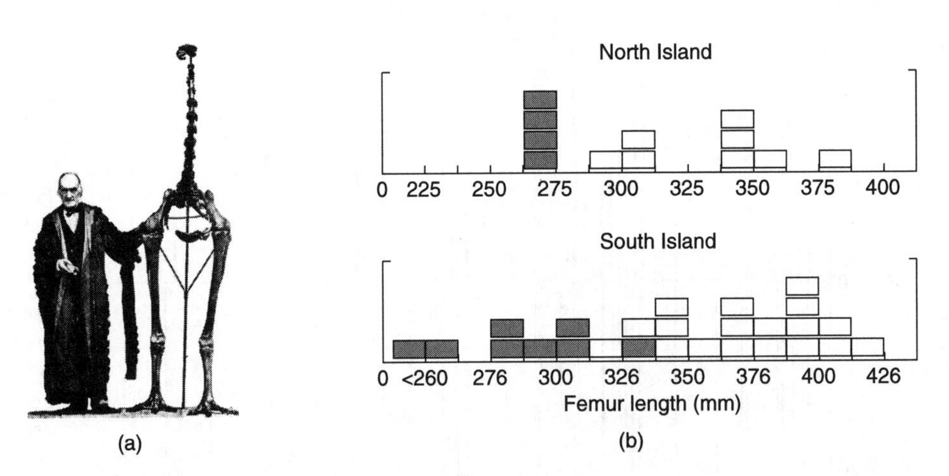

Figure 1.13. (a) Richard Owen, who predicted the existance of the moa in 1839 next to a reconstructed moa skeleton (Bunce *et al.*, 2003) (Photo ©George Bernard/Photo Researchers, Inc.) (b) The estimated size of extinct moa and their sex determined by sex-specific ancient DNA examination with males and females indicated by shaded and unshaded boxes, respectively. These moa were thought to consist of three species, but now the size variation has been found to be the result of sexual dimorphism on both the North and South Islands of New Zealand. (Courtesy of *Nature*, 425:2003, by Huynen, L., C.D. Millar, R.P. Scofield, and D.M. Lambert. Reprinted with permission of Nature Publishing Group.)

that females and males from the South Island were genetically similar and that females and males from the North Island were similar, but the two groups were different between the islands. Thus, it appears that there were only two *Dinornis* species, one on the South Island and one on the North Island and that both species were sexually dimorphic in size.

c. Visible Polymorphisms

The first genetic polymorphisms known were visible variants that affected color, shape, pattern, or other morphological aspects. Visible variants are known in virtually every kind of organism; examples include flower color variants in many plants, shell color and pattern variants in snails, and wing color variation in butterflies (Ford, 1971). Of course, not all visible variants are genetic, and of the genetic variants, the inheritance pattern of the polymorphism is sometimes poorly understood. That is, we often do not know the number of alleles, their dominance relationships, or even the number of genes involved. As a result, we concentrate on several examples in which the mode of inheritance is known and in which additional information exists on the temporal and/or spatial frequencies of these polymorphisms. Later, we discuss examples of color polymorphism in black bears (p. 83) and pocket mice (p. 521) where the molecular differences responsible for the color variation are known.

Color and pattern polymorphisms in the shells of several snail genera, particularly *Cepaea nemoralis*, have been extensively documented (p. 206). The African land snail, genus *Limicolaria*, also contains a number of species that exhibit polymorphisms in shell color and patterns (Owen, 1966). As in *Cepaea*, adjacent populations may vary considerably in the frequency of different forms involving variation in streaking and pigmentation. Fossil snails, 8000 to 10,000 years old, that appear to be *L. martensiana* have been found in areas of western Uganda that were covered by volcanic ash. These fossil snails have the same shell color and patterns as living populations in the same area. In fact, the frequencies of these types are quite similar to the fossil frequencies for some of the populations sampled (Table 1.6).

TABLE 1.6 The frequencies of color forms in a fossil population and three living populations of an African land snail. Sample size was greater than 800 in all cases (after Owen, 1966).

		Living		
Color form	*Fossil*	*1*	*2*	*3*
Streaked	0.610	0.546	0.406	0.337
Broken-streaked	0.052	0.089	0.042	0.098
Pallid 1	0.039	0.240	0.070	0.019
Pallid 2	0.283	0.119	0.432	0.373
Pallid 3	0.016	0.006	0.050	0.173

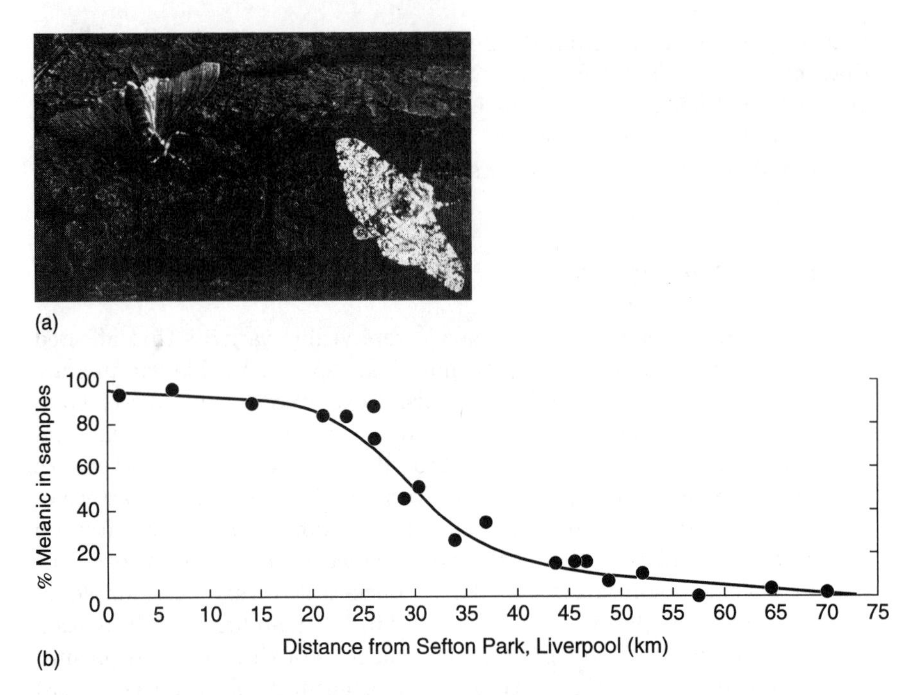

Figure 1.14. (a) Two morphs, carbonaria and typical, of the peppered moth resting on a dark tree trunk (from Kettlewell, 1973). (Courtesy of Bishop, J.A. 1973. An experimental study of the cline of industrial melanism in *Biston betularia* (L.) (Lepidoptera) between urban Liverpool and rural North Wales. *J. Anim. Ecol.* 41:209–243.) (b) The percent of melanics along a transect between Liverpool and North Wales (from Bishop, 1973).

The three contemporary populations given are all less than 30 km from the fossil site. This study suggests that these polymorphisms may be quite ancient and constant in frequencies.

Another organism for which polymorphic frequencies have been extensively investigated is the peppered moth, *Biston betularia* (p. 137). The frequencies of melanic, darkly pigmented forms of *B. betularia* have been observed in England for more than a century. The melanics originally were quite rare but increased in frequency so that in some urban populations they became nearly 100% of the population. The frequency differences of melanics in urban areas and nearby rural areas were quite striking. An example is given in Figure 1.14, where there were nearly 100% melanics in Liverpool and less than 10% melanics 50 km away in rural Wales.

A good example of a plant polymorphism, examined both in time and space, is a color variant in wild oats, *Avena fatua* (Jain, 1976). *A. fatua* is an annual grass introduced to the United States from Spain, and it is now one of the dominant grasses in many areas in California (see p. 254). In this study, the proportion of the gray lemma (a modified leaf enclosing the flower) phenotype caused by a single recessive gene (*bb*) was estimated

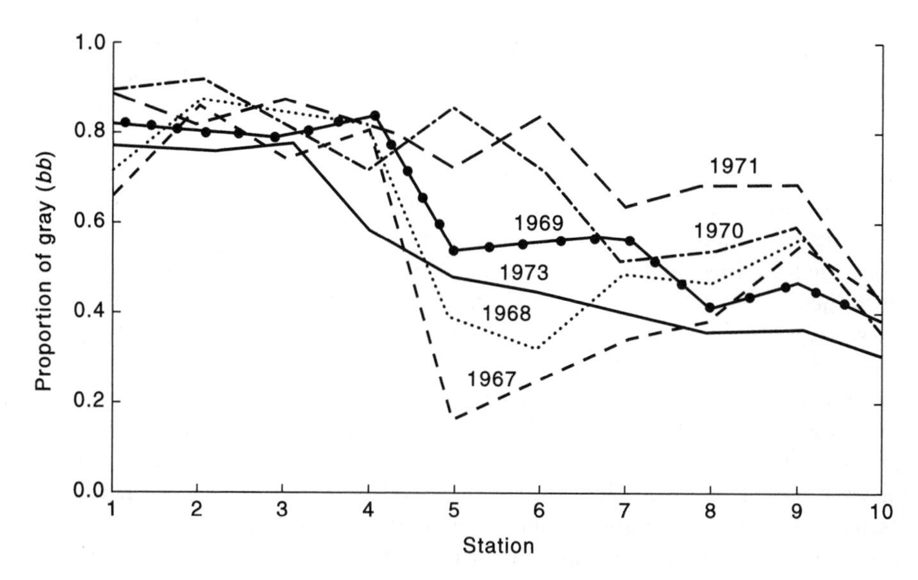

Figure 1.15. Variation in the frequency of the gray lemma phenotype in a population of the annual wild oats, *Avena fatua,* for six different generations along a clinal transect (from Jain, 1976).

for six different generations along a transect at Marysville, California. The 10 stations along the transect were only 3 meters apart, making the whole transect only 30 meters long. As indicated in Figure 1.15, there was substantial generation-to-generation variation at some sites, the proportion of gray lemma varying from less than 0.2 to greater than 0.8 at station 5. Furthermore, large spatial differences in genotype frequencies occur at the ends of the 30-meter transect, varying from a mean near 0.8 at station 2 to a mean near 0.4 at station 10, differences that are relatively constant over time.

d. Mutants, Lethals, and Fitness Modifiers

Another class of genetic variants consists of those that have demonstrable effects on fitness, usually detrimental in nature. These variants may be recognized as visible mutants by their phenotypic effects on other characters, or they may be known only because of their effect on fitness, as with many lethals. Lethals or deleterious variants may not be representative of the genetic variants in a population, both because they constitute extreme phenotypic consequences of mutation and because their negative effect on fitness leads to low frequencies. On the other hand, loci with variant forms that influence fitness, particularly those that have a positive effect, are "the stuff of evolution" even though they may be known only by their effect on fitness and cannot be characterized further.

Virtually every population that has been extensively sampled has morphological variants that are present in low frequencies and appear to have rather drastic and detrimental phenotypic effects, such as albinism in humans, rodents, frogs, and many other organisms and chlorophyll-deficiency mutants in many plants. Often these variants are recessive and are uncovered only after close inbreeding or as a result of the examination of a large sample. In a few species, data indicate the extent of this class of variation. In humans, extensive information exists because these variants are the causes of many diseases and approximately 12,000 different loci have been identified as causing human genetic disorders (Online Mendelian Inheritance in Man, http://www.ncbi.nlm.nih.gov/entrez/OMIM). Most of these variants are quite rare, with allele frequencies less than 1%, but as a group, they are an important source of genetic variation. Also, rather large differences exist among different populations in the frequencies of some of these diseases and their alleles. For example, a higher frequency of cystic fibrosis is found in the United Kingdom, sickle-cell allele in many countries of west Africa, and Tay-Sachs in Israel (Vogel and Motulsky, 1997). It is estimated that the average human is heterozygous for four to six such loci and that approximately 1% of newborn infants are homozygous for such a gene.

In *Drosophila,* several extensive surveys have been conducted to uncover the amount of recessive mutants in a population (see Lindsley and Zimm, 1992, for descriptions of the phenotypic effects of over 4000 loci in *D. melanogaster*). In these surveys, mated wild-caught females were placed in individual containers to lay eggs (lines of their progeny established in this manner are **isofemale lines**). The F_1 progeny were allowed to mate (brother–sister mating), and their progeny were scored for phenotypic abnormalities. If one of the original parents (the captured female or the male that mated with her) was heterozygous for a visible mutant, then 50% of the progeny of the cross would be heterozygous. Furthermore, 25% of the F_1 matings should be between two heterozygotes, and given Mendelian segregation, 0.0625 of the F_2 progeny should exhibit the visible mutant. For example, after examining 736 isofemale lines in *D. mulleri*, Spencer (1957) observed a considerable variety of mutants of the eye, bristle, body, and so on, even though he estimated from experiments with other species that he could identify only approximately one-sixth of the visible recessive mutants. In total, Spencer found 263 visible mutants and subsequently estimated that there were approximately 0.48 mutants/genome. This figure of approximately half a mutant per genome represents one mutant per individual and is the same order of magnitude as estimates in other *Drosophila* species (Lewontin, 1974).

This approach has been recently applied to wild-caught individuals of two fish species, bluefin killifish, and zebrafish (McCune *et al.*, 2002). In 43 F_1 crosses in the killifish, they found 39 recessive lethals (31 different ones) that could be divided into different categories based on the timing of their effect and influence on morphology. Similarly, in 86 F_1 crosses in the zebrafish, they found 59 recessive lethals (26 different ones), again with a variety of effects. Figure 1.16 shows the normal and mutant variants from

THE FAR SIDE® By GARY LARSON

"Hey! What's this *Drosophila melanogaster* doing in my soup?"

a single cross where an individual in Figure 1.16D is homozygous for two different recessive lethals. Overall, McCune *et al.* (2002) estimated that individual wild-caught killifish and zebrafish were heterozygous for 1.9 and 1.4 lethal alleles, respectively, similar to the estimates in *Drosophila* (Powell, 1997; see below). Some highly fecund species, such as oysters (Launey and Hedgecock, 2001) and pines (Remington and O'Malley, 2000), appear to have much higher high levels of lethals.

In *D. melanogaster, D. pseudoobscura,* and several other species of *Drosophila,* special techniques allow identification of lethal genes. Using these procedures, one can make a complete chromosome homozygous and then test for the presence of a lethal. The approach was first devised by H.J. Muller to study mutations and requires that the stock used have two different dominant visible markers, M_1 and M_2, on homologous chromosomes (these are usually lethal when homozygous) and an inversion(s) that "suppresses" crossing over on the chromosome under investigation. In addition, a useful aspect of *Drosophila* is that the males have virtually no crossing over, a necessary condition if wild chromosomes are going to be kept intact.

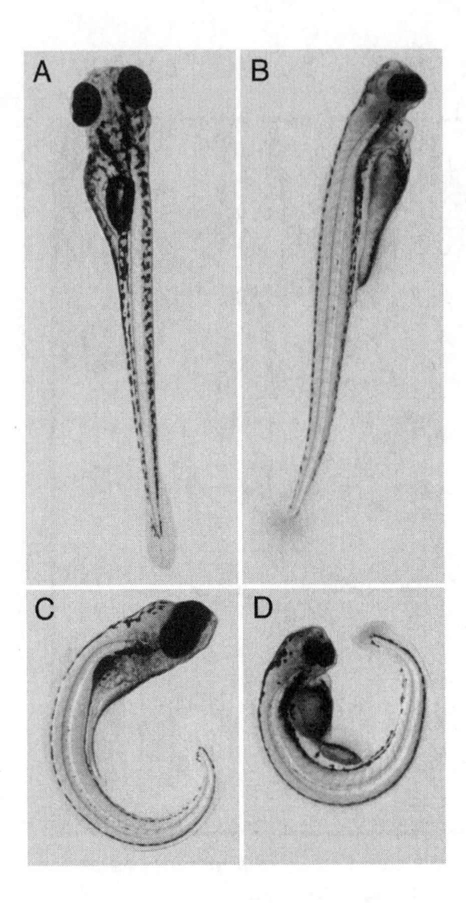

Figure 1.16. Normal (A) and recessive lethal phenotypes from a single F_1 sibship in the zebrafish. Fish (B) has the *pseudopunchout* phenotype with small eyes, knobby head, and no swim bladder. Fish (C) has the *spirograph* phenotype with curved spine, and fish (D) has both mutant phenotypes. (Reprinted Figure 2B with permission from McCune, A.R., R.C. Fuller, A.A. Aquilina, et al. 2002. A low genomic number of recessive lethals in natural populations of bluefin killifish and zebrafish. *Science*. 296:2398–2401. ©2004 AAAS.)

The technique is outlined in general in Figure 1.17. The parental cross is between a single wild-type male and females from the stock that has the dominant mutants and the inversion(s). The wild chromosomes of the male are labeled $+_1$ and $+_2$,and these are passed intact to the F_1 because there is no crossing over in males. Again, in the F_1, a single male, heterozygous for a wild chromosome and a stock chromosome, is backcrossed to stock females. In the F_2 progeny, the wild chromosome, $+_1$, is now in both the males and females. Crossing over between the wild chromosome and the mutant chromosome is avoided in the female because of the inversion in the M_1 balancer chromosome. The cross between heterozygous F_2 individuals gives three types of zygotes, $+_1/+_1$, $+_1/M_1$, and M_1/M_1, which are in proportions 0.25, 0.5, and 0.25, respectively (Table 1.7). Because M_1/M_1 is lethal, the proportion of wild-type adults, assuming that there is no lethal on the wild-type chromosome, is 0.333. If there is a lethal on $+_1$, then no $+_1/+_1$ individuals survive, and the proportion of wild-type progeny is 0.0. In this way, individual chromosomes can be categorized as either carrying or not carrying a lethal allele.

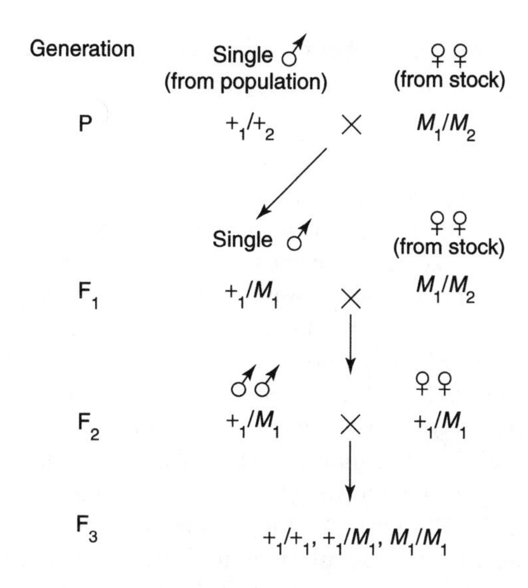

Figure 1.17. The general breeding technique used to make chromosomes homozygous in *Drosophila*. M_1 and M_2 indicate balancer chromosomes that carry different dominant marker genes and an inversion. The symbols $+_1$ and $+_2$ indicate two different wildtype chromosomes from the sampled population. In this example, chromosome $+_1$ is made homozygous.

TABLE 1.7 The zygotic proportions and adult phenotypes and proportions when a lethal is present or absent on the homozygous chromosome.

	$+_1/+_1$	$+_1/M_1$	M_1/M_1
Zygotes			
Proportions	0.25	0.5	0.25
Adults (with no lethal)			
Phenotype	wildtype	mutant	lethal
Proportions	0.333	0.667	0.0
Adults (with lethal)			
Phenotype	lethal	mutant	lethal
Proportions	0.0	1.0	0.0

It generally cannot be ascertained whether there is more than one lethal on a particular chromosome. However, if it is assumed that lethals are distributed independently at different loci, then the frequency of chromosomes with different numbers of lethal genes will follow a Poisson distribution (p. 23). If we let the frequency of chromosomes with lethals be X, then the proportion with no lethals is 1 - X. Assuming a Poisson distribution of lethals, the proportion of chromosomes with no lethals is equal to the first term of the Poisson distribution, or

$$1 - X = e^{-x}$$

where the x is the average number of lethals per chromosome. This equation can be solved for x so that

$$x = -ln(1 - X)$$

gives an estimate of the average number of lethals per chromosome.

Table 1.8 gives the results of a number of studies of the percent of lethal and semilethal second chromosomes and the expected number of lethals per chromosome in some populations of *D. melanogaster* (modified from Dobzhansky, 1970). There is considerable variation from site to site, with the maximum in Florida, where 61.3% of the chromosomes had at least one lethal or semilethal, and a minimum in a Korean survey, where only 11.2% of the chromosomes were lethal. Because the second chromosome is approximately 40% of the map length of *D. melanogaster,* the expected number of lethals per genome ranges from a minimum of 0.30 (Korea) to a maximum of 2.37 (Florida). Although this is a considerable amount of variation, the mean for the populations in Table 1.8 is approximately 1.0. This value is approximately twice that for visibles in *Drosophila*, as given above, but generally both are quite low.

TABLE 1.8 The percent of lethals or semilethals and the expected number of lethals per second chromosome in a number of populations of *D. melanogaster* (after Dobzhansky, 1970).

Population	*Number of chromosomes*	*Percent lethals or semilethals*	*x, Expected number of lethals per chromosome*
U.S.A.			
Florida	468	61.3	0.95
Massachusetts	2352	35.8	0.44
Ohio	177	49.7	0.69
Wisconsin	231	34.2	0.42
U.S.S.R.			
Caucasus	2971	15.6	0.17
Crimea	1630	24.8	0.29
Uman	2700	24.3	0.28
Korea	611	11.2	0.12
Japan	2773	21.5	0.24
Egypt	301	29.4	0.35
Italy	215	34.1	0.42
Israel	1222	34.7	0.43

Both visible mutants and lethals appear to be relatively infrequent and are probably not the major determinants of the considerable variation observed in fitnesses of different individuals. There must be, therefore, variants that individually affect fitness or the phenotype in more subtle ways that are the basis of evolutionary change. Such variants, termed *fitness modifiers*, can be studied in *Drosophila* in a way similar to that used to investigate lethals.

Remember that the proportion of wild-type individuals using the technique in Figure 1.17 was 0.0 when a lethal was present and 0.333 when no lethal was present. Although the proportion of wild-type individuals does clump around these values, other proportions are also observed, mostly between 0.0 and 0.333 but sometimes above 0.333. Such values indicate the presence of alleles on the chromosome that either lower the preadult viability below, or raise it above, that of the mutant $+_1/M_1$.

However, the dominant markers themselves may affect viability, thus invalidating a standard based on a comparison with $+_1/M_1$ individuals. What is needed is an evaluation of the viability of the $+_1/M_1$ relative to an average individual in the population. Such an individual would be heterozygous for wild-type chromosomes—for example, $+_i/+_j$. To obtain such individuals, the F_2 cross of Figure 1.17 can be modified to that shown in Figure 1.18. The proportion of wild-type individuals, heterozygotes in this case, is compared with the mutants, $+_i/M_1$ and $+_j/M_1$. The viability of the wild-type homozygote individuals is standardized by dividing by the average proportion of heterozygotes surviving. Therefore, a relative viability of 1.0 indicates that a particular chromosome, when homozygous, had the same viability as the average heterozygote combination.

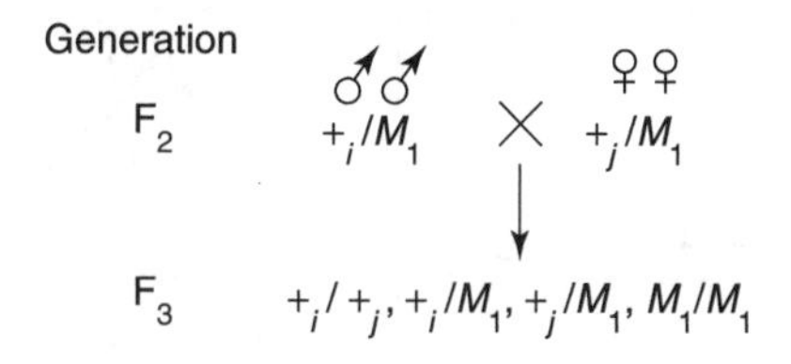

Figure 1.18. The modification of the cross given in Figure 1.17 for the F_2 generation so that wildtype chromosomal heterozygotes, $+_i/+_j$, are produced in the F_3 generation. The symbols $+_i$ and $+_j$ indicate two different wildtype chromosomes.

Standardized viabilities for a number of homozygotes can be plotted and are generally bimodal, as shown for the solid lines in Figure 1.19 for the surveys of second and third chromosomes by Mukai and Nagano (1983). In this study, 37% and 55% of the second and third chromosomes, respectively, were lethal, and most of the rest had relative viabilities in a broad distribution between 0.50 and 1.0. On the other hand, the heterozygote viabilities for both chromosomes were unimodal with a much narrower distribution. Such studies certainly indicate that there is considerable genetic variation modifying viability.

Recently, Kile *et al.* (2003) used the same general approach in mice to discover mutations from the general mutagen ENU. Instead of examining a whole chromosome, they used an inversion and a dominant marker on part of chromosome 11 to uncover recessive mutations in approximately 2% (about 700 genes) of the mouse genome. In 735 treated chromosomes examined, they found 88 mutations in this region, 55 of which were lethal. These mutations affected the skin, nervous system, head, fertility, and other traits.

One mutation that caused anemia was identified to be a single nucleotide change at a known candidate gene (see Figure 1.20). Such an approach could be expanded to document variants of smaller effects and to screen for variation in wild mouse chromosomes, as in *Drosophila.*

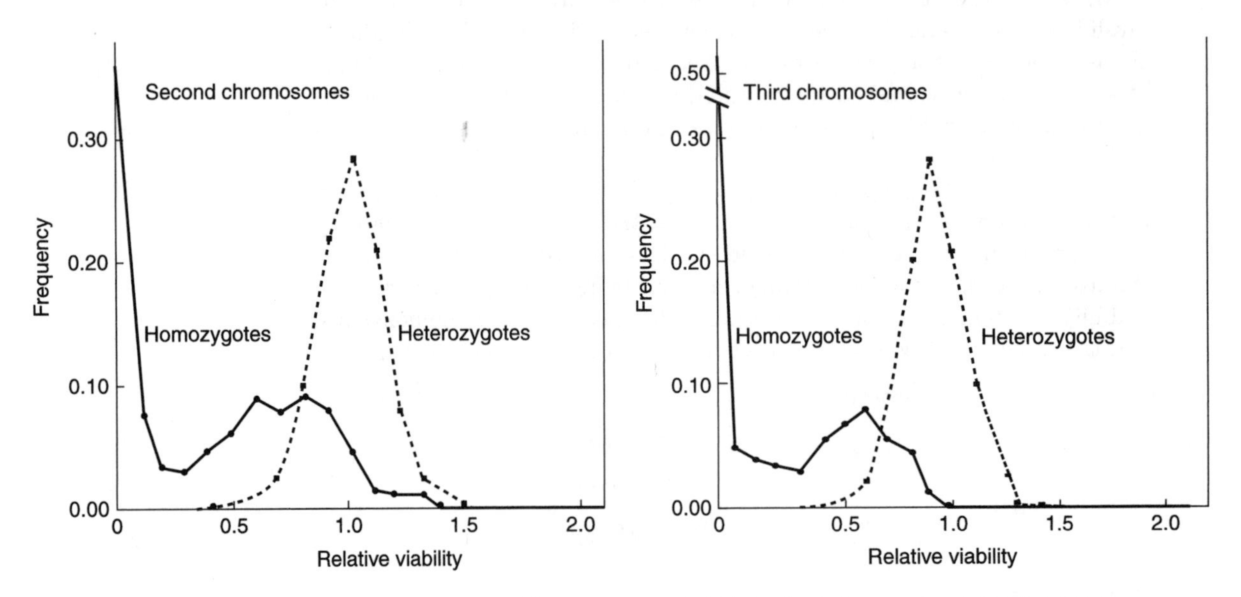

Figure 1.19. The distribution of homozygote and heterozygote viabilities for 475 second and 450 third chromosomes from Florida (Mukai and Nagano, 1983).

Figure 1.20. A litter of mice made homozygous for part of a chromosome in which the dark left two mice are normal and the pale right two mice are suffering from anemia. (Courtesy of *Nature*, 425:2003, by Kile, B.T., K.E. Hentges, A.T. Clark, et al. Reprinted with permission of Nature Publishing Group.)

e. Polygenic or Quantitative Traits

Many of the traits that appear to be of evolutionary importance, such as size, fitness components, and rate of growth, generally appear to be determined by many genes and not just a single locus. As a result, there is not the direct correspondence between particular phenotypes and genotypes that is apparent for the other genetic variants that we have discussed. These traits are either referred to as polygenic because many genes affect them or as quantitative because they are usually evaluated on a numerical scale. Detailed treatment of such traits is called **quantitative genetics**, and the books by Falconer and Mackay (1996) and Lynch and Walsh (1998) provide excellent coverage of the approaches and issues. Here we introduce quantitative traits to establish an appreciation for their role in evolution.

Several lines of evidence indicate the importance of genetic factors in polygenic or quantitative traits. First, when artificial selection is practiced over generations, the mean value of virtually any quantitative trait can be altered. Such a response to directional selection indicates an underlying genetic determination of the trait. Second, related individuals are generally more similar in phenotype than unrelated individuals. One can use this fact to estimate the amount of genetic variation for the trait under consideration. The third indication that genetic factors are involved in quantitative traits is the actual location on specific chromosomes of the genes that affect a particular trait. With the advent of detailed genetic maps in a number of organisms, it has become possible to locate genes affecting many different quantitative traits (p. 651).

Lewontin (1974) reviewed the type of traits that have been selected successfully in *Drosophila*. These include size, bristle number, developmental rate, fecundity, behavioral traits, mutant expression, environmental sensitivity, and resistance to insecticides. He concluded that "there appears to be no character—morphogenetic, behavioral, physiological, or cytological—that cannot be selected in *Drosophila*." Furthermore, it seems that virtually any population has enough genetic variants in it for response to selection for any trait. Of course, some traits respond much more quickly than others do, a difference that generally reflects the amount of genetic variation for that trait and the magnitude of effects at different genes. Traits that are closely related to fitness seem to be less likely to respond quickly to selection pressure than those not closely related to fitness. When individual genes have large effects, then the response to selection may be quite rapid.

Many of the early artificial selection experiments gave evidence of change in phenotypic values over 10 to 40 generations, often reaching a maximum or plateau before the termination of the experiment. There is evidence now that selection response may continue for many generations if there is an adequate initial sample of genetic variation, a reasonable population size during the selection process, and an effective selection scheme.

It has also become obvious that the input of new variation from mutation is significant in long term selection response.

One of the longest directional selection experiments is that for conducted for oil and protein percentage in maize at the University of Illinois. High and low directional selection have been carried out for 100 generations (Dudley and Lambert, 2004), and large changes in oil and protein percentage have been made (data for the high-oil and low-oil lines are given in Figure 1.21). The percentage of oil has increased over fourfold in the high line and continues to increase. The percentage of oil has been reduced to below 1% (obviously close to the lower selection boundary of 0%) in the low line, and the level is so low that it is difficult to measure it accurately. In generation 48, reverse selection lines were begun, and both the low selection from the high line and the high selection from the low line were successful in changing the mean values. These data, along with other estimates of genetic variability, indicate that selection progress took place in all lines over many generations, particularly in the high selection lines for oil and protein. This is particularly remarkable because the total gain from selection is far beyond the expectation from the distribution of oil and protein percentages in the original population. Although the major part of the response appears to be from favorable alleles with initially low frequency, it is possible that the input of advantageous variation from mutations may have made a substantial contribution to the extended response.

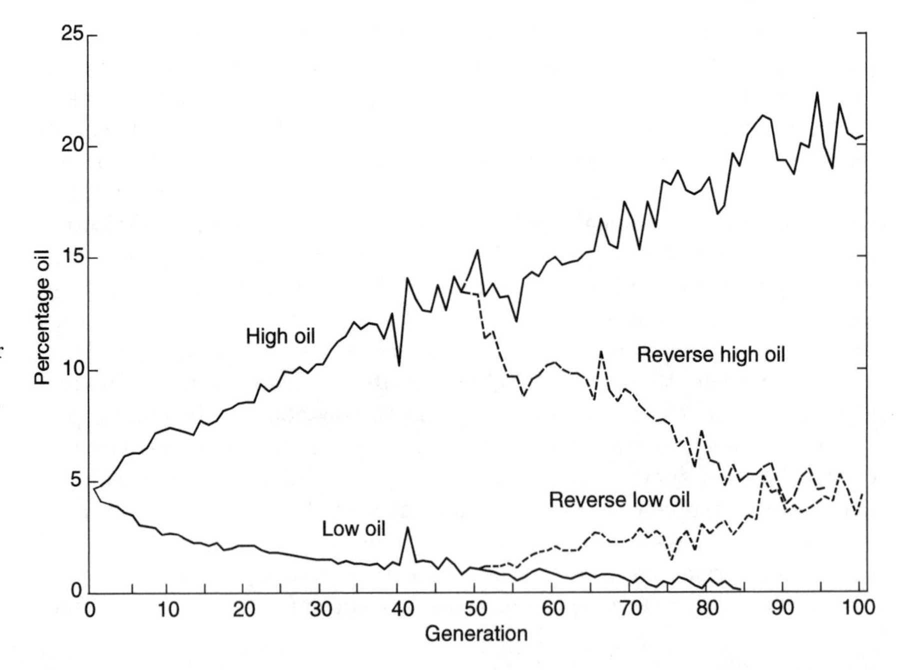

Figure 1.21. The percentage of oil in maize populations selected for 100 generations for either high oil content or low oil content (from Dudley and Lambert, 2004). Also given are selection responses in reverse selection lines for both the high and the low selection lines (broken lines).

For virtually any phenotypic character, the average pair of related individuals is more similar to each other than is a pair of unrelated individuals. Of course, some of this phenotypic similarity may be the result of a correlated environmental similarity for relatives, but much of the resemblance is the result of a common genetic heritage. It is often difficult to estimate the relative importance of environmental and genetic factors in determining the variation for quantitative traits for any particular situation, but by statistically examining phenotypic values for various types of relatives, one can obtain an indication of genetic determination. The most widely used measure to determine the importance of genetic factors is **heritability**, a quantity defined as ***the proportion of phenotypic variance that is genetically determined***. Heritability ranges from 0.0, when all of the phenotypic variance is caused environmentally, to 1.0, when all of it is the result of genetic factors. However, what is more interesting is the proportion of phenotypic variance that is the result of additive genetic variance—that is, the portion of the genetic variance that determines in large part the potential and rate of response to selection.

Heritability is a measure that is specific to a particular trait and organism and is also specific to a certain environment and population. These conditions are obvious restrictions because different genes are probably important for different traits and similar traits may be affected by different genes in different organisms. The amount of environmental variation is a function of the environment, and the amount of genetic variance is a function of the population. Heritability has been estimated in many organisms for a number of different traits, and the range of heritability values is quite large, from 0.1 for viability in chickens to 0.7 for plant height in corn. It appears that there is genetic variation for virtually any quantitative trait, although some of the lower estimates may not be significantly different from zero heritability values. Mousseau and Roff (1987) summarized data from 1120 estimates in wild outbred animal populations and gave the cumulative frequency distribution of four types of traits, those related to life history, behavior, physiology, and morphology (Figure 1.22). These results show that the heritability for traits related to life history have the lowest values and those related to morphology are the largest. The findings are generally consistent with the prediction from Fisher's fundamental theorem of natural selection (p. 133) that traits that have been under stronger selection would have lower heritability and genetic variation.

With the advent of new molecular techniques to locate polymorphic markers throughout the genome in most organisms, it is now possible to map the location of genes that affect specific quantitative traits, called **quantitative trait loci (QTLs)**. For example, Leips and Mackay (2002) located five QTLs that influence the life span in *D. melanogaster* (indicated by high likelihood values in Figure 1.23). Three of these loci appeared to be

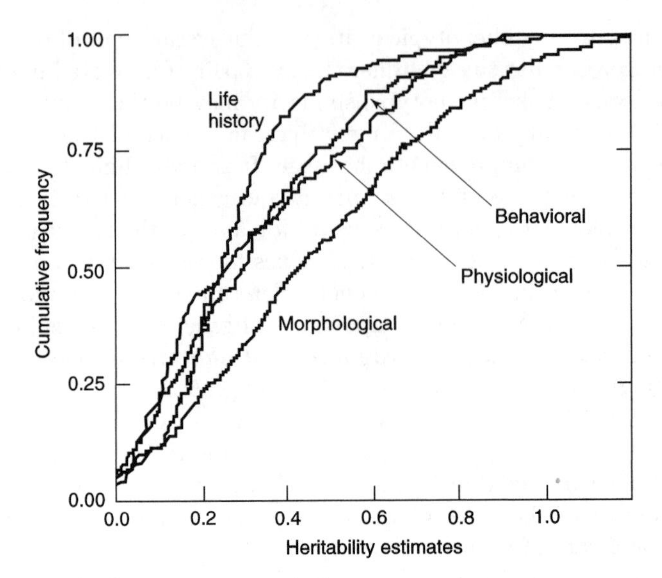

Figure 1.22. The cumulative frequency distributions of estimated heritability for four trait categories (from Mosseau and Roff, 1987).

male-specific (one on chromosome 2 and two on chromosome 3), one female-specific (at the right end of chromosome 3), and one affected life span in both sexes (also on chromosome 3). Using higher resolution techniques, the QTL on chromosome 2 was subsequently mapped to a specific cytogenetic interval. This region included a **candidate gene**—a gene recognized as likely to be related to the trait by either function or structure. The candidate gene in this case, the *Dopa decarboxylase* (*Ddc*) locus, catalyzes the final step in the synthesis of the neurotransmitters, dopamine, and serotonin. Further analysis demonstrated that 15% of the genetic contribution to variance in life span from chromosome 2 was the result of variation of three common SNPs at *Ddc* (De Luca *et al.*, 2003).

Crosses between related species or domesticated crops or animals and their wild progenitors and subsequent QTL analysis have identified genes that determine major morphological differences between species—for example, between maize and its wild progenitor teosinte (Doebley *et al.*, 1995). Recently, Van Laere *et al.* (2003) identified a variant that was responsible for approximately 15% to 20% of the phenotypic variation in muscle mass and back-fat thickness in crosses between European wild boars and domestic pigs. In earlier studies, a paternally expressed QTL had been mapped to the *IGF2* (insulin-like growth factor 2) region on pig chromosome 2. Van Laere *et al.* (2003) demonstrated that the QTL differences were caused by a single nucleotide substitution in intron 3 of the *IGF2* gene. It appears that this mutation in a noncoding region increases *IGF2* mRNA expression in muscle and the major differences in muscle mass and back fat. In other words, changes in regulatory regions may result in significant changes in phenotypic variation.

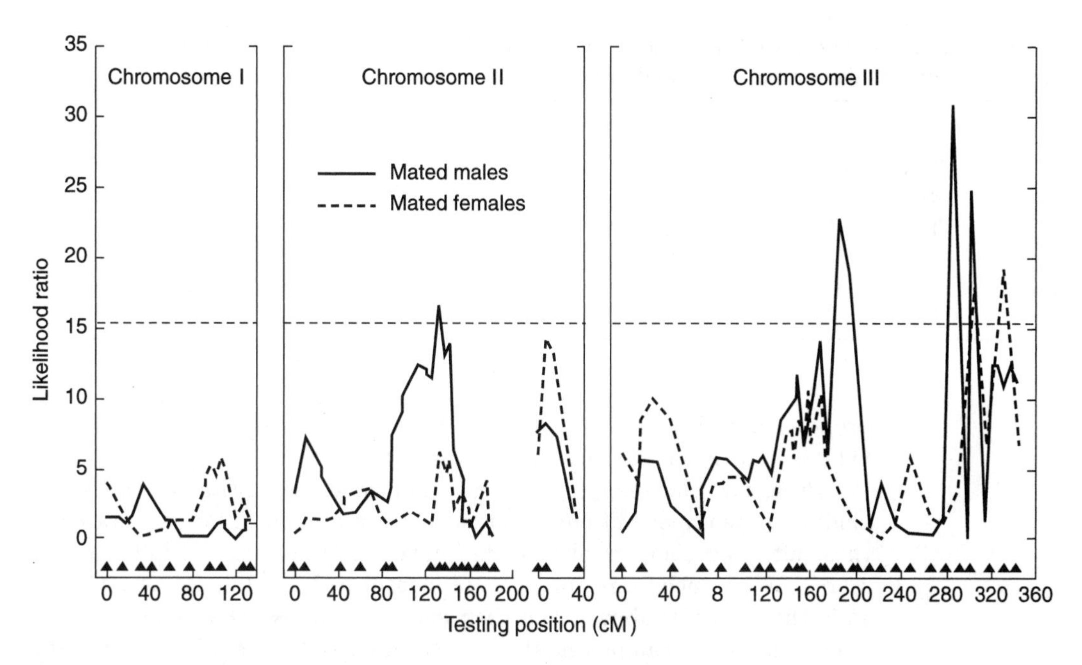

Figure 1.23. Genetic mapping results indicating the chromosomal positions of QTLs that contribute to the mean difference in life span for male and female *D. melanogaster*. The threshold significance, slightly above 15, is given by the broken horizontal line and is exceeded by five QTLs. The triangles on the horizontal axis indicate informative markers. (Courtesy of *Nature*, 34: 2003, by De Luca, M., N.V. Roshina, G.L. Gieger-Thornberry, *et al.* Reprinted with permission of Nature Publishing Group.)

In organisms without detailed genetic maps, it is much more difficult to map QTLs, and the location of candidate genes is generally not known; however, it is often still possible to map QTLs generally for complex trains in nonmodel organisms. For example, the Mexican axolotl, *Ambystoma mexicanum*, has a derived form of development, paedomorphosis, in which the animals do not develop into the normal terresterial form and adults have an entirely aquatic life cycle. On the other hand, the related salamander, *A. tigrinum*, has normal morphology and developes into the terrestrial morph. From crosses by Tompkins (1978), it appeared that the difference between these species is primarily due to a single recessive gene in the derived axolotl form. In an effort to locate the gene involved, Voss and Shaffer (1997) backcrossed F_1 metamorphs between the species to axolotls and examined segregation for 262 markers called amplified fragment length polymorphisms (AFLPs) and the state of metamorphosis. Table 1.9 summarizes their data for the three linked markers that showed a significant association with these morphologies. From statistical analysis (highest proportion of heterozygotes associated with terrestrial morphology and lowest

TABLE 1.9 Segregation for heterozygotes and homozygotes for three AFLP markers and for life cycle differences in *Ambystoma* salamanders (Voss and Shaffer, 1997)

Marker	*Form*	*Heterozygotes*	*Homozygotes*	*Proportion (heterozygotes)*
AFLP 11.7	Terrestrial	39	10	0.81
	Aquatic	9	22	0.31
AFLP 32.17	Terrrestrial	42	7	0.89
	Aquatic	5	26	0.21
AFLP 34.5	Terrestrial	41	8	0.85
	Aquatic	7	24	0.25

proportion with aquatic), it appears that the most likely position for the QTL is in a region near the center marker, *AFLP32.17*.

Presently, the mapping of genes influencing quantitative traits, including those important having important evolutionary characteristics and those affecting complex diseases in humans, is the object of intensive research. When such individual loci are identified that influence a trait, such as in the examples above, then their effects on fitness, their level of dominance, and so on, can potentially be measured. With this information, then many of the techniques of population genetics can be used to understand and predict the impact of evolutionary factors on the amount and pattern of genetic variation at these genes.

PROBLEMS

1. If a codon is CGC, what proportion of nucleotide changes in the third position result in a new amino acid? If the codon is AGG, what proportion of changes in the first position result in a new amino acid? If a codon is UUG, what proportion of changes in the third position result in a new amino acid?
2. Using the model for population growth in the text, what is R, given that the population numbers at times t and $t + 1$ are 60 and 90, respectively? What is the predicted population number at time $t + 3$?
3. Assume that the total body length in a group of jays was measured and there were 1, 12, 16, 27, 9, and 4 jays that were 25, 27, 28, 29, 30, and 32 cm in length, respectively. What is the mean length? What are the variance, standard deviation, coefficient of variation, and standard error of length calculated from these data?
4. Assume that the numbers of leopard frogs observed in a creek over five years were 105, 17, 266, 183, and 145. What are the arithmetic, geometric, and harmonic means for this data set?
5. Assume that the probability of a child being a boy is 0.52. What is the probability that a family would have three consecutive girls? What is the probability that a family would have two boys and two girls?

6. If the probability that a bomb would strike in a given area is 0.001 and 500 bombs are dropped, what is the probability that an area would not be struck? What is the probability that two bombs would strike the same area?
7. Using the sample matrix given on page 26, assume that a vector has the elements $y_1 = 0$, $y_2 = 1.0$, and $y_3 = 0$. What is the new vector if the matrix is multiplied by this vector? What is the new vector if the matrix is multiplied by the vector found in the first question?
8. Compare the balanced and classical views of the organization of the genome. Explain whether these views are relevant in understanding the amount of DNA variation.
9. Let us assume that you would like to document the amount of genetic variation in species and that you could examine both allozyme and DNA variation. Would you prefer to measure one type of genetic variation over the other? Why?
10. How would you experimentally investigate the basis of differentiation over space observed for the two allozymes in Figure 1.6? Assume that you have access to samples from past generations and that you can manipulate the populations in any way you wish and follow them in future generations.
11. From the *Adh* data given in Table 1.4, determine the proportion of sites, excluding insertions and deletions, that are variable within the S alleles (the top six). Determine the proportion of sites that are variable within the F alleles.
12. Think of an example of a visible genetic polymorphism other than those discussed in the text. Does the frequency of the polymorphism vary in space or time? How would you determine the cause of any spatial or temporal change, or of spatial or temporal stability, in the frequency of the polymorphism?
13. Judging on the basis of surveys of molecular variation, some organisms appear to have more genetic variation than others. Do you think this observation is significant? If so, explain why and design an experiment to support or falsify your hypothesis.
14. What proportion of wildtype offspring would you expect when there was a lethal on a chromosome that is made homozygous using the breeding technique in Figure 1.17? Trace this lethal on a diagram from the original wildtype male to the test progeny.
15. Construct a table listing the advantages and potential problems in using allozymes, DNA variants, visible polymorphisms, lethals, and polygenic traits to document the amount of genetic variation in a population.

2

Measures of Genetic Variation

> The frequencies with which the different genotypes occur define the gene ratio characteristic of the population, so that it is often convenient to consider a natural population not so much as an aggregate of living individuals as an aggregate of gene ratios. Such a change of viewpoint is similar to that familiar in the theory of gases, where the specification of the population of velocities is often more useful than that of a population of particles.
>
> Ronald Fisher (1953)

> The models employed in population genetics are mathematical. The model is always unsatisfactory in some respects. Inevitably, it is unable to reflect all the complexities of the true situation. On the other hand, it is usually true that the more closely the model is made to conform to nature the more unmanageable it becomes from the mathematical standpoint. If it is as complex as the true situation, it is not a model. We have to choose some sort of compromise between a model that is so crude as to be unrealistic or misleading and one that is incomprehensible or too complex to handle.
>
> James Crow and Motoo Kimura (1970)

To understand the influence of selection, inbreeding, genetic drift, gene flow, and mutation in population genetics, one must first be able to describe and quantify the amount of genetic variation in a population and the pattern of genetic variation among populations. In Chapter 1, we discussed examples of the large amount of genetic variation present in natural populations and detected in the form of different alleles, DNA sequence variation, or phenotypes. This chapter first introduces the relationship between allele and genotype frequencies and some techniques for estimating allele frequencies from observed genotype frequencies. Then we discuss how the variation found in DNA and amino acid sequences can be measured and how to compare allele frequencies and DNA sequences in different populations. In recent years, a number of software packages have become available to estimate many of the important parameters in population genetics and related topics (for a list, addresses, and some general comments about what they calculate and what tests they carry out, see Table 2.1). Also, the **Evolution Directory** (or **EvolDir**) (http://life.biology.mcmaster.ca/~brian/evoldir.html) is a source of comments about different estimation procedures in their archives and a place where one can ask other evolutionary geneticists for advice about different software packages and the best approach to use in a given situation.

However, before beginning this discussion, we must generally define the evolutionary or genetic connotation of the term population. As a simple ideal, a **population** is ***a group of interbreeding individuals that exist***

TABLE 2.1 A list of some of the more commonly used software packages used in population genetics and related topics, their websites (subject to change), and the analyses and tests that they carry out. All of these packages are free except PAUP.

Software Package	*Address*	*Analyses and Tests*
Arlequin	http://lgb.unige.ch/arlequin/	AMOVA, MSN, nucleotide diversity, mismatch distribution, linkage disequilibrium, Hardy–Weinberg, neutrality tests, Mantel test, pairwise population genetic distances
DnaSP (DNA sequence polymorphisms)	http://www.ub.es/dnasp/	Sequence variation within and between populations, linkage disequilibrium, recombination, gene flow, gene conversion, tests of neutrality
GDA (genetic data analysis)	http://lewis.eeb.uconn.edu/lewishome/software.html	Linkage disequilibrium, Hardy–Weinberg, genetic distances, and hierarchical F-statistics
Genepop	http://wbiomed.curtin.edu.au/genepop/	Linkage disequilibrium, Hardy–Weinberg, gene flow, F-statistics
LAMARC (likelihood analysis with metropolis algorithm using random coalescence)	http://evolution.genetics.washington.edu/lamarc.html	Maximum likelihood estimates of effective population size, gene flow, growth parameters, and recombination
MEGA (molecular evolutionary genetics analysis)	http://evolgen.biol.metro-u.ac.jp/MEGA	Pairwise distance matrices, nonsynonymous/synonymous ratios, neutrality tests, phylogenetic analyses
PAUP (phylogenetic analysis using parsimony)	http://paup.csit.fsu.edu/	Phylogenetic analysis using maximum likelihood, parsimony, and distance methods
PHYLIP (phylogeny inference package)	http://evolution.genetics.washington.edu/phylip.html	Phylogenetic analysis using maximum likelihood, parsimony, and distance methods
PowerMarker	http://152.14.14.48/	Linkage disequilibrium, Hardy–Weinberg, F-statistics, coancestry matrices, phylogenetic analysis, designed especially for SSR/SNP data analysis

together in time and space. Often it is assumed that a population is geographically well defined, although this may not always be true. In any case, this ideal and generally theoretical definition is very useful in developing the concepts of population genetics. Later, in Chapter 6, we discuss the concept of effective population size, which provides a more explicit definition of a population in evolutionary terms.

Let us go a step further in specifying this ideal population. Assume that there are two sexes and that the population consists of sexually mature individuals. If all possible matings between females and males are equal in probability (independent of distance between mates, type of genotype, age of individuals, and so on), then the population is called a **random-mating population**. Stated another way, if individuals are specified as to genotype or phenotype, then a random-mating population is a ***group in which the probability of a mating between individuals of particular genotypes or phenotypes is equal to the product of their individual frequencies in the population*** (sometimes a random-mating population is referred to as a panmictic population). This relationship is developed more fully below.

Many of the theoretical developments in population genetics assume a large, random-mating population that forms the **gene pool** from which the female and male gametes are drawn. In some real-life situations, such as dense populations of insects or outcrossing plants, this ideal may be nearly correct, but in many natural situations, it is not closely approximated. For example, there may not be random mating, as in self-fertilizing plants, or there may be small or isolated populations, as in rare or endangered species. In these cases, modifications of this theoretical ideal must be made. However, to develop the basic concepts of population genetics, we initially consider the ideal of a large, random-mating population and in later discussions consider the effects of nonrandom mating, small population size, or subdivided populations.

I. THE HARDY–WEINBERG PRINCIPLE

Although earlier studies concerned changes in traits of populations as the result of natural or artificial selection, population genetics as a field of study developed in the first decade of the 20th century. Shortly after the rediscovery of Gregor Mendel's work, an English mathematician named G.H. Hardy (1908) and a German physician named W. Weinberg (1908) formulated what has become known as the Hardy–Weinberg principle. This relationship is of basic importance to population genetics because it enables us to describe the genetic content in diploid populations in terms of allele, not genotype, frequencies. With the recent documentation of loci with many alleles or genes or genetic regions with many haplotypes, this principle has become very important.

The **Hardy–Weinberg principle** states that ***after one generation of random mating, single-locus genotype frequencies can be represented by a binomial (with two alleles) or multinomial (with multiple alleles) function of the allele frequencies***. This principle allows great simplification of the description of a population's genetic content by reducing the number of parameters that must be considered. Specifically, the Hardy–Weinberg principle allows us to consider only the frequencies of the n alleles at a locus to describe a population, instead of the frequencies of the $n(n+1)/2$ different diploid genotypes formed by the n alleles. For even a locus with five alleles, there is a threefold difference between 5 allele frequencies and 15 genotype frequencies. This simplification is particularly useful for a locus with two alleles because it allows us to follow changes in the frequency of one allele (the other allele frequency is its complement) instead of the frequency of two genotypes (the three genotype frequencies sum to unity, and thus, there are two independent ones).

Furthermore, in the absence of factors that change allele frequency (selection, genetic drift, gene flow, and mutation) and in the continued presence of random mating, the Hardy–Weinberg genotype proportions will not change over time. In the ensuing discussion of the Hardy–Weinberg principle, we consider the genotype and allele frequencies at a single locus. It is assumed that genes at other loci, the genetic background, do not influence genotype or allele frequencies. Problems associated with this assumption are considered in Chapter 10 when multilocus models are discussed.

a. Two Alleles

To illustrate the Hardy–Weinberg principle, assume that a population is segregating for two alleles, $A_1$and A_2, at the autosomal locus A in frequencies of p and q $(p + q = 1)$, respectively, and assume that female and male gametes unite at random to form zygotes. Random union of gametes, disassociated from parental genotypes, does not appear to be common in nature but may be visualized, for example, in marine organisms with pelagic gametes. Gametes uniting at random means that the frequency of a progeny zygote is equal to the product of the frequencies of its constituent gametes. This can be shown graphically, as in Figure 2.1, where the area of a unit square is divided into proportions representing the frequencies of different progeny genotypes. The unit length across the top of the square is divided into proportions p and q, representing the frequencies of the female gametes, A_1 and A_2, respectively. Likewise, the side of the square is divided into proportions representing male gametes. Therefore, the areas within the square represent the proportions of different progeny zygotes; for example, the upper left square has an area of p^2, the frequency of genotypes A_1A_1. Overall, the three progeny zygotes (A_1A_1, A_1A_2, A_2A_2)are formed in proportions (p^2, $2pq$, q^2).

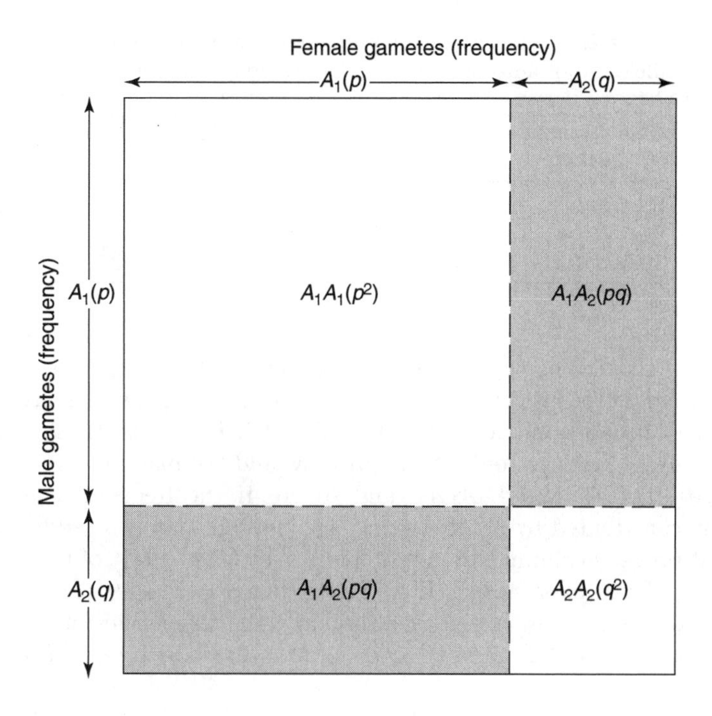

Figure 2.1. The Hardy–Weinberg proportions as generated from the random union of gametes using a unit square where the heterozygote area is shaded.

In most organisms, random union of gametes is quite unlikely because it is the parental genotypes that pair and then produce gametes that unite. Therefore, let us consider the situation in which reproductive individuals randomly pair to mate. If we consider a diploid organism, there are three possible genotypes in the population—A_1A_1, A_1A_2, and A_2A_2—that are present in frequencies P, H, and Q, respectively ($P+H+Q=1$). Because all of the alleles in the two homozygotes A_1A_1 and A_2A_2 are A_1 and A_2, respectively, and half the alleles in the heterozygote are A_1 and half are A_2, the allele frequencies in terms of the genotype frequencies are then

$$\begin{aligned} p &= P + \tfrac{1}{2}H \\ q &= Q + \tfrac{1}{2}H \end{aligned} \tag{2.1}$$

Let us assume that there is random mating in the population, which yields the nine possible combinations of matings between the male and female genotypes as given in Table 2.2. Only six of these need to be distinguished here because reciprocal matings, such as $A_1A_1 \times A_1A_2$ and $A_1A_2 \times A_1A_1$, where the first genotype is the female, have the same genetic consequences. The frequency of a particular mating type, given random mating, is equal to the product of the frequencies of the genotypes that constitute that mating type; for instance, the frequency of $A_1A_1 \times A_1A_1$ is P^2.

TABLE 2.2 The frequency of different mating types for two alleles at an autosomal locus when there is random mating.

Male genotypes (frequencies)	*Female genotypes (frequencies)*		
	$A_1A_1(P)$	$A_1A_2(H)$	$A_2A_2(Q)$
$A_1A_1(P)$	P^2	PH	PQ
$A_1A_2(H)$	PH	H^2	HQ
$A_2A_2(Q)$	PQ	HQ	Q^2

These six mating types, their frequencies, and the expected frequencies of their offspring genotypes, assuming that the gametes segregate in Mendelian proportions, are given in Table 2.3. For example, the mating $A_1A_1 \times A_1A_1$ produces only A_1A_1 progeny, and the mating $A_1A_1 \times A_1A_2$ produces $\frac{1}{2}A_1A_1$ and $\frac{1}{2}A_1A_2$, and so on. If the frequencies of A_1A_1 progeny contributed by all of the mating types are summed (adding down the first progeny column), then it is found that $(P + \frac{1}{2}H)^2$ of the progeny are A_1A_1. From expression 2.1, we know that $p = P + \frac{1}{2}H$, and, therefore, p^2 of the progeny are A_1A_1. Similarly, the frequencies of A_1A_2 and A_2A_2 progeny are $2(P + \frac{1}{2}H)(Q + \frac{1}{2}H) = 2pq$ and $(Q + \frac{1}{2}H)^2 = q^2$, respectively.

The crucial point here is that given any set of initial genotype frequencies (P, H, Q) after one generation of random mating, the genotype frequencies are in the proportions (p^2, $2pq$, q^2). For example, given the initial genotype frequencies (0.2, 0.4, 0.4), where from expression 2.1, $p = 0.4$ and $q = 0.6$, after one generation the genotype frequencies become $[(0.4)^2, 2(0.4)(0.6), (0.6)^2] = (0.16, 0.48, 0.36)$. Furthermore, the genotype frequencies will stay in these exact proportions generation after generation, given continued random mating. We have demonstrated that random union of gametes and random mating are equivalent in that they result in Hardy–Weinberg genotype proportions after one generation.

TABLE 2.3 Demonstration of the Hardy–Weinberg principle assuming random mating in the parents and Mendelian segregation to produce the progeny.

Mating type	*Frequency*	*Progeny*		
		A_1A_1	A_1A_2	A_2A_2
$A_1A_1 \times A_1A_1$	P^2	P^2	—	—
$A_1A_1 \times A_1A_2$	$2PH$	PH	PH	—
$A_1A_1 \times A_2A_2$	$2PQ$	—	$2PQ$	—
$A_1A_2 \times A_1A_2$	H^2	$\frac{1}{4}H^2$	$\frac{1}{2}H^2$	$\frac{1}{4}H^2$
$A_1A_2 \times A_2A_2$	$2HQ$	—	HQ	HQ
$A_2A_2 \times A_2A_2$	Q^2	—	—	Q^2
Total	1	$(P + \frac{1}{2}H)^2 = p^2$	$2(P + \frac{1}{2}H)(Q + \frac{1}{2}H) = 2pq$	$(Q + \frac{1}{2}H)^2 = q^2$

As discussed in later chapters, genotype proportions may deviate from Hardy–Weinberg expectations for several different reasons (see also Table 2.15). The most significant evolutionary factors are selection, inbreeding, and gene flow, and thus, it is often said that Hardy–Weinberg proportions are expected only in situations in which there is no selection, random mating, and no gene flow. In fact, small amounts of any one of these factors only slightly modify Hardy–Weinberg proportions (p. 96, Chapter 5), or they may in fact be present and the population may still be in Hardy–Weinberg proportions (p. 150). It also needs to be kept in mind, particularly for variants at microsatellite and other loci identified by molecular techniques, that deviations from Hardy–Weinberg proportions might result from technical problems.

Often the Hardy–Weinberg proportions are referred to as **Hardy–Weinberg equilibrium** (or **HWE**) frequencies; however, note that this is a special type of equilibrium (sometimes called a neutral equilibrium). If the genotype proportions are perturbed without changing allele frequencies, then they return to the same Hardy–Weinberg proportions after one generation of random mating. On the other hand, if the allele frequencies are changed, then the genotypes will be present in different Hardy–Weinberg proportions determined by the new allele frequencies.

Of course, different populations may have different allele frequencies. To illustrate the effect of different allele frequencies on genotype frequencies, Figure 2.2 gives the frequencies of the three different genotypes, assuming Hardy–Weinberg proportions, for the total range of allele frequencies. Note that the heterozygote is the most common genotype for

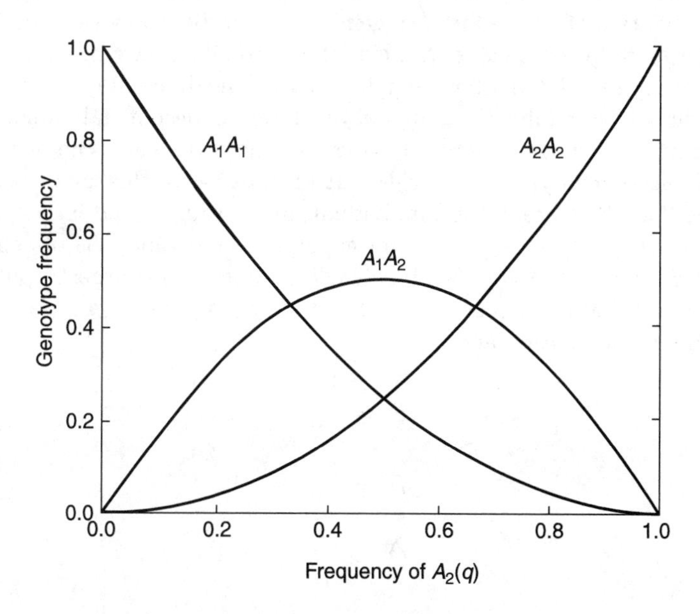

Figure 2.2. The relationship between the allele frequency and the three genotype frequencies for a population in Hardy–Weinberg proportions.

intermediate allele frequencies, whereas one of the homozygotes is the most common when the allele frequency is not intermediate. In fact, only 1/3 of the time, when q is between 1/3 and 2/3, is the heterozygote the most common genotype. When q is between 0 and 1/3, A_1A_1 is the most common, and when q is between 2/3 and 1, A_2A_2 is the most common. Because $p = 1 - q$, the horizontal axis also holds for the frequency of allele A_1 although the scale is reversed and goes from 1 to 0.

Figure 2.2 also illustrates that the **maximum heterozygote frequency** occurs when $q = 0.5$ ($p = 0.5$). This fact can be shown directly by setting the derivative of the Hardy–Weinberg heterozygosity, $2pq = 2q(1-q)$, equal to zero and solving for q or

$$\frac{d[(2q(1-q)]}{dq} = 2 - 4q = 0$$

with the result that $q = 0.5$.

In the previous discussion, we have assumed that generations are nonoverlapping—that the parents die after producing offspring and the offspring then become the next parental generation. This model describes the dynamics of growth of organisms such as annual plants or univoltine insects (insects with one generation per year). For many long-lived organisms, such as humans or trees, or even short-lived organisms, such as *Drosophila*, in which both birth and death may occur at nearly any time during much of the year, generations are generally not discrete but overlapping. When there are **overlapping generations**, Hardy–Weinberg proportions of the genotypes are still approached rapidly, the deviation from Hardy–Weinberg proportions only approximately 0.4 and 0.1 of the initial deviation after one and two generations, respectively (Moran, 1962).

The genotype or allele frequencies at a particular locus for a population are generally calculated after the individuals of the different genotypes have been counted as adults. In many ways, it would best if individuals were categorized as zygotes before selection, genetic drift, and gene flow could have changed genotype frequencies, but in most cases, this is not possible. Assume that there are N total individuals in the sample that is categorized with N_{11} of genotype A_1A_1, N_{12} of genotype A_1A_2, and N_{22} of genotype A_2A_2 ($N_{11} + N_{12} + N_{22} = N$). If this is a sample from a larger population, then the estimated frequencies of the three genotypes A_1A_1, A_1A_2, and A_2A_2 in the population are

$$\begin{aligned} \hat{P} &= \frac{N_{11}}{N} \\ \hat{H} &= \frac{N_{12}}{N} \\ \hat{Q} &= \frac{N_{22}}{N} \end{aligned} \tag{2.2}$$

respectively (estimates will be indicated by "hats" over symbols). $\hat{H}$ is an estimate of the observed heterozygosity, H_O, in the population (below we add a small sample size correction to this estimate). The estimated allele frequencies can also be calculated from the sample as

$$\hat{p} = \frac{N_{11} + {}^1\!/_2 N_{12}}{N}$$
$$\hat{q} = \frac{{}^1\!/_2 N_{12} + N_{22}}{N} \tag{2.3}$$

(see Example 2.1 for an illustration using the MN blood group locus in humans). This approach is the best way to estimate allele frequencies and is often referred to as the gene-counting technique (the result are also the maximum-likelihood estimates of the allele frequencies, see p. 79).

Example 2.1. The Hardy–Weinberg principle can be illustrated by the MN red blood cell locus in humans. For this gene, there are three genotypes, MM, MN, and NN, which, before DNA sequencing, were each distinguishable from the others by antigen–antibody reactions (see Blumenfeld and Huang, 1997; Daniels, 2002, for molecular details about the MN locus). Table 2.4 gives the number of MN genotypes for a sample of 1000 English blood donors. From these data, we find that the estimated frequency of allele M is $\hat{p} = 0.542$ and the estimated frequency of N is $\hat{q} = 1 - \hat{p} = 0.458$. The expected Hardy–Weinberg frequencies and the expected genotype numbers (the product of the expected frequencies and N) can then be calculated. The expected frequency of heterozygotes is $2\hat{p}\hat{q} = 0.496$, a value very close to the observed heterozygosity, $489/1000 = 0.489$. The observed numbers and the expected numbers are extremely close to each other (see Table 2.4). Later, a technique to test statistically whether a population is in Hardy–Weinberg proportions will be discussed.

TABLE 2.4 The observed numbers and the numbers expected from Hardy–Weinberg proportions for the MN blood group locus in an English sample of 1000 blood donors (from Cleghorn, 1960).

Phenotype	*Genotype*	*Observed number*	*Expected number*
M	MM	298	$\hat{p}^2 N = 294.3$
MN	MN	489	$2\hat{p}\hat{q}N = 496.4$
N	NN	213	$\hat{q}^2 N = 209.3$
Total		1000	1000

With all of this discussion about p and q, one might think that population genetics is the source of the cautionary phrase "mind your p's and q's." A more likely origin, however, is from British barkeepers who traditionally told their customers who had accumulated large bills on credit to "mind your pints and quarts"—p's and q's for short. It is also possible that the phrase originated in the difficulty typesetters had learning to tell p's and q's apart. The letter on each piece of handset type looked backward (in order to print forward), and, thus, it was very easy to confuse lowercase p's and q's.

b. Multiple Alleles

The Hardy–Weinberg principle can be extended to more than two alleles. This extension is particularly important because many loci, particularly those identified by molecular techniques such as microsatellites or genes known by their haplotype sequences, have been found to be segregating for more than two alleles or haplotypes. As a direct extension of the two-allele model, the frequency of an allele when there are multiple alleles is equal to the sum of the frequency of its homozygote plus half the frequency of each heterozygote that contains the allele. In order to use a general notation that is satisfactory for any number of alleles, let us assume that P_{ii} is equal to the frequency of genotype A_iA_i and P_{ij} is equal to the frequency of genotype A_iA_j.The frequency of allele A_i, p_i, is then

$$\begin{aligned} p_i &= \tfrac{1}{2}(P_{i1} + P_{i2} + \cdots + 2P_{ii} + \cdots + P_{ij} + \cdots + P_{in}) \\ &= P_{ii} + \tfrac{1}{2}\sum_{j=1}^{n} P_{ij} \end{aligned} \tag{2.4}$$

where $j \neq i$ and n is the number of alleles at the A locus.

Extending the Hardy–Weinberg principle to multiple alleles can be demonstrated by assuming either random union of gametes or random mating. Using either of the approaches, the expected genotype frequencies become p_i^2 for homozygotes and $2p_ip_j$ for heterozygotes. When there are Hardy–Weinberg proportions, the frequency of all heterozygotes combined (the Hardy–Weinberg or expected heterozygosity) can be written as one minus the expected proportion of homozygotes:

$$H_E = 1 - \sum_{i=1}^{n} p_i^2 \tag{2.5a}$$

As for two alleles, the maximum expected heterozygosity occurs when all alleles have equal frequencies. In this case, $p_i = 1/n$, and

$$H_E = 1 - \sum_{i=1}^{n} \left(\frac{1}{n}\right)^2$$
$$= 1 - n\left(\frac{1}{n}\right)^2$$
$$= \frac{n-1}{n} \qquad (2.5b)$$

For example, when there are three alleles in equal frequency ($p_1 = p_2 = p_3 = 1/3$), the heterozygosity is $2/3$. All other frequencies of three alleles give lower heterozygosity values. For example, $p_1 = 0.5$, $p_2 = 0.3$, and $p_3 = 0.2$ gives $H = 0.62$—not much less, but still smaller. If a number of alleles are in substantial frequency, nearly every individual in the population may be a heterozygote. For example, if there are 10 alleles in equal frequency (all $p_i = 0.1$) and Hardy–Weinberg proportions, then 0.9 of the genotypes would be heterozygotes, and only 0.1 would be homozygotes.

As for the two-allele case, the allele and genotype frequencies can be estimated from a population. Assume that a sample of N individuals is categorized and that N_{ii} and N_{ij} are the numbers of A_iA_i and A_iA_j genotypes observed, respectively. Therefore, the estimated allele frequency of allele A_i becomes

$$\hat{p}_i = \frac{N_{ii} + 1/2 \sum_{j=1}^{n} N_{ij}}{N} \qquad (2.5c)$$

where $j \neq i$ (see Example 2.2 for an illustration using a triallelic allozyme locus in *Daphnia*). Genotype frequencies can be estimated as in expression 2.2 and the observed heterozygosity estimated as

$$\hat{H}_O = \frac{\sum N_{ij}}{N} \qquad (2.5d)$$

where $i \neq j$.

Example 2.2. Hebert (1974) examined the frequencies of electrophoretic types at several polymorphic enzyme loci in *Daphnia magna* (water flea) from ponds near Cambridge, England. Some sample data from the malate dehydrogenase (*Mdh*) locus are given in Table 2.5, where S, M, and F are

TABLE 2.5 The observed numbers and the numbers expected from Hardy–Weinberg proportions for the *Mdh* locus in a *Daphnia* population (from Hebert, 1974).

Phenotype	*Genotype*	*Observed number*	*Expected number*
S	*SS*	3	2.4
SM	*SM*	8	10.9
SF	*SF*	19	17.4
M	*MM*	15	12.3
MF	*MF*	37	39.5
F	*FF*	32	31.5
Total		114	114

the alleles designated by Hebert. From the numbers of different electrophoretic types and expression 2.5*c* (assuming that alleles S, M, and F have allele frequencies p_1, p_2, and p_3, respectively), the estimates of allele frequencies are $\hat{p}_1 = 0.145$, $\hat{p}_2 = 0.329$, and $\hat{p}_3 = 0.526$. From these estimates, one can calculate the expected frequencies and numbers for the six different genotypes using the Hardy–Weinberg principle. Again, the observed numbers are very close to those expected assuming Hardy–Weinberg proportions. The expected heterozygosity using expression 2.5*a* is 0.594, whereas the observed heterozygosity is slightly lower at 0.561.

II. DIFFERENT FREQUENCIES BETWEEN THE SEXES

One of the implicit assumptions of the Hardy–Weinberg principle is that the parents of both sexes have the same genotype (allele) frequencies. However, in natural populations, female and male parents may have different genotype frequencies because of differential selection, gene flow, or sampling. In an experimental population, the female parents may be from one strain or population and male parents from another. Such a situation is particularly important when we are considering X-linked genes or haplodiploid organisms (such as Hymenoptera) where males have one copy of the gene and females have two copies.

a. Autosomal Genes

First, let us consider the effect of different sexual allele frequencies on an autosomal gene. Let p_f and p_m represent the frequencies of A_1 in females and males, respectively, and q_f and q_m the analogous frequencies of A_2. As a result, the frequencies of each type of zygote formed from a random union of gametes are

$$\begin{aligned} P &= p_f p_m \\ H &= p_f q_m + p_m q_f \\ Q &= q_f q_m \end{aligned} \tag{2.6a}$$

An example of these proportions is illustrated as a unit square in Figure 2.3. Notice that unlike Figure 2.1, the areas for homozygotes are not squares.

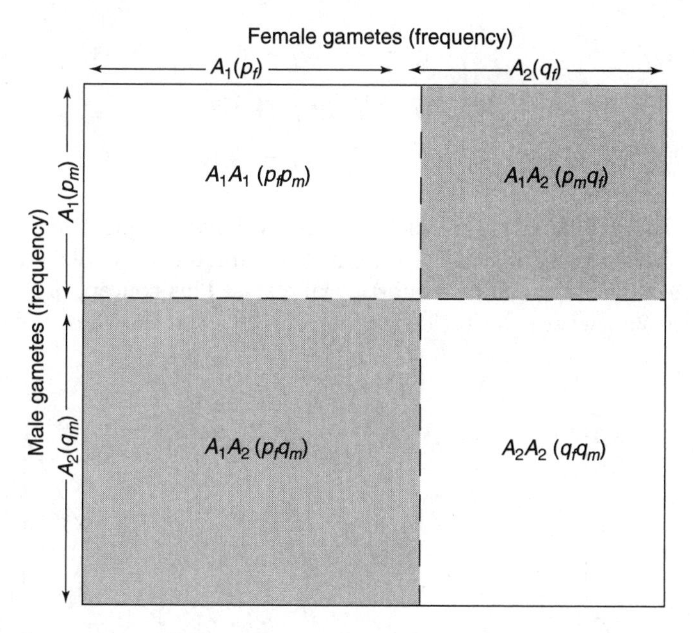

Figure 2.3. The proportions of zygotes formed from random union of gametes where female and male gametes differ in allele frequency and the heterozygotes area is shaded.

When the allele frequencies differ between the sexes, there will be an excess of heterozygotes and a deficiency of homozygotes over the Hardy–Weinberg proportions. This can be shown if we let the overall (mean) frequencies of the two alleles be $\bar{p} = 1/2(p_f + p_m)$ and $\bar{q} = 1/2(q_f + q_m)$ because half the genes are in females and half are in males. The amount of deviation from the Hardy–Weinberg proportion for genotype A_1A_1 is then (Robertson, 1965; Purser, 1966)

$$\begin{aligned} P - \bar{p}^2 &= p_f p_m - [1/2(p_f + p_m)]^2 \\ &= p_f p_m - 1/4(p_f^2 + 2p_f p_m + p_m^2) \\ &= -1/4(p_f^2 - 2p_f p_m + p_m^2) \\ &= -1/4(p_f - p_m)^2 \end{aligned}$$

There is always a deficiency of homozygotes because the square of the mean of two numbers is always greater than the product of the two numbers. Therefore, for this and the other genotypes, the deviations from Hardy–Weinberg proportions are

$$\begin{aligned} P - \bar{p}^2 &= -1/4(p_f - p_m)^2 \\ H - 2\overline{pq} &= 1/2(p_f - p_m)^2 \\ Q - \bar{q}^2 &= -1/4(p_f - p_m)^2 \end{aligned} \tag{2.6b}$$

For small differences in allele frequency between the two sexes, the increase in heterozygosity is minor, but if the differences in allele frequency are large, then there can be a substantial effect. This concept is illustrated in Figure 2.4, where the ratio of the observed over the Hardy–Weinberg

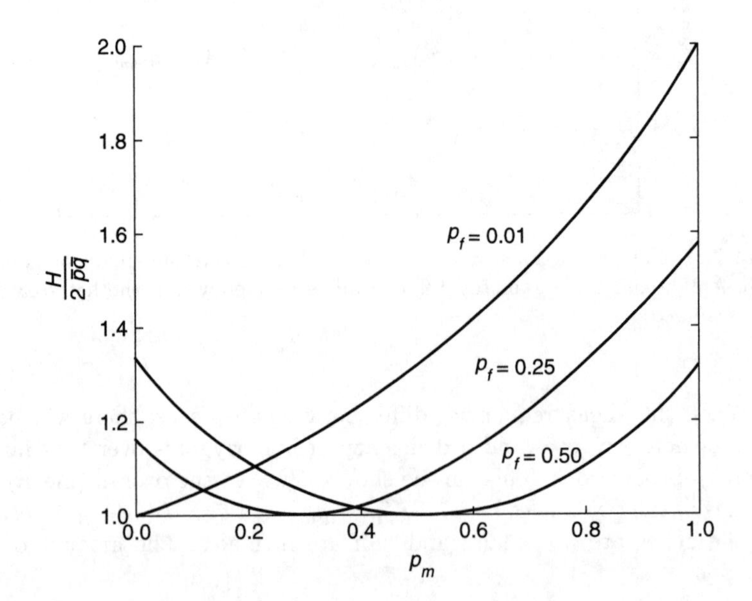

Figure 2.4. The ratio of the observed and Hardy–Weinberg heterozygous frequencies for different combinations of female and male allele frequencies.

heterozygosity, $H/(2\overline{pq})$, is plotted for different combinations of p_f and p_m. When there are multiple alleles and differences in allele frequencies between the sexes, the frequencies of all heterozygotes combined are elevated, and those of all homozygotes combined are decreased.

The deviation in genotype frequencies from what the Hardy–Weinberg principle leads us to expect lasts only one generation because in the progeny the genotype frequencies in both sexes are the same. In other words, the frequency of A_1 in progeny (p') of both sexes is

$$\begin{aligned} p' &= P + \tfrac{1}{2}H \\ &= p_m p_f + \tfrac{1}{2}(p_m q_f + p_f q_m) \\ &= \tfrac{1}{2}(p_f + p_m) \end{aligned}$$

Therefore, because $p' = p'_m = p'_f$, and assuming random union of gametes or random mating, the genotype frequencies in the next generation are in Hardy–Weinberg proportions.

b. X-Linked Genes or Genes in Haplo-Diploid Organisms

Genes on the X chromosome in mammals and other organisms have the same pattern of inheritance as do all genes in haplo-diploid organisms, such as Hymenoptera. In both situations, if there is an initial difference in allele frequencies in the two sexes, then equal allele frequencies in two sexes are achieved only over several generations, and the deviation in heterozygosity from Hardy–Weinberg expectations in the diploid females also disappears gradually with time. Here and later we assume that females are the homogametic (XX) sex and males the heterogametic (XY), although for some organisms, such as birds and Lepidoptera, the homogametic and heterogametic sexes are reversed. For alleles on the X chromosome, males are haploid (hemizygous), and heterozygosity occurs in only the females. In haplo-diploid organisms such as Hymenoptera, which have haploid males and diploid females, the following analysis holds for all loci.

For a gene polymorphic with two alleles, there are six mating types and five types of progeny, as listed in Table 2.6. The males receive all of their gametes from their mothers, and females receive half of their gametes from their mothers and half from their fathers. P_f, H_f, and Q_f refer to the frequencies of the diploid genotypes A_1A_1, A_1A_2, A_2A_2 in the females, and P_m and Q_m refer to the frequencies of the haploid genotypes A_1 and A_2 in the males; thus, the frequencies of A_2 in the two sexes are

$$\begin{aligned} q_f &= Q_f + \tfrac{1}{2}H_f \\ q_m &= Q_m \end{aligned} \qquad (2.7a)$$

TABLE 2.6 The genotype frequencies after one generation of random mating for an X-linked locus or a gene in a haplo-diploid organism.

Mating type			*Female offspring*			*Male offspring*	
♀	♂	*Frequency*	A_1A_1	A_1A_2	A_2A_2	A_1	A_2
A_1A_1	$\times A_1$	P_fP_m	P_fP_m	—	—	P_fP_m	—
A_1A_1	$\times A_2$	P_fQ_m	—	P_fQ_m	—	P_fQ_m	—
A_1A_2	$\times A_1$	H_fP_m	$\frac{1}{2}H_fP_m$	$\frac{1}{2}H_fP_m$	—	$\frac{1}{2}H_fP_m$	$\frac{1}{2}H_fP_m$
A_1A_2	$\times A_2$	H_fQ_m	—	$\frac{1}{2}H_fQ_m$	$\frac{1}{2}H_fQ_m$	$\frac{1}{2}H_fQ_m$	$\frac{1}{2}H_fQ_m$
A_2A_2	$\times A_1$	Q_fP_m	—	Q_fP_m	—	—	Q_fP_m
A_2A_2	$\times A_2$	Q_fQ_m	—	—	Q_fQ_m	—	Q_fQ_m
Total		1	p_fp_m	$p_fq_m + p_mq_f$	q_fq_m	p_f	q_f

If we assume equal numbers of the two sexes or equal contributions of the sexes to the next generation, then two-thirds of the alleles are in the females and one-third are in males; the mean allele frequency is

$$\bar{q} = {}^2\!/_3 q_f + {}^1\!/_3 q_m \tag{2.7b}$$

The genotype frequencies in the female progeny are calculated in the same manner as for an autosomal locus and are given at the bottom of Table 2.6. Therefore, the allele frequency of A_2 in the female progeny is

$$\begin{aligned} q'_f &= Q'_f + {}^1\!/_2 H'_f \\ &= q_fq_m + {}^1\!/_2(p_fq_m + p_mq_f) \\ &= {}^1\!/_2 q_f(p_m + q_m) + {}^1\!/_2 q_m(p_f + q_f) \\ &= {}^1\!/_2(q_f + q_m) \end{aligned} \tag{2.8a}$$

However, the allele frequency in the male progeny is equal to that of their female parents because they obtain all of their genes from their mothers. As a result, the frequency of A_2 in the next generation of males is simply

$$q'_m = q_f \tag{2.8b}$$

Let us examine how the allele frequencies in the two sexes change over time. First, expression 2.7b can be rearranged as $q_m = 3\bar{q} - 2q_f$. When we substitute this into expression 2.8a, the allele frequency in the female progeny becomes $q'_f = {}^1\!/_2(3\bar{q} - q_f)$. After rearranging this expression, we can show that the deviation of the female allele frequency from the mean allele frequency in the next generation is one-half that of the previous generation or

$$q'_f - \bar{q} = -{}^1\!/_2(q_f - \bar{q}) \tag{2.9a}$$

The deviation in the males is also halved because $q'_m = q_f$ but with a one-generation lag because of the inheritance pattern. As a result, the allele frequencies in the two sexes oscillate about the mean allele frequency, and the differences in allele frequency between the sexes are halved each generation, as shown in the example given in Figure 2.5. The oscillations dampen fairly quickly over time, and q_f and q_m both approach $\bar{q}$.

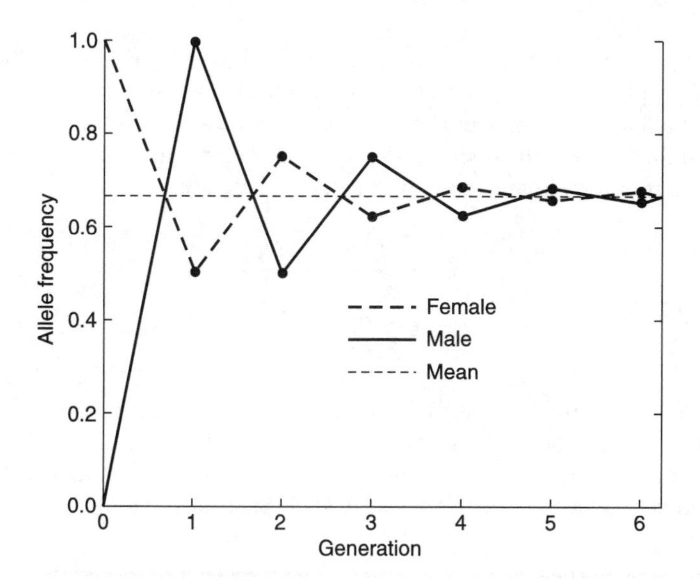

Figure 2.5. The female (q_f), male (q_m), and mean ($\bar{q}$) allele frequencies for an X-linked gene or a gene in a haplo-diploid organism, where $q_f = 1.0$ and $q_m = 0.0$ in the first generation.

If we let the deviation of the female allele frequency from the mean frequency in the zero and first generation be d_0 and d_1, respectively, then expression 2.9a becomes the recursive relationship

$$d_1 = -{}^1\!/_2 d_0$$

If the same process is repeated for more generations, the general relationship between the deviation in the tth generation and the initial generation is

$$d_t = (-{}^1\!/_2)^t d_0 \tag{2.9b}$$

Obviously, the deviation declines quickly because $(-{}^1\!/_2)^t$ closely approaches zero after only a few generations.

As the female and male frequencies approach the mean frequency, the female genotype frequencies also approach Hardy–Weinberg proportions. However, when the allele frequencies in the two sexes differ, there will also be a deviation from Hardy–Weinberg proportions in the females.

Following the same approach as for autosomal loci, the deviations from Hardy–Weinberg proportions are

$$\begin{aligned} P_f - \bar{p}^2 &= -1/9(4p_f - p_m)(p_f - p_m) \\ H_f - 2\bar{p}\bar{q} &= 1/9[(4p_f - p_m)(p_f - p_m) + (4q_f - q_m)(q_f - q_m)] \quad (2.9c) \\ Q_f - \bar{q}^2 &= -1/9(4q_f - q_m)(q_f - q_m) \end{aligned}$$

These deviations may be substantial because allele frequencies for X-linked or haplo-diploid genes can easily differ in the two sexes.

The allele frequencies in the two sexes can be estimated from the genotype numbers in a manner similar to that for an autosomal locus. Therefore, the estimated female and male frequencies of A_2 are

$$\begin{aligned} \hat{q}_f &= \frac{1/2 N_{12} + N_{22}}{N_f} \\ \hat{q}_m &= \frac{N_2}{N_m} \end{aligned} \qquad (2.9d)$$

where N_{12}, N_{22}, and N_f refer to the number of A_1A_2, the number of A_2A_2, and the total number of females, respectively, and N_2 and N_m refer to the number of A_2 males and the total number of males (see Example 2.3 about the X-linked gene that causes tortoise-shell pattern in cats).

Example 2.3. An interesting example of an X-linked gene is the *yellow* coat-color locus in cats, where the female heterozygote has a tortoise-shell or calico phenotype. Both the homozygote female and the hemizygote male for the yellow allele have a yellow phenotype. Some data for this locus collected on London cats sent to clinics for destruction are given in Table 2.7. From

TABLE 2.7 The observed numbers and the numbers expected from Hardy–Weinberg proportions in females for the *yellow* locus in a sample of London cats (data from Searle, 1949). + and y indicate the wildtype and mutant alleles, respectively.

			Number	
	Phenotype	*Genotype*	*Observed*	*Expected*
Females	Wildtype	$++$	277	273.2
	Tortoise shell	$+y$	54	61.4
	Yellow	yy	7	3.4
Males	Wildtype	$+$	311	—
	Yellow	y	42	—

these data, the estimated allele frequencies are $\hat{q}_f = 0.101$ and $\hat{q}_m = 0.119$, indicating that the male and female allele frequencies are very similar. However, using the estimated allele frequency in the females, we find that somewhat fewer heterozygotes were observed (0.160) than expected (0.182). This deficiency of heterozygotes may indicate that tortoise-shell cats were preferred and consequently were sent for destruction less frequently than cats with other phenotypes.

III. ESTIMATES OF ALLELE FREQUENCY

We have given estimates of the allele frequency for several cases in which we could identify the alleles in every individual in the sample. Even in these cases, the allele frequency estimates from the samples may not exactly reflect the allele frequency in the population if the sample is small. The accuracy of such an estimate depends on the sample size, and of course, the larger the sample, the better is the estimate of the frequency in the population. This assumes that the sample is drawn randomly from the population and is not biased in some manner. Thus, in the following discussion, the variance of different allele frequency estimates is given as well as the allele frequency estimate because the variance allows some statement to be made about the accuracy of the allele frequency estimates. The standard error of the mean for these estimates is the square root of the variance because the parameter being estimated, the allele frequency, is itself a mean. Given that the sample size is large enough (Weir, 1996), the 95% confidence interval around these estimates is approximately ± twice the standard error of the mean. The estimates and their variances given are the maximum-likelihood estimates of these parameters.

Potentially, there are a variety of ways to estimate allele frequencies (and other parameters), but the method of **maximum likelihood** (ML) (Fisher, 1958; Li, 1976; Felsenstein, 2001) exceeds all others in general applicability and widespread acceptance by statisticians. Assume that we want to estimate the allele frequency parameter, p, and are given some data. For a given value of p and the appropriate model, we can calculate the probability $\Pr(Data|p)$ (the vertical line stands for given) that the observed data would have occurred. The method of maximum likelihood is to ***vary p until we find the value that maximizes*** $\Pr(Data|p)$, ***the probability of the data, given p.***

Below we use as an example the multinomial probability (see p. 24) of obtaining the three genotypes, A_1A_1, A_1A_2, and A_2A_2, at a codominant, autosomal locus where the data are the numbers of the genotypes observed in a sample of size N—that is, N_{11}, N_{12}, and N_{22}. We want to compute $\Pr(Data|N_{11}, N_{12}, N_{22})$, assuming that the sample has been drawn from a population that has the genotype frequencies p^2, $2p(1-p)$, and $(1-p)^2$. Therefore, the **likelihood** (L) of the observed numbers is

$$L = \frac{N!}{N_{11}!N_{12}!N_{22}!}(p^2)^{N_{11}}[2p(1-p)]^{N_{12}}[(1-p)^2]^{N_{22}}$$

$$= \frac{N!}{N_{11}!N_{12}!N_{22}!}2^{N_{12}}p^{2N_{11}+N_{12}}(1-p)^{N_{12}+2N_{22}} \qquad (2.10a)$$

The value of p that maximizes this probability also maximizes the probability of its logarithm. Therefore, we can calculate the logarithm

$$\log(L) = (2N_{11} + N_{12})\log p + (N_{12} + 2N_{22})\log(1-p) + K$$

where K is a constant. We can find the value of p that maximizes this expression by taking its derivative, setting the derivative equal to zero, and solving for p:

$$\frac{d\log(L)}{dp} = \frac{2N_{11} + N_{12}}{p} - \frac{N_{12} + 2N_{22}}{1-p} = 0$$

so that

$$\hat{p} = \frac{N_{11} + {}^1\!/\!{}_2 N_{12}}{N} \qquad (2.10b)$$

as we gave in expression 2.3.

The negative reciprocal of the theoretical variance of a maximum likelihood estimate is equal to

$$-\frac{1}{V(\hat{p})} = \frac{d^2\log(L)}{dp^2} \qquad (2.10c)$$

or the second derivative of the log likelihood. If we take the second derivative, it becomes

$$\frac{d^2\log(L)}{dp^2} = -\frac{2N_{11} + N_{12}}{p^2} - \frac{N_{11} + 2N_{22}}{(1-p)^2}$$

Substituting p^2N for N_{11}, $2p(1 - p)N$ for N_{12}, and $(1 - p)^2N$ for N_{22}, then

$$\frac{d^2\log(L)}{dp^2} = -\frac{2N}{p} - \frac{2N}{(1-p)}$$

and

$$V(\hat{p}) = \frac{\hat{p}(1-\hat{p})}{2N} \qquad (2.10d)$$

which is the binomial variance.

The maximum likelihood method has several desirable properties. First, as the sample size increases, the estimate will approach the true value of p. Second, for a given sample size (provided that it is large), the variance of the estimate of p around the true value is less under the ML method than other approaches. Finally, although the ML estimate is not necessarily unbiased (i.e., the average estimate of p on repeated sampling may not be exactly p), the amount of bias does decline as the sample size increases.

In the allele frequency estimates given above, all alleles were assumed to be expressed or codominant in heterozygotes. There are many cases, as with metabolic disorders in humans and color polymorphisms in many organisms, in which the heterozygote is generally indistinguishable from one of the homozygotes. In these cases, it is usually assumed that the population is in Hardy–Weinberg proportions in order to estimate the allele frequencies.

The estimates of the allele frequency and their variances for two-allele systems are given in Table 2.8 and for some multiple-allele systems in Table 2.9. Only the derivation for codominance is given here, but a good discussion of the assumptions and the derivation of most of the estimates can be found in Li (1976). The methods of estimating allele frequency for autosomal loci with codominant gene action are given as (2) for the two-allele and n-allele cases in Tables 2.8 and 2.9, respectively. When there is codominance in females for X-linked genes or genes in haplo-diploid organisms, the estimates are given as (4) in Table 2.8. In addition, the estimates of allele frequency in a haploid organism are given as (1) in the two tables. These estimates are also appropriate for alleles at mtDNA, cpDNA, and Y chromosomes genes when N is the sample size of the appropriate sex.

TABLE 2.8 Estimates of allele frequency and their variances for two-allele genetic systems. For the estimate for organelle or Y-chromosome alleles, N is the sample size of the appropriate sex.

(1) Haploid, organelle, and Y chromosome	$\hat{q} = \frac{N_2}{N}$	$V(\hat{q}) = \frac{\hat{q}(1-\hat{q})}{N}$
(2) Codominance	$\hat{q} = \frac{\frac{1}{2}N_{12} + N_{22}}{N}$	$V(\hat{q}) = \frac{\hat{q}(1-\hat{q})}{2N}$
(3) Dominance	$\hat{q} = \left(\frac{N_{22}}{N}\right)^{1/2}$	$V(\hat{q}) = \frac{1-\hat{q}^2}{4N}$
(4) Codominance, X-linked or haplo-diploid	$\hat{q}_f = \frac{\frac{1}{2}N_{12} + N_{22}}{N_f}$	$V(\hat{q}_f) = \frac{\hat{q}_f(1-\hat{q}_f)}{2N_f}$
	$\hat{q}_m = \frac{N_2}{N_m}$	$V(\hat{q}_m) = \frac{\hat{q}_m(1-\hat{q}_m)}{N_m}$
	$\hat{\bar{q}} = \frac{\frac{1}{2}N_{12} + N_{22} + N_2}{2N_f + N_m}$	$V(\hat{\bar{q}}) = \frac{\hat{\bar{q}}(1-\hat{\bar{q}})}{2N_f + N_m}$

TABLE 2.9 Estimates of allele frequency and their variances for some multiple-allele systems.

System	Estimate	Variance
(1) Haploid, n alleles	$\hat{p}_i = \dfrac{N_i}{N}$	$V(\hat{p}_i) = \dfrac{\hat{p}_i(1-\hat{p}_i)}{N}$
(2) Codominance, n alleles	$\hat{p}_i = \dfrac{N_{ii} + \frac{1}{2}\sum_{j=1}^{n} N_{ij}}{N}$	$V(\hat{p}_i) = \dfrac{\hat{p}_i(1-\hat{p}_i)}{2N}$
(3) Dominant series, three alleles	$\hat{p}_1 = 1 - \left(\dfrac{N_{22}+N_{23}+N_{33}}{N}\right)^{1/2}$	$V(\hat{p}_1) = \dfrac{1-(\hat{p}_2+\hat{p}_3)^2}{4N}$
	$\hat{p}_2 = \left(\dfrac{N_{22}+N_{23}+N_{33}}{N}\right)^{1/2} - \left(\dfrac{N_{33}}{N}\right)^{1/2}$	$V(\hat{p}_2) = \dfrac{\hat{p}_2[2-\hat{p}_2(1-\hat{p}_1)]}{4N(1-\hat{p}_1)}$
	$\hat{p}_3 = \left(\dfrac{N_{33}}{N}\right)^{1/2}$	$V(\hat{p}_3) = \dfrac{1-\hat{p}_3^2}{4N}$
(4) Two alleles codominant, one recessive	$\hat{p}_1 = 1 - \left(\dfrac{N_{22}+N_{23}+N_{33}}{N}\right)^{1/2}$	$V(\hat{p}_1) = \dfrac{\hat{p}_1(1-\hat{p}_1)}{2N} + \dfrac{\hat{p}_1^2}{4N}$
	$\hat{p}_2 = 1 - \left(\dfrac{N_{11}+N_{13}+N_{33}}{N}\right)^{1/2}$	$V(\hat{p}_2) = \dfrac{\hat{p}_2(1-\hat{p}_2)}{2N} + \dfrac{\hat{p}_2^2}{4N}$
	$\hat{p}_3 = \left(\dfrac{N_{33}}{N}\right)^{1/2}$	$V(\hat{p}_3) = \dfrac{\hat{p}_3(1-\hat{p}_3)}{2N} + \dfrac{(1-\hat{p}_3)^2}{4N}$

The other genetic systems in Tables 2.8 and 2.9 depend on the population being close to Hardy–Weinberg proportions. For example, with two alleles and dominance, (3) in Table 2.8, it is assumed that the proportion of A_2A_2 genotypes in the population is

$$\frac{N_{22}}{N} = q^2$$

Therefore, an estimate of the frequency of A_2 is

$$\hat{q} = \left(\frac{N_{22}}{N}\right)^{1/2} \tag{2.11a}$$

(see Example 2.4 about color polymorphism in the Spirit or Kermode bear, a white form of the black bear).

Example 2.4. In the black bear, *Ursus americanus*, there is a striking white (not albino) color phase, called either the Kermode bear or the Spirit Bear, found in the rainforests along the north coast of British Columbia (Figure 2.6). The total number of Kermode bears is estimated to be 100 to 200, and they have been protected from hunting since 1925. Ritland *et al.* (2001) found that the Kermode form was the result of a recessive mutation caused by a single nucleotide replacement A to G at position 893 (A893G) in the

Figure 2.6. The white Kermode, or Spirit bear, next to a normally pigmented black bear. (©Charlie Russell/Peter Arnold, Inc.)

melanocortin 1 receptor gene (*mc1r*), which in turn results in a Tyr to Cys replacement at codon 298 (Y298C).

In a survey of 11 locations, three islands—Gribbell, Princess Royal, and Roderick—had the highest frequency of Kermode bears. If equation 2.11a is used to estimate the frequency of the white allele from the phenotype frequencies on these islands, then the overall frequency is 0.49 (Table 2.10a). However, Ritland *et al.* used their molecular findings to determine the genotypes of the 66 black bears (Table 2.10b). In fact, many fewer heterozygotes were observed than expected from Hardy–Weinberg proportions (24 vs. 38.3). When these genotype numbers are used in equation 2.3 to calculate allele frequencies, they are much lower (overall 0.38) than when the heterozygotes are not known. Also note that the standard errors of the allele frequency estimates are lower when the heterozygotes are known than when they are unknown, as expected from expression 2.11*b*.

Marshall and Ritland (2002) examined 10 microsatellite loci in bears from these islands and found a slight excess, not a deficit, in heterozygotes. As a result, it appears that the large deficiency of heterozygotes at the *mc1r* locus is the result of a factor intrinsic to the white color or the *mc1r* locus. Ritland *et al.* (2001) examined the potential impacts of positive-assortative mating, recent immigration of black homozygotes, and a fitness disadvantage to heterozygotes on a heterozygote deficit (see Table 2.15). They found that any one of these factors would have to be very large to cause the observed deficiency of heterozygotes; thus, either they appear to cause the deficit jointly, or some other unknown factor is the cause.

TABLE 2.10 (a) The observed numbers of black and white (Kermode) bears on three islands in British Columbia (Ritland *et al.*, 2001) and the calculated frequency of the recessive allele. (b) Using a molecular assay to determine heterozygosity and homozygosity of the black bears, the observed number of genotypes AA, AG, and GG (at nucleotide position 893) and the calculated frequency of the G allele.

(a)

		Phenotype		
Island	*Sample Size* (N)	*Black*	*White* (N_{22})	Allele frequency ($\hat{q}$) *(equation 2.11a)*
Gribbell	23	13	10	0.66 ± 0.08
Princess Royal	52	43	9	0.42 ± 0.06
Roderick	12	10	2	0.41 ± 0.13
Total	87	66	21	0.49 ± 0.05

(b)

		Genotype		
Island	$AA(N_{11})$	$AG(N_{12})$	$GG(N_{22})$	Allele frequency ($\hat{q}$) *(equation 2.3)*
Gribbell	7	6	10	0.56 ± 0.07
Princess Royal	26	17	9	0.33 ± 0.03
Roderick	9	1	2	0.21 ± 0.08
Total	42	24	21	0.38 ± 0.04

The variance of this allele frequency estimate is higher than when the heterozygote can be identified. To show the magnitude of this increase, the variance of the estimate can be rewritten as

$$\begin{aligned} V(\hat{q}) &= \frac{1-\hat{q}^2}{4N} \\ &= \frac{1-(1-\hat{p}^2-2\hat{p}\hat{q})}{4N} \\ &= \frac{\hat{q}(1-\hat{q})}{2N} + \frac{(1-\hat{q})^2}{4N} \end{aligned} \tag{2.11b}$$

The first term in expression 2.11*b* is equal to the variance for the two-allele codominance case and is the binomial variance. Therefore, the variance when the heterozygote is indistinguishable is increased by an amount equal to the second term. Obviously, this second term is largest when the frequency of A_2 is low and in fact is larger than the first term when $\hat{q} < 1/3$.

In metabolic diseases in humans, it is often important to know the proportion of heterozygotes, that is, **carriers** of the disease allele, in the population. The carriers for many diseases can now be detected using molecular techniques, but where they cannot be, an estimate of the proportion of carriers is

$$\hat{H} = 2\hat{p}\hat{q} \tag{2.11c}$$

where $\hat{q}$ is an estimate from expression 2.11*a* and $\hat{p} = 1 - \hat{q}$. It is often surprising to many nongeneticists how many individuals are carriers of recessive, disease alleles. For example, assume that only 1 in 10,000 individuals has a particular rare recessive disease such as albinism; then $\hat{q} = 0.01$ and $\hat{H}$ becomes 0.0198. In other words, nearly 2% of such a population are carriers for the albinism allele. In the Hopi Indians, where the frequency of albinism has been estimated to be approximately 1 in 200 (Woolf and Dukepoo, 1969), $\hat{q} = 0.0707$ and $\hat{H}$ becomes 0.131, or 13% of the individuals in the tribe are expected to be carriers for the albinism allele (see Example 4.1 for discussion of albinism in the Hopi Indians).

The multiple-allele systems given in Table 2.9 cover a large proportion of the known examples. In Table 2.11, the notation for some three-allele examples is given. The basis for the allele frequency estimates for the last two cases in Table 2.9, both of them three-allele examples, can be understood using the Hardy–Weinberg principle. For example, with

TABLE 2.11 Three-allele genetic systems and examples (ST = Standard, AR = Arrowhead, CH = Chiricahua, *F* = fast, *M* = medium, *S* = slow, 50, 52, and 54 indicate different length microsatellite locus alleles, him. = himalayan, carb. = carbonaria, ins. = insularia, typ. = typical).

Genotype	A_1A_1	A_1A_2	A_1A_3	A_2A_2	A_2A_3	A_3A_3
Number	N_{11}	N_{12}	N_{13}	N_{22}	N_{23}	N_{33}
Hardy–Weinberg frequency	p_1^2	$2p_1p_2$	$2p_1p_3$	p_2^2	$2p_2p_3$	p_3^2
Codominance						
(1) *D. pseudoobscura* inversions	ST/ST	ST/AR	ST/CH	AR/AR	AR/CH	CH/CH
(2) Electrophoretic alleles	*FF*	*FM*	*FS*	*MM*	*MS*	*SS*
(3) Microsatellite alleles	50/50	50/52	50/54	52/52	52/54	54/54
Dominant series						
(1) Rabbit coat color	full	full	full	him.	him.	albino
(2) *Biston betularia*	carb.	carb.	carb.	ins.	ins.	typ.
(3) *Uta stansburiana*	orange	orange	orange	yellow	yellow	blue
Two alleles codominant, one recessive						
(1) ABO	A	AB	A	B	B	O
(2) Electrophoretic series with null	F	FS	F	S	S	null
(3) Microsatellite locus with null	50	50/52	50	52	52	null

the dominant series, (3) in Table 2.9, where one allele is dominant over both other alleles and a second allele is dominant over the remaining allele, the Hardy–Weinberg proportion of A_3A_3 genotypes is

$$\frac{N_{33}}{N} = p_3^2$$

where N_{33} is the number of A_3A_3 genotypes and p_3 is the frequency of allele A_3, and, therefore,

$$\hat{p}_3 = \left(\frac{N_{33}}{N}\right)^{1/2} \tag{2.12a}$$

Also assuming Hardy–Weinberg proportions, the frequency of genotypes A_2A_2, A_2A_3, and A_3A_3 combined is

$$\begin{aligned}\frac{N_{22} + N_{23} + N_{33}}{N} &= p_2^2 + 2p_2p_3 + p_3^2 \\ &= (p_2 + p_3)^2\end{aligned}$$

and an estimate of the frequency of allele A_2 is

$$\hat{p}_2 = \left(\frac{N_{22} + N_{23} + N_{33}}{N}\right)^{1/2} - \left(\frac{N_{33}}{N}\right)^{1/2} \tag{2.12b}$$

The frequency of allele A_1 is then

$$\begin{aligned}\hat{p}_1 &= 1 - \left(\frac{N_{22} + N_{23} + N_{33}}{N}\right)^{1/2} \\ &= 1 - (p_2 + p_3)\end{aligned} \tag{2.12c}$$

(see Example 2.5 for an illustration using male throat color polymorphism in side-blotched lizards).

Example 2.5. Male side-blotched lizards, *Uta stansburiana*, have a striking throat color polymorphism (Sinervo and Lively, 1996). Males with orange throats are very aggressive and defend large territories. Males with yellow throat stripes are "sneakers" and do not defend territories, and males with blue throats are less aggressive and defend small territories. Sinervo and Lively (1996) followed a polymorphic population in California over 6 years and documented the change in phenotypic frequencies. They observed that changes in phenotype frequencies appeared to be driven by alternative male strategies by the different color morphs, resulting in fitnesses resembling the different strategies in the "rock-paper-scissors" game (see p. 226). There appear to be six distinguishable phenotypes, reflecting the six genotypes for a three-allele system with a codominant series of alleles (Sinervo, 2001). B. Sinervo has kindly provided me with the observed genotype numbers for males in the California population and below they are given for the years 1990 to 1995 (Table 2.12a).

The three alleles are designated as o (orange), y (yellow), and b (blue), and if a dominant series of alleles is assumed, o is dominant over y and b, y is dominant over b, and b is recessive to the other two alleles. Table 2.12b gives the estimated allele frequencies for the six years assuming a dominant series of alleles (equations 2.12a–2.12c) and codominance (equation 2.5c). In general, the estimated frequencies for the two different approaches are similar within the same year. Also the estimated frequencies for years 1992 to 1995 are fairly similar, but between 1991 and 1992, there was a substantial increase in the frequency of o and y and a consequent reduction in b. The standard errors for these estimates can be calculated from the equations in Table 2.9, and in general, they are slightly higher when a dominant series is assumed than when all genotypes are identified.

TABLE 2.12 (a) The phenotype and genotype numbers of males observed in a population of the side-blotched lizard over 6 years. (b) Allele frequencies calculated assuming either a dominant series of three alleles or codominance of the three alleles.

(a)

Phenotype		*Orange*			*Yellow*		*Blue*	
Genotype	*oo*	*oy*	*ob*	*yy*	*yb*	*bb*	*N*
1990	2	0	5	16	0	36	59
1991	4	2	5	5	3	43	62
1992	18	13	4	29	13	43	120
1993	2	10	14	32	5	42	105
1994	3	5	9	42	6	42	107
1995	5	0	12	34	7	48	106

(b)

	Dominant Series			*Codominant alleles*		
	p_o	p_y	p_b	p_o	p_y	p_b
1990	0.111	0.158	0.781	0.067	0.136	0.788
1991	0.093	0.074	0.833	0.121	0.105	0.774
1992	0.158	0.243	0.599	0.221	0.283	0.496
1993	0.133	0.235	0.632	0.133	0.248	0.619
1994	0.083	0.290	0.627	0.093	0.276	0.631
1995	0.084	0.243	0.673	0.103	0.226	0.670

Mixed dominance and codominance (case 4 in Table 2.9) can result in a situation in which there is a substantial number of recessive homozygotes, as for the *ABO* blood group locus, or the recessive allele can be low in frequency and no recessive homozygotes observed, as for an electrophoretic or microsatellite locus with two (or more) codominant alleles and one **null allele**, ***an allele that does not produce an identifiable product on the gel.*** For estimates of the *ABO* locus situation, the allele frequency estimates of alleles A_1 and A_3 are the same as for a dominant series. However, the estimate of the A_2 frequency uses the relationship based on Hardy–Weinberg proportions

$$\begin{aligned}\frac{N_{11}+N_{13}+N_{33}}{N} &= p_1^2 + 2p_1p_3 + p_3^2 \\ &= (p_1+p_3)^2\end{aligned}$$

Therefore, the frequency of allele A_2 can be estimated as one minus the square root of this expression, or

$$\hat{p}_2 = 1 - \left(\frac{N_{11}+N_{13}+N_{33}}{N}\right)^{1/2} \tag{2.13a}$$

However, this approach assumes that there are three parameters instead of two, and as a result, the sum of the frequency estimates for the three alleles generally does not equal unity unless the genotypes are very close to Hardy–Weinberg proportions. Adjusted estimates may be calculated (Bernstein, 1930) by letting the deviation from unity be

$$d = 1 - (\hat{p}_1 + \hat{p}_2 + \hat{p}_3)$$

and then obtaining new estimates of allele frequencies as

$$\begin{aligned}\hat{p}'_1 &= (1 + {}^1\!/\!_2 d)\hat{p}_1 \\ \hat{p}'_2 &= (1 + {}^1\!/\!_2 d)\hat{p}_2 \\ \hat{p}'_3 &= (1 + {}^1\!/\!_2 d)(\hat{p}_3 + {}^1\!/\!_2 d)\end{aligned} \tag{2.13b}$$

These values are practically equal to the maximum-likelihood solutions (Li, 1970) (see Example 2.6 for the calculation of the *ABO* blood group alleles in an African Pygmy population).

Example 2.6. For the *ABO* blood group in humans, alleles *A* and *B* are codominant, and both these alleles are dominant over the *O* allele (see Ogasawara *et al.*, 2001; Yip, 2002; and Daniels, 2002 for recent molecular research on the ABO system). In a sample from an African Pygmy population, there were 7 AB, 44 A, 27 B, and 88 O individuals (Cavalli-Sforza and Bodmer, 1971). The estimates of the frequencies for alleles A, B, and O and their standard errors from the expressions in Table 2.9 are found to be

$$\hat{p}_1 = 0.168 \pm 0.022$$

$$\hat{p}_2 = 0.108 \pm 0.018$$

$$\hat{p}_3 = 0.728 \pm 0.027$$

Because the frequencies do not add up to unity, a correction should be made. When the adjusted estimates in expression 2.13b are used, the allele frequency estimates for this example become

$$\hat{p}'_1 = 0.168 \pm 0.021$$

$$\hat{p}'_2 = 0.108 \pm 0.018$$

$$\hat{p}'_3 = 0.724 \pm 0.025$$

The standard errors for these corrected estimates are maximum-likelihood estimates as given by Li (1970).

For the situation in which the recessive homozygote is not present, the frequency of the null allele can be estimated as

$$\hat{p}_n = \frac{H_E - H_O}{1 + H_E} \tag{2.14}$$

where H_O is the observed heterozygosity as given in expression 2.5d and H_E is the expected heterozygosity as calculated using expression 2.5a for all the codominant alleles (Brookfield, 1996). The frequencies of the other alleles are then multiplied by $(1 - \hat{p}_n)$ to give allele frequency estimates that are close to the maximum likelihood estimates (see Example 2.7 for an application to a MHC locus in an endangered fish). Chakraborty *et al.* (1994) gave a somewhat different estimate for the frequency of a null allele (for a discussion of the differences in these two approaches, see Brookfield, 1996).

Example 2.7. Hedrick and Parker (1998a) examined the variation at a MHC locus in the endangered Gila topminnow, *Poeciliopsis occidentalis.* The sequences segregated consistently with Mendelian expectations in several family groups from Cienega Creek using SSCP (single-stranded conformation polymorphism) (Figure 2.7). However, in the sample from Sharp

Spring, which contained five different allele sequences, there was an observed deficiency of heterozygotes with $H_O = 0.475$ and $H_E = 0.729$. Microsatellite loci, examined in the same 40 fish (Parker *et al.*, 1999), were consistent with Hardy–Weinberg expectations.

Using expression 2.14, the estimate of the frequency of the null allele is 0.147 (Hedrick and Parker, 1998b). The expected number of null homozygotes is $\hat{p}_n^2 N$, so only 0.86 null homozygotes were expected (none were observed). Taking into account the estimated frequency of the null allele, the estimates of the frequencies for alleles *Pooc*-2, -3, -6, -8, and -9 become 0.277, 0.149, 0.043, 0.299, and 0.085, respectively. Alleles *Pooc*-8

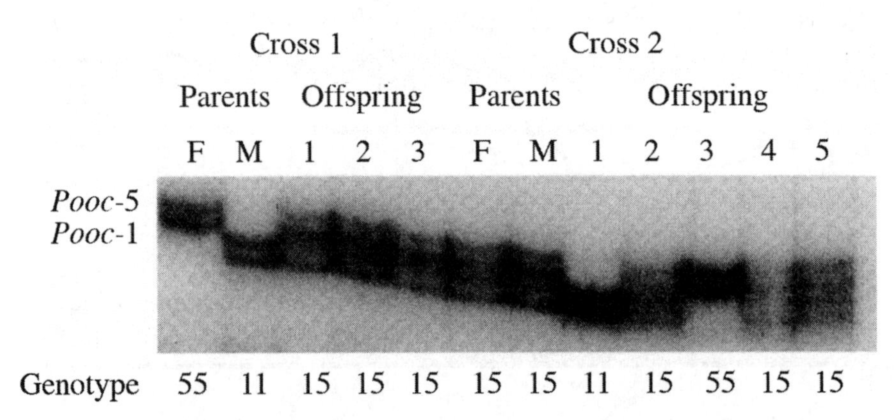

Figure 2.7. Segregation of *MHC* alleles as seen with SSCP in two crosses (from Hedrick and Parker, 1998a) where F and M are the female and male parents. In the first cross, between homozygotes *Pooc*-5/*Pooc*-5 and *Pooc*-1/*Pooc*-1, all offspring were heterozygotes *Pooc*-1/*Pooc*-5. In the second cross, between two *Pooc*-1/*Pooc*-5 heterozygotes, all three possible genotypes were seen in the five offspring.

and -9, which were found only in Sharp Spring, were the most divergent in the total study and have four and five amino acids, respectively, that are not found in any of the other alleles. In other words, it is not unlikely that an additional allele might be present in this population that was not amplified with PCR and therefore appeared as a null allele.

IV. TESTING HARDY–WEINBERG PROPORTIONS

In the examples with codominant alleles presented earlier, it was stated that the observed genotype numbers were quite close to the expected genotype numbers calculated from the Hardy–Weinberg principle. What is needed, however, is a statistical test that will allow us to decide whether the fit is "sufficiently" close to determine that the population is in the expected Hardy–Weinberg proportions. We use the χ^2 **(chi-square) test** for this purpose here because of its simplicity.

To determine whether the observed numbers are consistent with Hardy–Weinberg predictions, χ^2 can be calculated as

$$\chi^2 = \sum_{i=1}^{k} \frac{(O - E)^2}{E} \tag{2.15a}$$

where O and E are the observed and expected numbers of a particular type and k is the number of genotype classes. From the calculated value of the χ^2 and with the knowledge of another value, the degrees of freedom, we can obtain from a χ^2 table the ***probability that the observed numbers would deviate from the expected numbers as much or more by chance***.

The **degrees of freedom** used to determine the significance of a χ^2 value are ***equal to the number of phenotypic classes, k, minus one, and then minus the number of parameters estimated from the data***. One degree of freedom is always lost because there are only $k - 1$ independent genotype classes. That is, the numbers in all classes must sum to the total sample size, N, and given the numbers in all classes but one, the number in the last class is defined. Table 2.13 gives the degrees of freedom for several different genetic systems. In the systems with dominance or dominant series, there are no degrees of freedom left, meaning that a test of significance cannot be done.

TABLE 2.13 Computation of the degrees of freedom for a number of genetic systems.

Systems	*Number of phenotypic classes*	*Number of parameters*	*Degrees of freedom*
Two alleles			
Codominance	3	1	1
Dominance	2	1	0
Three alleles			
Codominance	6	2	3
Dominant series	3	2	0
ABO, null allele	4	2*	1
n alleles			
Codominance	$\frac{n(n+1)}{2}$	$n-1$	$\frac{n(n-1)}{2}$
Dominant series	n	$n-1$	0

*Because $p_1 = 1 - (p_2 + p_3)$, p_1 need not be estimated from the formula given in the text.

When the expected Hardy–Weinberg frequencies for the two-allele, codominance case are used, the χ^2 expression becomes

$$\chi^2 = \frac{(N_{11} - \hat{p}^2 N)^2}{\hat{p}^2 N} + \frac{(N_{12} - 2\hat{p}\hat{q}N)^2}{2\hat{p}\hat{q}N} + \frac{(N_{22} - \hat{q}^2 N)^2}{\hat{q}^2 N} \qquad (2.15b)$$

where the three terms represent the deviation from expected because of genotypes A_1A_1,A_1A_2, and A_2A_2, respectively. For more than two alleles,

$$\chi^2 = \sum_i \frac{(N_{ii} - \hat{p}_i^2 N)^2}{\hat{p}_i^2 N} + \sum_{i<j} \frac{(N_{ij} - 2\hat{p}_i\hat{p}_j N)^2}{2\hat{p}_i\hat{p}_j N} \qquad (2.15c)$$

Several other comments about the use of the χ^2 test are important here (and illustrate when an exact test, which does not have these problems, is preferable). First, even though the χ^2 test is generally conservative in detecting statistical significance, if the sample size is smaller than 50, the results should be interpreted cautiously. Also, as a general rule, the expected number in all classes should be greater than five. If the expected number in a class is less than five, then the expected and observed numbers for that class should be combined with an adjacent class (see Example 2.8 for calculation of χ^2 values from the data in Table 2.4 and Table 2.5).

Example 2.8. Let us use the data from previous examples to calculate χ^2 values. First, a χ^2 value can be calculated from the MN data given in Table 2.4. For these data, the χ^2 value is 0.22 with one degree of freedom, making the numbers consistent with Hardy–Weinberg expectations. To illustrate the problems encountered using χ^2 when there are low expected numbers in some classes, consider the *Daphnia* data given in Table 2.5. The expected numbers in the S class are only 2.4, so they should be combined with another class—the SM class in this case. Now there are five classes and two parameters estimated, so there are two degrees of freedom. The χ^2 value is equal to 1.23, also making these data consistent with Hardy–Weinberg expectations.

In small samples, the expected number of heterozygotes is increased slightly, and the expected number of homozygotes is decreased slightly. More accurate expectations of the expected genotype numbers (Levene, 1949) for homozygotes and heterozygotes are

$$E(Hom) = \frac{N\hat{p}_i(2N\hat{p}_i - 1)}{2N - 1} \qquad (2.16a)$$

and

$$E(Het) = \frac{4N^2\hat{p}_i\hat{p}_j}{2N-1} \qquad (2.16b)$$

This effect will generally be small but may be important in certain rare, endangered, or hard-to-obtain organisms where the sample size is less than 50. For example, if $N = 32$, $\hat{p}_1 = 0.25$, and $\hat{p}_2 = 0.75$, then the expected numbers of genotypes A_1A_1, A_1A_2, and A_2A_2 are 1.90, 12.19, and 17.90, respectively, rather than 2, 12, and 18, respectively, without the corrections.

When there are multiple alleles and the numbers in some of the genotype categories are small, then an exact probability test, or an **exact test** (Guo and Thompson, 1992; Rousset and Raymond, 1995, 1997; Weir, 1996), is a desirable approach. For small sample sizes, all possible combinations of genotypes can be generated and their probabilities calculated. However, as the sample size gets larger, the number of possible types of samples grows very quickly, and simulation is used to estimate the probability of obtaining a result as extreme, or more extreme, as that observed. Example 2.9 shows how **Monte Carlo** simulation, using a permutation procedure, can be used to estimate the exact probability.

Example 2.9. When there are multiple alleles (many genotypes), it is often useful to calculate the exact probability of observing the array by chance (many cells may have small numbers and/or expectations). To do this, one first calculates the probability of obtaining the given observed array of genotypes and then by repeated simulation determines how likely it is to observe an array this extreme by chance. There are two common ways to calculate such exact probabilities: the Monte Carlo and Markov Chain algorithms (Guo and Thompson, 1992; Rousset and Raymond, 1997).

To illustrate the Monte Carlo method, let us assume that we observed an array of genotypes where N_{ij} is the number of genotype A_iA_j and p_i is the observed frequency of A_i. Table 2.14 is an example observed array of 45 genotypes for four alleles with the expected number in parentheses (Louis and Dempster, 1987). The probability of the observed array is the multinomial probability

$$\Pr = \frac{N!}{\prod_{j \le i} N_{ij}!} \prod_{i=1}^{n} p_i^{2N_{ii}} \prod_{j<i}^{n} (2p_ip_j)^{N_{ij}}$$

In this case, the probability can be calculated as

$$\Pr = \frac{45!}{0!3!5!3!...2!}(0.122)^0(0.333)^2... + (0.081)^3(0.081)^5...$$

To carry out the Monte Carlo simulation, first order the 90 gametes as

$$\overset{11}{A_1A_1A_1\ldots}\quad \overset{30}{A_2A_2A_2\ldots}\quad \overset{30}{A_3A_3A_3\ldots}\quad \overset{19}{A_4A_4A_4\ldots}$$

Now for each draw, randomly permute the order of these gametes and consider each successive pair as a genotype. For example, a draw of the 90 ordered gametes might give genotypes

$$A_4A_2,\ A_3A_2,\ A_2A_1,\ A_4A_2,\ A_1A_3,\ A_3A_2,\ldots$$

Count up the number in each class and calculate the probability for this array with the formula above. Note that the allele frequencies are always the same (the order of gametes is just changed), and the only changes in each draw are the N_{ij} values. If the probability is less than that in the original observed array, add 1 to a counter. After 17,000 draws (this gives a probability value within 0.01 with 99% confidence) (Guo and Thompson, 1992), calculate the proportion that is less than the observed probability; this is the exact probability of a result more extreme than that observed. In this case, a Monte Carlo simulation gave $p = 0.017$, a statistically significant value—that is, the genotypes are significantly different from Hardy–Weinberg expectations.

We can also classify genotypes as heterozygotes or homozygotes, then we get

	Heterozygotes	*Homozygotes*
Observed	4	41
Expected	12.3	32.7

From this array, $\chi^2 = 7.7$ and $P < 0.01$. In other words, the explanation for the significance from the simulation is a deficiency of heterozygotes compared with that expected under Hardy–Weinberg.

TABLE 2.14 An array of observed numbers (expected numbers in parentheses) of genotypes where there are four alleles and $N = 45$ (Louis and Dempster, 1987).

	A_1	A_2	A_3	A_4	*Allele frequencies*
A_1	0 (0.62)	3 (3.71)	5 (3.71)	3 (2.35)	$p_1 = 11/90 = 0.122$
A_2		1 (4.89)	18 (10.11)	7 (6.40)	$p_2 = 30/90 = 0.333$
A_3			1 (4.89)	5 (6.40)	$p_3 = 30/90 = 0.333$
A_4				2 (1.92)	$p_4 = 19/90 = 0.211$

Another way to examine the difference from expected frequencies is to calculate the standardized deviation of the observed frequency from the Hardy–Weinberg expectation of the heterozygote (this is the **fixation index *F*** that is discussed in Chapter 5), which is

$$F = \frac{2pq - H}{2pq}$$
$$= 1 - \frac{H}{2pq} \tag{2.17a}$$

An excess in heterozygosity compared with Hardy–Weinberg expectations ($H > 2pq$) results in a negative F value, and a deficiency ($H < 2pq$) results in a positive F value. It can be shown that

$$\chi^2 = F^2 N \tag{2.17b}$$

We can then calculate χ^2 values for different F values and sample sizes as in Figure 2.8, where $p = q = 0.5$. For example, if $H = 0.45$, then the value of F is 0.1. With this deviation from the expected value, whether it be due to inbreeding, selection, or some other cause, a sample size of nearly 400 is necessary to detect this effect at the 5% level ($\alpha = 0.05$). The horizontal lines in Figure 2.8 indicate χ^2 values of 3.84 and 6.64, which are the 0.05 and 0.01 significance levels for χ^2 with one degree of freedom. To detect a value of $F = 0.05$ at the 0.05 level, a sample size of more than

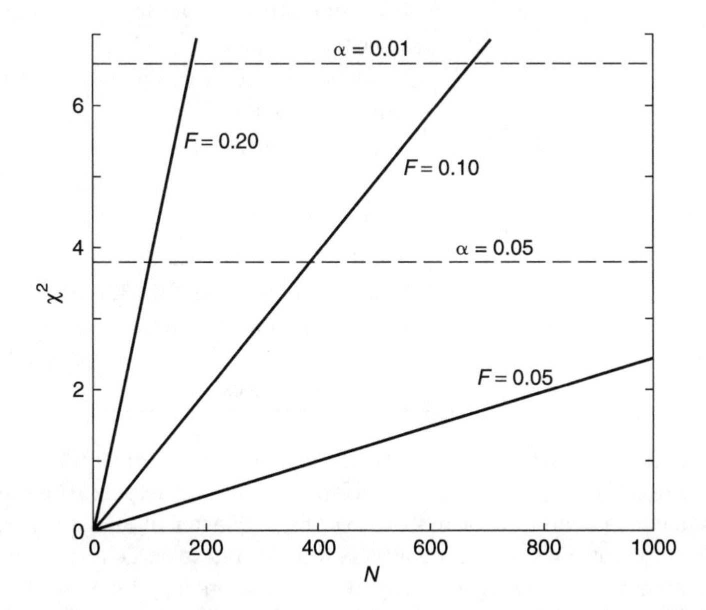

Figure 2.8. The χ^2 values expected for various sample sizes and deviations from the expected Hardy–Weinberg frequencies due to a systematic cause.

1500 is needed. On the other hand, if the sample size is large enough, then very small systematic deviations from Hardy–Weinberg proportions can be detected, even though these deviations may be of little or no biological consequence. In this case, it may be useful to determine the effect size, for example, the amount of selection that is necessary to cause a small but statistically significant deviation from Hardy–Weinberg proportions.

There are a number of potential reasons why there may be a deviation from Hardy–Weinberg proportions. Table 2.15 lists factors that may result in a relative decrease or increase in heterozygosity and the pages in the text where these factors are discussed. Some of these factors can have an impact in only certain situations; for example, the Wahlund effect occurs only when the groups that are being lumped together are different in allele frequency. It is often useful to determine how large a particular factor would have to be to result in an observed deviation from Hardy–Weinberg proportions because, in some instances, it may be unrealistic that a given factor could actually result in an observed effect.

TABLE 2.15 Some potential causes of deviations of heterozygotes relative to expected Hardy–Weinberg proportions and pages where these factors are discussed.

Effect on heterozygosity	*Cause*
Decrease	Selection against heterozygotes (p. 144)
	Inbreeding (p. 239)
	Positive-assortative mating
	Gene flow of zygotes (p. 482)
	Wahlund effect over space (p. 479) or time
	Null allele (p. 88)
	Allele dropout in old samples
Increase	Selection favoring heterozygotes (p. 139)
	Outbreeding
	Negative-assortative mating (p. 190)
	Gene flow of gametes (p. 482)
	Amplification artifact of new alleles
	Misclassification of alleles at different loci in multigene families

Some factors, such as genetic drift and mutation, are unlikely to result in a significant deviation from Hardy–Weinberg expectations. Other factors, such as gene flow or assortative mating, even though they may be present, may not have a statistically significant effect on genotype frequencies. Finally, in some cases, strong systematic factors may be present that do not result in a deviation from Hardy–Weinberg proportions. For example, strong selection may be present in the population and after selection the population may be still in Hardy–Weinberg proportions (p. 151). Workman (1969) provided a discussion of how various forces affecting genotype frequencies may cancel each other's effects and result in Hardy–Weinberg proportions.

V. MEASURES OF GENETIC VARIATION

In order to document the amount of genetic variation in a standardized way, we need measures that allow us to quantify this information. Heterozygosity remains the most used measure of variation and is desirable in a number of ways, but several other measures have been used to describe the genetic variation at a single locus or a number of similar loci.

a. Heterozygosity

The most widespread measure of genetic variation in a population is the amount of heterozygosity. Because individuals in diploid species are either heterozygous or homozygous at a given locus, this measure represents a biologically useful quantity. However, measures of heterozygosity are not very sensitive to additional variation because the upper limit, unity, is the same for any number of alleles. This limit makes it difficult to differentiate between populations for highly variable loci, such as microsatellites, where the heterozygosity may be 0.8 or higher.

As mentioned above, the expected Hardy–Weinberg heterozygosity of a population for a particular locus with n alleles can be calculated as

$$H_E = 1 - \sum_{i=1}^{n} p_i^2$$

which is one minus the Hardy–Weinberg homozygosity. Nei (1987) called this measure gene diversity and suggested that it is particularly useful because it is applicable for genes of different ploidy levels and in organisms with different reproductive systems. Using the estimated allele frequencies from expression 2.5c, an unbiased estimate of the expected heterozygosity at a locus, using a small samples size correction, is

$$\hat{H}_E = \frac{2N}{2N-1}\left(1 - \sum_{i=1}^{n} \hat{p}_i^2\right) \tag{2.18a}$$

(Nei and Roychoudhury, 1974) (this is the same correction used in expression 2.16b for the expected number of heterozygotes). Obviously, when $N > 50$, the bias in the heterozygosity estimate without the small sample size correction is low (see Example 2.10 to understand why this correction is necessary).

Example 2.10. Nei and Roychoudhury (1974) found that the observed heterozygosity in a sample of N individuals was expected to be

$$H = \frac{2N-1}{2N}\left(1 - \sum_{i=1}^{n} \hat{p}_i^2\right)$$

so that they recommended the use of the small sample size correction $2N/(2N-1)$ to compensate for the effect. However, why is the level of H in a sample lower than that in the population?

To illustrate, assume that the sample sizes are $N = 1$ and $N = 2$ and that the allele frequencies of A_1 and A_2 in the population from which the sample is drawn are $p_1 = 0.5$ and $p_2 = 0.5$. The three types of draws for $N = 1$ and the five types of draws for $N = 2$, their frequencies (obtained from binomial sampling), and heterozygosities are as given in Table 2.16. Overall, the observed heterozygosity is

$$H_O = \sum_i x_i H_i$$

that is, the sum of the products of the frequencies and the heterozygosities in the different samples. For $N = 1$, $H_O = 0.25(0) + 0.5\ (0.5) + 0.25(0) = 0.25$ and for $N = 2$, $H_O = 0.375$. If these values are multiplied by the small sample size correction $2N/(2N-1)$ given by Nei and Roychoudhury (1974), then they both are equal to 0.5, as is the heterozygosity in the population from which the samples were drawn.

TABLE 2.16 The different types of samples of sizes (a) $N = 1$ and (b) $N = 2$, their frequencies, and heterozygosities drawn from a population with alleles A_1 and A_2 with frequencies $p_1 = 0.5$ and $p_2 = 0.5$.

	Alleles	*Frequency* (x_i)	*Heterozygosity* ($H_i = 1 - \sum p_i^2$)
(a) $N = 1$	A_1, A_1	0.25	0
	A_1, A_2	0.5	0.5
	A_2, A_2	0.25	0
(b) $N = 2$	A_1, A_1, A_1, A_1	0.0625	0
	A_1, A_1, A_1, A_2	0.25	0.375
	A_1, A_1, A_2, A_2	0.375	0.5
	A_1, A_2, A_2, A_2	0.25	0.375
	A_2, A_2, A_2, A_2	0.0625	0

In most outbreeding populations, the observed heterozygosity is quite close to the theoretical heterozygosity. However, for populations in which genotype frequencies may not be close to Hardy–Weinberg proportions, such as those with high levels of self-fertilization, the observed heterozygosity may be calculated as

$$\hat{H}_O = \sum_{i<j}^{n} \hat{P}_{ij} \tag{2.18b}$$

where $\hat{P}_{ij}$ is the estimated frequency of genotype ij and $\hat{H}_O$ is the summation over the frequencies of all heterozygotes.

For populations with very high selfing and for haplotype (or allele) variation of mtDNA, cpDNA, or Y chromosomes, an unbiased estimate of the gene diversity is

$$\hat{H} = \frac{N}{N-1}\left(1 - \sum_{i=1}^{n} \hat{p}_i^2\right) \tag{2.18c}$$

where N is the number of individuals in a population with very high selfing or the number of the appropriate sex for the haplotype estimate (Nei, 1987). For the variance of these estimates of gene diversity and for tests of significance between populations, see Nei (1987).

For allozymes, microsatellites, SNPs, or other diploid loci detected using molecular techniques, one can obtain simultaneously information concerning the heterozygosity of a number of loci in many individuals in a population. For a given locus in a particular individual, there is either a heterozygous or homozygous state. Such observed heterozygosities can be represented in a matrix form as shown in Table 2.17, where H_{ij} represents

TABLE 2.17 Observed heterozygosity specified for particular loci and individuals.

	Locus							
Individual	1	2	3	$\cdots$	j	$\cdots$	m	
1	H_{11}	H_{12}	H_{13}	$\cdots$	H_{1j}	$\cdots$	H_{1m}	$H_{1\cdot}$
2	H_{21}				$\cdot$			$H_{2\cdot}$
3	H_{31}				$\cdot$			$H_{3\cdot}$
$\cdot$	$\cdot$				$\cdot$			$\cdot$
$\cdot$	$\cdot$				$\cdot$			$\cdot$
$\cdot$	$\cdot$				$\cdot$			$\cdot$
i	H_{i1}	$\cdot$	$\cdot$	$\cdots$	H_{ij}	$\cdots$	$\cdot$	$H_{i\cdot}$
$\cdot$	$\cdot$				$\cdot$			$\cdot$
$\cdot$	$\cdot$				$\cdot$			$\cdot$
$\cdot$	$\cdot$				$\cdot$			$\cdot$
$\cdot$	$\cdot$				$\cdot$			$\cdot$
N	H_{N1}				$\cdot$			$H_{N\cdot}$
	$H_{\cdot 1}$	$H_{\cdot 2}$	$H_{\cdot 3}$	$\cdots$	$H_{\cdot j}$	$\cdots$	$H_{\cdot m}$	$\overline{H}$

the value for the ith individual and the jth locus. The values in such a matrix are either zero (homozygotes) or unity (heterozygotes), and the estimated heterozygosities for each locus are the marginal values at the bottom of the array of heterozygosities given in the table. The marginal averages on the right side of the table are the heterozygosities for specific individuals over all loci examined. The estimated mean heterozygosity in the population over all loci is then

$$\hat{H} = \frac{1}{Nm}\sum_{i=1}^{N}\sum_{j=1}^{m} H_{ij} \tag{2.19a}$$

and the sampling variance is

$$V(\hat{H}) = \frac{\hat{H}(1-\hat{H})}{Nm} \tag{2.19b}$$

The sampling variance has two components: that caused by variation in heterozygosity among individuals and that caused by variation in heterozygosity among loci. These values may be quite different. For allozymes, individual heterozygosities generally form a fairly normal distribution and locus heterozygosities form a reverse J-shaped distribution. Figure 2.9 illustrates these distributions for 71 allozyme loci in humans (Harris and Hopkinson, 1972). The shaded area indicates the observed distribution of the heterozygosities of the 71 loci, with 51 monomorphic loci having a heterozygosity

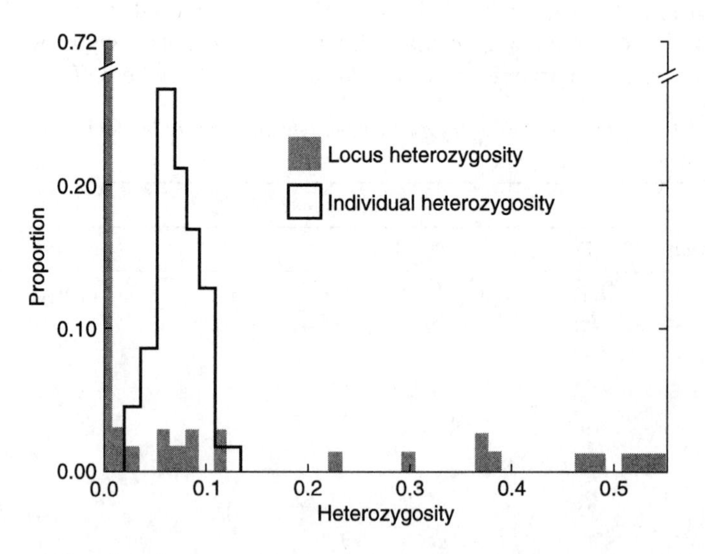

Figure 2.9. The distribution of heterozygosity for 71 allozyme loci in humans (from Harris and Hopkinson, 1972) and the distribution of heterozygosity for 71 individuals, generated by Monte Carlo simulation.

of 0.0, as given in Table 1.3. As a comparison, 71 hypothetical individuals, each with 71 loci, were randomly generated using Monte Carlo simulation. The distribution of heterozygosities of these 71 individuals is given as a solid line in Figure 2.9, and it is obvious that nearly all of these individuals had heterozygosities between 0.05 and 0.10. The mean heterozygosity for both these distributions is the same (0.067), but the variance in heterozygosity over individuals is approximately 3% of the variance in heterozygosity over loci.

If one wishes to obtain an estimate of the variation in heterozygosity among loci in order to evaluate the reliability of the heterozygosity estimate on these loci as a sample from all allozyme loci, then a variance using the horizontal marginal values of Table 2.17 would be appropriate. On the other hand, if one wishes to obtain an estimate of the variation in heterozygosity among individuals to evaluate the heterozygosity estimate on these individuals as a sample from the total population, then a variance using vertical marginal values would be appropriate. Nei (1987) provided variances for these components and suggested that given the choice to examine more loci or more individuals it is best to examine more loci because variation among loci is generally higher than among individuals. In fact, if only a few individuals are available, as in a rare species, then one can obtain a fairly good estimate of heterozygosity by examining a number of loci. However, it is doubtful that the interlocus variation will ever be reduced very much because of the real heterozygosity differences among allozyme loci (not just the result of sampling), resulting in extreme polymorphism for some loci and monomorphism for other loci.

Similar, but not as extreme, effects may also be seen for highly variable diploid loci, such as microsatellites. One should be aware that when using microsatellite loci, researchers often do not include invariant or low-variation loci, and they may even select for loci with high heterozygosity (Paetkau *et al.*, 1997) or a specific range of heterozygosity (Dib *et al.*, 1996) to obtain loci that are most useful for the purpose of their study. For example, the 5048 autosomal microsatellite loci, which had an average heterozygosity of 0.70, in humans reported by Dib *et al.* (1996) were selected for high (but generally less 0.8) heterozygosity (see Example 2.11 for some microsatellite data derived from bears).

Example 2.11. There are three species of bears in North America: the black bear (*Ursus americanus*), brown bear (*U. arctos*, grizzly bears), and polar bear (*U. maritimus*). Eight highly variable $(CA)_n$ microsatellites have been used to study the population genetics of the three species. Allele frequencies for one of these loci for several populations from each species are given in Table 2.18 (Paetkau *et al.*, 1997). The two most isolated populations, the black bear sample from Newfoundland Island (NI) and the brown bear sample from Kodiak Island (KI), have both the fewest

TABLE 2.18 The allele frequencies at microsatellite locus G10B in three species of bears from eight locations (Paetkau *et al.*, 1997). N is the sample size for each population, and the numbers in the left column designate different alleles and their nucleotide size. At the bottom are the number of alleles and expected heterozygosity for this locus and the average of these measures over eight $(CA)_n$ microsatellite loci.

Species	Black bear			Brown bear			Polar bear	
Location (N)	LM (32)	WS (116)	NI (32)	II (24)	FR (40)	KI (34)	NB (30)	WH (30)
Allele								
140	—	—	—	0.104	—	—	—	—
142	—	—	—	—	—	—	0.083	—
148	—	—	—	0.167	—	—	—	—
150	—	—	—	0.062	—	—	0.017	0.067
152	—	—	0.078	0.104	0.112	—	0.150	—
154	0.062	—	0.641	0.125	0.038	0.985	0.417	0.140
156	0.344	0.147	—	—	0.025	—	0.150	0.717
158	0.031	0.190	—	0.042	0.262	0.015	0.183	0.067
160	0.156	0.263	—	0.354	0.150	—	—	—
162	0.016	0.216	—	—	0.413	—	—	—
164	0.391	0.181	0.281	0.042	—	—	—	—
166	—	0.004	—	—	—	—	—	—
n	6	6	3	8	6	2	6	4
H_E	0.729	0.801	0.520	0.837	0.742	0.030	0.766	0.470
$\overline{n}$	8.75	9.50	3.00	6.63	6.50	2.13	6.38	5.38
$\overline{H}_E$	0.820	0.806	0.414	0.764	0.694	0.265	0.643	0.626

alleles and the lowest expected heterozygosity at this locus. None of the observed heterozygosities at this locus are different from Hardy–Weinberg expectations, and over all the 8 loci and 13 samples total, when corrected for multiple comparisons, there were no significant deviations from Hardy–Weinberg expectations (Paetkau *et al.*, 1997). At the bottom of the table, the average number of alleles and average expected heterozygosity are given for all 8 loci. Newfoundland and Kodiak Islands have both the lowest number of alleles, 3.00 and 2.13, and the lowest heterozygosites, 0.414 and 0.265. At the other extreme, the two continental black bear populations had the highest variation, with 8.75 and 9.50 alleles and heterozygosites of 0.820 and 0.806.

b. Other Measures

Ford (1940) defined a **genetic polymorphism** as "the occurrence together in the same habitat of two or more discontinuous forms in such proportion that the rarest of them cannot be maintained by recurrent mutation." Even though this definition is imprecise, it has had intuitive appeal because it appears to be based on population genetic theory. However, with the formulation of the neutrality hypothesis in which mutation (in combination with genetic drift) is a major force influencing genetic variation and with the discovery of highly variable loci, such as microsatellites, that may have quite high mutation rates, this definition is not generally appropriate. A more useful definition, given by Cavalli-Sforza and Bodmer (1971), is that ***"genetic polymorphism is the occurrence in the same population of two or more alleles at one locus, each with appreciable frequency."***

In a population survey, it is necessary to delineate what constitutes an appreciable frequency. A practical approach to defining polymorphism is to decide arbitrarily on a limit for the frequency of the most common allele; that is, polymorphic loci are those for which the frequency of the most common allele is smaller than 0.99 or smaller than 0.95. Both of these arbitrary cutoff points have been used, but 0.99 is used most frequently if the sample size is adequate (approximately 100 individuals or more). To estimate the **proportion of polymorphic loci** (***P***) for a population in which a number of loci have been examined, one must first count the number of polymorphic loci and then calculate the proportion that these loci represent of all the loci examined. In other words, the proportion of polymorphic loci is

$$\hat{P} = \frac{x}{m} \tag{2.20}$$

where x is the number of polymorphic loci in a sample of m loci. This measure is probably the most appropriate for allozyme loci and much less so for highly variable loci in which a high proportion of the loci is polymorphic in most populations.

SNPs are by definition nucleotide sites that are polymorphic. However, polymorphism for a SNP is generally a function of the sample size. For example, if 10 individuals are sequenced for SNPs ($2N = 20$), then the minimum frequency that could be discovered for a SNP is 0.05. Of course, a SNP may be discovered in one population and be monomorphic when examined in a second population. When SNPs are used for screening for disease genes in humans, only SNPs with the lower frequency variant above a given value, for example, 0.1 or 0.2, may be considered.

Another measure that is sometimes used is the **number of alleles**, n, a count of the number of alleles observed at a locus in a population (this measure is also called **allele diversity**, **allele richness**, or ***A***). However, this measure is often strongly influenced by the sample size so that comparison across populations with different sample sizes should be made cautiously. On p. 429 we discuss how the number of alleles is influenced by sample size under the neutral model. Petit *et al.* (1998) showed how rarefaction can be used to determine the expected number of alleles in samples of different sizes, and Leberg (2002) compared various approaches to correct estimates of A for differences in sample size.

The **effective number of alleles, n_e**, is also sometimes used to measure the amount of variation and is the inverse of the expected homozygosity

$$\hat{n}_e = \frac{1}{1 - \hat{H}} \tag{2.21}$$

To estimate this value accurately in small samples, see Nielsen *et al.* (2003).

VI. MEASURES OF NUCLEOTIDE AND AMINO ACID DIVERSITY

In recent years, nucleotide and amino acid sequences of individuals within populations have become available for a number of genes. Understanding the basis for the patterns of this variation within populations, between populations, and between species is the focus of molecular population genetics, a topic discussed throughout the book but emphasized in more detail in Chapter 8. Here we introduce two measures of nucleotide variation: the proportion nucleotide sites that are variable and the diversity (expected heterozygosity) at the nucleotide level. Similar measures can be used for determining the amount of amino acid variation. When we discuss molecular population genetics, we examine whether this variation is consistent with the predictions from the neutral model and other factors that may have an impact on sequence variation.

The simplest way to measure the amount of nucleotide variation is to determine the number of nucleotide sites that are variable in the sample of sites examined. If we examine a sample of DNA sequences and let the number of nucleotide sites that are different (segregating) be S and the total number of sites compared be N, the **proportion of nucleotide sites that differ in the population** is estimated by

$$\hat{p}_S = \frac{S}{N} \tag{2.22a}$$

and

$$V(\hat{p}_S) = \frac{\hat{p}(1 - \hat{p})}{N} \tag{2.22b}$$

This value is called the ***p* distance** for nucleotide sequences (Nei and Kumar, 2000).

The other general way to quantify the amount of nucleotide variation in a population is to determine the proportion of nucleotide differences between pairs of sequences and then weigh these differences by the frequencies of the sequences. If this is then summed over all of the possible pairs of sequences, then the **nucleotide diversity** becomes

$$\pi = \sum_{ij} p_i p_j \pi_{ij} \tag{2.23a}$$

where p_i is the frequency of sequence i and π_{ij} is the proportion of nucleotides that differs when the ith and jth DNA sequences are compared. Nucleotide diversity can be estimated as

$$\hat{\pi} = \frac{N}{N-1} \sum_{ij} \hat{p}_i \hat{p}_j \pi_{ij} \tag{2.23b}$$

where N is the number of sequences examined and $\hat{p}_i$ is the proportion of the ith sequence in the sample (see Example 2.12 for an application to the *Adh* data in *D. melanogaster*). Nei (1987) gave formulas for estimating the variance of $\hat{\pi}$, and Nei and Kumar (2000) suggested that resampling approaches, either the bootstrap or jackknife, may be useful to estimate the variance of this measure.

Example 2.12. One of the first estimates of nucleotide diversity was for the *Adh* gene in *D. melanogaster.* Kreitman (1983) reported 11 sequences of the *Adh* gene from samples from around the world. *Adh* is polymorphic for two allozyme alleles, F and S, and Kreitman (1983) sequenced six S allele samples and five F allele samples. Table 2.19 gives the proportion of nucleotide differences for all pairs of sequences. Three F sequences, Fr-F, Wa-F, Af-F, did not have any nucleotide differences over the 2379-nucleotide sequence (Af-F did differ at an insertion in the 3′ region) and are combined. As a result, the frequency is $p_i = 0.273$ for this sequence and 0.091 for the other eight sequences.

TABLE 2.19 Percent nucleotide difference ($\pi_{ij} \times 10^2$) between the 2379 nucleotides in the 11 sequences of the *Adh* locus in *D. melanogaster* (from Kreitman, 1983).

		S						*F*	
		Wa-S	*Fl-1S*	*Af-S*	*Fr-S*	*Fl-2S*	*Ja-S*	*Fl-F*	*(Fr-F, Wa-F, Af-F)*
	Fl-1S	0.13							
	Af-S	0.59	0.55						
S	Fr-S	0.67	0.63	0.25					
	Fl-2S	0.80	0.84	0.55	0.46				
	Ja-S	0.80	0.67	0.38	0.46	0.59			
	Fl-F	0.84	0.71	0.50	0.59	0.63	0.21		
F	(Fr, Wa, Af)	1.13	1.10	0.88	0.97	0.59	0.59	0.38	
	Ja-F	1.12	1.18	0.97	1.05	0.84	0.67	0.46	0.42

Surprisingly, the three samples with identical sequences were from geographically widespread sites in France, Washington (USA), and Burundi, Africa. The nucleotide diversity for the whole sample was 0.0065. If the sample is divided into the two allozyme categories, the mean diversity is 0.0056 for the six S sequences and 0.0029 for the five F sequences. Of course, three of the F sequences are identical. The average diversity for only the comparisons between the different allozyme alleles is much higher at 0.0084.

Similarly, the proportion of amino acid sites that are different (segregating) in the population can be estimated by

$$\hat{p}_{S.aa} = \frac{S}{N} \tag{2.24a}$$

where S indicates the number of amino acid sites that differ among sequences and N is the number of codons compared. Also, given the frequency of amino acid sequences for a given gene, then amino acid diversity can be calculated as

$$\pi' = \sum_{ij} p_i p_j \pi'_{ij} \tag{2.24b}$$

where π'_{ij} is the proportion of amino acids that differ when the ith and jth sequences are compared. When there is random mating, this is equal to amino acid heterozygosity. For most genes, the level of amino acid diversity appears to be quite low; for example, at the *D. melanogaster Adh* gene discussed in Example 2.12, only 1 of 255 sites in the whole gene has an amino acid difference. At the other extreme, for *MHC* genes, most of the nucleotide variation results in amino acid variation, as mentioned in Chapter 1 and as discussed further in Chapter 8.

VII. MEASURES OF GENETIC DISTANCE

Large studies of genetic variation often encompass many populations or species and a number of loci. A number of genetic similarity and distance measures have been proposed and used to evaluate the amount of variation shared among groups. These measures help to consolidate the data into manageable proportions and aid in visualizing general relationships among the groups. Although some information is lost when arrays of frequency data are reduced to a single value, patterns among populations obscured by the mass of numbers may become apparent by using genetic distance values. Distance measures are generally analogous to geometric distance; that is, zero distance is equivalent to no difference between the groups.

Similarities or differences in the type, amount, and pattern of genetic variation between populations can be the result of many factors. For example, if two populations are genetically similar, this could be either because (1) they recently separated into two populations, (2) gene flow occurred between them, (3) they were large populations (with little genetic drift), or (4) similar selection pressures affected loci similarly in both populations. Likewise, if two populations are different, then this could be because (1) they have been isolated for a long time and there has been no gene flow between them, (2) genetic drift has generated large differences, or (3) there are different selective pressures in the two populations. More than one—or possibly all—of these factors may be important in a particular situation. Furthermore, if these forces have been important in the past, it is often difficult to reconstruct historically their relative roles. We reserve until later discussion concerning the relative importance of these factors in affecting the amount of similarity or differences among populations.

Before discussing some approaches to measure genetic distance between populations, we should test the populations to determine whether the allele frequencies are significantly different. We can do this by using a χ^2 test for heterogeneity over populations where for two alleles

$$\chi^2 = \frac{2N\,V(\hat{p})}{\bar{p}\bar{q}} \tag{2.25a}$$

where N is the total sample size and $\bar{p}$ and $\bar{q}$ are the estimates of the average (weighted) allele frequencies for A_1 and A_2 in the total sample (Workman and Niswander, 1970). $V(\hat{p})$ is the weighted variance and is calculated as

$$V(\hat{p}) = \sum \frac{N_j}{N}\hat{p}_j^2 - \bar{p}^2$$

where N_j and p_j are the sample size and frequency of allele A_1 in population j. For n alleles, the χ^2 value is

$$\chi^2 = 2N\sum_{i=1}^{n} \frac{V(\hat{p}_i)}{\bar{p}_i} \tag{2.25b}$$

where $\bar{p}_i$ and $V(\hat{p}_i)$ are the estimated mean and variance of the frequency of allele i. A χ^2 value for m loci is the sum of the separate χ^2 values for each locus and has $(m-1)\ (n-1)$ degrees of freedom (see Example 2.13 below for a calculation of this χ^2 value).

A number of measures of genetic distance have been suggested over the past several decades. In practice, many of these measures are highly correlated, particularly when the differences between the populations are small, even though the measures are often based on different biological or mathematical assumptions. However, when the differences become larger, then there are often substantial differences between genetic distance measures on the same data set. This is particularly true when comparing genetic distance measures developed for microsatellite loci that assume particular modes of mutation and generally weight genetic distance by the square of the difference in the number of repeats for different alleles (Takezaki and Nei 1996; Goldstein and Pollock, 1997; Paetkau *et al.*, 1997). We do not present any of these measures here but suggest that they be interpreted cautiously and that a standardized measure (Goodman, 1997) may be preferable. Here we introduce the most commonly used genetic distance measure, the **standard genetic distance** of Nei (1972). This measure has the beneficial property that when there is neutrality (no differential selection) and genetic drift and all new mutations result in new alleles (the infinite-allele model) that it increases linearly with time (see below).

To calculate Nei's standard genetic distance for a single locus with n alleles, first calculate the genetic identity

$$I = \frac{J_{xy}}{(J_x J_y)^{1/2}} \tag{2.26a}$$

where

$$J_{xy} = \sum_{i=1}^{n} p_{i \cdot x} p_{i \cdot y}, \qquad J_x = \sum_{i=1}^{n} p_{i \cdot x}^2, \qquad J_y = \sum_{i=1}^{n} p_{i \cdot y}^2$$

and $p_{i \cdot x}$ and $p_{i \cdot y}$ are the frequencies of the ith allele in populations x and y. Note that I (and I', see below) has a range from zero, where no alleles are shared between the two populations, to unity, where the two populations have identical allele frequencies.

The genetic distance between the two populations is then defined as

$$\begin{aligned} D &= -\ln(I) \\ &= -\ln J_{xy} + 1/2 \ln J_y + 1/2 \ln J_x \end{aligned} \tag{2.26b}$$

One of the important aspects of this genetic distance measure is that, if there is no gene flow between the populations, it increases approximately linearly with time t as a function of the mutation rate u as

$$D = 2ut \tag{2.26c}$$

This is the relationship found for the molecular clock discussed on p. 417, that is, the genetic divergence between populations or species increases at a constant rate over time as a function of the mutation rate.

For multiple loci, J_{xy}, J_x, and J_y values are calculated by summing over alleles at all loci included in the study. The average value per locus is then calculated by dividing these sums by the number of loci. These average values, J'_{xy}, J'_x, and J'_y, are then used in the genetic identity formula above to calculate the identity I', and the distance becomes

$$D' = -\ln(I') \tag{2.26d}$$

D and D' have ranges from zero, for populations with identical allele frequencies, to infinity, for populations that do not share any alleles. The variance of D and D' can be calculated by the formulas in Nei (1987).

To calculate an unbiased estimate of D, a correction for the homozygosity estimates J_x and J_y is recommended but is probably necessary only when one of the sample sizes is less than 50. In this case, an unbiased estimate of J_x is

$$\hat{J}_x = \frac{2N_x \sum \hat{p}_i^2 - 1}{2N_x - 1} \tag{2.26e}$$

where N_x and p_i are the sample size and the frequency of the ith allele from population x (the unbiased estimate of J_y is analogous) (Nei, 1987). This estimate is one minus the estimate given for heterozygosity in expression 2.18a. However, this correction can give spurious results when the homozygosity is low, as for microsatellite loci, and the sample size is small. A simple example illustrating the calculation of the χ^2 and genetic distance values is given in Example 2.13.

Example 2.13. Human geneticists and anthropologists have surveyed many human populations for genetic variation at red blood cell, allozyme, and other variants (Nei and Roychoudhury, 1993; Cavalli-Sforza *et al.*, 1994). Table 2.20 gives an example of some of these data for two biallelic red blood cell loci, *Secretor* and *Lewis*, in Europeans and Africans. Using the sample sizes given, a χ^2 value for heterogeneity of these samples can be calculated. There is not significant heterogeneity for the *Secretor* locus, but there is both for the *Lewis* locus and for the two loci combined (values of $\chi^2 > 3.84$ are significant at the 0.05 level with one degree of freedom). We can also calculate genetic distance using the standard genetic distance of Nei given in the text. As is consistent with the χ^2 values, the genetic distance values for the *Secretor* locus are quite small, whereas for the *Lewis* locus they are quite large. The average genetic distance over the two loci is also rather large and is significantly greater than zero using the approach of Nei (1987).

TABLE 2.20 The allele frequencies for two human blood group loci in Europeans and Africans (from Cavalli-Sforza and Bodmer, 1971) with the χ^2 measure of heterogeneity, and Nei's standard genetic distance.

Allele	*Europeans* ($N = 1059$)		*Africans* ($N = 120$)
Secretor			
Se	0.523		0.573
se	0.477		0.427
Lewis			
Le	0.816		0.319
le	0.184		0.681
χ^2			
Secretor		$\chi^2 = 1.89$	
Lewis		$\chi^2 = 136$	
Total		$\chi^2 = 138$	
Genetic distance			
		$\hat{D}_N(\hat{I}_N)$	
Secretor		0.004 (0.996)	
Lewis		0.491 (0.612)	
Average		0.243 (0.784)	

The **precision**—that is, the reduction in the amount of variation around the estimate of the genetic distance between two populations—increases as more independent loci are examined. With the discovery and use of microsatellite loci with many alleles, similarly the precision of genetic distance estimation would be higher than for loci with only two or a few alleles. Kalinowski (2002) examined, using computer simulation, the effect of the number of loci and the number of alleles per locus on the precision of four commonly used genetic distance measures. Overall, he found that the level of precision was determined primarily by the number of independent alleles (a locus with three alleles would have two independent alleles). For example, samples with 20 independent biallelic loci or 5 independent loci with 5 alleles each would each have 20 independent alleles and similar levels of precision.

PROBLEMS

1. A survey of a SNP site segregating for nucleotides C and G in a chimpanzee population found the following numbers of the different genotypes:

C/C	C/G	G/G
11	42	64

What is the frequency of G and what are the expected numbers of the three genotypes?

2. A population survey for a microsatellite locus in a *Drosophila* population resulted in the following numbers of different genotypes:

A_1A_1	A_1A_2	A_1A_3	A_2A_2	A_2A_3	A_3A_3
8	38	121	27	252	401

Calculate the allele frequencies for all the alleles and test whether the genotype frequencies deviate significantly from Hardy–Weinberg proportions.

3. Assume that the frequencies of A_2 for an autosomal locus in males and females are 0.8 and 0.4, respectively. Calculate the genotype frequencies for the next two generations. Derive the last two equations in expression 2.6*b*.

4. Assume that the frequencies of A_1 for a locus in a haplo-diploid organism are 0.0 and 0.3 in males and females, respectively. What are the allele frequencies in the two sexes for the first four generations? Graph these results. In how many generations will the female frequency deviate from the mean frequency by less than 0.001?

5. For an allozyme survey in a plant species known to have self-fertilization, the following numbers of individuals were observed:

F	FS	S
25	14	21

Calculate the χ^2 value for these data using expression 2.15*b*. Calculate the χ^2 value using expected values from expressions 2.16*a* and 2.16*b*.

6. For a survey of a microsatellite locus in a snake population, the following numbers of individuals were observed:

Phenotype	120	120/122	122
Number	17	18	14

What is the estimated frequency of a null allele?

7. The number of newborn Americans of European ancestry who have the recessive autosomal disease cystic fibrosis is approximately 1 in 2500. What is the estimated frequency of this disease allele and what proportion of individuals would be heterozygous for this allele? In a random-mating population, what proportion of matings would be between two carriers?

8. A survey of a population of spotted owls found seven mtDNA haplotypes in the numbers 1, 4, 3, 21, 4, 1, and 7. What are the estimated frequencies of these haplotypes? What is the estimated mtDNA diversity using the small-sample-size correction?

9. In a sample of peppered moths from England after the frequency of melanics began to decline, there were 307 melanics, 5 insularia, and 448 typicals. What are the estimated frequencies for the three alleles?

10. Assume that the frequency of allele A_1 at a biallelic locus is 0.3. What observed heterozygosity would give an F value of -0.05? What sample size would be necessary to detect this effect at the 5% significance level?

11. Two populations of humans were sampled (x had 25 individuals and y had 50 individuals) and were found to have the following allele frequencies for two SNP sites:

	Site 1		Site 2	
	C	G	T	G
x	0.22	0.78	0.04	0.96
y	0.37	0.63	0.03	0.97

Calculate the χ^2 value to determine whether these populations are significantly different. Calculate D for each locus and over both loci.

12. Assume that the following mtDNA sequences were found in five different individuals:

AATCGAGACTTTAGC
ATTCCAGATTTAAGC
ATTCCAGATTTAAGC
AATCGAGACTTTAGC
TATCGAGACTATCCC

What are the estimates of p_n and π?

13. Access one of the software packages listed in Table 2.1, and calculate the answer for one of the questions above.

14. In a survey of *HLA-A* in the Havasupai, the following expected and observed numbers of homozygotes and heterozygotes were determined:

	Heterozygotes	Homozygotes
Observed	84	38
Expected	73.6	48.4

What is the probability of observing a difference this extreme? Assume that you are conducting a one-sided test to determine the probability of observing this many heterozygotes or more. Use the binomial formula with the expected proportions of the two classes calculated from the expected numbers above.

15. A survey of four sequences 900 bp long had the following numbers of pairwise differences between them: 1, 4, 2, 5, 6, and 10. What is the estimate of nucleotide diversity of this sample?

3

Selection: An Introduction

The importance of the great principle of Selection mainly lies in the power of selecting scarcely appreciable differences . . . which can be accumulated until the result is made manifest to the eyes of every beholder.

Charles Darwin (1859)

Selection is a wastebasket category that includes all causes of directed change in gene frequency that do not involve mutation or introduction from without.

Sewall Wright (1955)

Although there is no difficulty in theory in estimating fitnesses, in practice the difficulties are virtually insuperable. To the present moment no one has succeeded in measuring with any accuracy the net fitness of genotypes for any locus in any environment in nature.

Richard Lewontin (1974)

Although natural selection generally is believed to be the dominant force shaping the course of evolution, it may be complicated and is sometimes a difficult phenomenon to quantify. Most of the phenotypic characters that we associate with particular species and use in their description or classification—such as body shape and color in insects, flower morphology and color in plants, size and physiologic characteristics in mammals—are thought to be the end result of natural selection. In other words, these phenotypes were adaptive in a particular population and, therefore, increased in frequency so that these types are now characteristic of the species. Furthermore, these phenotypic characteristics are thought to reflect the genetic composition of a population or species. In some cases, this "adaptationist's view" may be a grossly oversimplified explanation of the evolution of a species because some characteristics may be due to historical, nonadaptive events (Lewontin, 1978).

Let us visualize the process of natural selection in a simple way (Figure 3.1). Imagine a population with a given amount of genetic (or genotypic)

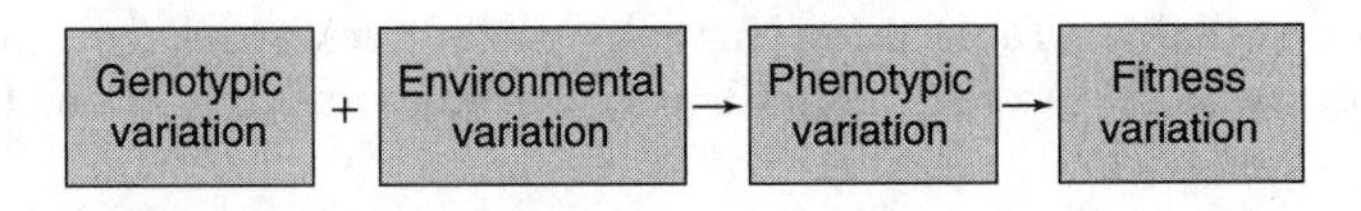

Figure 3.1. A schematic representation of the relationship among genetic (or genotypic), environmental, phenotypic, and fitness variation.

variation among individuals. This genetic variation, in combination with environmental components, determines the phenotypic variation for the ensemble of traits expressed by that particular organism among individuals. The amount and nature of the phenotypic variation in turn determine the extent of variation in survival, fecundity, mating ability, and other factors that ultimately determine whether the alleles of a particular individual will be passed on to future generations. The overall effect of these factors is termed the **relative fitness** of different individuals in the population and can be simply defined as ***the relative ability of different genotypes to pass on their alleles to future generations***. In the next chapter, we examine in more detail the different factors contributing to differential relative fitnesses. Evolution itself is the end result of the differential contribution of various genotypes to future generations.

In Chapter 1, we discussed in some detail the extent of genetic variation among individuals in a population. Environmental parameters have also been studied in detail by ecologists, climatologists, soil scientists, and other researchers. However, much remains unknown about the integration of these components. For example, the relative roles of genetic and environmental components in the determination of phenotypic variation in a population are often poorly understood. In addition, little is generally known about the relationship between phenotypic variants and relative fitness. The introductory characterization of the components important in natural selection presented here assumes that the different factors are independent and additive, assumptions that are not necessarily valid.

Because of the complexities of the process of natural selection, simplifying assumptions are usually made in studies of selection. This reductionist approach, which was initiated by Fisher, Haldane, and Wright, has been invaluable in establishing the feasibility and hence the potential significance of natural selection in evolution. Many of what are thought to be important biological complexities have since been incorporated into selection models to make them more realistic. This is our approach in examining selection. That is, we begin with simple models and examine the properties of these systems; then we relax the assumptions of the model one at a time to understand the effect of each assumption. For example, we use single-gene models until Chapter 10 because the predictions from these models can be generally used to understand situations in which two or more loci are important. Here we concentrate on the dynamics of genetic change caused by selection (the rate of change in allele or genotype frequencies) and the statics of genetic variation (those situations in which there is no change, or maintenance of genetic variation occurs, as the result of selection).

To visualize the basic effects of selection, and in later chapters, inbreeding, genetic drift, gene flow, and mutation and to teach about the impact of these factors, there are a number of free computer software packages available (Table 3.1). Virtually all of these packages cover the basic effects of selection, genetic drift, gene flow, and mutation. Populus

R. A. FISHER (1890–1962)

Sir Ronald Fisher is perhaps most widely known for his contributions to statistics. Throughout his lifetime, one of his major interests was evolutionary genetics, and his early book *The Genetical Theory of Natural Selection* (Fisher, 1930) was a landmark of synthesis of Darwinian selection and genetics. His contributions to genetics were theoretical (such as the development of concepts of adaptive selection and inbreeding) or statistical (such as the estimation of allele frequencies, selection intensity, or inbreeding coefficients). He also developed the "fundamental theorem of natural selection," part of an effort by him and others to provide a unifying conceptualization for evolution. Fisher was born and lived in England and was for most of his life a strong advocate of eugenics. His research papers have been published as a collection (Bennett, 1971–1974), and a biography details much of his life and contributions to genetics (Box, 1978; see also Crow, 1990). (©Photograph by Antony Barrington Brown. Reproduced with permission of the Fisher Memorial Trust.)

TABLE 3.1 A list of some of the software packages (and the developer) most commonly used in demonstrating the effects of population genetics factors to students with their websites (subject to change). All of these programs consider one-locus selection, genetic drift, gene flow, and mutation.

Software Package	*Website*
Populus (D. Alstad)	http://www.cbs.umn.edu/populus/
PopG (J. Felsenstein)	ftp://evolution.gs.washington.edu/pub/popgen/popg.html
AlleleA1 (J. Herron)	http://faculty.washington.edu/~herronjc/SoftwareFolder/ software.html.
EvoTutor (A. Lemmon)	http://www.evotutor.org/Software.html
PopGen (J. Aspi)	http://cc.oulu.fi/~jaspi/popgen/popgen.htm
WinPop (P. Nuin)	http://evol.biology.mcmaster.ca/paulo

(http://www.cbs.umn.edu/populus/) appears to be the most widely used. This package covers a number of more advanced topics, such as two-locus selection and selection in different environments. For more advanced understanding of these factors, it is often useful for students to use spreadsheets or programming to generate the effects of particular factors for themselves.

For several reasons, understanding and documenting the process of adaptive phenotypic change have been difficult. First, phenotypic change is usually gradual, taking place only over many generations. In a long-lived organism, it may take centuries for a change to become discernible, although still much faster than that observed in the paleontological record. However, examples are accumulating of rapid evolutionary change in response to strong ecological or environmental forces, both in response to natural forces and in response to human-caused environmental changes (Thompson, 1998; Hendry and Kinnison, 1999; Palumbi, 2001). Some documented selective changes to human environmental changes have been very rapid as in the evolution of pesticide resistance (see Example 3.1 for case studies in insects and rodents) and antibiotic resistance (Baquero and Blaszquez, 1997; Palumbi, 2001; Hughes, 2003). Second, the determination of many phenotypic traits is often quite complicated and may be influenced by a number of genes and complex environmental factors. In recent years, it has become possible to identify individual genes influencing quantitative traits, QTLs, so that understanding the genetic makeup of adaptive traits is feasible (Lynch and Walsh, 1998), and determining change in their allele frequencies is possible. Although understanding the process of natural selection continues to be difficult, recent advances are allowing us to understand in detail the components of evolutionary response.

Example 3.1. The use of chemicals to control pests has generally been successful for only a short time until the pests develop genetic resistance to the chemical (McKenzie, 1996). Synthetic organic pesticides were introduced on a large scale following World War II to control both agricultural and public health pests. At first these pesticides greatly increased crop production and reduced the incidence of many diseases. However, by the 1970s, most species of anopheline mosquitoes, which transmit malaria, were resistant to DDT, the house fly was resistant to many different insecticides, seven species of rodents were resistant to rodenticides, and a number of plants were resistant to herbicides. Since then, the number of species resistant to pesticides and the number of pesticides to which a given species is resistant have increased substantially.

As a result, new pesticides such as toxins derived from the cosmopolitan soil bacteria *Bacillus thuringiensis* (Bt) that are purported to not harm nontarget organisms or humans have become important in insect control. Bt genes that produce these toxins also have been introduced into a number of crop plants. Soon after Bt pesticides were widely used, resistance was reported (Gould *et al.*, 1997; Tabashnik *et al.*, 1997), and there appears to be a high potential for widespread resistance to both Bt toxins and Bt transgenic crops in many pest species (Ferré and Van Rie, 2002). For evolutionary genetics, the development of pesticide resistance has provided a number of case studies of rapid selective change—observations often not possible from natural selective forces. On the other hand, population genetic principles can provide a framework to evaluate the efficacy of various strategies to combat pesticide resistance (Vacher *et al.*, 2003) or to determine the potential for the introduction of resistant transgenes from genetically modified crops to wild relatives (Snow *et al.*, 2003).

The genetic basis of pesticide resistance may be the result of many genes, particularly in laboratory-selected strains, mutants at a single, or a few genes (Taylor and Feyereisen, 1996) or expansion of gene families (Ranson *et al.*, 2002). Resistance to some insecticides among mosquitoes that are vectors for diseases such as malaria (*Anopheles gambiae*) and West Nile virus (*Culex pipiens*) appears to be the result of a single-amino acid substitution. For example, a single nucleotide change, GGC (glycine) to AGC (serine) at codon position 119 in the gene for the enzyme acetylcholinesterase (*ace*-1), appears to result in insensitivity to organophosphates (Weill *et al.*, 2003). To determine the distribution of the G119S resistance mutation, Weill *et al.* (2003) examined 10 resistant strains and 19 susceptible strains of *C. pipiens* in two different subspecies (Table 3.2). All of

TABLE 3.2 Nucleotide polymorphism in exon 3 of gene *ace*-1 in *Culex pipiens* mosquitoes (Weill *et al.*, 2003). Samples are listed according to subspecies (*C. p. pipiens* and *C. p. quinquefasciatus*), presence of resistance (R) or susceptibility (S), and country of origin. Only polymorphic sites are indicated, and a dash indicates identity with the R consensus sequence for the subspecies (top line). The position of the G119S mutation at nucleotide position 739 is indicated by an asterisk.

			Nucleotide position																						
Subspecies	R or S	Country	450	453	471	498	528	564	573	603	651	691	696	714	732	*739	747	753	774	777	780	790	798	813	846
C. p. pipiens	R	Consensus	T	C	A	T	G	G	G	C	G	C	C	C	C	A	C	C	C	C	C	G	G	A	T
	R	Burkina Faso	-	-	-	-	-	-	-	-	-	-	-	-	-	-	-	-	-	-	-	-	-	-	-
	R	Zimbabwe	-	-	-	-	-	-	-	-	-	-	-	-	-	-	-	-	-	-	-	-	-	-	-
	R	Ivory Coast	-	-	-	-	-	-	-	-	-	-	-	-	-	-	-	-	-	-	-	-	-	-	-
	R	Mali	-	-	-	-	-	-	-	-	-	-	-	-	-	-	-	-	-	-	-	-	-	-	-
	R	Martinique	-	-	-	-	-	-	-	-	-	-	-	-	-	-	-	-	-	-	-	-	-	-	-
	R	Brazil	-	-	-	-	-	-	-	-	-	-	-	-	-	-	-	-	-	-	-	-	-	-	-
	S	United States	-	T	-	C	-	-	-	-	-	-	-	-	-	G	-	-	-	-	-	-	-	-	G
	S	United States	-	T	-	C	-	-	-	-	-	-	-	-	-	G	-	-	-	-	-	-	-	-	G
	S	United States	-	T	-	-	-	-	A	-	-	-	-	-	-	G	-	-	-	-	-	A	-	-	-
	S	United States	-	T	-	-	-	-	A	-	-	-	-	-	-	G	-	-	-	-	-	A	-	-	-
	S	China	-	T	C	C	-	-	-	-	-	-	-	-	-	G	-	-	T	-	-	-	-	-	-
	S	China	-	T	-	C	-	-	-	-	-	-	-	-	-	G	-	-	T	-	-	-	-	-	-
	S	Thailand	-	T	C	C	-	-	-	-	-	-	-	-	-	G	-	-	-	-	-	-	-	-	G
	S	India	-	T	-	C	-	A	-	-	-	-	-	-	-	G	-	-	-	-	-	-	-	-	G
	S	South Africa	-	T	C	C	-	-	-	-	-	-	-	-	-	G	-	-	-	-	-	-	-	-	G
	S	South Africa	-	T	-	-	-	-	A	-	-	-	-	-	-	G	-	-	T	-	-	-	-	-	-
	S	Ivory Coast	-	T	-	C	-	-	-	-	-	-	-	-	-	G	-	-	-	-	-	-	-	-	-
	S	Congo	-	T	C	C	-	-	-	-	-	-	-	-	-	G	-	-	T	-	-	-	-	-	G
	S	Brazil	-	T	-	C	-	-	-	-	-	-	-	-	-	G	T	-	-	-	-	-	-	-	G
	S	Polynesia	-	T	-	-	-	-	-	-	-	-	-	-	-	G	-	-	-	-	-	-	-	-	G
C. p. quinque	R	Consensus	A	T	A	C	G	G	A	C	C	A	G	T	T	A	C	T	C	T	T	G	G	G	T
	R	Tunisia	-	-	-	-	-	-	-	-	-	-	-	-	-	-	-	-	-	-	-	-	-	-	-
	R	Portugal	-	-	-	-	-	-	-	-	-	-	-	-	-	-	-	-	-	-	-	-	-	-	-
	R	Italy	-	-	-	-	-	-	-	-	-	-	-	-	-	-	-	-	-	-	-	-	-	-	-
	R	France	-	-	-	-	-	-	-	-	-	-	-	-	-	-	-	-	-	-	-	-	-	-	-
	S	Belgium	-	-	-	-	-	-	-	-	-	-	-	-	-	G	-	-	-	-	-	-	-	-	-
	S	Belgium	-	-	-	-	-	-	-	-	-	-	-	-	-	G	-	-	-	-	-	-	-	-	-
	S	Australia	-	-	-	-	-	-	-	A	-	-	-	-	-	G	-	-	-	-	-	-	-	-	-
	S	France	-	-	-	-	-	-	-	-	-	-	-	-	-	G	-	-	-	-	-	-	-	-	-
	S	Holland	-	-	-	-	A	-	-	A	-	-	-	-	C	G	-	-	-	C	-	-	A	-	-

the resistant strains had the same G119S mutation at nucleotide position 739. Furthermore, although 23 nucleotides were polymorphic in the exon sequenced, a different unique haplotype was associated with the resistance mutation in the two subspecies. This suggests that the same mutation arose independently on different haplotypes in the two subspecies. The complete lack of variation within samples for each of these two resistant haplotypes suggests that they have increased quite recently. In addition, Weill *et al.* showed that the mutated amino acid was in the active "gorge" of the enzyme and that the same catalytic properties and insecticidal sensitivity were present in transfected mosquitoes.

Instead of undergoing a qualitative change, some insects become resistant to insecticides by producing large amounts of detoxifying enzymes. For example, the overtranscription of a single allele at a single cytochrome P450 gene in *Drosophila melanogaster* has been shown to confer worldwide resistance to DDT (Daborn *et al.*, 2002). In this case, microarray analysis of all of the P450 genes identified the specific overtranscribed gene and subsequent genetic and sequence analysis demonstrated that the single resistance allele has resulted from the insertion of an *Accord* transposable element in the 5′ end of the gene.

The Norway or brown rat, *Rattus norvegicus*, a widespread agricultural and domestic pest, was controlled for about 15 years by the anitcoagulant warfarin. However, resistance to warfarin evolved quickly, and resistant rats have been found throughout the United States and Europe. Resistance appears to be the result of the dominant allele, R, at an autosomal locus. Most of the resistant animals are heterozygotes for the allele and for the wildtype, susceptible allele, S. However, RR homozygotes have low viability because of a 20-fold increase in vitamin K requirement. When warfarin is being applied, a net heterozygote advantage results, and the relative survivals of genotypes RR, RS, and SS are 0.37, 1.0, and 0.68, respectively (Greaves *et al.*, 1977). In one population, the frequency of the resistance allele was stable and polymorphic over nearly a decade, but when warfarin was no longer applied in another population, the frequency of the resistant allele declined very quickly, a result that illustrates the cost of maintaining the resistance allele.

Kohn *et al.* (2000, 2003) attempted to locate the warfarin resistance gene (Rw) and characterize variation at closely linked markers in German rat populations, which have varying levels of warfarin resistance (recently, Rost *et al.*, 2004, have identified a gene that causes warfarin resistance in rats and two related human disorders). It appears that there has been single and recent origin of warfarin resistance in these populations. Kohn *et al.* suggest, based on indirect genetic evidence, that heterozygote advantage is not present in the German populations, perhaps because these resistant rats do not suffer vitamin K deficiency, as documented in other populations (Smith *et al.*, 1993; Thijssen, 1995)—in other words, heterozygote advantage for warfarin resistance appears to present for some resistant alleles (populations) and not for others.

I. THE BASIC SELECTION MODEL

A number of assumptions are implicit in the basic selection model that we consider first. These assumptions are listed in Table 3.3 and are separated into those concerning the genetic system, those involving the mode of selection, and those involving other factors, such as inbreeding, genetic drift, and gene flow. Later in this chapter and in subsequent chapters, we relax most of these assumptions and evaluate their potential effects.

TABLE 3.3 A categorized list of assumptions in the basic selection model.

Genetic system
(a) Single, biallelic, autosomal locus
(b) Diploidy
(c) Random mating among individuals
Selection
(a) Selection identical in both sexes
(b) Selection occurs through differences in viability
(c) Constant value in space and time for each genotype
Other factors
(a) Nonoverlapping generations
(b) No inbreeding
(c) Infinite population size
(d) No gene flow
(e) No mutation

Let us begin by considering the situation in which the relative fitnesses of three possible genotypes, A_1A_1, A_1A_2, and A_2A_2, are designated as w_{11}, w_{12}, and w_{22}, respectively. Because we consider initially only viability selection, the relative fitness of an individual is its relative probability of survival. To illustrate what this means, let us assume that there are 100, 200, and 100 zygotes of the genotypes A_1A_1, A_1A_2, and A_2A_2, respectively. Because of various factors causing mortality in these individuals as they mature, only 80, 160, and 50 of the three genotypes survive to become adults. The proportions of the three genotypes that survive are then 0.8, 0.8, and 0.5. For heuristic reasons and algebraic simplicity, it is useful to standardize these survival values to obtain relative survival (fitness) values so that the largest is equal to 1. Therefore, the relative fitness values in this example become $0.8/0.8 = 1$, $0.8/0.8 = 1$, and $0.5/0.8 = 0.625$ for genotypes A_1A_1, A_1A_2, and A_2A_2. In this case, there is selection against the homozygote A_2A_2, and the other two genotypes, A_1A_1 andA_1A_2, have the same relative fitnesses. These relative fitness values are assumed here to be constant but, as discussed in the next chapter, may vary in response to different environments, the frequency of different genotypes, or other factors.

The genotype frequencies before selection are given in terms of allele frequencies assuming Hardy–Weinberg proportions (Table 3.4) where

TABLE 3.4 The frequency of genotypes before and after selection, assuming Hardy–Weinberg proportions before selection.

	Genotype			
	A_1A_1	A_1A_2	A_2A_2	*Total*
Relative fitness	w_{11}	w_{12}	w_{22}	—
Frequency before selection	p_0^2	$2p_0q_0$	q_0^2	1
Weighted contribution	$p_0^2w_{11}$	$2p_0q_0w_{12}$	$q_0^2w_{22}$	$\bar{w}$
Frequency after selection	$\frac{p_0^2w_{11}}{\bar{w}}$	$\frac{2p_0q_0w_{12}}{\bar{w}}$	$\frac{q_0^2w_{22}}{\bar{w}}$	1

p_0 and q_0 indicate the frequency of alleles A_1 and A_2 in generation zero (sometimes we drop the generational subscript for the allele frequency). The relative contribution of the three genotypes to the next generation is determined by the product of the relative fitness and the frequency before selection of that genotype. The **mean fitness** of the population, $\bar{w}$, is ***the sum of the relative contributions of the different genotypes***. Given Hardy–Weinberg proportions before selection, the mean fitness is

$$\bar{w} = p_0^2w_{11} + 2p_0q_0w_{12} + q_0^2w_{22} \tag{3.1}$$

If the relative contributions are standardized by the mean fitness value, then the frequencies of genotypes after selection can be obtained as shown in the bottom line of Table 3.4. The frequency of A_2 after selection (q_1) is equal to half of the frequency of the heterozygote, A_1A_2 (because only half the genes in A_1A_2 are A_2 alleles), plus the frequency of the homozygote A_2A_2, or

$$\begin{aligned} q_1 &= \frac{1}{2}\left(\frac{2p_0q_0w_{12}}{\bar{w}}\right) + \frac{q_0^2w_{22}}{\bar{w}} \\ &= \frac{p_0q_0w_{12} + q_0^2w_{22}}{\bar{w}} \end{aligned} \tag{3.2}$$

This expression, which we use to explore the effects of different types of selection below, shows that the frequency of A_2 after selection is a function of the allele frequencies before selection and the relative fitnesses of the genotypes.

If the amount of allele frequency change is defined as

$$\Delta q = q_1 - q_0$$

where the q_0 and q_1 are the frequencies of A_2 before and after selection (or, more specifically, the frequencies in the zero and first generations) and Δ is the Greek capital letter delta, then substituting for q_1 from expression 3.2 yields

$$\begin{aligned}\Delta q &= \frac{p_0 q_0 w_{12} + q_0^2 w_{22}}{\bar{w}} - q_0 \\ &= \frac{p_0 q_0 w_{12} + q_0^2 w_{22} - q_0 \bar{w}}{\bar{w}}\end{aligned}$$

If we substitute for $\bar{w}$ from expression 3.1 (remembering $q = 1 - p$), then we can simplify to get

$$\begin{aligned}\Delta q &= \frac{pqw_{12} + q^2 w_{22} - q(p^2 w_{11} + 2pqw_{12} + q^2 w_{22})}{\bar{w}} \\ &= \frac{q(pqw_{22} - pqw_{12} - p^2 w_{11} + p^2 w_{12})}{\bar{w}} \\ &= \frac{pq[q(w_{22} - w_{12}) - p(w_{11} - w_{12})]}{\bar{w}} \qquad (3.3a)\end{aligned}$$

This final expression is quite important, and we use it to demonstrate how different types of selection influence allele frequencies. If $p = 0$, $q = 0$, or the term in brackets equals 0, then there will be no change in allele frequency.

Another approach to examining the change in allele frequency illustrates how changes in the mean fitness are reflected as changes in allele frequency. Using the general expression for the mean fitness from expression 3.1, we find that the derivative with respect to q is

$$\begin{aligned}\frac{d\bar{w}}{dq} &= \frac{d}{dq}(p^2 w_{11} + 2pqw_{12} + q^2 w_{22}) \\ &= \frac{d}{dq}(w_{11} - 2qw_{11} + q^2 w_{11} + 2qw_{12} - 2q^2 w_{12} + q^2 w_{22}) \\ &= 2[-pw_{11} + (1 - 2q)w_{12} + qw_{22}] \\ &= 2[q(w_{22} - w_{12}) - p(w_{11} - w_{12})]\end{aligned}$$

The term in brackets in this last expression is a function of the fitness differences between the genotypes, weighted by allele frequencies, and is identical to the term in brackets in expression 3.3*a*. As a result, the expression for the change in frequency can be rewritten as

$$\Delta q = \frac{pq}{2\bar{w}} \frac{d\bar{w}}{dq} \qquad (3.3b)$$

Therefore, we can see that change in allele frequency is a function of the allele frequencies, the mean fitness, and the change in mean fitness with respect to allele frequency. Here, if $p = 0$, $q = 0$, or $d\bar{w}/dq = 0$, then the allele frequency will not change.

To explore how various fitness values can affect genetic change or stability, let us assume that the relative fitness of the heterozygote is unity. In other words, we standardize the fitnesses of the other two homozygotes against the heterozygote so that the fitnesses become w_{11}, 1, and w_{22} for genotypes A_1A_1, A_1A_2, and A_2A_2, respectively. If we assume that w_{11} and w_{22} can have different values, then four types of selection are possible (Figure 3.2). Two of these will result in directional selection (unshaded ar-

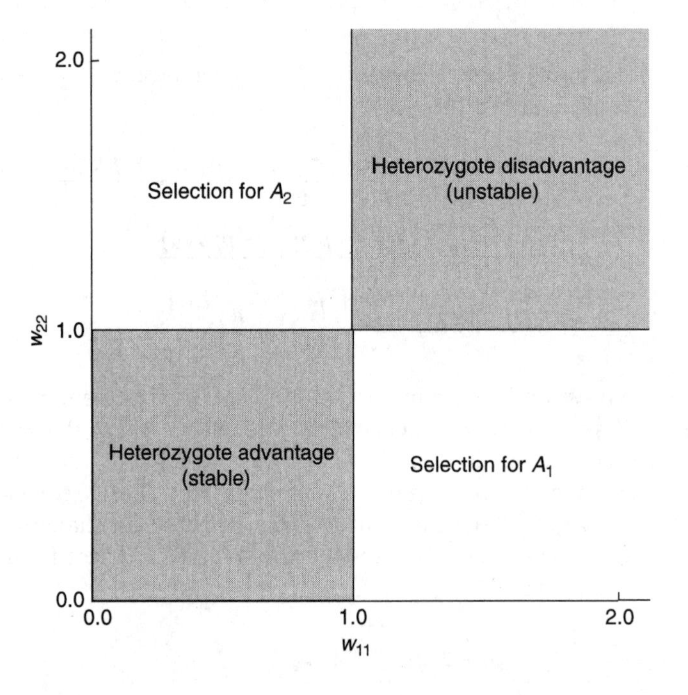

Figure 3.2. The different outcomes of selection when the fitnesses of the genotypes A_1A_1, A_1A_2, and A_2A_2 are w_{11}, 1, and w_{22}, respectively. The unshaded regions indicate fitness values that lead to directional selection, and the shaded regions indicate fitness values that give a stable equilibrium (heterozygote advantage) or an unstable equilibrium (heterozygote disadvantage).

eas). That is, if $w_{11} > 1.0 > w_{22}$, allele A_1 increases in frequency, and if $w_{11} < 1.0 < w_{22}$, allele A_2 increases. The two shaded areas lead to a stable equilibrium when the heterozygote has the highest fitness, $w_{11} < 1.0 > w_{22}$, and to an unstable equilibrium when the heterozygote has the lowest fitness, $w_{11} > 1.0 < w_{22}$. These fitness arrays for **heterozygous advantage** and **heterozygous disadvantage** are sometimes referred to as **overdominance** and **underdominance**, respectively, as originally used in plant and animal breeding. The only fitness region that leads to long-term polymorphism is the one in which there is a heterozygous advantage, whereas the other regions result in eventual fixation of either allele A_1 or A_2. However, even in these regions, allele frequency change may occur slowly for some fitness values, which makes it difficult in practice to actually differentiate between a stable polymorphism and a transient one.

At the point in which the two lines cross in the middle of the figure, all genotypes have the same relative fitness of unity. As we see in Chapter 8, having no selective differences between the three genotypes, or actually selective differences less important than the impact of genetic drift (a condition termed selective **neutrality**), plays a central role in molecular population genetics and evolution.

We now examine the properties of several fitness arrays with the specific fitness relationships summarized in Table 3.5. Two parameters are introduced here: s, the **selection coefficient** that ***measures the amount of selection against a homozygote***, and h, the level of **dominance**, which ***when multiplied by s measures the amount of selection against the heterozygote***. First, we consider the models of directional selection that lead to eventual fixation of one allele or the other. Then we consider selection that results in either a stable equilibrium—that is, the maintenance of both alleles in the population—or an unstable equilibrium, in which one allele or the other is lost, depending on the initial allele frequency.

TABLE 3.5 The fitness values for the different fitness relationships examined.

	Genotype		
	A_1A_1	A_1A_2	A_2A_2
General fitnesses	w_{11}	w_{12}	w_{22}
(a) Recessive lethal	1	1	0
(b) Detrimental alleles			
(1) Recessive	1	1	$1-s$
(2) Additive	1	$1-s/2$	$1-s$
(3) Dominant	1	$1-s$	$1-s$
(c) General dominance			
(1) Purifying selection	1	$1-hs$	$1-s$
(2) Adaptive Darwinian selection	$1+s$	$1+hs$	1
(d) Heterozygote advantage	$1-s_1$	1	$1-s_2$
(e) Heterozygote disadvantage	$1+s_1$	1	$1+s_2$

a. Recessive Lethal

Some alleles that cause diseases in humans, such as Tays-Sachs and cystic fibrosis, seem primarily to affect only the phenotype of the recessive homozygote, and before extensive medical treatment was available, they generally resulted in prereproductive death. Many chlorophyll mutants in plant species and many lethals in *Drosophila* also cause preadult mortality in homozygotes only. Presumably all species have some alleles that cause lethality as homozygotes before reproduction.

The relative fitnesses for a recessive lethal are represented by the fitness values in row (a) of Table 3.5. The fitness of one homozygote, A_1A_1, and that of the heterozygote are the same, but the other homozygote, A_2A_2, is lethal with a relative fitness of zero. Using equation 3.2 and substituting in these fitness values, we find that the allele frequency after selection is

$$q_1 = \frac{p_0q_0(1) + q_0^2(0)}{\bar{w}}$$

because

$$\begin{aligned}\bar{w} &= p_0^2(1) + 2pq(1) + q_0^2(0)\\ &= p_0^2 + 2p_0q_0\\ &= p_0(1+q_0)\end{aligned}$$

The allele frequency becomes

$$\begin{aligned}q_1 &= \frac{p_0q_0}{p_0(1+q_0)}\\ &= \frac{q_0}{1+q_0}\end{aligned} \tag{3.4a}$$

The change in allele frequency is

$$\begin{aligned}\Delta q &= q_1 - q_0\\ &= \frac{q_0}{1+q_0} - q_0\\ &= -\frac{q_0^2}{1+q_0}\end{aligned} \tag{3.4b}$$

For this array of fitness values, both the mean fitness and the change in allele frequency are functions of only the initial allele frequency. By examining the expression for mean fitness, one can see that it reaches a maximum of 1 when the population is fixed for $A_1 (q = 0)$. Furthermore, from expression 3.4*b*, it is clear that the change in frequency is largest when q is near 1.0 and is quite small when q is near 0.0. Example 3.2 shows an application of this theory to a *Drosophila* lethal. We use a number of *Drosophila* examples below to illustrate various evolutionary factors, both because *Drosophila* has played a central role in our achieving an understanding of these forces and because, as an important model organism in evolutionary genetics, it provides some of the clearest examples.

Example 3.2. When the frequency of a lethal is high, the allele frequency is reduced very quickly. An illustration is given in Figure 3.3 for the lethal *Glued* in *D. melanogaster*, a mutant that also reduces eye size and affects eye appearance in heterozygotes (Clegg *et al.*, 1976). The two replicate populations given in Figure 3.3 were initiated with all heterozygotes so that $q_0 = 0.5$ and then were followed for several generations. Note that the general trend observed is close to that expected for the first few generations but that there is a faster decline than expected in the later generations. Further investigation showed that selection against the heterozygote resulted in this faster decline.

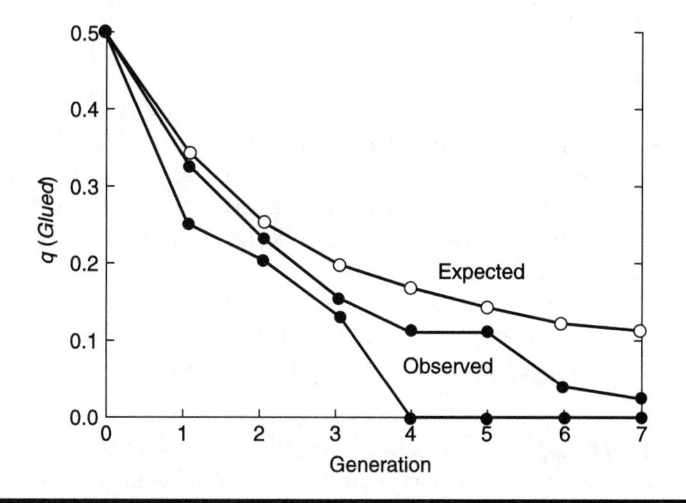

Figure 3.3. The expected decline in the frequency of the lethal *Glued* (open circles) and that observed in two experimental replicates (closed circles).

The relationship given in expression 3.4*a* is a recursive relationship; that is, the allele frequency at any time $t + 1$ is a function of the frequency at time t or

$$q_{t+1} = \frac{q_t}{1 + q_t}$$

Therefore, the allele frequency in the second generation is a function of the frequency in the first generation and is

$$q_2 = \frac{q_1}{1 + q_1}$$

If we substitute the value of q_1 from expression 3.4*a* in this expression, it becomes

$$q_2 = \frac{q_0}{1 + 2q_0}$$

This relationship can be generalized to give the frequency in generation t as a function of the frequency in generation 0 or

$$q_t = \frac{q_0}{1 + tq_0} \tag{3.5a}$$

The maximum allele frequency possible is ½ (all heterozygotes), and with this starting point, the allele frequency over generations 0, 1, 2, 3, 4 . . . forms the progression 1/2, 1/3, 1/4, 1/5, 1/6. . . .

Furthermore, this expression can be solved for the number of generations t as

$$t = \frac{1}{q_t} - \frac{1}{q_0} \tag{3.5b}$$

These last two expressions allow us to ask two different questions about the dynamics of allele frequency change of recessive lethals. How much allele frequency change occurs over t generations, given an initial frequency for the lethal? How many generations will it take for the allele frequency to change by a specific amount?

Table 3.6 gives the values calculated from expression 3.5b for a few examples. When the frequency of the lethal is high, it is reduced very quickly—for instance, from 0.5 to 0.1 in 8 generations. However, when

TABLE 3.6 The number of generations (t) needed to reduce the allele frequency from an initial value of q_0 to q_t for a recessive lethal.

q_0	q_t	t
0.5	0.25	2
	0.1	8
	0.01	98
0.1	0.05	10
	0.01	90
	0.001	990
0.01	0.005	100
	0.001	900
	0.0001	9900

the lethal frequency initially is low, it is reduced slowly; for example, a change of frequency from 0.01 to 0.005—a halving of the frequency—takes 100 generations. The slow change at low allele frequencies occurs because most of the lethal alleles are "hidden" in heterozygotes and not subject to differential selection. To illustrate this phenomenon, we can calculate the proportion of A_2 alleles in the heterozygotes relative to homozygotes (remembering that only half the alleles in the heterozygote are A_2) as

$$\frac{pq}{q^2} = \frac{p}{q}$$

When the frequency of A_2 is low, this ratio becomes quite large. For example, when $q = 0.01$, the ratio is 99, meaning that for every A_2 allele in a homozygote undergoing selection, there are 99 masked by dominance in heterozygotes and not subjected to selection. It is obvious from this discussion that it is quite difficult to eliminate recessives completely from a population, and efforts to reduce the frequency of recessives from a low level to a lower level, unless heterozygotes can be identified, will be largely ineffective.

b. Selection Against Recessives

In many instances, there is not complete selection against a homozygote, and thus, the relative fitness of the homozygote is only partially reduced compared with the other genotypes. For many human genetic diseases, such as albinism and sickle cell anemia, homozygous recessive individuals can survive and produce progeny, although the probability of this occurring generally is reduced compared with that of other individuals. In *Drosophila*, mice, corn, and other organisms that have been investigated in detail genetically, there are many examples of recessive morphologic mutants that reduce the fitness of homozygotes but do not cause lethality.

To represent the relative fitnesses for this type of selection in which the recessive is **detrimental** or **deleterious**, we can use the fitness values in row (b1) of Table 3.5. In other words, the relative fitnesses of A_1A_1 and A_1A_2 are the same, and the fitness of A_2A_2 is less by an amount s, the **selective disadvantage** or **selection coefficient** of the homozygote. The selective disadvantage has a maximum value of unity, in which case there is a recessive lethal, as discussed above, and a minimum value of zero where there is selective neutrality. (Theodosius Dobzhansky facetiously commented that a value of s greater than one was a "fate worse than death.") When these fitness values are substituted in equation 3.2, then the frequency of A_2 after selection becomes

$$q_1 = \frac{q_0(1 - sq_0)}{1 - sq_0^2} \tag{3.6a}$$

The term in the denominator here and other similar equations below is equal to $\bar{w}$. The change in the frequency of A_2 from expression 3.3a is

$$\Delta q = -\frac{sq^2(1 - q)}{1 - sq^2} \tag{3.6b}$$

Both the mean fitness and the change in allele frequency are functions of the allele frequency and the selection coefficient. Figure 3.4 illustrates how these values change for different frequencies when the selection coefficient is 0.2 against a recessive (solid lines). As for a recessive lethal, the mean fitness (in the top half of Figure 3.4) reaches a maximum of 1.0 when the population is fixed for A_1 and a minimum (of 1 - $s = 0.8$ here) when fixed for A_2. However, the change (reduction) in frequency is greatest for an intermediate allele frequency. For the selection coefficient used in Figure 3.4, the maximum change in allele frequency occurs approximately at a frequency of 0.69.

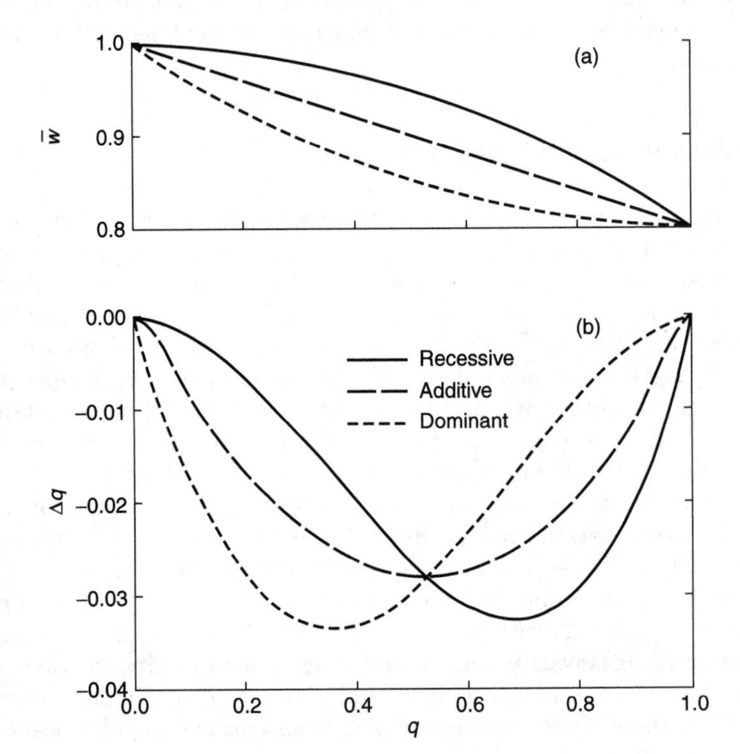

Figure 3.4. The mean fitness (a) and change in allele frequency (b) for selection against a recessive, additive, or dominant allele with $s = 0.2$.

When the selection coefficient is small, then the maximum frequency change occurs when $q = 2/3$. One can demonstrate this by taking the derivative of expression 3.6b, setting it equal to zero, and solving for q. To do this, assume that s is near zero so that $\overline{w} = 1 - sq^2 \approx 1$. Then

$$\Delta q \approx -sq^2 + sq^3$$

and

$$\frac{d(\Delta q)}{dq} = -2q + 3q^2 = 0$$

When this is solved for the allele frequency, $q = 2/3$ when Δq is at a maximum (actually a minimum in this case). In Figure 3.4, where $s = 0.2$, the maximum Δq was close to this frequency. As we found previously for recessive lethals, the change in allele frequency is low when q approaches zero. This is again the result of dominance "hiding" A_2 alleles from selection in the heterozygotes.

There is no exact solution to the recursion relationship given in expression 3.6a. As a result, the following approach can be used to determine the number of generations that it takes for the allele frequency to change by a certain amount. First, assume that the derivative of q with respect to time is approximately equal to Δq or

$$\frac{dq}{dt} \approx \Delta q$$

We can then separate the parts of this differential so that

$$dt \approx \frac{dq}{\Delta q}$$

By integrating both sides of this equation from 0 to t generations, then

$$\int_0^t dt = \int_{q_0}^{q_t} \frac{dq}{\Delta q}$$

and

$$t = \int_{q_0}^{q_t} \frac{dq}{\Delta q}$$

The same assumption can be made about Δq—that is

$$\Delta q \approx -sq_0^2(1 - q_0)$$

so that

$$t = \int_{q_0}^{q_t} \frac{dq}{sq^2(1 - q)}$$

This can be solved to give

$$t = \frac{1}{s}\left[\frac{q_0 - q_t}{q_0 q_t} + \ln \frac{q_0(1 - q_t)}{q_t(1 - q_0)}\right] \qquad (3.6c)$$

Using this expression, we can calculate the number of generations for a given amount of allele frequency change. For example, it takes approximately 58 generations for the allele frequency to decrease from 0.1 to 0.05 when $s = 0.2$. The results are quite close to the values obtained by iterating expression 3.6a when s is not too large. The pattern of change of frequency over time for a recessive detrimental is similar to that for a recessive lethal, although the time to change is longer by the proportion $1/s$. To appreciate this relationship, the first term in the brackets in expression 3.6c is equivalent to expression 3.5b. In addition, expression 3.6c gives the important general result that the time for a given amount of allele change is an inverse function of the amount of selection.

THE FAR SIDE® By GARY LARSON

Natural selection at work

c. Intermediate Dominance (Additivity)

Studies on quantitative traits (Falconer and MacKay, 1996; Lynch and Walsh, 1998) indicate that the heterozygote often may be nearly intermediate in value between the two homozygotes. When ***the value of the***

heterozygote is exactly midway between the two homozygotes, this is called **additivity** or **additive gene action**. This term is based on the fact that as each allele of a particular type is added, there is a given increment in the phenotypic value being measured. This model is illustrated by the relative fitness values in row (b2) of Table 3.5, where the fitness of the homozygotes, A_1A_1 and A_2A_2, are 1 and $1 - s$ and the fitness of the heterozygote, A_1A_2, is $1 - s/2$.

With this fitness array, the allele frequency after selection from expression 3.2 becomes

$$q_1 = \frac{q_0\left[1 - \frac{s}{2}(1 + q_0)\right]}{1 - sq_0} \tag{3.7a}$$

and the change in the frequency of A_2 is

$$\Delta q = -\frac{sq(1-q)}{2(1-sq)} \tag{3.7b}$$

As for a recessive detrimental, the mean fitness is a function of the selection coefficient and the allele frequency and again is at a maximum when A_2 is not present and a minimum when A_2 is fixed (Figure 3.4, long, broken line). The maximum change in allele frequency is when the two alleles are equally frequent—that is, when $p = q = 0.5$. This is shown graphically in Figure 3.4 and can also be shown analytically by assuming that s is small so that

$$\Delta q \approx -\frac{s}{2}q(1-q) \tag{3.7c}$$

Taking the derivative of this expression, setting it equal to zero, and solving for q demonstrate that the maximum change in allele frequency is for $q = 0.5$.

Again, there is no exact solution to the recursion relation given in expression 3.7a, and thus, we can use the same approach as that used for recessive detrimentals to calculate the change over time. Assume that the change in allele frequency is approximated as above, then

$$dt \approx \frac{dq}{\Delta q} = \frac{2dq}{-sq_0(1-q_0)}$$

If we integrate both sides of this expression,

$$t = \frac{2}{s}\left[\ln\frac{q_0(1-q_t)}{q_t(1-q_0)}\right] \tag{3.8}$$

The number of generations necessary to change the allele frequency by specific amounts when $s = 0.1$ is given in Table 3.7. For example, when

TABLE 3.7 The number of generations (t) needed to reduce the allele frequency from an initial value of q_0 to q_t for intermediate dominance (additivity) when $s = 0.1$.

q_0	q_t	t
0.9	0.5	44
	0.1	89
	0.01	136
0.5	0.25	22
	0.1	44
	0.01	92
0.1	0.05	15
	0.01	48
	0.001	194

$q_0 = 0.9$, the allele frequency is reduced to 0.1 in approximately 89 generations. The term in brackets here is the same as the second term in brackets in expression 3.6c. The reduction in allele frequency when the initial frequency is low is generally much faster than for a recessive lethal, even though s is an order of magnitude lower in Table 3.6. For example, the numbers of generations that it takes to decrease the allele frequency from 0.1 to 0.001 are 194 generations for an additive allele and 990 generations for a lethal allele. This difference occurs because at low frequencies most of the alleles are in heterozygotes, and there is selection against a detrimental additive allele in heterozygotes, whereas a recessive lethal is hidden from selection in heterozygotes.

The dynamics of allele frequency change for **gametic selection** or **selection in haploids** are approximated by that in the additive gene model. If the fitnesses of the A_1 and A_2 gametes (or genotypes) are 1 and 1 - s and their frequencies are p and q, then the frequency of A_2 gametes after selection is (Table 3.8)

$$q_1 = \frac{q_0(1-s)}{1-sq_0} \tag{3.9a}$$

TABLE 3.8 The frequency of gametes or haploid genotypes before and after selection.

	Genotype		
	A_1	A_2	*Total*
Fitness	1	$1-s$	—
Frequency before selection	p_0	q_0	1
Weighted contribution	p_0	$q_0(1-s)$	$1-sq_0$
Frequency after selection	$\frac{p_0}{1-sq_0}$	$\frac{q_0(1-s)}{1-sq_0}$	1

and the change in the frequency of A_2 is

$$\Delta q = -\frac{sq(1-q)}{1-sq} \tag{3.9b}$$

The numerator of this expression is the same as in expression 3.7*b*, but the denominator differs by a factor of 2 (here s indicates the selective difference between genotypes that differ by one allele, A_1 versus A_2, whereas above s was the selective difference between genotypes that differed by two alleles, A_1A_1 versus A_2A_2). Because the relationship of mean fitness and change in allele frequency is the same as for additivity, expression 3.8 can be used to predict the number of generations for a certain allele frequency change due to gametic selection if the result is halved.

The haploid selection model is the simplest of all selection models and is therefore useful to illustrate the **fundamental theorem of natural selection**. First, we must realize that for differential selection to occur in a population there must be genetic variation in fitness. Conversely, without genetic variation in fitness, there is no potential for significant evolutionary change. More specifically, Fisher (1930) showed that ***the rate of increase in fitness of any organism at any time is equal to its genetic variance in fitness at that time***, a concept he termed the fundamental theorem of natural selection. Although the theorem applies exactly to only simple genetic models (Frank and Slatkin, 1992; Burt, 1995), it is useful in heuristic terms and to obtain a general estimate of the amount of present or past selection. For some recent exposition and perspective, see Crow (2002) and Edwards (2002).

To illustrate the basis of this concept, let us assume the simplest form of inheritance—inheritance in which there is only asexual reproduction so that each offspring is genetically identical to its parent. Let the frequency and the relative fitness of the ith genotype be p_i and w_i, respectively, so that the mean fitness is

$$\overline{w} = \sum_{i=1}^{k} p_i w_i$$

The frequency of genotype i after selection is

$$p_i' = \frac{p_i w_i}{\overline{w}}$$

where the prime indicates the frequency after selection (in the next generation). The mean fitness in the next generation is

$$\begin{aligned}\overline{w}' &= \sum_{i=1}^{k} p_i' w_i \\ &= \frac{1}{\overline{w}} \sum_{i=1}^{k} p_i w_i^2\end{aligned}$$

Let us define the relative increment in fitness in a generation as

$$\begin{aligned}I &= \frac{\overline{w}' - \overline{w}}{\overline{w}} \\ &= \frac{1}{\overline{w}}\left(\frac{1}{\overline{w}}\sum_{i=1}^{k} p_i w_i^2 - \overline{w}\right) \\ &= \frac{V_w}{\overline{w}^2}\end{aligned}$$

or the variance in fitness in the parental generation divided by the square of the mean fitness. Of course, the fitnesses can be standardized so that $\overline{w} = 1$, which makes

$$I = V_w \tag{3.10}$$

In other words, the change (increment) in fitness is equal to the genetic variation in fitness, as Fisher stated in the fundamental theorem of natural selection. This model and derivation assume that all differences in fitness are genetically determined and that it is a measure of the rate of evolution with this simple genetic model.

d. Selection Against Dominants

Selection may act against the dominant phenotype in some situations. For example, a dominant color morph may not provide protective coloration in some environments, or a recessive mutant may be advantageous in a particular situation. In this case, the relative fitness values given in row (b3) of Table 3.5 can be used. Thus, the allele frequency after selection becomes

$$q_1 = \frac{q_0(1-s)}{1 - sq_0(2-q_0)} \tag{3.11a}$$

and the change in the frequency of A_2 is

$$\Delta q = -\frac{sq(1-q)^2}{1-sq(2-q)} \tag{3.11b}$$

The mean fitness and change in allele frequency for selection against dominants with $s = 0.2$ are shown in Figure 3.4 as short, broken lines.

Complementary to that for a recessive detrimental, the maximum allele frequency change for this type of selection is when $q = 1/3$ if s is small. Also, the time for a certain change in allele frequency is approximately

$$t = \frac{1}{s}\left[\frac{q_0 - q_t}{(1-q_0)(1-q_t)} - \ln\frac{(1-q_0)q_t}{(1-q_t)q_0}\right] \tag{3.11c}$$

using the approach outlined above.

e. General Dominance

Presumably there are loci that have virtually every possible level of dominance with the value of the heterozygote still remaining between the two homozygotes. A general fitness array such as that given in row (c1) of Table 3.5 allows consideration of different levels of dominance. A new parameter to denote the **level of dominance**, h, is then used. If $h = 0$, then the fitness array becomes that for the detrimental recessive model, whereas if $h = 0.5$, it becomes the additive model. In this general model, the allele frequency after selection becomes

$$q_1 = \frac{q_0[1 - s(hp_0 + q_0)]}{1 - 2hsp_0q_0 - sq_0^2} \tag{3.12a}$$

and the change in the frequency of A_2 is

$$\Delta q = -\frac{spq[h-(2h-1)q]}{1-2hspq-sq^2} \tag{3.12b}$$

Assuming that $0 \leq h \leq 1$, the maximum fitness occurs when A_1 is fixed. However, the maximum change in allele frequency is a function of the level of dominance, as we have seen from the discussion of the previous models. The number of generations required to change allele frequency by a certain amount can be determined numerically, if s is not too large, by evaluating

$$t = \int_{q_0}^{q_t} \frac{dq}{\Delta q}$$

where Δq is given by expression 3.12b.

We have assumed that selection was against genotypes containing allele A_2, thereby reducing its frequency. This can be thought of as **purifying selection**, ***selection that reduces the frequency of detrimental alleles in a population*** (Lewontin, 1974). New mutants often have detrimental effects and purifying selection keeps them at low frequencies.

However, some mutant alleles, or alleles introduced by gene flow, may have an adaptive advantage. In this case, directional genetic change may allow a population to adapt to its environment, and new, better adapted alleles may be replacing old, less adapted alleles. Such ***selection for alleles that are advantageous in the present environment*** is often called **adaptive Darwinian selection** or **positive Darwinian selection**. The dynamics of genetic change are the same for purifying and adaptive Darwinian selection, but it is convenient to give the relative fitnesses for the latter view as in row (c2) of Table 3.5, where the homozygote for the new allele has a fitness advantage of s. Again, h indicates the level of dominance, but notice that the scale is reversed from row (c1). For example, when $h = 0$ for row (c1), the heterozygote has the same fitness as A_1A_1, and when $h = 0$ for (c2), the heterozygote has the same fitness as A_2A_2.

The dynamics of adaptive Darwinian selection vary as a function of dominance, as does that of purifying selection. However, the general emphasis in considering adaptive Darwinian selection is on the increase of allele A_1 from some low frequency, assuming that it was introduced by mutation or gene flow, to a high frequency. With purifying selection, the focus is on the reduction of the frequency of detrimental allele A_2 from a low level to a lower frequency still.

To evaluate the effect of adaptive Darwinian selection, the fitness values in row (c2) can be substituted into expression 3.5a so that the change in allele frequency becomes

$$\Delta q = -\frac{spq[h + p(1 - 2h)]}{1 + 2hspq + sp^2} \tag{3.13}$$

where the denominator is the mean fitness. As an illustration of the effect of different levels of dominance on the rate of increase of a favorable allele, let us examine the allele frequency over time for an allele with a low initial frequency. An example is given in Figure 3.5 where $p_0 = 0.01$ and $s = 0.1$. The initial rise in allele frequency given (Figure 3.5b) is fastest when the allele is dominant, although allele frequency also changes rather quickly when the allele is additive. At a frequency of about 0.85 and after approximately 70 generations, the frequency of the additive allele surpasses that of the dominant. When the favorable allele is recessive, the initial increase is quite slow, and it takes more than 500 generations to increase its frequency above a value of 0.1.

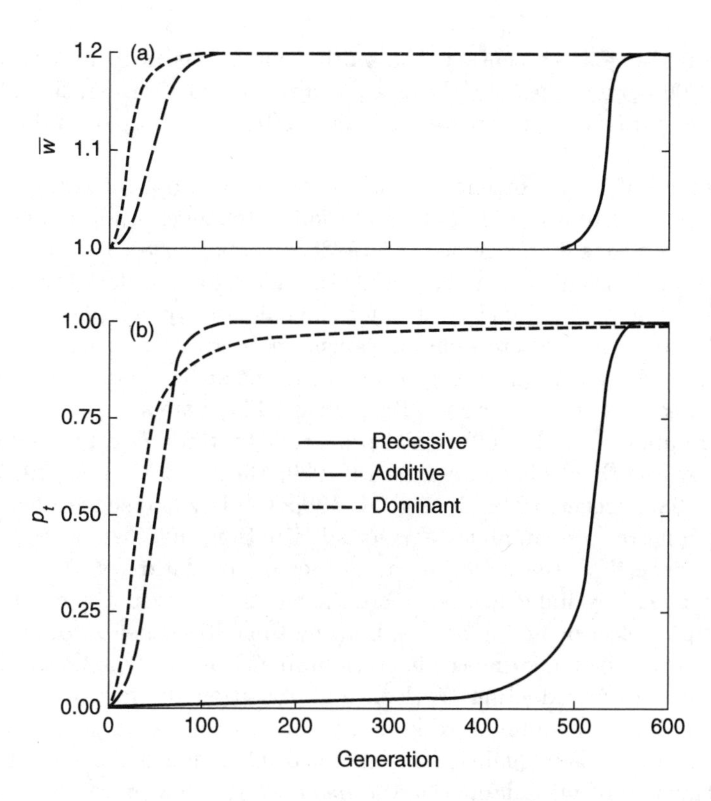

Figure 3.5. The mean fitness (a) and allele frequency (b) over time when adaptive Darwinian selection favors a recessive, additive, or dominant allele, where $p_0 = 0.01$ and $s = 0.1$.

The change in mean fitness is similar to the change in allele frequency for the three levels of dominance. With dominance, the increase in fitness is nearly immediate; for additivity, it is somewhat delayed, and for recessive gene action, it only rises after a long time. The fast response for the dominance model occurs because genotypes with only one dose of A_1, the heterozygotes, have the maximum fitness possible. From this example, it is obvious that adaptive Darwinian selection increases allele frequency most rapidly when there is a high degree of dominance for the favored allele and increases it more slowly when the favored allele is recessive (see Example 3.3 for a discussion the increase and recent decrease in melanism in peppered moths).

Example 3.3. An example of a genetic polymorphism that appears to have undergone adaptive Darwinian selection is industrial melanism in the peppered moth, *Biston betularia.* Industrial melanism refers primarily to black or dark forms of normally light-colored cryptic moths (or other insects) that have spread into areas subject to industrial pollution (see Figure 1.12a). Industrial melanics were first noticed in England in the mid-nineteenth century, in frequencies that were initially low but increased rapidly so that by the mid-twentieth century, the moth populations in many areas were nearly 100 percent melanic. However, at a particular location, the increase from a low frequency to greater than 90% often took only several decades (see Berry, 1990).

Because this phenomenon is such a striking example of both morphological polymorphism and rapid evolutionary change, studies concerning it have been extensive (Kettlewell, 1973; Majerus, 1998). Genetic control of industrial melanism is unifactorial, the allele for the dark, or melanic, phenotype being dominant to that for the pale, or typical, phenotype. Although the gene that determines melanism in *B. betularia* has not been yet identified, characterizing the genes causing melanism in *Drosophila*, and other insects, is on the horizon (True, 2003). The main selective agent acting here appears to be differential predation by birds (see Example 4.6). Some reviews (Brakefield, 1987; Berry, 1990; Mani and Majerus, 1993; Majerus, 1998; Sargent, 1998; Cook *et al.*, 1999; Cook, 2000) suggest that such an explanation oversimplifies the actual situation and that other factors, such as gene flow, site selection, and other nonpredation selective effects, may significantly influence melanic frequencies. However, the overall case for natural selection acting on the rise (and the fall, see below) of melanism frequencies in peppered moths is indisputable (Coyne, 2002; Grant, 2002).

Since the introduction of clean air legislation in the 1960s in England, the levels of sulfur dioxide and suspended particulates, two important measures of air quality, have declined. Over much the same period, the frequency of melanic moths also has declined. For example, at Caldy Common, melanic frequency declined from greater than 90% in 1960 to below 20% by 1995 (Fig. 3.6) (Grant *et al.*, 1996). When the frequency of melanics increased in the nineteenth and early twentieth centuries, the dominant melanic allele had a selective advantage, whereas the recessive typical has been selectively favored since the increase in air quality. As a

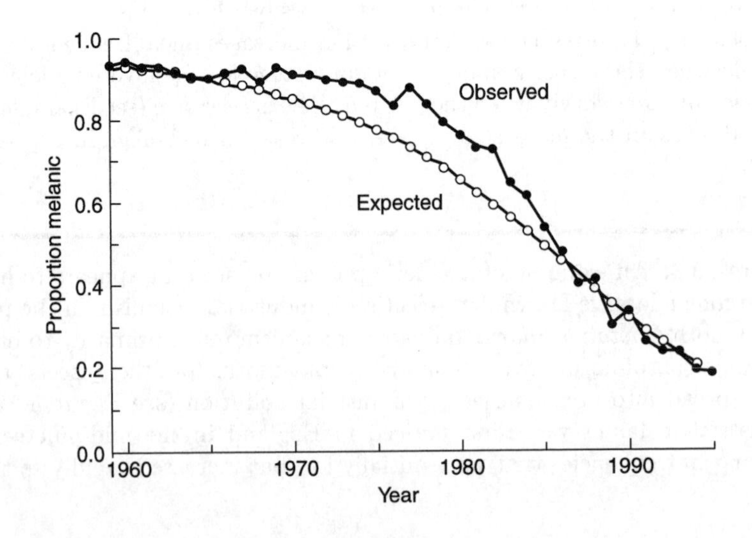

Figure 3.6. The observed decline in the frequency of melanics over 26 years at Caldy Common in England (closed circles). The open circles give the predicted decline when $s = 0.153$ against the melanics (after Grant *et al.*, 1996).

result, the frequency of the melanic phenotype initially declines slowly and then decreases more quickly as the frequency of the typical phenotype increases. Also given in Figure 3.6 is the expected decrease of the melanic phenotype, assuming that the fitnesses of MM (melanic), Mm (melanic), and mm (typical) are 0.847, 0.847, and 1 (Grant *et al.*, 1996). Obviously, this selective difference closely approximates the observed decline at Caldy Common. The decline in melanic frequency in the Manchester area was much more rapid, declining over 90% in the early 1980s to less than 10% in the late 1990s (Cook *et al.*, 1999). In this case, the estimated selection coefficient was much higher at $s = 0.45$.

B. betularia is also found in North America, and in some areas the frequency of melanics has been quite high. For example, at a site at the George Reserve in Michigan, the phenotypic frequency of melanics was nearly 90% around 1960 (Grant *et al.*, 1996). In the 1960s, clean air legislation also began to result in better air quality in North America. When the George Reserve was censused again in 1995 and 2001, the frequency of melanics had dropped to 20% and 5.3%, respectively (Grant and Wiseman, 2002). Presumably, similar reversals in selection pressures occurring on both continents were responsible for this remarkable parallel decline. One difference between the two populations is that the English population is univoltine (one generation per year), whereas the Michigan population is bivoltine (two generations per year). In other words, for the Michigan population to have experienced a decline in frequency similar to that of the English population in twice as many generations, the selection pressures in Michigan would only have to have been approximately half as strong as in England.

---

f. Heterozygote Advantage (Overdominance)

All of the selection models that we have discussed lead to the eventual fixation of one allele or the other and, as a result, reduce the genetic variation in the population. On the other hand, when the heterozygote has a higher fitness than both homozygotes, then two alleles can be maintained in the population. There are some documented cases of strong heterozygote advantage, such as sickle-cell anemia (p. 156), warfarin resistance in Norway rats (p. 118, Greaves *et al.*, 1977), transferrin genotypes in pigeons (p. 232, Frelinger, 1972), and prion disease (p. 149, Mead *et al.*, 2003), although there is disagreement about the overall proportion of genes with this type of fitness array (Chapter 8). In the next chapter, we examine some other types of selection that can also result in the maintenance of a genetic polymorphism.

To investigate the heterozygote advantage (or overdominance) model, we use the fitness array in row (d) of Table 3.5, where the heterozygote, A_1A_2, has the maximum relative fitness, 1, and the fitnesses of the homozygotes are $1-s_1$ and $1-s_2$, where $\boldsymbol{s_1}$ **and** $\boldsymbol{s_2}$ **are the selective disadvantages of the homozygotes** $\boldsymbol{A_1A_1}$ **and** $\boldsymbol{A_2A_2}$, respectively. When we put these fitness values into expression 3.2, the frequency of allele A_2 after selection is

$$q_1 = \frac{q_0 - s_2 q_0^2}{1 - s_1 p_0^2 - s_2 q_0^2} \tag{3.14a}$$

and the change in the frequency of A_2 is

$$\Delta q = \frac{pq(s_1 p - s_2 q)}{1 - s_1 p^2 - s_2 q^2} \tag{3.14b}$$

In order to maintain both alleles in the population, Δq must be zero for some value of q between 0 and 1. The point where this occurs is called the **equilibrium frequency** of allele A_2, ***the allele frequency for which the change in frequency is zero***. What are termed trivial equilibria in expression 3.14b occur when Δq is zero because either $p = 0$ or $q = 0$. There is, of course, no change in allele frequency in these cases because only one allele is present.

Another value of q for which expression 3.14*b* is equal to zero occurs when the term in parentheses in the numerator

$$s_1 p - s_2 q = 0$$

Solving this equation makes it apparent that equilibrium occurs when

$$q_e = \frac{s_1}{s_1 + s_2} \tag{3.15}$$

where q_e signifies the **equilibrium frequency of allele A_2**. In other words, at this allele frequency, there is no change in allele frequency, and both alleles are maintained in the population. The equilibrium is a function of only the selection coefficients for the two homozygotes.

The properties of the equilibrium can be seen in another way by expressing the deviation in allele frequency from the equilibrium as (Li, 1955)

$$q - q_e = q - \frac{s_1}{s_1 + s_2}$$
$$(s_1 + s_2)(q - q_e) = q(s_1 + s_2) - s_1$$

The term in parentheses in the numerator of expression 3.14*b* is equal to $s_1 - q(s_1 + s_2)$. Therefore, expression 3.14b becomes

$$\Delta q = \frac{-pq(s_1 + s_2)(q - q_e)}{\overline{w}}$$

It is obvious from this formulation that when $q > q_e$, Δq is negative, and when $q < q_e$, Δq is positive. In other words, after a perturbation away from the stable equilibrium, the allele frequency will change back toward the equilibrium frequency. Because of these properties, the equilibrium is called **stable** (see Example 3.4 for an illustration of these dynamics from a *Drosophila* experiment).

The mean fitness and the change in allele frequency are both functions of the allele frequency and the selection coefficients. Two examples are given in Figure 3.7, where selection results in either a stable equilibrium

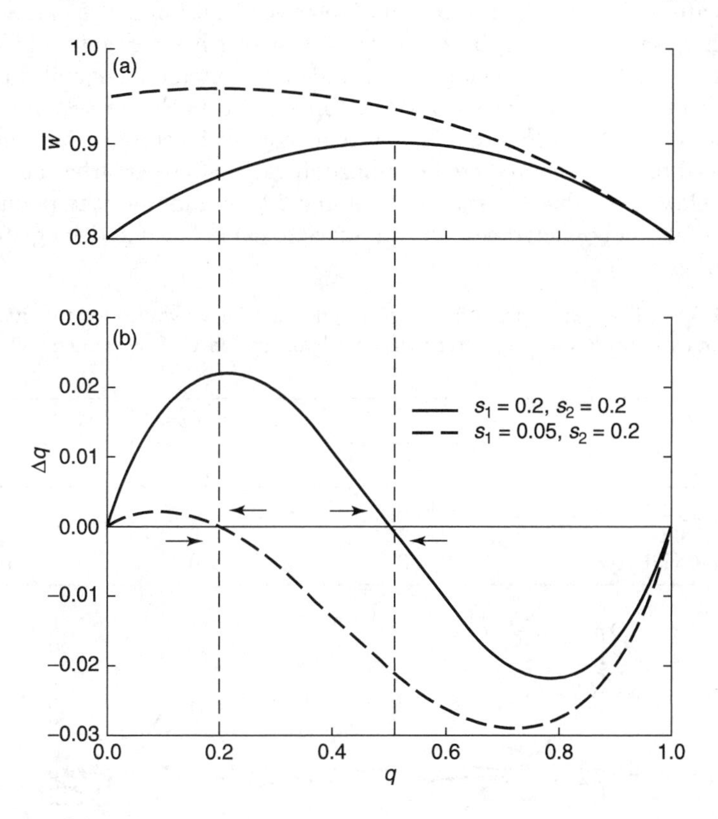

Figure 3.7. The mean fitness (a) and change in allele frequency (b) when selection favors the heterozygote. The vertical broken lines indicate the stable equilibrium frequency for the two sets of fitness values.

of A_2 at either 0.5 (solid lines) or 0.2 (broken lines). For the Δq curve when $q_e = 0.5$ and if $0.0 < q < 0.5$, the change in allele frequency is positive, with the largest Δq of slightly greater than 0.02 just above 0.2. Likewise, when $0.5 < q < 1.0$, the change in allele frequency is negative with the largest negative Δq of -0.02 just below 0.8. On the other hand, the Δq curve when $q_e = 0.2$ is quite asymmetrical, the magnitude of the Δq curve below the equilibrium being quite small and the magnitude above the equilibrium being quite large. The slope for both Δq curves is negative at the equilibrium. A negative slope at the equilibrium indicates that the equilibrium is stable. The magnitude of the slope tells how quickly the allele frequency will return to the equilibrium after a small perturbation away from it.

Example 3.4. An experimental examination of the effect of heterozygote advantage was given by Prout (1971) using mutants on the fourth chromosome in *D. melanogaster.* This chromosome is less than 1 map unit long, and there is little known recombination. Therefore, using recessive eye and bristle mutants, *eyeless* (ey) and *shaven* (sv), respectively, Prout constructed populations that had the genotypes and phenotypes given in Table 3.9. Estimates of two fitness components, relative virility in males and average viability (see Table 3.9), both indicated a strong heterozygote advantage. When all fitness components were included, the predicted equilibrium frequency for chomosome $ey+$ was 0.604 (broken line in Figure 3.8). As shown by the solid lines in Figure 3.8, the frequency of the chromosomes quickly returned to near the predicted equilibrium value after perturbations above and below this value in generations 0 and 12. In addition, the population started at the predicted equilibrium stayed near this value throughout the experiment.

TABLE 3.9 The genotypes, phenotypes and the relative values for two fitness components, both showing heterozygote advantage, in the experiment of Prout (1971).

	Genotype		
	$+sv/+sv$	$ey+/+sv$	$ey+/ey+$
Phenotype	shaven	wildtype	eyeless
Male virility	0.047	1.0	0.247
Average viability	0.856	1.0	0.852

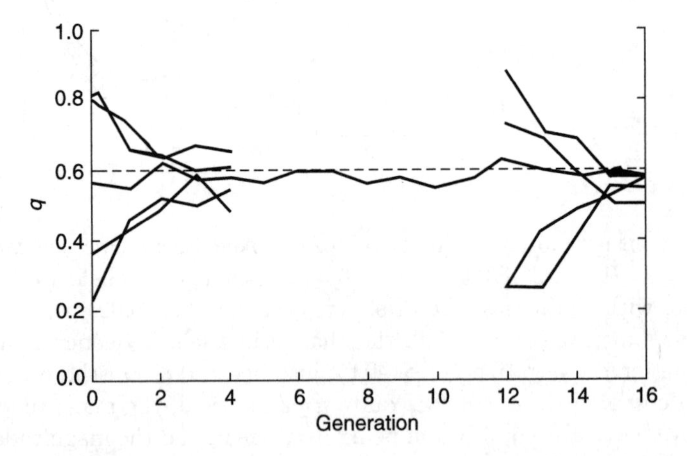

Figure 3.8. The frequency of chromosome $ey+$ after perturbations in generations 0 and 12. The broken line indicates the theoretical equilibrium of 0.604, given the estimated heterozygote advantage, and the solid lines indicate different experimental populations (from Prout, 1971).

The mean fitness reaches a maximum at the allele frequency equilibrium. This can be seen in Figure 3.7 and can be shown analytically by taking the derivative of $\overline{w}$ from the denominator of expression 3.14*b*, setting it equal to zero, and solving for q as

$$\frac{d\overline{w}}{dq} = 2s_1 - 2s_1q - 2s_2q = 0$$

so that

$$q = \frac{s_1}{s_1 + s_2}$$

In other words, allele frequency changes so that fitness is increased until it reaches a maximum value at the stable equilibrium. At the equilibrium, the allele frequency stops changing, and the fitness remains at the maximum. Substituting the allele frequencies at equilibrium $[p_e = s_2/\,(s_1 + s_2)]$ into the expression for mean fitness reveals that the mean fitness at the equilibrium is

$$\begin{aligned}\overline{w}_e &= 1 - s_1\left(\frac{s_2}{s_1 + s_2}\right)^2 - s_2\left(\frac{s_1}{s_1 + s_2}\right)^2 \\ &= 1 - \frac{s_1s_2(s_1 + s_2)}{(s_1 + s_2)^2} \\ &= 1 - \frac{s_1s_2}{s_1 + s_2} \qquad (3.16)\end{aligned}$$

In other words, the mean fitness is less than that of the genotype with the highest fitness (the heterozygote has a relative fitness of unity) by an amount $s_1s_2/(s_1 + s_2)$.

Beginning in the 1940s, there was concern that radiation and fallout from nuclear weapons would cause excessive amounts of detrimental mutations. Muller (1950), in particular, focused attention on the potential burden caused by the addition of these mutations to the human gene pool. To evaluate this effect, he introduced the concept of **genetic load**, which is ***the reduction in fitness from the maximum possible in a population***. The principle factors thought to cause genetic load are the presence of recessive detrimental alleles maintained by a selection–mutation balance (Chapter 7) and the segregation of homozygotes when there is a heterozygote advantage (Wallace, 1970; Crow, 1986).

For the heterozygote advantage model, the heterozygote has the maximum fitness of unity, and, thus, the genetic load is

$$L = 1 - \overline{w}$$

If we substitute the mean fitness value at equilibrium from expression 3.16, the load becomes

$$L = \frac{s_1 s_2}{s_1 + s_2} \tag{3.17}$$

and is a function only of the selection coefficients. The genetic load in this case is greatest when s_1 and s_2 are both large and similar in magnitude.

g. Heterozygote Disadvantage (Underdominance)

In some cases, the heterozygote may have a fitness less than that of either homozygote. Such a fitness array may occur in hybrids between species or subspecies, although many genes are probably involved in these cases, and a single-gene model may not always be appropriate. Other examples of low heterozygote fitness involve individuals that are heterozygous for different chromosomes, such as a translocation or an inversion. Often chromosomal heterozygotes produce unbalanced gametes with low viability, thereby reducing the fitness of the heterozygous parent compared with chromosomal homozygotes. The heterozygote disadvantage model may also be useful in genetic control of pest populations (Foster *et al.*, 1972; Whitten, 1979; see Example 3.5). On p. 180 we discuss Rh incompatibility in humans, caused by an interaction between mother and fetus such that selection is against heterozygotes from incompatible matings.

To examine the heterozygote disadvantage (or underdominance) model, let us use the fitness array in row (e) of Table 3.5, where the heterozygote has a fitness of 1 and the homozygotes A_1A_1 and A_2A_2 have fitnesses of $1 + s_1$ and $1 + s_2$, respectively. (In some cases, this array is standardized so that the fittest homozygote has a relative fitness of 1.) With this array, which is an extension of the heterozygote advantage array, the allele frequency after selection becomes

$$q_1 = \frac{q_0 + s_2 q_0^2}{1 + s_1 p_0^2 + s_2 q_0^2} \tag{3.18a}$$

and the change in the frequency of A_2 is

$$\Delta q = \frac{pq(s_2 q - s_1 p)}{1 + s_1 p^2 + s_2 q^2} \tag{3.18b}$$

Again, this last equation can be shown to have a nontrivial equilibrium at

$$q_e = \frac{s_1}{s_1 + s_2} \tag{3.18c}$$

In this case, however, Δq is positive above the equilibrium frequency and negative below it. As a result, this equilibrium is **unstable**, and with a slight perturbation away from it, the allele frequency continues to move away (see Example 3.5 for an application using chromosomal variation in *Drosophila*).

Example 3.5. Foster *et al.* (1972) provided an experimental examination of the change in allele frequency when a heterozygote disadvantage exists. Their example is the basis of a proposed insect control program using compound or translocated chromosomes but using *D. melanogaster* as the model organism. Because translocation heterozygotes often produce unbalanced gametes and thereby have lower fertility, their fitness is much less than for translocation homozygotes. Figure 3.9a gives an example where the unstable equilibrium is at 0.5; in Figure 3.9b, it is around 0.9. In a matter of a few generations, all the replicates were fixed either for one chromosome type or for the other. The chromosome type for which individual replicates became fixed depended on whether the chance change in the early generations resulted in frequencies above or below the unstable equilibrium.

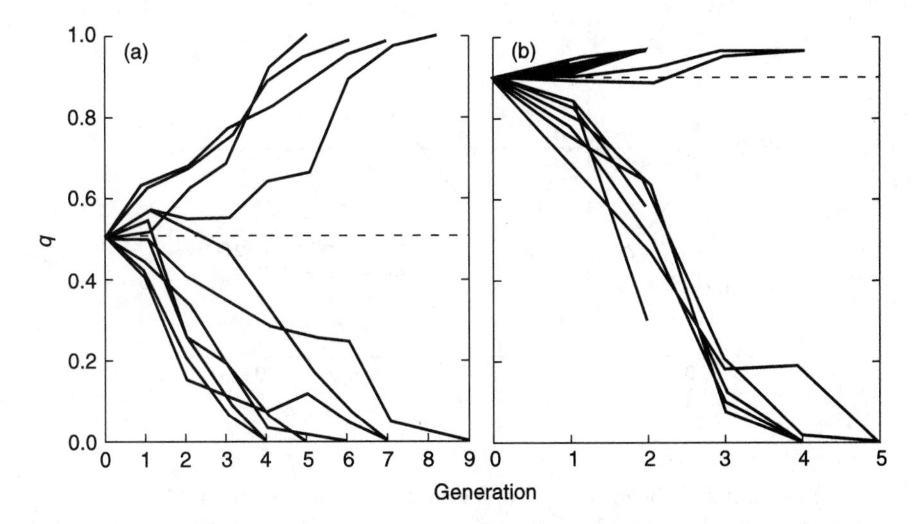

Figure 3.9. The change in chromosomal frequency over time when the heterozygote has a disadvantage and the unstable equilibrium (indicated by the broken lines) is at 0.5 (a) or 0.9 (b) for translocations in *D. melanogaster* (from Foster *et al.*, 1972). The solid lines indicate different experimental replicates.

Using an approach analogous to that used for heterozygote advantage, we can modify expression 3.18*b* to give

$$\Delta q = \frac{pq(s_1 + s_2)(q - q_e)}{\overline{w}}$$

In this case, however, when $q > q_e$, Δq is positive, and when $q < q_e$, Δq is negative, which indicates that the allele frequency moves away from the equilibrium.

Again, the mean fitness and the change in allele frequency are functions of the allele frequency and the selection coefficients. Two examples are given in Figure 3.10 where selection results in an unstable equilibrium at 0.5 (solid

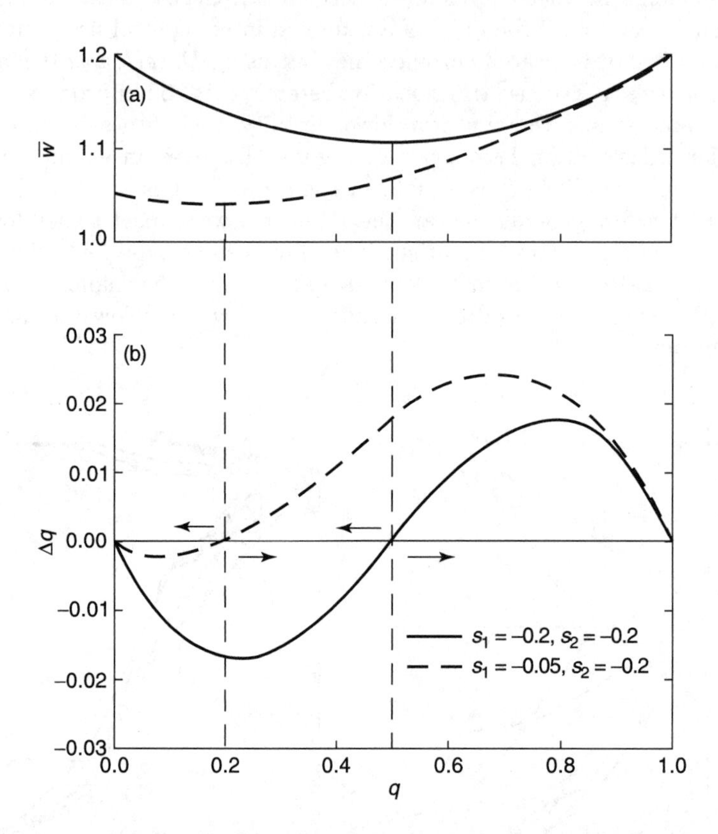

Figure 3.10. The mean fitness (a) and change in allele frequency (b) when selection is against the heterozygote. The vertical broken lines indicate the unstable equilibrium frequency for the two sets of fitness values.

lines) or at 0.2 (broken lines). In the first example, when the frequency of A_2 is smaller than 0.5, Δq is negative, and when it is greater than 0.5, Δq is positive. Therefore, all populations with this fitness array will eventually become fixed for A_1 if $q_0 < 0.5$ or A_2 if $q_0 > 0.5$. The mean fitness is at a minimum at the unstable equilibrium so that as the frequency changes, either increasing A_1 or A_2 in frequency (depending on what q_0 is), the fitness also increases. However, in the second example, if $q_0 < 0.2$, fitness cannot reach the maximum value of $1 + s_2$ because the initial frequency is on the wrong side of the unstable equilibrium.

II. ESTIMATION OF SELECTION

Relative fitness is simply defined as the ***ability of a particular genotype, relative to that of other genotypes, to pass on alleles to the next or future generations***. Although this at first appears to be a simple concept, measuring the relative fitness values for different genotypes is often difficult (Hedrick and Murray, 1983; Lewontin, 1974). Generally, many estimation techniques are most easily applied to laboratory populations or well-understood natural populations in which other factors may be controlled or accounted for. The use of approaches based on DNA sequence data, which we discuss in Chapter 8, provides another manner in which to estimate differential selection.

Dobzhansky (1955) suggested that work in population genetics was "severely handicapped by lack of reliable methods of comparing the fitnesses of populations." In population genetics, the mean fitness of a population is defined as the weighted mean of the relative fitnesses of its genotypes. However, population fitness in this sense is of limited value for comparisons among populations because it is an average of relative values measured within a population. In trying to focus on what is evolutionarily important, Dobzhansky (1970) defined the absolute measure **adaptedness** as ***the ability of a population, organism, or genotype to survive and reproduce in a particular environment***. In this context, population ecology measures, such as the intrinsic rate of increase and the carrying capacity in particular environments, are useful. However, it is often simpler to obtain two related measures: productivity, the number or biomass of adults produced in the population per time unit, and population size, the average biomass or number of individuals living in the population. Although we concentrate on estimating relative fitness differences among genotypes here, it is important to remember this broader perspective.

a. Viability Selection

Let us begin discussing estimation of selection for different genotypes by considering preadult viability selection. Estimates of viability selection allow a focus on one aspect of the life cycle that is often thought to be important. However, such an emphasis may lead one to overlook effects on other fitness components (see discussion in Chapter 4) that are basic to understanding the overall fitness differences among the genotypes. For example, there is no guarantee that if a particular genotype is favored in viability selection that it will also have an advantage for other fitness components. In fact, sexual and fecundity selection are often found to be quite important, which indicates that these components should be carefully evaluated along with viability selection.

If possible, it is important to estimate all of the components of selection, either in a direct or an indirect manner, to understand the impact of selection (Prout, 1965). Christiansen and Frydenberg (1973) developed an approach that contains a series of specific null hypotheses that can be tested to evaluate the extent of different types of selection. The approach is designed to measure the components of selection for a monogamous population with discrete generations in which sampling of the genotypes takes place over a one-generation transition from zygotes to zygotes (see Chistiansen *et al.*, 1973, for an application of this approach to the eelpout fish).

To measure the amount of selection that occurs in preadult viability, one can generally use the ratio of the genotype frequencies in an early stage of census to the genotype frequencies of adults. Let us assume that P_{ij} and P'_{ij} are the genotype frequencies before and after viability selection and that the relative viabilities of genotypes A_1A_1, A_1A_2, and A_2A_2 are v_{11}, 1, and v_{22}, respectively. Therefore, the genotype frequencies after selection are

$$\begin{aligned} P'_{11} &= \frac{P_{11}v_{11}}{\overline{v}} \\ P'_{12} &= \frac{P_{12}}{\overline{v}} \\ P'_{22} &= \frac{P_{22}v_{22}}{\overline{v}} \end{aligned} \tag{3.19a}$$

where $\overline{v} = P_{11}v_{11} + P_{12} + P_{22}v_{22}$. The ratios of the genotype frequencies of the homozygotes over the heterozygote are

$$\begin{aligned} \frac{P'_{11}}{P'_{12}} &= \frac{P_{11}v_{11}}{P_{12}} \\ \frac{P'_{22}}{P'_{12}} &= \frac{P_{22}v_{22}}{P_{12}} \end{aligned}$$

These two equations can be solved for estimates of the relative viability of the homozygotes as

$$\begin{aligned} \hat{v}_{11} &= \frac{P'_{11}P_{12}}{P'_{12}P_{11}} \\ \hat{v}_{22} &= \frac{P'_{22}P_{12}}{P'_{12}P_{22}} \end{aligned} \tag{3.19b}$$

(see Example 3.6 for an application in the prion disease Kuru).

Example 3.6. Kuru is a prion disease found mainly in the Fore tribe of the New Guinea highlands (other prion diseases are Creutzfeldt-Jacob disease [CJD] in humans, bovine spongiform encephalopathy [BSE] in cattle and other animals, and scrapie in sheep). The Fore women and children of both sexes had a tradition of eating their deceased relatives at mortuary feasts. Kuru killed approximately 1% of the population annually, but a ban on cannibalism in the 1950s resulted in a decline in kuru incidence.

The polymorphism at codon position 129 in the prion protein gene (*PRNP*) influences the susceptibility for human prion diseases (Palmer *et al.*, 1991). Recently, Mead *et al.* (2003) examined variability at codon 129 and other variants in and near the *PRNP* gene in a sample of 30 women age 50 and older and who had a history of multiple mortuary feasts (Table 3.10). The frequency of the two sequences that had M (methionine) or V (valine) at codon position 129 was nearly equal ($q_V = 0.484$). However, there were only 7 homozygotes, significantly below the 15.0 expected under Hardy–Weinberg. In another sample of 140 unexposed Fore individuals, the genotypes were in Hardy–Weinberg proportions.

Using equation 3.19b and assuming that these two groups indicate the genotype frequencies before and after selection, the viability of genotype MM relative to genotype MV (1.0) can be estimated as

$$v_{MM} = \frac{P'_{MM}P_{MV}}{P'_{MV}P_{MM}} = \frac{(0.133)(0.514)}{(0.767)(0.221)} = 0.303$$

and the viability of genotype VV relative to genotype MV can be estimated as

$$v_{MM} = \frac{P'_{VV}P_{MV}}{P'_{MV}P_{VV}} = \frac{(0.100)(0.514)}{(0.767)(0.264)} = 0.254$$

In other words, the relative viabilities of the genotypes MM, MV, and VV are 0.303, 1.0, and 0.254, respectively, a very strong heterozygote advantage (Hedrick, 2003a). These groups are not really the before and after selection

TABLE 3.10 The genotype frequencies in two groups of Fore people, 140 that were never exposed to mortuary feasts and 30 women over age 50 that were exposed to multiple feasts. The genotypes are variants at codon position 129 of the *PRNP* gene where M and V indicate amino acids methionine and valine.

		Genotype			
	Sample size	MM	MV	VV	q_V
Unexposed to feasts (before selection)	140	0.221	0.514	0.264	0.521
Exposed to feasts (after selection)	30	0.133	0.767	0.100	0.484
Estimated viabilities		0.303	1.000	0.254	

cohorts that this estimation procedure is designed for, but unselected and selected samples from the same population.

Because adult males participated little at feasts, this heterozygote advantage acts primarily in females. Therefore, the average selection coefficient ($s = 1 - v$) against MM homozygotes is approximately $\overline{s}_{MM} = (1 - 0.303)/2 = 0.348$ and against VV homozygotes is $\overline{s}_{VV} = (1 - 0.254)/2 = 0.373$. The expected equilibrium frequency of the V allele is therefore $q_V = \overline{s}_{MM}/(\overline{s}_{MM} + \overline{s}_{VV}) = 0.483$, not very different from the observed frequency of 0.55 reported by Mead *et al.* Although it is not known whether selection has been this strong in previous generations, the strength of balancing selection in this one generation appears to be the strongest yet documented in any human population.

---

b. Genotype Frequencies and Some Caveats

An approach that at first appears to be an obvious one to detect selection is the examination of genotype frequencies to determine deviations from Hardy–Weinberg proportions. However, using observed genotype proportions to estimate selection is fraught with a number of difficulties. First, genotype proportions are affected by a number of factors besides selection (Workman, 1969). For example, inbreeding, an effect that can obscure the impact of selection (Chapter 5), may also influence the observed genotype proportions.

Second, even though selection is occurring, it may not result in a deviation from Hardy–Weinberg proportions (Lewontin and Cockerham, 1959). This can be demonstrated using the fixation index, a measure based on observed and expected heterozygosities that are zero if there are Hardy–Weinberg proportions (see p. 95). Let us assume that we are examining adult frequencies after viability selection (here indicated by primes) so that fixation index is

$$F = 1 - \frac{P'_{12}}{2p'q'}$$

Assuming that the relative viabilities of genotypes A_1A_1, A_1A_2, and A_2A_2 are w_{11}, w_{12}, and w_{22}, respectively, we have $P'_{12} = 2pqw_{12}/\overline{w}$, $p' = (p^2w_{11} + pqw_{12})/\overline{w}$, and $\overline{w} = p^2w_{11} + 2pqw_{12} + q^2w_{22}$, and then

$$\begin{aligned} F &= 1 - \frac{2pqw_{12}\overline{w}}{(p^2w_{11} + pqw_{12})(pqw_{12} + q^2w_{22})} \\ &= \frac{pq(w_{11}w_{22} - w_{12}^2)}{(p^2w_{11} + pqw_{12})(pqw_{12} + q^2w_{22})} \end{aligned} \tag{3.20}$$

Obviously, the condition necessary for $F = 0$ and Hardy–Weinberg proportions is that $w_{11}w_{22} = w_{12}^2$. This is of course true if there is no selection present (all fitnesses are 1), but it can also occur when there is strong selection and the relative fitnesses of the genotypes form a geometric progression—for example, 1, $1 - s$, $(1 - s)^2$ for genotypes A_1A_1, A_1A_2, and A_2A_2, respectively. To illustrate this point, assume that w_{11}, w_{12}, and w_{22} are 1.0, 0.5, and 0.25, respectively, and that $p = q = 0.5$. After selection, the genotype proportions are $4/9$, $4/9$, and $1/9$; Hardy–Weinberg proportions for $q = 1/3$ and $F = 0$, even though there has been strong directional selection.

Finally, as mentioned above, selection may include selection other than differential viability of zygotes, the type of selection assumed when observed genotype proportions are used. For example, selection in the previous generation resulting in different allele frequencies in the gametes of the two sexes can result in deviations from Hardy–Weinberg proportions (see p.). Second, strong selection favoring the heterozygote in fecundity in only one sex may result in small or no deviations from Hardy–Weinberg proportions (Frelinger and Crow, 1973). As an illustration, Table 3.11 gives the offspring proportions when the homozygous females have zero fecundity. As a result, the progeny frequencies are $\frac{1}{2}p$, $\frac{1}{2}$, and $\frac{1}{2}q$ for genotypes A_1A_1, A_1A_2, and A_2A_2, respectively. When $p = q = 0.5$, the equilibrium in this case, there is no deviation from Hardy–Weinberg proportions even though there is strong selection. If the population is not at equilibrium, then there is a small excess of heterozygotes, as compared with Hardy–Weinberg expectations, because of differential allele frequencies in the two sexes.

TABLE 3.11 The frequency of offspring when only the A_1A_2 genotype is capable of producing offspring. That is, there is complete fecundity (or viability) selection against homozygous females.

Mating type ♀ × ♂	*Offspring*		
	A_1A_1	A_1A_2	A_2A_2
$A_1A_2 \times A_1A_1$	$\frac{1}{2}p^2$	$\frac{1}{2}p^2$	—
$A_1A_2 \times A_1A_2$	$\frac{1}{2}pq$	pq	$\frac{1}{2}pq$
$A_1A_2 \times A_2A_2$	—	$\frac{1}{2}q^2$	$\frac{1}{2}q^2$
Total	$\frac{1}{2}p$	$\frac{1}{2}$	$\frac{1}{2}q$

Let us temporarily ignore these problems associated with using observed genotype frequencies. If the allele frequencies are at equilibrium in the population, then the deviations from Hardy–Weinberg proportions are adequate for estimating the relative viabilities of the two homozygotes (Lewontin and Cockerham, 1959; see also Brown, 1970). However, the power of such a test, the probability that the null hypothesis (no viability differences) will be rejected when there is a selective difference, is generally low. The power of the test, $1 - \beta$ (see p. 24), is

$$1 - \beta = F^2 N \tag{3.21}$$

where F is as given by expression 3.20 and N is the sample size. For example, when w_{l1}, w_{l2}, and w_{22} are 0.8, 1.0, and 0.8 ($s = 0.2$), the sample size necessary to detect selection 90% of the time is 851.

c. Deviations from Mendelian Expectations

Some of the problems mentioned above could be avoided by estimating the relative survival rates for a cross in which the zygotic proportions are known and the adult numbers are subsequently counted. For example, when equal proportions of zygotes are expected—from a cross between A_1A_1 and A_1A_2—a nearly unbiased estimate of the relative viability of A_1A_1 is

$$\hat{v}_{11} = \frac{N_{11}}{N_{12} + 1} \tag{3.22a}$$

where N_{11} and N_{12} are the observed numbers of the A_1A_1 and A_1A_2 adults, respectively (Haldane, 1956). The estimate of the selection coefficient against A_1A_1 is $s = 1 - v_{11}$. The approximate sampling variance of this estimate is

$$V(\hat{v}_{11}) \approx \frac{v_{11}\,(1 + v_{11})^2}{N}$$

where N is the total number of adults counted (see Example 3.7 estimating selection for melanic and typical peppered moths where equal numbers of the two types where placed on different backgrounds).

If there are twice as many zygotes of one type produced as another, such as the number of A_1A_2 genotypes compared with A_1A_1 from $A_1A_2 \times A_1A_2$ cross, then an almost unbiased estimate of the relative viability of A_1A_1 is

$$\hat{v}_{11} = \frac{2N_{11}}{N_{12} + 1} \tag{3.22b}$$

Example 3.7. Melanism in the peppered moth, *Biston betularia*, is often used as an example of adaptation to a changing environment. The major selective agent is thought to be predation by birds that favors melanics in polluted areas and favors the typical form in unpolluted areas. This hypothesis has been tested by capture-recapture experiments in both polluted and unpolluted environments. It was found that typicals had approximately twice the survival of melanics in an unpolluted environment (Kettlewell, 1973).

Another approach was used by Clarke and Sheppard (1966). They exposed dead moths to predation on either dark or pale backgrounds (Table 3.12). Their results are entirely consistent with the view that protective camouflage is important in the survival of the two forms of the moth. Relative viabilities can be calculated using expression 3.22*a*. For example, typicals have a relative survival of only 0.672 ± 0.139 on a dark background, and melanics have a relative survival of only 0.727 ± 0.197 on a pale background. As we mentioned on p. 138, recent reviews suggest that other evolutionary factors may also be of great significance.

TABLE 3.12 The "survival" of dead melanic or typical moths placed on either dark or pale backgrounds (Clarke and Sheppard, 1966).

	Melanic			*Typical*		
Background	*Exposed*	*Survived*	*Proportion*	*Exposed*	*Survived*	*Proportion*
Dark	70	58	0.83	70	39	0.56
Pale	40	24	0.60	40	32	0.80

The MHC genes are part of immune systems in vertebrates and differential selection through resistance to pathogens is widely thought to be the basis of their high genetic variation (Edwards and Hedrick, 1998). To test this in the endangered winter-run Chinook salmon, Arkush *et al.* (2002) exposed captive-bred families from known MHC genetic matings (one parent was heterozygous and the other homozygous) to three different important salmonid pathogens. For the virus IHNV, Table 3.13 gives the number

TABLE 3.13 The number of MHC heterozygotes and homozygotes exposed and surviving the virus IHNV in 10 segregating families (Arkush *et al.*, 2002).

	Heterozygotes			*Homozygotes*		
Family	*Exposed*	*Survived*	*Proportion*	*Exposed*	*Survived*	*Proportion*
1	29	27	0.93	32	25	0.78
3	32	24	0.75	27	21	0.78
7	31	24	0.77	29	17	0.59
8	27	24	0.89	33	25	0.86
9	25	18	0.72	35	24	0.69
10	24	21	0.88	35	28	0.80
11	38	30	0.79	32	19	0.83
12	30	23	0.77	30	24	0.80
13	33	25	0.76	27	16	0.59
14	30	29	0.97	34	29	0.85
Total	299	245	0.819	305	228	0.748

exposed and surviving from each segregating family. The overall difference in survival between heterozygotes and homozygotes is significantly different at the 0.022 level (Arkush *et al.*, 2002). We can estimate the relative fitnesses using expression 4.8*b* and the fitness of the homozygotes relative the heterozygotes is $\hat{v}_{\text{hom}} = (0.482)(0.495)/(0.518)(0.505) = 0.912$, making $s = 0.088$. If we use expression 4.9*a*, which ignores the known small difference in the numbers of heterozygotes and homozygotes exposed, then $\hat{v}_{\text{hom}} = 228/(245 + 1) = 0.927$ and $s = 0.073$, a slightly smaller selection coefficient.

Likewise, from a cross of $A_iA_j \times A_iA_k$, then there are three times as many heterozygotes expected as homozygotes, viability can be estimated as

$$\hat{v}_{ii} = \frac{3N_{ii}}{N - N_{ii} + 1} \tag{3.22c}$$

and the amount of selection against homozygotes is $s = 1 - v_{ii}$ (see Example 3.8, which gives selection estimates for HLA genes from genetic segregation in South American Indian families).

Example 3.8. There is substantial evidence that there has been strong balancing selection on the human major histocompatibility complex (*HLA*) genes in South Amerindians (Black and Salzano, 1981; Parham and Ohta, 1996). Perhaps the best way to demonstrate that balancing selection in the present generation is to examine the proportions of homozygous and heterozygous progeny produced from parents of known genotypes. For most human populations, this is not feasible because of the high polymorphism at most *HLA* loci, where it is unusual for parents to share an allele. However, in many South Amerindians, polymorphism is much less extreme, and parents who share one or more alleles are common.

Table 3.14 gives the numbers of heterozygous and homozygous progeny from South Amerindian matings that were between identical heterozygotes or shared one allele for the two loci *HLA-A* and *HLA-B* (Black and Hedrick, 1997). In all four combinations, there was an excess of heterozygous progeny over Mendelian expectations, and for the three statistically significant comparisons, the average estimate of selection against homozygotes, 0.5, was quite large. The proportion of heterozygous progeny in older age classes was not larger than in younger age classes, thereby not generally supporting the importance of infectious disease as the selective agent. However, for both loci, there was no evidence of a reduced number of homozygotes for the mating type in which the female was homozygous for an allele shared by her mate (data not shown). Because the reciprocal mating did show significant selection against homozygotes, selection based on a maternal–fetal interaction that results in heterozygous advantage appears likely (see p. 182).

TABLE 3.14 Heterozygous and homozygous progeny for two genes from South Amerindian matings (a) between identical heterozygotes, $A_iA_j \times A_iA_j$, and (b) where allele A_i is shared in the parents, $A_iA_j \times A_iA_k$; s is the estimate of selection against homozygotes (Black and Hedrick, 1997).

	Heterozygotes		*Homozygotes*			
	Observed	*Expected*	*Observed*	*Expected*	χ^2	s
(a)						
HLA-A	56	40.0	29	40.0	12.8***	0.58
HLA-B	55	41.5	28	41.5	8.8**	0.50
(b)						
HLA-A	169	151.5	33	50.5	8.1**	0.42
HLA-B	140	132.8	37	44.2	1.6	0.21

**$P < 0.01$, ***$P < 0.001$.

d. Relative Risk

For many diseases in humans, large epidemiological studies evaluate the **relative risk (*RR*)** of different types of individuals getting the disease (Agresti, 1996). A general approach to determine the risk of individuals with a given genotype getting a disease, relative to that in the rest of the population, is to calculate the RR as

$$RR = \frac{f_d(1 - f_c)}{f_c(1 - f_d)} \tag{3.23}$$

where f_c and f_d are the frequencies of the genotype in control and diseased groups, respectively. If a genotype reduces susceptibility to a disease, then the frequency of the genotype in individuals with the disease is lower than in the control group—that is, $RR < 1$. On the other hand, if the genotype increases susceptibility to the disease, then the frequency of the genotype in the disease group is higher than in the control group and $RR > 1$.

In order to use RR values for different genotypes in population genetics, they need to be translated into relative fitnesses for the different genotypes (Hedrick, 2004). The selection coefficient for individuals with a given genotype that confers resistance to a disease can be calculated as

$$s = m(1 - RR) \tag{3.24a}$$

where m is mortality rate for the subjects infected with the disease, independent of genotype. In this case, the relative fitnesses of the genotype being examined and that in the rest of the population are 1 and $1 - s$, respectively. However, to obtain the positive selective effect of the genotype compared with the rest of the population, the relative fitnesses need to be scaled so that the fitnesses of the genotype and the rest of the population become $1/(1 - s)$ and 1, respectively. For example, if $RR = 0$ (complete resistance of the genotype) and $m = 0.5$ (50% mortality of infected individuals), then $s = 0.5$, and the fitness of the resistant genotype, relative to the rest of the population, is 2.0.

To determine the relative fitness of genotypes when $RR > 1$, the selective effect (which by definition is bounded by 0 and 1) needs to be calculated in a different way. In this case, we can estimate the selection coefficient as

$$s = m\left(1 - \frac{1}{RR}\right) \qquad (3.24b)$$

The fitnesses of the susceptible genotype and the rest of the population become $1 - s$ and 1, respectively. For example, if $1/RR = 0$ (complete susceptibility of the genotype) and $m = 0.5$, then $s = 0.5$, and the relative fitness of the susceptible genotype is 0.5. (see Example 3.9 for an application of this approach to estimating fitness of hemoglobin genotypes in the presence of malaria).

Example 3.9. The normal (A) and the sickle-cell (S) alleles at the human hemoglobin β locus, which differ by a single amino acid at position 6, glutamic acid (E) to valine (V), E6V (see p. 546), have long provided a classic example of balanced polymorphism. AS heterozygotes have greater resistance to malaria than homozygotes AA and SS, and in addition, SS homozygotes suffer from sickle-cell anemia. It has been known that a third allele C at this locus, which has a different single amino acid substitution at position 6, glutamic acid to lysine (E6K), is in substantial frequency in several west African populations. Recently, allele C has been shown to confer higher protection to malaria than allele S from Burkina Faso, West Africa (Modiano *et al.*, 2001). This epidemiological finding supports earlier research suggesting that red blood cells in CC individuals are unsuitable hosts for the malarial parasite. In addition, it appears that CC homozygotes have only limited, if any, costs from anemia.

Various procedures for estimation of the relative fitness of genotypes at the hemoglobin β locus in populations polymorphic for alleles A, S, and C have been used (Allison, 1956; Cavalli-Sforza and Bodmer, 1971). Estimates in the past of the relative fitnesses for genotypes with the C allele—that is, AC, CC, and CS—have not been very accurate, mainly because the sample sizes for these genotypes were often small. The new very large data set from

Burkina Faso with substantial numbers of these genotypes overcomes this problem. The *RR* values were statistically significant for the hemoglobin genotypes *AA*, *AC*, *AS*, and *CC* (*AC*, *AS*, and *CC* showed relative resistance *AA* showed relative susceptibility). Table 3.15 gives the frequencies of these four genotypes in the control (healthy subjects) and disease (malaria patients) groups and the resulting *RR* values. The *RR* values for the two other genotypes, *SC* and *SS*, were not statistically significant, primarily because there were few individuals of these genotypes in either the control or disease groups (however, see p. 162).

Using the approach in expression 3.24*b* for genotypes *AC*, *AS*, and *CC* and the approach in expression 3.24*a* for *AA*, Table 3.15 gives the relative fitness for the six genotypes, assuming that $m = 0.1$ (the mortality rate for hospitalized patients from the World Health Organization; WHO, 1998). Also given are the fitnesses relative to the genotype with highest fitness, *CC*. From these calculations, it appears that *CC* has approximately a 2% higher fitness than the *AS* heterozygotes. In Example 3.10. (p. 161), we use these fitness values to predict genetic change in this population.

TABLE 3.15 The frequency of malaria in healthy (control) subjects (f_c), malaria (diseased) patients (f_d), and the *RR* of malaria for the four genotypes where the *RR* is statistically significant (Modiano *et al.*, 2001). Below are given the estimated relative fitness values and their values relative that of genotype *CC*.

	Genotype			
Sample	*AA*	*AC*	*AS*	*CC*
Healthy subjects (f_c)	0.6641	0.2172	0.0954	0.0165
Malaria patients (f_d)	0.8036	0.1641	0.0275	0.0012
RR	2.070	0.7075	0.2681	0.0715
Fitness	0.948	1.030	1.079	1.102
Fitness (relative to *CC*)	0.861	0.935	0.979	1

III. RELAXING SOME GENETIC ASSUMPTIONS

Intrinsic to the basic selection model developed above are a number of assumptions regarding the genetic system. We relax two of these assumptions here. First, we discuss selection models in which the genes have multiple alleles. Then we discuss selection for X-linked genes or for genes that are in a haplo-diploid organism. These extensions, which are quite important biologically, significantly increase the complexity of the system and, as a result, can provide results quite different from those of the basic selection model discussed above. Consideration of multilocus selection, another extension of the simple genetic models of selection, is discussed in Chapter 10.

a. Multiple Alleles

As indicated in Chapter 1, some loci may have a number of alleles. Although SNPs generally only have two alleles and allozyme loci generally do not have more than three or four alleles, other loci, such as microsatellite and MHC, may have many different alleles. As we see below, the conditions for maintaining many alleles by selection favoring heterozygotes become quite complicated and restrictive (Lewontin *et al.*, 1978).

The basic selection model for two alleles that we discussed in the last chapter can be extended to multiple alleles. Assuming that the frequency of the ith allele is p_i and Hardy–Weinberg proportions before selection, the change in allele frequency after selection is

$$\Delta p_i = \frac{p_i(\overline{w}_i - \overline{w})}{\overline{w}} \tag{3.25}$$

where the average fitness of individuals with allele A_i is

$$\overline{w}_i = \sum_{j=1}^{n} p_j w_{ij}$$

and the mean fitness of the population is

$$\overline{w} = \sum_{i=1}^{n}\sum_{j=1}^{n} p_i p_j w_{ij}$$

To calculate changes in allele frequencies over generations, expression 3.25 can be iterated.

With more than two alleles undergoing selection, the changes in allele frequency may not change monotonically. An illustration can be given from the fitness array (a) in Table 3.16. In this three-allele system, both alleles

TABLE 3.16 The relative fitnesses for genotypes in a three-allele system for (a) a directional selection model, those estimated for the genotypes in the hemoglobin β system (b) from Allison (1956), and (c) from Hedrick (2004).

	Genotype and fitness					
(a) Directional selection	A_1A_1 $1+s$	A_1A_2 $1+s$	A_1A_3 $1+s$	A_2A_2 $1+s$	A_2A_3 1	A_3A_3 1
Hemoglobin β	AA	AS	AC	SS	SC	CC
(b) Allison (1956)	0.858	1.000	0.969	0.169	0.358	0.483
(c) Hedrick (2003)	0.861	0.979	0.935	0.109	0.498	1.000

A_1 and A_2 have the same selective advantage in the homozygote; however, A_1 is dominant over A_3, and A_2 is recessive to A_3. James (1965) suggested these fitness values for different alleles in industrial melanism in peppered moths in which A_1 is *melanic*, A_2 is *insularia*, and A_3 is *typical*. If it is assumed that initially $p_1 = 0.01$, $p_2 = 0.2$, and $p_3 = 0.79$, then the allele frequencies change as indicated in Figure 3.11. In the early generations, both favorable alleles increase, but as A_1 reaches a frequency near the others after approximately 25 generations, the frequency of A_2 stops increasing and eventually declines somewhat. This effect is due to the relative disadvantage incurred by the A_2 allele when a heterozygote with A_3, a factor that does not affect the A_1 allele. A_1 continues to increase until $A_3$is nearly eliminated and then stops increasing because there is no differential selection when only alleles A_1 and A_2 are present.

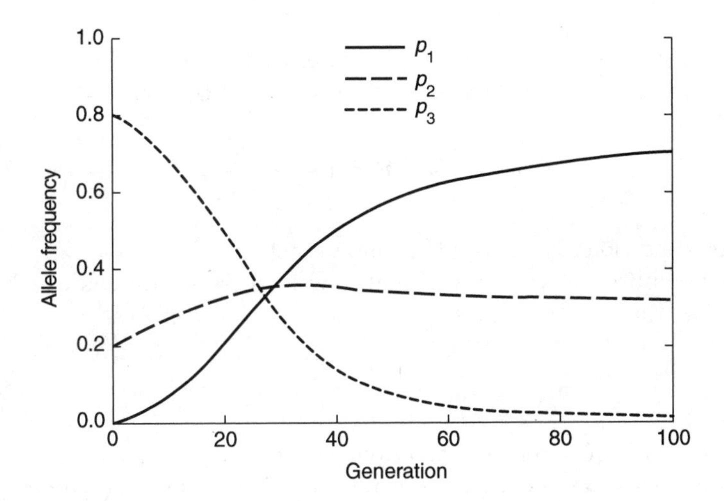

Figure 3.11. The change in allele frequency over time for a three-allele situation where both A_1 with a frequency of p_1 and A_2 with a frequency of p_2 are favorable alleles, but A_1 is dominant and A_2 is recessive to A_3 (row a of Table 3.16).

Another way to specify the conditions for an equilibrium for the heterozygote advantage model with two alleles is to show that

$$\overline{w}_1 - \overline{w} = 0 \tag{3.26a}$$

and

$$\overline{w}_2 - \overline{w} = 0$$

In other words, the average fitness for both alleles is equal to the overall mean fitness. To illustrate these conditions for two alleles, A_1 and A_2, with frequencies p_1 and p_2, assume that the fitness of genotypes A_1A_1, A_1A_2, and A_2A_2 are $1 - s_1$, 1, and $1 - s_2$, respectively. Therefore, using the expressions above

$$\overline{w}_1 = 1 - s_1 p_1$$

$$\overline{w}_2 = 1 - s_2 p_2$$

$$\overline{w} = 1 - s_1 p_1 - s_{2p_2}$$

If the alleles are at the stable equilibrium, that is, $p_1 = s_2/(s_1 + s_2)$ and $p_2 = s_1/(s_1 + s_2)$, then these three expressions all become

$$\overline{w}_1 = \overline{w}_2 = \overline{w} = 1 - \frac{s_1 s_2}{s_1 + s_2}$$

and the conditions for an equilibrium are met.

For equilibrium with n different alleles, these conditions can be expanded so that we must have

$$\overline{w}_i - \overline{w} = 0 \tag{3.26b}$$

for all n alleles. However, the conditions for a stable polymorphism for three or more alleles are not a simple extension of the two-allele heterozygote advantage model. Several different researchers have derived the details of these conditions, and Mandel (1970) showed their results to be equivalent.

For example, the conditions for the maintenance of three alleles and the equilibrium frequencies are a function of several expressions. Let us define these expressions as

$$x_1 = w_{23}^2 - w_{22}w_{33}$$

$$x_2 = w_{13}^2 - w_{11}w_{33}$$

$$x_3 = w_{12}^2 - w_{11}w_{22}$$

$$y_1 = w_{12}w_{13} - w_{11}w_{23}$$

$$y_2 = w_{23}w_{12} - w_{22}w_{13}$$

$$y_3 = w_{23}w_{13} - w_{33}w_{12}$$

$$z_1 = y_2 + y_3 - x_1$$

$$z_2 = y_1 + y_3 - x_2$$

$$z_3 = y_1 + y_2 - x_3$$

With these expressions, a three-allele stable equilibrium exists if x_1, x_2, x_3, z_1, z_2, and z_3 are all > 0. The equilibrium frequencies for the three alleles are

$$\begin{aligned} p_{1(e)} &= \frac{z_1}{z_1 + z_2 + z_3} \\ p_{2(e)} &= \frac{z_2}{z_1 + z_2 + z_3} \\ p_{3(e)} &= \frac{z_3}{z_1 + z_2 + z_3} \end{aligned} \qquad (3.27)$$

In general, the condition for a three-allele polymorphism is that the average viability of the heterozygotes must be greater than the average viability of the homozygotes. There are, however, three other particular cases that may give a three-allele equilibrium. None of these is sufficient by itself for a stable polymorphism; even given such viability relationships, some actual values may not satisfy the conditions given above. These viability relationships are as follows:

- All heterozygotes are more fit than all homozygotes; that is, w_{12}, w_{13}, $w_{23} > w_{11}$, w_{22}, w_{33}.
- One heterozygote is less fit than one of its associated homozygotes; for example, w_{12}, $w_{13} > w_{33} > w_{23} > w_{11}, w_{22}$.
- One heterozygote is less fit than the nonassociated homozygote; for example, $w_{12}, w_{13} > w_{11} > w_{23} > w_{22}$, w_{33}.
- One heterozygote is less fit than two homozygotes, the nonassociated homozygote, and one of the associated homozygotes; for example, $w_{12}, w_{13} > w_{11}$, $w_{33} > w_{23} > w_{22}$.

An example of this last case is given in Example 3.10 for β-chain hemoglobin alleles using fitness array (b) in Table 3.16. In all four cases above, at least two heterozygotes must have higher viability than all of the homozygotes. Additionally, no heterozygote has a lower viability than all of the homozygotes. Example 3.10 also shows the predicted elimination of the sickle-cell allele with another fitness array in which genotype *CC*, not *AS*, has the highest fitness.

Example 3.10. As we discussed in Example 3.9, there are three alleles; *A*, *S*, and *C* at the hemoglobin β locus in a number of African populations. Relative-fitness values have been calculated for the six genotypes resulting from these three alleles by Allison (1956) based on differential viability (see row b in Table 3.16). Note that two heterozygotes, *AS* and *AC*, have the highest relative-fitness values and that two of the homozygotes, *AA* and *CC*, have fitness values higher than the other heterozygote, *SC*.

Using these fitness values, the equilibrium frequencies of A, S, and C are $p_1 = 0.83$, $p_2 = 0.07$, and $p_3 = 0.10$, respectively, using expression 3.27. The average relative fitnesses for each allele and the mean fitness are all 0.897 at equilibrium satisfying equation 3.26b.

If the C allele is not present, as in many populations, alleles A and S together constitute a balanced polymorphic system with frequencies of 0.854 and 0.146, respectively. Figure 3.12 shows the change in allele frequencies over time if the C allele is introduced into such a population at a frequency of 0.01 ($p_1 = 0.844$, $p_2 = 0.146$). Although allele C can enter the system as shown in Figure 3.12, an extremely long time is necessary for it to increase in frequency. In fact, it takes approximately 5000 generations to

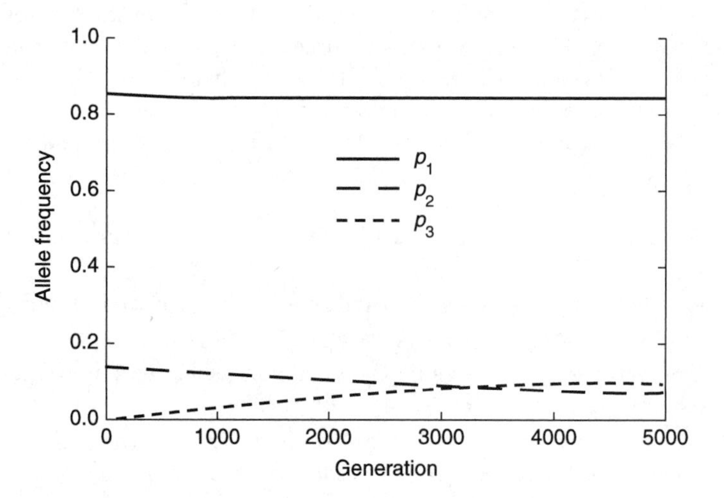

Figure 3.12. The theoretical change in allele frequency over time for β-chain variants using the relative fitness values in row (b) of Table 4.16.

approach its equilibrium of 0.10. This slow change is due to the miniscule difference between the mean fitness for the two-allele equilibrium and that for the three-allele equilibrium: less than a 0.01 percent difference.

The low observed frequencies of SC and SS in the study of Modiano *et al.* (2001) appear to be the result of high mortality (short life expectancy) caused by anemia and related complications in individuals with these genotypes. In this case, the relative fitness can be estimated as $1 - s = N_O/N_E$, where N_O and N_E are the observed and expected numbers (using Hardy–Weinberg proportions) of the genotype. Modiano *et al.* (2001) stated that in the control group for genotype SC, N_O and N_E were 23 and 46.2, respectively, and for SS, N_O and N_E were 1 and 9.2, respectively. The estimated fitnesses are then 0.498 and 0.109 for genotypes SC and SS. Using the relative fitnesses for the other genotypes given in Example 3.9, we have the array of fitnesses given in row (c) of Table 3.16 for this population.

Unlike other fitness arrays (such as row b of Table 3.16), these fitnesses give no stable or unstable three-allele equilibria. If C is introduced by mutation or gene flow, whether S is present or not, then C will always increase, and eventually the population will become fixed for C. This outcome occurs primarily because genotype CC has the highest estimated relative fitness of any genotypes. However, the rate of increase of C is a function of the frequency of S. If S is not present, C is introduced by mutation or gene flow at a low frequency, and then C quickly increases to a substantial frequency in a few generations (from 0.01 to 0.5 in approximately 60 generations) (Figure 3.13). If both C and S are introduced simultaneously at low frequencies, they both initially increase (S actually faster than C). If A and S are at their predicted equilibrium frequencies, then the initial increase of C from a low frequency is greatly slowed.

Modiano *et al.* (2001) estimated that the frequencies of A, S, and C in their sample from Burkina Faso were 0.820, 0.051, and 0.128, respectively. In the simulations in Figure 3.13, frequencies of S and C very similar to those observed (indicated by solid squares) were seen in generation 59 (when both alleles were simultaneously introduced at low frequencies) and in generation 139 (when S was initially at equilibrium with A and C was introduced at low frequency). Therefore, it appears that C has been slowly increasing in this population because of the presence of S and that now it is poised to increase rapidly to a high frequency within the next 50 generations, eliminating allele S and sickle cell anemia, and going to fixation.

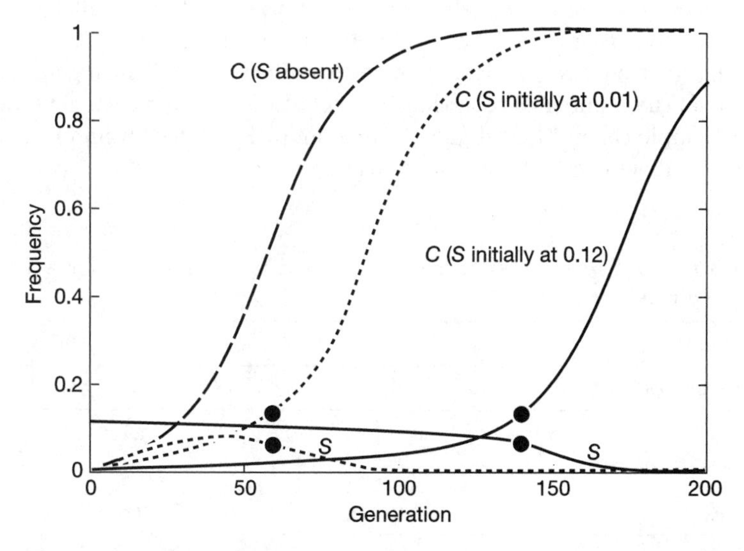

Figure 3.13. The increase in frequency of allele C when it begins at a frequency of 0.01 (long, broken line), when S also begins at a frequency of 0.01 (short, broken line), and when S begins at its equilibrium frequency of 0.12 (solid line) (Hedrick, 2004). The change in frequency of S is also given for the last two situations. The solid circles indicate the frequencies in generations closest in frequency to that observed in the Burkina Faso sample.

Three-allele polymorphic systems may evolve from two-allele polymorphisms. The conditions for a third allele to increase in a population at equilibrium for two alleles are related to the equilibrium condition. That is, A_1 will increase in the presence of A_2 and A_3 if $z_1 > 0$; A_2 will increase in the presence of A_1 and A_3 if $z_2 > 0$, and A_3 will increase in the presence of A_1 and A_2 if $z_3 > 0$. In other words, for a stable three-allele polymorphism to occur, each of the alleles must be able to increase from a low frequency, or invade, in the presence of the other two alleles.

b. X-Linked Genes or Genes in Haplo-Diploid Organisms

As we saw in Chapter 2, for X-linked genes or genes in haplo-diploid organisms, males are haploid and females diploid (for a review of their population genetics, see Hedrick and Parker, 1997). As a result, selection may greatly differ in the two sexes, acting in a haploid manner for males and a diploid manner for females. First, let us consider the situation of a lethal—that is, where genotypes A_2A_2 in the females and A_2 in the males are lethal. In this case, no A_2A_2 females will be produced in the progeny because such an outcome would necessitate an A_2 male reproducing in the previous generation. In other words, the only adults to have the A_2 allele will be A_1A_2 females. Because half the A_2 alleles are given each generation to sons that die, the frequency of A_1A_2 females is reduced by half each generation. Of course, the number of sons that die will also be halved each generation. (These simple calculations do not include mutation, which would introduce the lethal allele back into the population.)

TABLE 3.17 Different fitness arrays used as examples of selection for X-linked or haplo-diploid genes.

	Females			*Males*	
Fitness arrays	A_1A_1	A_1A_2	A_2A_2	A_1	A_2
(a)	w_{11}	w_{12}	w_{22}	w_1	w_2
(b)	1	1	$1 - s_f$	1	$1 - s_m$
(c)	$1 - s_1$	1	$1 - s_2$	$1 + s_m$	$1 - s_m$
(d)	1.5	1.0	0.5	0.5	1.5

Let us now consider a general array of fitnesses values in row (a) of Table 3.17, where the change in allele frequency caused by selection can be calculated from Table 3.18. The genotype frequencies before selection are the ones from Table 2.6, which were used previously to examine frequency change for an X-linked or haplo-diploid allele with no selection. The allele

TABLE 3.18 The frequency of female and male genotypes before and after selection for an X-linked gene or for a gene in a haplo-diploid organism.

	Female genotype			
	A_1A_1	A_1A_2	A_2A_2	*Total*
Fitness	w_{11}	w_{12}	w_{22}	—
Frequency before selection	$p_f p_m$	$p_f q_m + p_m q_f$	$q_f q_m$	1
Weighted contribution	$p_f p_m w_{11}$	$(p_f q_m + p_m q_f)w_{12}$	$q_f q_m w_{22}$	$\overline{w}_f$
Frequency after selection	$\frac{p_f p_m w_{11}}{\overline{w}_f}$	$\frac{(p_f q_m + p_m q_f)w_{12}}{\overline{w}_f}$	$\frac{q_f q_m w_{22}}{\overline{w}_f}$	1
	Male genotype			
	A_1	A_1	*Total*	
Fitness	w_1	w_2	—	
Frequency before selection	p_f	q_f	1	
Weighted contribution	$p_f w_1$	$q_f w_2$	$\overline{w}_m$	
Frequency after selection	$\frac{p_f w_1}{\overline{w}_m}$	$\frac{q_f w_2}{\overline{w}_m}$	1	

frequencies after selection in females and males are

$$q_f' = \frac{1/2(p_f q_m + p_m q_f)w_{12} + q_f q_m w_{22}}{\overline{w}_f} \tag{3.28a}$$

and

$$q_m' = \frac{q_f w_2}{\overline{w}_m} \tag{3.28b}$$

where

$$\overline{w}_f = p_f p_m w_{11} + (p_f q_m + p_m q_f)w_{12} + q_f q_m w_{22}$$

and

$$\overline{w}_m = p_f w_1 + q_f w_2$$

where the subscripts f and m indicate females and males, respectively.

If there is selection only against the recessive A_2A_2 females and the haploid A_2 males, the fitnesses of row (b) of Table 3.17 can then be used. Expressions 3.28a and 3.28b can be written as

$$q_f' = \frac{1/2(p_f q_m + p_m q_f) + (1 - s_f)q_f q_m}{1 - s_f q_f q_m}$$

and

$$q'_m = \frac{q_f(1 - s_m)}{1 - s_m q_f}$$

For the lethal case as discussed above, $s_f = s_m = 1$ so that $q'_m = 0$ and $q'_f = \frac{1}{2}q_f$.

The change in allele frequency is, for the two sexes,

$$\Delta q_f = -\frac{\frac{1}{2}p_f q_m - q_f(\frac{1}{2}p_m + s_f q_m p_f)}{1 - s_f q_f q_m}$$

and

$$\Delta q_m = -\frac{s_m q_f (1 - q_m) + q_f - q_m}{1 - s_m q_f}$$

Because inheritance of these genes is different for the two sexes and selection is among diploids in females and haploids in males, even if the allele frequencies are initially the same, they generally quickly become unequal. Expressions 3.28*a* and 3.28*b* can be similarly modified with other fitness arrays to calculate allele frequency change (see Example 3.11 for an application to the *white* mutant in *Drosophila*).

Example 3.11. An example of X-linked selection is that in *D. melanogaster* for the recessive eye-color mutant *white* (Hedrick, 1976). In this experiment, selection primarily acted against white males that had lowered mating ability, although recessive white females had a somewhat lowered fertility. When the initial allele frequencies were the same in the two sexes, selection smoothly eliminated the mutants, as shown in Figure 3.14a. However, when the initial allele frequencies were different in the two sexes, male and female allele frequencies oscillated for several generations before the eventual elimination of the mutant. Note that the male frequencies lag behind the female frequencies after the oscillations have ceased. In addition, the theoretical predictions (open circles), using a model that incorporated the estimated fitness differences, were very similar to the experimental results.

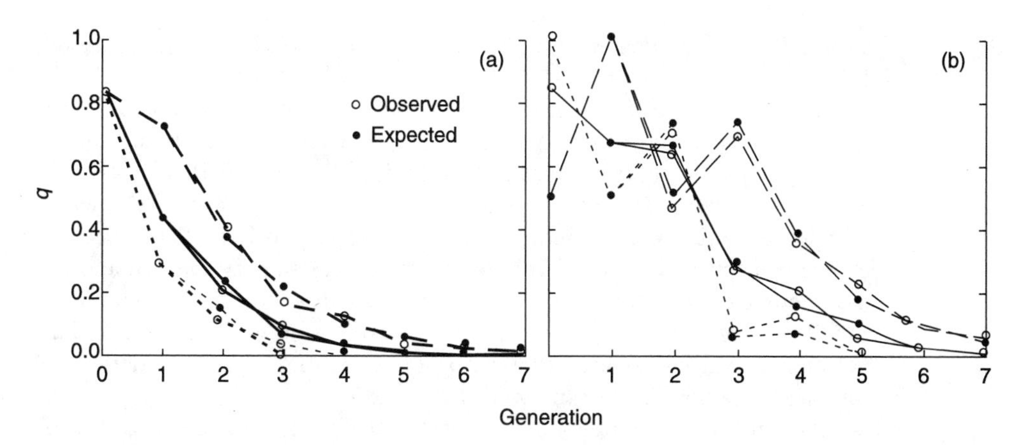

Figure 3.14. The change in frequency for the X-linked mutant *white* when (a) there is equal initial frequency in the two sexes and (b) when the initial frequencies are different (from Hedrick, 1976). The closed and open circles indicate experimental results and theoretical predictions, respectively. The solid, the short broken, and the long broken lines refer to the mean, female, and male allele frequencies, respectively.

A stable polymorphism is also possible for an X-linked or haplodiploid gene. However, the conditions are somewhat more stringent than for diploids because the males are haploids and therefore cannot exhibit a heterozygous advantage. Given the fitnesses in row (a) of Table 3.17, the conditions for a stable polymorphism are that

$$w_1 w_{11} < 1/2\,(w_1 + w_2)\, w_{12} > w_2 w_{22} \tag{3.29a}$$

The fitness values can also be written as in row (c) of Table 3.17. If we use these values, the conditions for equilibrium become

$$(1 + s_m)\;(1 - s_1) < 1 > (1 - s_m)\;(1 - s_2) \tag{3.29b}$$

In other words, a balanced polymorphism is possible either if there is heterozygous advantage in the females and not very strong selection in the males or with selection of similar magnitude acting in opposite directions in the two sexes. To illustrate the restrictiveness of these conditions, let $s_m > 0$. For a stable polymorphism, then

$$s_1 > \frac{s_m}{1 + s_m} \qquad \text{and} \qquad s_2 < \frac{s_m}{s_m - 1}$$

whereas for the diploid case, s_1 and s_2 only need to be greater than zero. The region in which there is a stable polymorphism is only approximately

58% that of the analogous diploid case (Pamilo, 1979; Hedrick and Parker, 1997).

The fire ant, *Solenopsis invicta*, which has been introduced to the United States from South America, has both colonies with one queen (monogyne) and multiple queens (polygyne) (Keller and Ross, 1999). In polygyne colonies, all queens are heterozygous *Bb* for the gene *Gp-9*, which encodes for a pheromone-binding protein (Krieger and Ross, 2002) because homozygous *BB* queens are destroyed by workers as they mature and homozygous *bb* queens are lethal. Obviously, because $s_1 = s_2 = 1$, then the conditions for a stable polymorphism in expression 3.29*b* are met. This example also demonstrates that variation at a single gene can determine complex behaviors important in social evolution.

An important example of selection for an X-linked variant is at the *G-6PD* locus for malarial resistance in humans (Ruwende *et al.*, 1995; Tishkoff *et al.*, 2001; Saunders *et al.*, 2002; Verrelli *et al.*, 2002). There are several different *G-6PD* alleles that appear to be involved in malarial resistance, but in the African populations examined by Ruwende *et al.* (1995), the common resistance allele was *A-*. They estimated the *RR* of both females heterozygous for the ancestral allele *B*, that is, genotype *BA-*, and male *A-* hemizygotes having malaria as 0.54 and 0.42, both significantly less than one. Using the approach in equation 3.24a, and assuming that mortality m from malaria is 0.1, then the estimates of fitness for *BA-* and *A-* hemizygotes are 1.046 and 1.058, respectively. Standardizing these values, then $w_{BB} = 0.956$, $w_{BA-} = 1$, $w_B = 0.945$, and $w_{A-} = 1$. Using these values in the polymorphism conditions in expression 3.29*a*, if the fitness of the *A-A-* homozygotes (w_{22}), which have a deficiency of the enzyme *G-6PD*, is less than 0.972, then a stable polymorphism is expected. However, the relative values of the fitnesses and whether there is a stable polymorphism are not clear (see discussion in Saunders *et al.*, 2002).

If the conditions for a stable polymorphism are met, then the equilibrium frequency of allele A_2 in males is

$$p_{m(e)} = \frac{s_2 + s_m + s_m^2(1 - s_2)}{s_1 + s_2 + s_m^2(2 - s_1 - s_2)} \tag{3.30a}$$

and that in females is

$$p_{f(e)} = \frac{s_2 + s_m - s_2 s_m}{s_1 + s_2 + s_m(s_1 - s_2)} \tag{3.30b}$$

When s_1, s_2, and s_m are small (products involving them are quite small), then the equilibrium allele frequencies in both sexes are similar, and approximately

$$P_{m(e)} \approx P_{f(e)} \approx \frac{s_2 + s_m}{s_1 + s_2} \tag{3.30c}$$

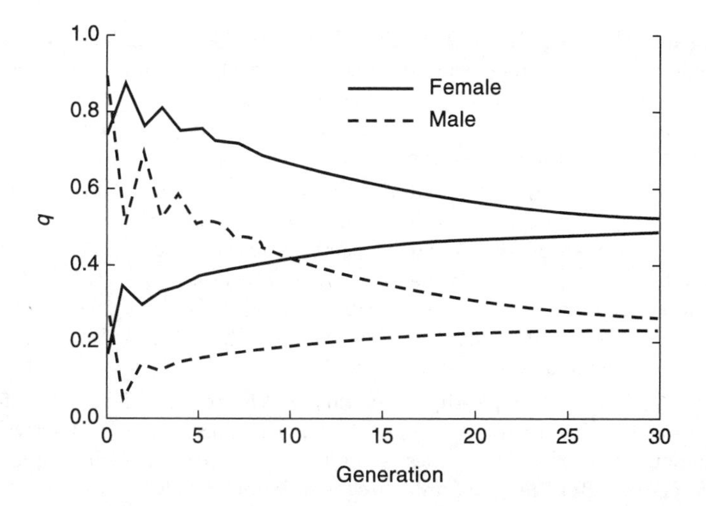

Figure 3.15. The change in allele frequency over time from two starting frequencies for an X-linked or haplo-diploid gene where there is a stable equilibrium for males at 0.25 and females at 0.5.

Figure 3.15 shows an example of the change in allele frequency for a situation in which there is a balanced polymorphism caused by differential selection in the two sexes. In this case, the fitnesses are as given in row (d) of Table 3.17 so that $s_1 = -0.5$, $s_2 = 0.5$, and $s_m = -0.5$. The equilibrium allele frequencies in the males and females from expressions 3.30*a* and 3.30*b* are 0.25 and 0.5, respectively—quite different equilibrium values for the two sexes. Two initial sets of allele frequencies are used in Figure 3.15, one above the equilibrium ($\overline{p} = 0.8$, $p_m = 0.9$, $p_f = 0.75$) and one below the equilibrium ($\overline{p} = 0.2$, $p_m = 0.3$, $p_f = 0.15$). The initial differences in allele frequencies between the sexes result in oscillations in the early generations and then a steady asymptotic approach to the equilibrium values, from both above and below the equilibrium.

PROBLEMS

1. Compare the change in allele frequency expected for selection against a recessive using expressions 3.6*a* and 3.6*c*. (One approach is to calculate q_t for a specific s value using expression 3.6*a* and then to substitute q_t in 3.6*c* to calculate t.) How good is the approximation in expression 3.6*c*?
2. Show that the expression for change in allele frequency for general dominance (expression 3.12*a*) becomes the specific one discussed previously when $h=$ 0.0, 0.5, and 1.0. Derive expression 3.13 for allele frequency change when there is adaptive Darwinian selection.
3. The relative fitnesses in a rat population for a gene conferring warfarin resistance have been estimated to be 0.37, 1.0, and 0.68 for genotypes RR, RS, and SS, respectively (see discussion in Example 3.1).What are the equilibrium frequency and the eventual allele frequency for populations beginning with the frequency of R equal to 0.0, 0.3, 0.7, and 1.0? Plot the mean fitness and the Δq values for the complete range of allele frequencies.

4. Assume that the fitnesses of three genotypes for a biallele locus form a geometric progression—that is, 1, 1 - s, and $(1 - s)^2$ for genotypes A_1A_1, A_1A_2, and A_2A_2, respectively. What is the frequency of A_2 after selection for this fitness array? Give this expression in its simplest form and compare it to that for haploid selection.
5. The relative fitnesses found for a chromosomal variant and the ancestral chromosomal type are 1.0, 0.5, and 0.8 for chromosomal types A_1A_1, A_1A_2, and A_2A_2, respectively. What are the relative fitnesses when the heterozygote has a fitness of 1.0? What are the equilibrium frequency and the eventual allele frequencies for populations beginning with the frequency of A_2 equal to 0.0, 0.3, 0.7, and 1.0? Plot the mean fitness and Δq values for the complete range of allele frequencies.
6. Assume that in a haplo-diploid organism, the relative fitnesses in females are 1.0, 1.0, and 0.8 for genotypes A_1A_1, A_1A_2, and A_2A_2, respectively, and 1.0 and 0.8 for male genotypes A_1 and A_2.What are the allele frequencies in both sexes after one, two, and three generations when $q_m = q_f = 0.2$ in the parental generation?
7. Calculate the expected equilibrium frequencies for the A, S, and C alleles using the fitnesses given in Table 3.16 from Allison (1956).
8. Why do you think there are only a few good examples of heterozygote advantage? Do you have any suggestions that may lead to the discovery of more examples?
9. Why do you think the major examples of heterozygote disadvantage are for chromosomal variants? Can you think of a mechanism that may result in heterozygote disadvantage at the gene level?
10. What is the expected frequency, in an experimental situation after 5 generations, of a lethal that has an initial frequency of 0.5? If, after 5 generations, the allele frequency was observed to be 0.28, what value of s, for a completely recessive gene, would explain the decrease in allele frequency?
11. Explain why the greatest expected change in allele frequency per generation for a recessive allele occurs around a frequency of 2/3. Explain why the greatest expected change in allele frequency per generation for a dominant allele occurs around a frequency of 1/3.
12. What are the implications of Fisher's fundamental theorem of natural selection? When could the genetic variance in fitness be equal to 0?
13. Assume that the fitnesses are 0.9, 1.0, and 0.6 for a locus. Calculate the mean fitness when $q = 0.4$. Is the mean fitness higher for any other allele frequency? Why?
14. Assume that the initial and final allele frequencies before and after selection are 0.2 and 0.01. How many generations does it take for this amount of change when there is a recessive lethal, selection against a recessive with $s = 0.2$, and additive selection with $s = 0.1$?
15. There are four different viability relationships that can satisfy the conditions for maintenance of three alleles. Give an example of the six viabilities for each that satisfies the conditions given in the text.
16. Using the genotype frequencies in two different generations may lead to spurious frequency-dependent selection. To illustrate, assume that the relative

fitnesses of genotypes A_1A_1, A_1A_2, and A_2A_2 are 0.9, 1.0, and 0.9, respectively, and that before selection $q = 0.1$, 0.25, or 0.9 (and initially Hardy–Weinberg proportions exist). Use expression 3.19*b* but with the genotype frequencies before selection. In the generation after selection, use the values calculated as in Table 3.3. Interpret your results.

17. Assume that mating of $A_1A_2 \times A_1A_2$ produced 40 A_1A_1, 100 A_1A_2, and 45 A_2A_2 individuals, counted at a pre-adult stage. What are the unbiased estimates of the relative viabilities of the two homozygotes?

4

Selection: Advanced Topics

> Haldane, Wright, and Fisher are the pioneers of population genetics whose main research equipment was paper and ink rather than microscopes, experimental fields, *Drosophila* bottles, or mouse cages. Theirs is theoretical biology at its best, and it has provided a guiding light for rigorous quantitative experimentation and observation...A body of evidence is now in existence to test the validity of some of the mathematical deductions...It is probably fair to say that accumulation of observational and experimental data is the order of the day...
>
> Theodosius Dobzhansky (1955)

> And NUH is the letter I use to spell Nutches
> Who live in small caves, known as Nitches, for hutches.
> These Nutches have troubles, the biggest of which is
> The fact there are many more Nutches than Nitches.
> Each Nutch in its Nitch knows that some other Nutch
> Would like to move into his Nitch very much.
> So each Nutch in a Nitch has to watch that small Nitch
> Or Nutches who haven't got Nitches will snitch.
>
> Dr. Seuss

It is obvious to any biologist that selection is the major mechanism that generates adaptive evolution. That is, the phenotypic results of evolution in terms of adaptive behavior, morphology, and physiology are familiar to everyone who has carefully observed organisms. Although both the genetic basis of adaptive change is complex and providing a general framework for selective challenges is difficult, only by attempting to organize and evaluate different types of selection is it possible to understand some basic evolutionary principles. This chapter introduces and integrates a number of detailed topics on selection so that readers will develop both an appreciation for the effects of various modes of selection and an understanding of their significance in evolutionary genetics.

In Chapter 3, we assumed that selection was occurring because of differential viability of various genotypes. Selection can also result when genotypes have differences in fecundity, mating, or gametic production as well as in viability. We next discuss how selection can occur for these other components of fitness and then present some examples of the different forms selection may take, such as meiotic drive, self-incompatibility systems, and negative-assortative mating. Finally, we consider how selection may vary in different environments, whether it is affected by physical aspects of the environment (such as temperature or rainfall) that vary in time or space, or biotic aspects of the environment, including both the species being investigated and other species interacting with it (such as hosts or parasites). From this account, it should be apparent that selection is a complicated phenomenon and that the use of selection models in predicting genetic change or understanding the maintenance of genetic variation requires thoughtful application.

Differential selection may occur at any or all of the different life stages of an organism, such as the gametic, juvenile, or adult stages. The aspects of the life cycle of an organism can be organized to reflect selection at different stages. For example, Bundgaard and Christiansen (1972) divided the total amount of selection in a particular population into four major selection or **fitness components**: viability (zygotic), sexual, gametic, and fecundity. The same general scheme can be used for insects, mammals, plants, or other organisms, with an emphasis on the aspects that seem most important in the species in question.

Figure 4.1 is a diagram outlining this division of selection into components though a complete generation cycle. Of these different selection components, selection affecting viability or survival has often been emphasized in population genetics, probably in part because it is easiest to quantify and to use in theoretical modeling. However, it appears from a number of studies that sexual, gametic, and fecundity selection may be as important as viability differences in many situations. First, we present

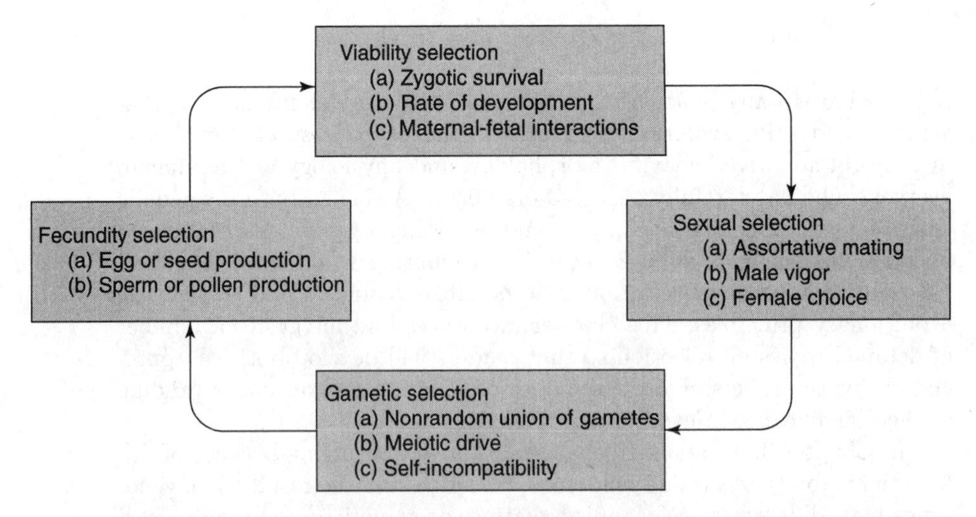

Figure 4.1. The division of selection into components, with examples of types of selection in each.

a brief discussion of how differential selection may occur through each of these fitness components. Then we discuss in the following sections some specific models of selection that result when there is selection for these different components.

The generation starts with a given composition of zygotes at the top of the diagram, which is then modified by differential survival in various life stages or **viability selection**. This selection includes survival of different preadult stages. In holometabolous insects there is differential hatchability (egg survival), larval survival, pupal survival, and pre-mating adult survival. In plants, there are analogous stages, such as differential germination of seeds, seedling survival, and prereproductive survival. In addition, the differential rate of development or time spent in these stages can influence the proportions of different genotypes at later life stages. In addition, in live-bearing organisms, there may be **maternal–fetal interactions** that can influence the viability of the offspring.

In sexually reproducing organisms, **sexual selection** occurs when the genotype proportions of reproductively mature females and males are not the same as the actual genotype or mating proportions among successful mates. For example, some female and male genotype combinations may be favored by assortative mating either between like or different types, males may have differential vigor in mating, or females may have differential receptivity to males because of female choice. Although such selection is probably most important in animals, it may also occur in plants. For example, an insect or animal pollinator may selectively visit plants of particular genotypes or phenotypes.

In some cases, **gametic selection** occurs because the gametes that are produced from heterozygotes are not in equal proportions, as from **meiotic drive (segregation distortion)**, or there may be nonrandom union of gametes with respect to the various alleles. In these cases, zygotic proportions may occur that do not reflect the allele frequencies in the parental genotypes, given Mendelian segregation and random union of gametes. Of particular importance in a number of plants are factors that inhibit fertilization by particular types of pollen and generally are mechanisms for reducing self-fertilization. These self-incompatibility systems are usually a function of either the pollen genotype or the genotype of the pollen parent.

If the contribution to the zygote pool for the next generation varies with particular male–female mating types, then there is the potential for selection via egg, seed, sperm, or pollen production called **fecundity selection**. Generally, such selection is the result of differences in the gametic potential of females because sperm and pollen production are often not limiting. Furthermore, the schedule of egg or seed production over time may be an important factor in determining the contribution to the next generation. We do not discuss fecundity selection here, mainly because differential fecundity among genotypes often results in changes in allele frequencies similar to those resulting from viability selection (Penrose, 1949; Bodmer,

1965). For example, when fecundity selection is assumed to be the same in the two sexes and there is no interaction between the sexes, then fecundity selection reduces to a viability selection model in which selection is the same in both sexes. However, one must be careful how one does (or does not) incorporate Mendelian segregation when considering fecundity selection in other situations. For example, in the model developed by Frelinger and Crow (1973) for the data in Example 4.14 (it documents that female pigeons heterozygous for a transferrin locus have higher fecundity in the presence of pathogens), selection in only one sex makes the model different from a viability model.

Sometimes when multiple selection components are influenced by variation at a single locus, there may be **antagonistic pleiotropy**, in which there is a negative correlation between different selection components (e.g., one allele at a given locus results in low viability and high fecundity, and another allele results in low fecundity and high viability). Such fitness arrays have been suggested as a mechanism for maintaining genetic variation and are thought to be important in both life history theory and the evolution of senescence.

I. VIABILITY SELECTION

Let us begin by considering several extensions of the basic viability selection model discussed in the last chapter. We discuss first the effect of survival variation in different life stages and then the effect of zygotic selection that differs in the two sexes. We then consider the effects of maternal–fetal interaction on viability. This type of selection may result in either an unstable or stable equilibrium, depending on the fitness values. Finally, we consider the conditions for maintenance of a balanced polymorphism from antagonistic pleiotropy.

a. Viability Differences in Different Life Stages

When premating viability (survival) of different zygotes varies in different life stages, then the total effect can be collapsed into one value. To understand this notion in its simplest form, assume first that there is a group of zygotes of a given genotype, next that a proportion of these individuals survive the larval (or seedling) stage and, finally, that a proportion of the remaining individuals survive the preadult stage. The overall proportion of survivors is the product of the proportions of survivors for the different life stages; in other words, the survival values for the different life stages collapse multiplicatively into one value.

The collapse of viability values of different life stages is shown in Table 4.1 for an example in plants where it is assumed that survival differences

TABLE 4.1 Different viability in various life stages of a plant and the collapsed overall survival values for each genotype.

	Genotype		
Life stage	A_1A_1	A_1A_2	A_2A_2
Germination proportion (seed survival)	w_{11g}	w_{12g}	w_{22g}
Seedling survival	w_{11s}	w_{12s}	w_{22s}
Remaining pre-adult survival	w_{11p}	w_{12p}	w_{22p}
Overall survival or Relative viability	$w_{11g}w_{11s}w_{11p}$ w_{11}	$w_{12g}w_{12s}w_{12p}$ w_{12}	$w_{22g}w_{22s}w_{22p}$ w_{22}

exist in three life stages. In this case, $w_{ij \cdot g}$, $w_{ij \cdot s}$, and $w_{ij \cdot p}$ are the proportions of plants of genotype ij that germinate, survive as seedlings, and survive as preadults, respectively. The overall survival for a given genotype is the product of these three survival values. Remember from the last chapter that the condition for a stable polymorphism with viability selection when there are two alleles is heterozygote advantage. We can either examine the products of these viability values or standardize the products so that the largest overall survival has a value of 1. In other words, using either the overall collapsed values or standardized values, for a stable polymorphism, $w_{11} < w_{12} > w_{22}$. Obviously, just because there is strong differential survival in some life stages, it is far from a foregone conclusion that there will be heterozygote advantage. With differential viability selection in different life stages, genotype proportions and allele frequencies may vary in different life stages, reflecting the cumulative selection to that point. However, one must be very careful in using genotype proportions to estimate the extent of selection (Prout, 1969).

b. Viability Differences in the Sexes

In many organisms, environmental factors may vary for individuals of the two sexes. For species that are sexually dimorphic, such as many birds and mammals, it is presumed that selection pressures have been significantly different enough in the two sexes to lead to differences in morphology, behavior, or other phenotypic characters. Although sexual dimorphism is often attributed to sexual selection (see p. 188), there are a number of examples where the two sexes of a species may have different viability values (Selander, 1966; references in Hedrick, 1993). To illustrate the effect of such selection, let us assume that there are selective differences in the two sexes at a single locus as given in Table 4.2a, where w_{f11}, w_{f12}, and w_{f22} are the relative fitnesses of genotypes A_1A_1, A_1A_2, and A_2A_2 in the females and w_{m11}, w_{m12}, and w_{m22} are the analogous fitnesses in the males.

TABLE 4.2 The relative fitnesses and change in genotype frequencies when there is differential selection between the sexes in general (a) and relative fitnesses for additive and opposite selection (b).

	Genotypes		
	A_1A_1	A_1A_2	A_2A_2
(a) **Relative fitness**			
Females	w_{f11}	w_{f12}	w_{f22}
Males	w_{m11}	w_{m12}	w_{m22}
Frequency before selection	$p_m p_f$	$p_m q_f + p_f q_m$	$q_m q_f$
Frequency after selection			
Females	$\frac{w_{f11}p_m p_f}{\bar{w}_f}$	$\frac{w_{f12}(p_m q_f + p_f q_m)}{\bar{w}_f}$	$\frac{w_{f22}q_m q_f}{\bar{w}_f}$
Males	$\frac{w_{m11}p_m p_f}{\bar{w}_m}$	$\frac{w_{m12}(p_m q_f + p_f q_m)}{\bar{w}_m}$	$\frac{w_{m22}q_m q_f}{\bar{w}_m}$
(b) **Relative fitness**			
Females	$1 - s_f$	$1 - \frac{1}{2}s_f$	1
Males	1	$1 - \frac{1}{2}s_m$	$1 - s_m$

Assuming that q_f and q_m are the frequencies of A_2 in the females and males, respectively, the mean fitnesses in the two sexes are

$$\bar{w}_f = w_{f11}p_m p_f + w_{f12}\left(p_m q_f + p_f q_m\right) + w_{f22}q_m q_f$$
$$\bar{w}_m = w_{m11}p_m p_f + w_{m12}\left(p_m q_f + p_f q_m\right) + w_{m22}q_m q_f$$

The change in the frequency of A_2 in the two sexes is (see Table 4.2a)

$$\Delta q_f = \frac{1/2 w_{f12}\left(p_m q_f + p_f q_m\right) + w_{f22}q_m q_f - \bar{w}_f q_f}{\bar{w}_f} \qquad (4.1a)$$

$$\Delta q_m = \frac{1/2 w_{m12}\left(p_m q_f + p_f q_m\right) + w_{m22}q_m q_f - \bar{w}_m q_m}{\bar{w}_m}. \qquad (4.1b)$$

Selection may take a number of different patterns in the two sexes. For example, directional selection may occur in different magnitudes but in the same direction in the two sexes. However, it is particularly interesting when directional selection occurs in opposite directions in the two sexes and leads to a stable polymorphism. In this case, for an equilibrium to exist, both expressions 4.1a and 4.1b must be zero for some polymorphic frequency.

Let us consider the situation in which selection occurs in opposite directions in the two sexes and heterozygotes are intermediate. These fitness values are given in Table 4.2b, where s_f and s_m are the selection coefficients in the A_1A_1 females and A_2A_2 males, respectively. In this case, there is a stable equilibrium, which for females and males is

$$q_f = \frac{s_f - 1}{s_f} + \left(\frac{s_m s_f - s_m - s_f + 2}{2 s_m s_f}\right)^{1/2} \tag{4.2a}$$

$$q_m = \frac{1}{s_m} - \left(\frac{s_m s_f - s_m - s_f + 2}{2 s_m s_f}\right)^{1/2} \tag{4.2b}$$

(Kidwell *et al.*, 1977). This equilibrium exists only when

$$\frac{s_m}{1 - s_m} > s_f > \frac{s_m}{1 + s_m} \tag{4.2c}$$

The region of the equilibrium can be shown graphically, as in the shaded region in Figure 4.2. Obviously, a fairly large range of selection values will result in a stable polymorphism. However, when selection is not large, the conditions for a polymorphism become quite restrictive. For example, if s_m is 0.1, then s_f must be between 0.091 and 0.111 for a stable polymorphism. It is noteworthy that this region is similar to the region of equilibria for the model for differential selection in two niches, which we discuss later. However, for selection differing between the sexes, the allele frequencies in the two sexes can be quite different while remaining the same for the two-niche model. Interestingly, sex-dependent habitat selection, even without differential selection in the two sexes, may enhance the conditions for the maintenance of polymorphism (Hedrick, 1993).

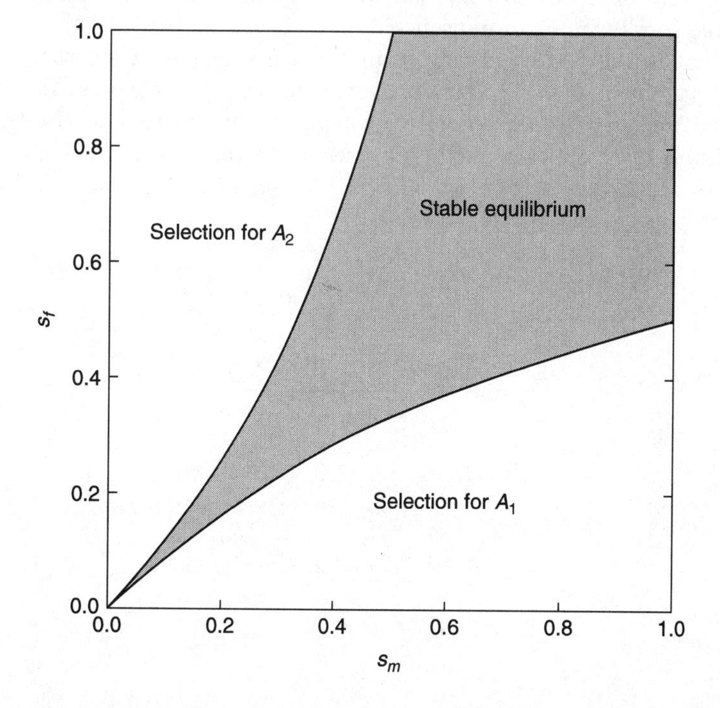

Figure 4.2. The region of a stable equilibrium (shaded area) when there is additive and opposite viability selection in the two sexes. The terms s_f and s_m indicate the selection against the female A_1A_1 and male A_2A_2 genotypes, respectively.

c. Maternal-Fetal Interactions

In mammals, there are interactions between a mother and her fetus that influence the survival of offspring of particular genotypes (similar interactions are possible in other live-bearing organisms). The best understood maternal–fetal incompatibility system is the Rh (rhesus) blood group system on the short arm of chromosome 1 in humans. The Rh system is the most polymorphic of the human blood groups and is composed of two closely linked, very similar genes: *RHCE* and *RHD* (extensive detail is known about the molecular basis and evolution of the system; Avent and Reid, 2000; Okuda *et al.*, 2000; Daniels, 2002). For our discussion here, if there is incompatibility between the mother and fetus, then fetal mortality may occur because maternal antibody production destroys red blood cells of the fetus. This interaction occurs when the fetus is Rh positive (genotypes RR or Rr) and the mother is Rh negative (genotype rr), where allele R indicates the presence of antigen D from gene *RHD* and allele r the absence of antigen D.

Because the R allele is dominant, incompatible combinations can only occur when the mother is Rh negative and father is Rh positive, that is, matings $rr \times RR$ and $rr \times Rr$ (Table 4.3a). Approximately 10% to 15% of the matings in most Caucasian populations are incompatible. In the mating $rr \times RR$, all offspring are incompatible with the mother, and thus, selection against these infants will be somewhat greater than from $rr \times Rr$, where only half of the progeny are incompatible. Selection is only occurring against heterozygotes so that one would predict that there might be an unstable equilibrium as we discussed on p. 145.

What is the predicted change in allele frequencies when there is Rh incompatibility? Using the mating frequencies in Table 4.3a, assuming Mendelian segregation, and selection against Rr offspring in the two incompatible matings, then the frequencies of the three types of offspring are as given at the bottom of Table 4.3a. The change in frequency of the r allele in one generation becomes

$$\Delta q = \frac{pq - {}^1\!/\!_2 pq^2(s_1 p + s_2 q) + q^2}{\bar{w}}$$

$$= \frac{pq^2(s_1 p + s_2 q)(q - {}^1\!/\!_2)}{\bar{w}} \tag{4.3}$$

where

$$\bar{w} = 1 - pq^2(s_1 p + s_2 q)$$

TABLE 4.3 (a) The mating types and their frequencies and the relative fitness of the offspring genotypes for the Rh maternal–fetal incompatibility system (the asterisk indicates the incompatible matings). The relative fitnesses of the offspring genotypes for *HLA* genes when there are maternal–fetal interactions resulting in (b) higher fitness when the fetus has an allele different from the mother and (c) low fitness of homozygotes from heterozygous mothers.

(a) Rh

		Offspring		
Female × *Male*	*Frequency*	*RR*	*Rr*	*rr*
$RR \times RR$	p^4	1	—	—
$RR \times Rr$	$2p^3q$	1	1	—
$RR \times$ rr	p^2q^2	—	1	—
$Rr \times RR$	$2p^3q$	1	1	—
$Rr \times Rr$	$4p^2q^2$	1	1	1
$Rr \times rr$	$2pq^3$	—	1	1
$rr \times RR*$	p^2q^2	—	$1 - s_1$	—
$rr \times Rr*$	$2pq^3$	—	$1 - s_2$	1
$rr \times rr$	q^4	—	—	1
		$\frac{p^2}{\overline{w}}$	$\frac{2pq - pq^2(s_1p + s_2q)}{\overline{w}}$	$\frac{q^2}{\overline{w}}$

	(b) *HLA*			(c) *HLA*		
Female × *Male*	A_1A_1	A_1A_2	A_2A_2	A_1A_1	A_1A_2	A_2A_2
$A_1A_1 \times A_1A_1$	$1 - s$	—	—	1	—	—
$A_1A_1 \times A_1A_2$	$1 - s$	1	—	1	—	—
$A_1A_1 \times A_2A_2$	—	1	—	—	1	—
$A_1A_2 \times A_1A_1$	$1 - s$	$1 - s$	—	$1 - s$	1	—
$A_1A_2 \times A_1A_2$	$1 - s$	$1 - s$	$1 - s$	$1 - s$	1	$1 - s$
$A_1A_2 \times A_2A_2$	—	$1 - s$	$1 - s$	—	1	$1 - s$
$A_2A_2 \times A_1A_1$	—	1	—	—	1	—
$A_2A_2 \times A_1A_2$	—	1	$1 - s$	—	1	1
$A_2A_2 \times A_2A_2$	—	—	$1 - s$	—	—	1

Because s_1 and $s_2 > 0$, then from expression 4.3, the only polymorphic equilibrium occurs when $q = 0.5$. However, this is an unstable equilibrium because Δq is positive when $q > 0.5$ and Δq is negative when $q < 0.5$. As a result, one would predict that a population would become homozygous for whichever allele was initially most common.

The ancestral Rh haplotype appears to be *Dce* (or *R*), and the *dce* (*r*) was generated by deletion or inactivation of the *D* gene with the other six haplotypes hypothesized to have been generated by recombination, gene conversion, and mutation (Daniels, 2002). Most human populations have high frequencies of the *R* haplotypes, but Europeans, particularly Basques, who live in the Pyrenees Mountains between Spain and France, have a high frequency of *dce*. Given the unstable equilibrium predicted above, one would expect that selection would have kept the frequency of r low. To explain the high frequency of r in Europeans, Cavalli-Sforza *et al.* (1994) suggested that the Basques are an ancient European population that was predominately r. For example, the estimate of r was 0.65 in one sample of Basques (Mourant *et al.*, 1976). Subsequently, settlers that were predominately R mixed with them and somewhat lowered the frequency of r. Of course, the prediction is still that in populations with r above 0.5 (only the Basques), r would increase and the population would become Rh-, whereas in the rest of the world where r is below 0.5, r would decrease and become Rh+.

As we mentioned on p. 40, the human MHC (called *HLA*) is the most polymorphic region in the human genome. One mode of balancing selection proposed to contribute to this polymorphism is maternal–fetal interaction that results in a net heterozygote advantage (Clarke and Kirby, 1966). This hypothesis suggests that a fetus with an antigen not present in its mother may have a higher survival than a fetus sharing antigens with its mother. Supporting evidence for this scenario comes from human couples that have a history of spontaneous abortions who are more likely to share antigens at *HLA* loci than control couples (Thomas *et al.*, 1985).

For such a model of selection, Table 4.3b gives the fitnesses of offspring from different matings (Hedrick and Thomson, 1988). What is important here in determining fitness is the comparison of the offspring genotypes and those of their mother. For example, in the first-row mating, $A_1A_1 \times A_1A_1$, the offspring A_1A_1 share both alleles with their mother and, therefore, have the reduced relative viability $1-s$. In the second-row mating, $A_1A_1 \times A_1A_2$, half of the offspring A_1A_2 have an allele different from their mother and consequently have the higher relative viability of 1. Overall, some heterozygotes are not selected against, whereas all homozygotes have a selective disadvantage, suggesting that there would be a net heterozygote advantage. Using the same approach as we did for Rh incompatibility but with the fitnesses in Table 4.3b, the expected change in allele frequency is

$$\Delta q = \frac{sp_1(1-p_1)(1/2-p_1)}{\bar{w}} \tag{4.4}$$

where

$$\bar{w} = 1 - s(1 - p_1p_2)$$

Here the only polymorphic equilibrium occurs when $p_1 = 0.5$, and this is a stable equilibrium because Δq is positive when $q < 0.5$ and Δq is negative when $q > 0.5$.

In examining a large sample of parent-offspring data from *HLA* loci in South Amerindians, Black and Hedrick (1997) found that there was a large excess of heterozygous progeny when the mother was heterozygous (491 heterozygotes observed and 418.25 expected) and no excess of heterozygous progeny when the mother was homozygous (75 heterozygotes observed and 71.5 expected). This was explained by another array of maternal–fetal interactions, and Table 4.3c gives the relative viabilities consistent with these observations. Using the same approach as above, but using genotype frequencies (H is the frequency of heterozygotes), Hedrick (1997) found that

$$\bar{w} = 1 - \frac{sH}{2} \tag{4.5a}$$

and the expected change in the frequency of A_2 becomes

$$\Delta q = \frac{q(q - wH/2) + pq - q\bar{w}}{\bar{w}} = 0 \tag{4.5b}$$

Quite surprisingly, there is no change in allele frequency for any value of q, as also found for neutral variants in which all genotypes have the same relative fitness (Figure 4.3).

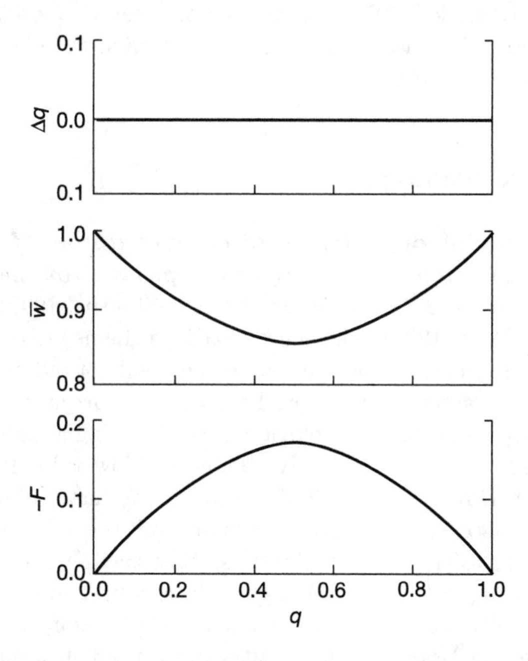

Figure 4.3. From top to bottom, the change in allele frequency (Δq), the mean fitness ($\bar{w}$), and the excess of heterozygotes (-F) as a function of allele frequency for the selection model in Table 4.3c when $s = 0.5$.

If we solve for equilibrium frequency of heterozygosity, then

$$H_e = \frac{2pq}{1 - sH_e/2}$$
$$= \frac{1 - (1 - 4spq)^{1/2}}{s} \tag{4.5c}$$

Using this value, we can calculate the mean fitness and heterozygosity at equilibrium (Figure 4.3). Here the mean fitness is a maximum at 0 and 1 and a minimum at 0.5, reminiscent of heterozygote disadvantage and an unstable equilibrium at 0.5. The heterozygosity is a maximum at 0.5 (as measured by the fixation index F, expression 2.17a), reminiscent of selection favoring heterozygotes.

Although it is often possible to predict the results of particular models logically, in this case, a single model simultaneously gives results expected from neutrality, heterozygote disadvantage, and heterozygote advantage. How can this be? Heterozygous females have lower fitness than homozygous females because of the lowered average fitness of their offspring. However, they compensate precisely for this by the relatively higher fitness of offspring that are exactly like themselves (heterozygotes), thus explaining the neutrality equilibrium. The relatively higher fitness of their heterozygous offspring results in an overall excess of heterozygous offspring. Overall, this example suggests that even for relatively simple models, predictions may not be intuitive and that great caution should be used when basing conclusions on what appears to be common sense. Wade (2000) interpreted the findings of Hedrick (1997) as a balance between two levels of selection, within and among families, and suggested a similarity to maternal effect selfish genes in several organisms.

d. Antagonistic Pleiotropy

Antagonistic pleiotropy, ***the negative correlation of two components of fitness, such as viability and reproduction***, has been widely suggested to be relevant in life history theory (Stearns, 1992) and the evolution of aging (Rose, 1991). The genetic basis of the negative correlation of two characters is generally thought to be the result of individual genes influencing both characters, pleiotropy. The negative correlation is thought to be the result of a metabolic or ecological cost of having a high value for one fitness component that subsequently results in a low value for another fitness component. However, the conditions for maintenance of polymorphism at a gene by antagonistic pleiotropy often are quite restrictive (Rose, 1982; Curtsinger *et al.*, 1994; Hedrick, 1999a) so that the general significance in evolutionary ecology is unclear. Below we discuss first some simple models of antagonistic pleiotropy and then show an application to this approach for albinism in the Hopi Indians, a situation in which a single gene has been suggested to influence two components of fitness, namely viability and male-mating success.

Selection may act in different directions on different fitness components over the lifetime of an organism; however, the net effect is often an advantage of one allele over another. We discussed this on p. 176 when we assumed that viability could act differentially and multiplicatively on different life stages. Similarly, we can assume that high reproductive effort, such as high fecundity in females or high mating success in males, may result in lowered subsequent viability, or high effort expended in survival may result in subsequent lower reproductive effort or success. In the simplest two-allele situation, if we assume that the reproductive and viability values of genotype A_iA_j are f_{ij} and v_{ij}, and that they are multiplicative in their effects, then the conditions for a stable polymorphism are

$$f_{11}v_{11} < f_{12}v_{12} > f_{22}v_{22} \tag{4.6a}$$

(Rose, 1982).

Let us give these values in terms of the level of selection against the homozygotes and dominance so that the relative reproductive values are $1-s_1$, $1-hs_1$, and 1 and the relative viability values are 1, $1-(1-h)s_2$, and $1-s_2$ for genotypes A_1A_1, A_1A_2, and A_2A_2, respectively. We have assumed the same level of dominance for both traits, called parallel dominance by Curtsinger *et al.* (1994). In this case, equation 4.6a becomes

$$(1-s_1) < (1-hs_1)[1-(1-h)s_2] > (1-s_2) \tag{4.6b}$$

From this, the conditions for the maintenance of polymorphism can also be given as

$$\frac{s_2}{1+hs_2} < s_1 > \frac{s_2}{1-(1-h)s_2}$$

Figure 4.4 gives the conditions for a polymorphism, as a function of the overall fitness of the two homozygotes, when $h = 0.5$ (Hedrick, 1999a). A polymorphism may be maintained in this case, but the conditions are

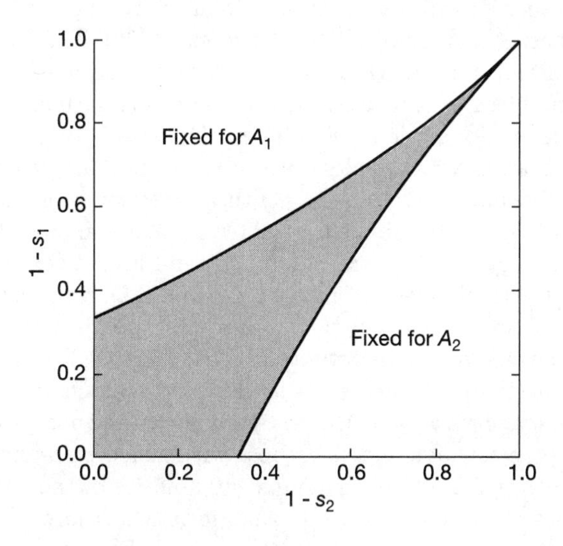

Figure 4.4. The region of polymorphism for antagonistic pleiotropy (shaded) when the relative viability of genotype A_2A_2 is $1-s_2$ and the relative reproduction of genotype A_1A_1 is $1-s_1$ and the level of dominance is $h = 0.5$.

somewhat restrictive, particularly when the amount of selection is low ($1 - s_1$ and $1 - s_2$ approach 1). For example, if $h= 0.5$ and $s_1 =0.1$, then s_2 must be between 0.0952 and 0.1053, a range of only 0.0101. Only when there are large selective differences, which if s_1 approaches 1 implies sterility or inability to mate for genotypes A_1A_1, or if s_2 approaches 1 implies lethality for A_2A_2, are the conditions significantly broadened.

Hedrick (1999a) also examined the impact of a number of other factors on the maintenance of a polymorphism by antagonistic pleiotropy and concluded that the conditions were even more restrictive when there was sex-limited selection; for example, the reproductive effects are only in one sex such as female fecundity or male-mating success. In addition, inbreeding or finite population size also reduces the likelihood of a balanced polymorphism. Curtsinger *et al.* (1994) suggested that the only major exception to these restrictive conditions is when there is a reversal of dominance so that the heterozygote is closest to fitness to the favored homozygote for both traits. However, they suggested that the likelihood that dominance is commonly reversed for pleiotropic traits at a given gene is quite low. When the extent of selection is high, then the conditions for maintenance of polymorphism are also broader. For example, Hedrick (2001) showed that there are broad conditions for a transgene to invade a natural population when it has mating advantage, even though it has a viability disadvantage. A similar model is given in Example 4.1, which investigates the potential influence of male-mating advantage for individuals with albinism influencing the incidence of this recessive, detrimental disease in the Hopi Indians.

Example 4.1. An often-cited example of pleiotropic effects of viability and male-mating success in humans is for albinism in Hopi Indians, a small tribe in northern Arizona (Woolf and Dukepoo, 1969). The number of individuals with albinism in the Hopis was estimated to be 26 of approximately 5,000, or 1 in 200, approximately two orders of magnitude higher than is observed in most populations. Woolf and Dukepoo (1969) thought that cultural selection, in which males with albinism had a mating advantage, influenced the incidence of albinism. Cultural selection could have occurred because Hopi males with albinism were traditionally allowed to remain in the village, thereby avoiding bright sunlight and its detrimental effects on them, while other males were farming in the fields. During this time, Woolf and Dukepoo (1969) suggested that males with albinism "had ample opportunity to engage in sexual activity."

To examine the joint effects of viability and male-mating selection on the albinism gene, first assume that individuals with genotypes AA and Aa have normal pigmentation and that those with genotype aa have albinism. Then assume that the relative male-mating abilities of the genotypes AA, Aa, and aa are 1, 1, and $1 + m$ so that m is a measure of the possible additional mating success of males with albinism. Finally, assume that the relative viabilities of genotypes AA, Aa, and aa are 1, 1, and $1 - s$, where

s is the selective disadvantage in survival of individuals with albinism of both sexes. With this model and assuming no mutation for the moment, the allele for albinism will increase in frequency when $m > 2s/(1 - s)$ (Hedrick, 2003b). In other words, the effect of male-mating success needs to more than twice the effect of viability selection, mainly because the mating difference occurs only in one sex and the viability difference in both sexes. For example, if $s = 0.5$, about the viability selection against albinism estimated in other populations, then $m > 2$. This means that the mating success of males with albinism would need to be three times that of a male without albinism to increase the frequency of the a allele. Such a high mating advantage seems quite unlikely, particularly because none of the Hopi males with albinism were married.

A more complete picture of the situation is obtained if mutation from allele A to a is assumed to occur at a rate $u = 0.000025$ (see discussion of mutation in Chapter 7) where the equilibrium incidence of albinism are examined as a function of the male-mating advantage (Figure 4.5). First, can male-mating advantage maintain the observed incidence of 1 in 200 individuals having albinism? The rightmost line in Figure 4.5 gives the equilibrium incidence of individuals with the albinism genotype when $s = 0.5$. To maintain an incidence of 0.005 for albinism, the value of m must be 1.987, almost 2, the value without mutation. Let us assume that viability selection is not as strong against albinism, either because of cultural selection increasing survival or that the type of albinism in the Hopis is milder than the common type in other populations (Hedrick, 2003b). For example, if $s = 0.3$, then m needs only to be 0.845 to increase the incidence to 0.005. Therefore, with lower viability selection, the necessary extent of mating advantage of males with albinism is reduced, and it could potentially play a role in the maintenance of albinism.

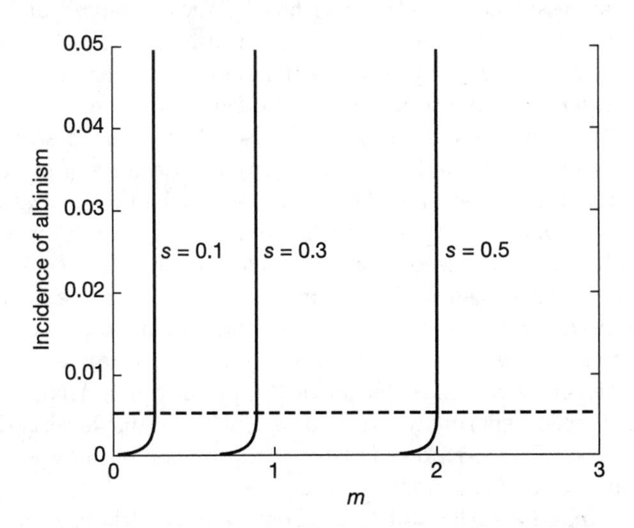

Figure 4.5. The equilibrium incidence of albinism as a function of mating advantage m for males with albinism and three values of selective disadvantage s in viability (Hedrick, 2003b). The broken line at 0.005 indicates the estimated incidence of Hopi Indians with albinism.

II. SEXUAL SELECTION

Sexual selection occurs in many ways, some of which result in selection favoring a particular allele and others of which result in polymorphism of two or more alleles. Nonrandom mating in the form of **positive-assortative mating**, in which ***similar individuals mate more often than randomly expected***, generally results in no change in allele frequency but an increase in homozygosity, just as for inbreeding (Chapter 5). For **negative-assortative mating** where ***individuals with unlike genotypes or phenotypes tend to mate with each other***, there is generally a change in allele frequency resulting in either fixation of a particular allele or a stable polymorphism.

Before we discuss two population genetic models of nonrandom mating, let us mention the basis of sexual selection in a more general context. It has been long recognized that mate success in many species is strongly affected by **male competition** for mates and **female choice** of mates (Darwin, 1871; for a recent thorough discussion, see Shuster and Wade, 2003). In most of these cases, the males in many species of birds, mammals, amphibians, fishes, and insects have extreme morphological or behavioral traits that appear to have evolved because they enhance the individual's odds either of winning in male–male competition to obtain matings with females or of attracting or persuading females to mate with him. In other words, variation in male phenotypic traits is assumed to indicate differences in male genetic quality (Wilkinson *et al.*, 1998). Examples of extreme male morphological traits are the large horns in male bighorn sheep and the immense size of male elephant seals. Examples of extreme behavioral traits include the construction of elaborate bowers (arenas) by male bowerbirds and the loud and constant calling by male frogs. The most extreme male individuals are assumed to be those with the highest fitness, which accounts for directional selection on these traits. We discuss male competition for mates among three color types in the side-blotched lizard on p. 227 that appears to resemble the interactions in the game "rock–scissors–paper."

In addition, male competition may also occur after mating through **sperm competition** in multiply mated females. Because of mechanical or biochemical factors, sometimes the first male may be the successful father, for example, when a vaginal plug excludes successful mating by subsequent males. In other cases, the second or later males may be the successful father, for example, when the second male removes the sperm in the female reproductive from earlier males. Clark *et al.* (1995) examined genetic variation for sperm competition in *D. melanogaster* for both **defense**, when sperm from an earlier male resists replacement by sperm from a subsequent male, and **offense**, when sperm from a subsequent male displaces sperm from an earlier male. They found extensive variation for both types of sperm competition spread throughout the genome.

Females may have the ability to choose mates with extreme morphological or behavioral characteristics over less extreme males. For female preference to result in selection favoring extreme phenotypes, females must be able to perceive accurately differences in male size, shape, sound, and so on. In other words, selection in the females is also directional but for a different trait—one that is dependent on accurate discrimination between males. It is generally thought that females choose mates that can provide important resources to them. For example, females may choose males with better territories that would provide more food for her offspring. Female insects may choose males that offer better nuptial gifts, generally food that she eats during mating. Another hypothesis explaining female choice is that females choose males with **good genes** (Andersson, 1994). For example, more brightly colored males may have greater genetic resistance to parasites (Hamilton and Zuk, 1982), a trait that then would be passed on to the female's offspring.

If directional selection occurs in both sexes (or even just in males, with females always choosing extreme mates), then theoretically, runaway sexual selection could result with more and more extreme males being chosen by females (Fisher, 1930; Lande and Arnold, 1985; Kirkpatrick and Barton, 1997). However, some of the assumptions of these models have been challenged (e.g., Holland and Rice, 1998; Qvarnstrom and Forsgren, 1998; Widemo and Saether, 1999), and the traits involved in male-mating success and female choice may be under strong selection in other ways (Ryan, 1998). For example, extreme male calling behavior in frogs may result in greater mating success but may also result in higher predation from bats. Because of higher mortality for extreme traits, the response to sexual selection may reach a limit beyond which the cost in male viability is greater than the advantage in male mating. In a series of experiments in nature, Endler (1980) showed that the patterns of spot coloration on the sides of male guppies were determined by a balance between sexual selection and natural selection. That is, it was advantageous in mate success for male guppies to have more colorful spots, but this was counteracted by the viability disadvantage in the presence of predatory fish to have this conspicuous coloration.

Genes that have opposite selective effects in females and males have been called **sexually antagonistic genes** (Rice, 1992; Rice and Chippindale, 2001a). On p. 177, we discussed a theoretical situation in which viability selection occurs in opposite directions in males and females, but perhaps more likely is where selection may favor given genotypes in one sex for one component of fitness, for example, mating success, and may favor other genotypes in the other sex for a different component of fitness, for example, viability. Because such an antagonism here is assumed to be sex specific, it is somewhat more restrictive than antagonistic pleiotropy, which we discussed on p. 184. In addition, the genes involved in determining the traits in the two sexes may be different and not the pleiotropic effects of alleles at a single gene. In this case, there may be a **Red Queen** process (Van Valen, 1973) or an **evolutionary arms race** between the two sexes in which they both change evolutionarily in response to the other, but there is no net overall change.

For example, there may be selection on a trait to increase female reproductive success that consequently reduces male viability. To counter this, there may subsequent selection to increase male viability, but this may then reduce female reproductive success. Rice and Chippendale (2001a) gave such a scenario based on human hip width, in which they assumed that larger hip width is advantageous for female reproduction but disadvantageous for male viability. They suggested that there is sexual antagonism in fitness for various mutants until a variant is generated that allows each sex to evolve hip size that is reflective of its gender-specific fitness optimum.

Although it is often said that in human populations "opposites attract," the correlation between human mates for almost any trait is positive (except sex). In fact, it appears that in general in most species, the phenotypes—and presumably the genotypes—of mates are positively correlated. A major exception to this is in plants where successful fertilization occurs only between individuals with different flower types. Although less generally accepted, another example is in populations in which rare males (or females) have a mating advantage. Finally, some reports suggest that negative-assortative mating in mammals may be based on MHC differences. We discuss and give models that can be used for these three types of negative-assortative mating below.

a. Negative-Assortative Mating

First, let us examine a simple model with negative-assortative mating. Assume that allele A_1 is dominant so that A_1A_1 and A_1A_2 have the same phenotype and that the frequency of the dominant phenotype is $P + H$. The mating pattern is such that mating results in fewer like $\times$ like (dominant $\times$ dominant and recessive $\times$ recessive) matings than expected from random-mating proportions (see Example 4.2 for an illustration of negative-assortative mating of different flower types in plants). Let us assume (after Li, 1976) that a proportion $(1 - R)$ of the population mates at random and the remainder, R, mate assortatively, having dominant $\times$ recessive matings (Table 4.4).

TABLE 4.4 The frequency of mating types and the proportion of progeny produced for a general negative-assortative mating model.

		Progeny		
Mating type	*Frequency*	A_1A_1	A_1A_2	A_2A_2
$A_1- \times A_1-$	$(P+H)^2(1-R)$	$p^2(1-R)$	$pH(1-R)$	$\frac{1}{4}H^2(1-R)$
$A_1- \times A_2A_2$	$2(P+H)Q(1-R)+R$	—	$2pQ(1-R)+\dfrac{pR}{P+H}$	$HQ(1-R)+\dfrac{HR}{2(P+H)}$
$A_2A_2 \times A_2A_2$	$Q^2(1-R)$	—	—	$Q^2(1-R)$
	1	$p^2(1-R)$	$2pq(1-R)+\dfrac{pR}{P+H}$	$q^2(1-R)+\dfrac{HR}{2(P+H)}$

Example 4.2. A number of plants have two flower types that differ in the length of the style and consequently result in obligate outcrossing between the different flower types (Barrett, 1990). Probably the most thoroughly examined such species is the primrose, *Primula vulgaris*, which commonly has the two flower types pin and thrum (see Figure 4.6). There appear to be different alleles at several closely linked genes that affect the positions of the anthers, style length, and other characteristics in this species (Dowrick, 1956). These alleles are associated into either the dominant thrum "allele" (S) or the recessive pin "allele" (s).

The primrose is pollinated by insects, and the position of the anthers and that of the stigma are such that virtually all pollination is between different flower types. In other words, insects generally transfer pollen only from a thrum flower with high anthers to a pin flower with a high stigma or from a pin flower with low anthers to a thrum flower with a low stigma. Furthermore, even if pin pollen is placed on a pin stigma or if thrum pollen is placed on a thrum stigma, very little seed is generally produced. In a typical example of a survey of an English population of primroses, there were 1553 thrum (SS or Ss) and 1827 pin (ss) plants (Ford, 1971). Natural populations generally have an excess of pins, as in this case, although an excess of thrums has occasionally been detected. The difference from equal proportions of pins and thrums, the proportions expected if there were only outcrossing, seems to occur because there is some self-fertilization of pins when no thrum pollen is available.

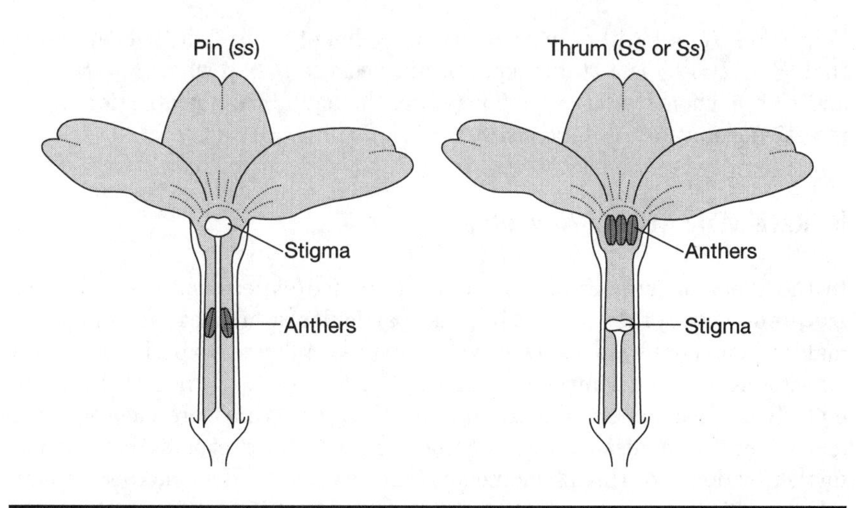

Figure 4.6. Diagrams of the structure of the pin and thrum flowers in the primrose, indicating the relative positions of the stigma and anthers in each.

Let us first examine the progeny frequencies for the dominant × dominant mating. The frequency of this mating is $(P + H)^2$ $(1 - R)$ because all such matings are in the random-mating proportion of the population. Among individuals with the dominant phenotype, the frequency of A_1 is

$P + \frac{1}{2}H = p$ so that the frequency of A_1A_1 progeny from dominant $\times$ dominant matings is $p^2(1 - R)$. Likewise, the frequency of A_2 among dominant individuals is $\frac{1}{2}H$ so that the proportion of A_1A_2 and A_2A_2 offspring are $pH\ (1-R)$ and $\frac{1}{4}H^2(1 - R)$, respectively. The progeny from the other random mating types can be calculated in a similar manner.

Now let us examine the progeny proportions from the negative-assortative mating proportion, R. These matings consist only of the dominant $\times$ recessive mating, a mating that results in either A_1A_2 or A_2A_2 progeny. The $A_1A_1 \times A_2A_2$ mating will give all heterozygotes, and half the progeny from the $A_1A_2 \times A_2A_2$ matings will be heterozygotes. Therefore, the amount of heterozygous progeny from the negative-assortative mating proportion is

$$R\left[\frac{P}{P+H} + \frac{H}{2(P+H)}\right] = \frac{pR}{P+H}$$

The frequency of progeny from all the matings is summarized in Table 4.4.

Using these values, we find that the frequency of A_1 in the next generation becomes

$$\begin{aligned} p' &= p^2(1-R) + pq(1-R) + \frac{pR}{2(P+H)} \\ &= p\left[1 - R + \frac{R}{2(P+H)}\right] \end{aligned} \tag{4.7}$$

When $P + H = \frac{1}{2}$, then $p' = p$, for any value of R. In fact, it is apparent that $P + H = \frac{1}{2}$ is a stable equilibrium because $p' > p$ when $P + H < \frac{1}{2}$ and $p' < p$ when $P + H > \frac{1}{2}$. Therefore, the equilibrium proportion of the recessive genotype A_2A_2 is also $\frac{1}{2}$.

b. Rare-Male Mating Advantage

In the 1960s, a number of researchers obtained experimental support for **frequency-dependent mating**, in particular a mating advantage to males when there were the rare type (Example 4.3 gives experimental data for two chromosomal inversions in *D. pseudoobscura*). When this type of male became common, the mating advantage was not present, or these males were at a disadvantage. Although some later studies have found further evidence of this phenomenon (Salceda and Anderson, 1988), many other studies were often not able to reproduce these effects. Partridge (1988), in a rigorous review, suggested that many of the rare-male mating studies had problems related to observer bias, experimental design, and data analysis (see also Knoppien, 1985). Some recent studies have again suggested the importance of the rare-male effect in both guppies (Hughes *et al.*, 1999) and house wrens (Masters *et al.*, 2003).

Example 4.3. It has been suggested that mating success in males may be a function of the frequencies of different males in a population. For example, in *D. pseudoobscura*, a number of studies indicated that males with an uncommon genotype have an advantage in mating, often called rare-type male advantage. To examine this phenomenon, one would place a number of males of two types in a container and record their mating success. The proportions of the two types are generally varied over a range, say from 10 to 90%, while the total number of males is kept constant. Data from an experiment using different inversion types in *D. pseudoobscura* are given in Figure 4.7. Note that the output ratio is much lower than the input ratio when the input ratio is high, and vice versa. This is in contrast to the expectation of equal input and output ratios for no frequency dependence. The slope of these ratios is significantly smaller than unity and crosses the no-frequency-dependence line at an input ratio of about 0.667 as indicated where the lines cross. As a result, Ayala (1972) infers that there is a stable equilibrium around 0.4.

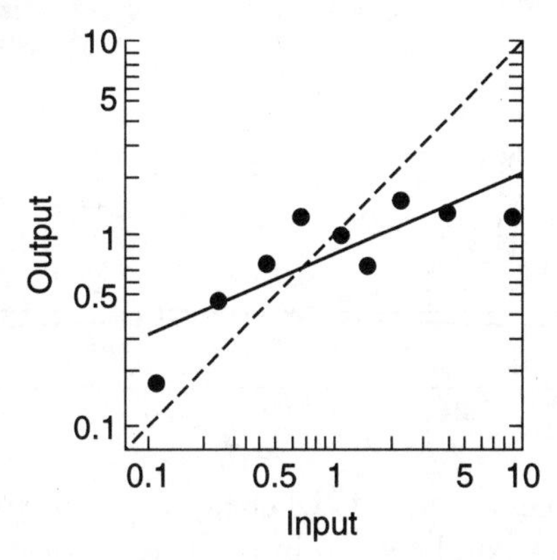

Figure 4.7. The slope (solid line) of the logarithm of the ratio of males that mated (output) to those present (input) for two inversions (Ehrman, 1967). The broken line indicates equal input and output ratios and where the lines cross indicates the putative equilibrium.

Let us examine a frequency-dependent mating model used by Anderson (1969) to describe rare-male advantage data in *Drosophila* where dominance is assumed. In this scheme, designed specifically to give a large mating advantage to rare individuals, the relative mating success of males with genotype ij is

$$w_{ij} = 1 + \frac{s_{ij}}{P_{ij}} \tag{4.8a}$$

where P_{ij} and s_{ij} are the frequency and the selection coefficient associated with genotype ij. For complete dominance,

$$w_{11} = w_{12} = 1 + \frac{s_1}{1 - q^2}$$

$$w_{22} = 1 + \frac{s_2}{q^2}$$

where s_1 and s_2 are the selection coefficients associated with the dominant and recessive phenotypes, respectively. In this case, the mating success of a genotype increases quickly if it is rare, as shown in Figure 4.8 as a function

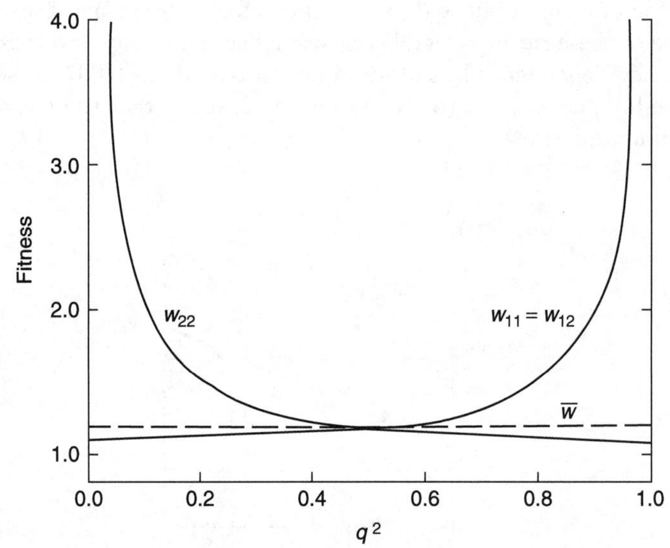

Figure 4.8. The relative fitnesses for a frequency-dependent model as a function of the frequency of genotype A_2A_2.

of the genotype frequency of A_2A_2 where s_1 and s_2 are both 0.1. The stable equilibrium frequency can be given in terms of $s_1$and s_2 as

$$q_e = \left(\frac{s_2}{s_1 + s_2}\right)^{1/2} \tag{4.8b}$$

and the mean fitness is

$$\bar{w} = 1 + s_1 + s_2$$

Several interesting properties found in frequency-dependent models that are different from the heterozygote advantage model are illustrated here. First, at the allele frequency equilibrium, all of the genotypes have the same fitness. This can be seen in the numerical example in Figure 4.8

and can be shown in general by substituting the equilibrium frequency into the fitness expressions. Second, the mean fitness is not maximized at the equilibrium because the mean fitness at the equilibrium is independent of the allele frequency. Finally, there is no marginal (overall) heterozygous advantage in this model because the fitness of the homozygote A_1A_1 is always the same as that of the heterozygote.

c. Female Choice, MHC, and Negative Assortative Mating

One of the main examples of selection by females for specific good genes in males is selection for genes in the MHC complex. Because the fundamental function of the MHC is to recognize pathogens and initiate an immune response, mating with males having MHC alleles that increase the pathogen resistance of offspring is consistent with the good genes selection scenario. It has been shown that mice can detect MHC differences in other mice, and experiments have shown female choice of males that differ in their MHC type (Egid and Brown, 1989). How can such female choice of male MHC type be modeled?

Let us assume that females preferentially mate males that differ genetically from themselves. For example, assume that a female is homozygous at a MHC gene, for example, A_1A_1, and males are identical at 2, 1, or 0 alleles, for example, genotypes A_1A_1, A_1A_2, or A_2A_2, then the relative preferences of males by these females are $1 - s$, $1 - hs$, and 1, respectively (Hedrick, 1992). If there are only two alleles, then the number of alleles that males differ from females, and the consequent mating preferences are as given in Table 4.5, where P_{ij} is the frequency of genotype A_iA_j and w_{ij} is a standardization so that the sum of the mating frequencies for each female genotype is equal to her frequency. With only two alleles, the heterozygous females mate at random with the three types of males (they all have the same low preference) because none of them has alleles different from the female.

From examining Table 4.5, it is obvious that there is an excess of matings in which males differ from females at two alleles, that is, matings $A_1A_1 \times A_2A_2$ and $A_2A_2 \times A_1A_1$. This has two effects: it results in a stable polymorphism and an excess of heterozygotes over that expected under Hardy–Weinberg proportions. For example, if $h = 0.5$, then the expected change in allele frequency for $s = 0.5$ in this model is virtually identical to heterozygote advantage when $s = 0.186$ (Hedrick, 1992). The smaller effect for the negative assortative mating model occurs because mate preference is only in females, and there is no mate preference when females are heterozygotes. Similarly, female preference for matings that are homozygous for different alleles produces an excess of heterozygotes in their progeny.

TABLE 4.5 The mating types for a MHC gene, giving the number of alleles that males differ from females, relative mating preference, and the relative mating frequencies

Mating type *Female × male*	*Number of different alleles*	*Mating preference*	*Mating frequencies*
$A_1A_1 \times A_1A_1$	0	$1-s$	$P_{11}^2(1-s)/w_{11}$
$A_1A_1 \times A_1A_2$	1	$1-hs$	$P_{11}P_{12}(1-hs)/w_{11}$
$A_1A_1 \times A_2A_2$	2	1	$P_{11}P_{22}/w_{11}$
$A_1A_2 \times A_1A_1$	0	$1-s$	$P_{11}P_{12}(1-s)/w_{12}$
$A_1A_2 \times A_1A_2$	0	$1-s$	$P_{12}^2(1-s)/w_{12}$
$A_1A_2 \times A_2A_2$	0	$1-s$	$P_{12}P_{22}(1-hs)/w_{12}$
$A_2A_2 \times A_1A_1$	2	1	$P_{11}P_{22}/w_{22}$
$A_2A_2 \times A_1A_2$	1	$1-hs$	$P_{12}P_{22}(1-hs)/w_{22}$
$A_2A_2 \times A_2A_2$	0	$1-s$	$P_{22}^2(1-s)/w_{22}$

The overall support for MHC-based, negative assortative mate choice is mixed and contentious (Hughes and Hughes, 1995; Penn, 2002). The most widely known study is the "t-shirt study" in which female Swiss university students ranked the smell of t-shirts worn by male students on characteristics such as pleasantness (Wedekind *et al.*, 1995). The findings of this study suggested that females preferred the odor of males that differed at MHC genes, except when they were on birth control pills, in which case they preferred males that were similar at MHC genes! The generality of these findings have been questioned (Hedrick and Loeschcke, 1996), and in fact, Hedrick and Black (1997) did not find MHC preference in Amerindians and Jacob *et al.* (2002) have suggested that Hutterite (an isolated religious group) females prefer MHC-similar males. The primary function of MHC in mate choice in mice has also been questioned because of the demonstrated role of other molecules, the major urinary proteins (MUPs) (Hurst *et al.*, 2001). This protein, found in only urine and in concentrations a million times greater than MHC, appears fundamental to individual recognition in mice.

III. GAMETIC SELECTION

In the last chapter, we considered a general model of gametic selection in which differential viability of gametes existed, and we showed that such selection is related to zygotic selection when the heterozygote is exactly intermediate. Selection involving differential viability of gametes is probably most common in plants. In animals, the diploid genotype of the male parent determines the protein constitution of the sperm cell, and in general, the haploid genome of the sperm is not used in either RNA or protein production. Here we consider two particular types of selection: meiotic drive (segregation distortion) and self-incompatibility systems that involve differential viability of gametes.

a. Meiotic Drive or Segregation Distortion

There are several well-known genetic systems in animals in which ***heterozygous individuals do not produce equal proportions of their two different alleles in gametes, as predicted by Mendelian segregation***. This phenomenon is generally called **meiotic drive** or **segregation distortion** and is an example of ultraselfish genes (Crow, 1988)—genes that interfere with the function of other genes and thereby increase their own frequency. In heterozygotes, the selfish chromosome interacts with the normal chromosome to either destroy their gametes or make them nonfunctional. In three well-studied examples—the t alleles in the house mouse (Lyon, 2003; see Example 4.4), *SD* (*segregation distorter*) alleles in *D. melanogaster* (Merrill *et al.*, 1999), and *SR* (*sex-ratio*) in *D. pseudoobscura* and other *Drosophila* species (Jaenike, 2001; Derome *et al.*, 2004)—the distortion from normal segregation proportions takes place only in males (see also Wilkinson and Fry, 2001; Taylor and Ingvarsson, 2003), and in general, such meiotic deviations usually take place in only one sex. However, because meiotic drive results in a large excess of the driven allele, to maintain a polymorphism there also needs to be countervailing selection that favors the wild-type allele. To illustrate the selective mechanisms involved in meiotic drive systems, we examine a simple model suggested for the maintenance of the t allele in the house mouse.

To understand the effect of meiotic drive on allele frequencies, assume that the male heterozygote produces a proportion k of the driven allele A_2 and $1 - k$ of wild-type allele A_1. Therefore, the allele frequency of A_2 after one generation, assuming half of the alleles come from females with normal segregation and half from males with meiotic drive, is

$$\begin{aligned} q_1 &= {}^1\!/\!_2\left(p_0q_0 + q_0^2\right) + {}^1\!/\!_2\left(2kp_0q_0 + q_0^2\right) \\ &= q_0[p_0\left(k + {}^1\!/\!_2\right) + q_0] \end{aligned}$$

The change in allele frequency is

$$\begin{aligned} \Delta q &= q_1 - q_0 \\ &= pqk* \end{aligned} \tag{4.9}$$

where $k^* = k - {}^1\!/\!_2$, the deviation from normal segregation. This expression is identical to that caused by gametic selection in Chapter 3 (expression

3.7c) if k^* is equal to the selection coefficient between the two gametic types. In this case, if $k* > 0$, then A_2 will asymptotically approach fixation.

However, for the t, SD, and SR alleles, and perhaps in other cases, there is counterbalancing selection in another component of fitness; thus, A_2 does not go to fixation. For example, in the t allele system, either homozygous tt individuals are generally lethal or tt males are sterile. Let us examine the situation in which tt individuals are lethal, a case in which there are only four matings, as indicated in Table 4.6 (Bruck, 1957). The frequency

TABLE 4.6 The matings and frequency of progeny (before selection) for the t allele.

Mating ♀ × ♂	*Frequency*	*Progeny (before selection)* ++	+*t*	*tt*
++ × ++	P_{11}^2	P_{11}^2	—	—
+*t* × ++	$P_{11}P_{12}$	$\frac{1}{2}P_{11}P_{12}$	$\frac{1}{2}P_{11}P_{12}$	—
++ × +*t*	$P_{11}P_{12}$	$(1-k)P_{11}P_{12}$	$kP_{11}P_{12}$	—
+*t* × +*t*	P_{12}^2	$\frac{1}{2}(1-k)P_{12}^2$	$\frac{1}{2}P_{12}^2$	$\frac{1}{2}kP_{12}^2$

of zygotes produced from these mating types before selection is the result of Mendelian segregation in heterozygous females and k and $1-k$, t and + alleles, respectively, in heterozygous males. The frequency of the t allele in the next generation, after selection, is

$$q' = \frac{1/2\left[P_{11}P_{12}\,(1/2 + k) + 1/2P_{12}^2\right]}{P_{11}^2 + 2P_{11}P_{12} + 1/2P_{12}^2\,(2-k)}$$

Because only ++ and +t individuals survive, all of the t alleles are in heterozygotes. In the previous generation, $P_{12} = 2q'$ and $P_{11} = 1 - q'$ for the same reasons. When we substitute these values in the previous expression, it becomes

$$q' = \frac{q\,(2k - 4kq + 1)}{2\,(1 - 2kq^2)} \tag{4.10a}$$

At the equilibrium $q' = q$, this expression becomes the quadratic

$$4kq^2 - 4kq + 2k - 1 = 0$$

with the solution

$$q_e = 1/2 - \frac{[k\,(1-k)]^{1/2}}{2k} \tag{4.10b}$$

For an application of this formula to wild populations of the house mouse, see Example 4.4.

Example 4.4. Many wild populations of the house mouse, *Mus musculus*, are polymorphic for alleles at the t locus, known to be on chromosome 17 (for genetic details, see Lyon, 2003; Dod *et al.*, 2003). Most of these alleles are lethal when homozygous, but heterozygous males generally produce a large majority of sperm with the t allele. Lewontin and Dunn (1960) reported the estimated proportion of t-carrying sperm, k, in laboratory crosses for 19 newly arisen t mutants and for 16 t alleles from wild populations (Figure 4.9). The distribution of k is quite different for these two groups of alleles. Heterozygotes for the wild t alleles have an average of $k = 0.952$, and all the wild alleles have a segregation ratio greater than 0.85. On the other hand, 8 of the 19 mutants do not have k values significantly different from 0.5, although four have high k values and several have k values less than 0.5.

If we use $k = 0.95$, the average estimate for wild t alleles, in expression 4.10*b*, then $q_e = 0{:}385$. However, this predicted frequency is much higher than is generally observed in natural populations. For example, a survey

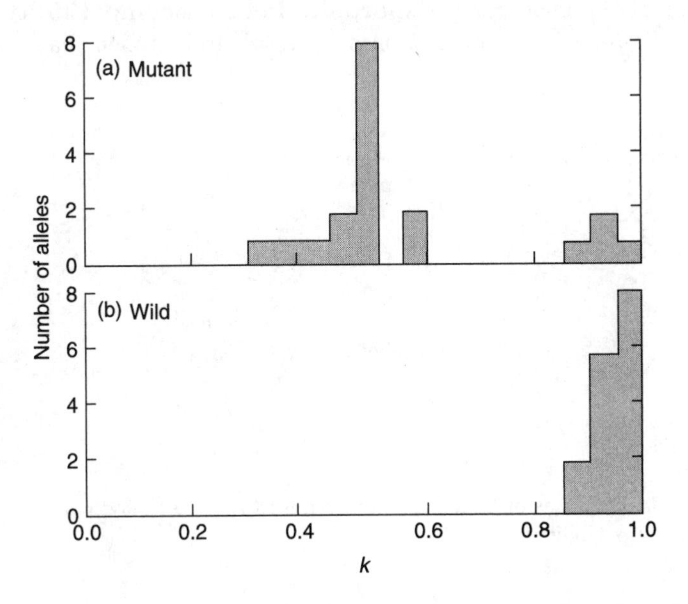

Figure 4.9. The distribution of segregation (k), (a) in mutant t alleles and (b) in wild t alleles (Lewontin and Dunn, 1960).

of 3263 mice in 63 populations, using molecular techniques to detect t alleles, found an average t allele frequency of only 0.062 (Ardlie and Silver, 1998). A number of researchers have tried to determine what factors are responsible for the large difference between the theoretical prediction and actual observations. Various combinations of genetic drift in small populations, extinction of populations, and gene flow between populations can result in reduction of the t allele frequency (Lewontin, 1968; Levin *et al.*, 1969; Nunney and Baker, 1993). Durand *et al.* (1997) demonstrated that the amount of gene flow required in such models falls in a very restrictive range, and they suggest that selection against heterozygotes and variation in the amount of meiotic drive may be important in lowering the t allele frequency.

---

b. Self-Incompatibility Alleles

In a number of plants, self-fertilization is prevented by **self-incompatibility**, resulting in the ***absence of germination or growth of pollen on the stigma of flowers from the same or genetically similar plants***. Self-incompatibility and heterostyly (Example 4.2) are two mechanisms in plants that prevent inbreeding (for evolutionary background, see Charlesworth and Charlesworth, 1987; Barrett, 2002). There are two basic types of self-incompatibility: that related to the genotype of the pollen (**gametophytic self-incompatibility, GSI**) and that resulting from the genotype of the male parent (**sporophytic self-incompatibility, SSI**). In GSI (Figure 4.10a), fertilization must result from pollen that has a dif-

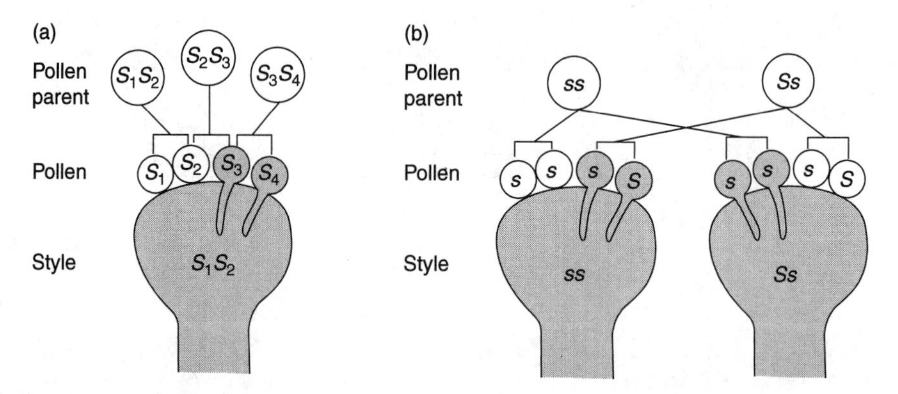

Figure 4.10. Diagrams of pollen tube growth for (a) gametophytic and (b) sporophytic incompatibility.

ferent allele from the female parent; thus, for example, a S_1S_2 plant cannot be fertilized by S_1 or S_2 pollen and must be pollinated by S_3, S_4, or some other pollen. This of course leads to obligate heterozygosity at the locus determining pollen incompatibility. For SSI, the genotype of the pollen parent determines the ability to grow on the stigma, and some pollen can grow on a plant containing the same allele. For example, s pollen can grow on an ss stigma if it comes from an Ss pollen parent (Figure 4.10b).

Let us consider a model that describes GSI when there are three alleles. Table 4.7 gives the relationship between the genotype frequencies

TABLE 4.7 The frequency of progeny in a GSI system with three alleles such that pollen will not grow on a plant with one of the same self-incompatibility alleles.

Female parent	*Pollen*	*Frequency*	*Progeny*		
			S_1S_2	S_1S_3	S_2S_3
S_1S_2	S_3	P_{12}	—	$\frac{1}{2}P_{12}$	$\frac{1}{2}P_{12}$
S_1S_3	S_2	P_{13}	$\frac{1}{2}P_{13}$	—	$\frac{1}{2}P_{13}$
S_2S_3	S_1	P_{23}	$\frac{1}{2}P_{23}$	$\frac{1}{2}P_{23}$	—
			$\frac{1}{2}(1 - P_{12})$	$\frac{1}{2}(1 - P_{13})$	$\frac{1}{2}(1 - P_{23})$

(remember that only heterozygotes can be formed, and thus, $P_{12} + P_{13} + P_{23} = 1$) in different generations, which for the three genotypes are

$$\begin{aligned} P'_{12} &= 1/2\,(1 - P_{12}) \\ P'_{13} &= 1/2\,(1 - P_{13}) \\ P'_{23} &= 1/2\,(1 - P_{23}) \end{aligned} \tag{4.11a}$$

The change in genotype frequency for S_1S_2 is

$$\begin{aligned} \Delta P_{12} &= P'_{12} - P_{12} \\ &= 1/2\,(1 - 3P_{12}) \end{aligned}$$

If we set $\Delta P_{12} = 0$, then the equilibrium frequency for S_1S_2 is

$$P_{12(e)} = 1/3 \tag{4.11b}$$

The equilibrium values for genotypes S_1S_3 and S_2S_3 are also $1/3$.

The allele frequency values can be calculated, and for allele S_1

$$\begin{aligned} p'_1 &= 1/2 P_{12} + 1/2 P_{13} \\ &= 1/2\,(1 - P_{23}) \\ &= P'_{23} \end{aligned}$$

which is the proportion of S_2S_3 zygotes in the next generation. The other allele frequencies can be found in a similar manner so that $p'_2 = P'_{13}$ and $p'_3 = P'_{12}$. In other words, each allele frequencies is equal to the frequency of the heterozygote that does not contain that allele in the next generation. Furthermore, the equilibrium frequencies of all of the alleles are also equal to one third. This equilibrium is reached rather quickly, even if the initial frequencies are not close to the equilibrium values, because of the occurrence of strong selection among the mating types. The example in

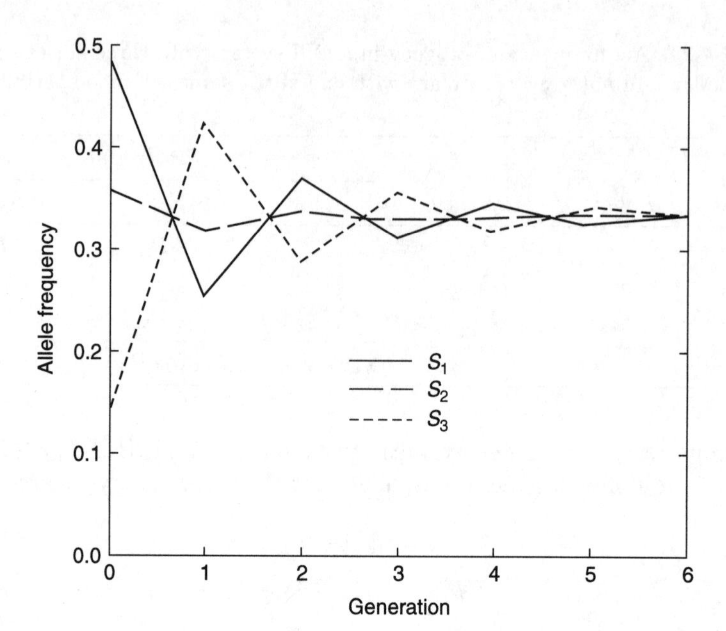

Figure 4.11. The frequency of three self-incompatibility alleles over time when the initial frequencies of genotypes S_1S_2, S_1S_3, S_2S_3 were 0.7, 0.28, and 0.02, respectively.

Figure 4.11 shows that there may be large oscillations in allele frequency before the equilibrium value is reached. If an allele is rare, it will have a great advantage in pollination success and will increase in frequency because pollen containing the rare allele will seldom encounter maternal genotypes with the same allele.

In a result analogous to the three-allele GSI example, if there are n alleles, then the equilibrium frequency of all alleles is $1/n$. Furthermore, there are $n(n-1)/2$ heterozygotes, and the equilibrium frequency of all these genotypes is $2/[n(n-1)]$. Theoretical investigations of GSI systems (Wright, 1965a; Yokoyama and Hetherington, 1982) find that the theory that includes many of the joint effects of selection, finite population size, population structure, and mutation is generally consistent with observed numbers of alleles in populations (see Example 4.5, which presents some of the data from the classic study in an *Oenothera* species). However, accounting for all of the potential evolutionary effects, such as other types of selection and subdivided populations, becomes quite complicated (Schierup, 1998; Vekemans *et al.*, 1998).

Example 4.5. *Oenothera organensis* is a rare perennial that is endemic to the Organ Mountains of southern New Mexico. The total number of plants in the species is 5000 or less (Levin *et al.*, 1979), existing in canyons above 2000 meters. Emerson (1939) identified self-incompatibility alleles from this species with GSI by observing pollen tube growth in more than 3000 crosses. He found 34 different alleles in a sample of 134 plants (at least another 11 alleles have since been identified). The distribution of these alleles is given in Figure 4.12, where the frequency of alleles is shown on the horizontal

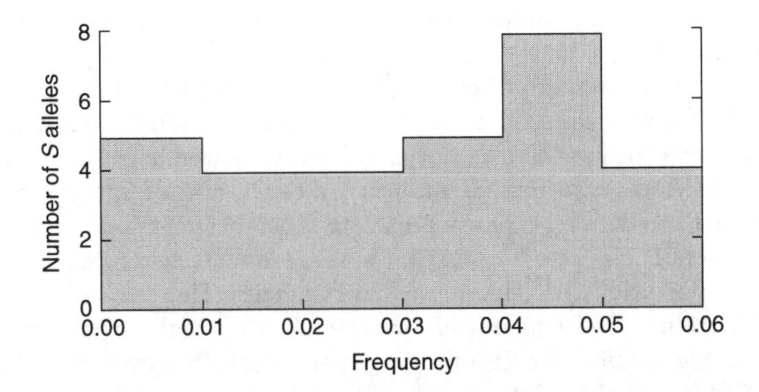

Figure 4.12. The frequencies of 34 different self-incompatibility alleles in *Oenothera organensis* (after Emerson, 1939).

axis. For example, five different S alleles (S_9, S_{15}, S_{16}, S_{35}, and S_{37}) were observed four times in the sample, for a frequency of $4/134 = 0.0299$. If the 34 alleles were at equilibrium, then the expected frequency would be 0.0294. In this case, a great deal of variation in frequency exists among the alleles, a number of alleles having frequencies higher than the equilibrium and a number lower. Such variation is not unexpected and probably results both from the effects of genetic drift and from the small sample size relative to the number of alleles.

Mable *et al.* (2003) estimated the numbers and frequencies of self-incompatibility alleles in the SSI system of a natural population of *Arabidopis lyrata.* Using a combination of pollination studies and molecular techniques, they identified 11 putative alleles in a sample of 20 individuals. For various reasons, they were not able to identify all of the alleles in the sample but estimate that there are between 13 and 16 total. The allele frequencies were more uneven than in the *Oenothera* example above, with one allele at a frequency of 0.325, another at 0.125, two at 0.075, four at 0.05, and three at 0.025 (0.125 of the alleles could not be identified). The most common allele does not appear to be dominant, consistent with the general prediction that recessive alleles in SSI systems would be in higher frequency.

When there is SSI and dominance of one allele over the other as shown in Figure 4.10b, then there are two successful mating types, $Ss \times ss$ and $ss \times Ss$, where the female genotype is given first. Both of these mating types produce equal proportions of Ss and ss progeny. Assuming that the frequencies of Ss and ss are H and Q, then in one generation, it is expected from any starting point that $H = Q = 0.5$. As a result, the frequency of the dominant allele S is 0.25, lower than the expected frequency of 0.75 for the recessive allele s. The general prediction of higher frequency of recessive alleles in a SSI system was first made by Bateman (1952), and theoretical studies have examined the impact of other patterns of dominance on allele frequencies and maintenance of polymorphism (Schierup *et al.*, 1997; Uyenoyama, 2000).

A review of the number of alleles at self-incompatibility loci in different species (Lawrence, 2000) showed that there are a substantial number of alleles in some species, such as clover species that often have more than 100 alleles in a population, but the majority of other studied species generally have from 12 to 45 alleles per population. Lawrence (2000) also concluded that the frequencies of alleles in GSI systems are consistent with the theoretical expectations of generally equal frequency. He also suggested that for SSI systems, the frequency of recessive alleles is higher than other alleles, again consistent with theoretical expectations (however, see Mable *et al.*, 2003; some of their data are presented in Example 4.5). Sequence data from self-incompatibility alleles have demonstrated that many alleles are old, predating speciation (Ioerger *et al.*, 1990), an observation consistent with long-term and strong balancing selection. Recent work has provided details about the molecular basis of self-incompatibility systems (Nasrallah, 2002) and both examination of observed variation (Richman and Kohn, 1999; Charlesworth *et al.*, 2003) and extension of theoretical considerations (Uyenoyama, 2003).

IV. ECOLOGICAL GENETICS AND BALANCING SELECTION

Starting in the 1960s, increased biological realism was advocated in population genetics, based in part on the inclusion of meaningful ecological factors. The hope was that a broader discipline of population biology would be created through the application of the concepts of population ecology to population genetics (and vice versa). Such an expansion implies that relative fitness values depend not only on a particular genotype but also on the environment in which the genotype exists. For example, in one environment, the relative fitness of a given genotype may be high, whereas in another environment, it may be low. Environmental factors that potentially affect relative fitness are physical factors such as temperature, moisture, and soil type; biotic factors in other species such as interspecific competitors, predators or prey, and hosts or parasites; and the composition of the population being considered in terms of population numbers, age distribution, and genotype proportions. Both the contemporary disciplines of ecological genetics and evolutionary ecology include elements of both ecology and population genetics or evolution.

As a first step in introducing such ecological effects, we assume that selective values are environmentally dependent and that the environment may vary over time or in space. Such an analysis is particularly appropriate in an examination of physical environment factors. Second, we consider the selective effects of different genotype frequencies within a population, that is, frequency-dependent selection. The selective factors causing differential selection in this case may be other organisms such as predators, prey, hosts, or parasites. The effects of population numbers of the same species can be examined by using an extension of the logistic equation from population ecology that allows density-dependent selection (see Charlesworth, 1971; Roughgarden, 1971). For data demonstrating a density-dependent heterozygote advantage in a marine copepod, see Example 4.6. The effects of age-specific mortality and fecundity can be examined using matrices giving fecundity and survival values for different genotypes (see Anderson and King, 1970; Charlesworth, 1994).

Example 4.6. A number of marine copepods exhibit polymorphisms of color patterns in natural populations. Battaglia (1958) examined the effects of crowding on a polymorphism in the pigment pattern of the copepod *Tisbe reticulata* from the brackish waters of the Venice lagoon. In this population, there are three common phenotypes that result from the genotypes V^vV^v, V^vV^m, and V^mV^m. In one experiment, Battaglia crossed $V^vV^m \times V^vV^m$ and raised the progeny under three different levels of crowding (Table 4.8). Under low crowding, the progeny are produced in close to a 1:2:1 ratio, although there is an excess of heterozygotes. Under medium and high crowding, however, there is a large excess of heterozygotes. Using expression 3.22*b*, we can obtain estimates of the viabilities of the homozygotes relative to that of the heterozygote. For high crowding, the relative viabilities of the homozygotes are only 0.66 and 0.61 that of the heterozygote. This appears to be a good example of density-dependent viability where heterozygote advantage increases with higher density.

TABLE 4.8 The number of progeny, estimated viabilities (in parentheses), and predicted equilibrium frequencies for allele V^m of a marine copepod raised at three levels of crowding (Battaglia, 1958).

Culture condition	N	V^vV^v	V^vV^m	V^mV^m	q_e
Low crowding	3839	904 (0.89)	2023 (1.0)	912 (0.90)	0.474
Medium crowding	1743	343 (0.68)	1015 (1.0)	385 (0.76)	0.571
High crowding	1751	353 (0.66)	1069 (1.0)	329 (0.61)	0.466

A number of approaches and studies have shown apparent associations between genetic variation and environmental factors (for reviews, see Hedrick *et al.*, 1976; Hedrick, 1986a). Of course, such associations do not mean that the genetic pattern is caused by the environmental factor unless additional evidence supports a cause–effect relationship. However, some examples do have additional supportive evidence consistent with the hypothesis that different selective pressures are actually occurring in different environments. One of the most thoroughly investigated polymorphisms is that of the color and banding patterns in the shell of the snail *Cepaea nemoralis* (Example 4.7), but even in this case, the genetic–environmental tie is not fully understood.

Example 4.7. In the land snail *Cepaea nemoralis*, a series of Mendelian polymorphisms affecting shell color and banding pattern (see Figure 4.13) appear to be maintained by environmental heterogeneity in different areas (see Jones *et al.*, 1977; Cook, 1998, for reviews). A number of studies have

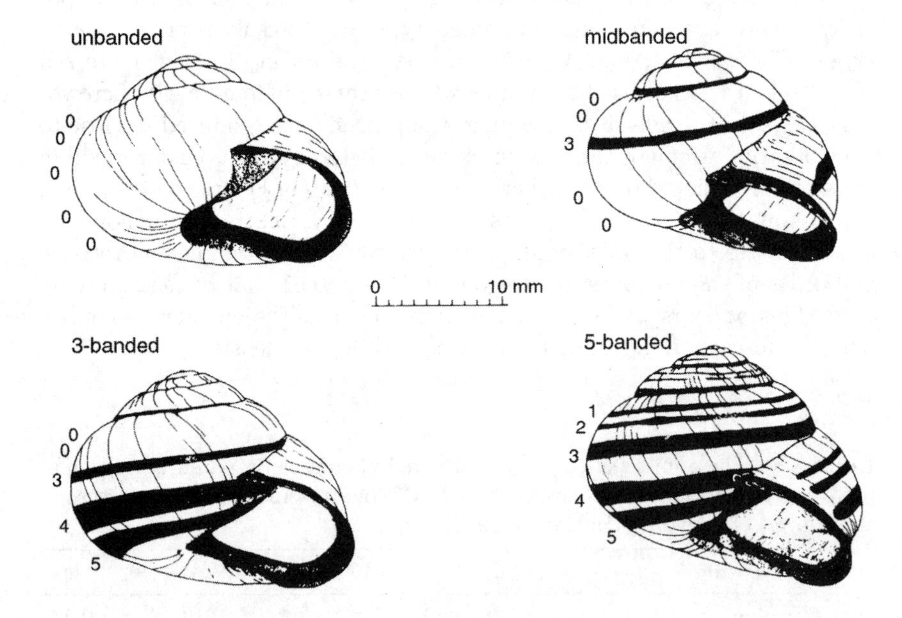

Figure 4.13. The shell banding patterns observed in the snail *Cepaea nemoralis* (after Jones *et al.*, 1977).

indicated that the polymorphism is maintained by differential predation by birds, particularly thrushes, of snails whose shell color and banding pattern do not appropriately match the background of their habitat (Cain and Sheppard, 1954; Clarke, 1960). For example, brown, unbanded snails appear protectively camouflaged on dark soil, whereas yellow, banded types blend visually with the stems of many plants.

However, selective predation does not appear to explain completely the presence of these polymorphisms (Cook, 1998). In some regions, morph frequencies are uniform in spite of apparent environmental heterogeneity, and other areas exhibit steep clines in the absence of any discernible environmental gradient. It has been suggested that these phenomena, called **area effects**, are also due to environmental heterogeneity in climatic factors that affect the various morphs differentially (Cain and Currey, 1963). The available evidence indicates that the maintenance of shell polymorphisms in *C. nemoralis* is due to a complex interaction of environmental heterogeneity and genetic factors and that the relative importance of different factors varies from locality to locality (Davison and Clarke, 2000). These general conclusions are also supported by several long-term studies of *C. nemoralis* populations (Cain *et al.*, 1990; Cook and Pettitt, 1998; Cowie and Jones, 1998).

Interestingly, *C. nemoralis* has been accidentally introduced to several areas of eastern North America, apparently as a stowaway on vegetation imported from Europe. Brussard (1975) examined a number of these colonies for color and banding patterns, as well as for several allozyme loci. He concluded that the colonies near Lexington, Virginia, originated from an importation from Italy, a conclusion consistent with historical information, and that the other populations in the northern United States and southern Canada originated from northern Europe. As a result, there was general concordance with the climatic hypothesis, but strong founder effects were observed in several populations. For example, all snails from Harrisonburg, Virginia, were banded pink, and all snails from London, Ontario, were banded yellow.

In the classic, two-allele, constant-fitness model presented in Chapter 3, only when the fitnesses of the homozygotes are less than that of the heterozygote is there a stable equilibrium. In general, such selection models that may lead to the maintenance of a polymorphism are called **balancing-selection models**. Balancing-selection models include the heterozygous advantage model of Chapter 3, the sex-dependent viability selection, self-incompatibility, and the negative assortative mating models discussed above, and variable selection models we discuss here.

A number of biological and environmental factors may cause fitnesses to vary, with the potential result that the relative fitness of a particular genotype may be high in one environment and low in another. If we apply the term environments generally to the factors that affect relative fitness, then over a series of environments the fitnesses may vary as illustrated in Table 4.9. In this example, genotype A_1A_1 has the highest fitness in environment 1 and the lowest fitness in environments 2 and 4. If the average fitness for each genotype over all environments is evaluated, the average fitness of the heterozygote is greater than that of either homozygote, even

TABLE 4.9 Variation in fitness as the result of different environments leading to a marginal heterozygote advantage. In other words, the arithmetic mean of the heterozygote is higher than that of the homozygotes.

Environment	A_1A_1	A_1A_2	A_2A_2
1	$1+s$	1	$1-s$
2	$1-s$	1	1
3	1	1	$1-s$
4	$1-s$	1	$1+s$
Average	$1-s/4$	$< 1 >$	$1-s/4$

though the heterozygote by itself does not have the highest fitness in a single environment.

This phenomenon may be called **marginal heterozygote advantage** (or marginal overdominance) and is useful in visualizing how variable selection may lead to a net heterozygous advantage. However, in this formulation, dominance is also varying with different environments. For example, in environment 2, A_1A_2 has the same fitness as A_2A_2, and in environment 3, it has the same fitness as A_1A_1. This complete reversal of dominance (h changing from 1 to 0) is probably an uncommon phenomenon in nature. In addition, the arithmetic mean may not be the appropriate value to use in judging heterozygous advantage, and as we shall see, the geometric or harmonic mean may be appropriate in some situations.

It is important to realize that variation in selection does not necessarily lead to the maintenance of a polymorphism (Prout, 1967; Lewontin, 1974). Consider an example in which differential selection affecting viability occurs in the life cycle of an organism and where there are just two morphs: the dominant morph (consisting of one homozygote and the heterozygote) and the recessive morph (Table 4.10). Assume that the relative viability is higher for the recessive morph in the larval (or seedling) stage ($v_L > 1$) but that the recessive morph has a lower premating adult survival ($v_A < 1$). On the surface, it appears that balancing selection would be operating and that selection would maintain a polymorphism.

TABLE 4.10 An example of differential viability in two life stages that collapses into a single viability measure.

	A_1A_1	A_1A_2	A_2A_2
Larval (or seedling) viability	1	1	v_L
Premating, adult viability	1	1	v_A
Overall viability	1	1	v_Lv_A

However, these fitness components "collapse" (see the discussion on p. 176) so that the overall survival of the morphs is the product of the relative viability in the different life stages, v_Lv_A. Because of this property, assuming that there are no other fitness differences for these genotypes, the overall relative fitness of the dominant form is either less than, equal to, or greater than the fitness of the recessive morph. When the overall relative fitness of the dominant morph is greater than the recessive, the A_1 allele will eventually become fixed, whereas if it is lower, the A_2 allele will eventually become fixed. Only in the unlikely case in which the relative fitnesses are exactly equal will the two morphs remain in the population because they are neutral with respect to each other. This logic would predict that it would be difficult to maintain a polymorphism in a haploid organism, but Example 4.8 gives an experimental example of a stable polymorphism in a bacterium.

Example 4.8. The common bacterium *Pseudomonas fluorescens* evolves rapidly under novel environmental conditions and generates a variety of mutants. In a homogeneous environment (constantly shaken microcosm) the original morph, smooth, is the only detectable one. On the other hand, in a spatially heterogeneous environment (no shaking in microcosm) at least three morphs, smooth, wrinkly-spreader, and fuzzy-spreader, are found in substantial frequency in all replicate microcosms (Rainey and Travisano, 1998) (Figure 4.14). To examine whether spatial environmental heterogeneity is the selective force influencing genetic variation, Rainey and Travisano transferred cultures from the heterogeneous environment to the homogeneous environment. In just 2 weeks, the cultures declined in diversity to about one-quarter that in the heterogeneous environment.

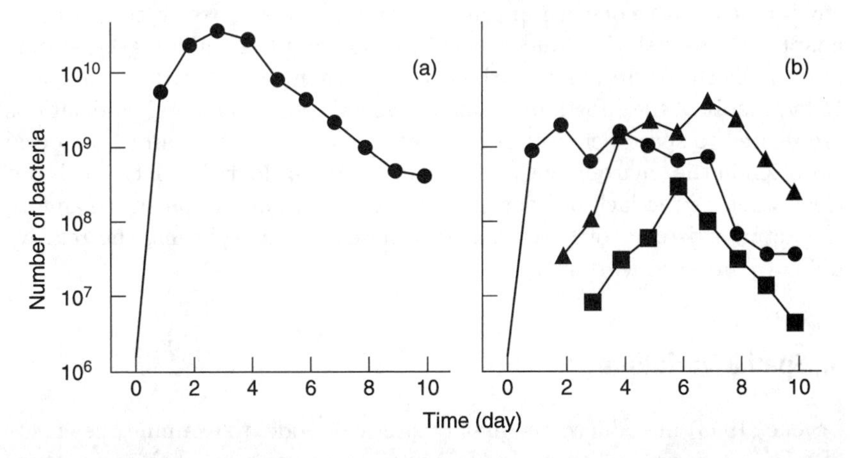

Figure 4.14. Homogeneous (a) and heterogeneous (b) microcosms were founded with the smooth morph (circles). The heterogeneous microscosms also had the wrinkly-spreader (triangles) and fuzzy-spreader (squares) morphs generated by mutation (after Rainey and Travisano, 1998).

Transfer of one sent back to the heterogeneous environment from the homogeneous quickly restored the diversity level.

For a new mutant to invade a population, it must have a fitness advantage when rare. To determine whether the variants have this property, Rainey and Travisano measured the competitive abilities of each of the three morphs when rare in pairwise tests against the other morphs. For five of the six combinations, the rare morphs had significantly higher fitness values than the common morph. The fitness differences between the ancestral (smooth) and both of the derived morphs provide evidence supporting a stable polymorphism. The ecological mechanisms maintaining diversity are not completely understood, but the wrinkly-spreader morphs adhere to each other and form, on the air–broth surface of the microcosm, mats that allow better access to oxygen and nutrients. However, when the wrinkly-spreader morph becomes too common, the mat sinks, eliminating this advantage.

Before we examine some specific models of environmental heterogeneity, let us discuss an intuitive explanation for the basis of the difference between the effects of spatial and temporal variation in the environment on genetic variation. If the environment varies over time—for example, one year is wet and the next is dry—then every individual must endure every different environment whether they are genetically adapted to it or not. However, if the environment varies over space—for example, one area is wet and another dry—then only part of the population encounters a particular environment at a given time. In other words, some genotypes and a proportion of their descendants that are well adapted in a particular environment may not even encounter environments to which they are not adapted. Intuitively, it would appear more difficult to maintain genetic variation when the environment varies over time than when it varies over space. In fact, most of the substantive cases of genetic–environmental associations are related to spatial environmental heterogeneity rather than to temporal variation in the environment (Hedrick *et al.*, 1976; Hedrick, 1986a). It is not clear whether the lack of known temporal-based polymorphisms is due to the actual existence of fewer temporal-based polymorphisms, the relative difficulty in documenting them, or both.

a. Spatial Variation

Levene (1953) introduced the first theoretical model to examine the effects of spatial variation in fitness. In his model, it is assumed that the area where the population exists is subdivided into different environmental niches and that "after fertilization the zygotes settle down at random in large numbers into each of the niches. There is then differential mortality ending with a fixed number of individuals in each niche." Implicit in this formulation is that there is complete mixing of adults before mating from all of the niches each generation. A general diagram of this model for a complete generation is given in Figure 4.15, where it is assumed that there are two or more different niches.

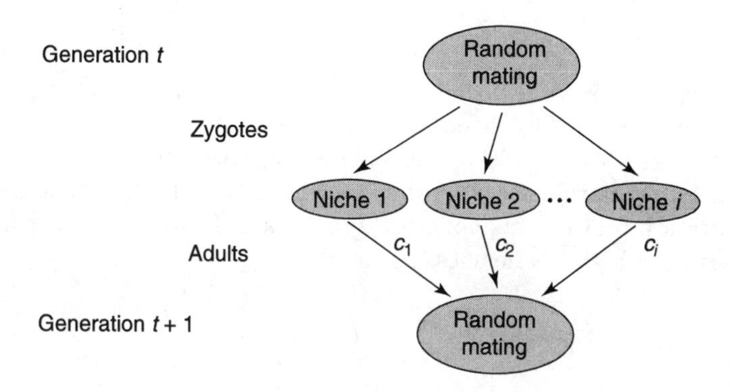

Figure 4.15. The selection model for variable selection in space.

Let us demonstrate how the conditions for a polymorphism in this case are derived; it is a technique that can be used in other situations. Assume that we are examining a biallele locus and that the fitnesses of genotypes A_1A_1, A_1A_2, and A_2A_2 are $w_{11 \cdot i}$, 1, and $w_{22 \cdot i}$, respectively, in the ith niche. We can calculate the change in allele frequency in the ith niche by substituting these fitness values into expression 3.3a so that

$$\Delta q_i = \frac{pq[p(1 - w_{11 \cdot i}) - q(1 - w_{22 \cdot i})]}{\bar{w}_i}$$

where $\bar{w}_i = w_{11 \cdot i}p^2 + 2pq + w_{22 \cdot i}q^2$. Let us assume that there are m niches and that c_i is the proportion of individuals from niche i so that

$$\sum_{i=1}^{m} c_i = 1$$

The change in allele frequency over all m niches is then

$$\begin{aligned} \Delta q &= \sum_{i=1}^{m} c_i \Delta q_i \\ &= pq \sum c_i \left[\frac{p(1 - w_{11 \cdot i}) - q(1 - w_{22 \cdot i})}{\bar{w}_i} \right] \end{aligned} \tag{4.12a}$$

or the weighted proportion of genetic change from the different niches.

In order to establish the conditions for a stable polymorphism, we can define a new function, $h(q)$, such that

$$h(q) = \frac{\Delta q}{pq}$$
$$= \sum c_i \left[\frac{p(1 - w_{11 \cdot i}) - q(1 - w_{22 \cdot i})}{\bar{w}_i} \right]$$

By dividing by pq, we now have a function that is nonzero for $p = 0$ or $q = 0$. The function $h(q)$ is continuous between 0 and 1 so that if it can be shown that $h(0)$ is positive and $h(1)$ is negative, there must be at least one stable, polymorphic equilibrium (Levene, 1953). This is analogous to a demonstration that the frequency of A_2 increases from low frequencies and decreases from high frequencies. For $q = 0$,

$$h(0) = \sum c_i \left(\frac{1 - w_{11 \cdot i}}{w_{11 \cdot i}} \right)$$

Because this expression must be positive for A_2 to increase from a low frequency, the condition for a stable equilibrium is that

$$\sum c_i \left(\frac{1 - w_{11 \cdot i}}{w_{11 \cdot i}} \right) > 0$$

or

$$\frac{1}{\sum c_i \frac{1}{w_{11 \cdot i}}} < 1 \qquad (4.12b)$$

The left side of this expression is equal to the harmonic mean of the relative fitness of genotype A_1A_1 over all niches. Similarly, for $h(1)$ to be negative, it can be shown that

$$\frac{1}{\sum c_i \frac{1}{w_{22 \cdot i}}} < 1 \qquad (4.12c)$$

or that this condition for stability is that the harmonic mean of the relative fitness of A_2A_2 must also be less than unity. Therefore, for a stable polymorphism, the harmonic mean fitness of both homozygotes must be less than that of the heterozygote. The value of the harmonic mean such that there is a stable polymorphism is the result of a balance between the fitnesses and the proportionate contributions from each niche. For a balanced polymorphism, the fitnesses and contributions must be somewhat finely balanced for the harmonic means to be less than unity, particularly if there is weak selection.

To illustrate these conditions, let us assume that there are only two niches and that there is additive and directional selection in niche 1 favoring allele A_1. That is, the fitness values are $1 + s$, 1, and $1 - s$, and there is equal, but opposite, directional selection in niche 2 favoring A_2, that is, fitness values of $1 - s$, 1, and $1 + s$. If the two niches are equally frequent, then $c_1 = c_2 = 0.5$ and the left-hand sides of equations 4.12*b* and 4.12*c* become $1 - s^2$, and the conditions for the polymorphism are met. However, as one niche becomes more common, the overall allele frequency reflects selection in that niche until there is no longer a stable polymorphism. In this case, when $s = 0.1$, then c_1 must be between 0.45 and 0.55 for a stable polymorphism.

The Levene model requires that there be some type of marginal heterozygote advantage—that is, the harmonic mean heterozygote advantage. However, in some situations, a balanced polymorphism can occur even when the heterozygote and one of the homozygotes have the same relative fitnesses in all environments. In other words only one genotype—for example, A_2A_2—varies in relative fitness over environments so that the fitnesses of A_1A_1, A_1A_2, and A_2A_2 in the ith niche are 1, 1, and $w_{22 \cdot i}$. This model has been termed the **absolute dominance** model (Prout, 1968) and may be appropriate for a number of dominant morphological polymorphisms such as those involving mimicry or protective coloration.

For absolute dominance, the conditions for a polymorphism when fitness varies in different niches are

$$\frac{1}{\sum c_i \dfrac{1}{w_{22 \cdot i}}} < 1 < \sum c_i w_{22 \cdot i} \tag{4.13}$$

(Prout, 1968). In other words, the harmonic mean of the variable genotype must be smaller than unity and the arithmetic mean greater than unity. These conditions are given in Figure 4.16, where two environments are in equal proportions, $c_1 = c_2 = 0.5$ (Hedrick *et al.*, 1976). When the amount of selection is weak, the range of values that can give a polymorphism is small. For example, when the relative fitness of the A_2A_2 homozygote in environment 1 is 0.9, then the fitness in the other environment must be between 1.1 and 1.125 for a stable polymorphism.

If there is genotype-specific **habitat selection** such that individuals prefer niches in which they have higher fitness, then the conditions for a polymorphism are broader. The extent of habitat selection can be quantified using the standardized values in Table 4.11 where h is the proportion of homozygotes that choose the niche where they have the higher fitness (Hedrick, 1990a). For example, if A_1A_1 has a higher fitness in niche 1 and

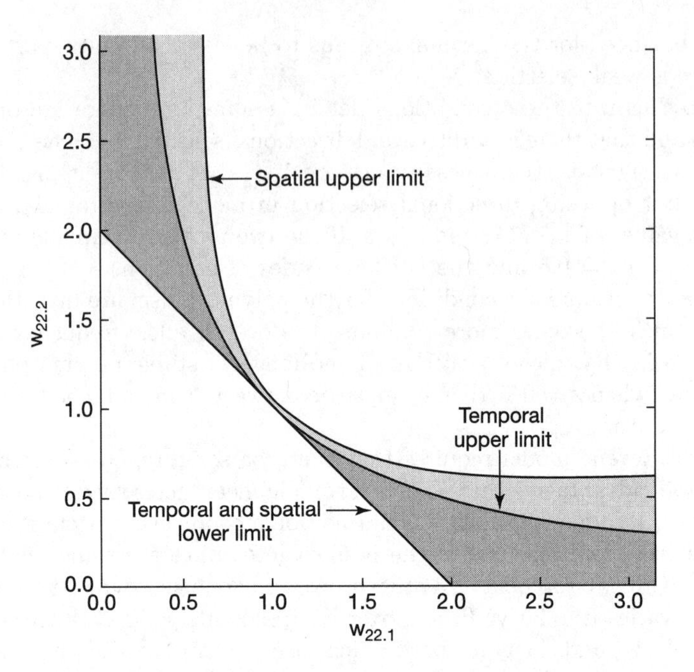

Figure 4.16. The regions for a stable equilibrium (shaded area) when the relative fitnesses of genotypes A_1A_1, A_1A_2, and A_2A_2 are 1, 1, and $w_{22 \cdot 1}$ in environment 1 and 1, 1, and $w_{22 \cdot 2}$ in environment 2. The light shaded area is the region where spatial variation gives stability and temporal variation does not.

TABLE 4.11 Genotypic habitat selection, where h is the proportion of homozygotes that choose the niche to which they are most adapted.

Niche	A_1A_1	A_1A_2	A_2A_2
1	$\dfrac{hc_1}{hc_1 + (1-h)c_2}$	c_1	$\dfrac{(1-h)c_1}{(1-h)c_1 + hc_2}$
2	$\dfrac{(1-h)c_2}{hc_2 + (1-h)c_2}$	c_2	$\dfrac{hc_2}{(1-h)c_1 + hc_2}$

A_2A_2 a higher fitness in niche 2 and the difference in fitness between the homozygotes within a niche is s (and the heterozygote is exactly intermediate), then the conditions for a polymorphism are given in Figure 4.17. As a comparison, the conditions with the Levene model in which there is no habitat selection but differential fitnesses in the two environments are also given. Even as here when there is only moderate habitat selection ($h = 0.625$) for all proportions of the two environments, the conditions are greatly broadened compared with the Levene conditions. In fact, the estimate of average h (Hedrick, 1990a) using the *Drosophila* data of Jaenike (1985) for mushroom versus tomato is 0.745 and that of Hoffman and O'Donnell (1990) for lemon versus orange is 0.626, suggesting that this level of habitat selection is not extreme. Smith (1993) determined that a single gene appears responsible for the polymorphism in bill size in the African finch *Pyrenestes*. In this case, bill size and feeding preference are correlated, suggesting that the maintenance of genetic variation in this random-mating population is related to habitat selection.

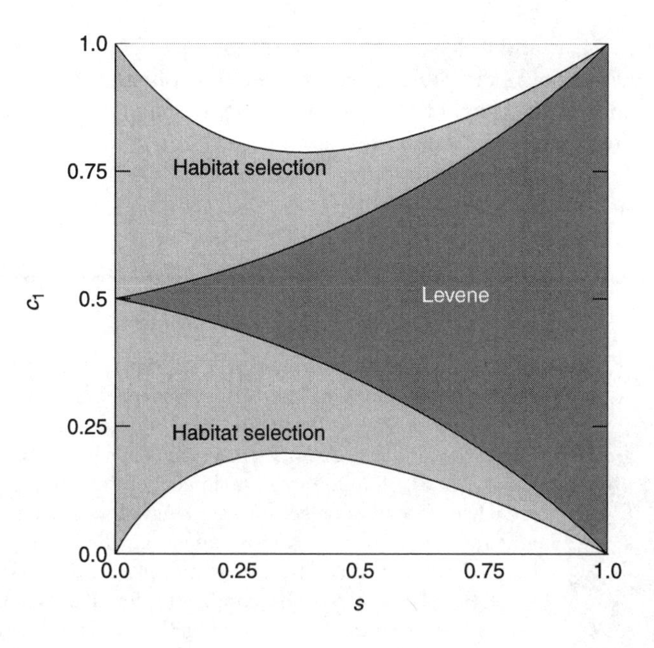

Figure 4.17. The region of a stable polymorphism for the Levene model, in which there is no habitat selection (dark shaded) and where there is habitat selection with $h = 0.625$ (all shaded area between outside curved lines), where c_1 is the proportion of niche 1 and s is the selective difference between the two homozygotes.

b. Temporal Variation

As suggested earlier, the conditions for a stable polymorphism when there is temporal variation in the environment are somewhat more restrictive than for spatial variation. The conditions for a stable polymorphism when fitnesses vary in different generations were first derived by Haldane and Jayakar (1963), who showed that, assuming that the relative fitnesses of genotypes A_1A_1, A_1A_2, and A_2A_2 are $w_{11\cdot i}$, 1, and $w_{22.i}$ in the ith generation, a stable polymorphism can exist when

$$\left(\prod_{i=1}^{n} w_{11\cdot i}\right)^{1/n} < 1 > \left(\prod_{i=1}^{n} w_{22\cdot i}\right)^{1/n} \tag{4.14a}$$

where n is the number of generations. In other words, for a stable polymorphism with temporal variation in fitness, the geometric means of both homozygotes must be smaller than the geometric mean of the heterozygote. Because the geometric mean is always larger than the harmonic mean, the conditions for a polymorphism for temporal variation are more restrictive than for spatial variation.

To illustrate these conditions, let us again assume that there are only two different environments, such as hot and cold seasons, and in different generations, the population exists in these different environments. Allele A_1 is favored in environment 1 with fitness values $1 + s$, 1, and $1 - s$ for genotypes A_1A_1, A_1A_2, and A_2A_2, respectively, and there is equal, but opposite, directional selection in environment 2 favoring A_2—that is, fitness values of $1 - s$, 1, and $1 + s$. When equal numbers of generations are spent in each environment, the conditions for a polymorphism are met. However, if slightly more time is spent in one environment, then one of the geometric means is greater than 1, and directional selection occurs. If $s = 0.1$, then for a stable polymorphism, between 0.475 and 0.525 of the time must be spent in each environment.

J. B. S. HALDANE (1882–1964)

J. B. S. Haldane, in addition to formulating (along with Fisher and Wright) much of the basis of theoretical population genetics, was also a well-known biochemist and prolific political writer (see Clark, 1968, for a fascinating biography; see also Crow, 1992). Haldane was particularly interested in the mathematical theory of selection at a single locus. Using selection and mutation as joint factors affecting allelic frequencies, he derived the equilibrium that results from this balance of factors. He also derived the conditions for a polymorphism in a number of situations, including that for an X-linked locus and for variable selection over time (see Dronamraju, 1990, for selected papers). One of his very insightful articles was "A defense of beanbag genetics" (Haldane, 1964), a response to criticism by Mayr (1963) of the mathematical school of evolutionary biology. In 1957, he left England for India, where he had an important influence on a number of biologists. After he was diagnosed with cancer (of which he later died), he wrote the poem "Cancer's a Funny Thing," which begins

> I wish I had the voice of Homer
> To sing of rectal carcinoma,
> Which kill a lot more chaps, in fact,
> Than were bumped off when Troy was sacked.

(Photo ©American Philosophical Society.)

The conditions for a stable polymorphism with absolute dominance—when only the relative fitness of A_2A_2 varies over time—are also quite restrictive:

$$\left(\prod_{i=1}^{n} w_{22\cdot i}\right)^{1/n} < 1 < \frac{1}{n}\sum_{i=1}^{n} w_{22\cdot i} \qquad (4.14b)$$

In this case, the geometric mean of the variable genotype must be smaller than unity and the arithmetic mean greater than unity. These conditions are given in Figure 4.16 for two environments that occur in equal numbers of generations. The light shading indicates the region in which there is a stable polymorphism for spatial variation but not for temporal. When selection is weak, the range of values that give a polymorphism is very small. For example, when the relative fitness of A_2A_2 in environment 1 is 0.9, then the fitness in the other environment must be between 1.1 and 1.111. An exception to these restrictive conditions for maintenance of polymorphism with temporal environmental variation occurs when there are generations with diapause, as in insects or other animals, or seed dormancy, as in plants (Ellner and Hairston, 1994; Hedrick, 1995a; Turelli *et al.*, 2001). In this case, genotypes may escape environments in which they have low fitness because they are in a stage that is not active. Example 4.9 discusses a population of a small desert annual plant, polymorphic for flower color, in which the conditions for a stable polymorphism resulting from variable temporal selection are met.

Example 4.9. Flower color variation in a small annual plant of the Mohave Desert, *Linanthus parryae*, was the focus of debate among early evolutionary geneticists about what factors were important in determining the pattern of genetic variation. Epling and Dobzhansky (1942) and Wright (1943a) suggested that the pattern of flower color polymorphism observed over populations was consistent with genetic drift. Many of the populations surveyed by Epling and Dobzhansky (1942) were monomorphic—78% all recessive white and 10% all dominant blue—with the only remaining approximately 12% polymorphic. This distribution was consistent with the predictions of genetic drift and isolation by distance; that is, most small populations were fixed for one or another allele in different populations. However, Epling *et al.* (1960) later rejected this conclusion and proposed that natural selection maintained this flower color polymorphism.

To re-examine this classic evolutionary example, Schemske and Bierzychudek (2001) studied a polymorphic population at Pearblossom over 11 years (in 1990, 1994, 1996, and 1997, the densities were too low to carry out the study). The frequency of plants at this site with blue flowers was fairly stable over this period, ranging from 9% to 15.9% (Table 4.12); however, the density variation was extreme from no plants in 1990, 1996, and 1997 to 66.5 plants/m^2 in 1995. For the 7 years with substantial densities, the number of seeds per plant was determined for both blue- and white-flowered plants. For the 4 years with lower seed numbers, blue-flowered plants had many more seeds than white-flowered plants, whereas for the 3 years with higher seed numbers, the opposite was true. Such high selective differences observed between genotypes are counter to what would be expected for neutral alleles dominated by genetic drift.

Turelli *et al.* (2001) developed theory to examine this situation and determined the conditions for a stable polymorphism with variable selection and variable contributions to the seed bank (*L. parryae* seeds may remain viable in the soil for at least 7 years). They showed that a genotype whose arithmetic and geometric means are both less than one can persist if its relative fitness is greater than one in years with high reproduction (high contribution to the seed bank). Remember from our discussion above that without a seed bank, the arithmetic mean of the variable genotype must be > 1 and geometric < 1. Here the arithmetic and geometric means can be calculated from the blue/white ratios given in Table 4.12 and are 1.34 and 1.25 for the blue-flowered plants and 0.85 and 0.80 for the white-flowered plants (white/blue ratios need to be used for calculating the values for the white plants). The contributions to the seed bank can be calculated as the mean of the product of the ratios and the seed bank contributions in the right-hand column of Table 4.12. The values for blue- and white-flowered plants are 0.92 and 1.10, respectively. In other words, the lower fitness value of the white-flowered plants is balanced by a higher contribution of white flowered plants to the seed bank.

TABLE 4.12 The frequency of plants with blue flowers at Pearblossom over the study, the number of seeds/plant in the different years (Schemske and Bierzychudek, 2001), and the estimated contributions to the seed bank (Turelli *et al.*, 2001).

Year	*Blue frequency*	*Density (plants/m²)*	*Number of seeds/plant* Blue	White	Blue/white	*Contribution to seed bank*
1988	0.135	18.4	30.45	19.67	1.55***	0.269
1989	0.137	1.6	5.00	3.83	1.31	0.004
1991	0.159	5.8	260.13	302.42	0.86*	1.190
1992	0.147	21.8	18.58	13.28	1.40*	0.212
1993	0.104	12.8	2.77	1.12	2.47***	0.012
1995	0.116	66.5	65.62	73.35	0.89***	3.341
1998	0.090	22.5	108.87	125.81	0.87*	1.972

* $P < 0.05$ and *** $P < 0.001$ and significant after Bonferroni correction.

Even when there is a stable polymorphism, the change in allele frequency toward the equilibrium may be quite slow. An example is given in Figure 4.18, where the fitness of the variable homozygote is 0.9 in environment 1 and 1.108 in environment 2. It is assumed that there is an environmental switch every generation (see discussion of environmental patterns below) and that the equilibrium frequencies are 0.484 and 0.497 in the two different environments. After 5,000 generations the frequency has still not reached the equilibrium value when approaching the equilibrium from a low initial frequency.

Interestingly, the conditions for a stable polymorphism when there is temporal variation in the environment are independent of the temporal pattern of environmental variation (Haldane and Jayakar, 1963). However, such different environmental patterns may lead to quite different distributions of allele frequencies over populations. The simplest way to imagine

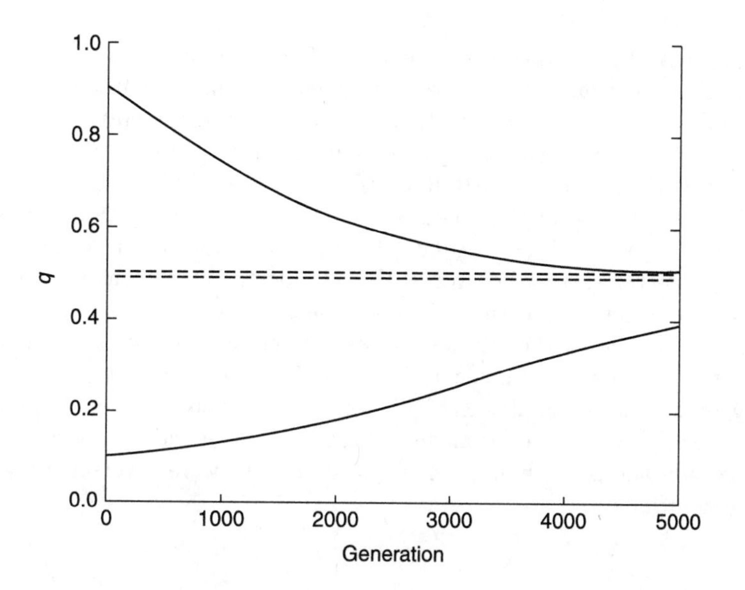

Figure 4.18. The allele frequency over time when there is a stable polymorphism resulting from temporal variation in selection for one homozygote. The two broken lines indicate the equilibrium values in the two different environments.

such a sequence is to consider two environments, such as wet and dry years, and to define the probabilities of a transition from one type of environment to another. Let us say that the probability of staying in environment 1 in generation $t+1$ given environment 1 in generation t is α and that the probability of staying in environment 2 in generation $t+1$ given environment 2 in generation t is β. Because the probability of a different environment is the complement of α and β for the two environments, the transition between environments can be specified as in Table 4.13a. The correlation (r) between subsequent environments is

$$r = \alpha + \beta - 1 \tag{4.15}$$

TABLE 4.13 A general model (a) of stochastic environmental variation between two environments. Below in (b) are three specific patterns that show different amounts of autocorrelation, r, between subsequent environments.

(a)		Generation t			
Generation $t+1$		*Environment 1*		*Environment 2*	
Environment 1		α		$1-\beta$	
Environment 2		$1-\alpha$		β	
		$r = \alpha + \beta - 1$			
(b)					
0.5	0.5	0.9	0.1	0.1	0.9
0.5	0.5	0.1	0.9	0.9	0.1
$r = 0.0$		$r = 0.8$		$r = -0.8$	

and is referred to as the autocorrelation. Therefore, if $\alpha = \beta = 0.0$, which indicates an environmental switch every generation, the autocorrelation is –1.0. Likewise, if $\alpha = \beta = 1.0$, there is no change in the environment and $r = 1.0$. Three environmental patterns are given in Table 4.13b, where the autocorrelation values are 0.0, 0.8, and –0.8, respectively.

Let us assume that the fitnesses in environment 1 are 1, 0.9, and 0.8 and that in environment 2 they are 0.8, 0.9, and 1 for genotypes A_1A_1, A_1A_2 and A_2A_2, respectively. In other words, there is additivity in both environments and equal selection favoring different alleles. A typical pattern of allele frequency change for these fitness values is given in Figure 4.19 for the three environmental patterns given in Table 4.13b (these data were generated using Monte Carlo simulation). When $r = -0.8$, there is an environmental switch nearly every generation, which results in a reversal in allele frequency change. On the other hand, when $r = 0.8$, there were reversals only in a few generations.

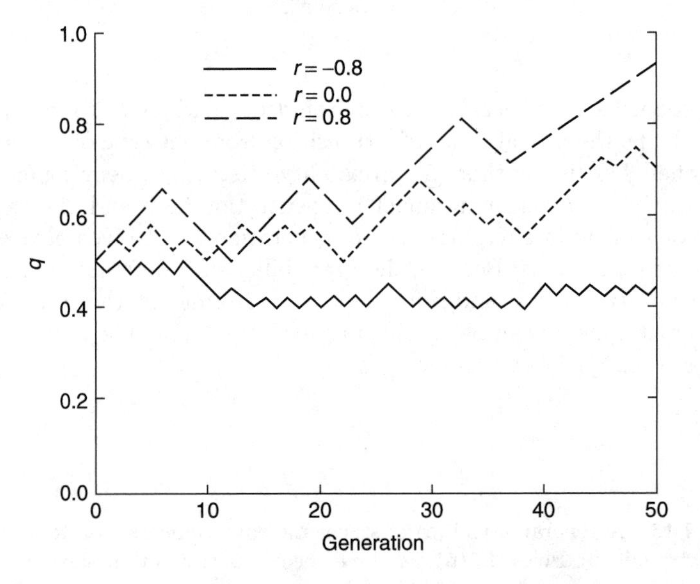

Figure 4.19. The frequency of A_2 over time when there is selection in two different environments and the autocorrelation r between different environments is –0.8, 0.0, or 0.8.

As a result of these different environmental patterns, allele frequencies tend to stay near their starting value or tend to move away. For example, when $r = -0.8$, the frequencies remain near the starting frequency of 0.5 for a number of generations. When $r = 0.8$, many of the populations soon have frequencies below 0.1 or above 0.9. Obviously, the pattern of environmental change can have substantial effects on the allele frequency distribution and can lead to frequencies near 0.0 or 1.0 in some populations even though a stable polymorphism is predicted. Of course, because the population being considered here is infinite, fixation cannot occur. Instead, the term **pseudofixation** is used to describe situations in which a population reaches allele frequencies very near 0.0 or very near 1.0. In this case, the level of pseudofixation increases with the level of autocorrelation.

c. Frequency-Dependent Selection

In a variety of organisms, including insects, birds, and crop plants, it appears that relative fitness values may be a function of the frequencies of different genotypes in a population (for a review, see Clarke *et al.*, 1988). Generally, models of frequency-dependent selection that lead to a balanced polymorphism have been of major interest where the relative fitnesses of genotypes are usually some inverse function of their frequencies in the population, sometimes called negative frequency-dependent selection (see p. 192 and Wright, 1969). For example, if an allele is uncommon, the homozygote for that allele would have a higher fitness than other genotypes because of an advantage in viability or mating. Frequency-dependent selection may also lead to a faster fixation of a particular allele. For example, a rare allele may have a greater disadvantage when uncommon, as when a rare flower type is not recognized by a pollinator or an unusual color morph is picked out of a school of fish by a predator. This greater advantage to the common type is sometimes called positive frequency-dependent selection.

Frequency-dependent selection implies that the relative fitnesses of particular genotypes are a function of the frequency of other genotypes in the population. Selection of this type has been most frequently observed in viability and sexual selection and is often mediated by some behavioral factor, such as a predator preferentially killing particular genotypes (see Example 4.10 for frequency-dependent selection for "handedness" in a scale-eating fish) or particular male genotypes being favored by females for mating (see Example 4.3). In some cases, the basis for frequency-dependent selection may be physiological, as in the interactions among plants or insect larvae relating either to differential use of nutrients or differential production of toxins.

Example 4.10. Hori (1993) presented an intriguing example of frequency-dependent selection in an unusual scale-eating cichlid fish, *Perissodus microlepis*, from Lake Tanganyika. The scale-eaters are specialized to feed on the scales of other living fish, approaching their prey from behind to snatch off several scales from the flank of the prey. The mouth of a given cichlid either opens to the right or opens to the left because of an asymmetrical joint in the jaw (Figure 4.20). The right- or left-handedness of the mouth appears to be determined by segregation at a single gene with the right-handed allele dominant over the left-handed allele.

Figure 4.20. The handedness of the mouth opening of a Lake Tanganyikan scale-eating cichlid. The top fish is right-handed, and the lower fish is left-handed. (Reprinted Figure 1 with permission from Hori, M. 1993. Frequency-dependent natural selection in the handedness of scale-eating cichlid fish. *Science*. 260:216–219. ©2004 AAAS.)

In several experiments, Hori (1993) found complete correspondence between the handedness of the cichlid and the flank of the prey attacked. That is, right-handed fish always attacked the left flank of the prey, and vice versa. However, the prey fish are alert to the scale-eaters, and only about 20% of attacks are successful. A survey of the flanks of prey species demonstrated that in a year when the right-handed cichlid was more common, there were more scars on the left sides and that in another year, when the left-handed cichlid was more common, there were more scars on the right sides.

Overall, there appears to be strong frequency-dependent selection such that if one of the handedness forms becomes more common, the prey species will guard against attack from that side, consequently reducing success of that type. As a result, at equilibrium, selection should result in equal frequencies of the two morphs, which is in fact exactly what is observed (Figure 4.21). Although the average frequency of each handedness type is not different from 0.5, there does appear to be a cycle with amplitude of about 0.15 approximately every five years. It is not clear whether the cycle is related to a time lag because *P. microlepis* starts feeding on scales at two years of age or because the response of the prey species to attack is not exactly frequency-dependent. In most cases of frequency-dependent selection, variation in a prey is maintained by predation, but here the morphological variation in a predator is maintained by its prey.

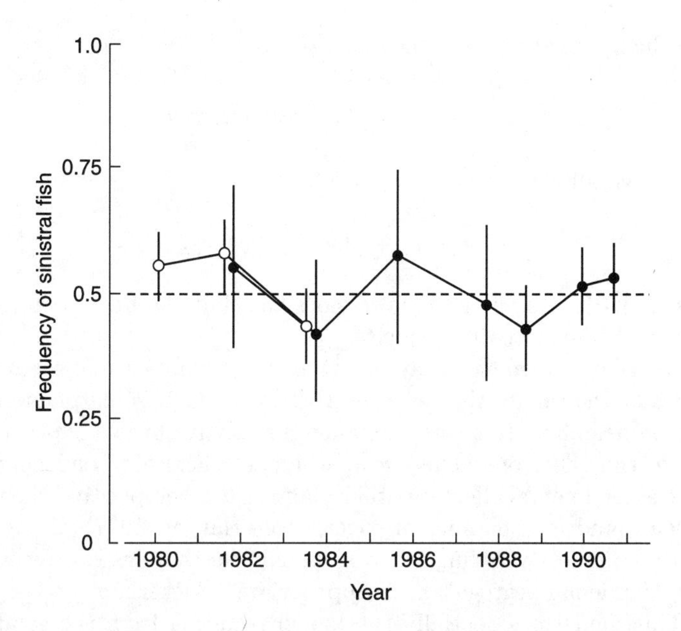

Figure 4.21. The frequency of sinistral (left-handed) fish over an 11-year period at two different sites, with the 95% confidence interval for each sample (after Hori, 1993).

One way to envision frequency-dependent viability is to assume that viability is specifically affected by the genotype of the organisms that are encountered or contacted and that viability is differentially affected by these different genotypes (Huang *et al.*, 1971; Cockerham *et al.*, 1972; Asmussen *et al.*, 2004). For example, if the relative fitness of an individual is inhibited more by others with the same genotype than those of different genotypes, then a stable polymorphism may result. If there are two alleles, this results in nine specific conditional fitness values because each genotype and its specific association with each other genotype must be considered; for example, $w_{11 \cdot 22}$ is the fitness of genotype A_1A_1 when in association only with individuals of genotype A_2A_2.

Assuming that the relative fitnesses are a function of the frequencies of the other genotypes in the population and Hardy–Weinberg proportions, the resulting fitness values are

$$\bar{w}_{11} = p^2 w_{11 \cdot 11} + 2pq w_{11 \cdot 12} + q^2 w_{11 \cdot 22}$$
$$\bar{w}_{12} = p^2 w_{12 \cdot 11} + 2pq w_{12 \cdot 12} + q^2 w_{12 \cdot 22}$$
$$\bar{w}_{22} = p^2 w_{22 \cdot 11} + 2pq w_{22 \cdot 12} + q^2 w_{22 \cdot 22}$$

The frequency of A_2 after selection is

$$q' = \frac{q\,(p\bar{w}_{12} + q\bar{w}_{22})}{\bar{w}}$$

and the change in allele frequency is

$$\Delta q = \frac{pq}{\bar{w}}[q(\bar{w}_{22} - \bar{w}_{12}) - p(\bar{w}_{11} - \bar{w}_{12})] \tag{4.16a}$$

where the overall mean fitness is

$$\bar{w} = p^2\bar{w}_{11} + 2pq\bar{w}_{12} + q^2\bar{w}_{22}$$

Expression 4.16*a* is similar to expression 3.3a, but the fitness values are a function of the genotype frequencies.

There is some evidence that the fitnesses of plants are affected by the genotypes of the plants that surround them. In fact, some plants are so-called good neighbors (i.e., they enhance the relative fitness of other plants relative to the effect on themselves), whereas others are "bad neighbors" and have a detrimental effect on other plants. Such competitive phenomena have been found in a number of species (see Harper, 1977). If the plants are nearly completely selfing, then a model like that in expression 4.16*a* without Mendelian segregation is appropriate (Cockerham and Burrows, 1971). This model is essentially the model resulting from the conditional fitnesses given above without the Mendelian component and is similar to an interspecific competition model in which the equilibria are solely the function of interspecific competition coefficients. Such a model was used by Allard and Adams (1969) in conjunction with relative yield estimates in pure and mixed stands of barley to predict the change in proportions of four different varieties. Although the predicted results were generally consistent with the pattern of change observed over 16 generations, the rate of change observed was even faster than that predicted. Example 4.11 gives some experimental data on an orchid with a flower polymorphism showing negative frequency-dependence and how to use the previous equations.

Example 4.11. Although pollinators are thought to favor common flower colors, resulting in positive frequency-dependent selection and eventually fixation, there is a counterexample in which naïve pollinators favor the rare flower types (Gigord *et al.*, 2001). The orchid *Dactylorhiza sambucina* is polymorphic for striking yellow- and purple-flowered individuals throughout Europe, and these orchids are known to not produce a reward, either in the form of nectar or pollen, for pollinators. Newly emerged, naïve pollinators tend to change flower color when they encounter a rewardless flower. Because they encounter more of the common type of flower, they overvisit the rarer color types and result in higher fitness for the rare flower types, a negative frequency dependence.

To determine whether field data are consistent with this hypothesis, Gigord *et al.* (2001) put out arrays of *D. sambucina* and varied the proportion of yellow plants between 0.1, 0.3, 0.5, 0.7 and 0.9. They then measured

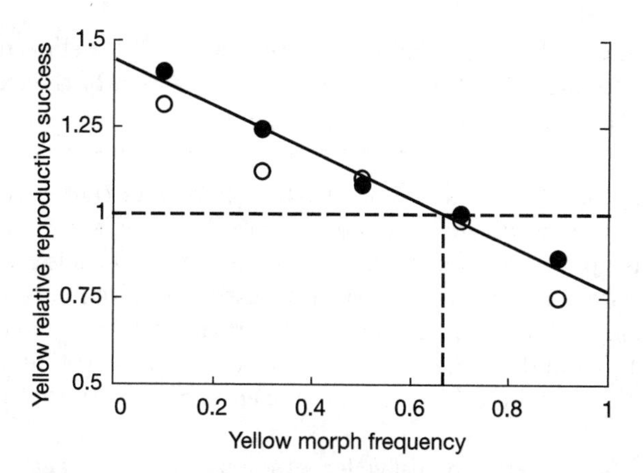

Figure 4.22. Measures of male (closed circles and solid line) and female (open circles) relative reproductive success for yellow-flowered orchids as a function of the frequency of yellow flowers in experimental populations (from Gigord *et al.*, 2001).

a male fitness component, the number of pollinia removed by the pollinators, and a female fitness component, the amount of fruit set, on yellow- and purple-flowered plants. Figure 4.22 shows that for both male and female components there was negative frequency dependence; that is, the estimate of fitness for yellow-flowered plants decreased as the frequency of the yellow flowers increased. By the point where the data intercept the line at 1, one would predict that there would be an equilibrium between 0.6 and 0.7 (vertical broken line).

These data can be used in the expressions given above to determine the frequency-dependent fitnesses and predict changes in allele frequency. Using y to indicate yellow plants, p to indicate purple plants, and q to indicate the frequency of yellow plants and assuming Hardy–Weinberg proportions and dominance of purple, then

$$\bar{w}_p = q^2 w_{p \cdot y} + (1 - q^2) w_{p.p}$$

$$\bar{w}_y = q^2 w_{y \cdot y} + (1 - q^2) w_{y \cdot p}$$

The frequency of yellow after selection would be

$$q' = \frac{q(1-q)\bar{w}_p + q^2 \bar{w}_y}{\bar{w}}$$

where

$$\bar{w} = (1 - q^2)\bar{w}_p + q^2 \bar{w}_y$$

From the data in Gigord *et al.* (2001), on average, then approximately $w_{y.y} = 0.8$, $w_{y.p} = 1.45$, $w_{p.y} = 1.2$, and $w_{p.p} = 0.55$. Using the expression above and these values, the equilibrium frequency of the yellow morph (q^2) is 0.69. The frequency of the yellow morph in the area of the experiments is 0.69, the same as this predicted equilibrium.

An array of fitness interactions that gives results that may be significant in maintaining diversity within and over species is one that satisfies the relationships in the game **rock–scissors–paper**. This is a classic nontransitive system in which rock crushes scissors, scissors cuts paper, and paper covers rock (if it was transitive, then rock would win in competition with paper). To understand this situation, let us assume that there is an array of fitnesses as in Table 4.14a, and then using an approach similar to

TABLE 4.14 (a) The array of fitness for a rock–paper–scissors game where fitness $w_{i.j}$ is for type i in the row and j is competing type given in the columns and (b) gives the general fitness relationships. Specific arrays of values for (c) the different colors, orange, yellow, and blue of male side-blotched lizards and (d) the strains of *Esherichia coli*, C (colicinogenic), S (sensitive), and R (resistant).

(a) *Competing with*

Fitness of	Rock	Scissors	Paper
Rock	$w_{R.R}$	$w_{R.S}$	$w_{R.P}$
Scissors	$w_{S.R}$	$w_{S.S}$	$w_{S.P}$
Paper	$w_{P.R}$	$w_{P.S}$	$w_{P.P}$

(b)

	R	S	P
R	1	> 1	< 1
S	< 1	1	> 1
P	>1	< 1	1

(c)

	Orange	Blue	Yellow
Orange	1.0	4.31	0.4
Blue	0.2	1.0	1.8
Yellow	1.76	1.00	1.0

(d)

	C	S	R
C	1.0	$\rightarrow \infty(10)$	0.52
S	$\rightarrow 0(0.1)$	1.0	1.69
R	1.91	0.59	1.0

that in equation 4.16a, the mean fitnesses for rock (R), scissors (S), and paper (P) are

$$\bar{w}_R = p_R w_{R.R} + p_S w_{R.S} + p_P w_{R.P}$$

$$\bar{w}_S = p_R w_{S.R} + p_S w_{S.S} + p_P w_{S.P}$$

$$\bar{w}_P = p_R w_{P.R} + p_S w_{P.S} + p_P w_{P.P}$$

For simplicity, let us assume that the types are pure breeding (no Mendelism). Then the frequencies of R, S, and P after selection are

$$p'_R = \frac{p_R \bar{w}_R}{\bar{w}}$$

$$p'_S = \frac{p_S \bar{w}_S}{\bar{w}} \qquad (4.16b)$$

$$p'_P = \frac{p_P \bar{w}_P}{\bar{w}}$$

where the overall mean fitness is

$$\bar{w} = p_R \bar{w}_R + p_S \bar{w}_S + p_P \bar{w}_P$$

The behavior of this system is complicated, but some insight is obtained from theoretical examples. For example, if in Table 4.14b the > 1 and < 1 categories are set to 2 and 0.5, respectively, then the three types approach an equilibrium of 1/3 from all starting points. On the other hand, if the > 1 and < 1 categories are set to 1.5 and 0.5, then there are cycles of approximately 20 generations going to nearly 100% rock, then nearly 100% scissors, and finally nearly 100% paper. In addition, the data from estimates in side-blotched lizards and *E. coli* can be used to predict the effects of these interactions (see Example 4.12).

Example 4.12. Two examples of the nontransitive rock–paper–scissors game have been described. First, in the side-blotched lizard (see also Example 2.5), males are orange, blue, or yellow, and the relative fitnesses of the interactions between males is given in Table 4.13c. Here the dominant orange male defeats the mate-guarding strategy of the blue males (4.31), the mate guarding strategy of the blue males defeats the sneaker yellow males (1.8), and the sneaker yellow males defeats the dominant orange males (1.91). If this array is used in equation 4.16*b*, then there is a stable cycle (Figure 4.23) with about 15-year cycles as observed in the simulations of Sinervo and Lively (1996).

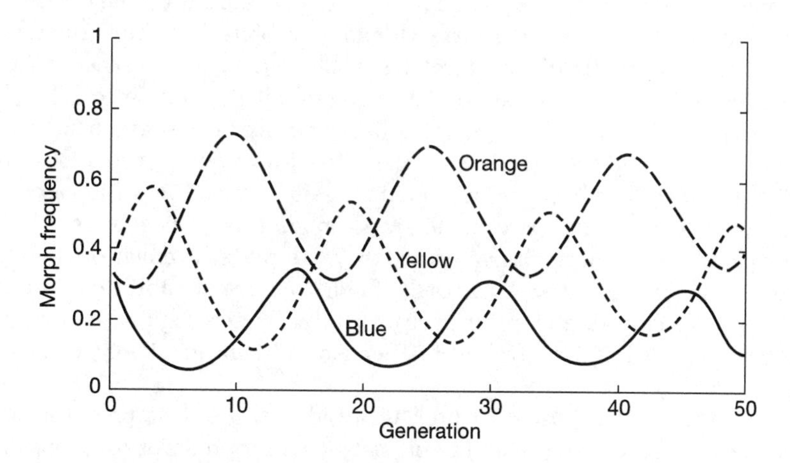

Figure 4.23. Predicted cycles in the frequency of the three male color morphs of side-blotched lizards; the dominant orange, the mate-guarding blue, and sneaker yellow when using the fitnesses in Table 4.14c. (after Sinervo and Lively, 1996).

Kerr *et al.* (2002) described an example in a model system of *E. coli* in which one bacterial form, colicinogeneic (C), produces the toxin colicin; another form, sensitive (S), is killed by colicin, and the third form is resistant (R) to colicin (see also Kirkup and Riley, 2004). The interactions between the forms are nontransitive because C displaces S because it kills it, S displaces R because S has a growth rate advantage, and R displaces C because R has a growth rate advantage (see fitness interactions in Table 4.14d). If we use the fitness values in Table 4.14d and substitute 10 for ∞ for $w_{C.S}$ and 0.1 for 0 for $w_{S.C}$, then using equations 4.16b, we obtain population cycles that are about 24 generations long.

d. Host–Pathogen Interactions

When different species interact, then selective changes may occur as a result of the interaction. In fact, the concept of **coevolution** suggests that species that are either host and pathogen (or parasite), predator and prey, competitors, or mutualists may evolve in response to each other (Thompson, 1994). In particular, the interaction of host and their pathogens and predators and their prey may result in frequency-dependent selection (Ebert, 1998). For example, a new genetic type of the pathogen may have high reproduction when it is rare because most of the host population is susceptible. However, as this pathogen type increases in frequency, resistant host types may also increase, thereby reducing the reproduction rate of the pathogen, or vice versa, a new host type may be resistant to the common type of pathogen. By the time the host type becomes common because of its selective advantage, the pathogen may have had time to evolve further virulence, and the new host type is no longer as successful.

A classic example of host–pathogen coevolution was documented in European rabbits that were introduced to Australia from England (Fenner and Fantini, 1999). Myxoma virus (found in rabbits from the Americas) was introduced in 1950 to control the rabbits that had reached plague proportions. The virus initially reduced the rabbit population to 1% of its former level. Then the virus evolved lower virulence, and subsequently, the rabbit evolved higher resistance to the virus. However, in the 1980s and 1990s, the level of virulence of myxoma again changed and increased to high levels (Fenner and Fantini, 1999). Although rabbits now appear to be controlled in some agricultural areas, they are pests in much of the arid inland areas of Australia. As a result, further efforts at rabbit control are still being developed, and another virus (rabbit hemorrhagic disease virus) has been released as a control agent (Fenner and Fantini, 1999; Cooke and Fenner, 2002).

The ability of a host or a prey to avoid pathogens or predators, respectively, and likewise the ability of pathogen or a predator to use or find a host or prey, respectively, appears to be under genetic control in many cases. Although the genetic determination of host–pathogen or predator–prey traits may be determined by a number of genes, there are a number of examples in which these traits are determined by only a few genes. For example, mimicry, a form of predator avoidance, may be the result of only a few genes (Joron and Mallet, 1998), and host resistance and pathogen virulence may be the result of a small number of genes. In many plants, there seems to be a widespread matching of host alleles conferring resistance to specific strains of pathogen, resulting a **gene-for-gene** change in both species (Thompson and Burdon, 1992; Bergelson *et al.*, 2001; de Meaux and Mitchell-Olds, 2003). Example 4.13 shows an example from the first gene-for-gene study in rust resistance in flax and the correspondence of alleles at an *R* gene and resistance in *Arabidopsis*.

Example 4.13. The first definitive studies on the genetics of plant resistance that showed the gene-for-gene pattern of resistance was in a rust disease on flax (Flor, 1956). For example, two varieties of flax, Ottawa and Bombay, were tested for their susceptibility (S) or resistance (R) to two rust strains, 22 and 24. As shown in Table 4.15, Ottawa is resistant to rust strain 24, and Bombay is resistant to rust strain 22. When the flax varieties are crossed, the F_1 is resistant to both rust strains, demonstrating that the alleles conferring resistance are dominant over the alleles allowing susceptibility. A cross of F_1 individuals results in segregation of four different types of individuals, those resistant to both rust strains, those resistant to 24 and not 22, those resistant to 22 and not 24, and those susceptible to both. Table 4.15 gives the numbers of these classes observed in an experimental cross and the expected number based on a 9:3:3:1 ratio of the segregation of two independent dominant genes; the observed numbers are not significantly different from the expected ones. These findings demonstrate that resistance to rust strain 22 is determined by a dominant allele at one gene and that resistance to rust strain 24 is determined by a dominant allele at another unlinked gene.

In recent years, there has been extensive research examining *R* (disease resistance) genes in a number of plant species. For example, Stahl *et al.* (1999) showed that resistance and susceptibility in the highly selfing, plant *Arabidopsis thaliana* to the bacterial pathogen *Pseudomonas* is the result of molecular variation at the gene *Rpm1*. In a worldwide survey of *A. thaliana* accessions, they found that all susceptible plants were genotype *rr* and all resistance plants were *RR*, except for one resistant accession from Kazakhstan that was heterozygous *Rr*. The overall frequency of the resistant *R* allele was estimated to be 0.52 throughout the range of the species. A fitness cost for the presence of this resistance allele in the absence of pathogens has been recently shown (Tian *et al.*, 2003), providing a counter to the advantage of this allele for disease resistance and an explanation for the polymorphism of this allele (however, see Korves and Bergelson, 2004). In general, to maintain polymorphism in a gene-for-gene system it is thought that there needs to be a cost to the resistant host allele and a cost for the virulent pathogen allele or they will go to fixation.

TABLE 4.15 The resistance (R) or susceptibility (S) of two flax varieties (Ottawa and Bombay), their F_1 progeny, and their F_2 progeny to two strains of rust (22 and 24) (Flor, 1956).

	Flax variety or cross						
Rust strain	Ottawa	Bombay	F_1	F_2			
22	S	R	R	R	S	R	S
24	R	S	R	R	R	S	S
Observed				109	36	36	12
Expected				108.6	36.2	36.2	12.1

We can understand the genetic interaction of a host and a pathogen using a simple **matching-allele model** (sensu Frank, 1994) in which both species are polymorphic for two alleles (for recent discussion of genetically defined host-pathogen models, see Agrawal and Lively, 2002; Dybdahl and Storfer, 2003). Let us assume that the diploid host has alleles A and a that influence resistance, with frequencies p and $1 - p$, and that the haploid pathogen has alleles B and b, with frequencies q and $1 - q$, that influence virulence. There are six different host fitnesses, the combinations of the three host genotypes infected with the two different pathogen genotypes (Table 4.16a). For example, host genotype AA has the highest fitness when infected with pathogen genotype B, but pathogen genotype B has the lowest fitness on host genotype AA.

TABLE 4.16 The relative fitnesses (a) of a host with diploid genotypes AA, Aa, and aa when infected with a pathogen with haploid genotypes B and b and (b) the relative fitnesses of the pathogen genotypes on the different host genotypes.

		(a) *Fitness of host*			(b) *Fitness of pathogen*		
		Host			*Host*		
		AA	Aa	aa	AA	Aa	aa
Pathogen	B	1	$1 - hs_h$	$1 - s_h$	$1 - s_p$	$1 - hs_p$	1
	b	$1 - s_h$	$1 - hs_h$	1	1	$1 - hs_p$	$1 - s_p$

The overall fitnesses of the host genotypes are then the sum of the products of these fitnesses and the frequencies of the pathogen genotypes as

$$
\begin{aligned}
w_{AA} &= 1 - s_h(1 - q) \\
w_{Aa} &= 1 - hs_h \qquad (4.17a) \\
w_{aa} &= 1 - s_h q
\end{aligned}
$$

Similarly, the fitnesses of the pathogen genotypes on different hosts are given in Table 4.16b, and the overall fitnesses of the pathogen genotypes are the sum of the products of the pathogen fitnesses and the frequencies of the host genotypes as

$$\begin{aligned} w_B &= 1 - s_p p[p + 2(1-p)h] \\ w_b &= 1 - s_p(1-p)(2hp + 1 - p) \end{aligned} \tag{4.17b}$$

The frequency of host allele A in the next generation is then

$$p' = \frac{p^2 w_{AA} + p(1-p)w_{Aa}}{\bar{w}_h} \tag{4.17c}$$

where

$$\bar{w}_h = p^2 w_{AA} + 2p(1-p)w_{Aa} + (1-p)^2 w_{aa}$$

and the frequency of the pathogen allele B in next generation is then

$$q' = \frac{q w_B + (1-q)w_b}{\bar{w}_p} \tag{4.17d}$$

where

$$\bar{w}_p = q w_B + (1-q)w_b$$

Even for this rather simple system, the genetic changes are somewhat complicated, and thus, the best approach is to discuss examples illustrating some of the different outcomes. First, let us assume that the selection coefficients are the same in both species, which means that $s_h = s_p/2$ because the host is diploid and the pathogen is haploid. Then, let us assume that $h = 0$; that is, one copy of an allele gives the host the same increased resistance as does two copies, and the virulence of the pathogen is as high when there are one or two copies of the host allele that increases virulence. In this case, the host fitnesses from equation 4.17a for AA, Aa, and aa become $1 - s_h(1-q)$, 1, and $1 - s_h q$, so that there is a heterozygote advantage for all polymorphic frequencies and a stable equilibrium is predicted. By iterating the equations 4.17c and 4.17d for this situation, the frequency of A and B always go to a stable equilibrium at 0.5 from any starting point.

Now let us assume that $h = 0.5$; host heterozygotes have an intermediate effect in both resistance and virulence, and let $s_h = 0.2$. In this case, when we iterate equations 4.17c and 4.17d, the frequencies of the alleles in the two species oscillate up and down with the frequency of the A allele in the host following the B allele in the pathogen by approximately 50 generations (Figure 4.24). Finally, let us assume that $h = 0$ so that the host fitnesses for AA, Aa, and aa become $1 - s_h(1-q)$, $1 - s_h$, and, $1 - s_h q$. In

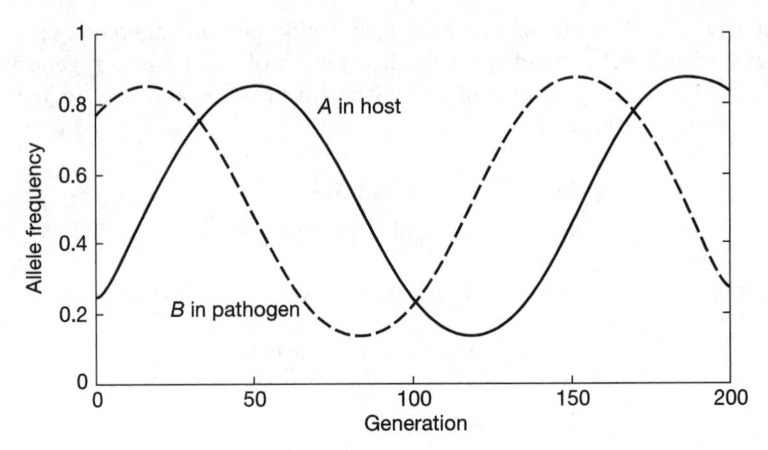

Figure 4.24. The change in the frequency of host allele A (solid line) and pathogen allele B (broken line) over time when there is additivity ($h = 0$) in host resistance and virulence on different hosts.

this case, there is a heterozygote disadvantage for all polymorphic frequencies, and an unstable equilibrium is predicted. In fact, for different starting points, there are different outcomes that appear to happen in sequence. For example, if the host allele a goes to fixation, then the pathogen allele B goes to fixation, and vice versa, if the host allele A goes to fixation, then the pathogen allele b goes to fixation. The fixation of allele frequencies in the pathogen result in the maximization of mean fitness in the pathogen, given the allele frequencies in the host.

Another type of model that has great contemporary relevance is one that shows how genetic variation may be important in determining resistance to epidemic infectious diseases of the past and present. Hedrick (2002a) suggested a selection model for maintaining MHC variation in which a given allele confers resistance to a given pathogen and there is variable presence or absence of the pathogen (selection) over time. In this instance, pathogen variation may be between the same or different species, and host resistance is due to multiple alleles at a given gene. It also has been suggested that various strains of a given viral pathogen even within a host may be different enough that different alleles may confer resistance (Thurz *et al.* 1997; Carrington *et al.*, 1999) (see Example 4.14, which describes the effect of transferrin genotypes on pathogen resistance in pigeons, another system that may be of general importance in the genetics of host–pathogen interactions).

Example 4.14. One of the best examples of both heterozygote advantage and differential host resistance to pathogens was provided by Frelinger (1972). He demonstrated that female pigeons heterozygous at the non–heme-binding protein transferrin for the two common variants, Tf^A and Tf^B, had higher egg hatchability than either of the two homozygotes. The data in Table 4.17 show that the hatchability of eggs was 67% from heterozygous females and only 46% and 52% from the two homozygotes. The selection against the homozygotes gives estimates of $s_1 = 0.31$ and

$s_2 = 0.22$ and a predicted equilibrium of Tf^A of 0.58 (calculation of the equilibrium frequency is the same as for heterozygote advantage; Frelinger and Crow, 1973). In three different populations, the frequency of Tf^A was 0.38, 0.52, and 0.59, close to the predictions from the hatchability experiment.

To understand the mechanism of heterozygote advantage, Frelinger used yeast as a microbial assay and determined its growth in a medium with egg white or purified transferrin from the different genotypes. In both experiments, yeast growth was much more inhibited by media using extracts from heterozygotes than from homozygotes (Figure 4.25). Transferrin is known to inhibit growth in a wide variety of iron-dependent microorganisms, and these data suggest that the two common transferrin alleles somehow complement each other to enhance protection from microbial infection in eggs (and probably in young birds that still have maternal transferrin).

TABLE 4.17 The hatchability of eggs from female pigeons with the common transferrin genotypes Tf^ATf^A, Tf^ATf^B, and Tf^BTf^B, and the resulting relative fitnesses.

	Female genotype		
	Tf^ATf^A	Tf^ATf^B	Tf^BTf^B
Number of eggs laid	128	267	144
Number of eggs hatched	59	180	75
Proportion hatched	0.46	0.67	0.52
Relative fitness	0.69	1.0	0.78

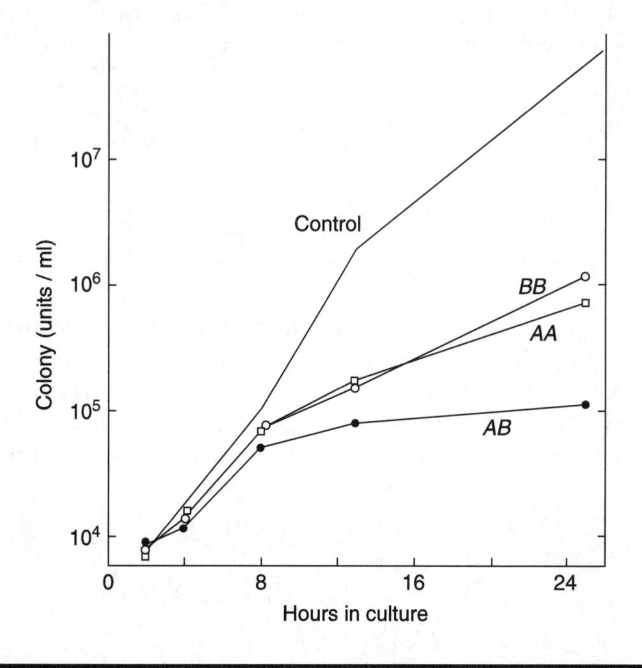

Figure 4.25. The growth of yeast in media with purified transferrin from pigeon genotypes $Tf^ATf^A(AA)$, $Tf^ATf^B(AB)$, and $Tf^BTf^B(BB)$ and a control with no transferrin.

Let us assume that in a population the presence of different infectious diseases varies over time, with particular pathogens (or pathogen types) being present in some generations and absent in others (we are not following the evolution of the pathogens here). If we assume that different alleles confer different resistance to these pathogens, then we can examine the conditions under which genetic variation can be maintained. Let us initially assume that there are two alleles, A_1 and A_2, that result in resistance to pathogens 1 and 2, respectively. The pathogens could be two different species or strains of the same species that are genetically different enough for differential recognition. If we assume that pathogens 1 and 2 are present in proportions e_1 and e_2 and that the presence of the pathogens is independent of each other, then the frequencies of the four environments and the fitnesses of the three genotypes in these environments are given in Table 4.18 (Hedrick, 2002a). In a proportion e_1e_2 of the generations, both pathogens are present. For example, here the selective disadvantages of homozygotes (and heterozygotes, see below) that do not have allele A_1 and, therefore, do not have resistance to pathogen 1, is s_1. The level of dominance in heterozygotes that have one copy of the allele A_1 that confers resistance to pathogen 1 is h_1. Dominance for resistance may not be complete; that is, the dominance levels in the two environments, h_1 and h_2, may be greater than 0.

TABLE 4.18 The relative fitnesses of the three genotypes when pathogens 1 and 2 are both absent (−−), only pathogen 1 is present (+−), only pathogen 2 is present (−+), or both are present (++) and the frequencies of these pathogen environments, assuming that the presence of the pathogens are independent.

Pathogen		*Fitness of host*		
1 2	Frequency	A_1A_1	A_1A_2	A_2A_2
−−	$(1-e_1)(1-e_2)$	1	1	1
+−	$e_1(1-e_2)$	1	$1-h_1s_1$	$1-s_1$
−+	$(1-e_1)e_2$	$1-s_2$	$1-h_2s_2$	1
++	e_1e_2	$1-s_2$	$(1-h_1s_1)(1-h_2s_2)$	$1-s_1$

As we discussed above, Haldane and Jayakar (1963) showed that the conditions for a stable polymorphism in a temporally variable environment are that the geometric mean fitness over environments for the heterozygote must be larger than the geometric mean of the two homozygotes. In the two-allele example given in Table 4.18, the geometric means of the fitnesses for genotypes A_1A_1, A_1A_2, and A_2A_2 are

$$\bar{w}_{11} = (w_{11.1})^{e_1}(w_{11.2})^{e_2} = (1-s_2)^{e_2}$$
$$\bar{w}_{12} = (w_{12.1})^{e_1}(w_{12.2})^{e_2} = (1-h_1s_1)^{e_1}(1-h_2s_2)^{e_2} \qquad (4.18a)$$
$$\bar{w}_{22} = (w_{22.1})^{e_1}(w_{22.2})^{e_2} = (1-s_1)^{e_1}$$

where $w_{ij.k}$ indicates the fitness of genotype ij in the presence of pathogen k. Therefore, the conditions for a polymorphism are

$$(1 - s_2)^{e_2} < (1 - h_1 s_1)^{e_1}(1 - h_2 s_2)^{e_2} > (1 - s_1)^{e_1} \tag{4.18b}$$

If $h_1 = h_2 = 0$, that is, complete dominance for resistance, then these conditions become

$$(1 - s_2)^{e_2} < 1 > (1 - s_1)^{e_1} \tag{4.18c}$$

showing that in this case as long as s_1 and s_2 are greater than 0, then there should be a stable polymorphism. If the pathogens are present every generation, then $e_1 = e_2 = 1$, and the conditions are equivalent to that for the heterozygous advantage model.

Turelli (1981) has shown that the conditions of maintenance for multiple alleles in a randomly fluctuating environment are like those for multiple-allele selection in constant environments, as we discussed on p. 160, except that the relevant fitnesses are replaced by the geometric mean fitness of the genotypes over time. If we assume that all k pathogens are equivalent so that all $s_k = s$, all $e_k = e$, and all $h_k = 0$ and that there are n alleles, then the fitness for the homozygotes become becomes $(1-s)^{(n-1)e}$ and the fitness for the heterozygotes becomes $(1 - s)^{(n-2)e}$. Notice that as there are more alleles and more pathogens, the fitnesses become lower, but the fitness of the heterozygotes relative to the homozygotes remains the same, that is, 1 and $(1 - s)^e$, respectively.

PROBLEMS

1. For a locus in a given organism, go through all of the selection components, suggest which ones may and which may not be important, and design experiments to examine your hypotheses.
2. The relative fitnesses of A_1A_1 and A_2A_2 homozygotes in males are 1.0 and 0.89 and in females they are 0.9 and 1.0, respectively. If heterozygotes have intermediate fitnesses, what are the equilibrium frequencies in both sexes? Assuming the same selection in females, determine what range of s_m values will give a stable equilibrium.
3. Show that the mean fitness and equilibrium frequencies of the complete dominance frequency-dependent model discussed in the text are given by expression 4.8b.
4. Do you think there are many examples of meiotic drive or related phenomena that have not been documented? How would you conduct a survey to determine the general importance of meiotic drive?
5. Assume that the initial frequencies at a self-incompatibility locus are 0.2, 0.5, and 0.3 for S_1S_2, S_1S_3, and S_2S_3, respectively. What is the frequency of S_1 initially and after one and two generations?

6. Discuss the factors that may contribute to variation in the frequencies of self-compatibility alleles given in Figure 4.12. How would you evaluate the relative importance of these different factors?
7. Go to an original journal article that discusses genetics and sexual selection. Would the type of sexual selection discussed result in a genetic polymorphism? Discuss and support your answer.
8. Give a numerical example where fitness variation in space yields a stable polymorphism but the same fitness values varying over time do not give a stable polymorphism (assume only two environments). Show why this is true.
9. Assume that there is fitness variation over space, that in one environment the fitnesses are 1, 1, and 0.8, and that in the other environment the fitnesses are 1, 1, and 1.2. What is the range of c_1 values that will allow a stable polymorphism? Graph the equilibrium frequency over this range.
10. Assume that the environment varies temporally between a spring and a fall generation. If the fitnesses in the spring and fall generations are 1, 1, and 1.22 and 1, 1, and 0.8, respectively, what are the equilibrium frequencies in the two environments?
11. For a given known polymorphism, design experiments that would investigate selective differences based on the physical environment, the frequency of the morphs, and the population density.
12. Why do you think there appear to be more polymorphisms based on spatial variation in the environment than on temporal variation?
13. The data in Examples 3.8, 4.3, and 4.6 are consistent with explanations of balancing selection. Pick one of these examples and design experiments to investigate the selective mechanisms in this example.
14. Example 4.14 demonstrates that transferrin may play an important role in pathogen resistance. How would you examine this situation in more detail, both empirically and experimentally, to determine whether heterozygote advantage is a general phenomenon for transferrin loci?
15. Do you think that the rock–paper–scissors relationship is a common one in natural populations? Why?

5

Inbreeding

Down, July 17, 1870

My dear Lubbock,

As I hear that the Census will be brought before the House to-morrow, I write to say how much I hope that you will express your opinion on the desirability of queries in relation to consanguineous marriages being inserted. As you are aware, I have made experiments on the subject during several years; and it is my clear conviction that there is now ample evidence of the existence of a great physiological law, rendering an enquiry with reference to mankind of much importance. In England and many parts of Europe the marriages of cousins are objected to from their supposed injurious consequences: but this belief rests on no direct evidence. It is therefore manifestly desirable that the belief should be either proved false, or should be confirmed, so that in this latter case the marriages of cousins might be discouraged. . . .

It is, moreover, much to be wished that the truth of the often repeated assertion that consanguineous marriages lead to deafness and dumbness, blindness, &c. should be ascertained: and all such assertions could be easily tested by the returns from a single census.

Believe me,
Yours very sincerely,

Charles Darwin

My mother says that shortly after I was born, ants attacked me in my cradle. The place was Cairo, so I like to guess the ant was "Pharaoh's," *Monomorium pharaonis*. . . . I like the idea because, beside being a notorious lover of sweet things, this tiny yellow ant is noted for indiscriminate inbreeding within the nest, including brother with sister.

William Hamilton (1993)

We have been considering a population in which there was random mating, a condition that appears to be nearly met for many organisms. There are obvious exceptions, such as populations of highly self-fertilizing plants and some invertebrates, some Hymenoptera that have brother–sister matings, and human populations characterized by various amounts of **consanguinity**, or mating between relatives. All of these situations of nonrandom mating have long fascinated scientists, and they still are the focus of extensive research because of the often different or surprising conclusions found in such systems.

Generally, the greatest extent of inbreeding is found in plants because of the high amount of **self-fertilization**, where the ***female and male gametes come from the same individual***, in some species. Inbreeding that results from matings between different, related individuals in plants is often called **biparental inbreeding** to distinguish it from self-fertilization. Plant mating systems have traditionally been divided into three basic types: predominantly outcrossing, mixed self-fertilizing and outcrossing, and predominantly self-fertilizing. Only for predominantly outcrossing species are the basic conclusions from a random-mating population appropriate, and even in this case, when there is selection, having some inbreeding may result in quite different consequences than having no inbreeding. In some plant populations, there appears to be substantial biparental inbreeding, and thus, the above mating system categories need to be expanded to include different types of inbreeding besides selfing.

At the end of the chapter, we also give an introduction to nonrandom mating with respect to phenotype as it occurs in insects, birds, humans, and other organisms. Some of these mating patterns, such as positive-assortative mating, affect only genotype frequencies and not allele frequencies, as do the various patterns of inbreeding. However, most mating patterns, as we saw in Chapter 4, involve selective mating that can lead to changes in allele frequencies and result in either a stable polymorphism or fixation of one allele or another.

I. INBREEDING

Nonrandom mating with respect to genotype occurs in populations in which ***the mating individuals are more closely or less closely related than those drawn by chance from the population***. The results of these two types of matings are called **inbreeding** and **outbreeding**, respectively. Neither inbreeding nor outbreeding by itself causes a change in allele frequency, but both cause a reorganization of the alleles into genotypes. In a population that is inbred, the frequency of homozygotes is increased, and the frequency of heterozygotes is reduced relative to random-mating (Hardy–Weinberg) proportions. With outbreeding, the opposite occurs, and the frequency of heterozygotes is increased and that of homozygotes reduced relative to random-mating proportions. Of course, within a population with no structure, the impact of outbreeding cannot be very great. Only when individuals preferentially mate with individuals from another population (or subpopulation) can there be a large effect of outbreeding on genotype frequencies.

Several other important generalities should be noted. First, the ***genotype changes caused by inbreeding affect all loci in the genome.*** Genetic drift and gene flow also influence all loci, but selection and mutation influence only single loci (selection at a locus also may potentially influence loci closely linked to it or interacting with it). Second, the ***effect on genotype frequencies may be quite ephemeral*** if the mating system changes. For example, the high frequency of homozygotes resulting from self-fertilization can be eliminated completely in one generation of random mating. Finally, ***inbreeding and genetic drift appear to have similar overall effects on heterozygosity, but when examining a given locus within a population, the predicted effect is different.*** Inbreeding (in a large population) can result in deficiency of heterozygotes with no change in allele frequency for a given locus within a population. On the other hand, genetic drift may cause a change in allele frequency but generally no deficiency of heterozygotes within a population. Only when averaged over replicate populations for a given locus or averaged over loci within a given population does genetic drift result in a deficiency of heterozygotes and no change in allele frequency (see Chapter 6).

First, let us define the **coefficient of inbreeding**, f, as ***the probability that the two homologous alleles in an individual are* identical by descent** (**IBD**)—that is, that they are derived from one particular allele possessed by a common ancestor. To illustrate this concept, let us imagine that to form a diploid individual, we draw two random alleles (with replacement) from the alleles available in the gene pool. After the first allele is drawn, then there is a probability f that the second homologous allele to be drawn will be the same allele as the first one. If the probability that the first allele is A_1 is p and the probability that the second allele is A_1 is f, then the probability of getting two A_1 alleles IBD is pf.

However, there is also a probability that both alleles may be A_1, but not IBD. This is often termed **identity in state**, which includes homozygotes with two alleles that appear to be identical as, for example, allozyme, microsatellite, or SNP variants, or perhaps even in DNA or amino acid sequence, but that did not descend (at least in the very recent past) from the same ancestral allele. The probability of identity in state is $p^2(1-f)$, where p^2 is the probability of drawing two consecutive A_1 alleles and $1-f$ is the probability that the second allele is not IBD to the first. Therefore, the frequency of genotype A_1A_1 is

$$\begin{aligned} P &= pf + p^2(1-f) \\ &= p^2 + fpq \end{aligned}$$

This can be understood visually from examining the unit square in Figure 5.1. First, the top of the square, indicating the first allele in the diploid

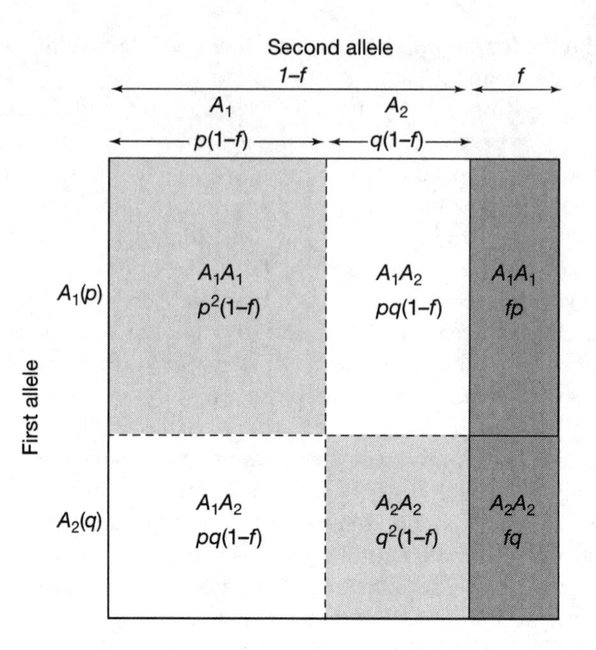

Figure 5.1. The proportions of homozygous genotypes that are the result of inbreeding or IBD (darkly shaded area) and homozygotes that are the result of identity in state (lightly shaded area) using a unit square.

genotype, is divided into the inbred (f) and noninbred ($1 - f$) components. Then the noninbred component for the first allele is divided into parts that represent the frequencies of the two alleles, A_1 and A_2. Next, the second allele is indicated on the vertical axis and is divided into parts that represent the frequencies of the two alleles, A_1 and A_2. Now if we multiply the values on the two axes, the areas within the square are divided into parts that represent the frequencies of the different genotypes. For example, the proportions of genotypes IBD are fp for A_1A_1 and fq for A_2A_2. Likewise, the proportions of genotypes identical in state are $p^2(1 - f)$ for A_1A_1 and $q^2(1 - f)$ for A_2A_2.

A general formulation of the proportion of the three genotypes with inbreeding level f is

$$\begin{aligned} P &= p^2 + fpq \\ H &= 2pq - 2fpq \\ Q &= q^2 + fpq \end{aligned} \tag{5.1}$$

(Wright, 1931) where the first term is the Hardy–Weinberg proportion and the second is the deviations from that value. The size and sign of the coefficient of inbreeding reflect the deviation from Hardy–Weinberg proportions of the genotypes such that when f is zero, the zygotes are in Hardy–Weinberg proportions, and when f is positive, there is a deficiency of heterozygotes. Because f as we have defined it here is a probability, it has a range from 0 to 1. If $f = 1$, its maximum value when there is inbreeding, then

$$P = p$$
$$H = 0$$
$$Q = q$$

In other words, only homozygotes are present, and they occur in the proportions of the allele frequencies.

One way to visualize the effect of inbreeding on the proportions given in expression 5.1 is to imagine that the population is separated into a random-mating proportion $(1 - f)$ and an entirely inbred proportion (f). Because the frequencies of heterozygotes in these two proportions are $2pq$ and 0, the overall frequency of the heterozygote becomes

$$\begin{aligned} H &= 2pq(1 - f) + 0(f) \\ &= 2pq - 2fpq \end{aligned}$$

Likewise, the frequency of genotype A_1A_1 is

$$\begin{aligned} P &= p^2(1 - f) + pf \\ &= p^2 + fpq \end{aligned}$$

because the frequency of A_1A_1 genotypes in the completely inbred proportion is p.

Expression 5.1 for the frequency of heterozygotes can be rewritten as

$$H = 2pq(1 - f) \tag{5.2a}$$

If we assume that the level of heterozygosity in nonbred individuals is H_0 and the amount of heterozygosity in individuals with inbreeding f is H_f, then

$$H_f = H_0(1 - f) \tag{5.2b}$$

Example 5.1 shows how this prediction is consistent with the observed heterozygosity for microsatellite loci in wolves with different known levels of inbreeding.

Example 5.1. Scandinavian gray wolves were thought to be nearly extinct in the wild, and a captive population was started, primarily in Swedish zoos. This population was initiated with four founders in the 1950s and 1960s, two from Sweden and two from Finland, and thus, there were rather high

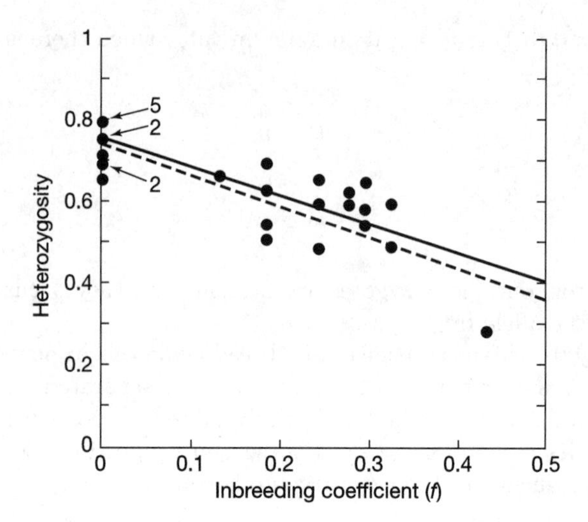

Figure 5.2. The heterozygosity at 29 microsatellite loci for 29 captive Scandinavian wolves with known inbreeding coefficients (from Ellegren, 1999). The solid line is the regression of the observed individual heterozygosity on f, and the broken line is the expected relationship (numbers indicate multiple individuals with identical values).

levels of inbreeding in subsequent generations. Starting in 1980, four more founders were added, two each from Russia and Estonia, to reduce the level of inbreeding (and inbreeding depression) (Laikre and Ryman, 1991) and to add more genetic variation.

Although population genetics theory predicts that the level of molecular genetic variation in individuals with different inbreeding level should follow the pattern predicted by expression 5.2b, only a few situations have examined this relationship (for a related example, see Cunningham *et al.*, 2001). Ellegren (1999) determined the heterozygosity for 29 microsatellite loci of 29 wolves in this captive wolf population with different levels of inbreeding as estimated by pedigree analysis (see p. 264). As illustrated in Figure 5.2, the mean heterozygosity for individual noninbred wolves (f = 0.0) was approximately 0.75 and decreased to 0.28 for the animal with an inbreeding coefficient of 0.42. The linear regression of the observed values explains 67% of the variation among individuals. If the observed heterozygosity for animals with different levels of inbreeding is compared with the expected decline using expression 5.2b with $H_0 = 0.75$, the observed pattern is not significantly different from that expected (Ellegren, 1999), confirming that the observed values are consistent with theoretical predictions.

Expression 5.2a can be solved for f as

$$f = 1 - \frac{H}{2pq} \tag{5.2c}$$

From this equation, it can be seen that f is a function of the ratio of the observed heterozygosity H over the Hardy–Weinberg or expected (with no inbreeding) heterozygosity. With inbreeding, $H < 2pq$, and therefore, $f > 0$.

For any value of f, the allele frequency remains the same. That is, by substituting the genotype frequencies as given in expression 5.1 in the equation for calculating the frequency of A_1, we have

$$\begin{aligned} p &= P + {}^1\!/_2 H \\ &= (p^2 + fpq) + {}^1\!/_2(2pq - 2fpq) \\ &= p^2 + pq = p \end{aligned}$$

This is a demonstration of the fact that ***inbreeding affects only genotype and not allele frequencies***.

Inbreeding can have dramatic effects on the relative genotype frequencies for genes exhibiting a dominant–recessive relationship. Most importantly, rare recessive genotypes become much more common. Data to illustrate this concept are given in Figure 5.3, where the frequencies of several recessive genetic conditions are given for affected progeny from marriages between first cousins ($f = {}^1\!/_{16}$, as we see below) and the general population (f assumed to very close to 0). These frequencies should be compared with the proportion of first-cousin marriages in the total population. For example, in this sample, approximately 5% of the Japanese marriages were between first cousins, but the proportion of first-cousin marriages among families with offspring having albinism was 10-fold higher, approximately 0.56. In the European sample, only approximately 2% of the marriages were between first cousins, but 0.2 of the marriages that had offspring with albinism were among first cousins.

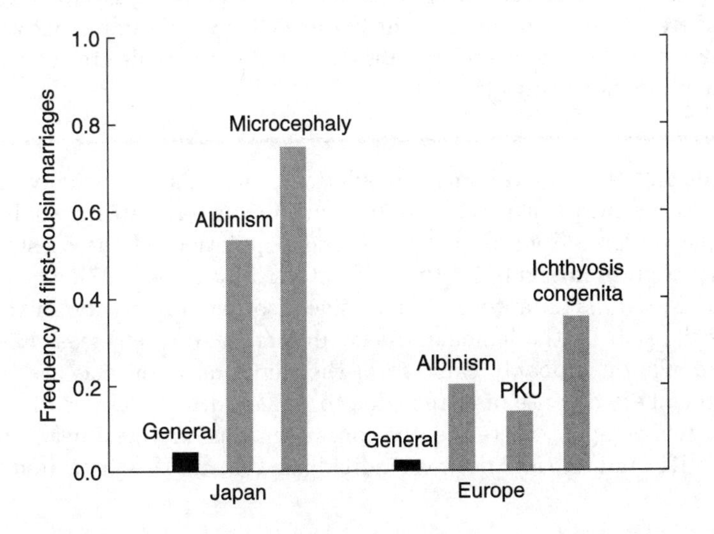

Figure 5.3. The frequency of first-cousin marriages in the parents of individuals affected with several recessive disorders and the frequency in the general population in both Japan and Europe (after Bodmer and Cavalli-Sforza, 1976).

TABLE 5.1 The ratio of the proportion of recessives with a given inbreeding coefficient (f) to the Hardy–Weinberg proportion of recessives for several allele frequencies.

	f		
q	$\frac{1}{32}$	$\frac{1}{16}$	$\frac{1}{8}$
0.001	32.2	63.4	125.9
0.025	13.5	25.9	50.9
0.01	4.1	7.2	13.4
0.1	1.3	1.6	2.1
0.5	1.0	1.0	1.1

To demonstrate algebraically the genetic explanation for these observations, we can examine the incidence of recessive genetic diseases among individuals from consanguineous matings. Inbreeding has the effect of increasing the frequency of homozygous recessives, which in this case increases the incidence of the recessive disease. Let us compare the proportion of recessive homozygotes for a given inbreeding coefficient (Q_f) to that in a noninbred population ($Q = q^2$). The ratio of these two quantities is

$$\frac{Q_f}{Q} = \frac{q^2 + fpq}{q^2}$$

$$= 1 + \frac{fp}{q} \tag{5.3}$$

which is greater than unity for any allele frequency. The ratio is very large for low allele frequencies, as is generally seen for recessive diseases, and increases with the level of inbreeding. The ratio is given for three levels of inbreeding and several allele frequencies in Table 5.1. For example, when $f = 1/16$ and $q = 0.01$, there are more than seven times as many affected individuals in the inbred group as in the random-mating group, similar to the differences seen in the data in Figure 5.3. (See Example 5.2 about a rare skin disease in which many of the affected individuals are the result of consanguineous matings.)

Example 5.2. In isolated human populations, often there are many consanguineous matings because of the limited number of mates available. An example pedigree from an isolated island population off the coast of Yugoslavia is given in Figure 5.4 (from Vogel and Motulsky, 1997). The disease Mal de Meleda is an autosomal recessive disorder that results in thickening of the skin on the hands and feet. In fact, all of the disease alleles in this population probably came from the same mutation and are in high frequency here because of chance due to genetic drift (Chapter 6).

In this pedigree, there are two consanguineous matings, between individuals IIIa and IIIb and between individuals IVa and IVb, indicated by the double lines connecting these mates. First cousins IIIa and IIIb must both be heterozygotes, because two of their seven offspring were affected. The other consanguineous mating is between IVa, apparently a heterozygote, and a diseased homozygote IVb; it resulted in two of the three progeny being affected. Although this disease is virtually unknown in other populations, its commonness in this population results in many heterozygotes.

Recently, different independent mutations in Croatian, Algerian, Palestinian, Turkish, and German families causing Mal de Meleda have been described (Eckl *et al.*, 2003; Fischer *et al.*, 2001). Most of these mutations are single nucleotide changes in the gene *LY6LS*, and they generally result in single amino acid changes in the SLURP-1 protein encoded by the *LY6LS* gene.

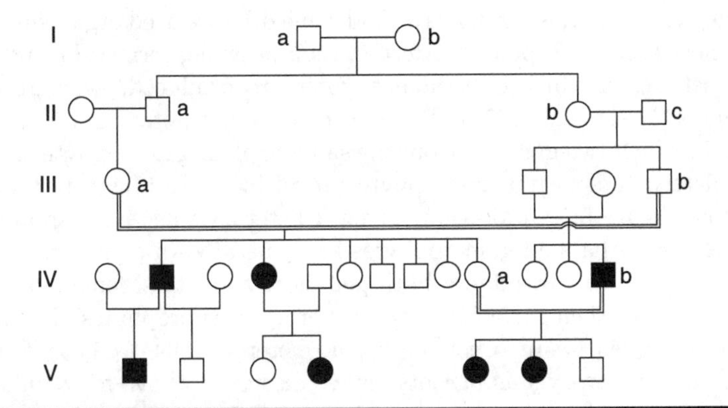

Figure 5.4. A pedigree with two consanguineous matings and a high incidence of a rare recessive disease (from Vogel and Motulsky, 1997). As is the convention in human genetics, circles and squares indicate females and males, respectively, filled symbols indicate affected individuals, generations go downward, and consanguineous matings are indicated by double lines.

a. Self-Fertilization

The most extreme type of inbreeding generally found, assuming there is sexual reproduction, is self-fertilization (see the discussion of the most extreme form of inbreeding, intragametophytic selfing, in Example 5.3). With complete self-fertilization, a population is divided into a series of lines that quickly become highly homozygous. As shown in Table 5.2, with only self-fertilization, there are just three mating types, and they occur in the relative proportions of the genotypes in the population. Allowing for segregation in the heterozygote, the proportions of the genotypes in the progeny are

$$P_1 = P_0 + {}^1\!/_4 H_0$$

$$H_1 = {}^1\!/_2 H_0$$

$$Q_1 = Q_0 + {}^1\!/_4 H_0$$

TABLE 5.2 The mating types and frequency of progeny when there is complete self-fertilization.

		Progeny		
Mating type	*Frequency*	A_1A_1	A_1A_2	A_2A_2
$A_1A_1 \times A_1A_1$	P_0	P_0	—	—
$A_1A_2 \times A_1A_2$	H_0	$\frac{1}{4}H_0$	$\frac{1}{2}H_0$	$\frac{1}{4}H_0$
$A_2A_2 \times A_2A_2$	Q_0	—	—	Q_0
Total	1	$P_0 + \frac{1}{4}H_0$	$\frac{1}{2}H_0$	$Q_0 + \frac{1}{4}H_0$

Example 5.3. In homosporous ferns and some other related organisms, there is the most extreme type of self-fertilization possible, termed intragametophytic selfing, in which only homozygotes are produced, even from heterozygotes (Klekowski, 1979). These ferns have the potential for a mating system in which two gametes from the same haploid gametophyte can form a completely homozygous sporophyte (at all loci). To illustrate for a single locus, Figure 5.5 shows that, given a heterozygous A_1A_2 sporophyte, either A_1 or A_2 haploid gametophytes are generated in equal proportions. A gametophyte then produces both genetically identical female and male gametes, which then form homozygous progeny sporophytes that are either A_1A_1 or A_2A_2. In other words, no heterozygous A_1A_2 zygotes are formed from heterozygous parents, as we discussed above for regular self-fertilization, and in one generation of complete intragametophytic selfing, the population becomes 100% homozygous (McCauley *et al.*, 1985; Hedrick, 1987a).

The alpine lady-fern, *Athyrium distentifolium*, occurs throughout the northern hemisphere but is confined to a few locations in the Scottish highlands in Great Britain (McHaffie *et al.*, 2001). In some of these populations, a distinct taxon, *A. distentifolium* var. *flexile*, is found that has a very different sporophyte appearance with smaller, narrower, and morphologically different fronds. *Flexile* has been considered a separate species, *A. flexile*, and has been protected because of its rare status.

To investigate the inheritance of the sporophyte morphology, McHaffie *et al.* (2001) isolated individual gametophytes from plants with known morphology, allowed intragametophytic selfing, and subsequently determined the sporophyte morphology of the progeny. First, all *flexile* plants produced only *flexile* offspring. Second, the *distentifolium* plants produced either all *distentifolium* plants or *distentifolium* and *flexile* plants in equal proportions. Equal proportions of the two types of offspring are expected from heterozygous plants undergoing intragametophytic selfing. These and other results are consistent with hypothesis that the morphological differences between *distentifolium* and *flexile* are the result of a single gene polymorphic for two alleles, the dominant A^D (*distentifolium*) and the

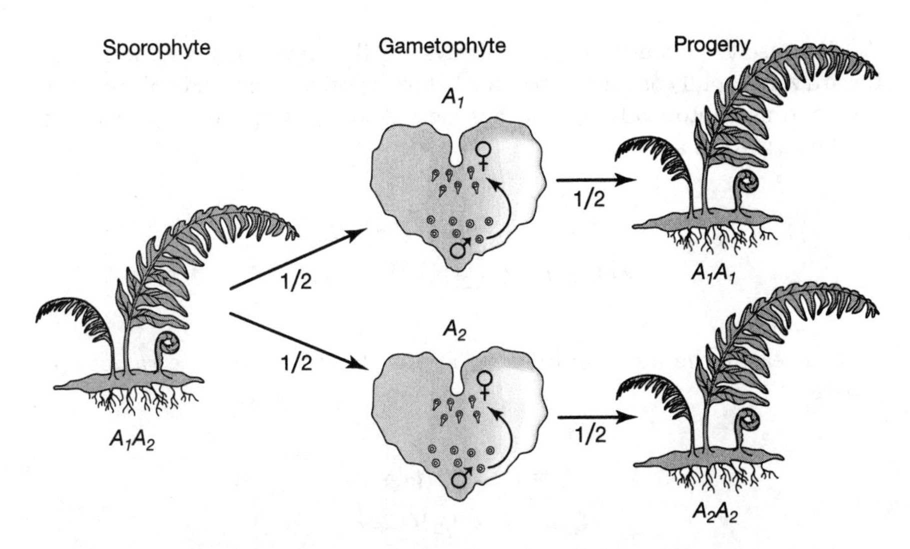

Figure 5.5. A diagrammatic representation of intragametophytic selfing in a fern for a single locus in which the heterozygote sporophyte A_1A_2 produces equal proportions of haploid gametophytes with genotypes A_1 and A_2. Each gametophyte then produces female and male gametes, which after fertilization produce equal proportions of either A_1A_1 or A_2A_2 progeny sporophytes.

recessive A^F (*flexile*). Furthermore, McHaffie *et al.* (2001) also found that this gene had major effects on the performance, ecology, and distribution of the plants and consequently made a case that conserving this genetic variation would help safeguard ecological and evolutionary processes in this species.

The frequency of allele A_1 in the next generation is, therefore,

$$\begin{aligned} p_1 &= P_1 + \tfrac{1}{2}H_1 \\ &= (P_0 + \tfrac{1}{4}H_0) + \tfrac{1}{4}H_0 \\ &= p_0 \end{aligned}$$

which demonstrates that there is no overall change in allele frequency caused by self-fertilization.

However, the genotype proportions are greatly changed. The relationship for heterozygosity in two succeeding generations is a general one

$$H_{t+1} = \tfrac{1}{2}H_t$$

where H_t is the heterozygosity in generation t. This recursion relationship can be generalized for any number of generations so that

$$H_t = (\tfrac{1}{2})^t H_0 \tag{5.4}$$

Because the quantity $(1/2)^t$ is always smaller than unity and decreases toward zero rapidly as t increases, the heterozygosity is reduced quickly over time and asymptotically approaches zero. Analogously, the frequencies of the homozygotes are

$$P_t = P_0 + 1/2[1 - (1/2)^t]H_0$$
$$Q_t = Q_0 + 1/2[1 - (1/2)^t]H_0$$

As t increases, the terms in brackets approach unity and these values approach

$$P_t = P_0 + 1/2 H_0 = p_0$$
$$Q_t = Q_0 + 1/2 H_0 = q_0$$

Therefore, the frequency of the homozygotes eventually will become the frequency of their respective alleles in the initial generation. A numerical example is given in Figure 5.6 when $H_0 = 0.5$, and there are initially Hardy–Weinberg proportions. Reduction in heterozygosity is rapid and is reduced to less than 1% after six generations. These results are true when there multiple alleles as well and, for example, equation 5.4 describes the decay of heterozygosity caused by complete self-fertilization for any number of alleles (Li, 1976).

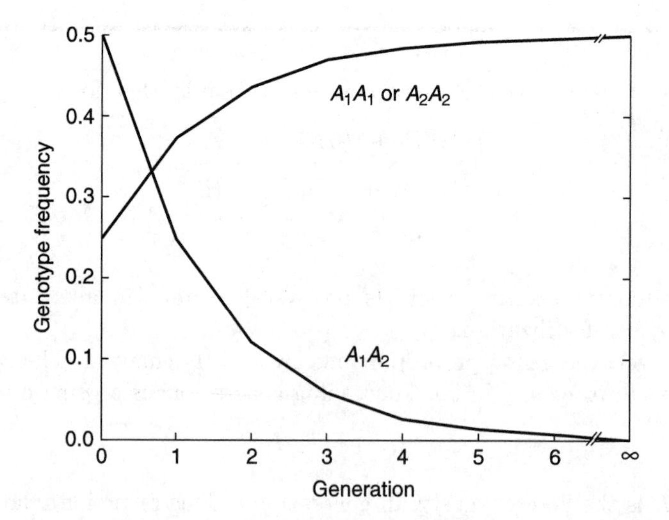

Figure 5.6. The genotype frequencies over time when there is complete self-fertilization and Hardy–Weinberg proportions existed initially with $p_0 = q_0 = 0.5$.

b. Partial Self-Fertilization

For many organisms in which self-fertilization occurs, there is some degree of outcrossing (random mating) as well. In such mixed-mating systems, the simplest procedure is to designate a proportion, S, of the progeny produced by self-fertilization with the remaining proportion, T, being produced by outcrossing ($S+T=1$). Therefore, in order to determine the genotype frequencies after one generation, we can sum the frequencies of the genotypes produced by selfing and outcrossing as in Table 5.3 so that

$$\begin{aligned} P_1 &= Tp_0^2 + S(P_0 + {}^1\!/\!_4H_0) \\ H_1 &= 2Tp_0q_0 + {}^1\!/\!_2SH_0 \\ Q_1 &= Tq_0^2 + S(Q_0 + {}^1\!/\!_4H_0) \end{aligned} \tag{5.5a}$$

where the first term indicates the amount of progeny produced by outcrossing and the second term the amount by self-fertilization.

Again, it can be shown that the frequency of allele A_1 in the next generation does not change; that is,

$$\begin{aligned} p_1 &= P_1 + {}^1\!/\!_2H_1 \\ &= Tp_0^2 + S(P_0 + {}^1\!/\!_4H_0) + {}^1\!/\!_2(2Tp_0q_0 + {}^1\!/\!_2SH_0) \\ &= Tp_0(p_0 + q_0) + S(P_0 + {}^1\!/\!_2H_0) \\ &= p_0 \end{aligned}$$

TABLE 5.3 The mating types and frequencies of progeny when there is a proportion S of the progeny produced by self-fertilization and a proportion T produced by outcrossing.

		Progeny		
Mating type	*Frequency*	A_1A_1	A_1A_2	A_2A_2
Selfing (S)				
$A_1A_1 \times A_1A_1$	SP	SP	—	—
$A_1A_2 \times A_1A_2$	SH	$\frac{1}{4}SH$	$\frac{1}{2}SH$	$\frac{1}{4}SH$
$A_2A_2 \times A_2A_2$	SQ	—	—	SQ
Outcrossing (T)				
$A_1A_1 \times A_1A_1$	TP^2	TP^2	—	—
$A_1A_1 \times A_1A_2$	$T2PH$	TPH	TPH	—
$A_1A_1 \times A_2A_2$	$T2PQ$	—	$T2PQ$	—
$A_1A_2 \times A_1A_2$	TH^2	$\frac{1}{4}TH^2$	$\frac{1}{2}TH^2$	$\frac{1}{4}TH^2$
$A_1A_2 \times A_2A_2$	$T2HQ$	—	THQ	THQ
$A_2A_2 \times A_2A_2$	TQ^2	—	—	TQ^2
Total	1	$Tp^2 + S(P + \frac{1}{4}H)$	$2Tpq + \frac{1}{2}SH$	$Tq^2 + S(Q + \frac{1}{4}H)$

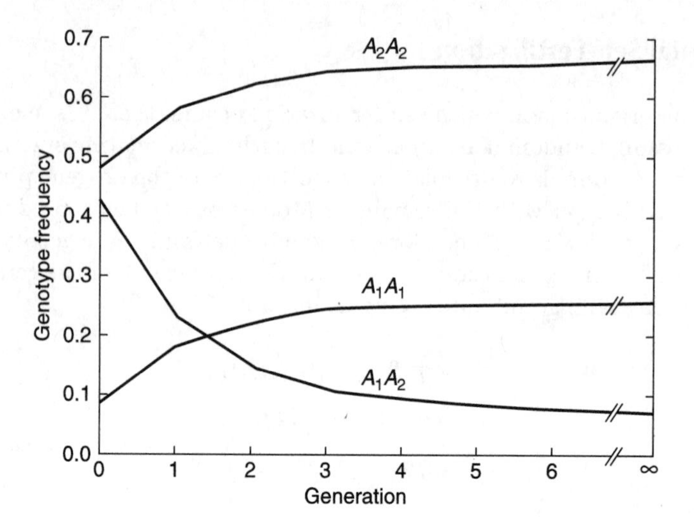

Figure 5.7. The genotype frequencies over time for an example with partial selfing ($S = 0.9$) where Hardy–Weinberg proportions existed initially with $q_0 = 0.7$.

The genotype frequencies do change over time, but in contrast to complete self-fertilization, the heterozygosity does not asymptotically approach zero. With partial self-fertilization, the random-mating (outcrossing) proportion of the population reconstitutes heterozygotes each generation so that their frequency cannot be reduced to zero. To illustrate, Figure 5.7 gives the change in genotype frequencies for an example with $S = 0.9$, $q_0 = 0.7$, and initial Hardy–Weinberg proportions. The frequencies of all three of these genotypes change rather quickly and approach their asymptotic values in a few generations.

The asymptotic values are the **inbreeding equilibrium** proportions for a given amount of self-fertilization and a given allele frequency. These equilibrium proportions can be derived from the recursion relationships given in expression 5.5*a*. These recursion relationships hold for any adjacent pair of generations so that

$$H_{t+1} = 2Tpq + {}^1\!/_2 SH_t$$

where the subscripts of the allele frequencies have been dropped because there is no change in allele frequency. At equilibrium, there is no change in heterozygosity, and thus, $H_{t+1} = H_t$. Therefore, by setting $H_{t+1} = H_t = H_e$ where H_e is the equilibrium heterozygosity

$$\begin{aligned} H_e &= 2Tpq + {}^1\!/_2 SH_e \\ &= \frac{4pq(1 - S)}{2 - S} \end{aligned}$$

We can take the same approach for the homozygote A_1A_1, and thus, if we substitute the equilibrium heterozygosity from above

$$\begin{aligned} P_e &= Tp^2 + S(P_e + {}^1\!/_4 H_e) \\ &= Tp^2 + SP_e + {}^1\!/_4 S\left(\frac{4pq(1-S)}{2-S}\right) \\ &= p^2 + \frac{Spq}{2-S} \end{aligned}$$

The same procedure can be used for A_2A_2 so that the inbreeding equilibrium proportions for the three genotypes are

$$\begin{aligned} P_e &= p^2 + \frac{Spq}{2-S} \\ H_e &= \frac{4pq(1-S)}{2-S} \\ Q_e &= q^2 + \frac{Spq}{2-S} \end{aligned} \tag{5.5b}$$

After a number of generations, no matter what the initial genotype proportions, the genotype proportions approach these equilibrium values, as determined by the allele frequencies and the proportion of self-fertilization, indicating that this is a **stable equilibrium**. For example, the genotype frequencies for the example in Figure 5.7 eventually become 0.262, 0.076, and 0.662 for genotypes A_1A_1, A_1A_2, and A_2A_2, respectively, as they would from any starting point where $q_0 = 0.7$ and $S = 0.9$.

To illustrate the effects of the level of self-fertilization S on heterozygosity, Figure 5.8 gives the equilibrium heterozygosity frequency for several allele frequencies and different levels of S. When $S = 0$, there is Hardy–Weinberg heterozygosity, and when $S = 1$, the heterozygosity is zero. Naturally, the greatest effect on reducing the proportion of heterozygotes occurs when S is high.

Recursion relationships for the amount of inbreeding caused by complete selfing and partial selfing can also be used to describe the changes in genotype frequencies. The recursion relationship for complete selfing is

$$f_{t+1} = {}^1\!/_2 + {}^1\!/_2 f_t$$

and the solution for this is

$$f_t = 1 - ({}^1\!/_2)^t(1 - f_0) \tag{5.6a}$$

As t becomes large, the inbreeding coefficient approaches unity as expected.

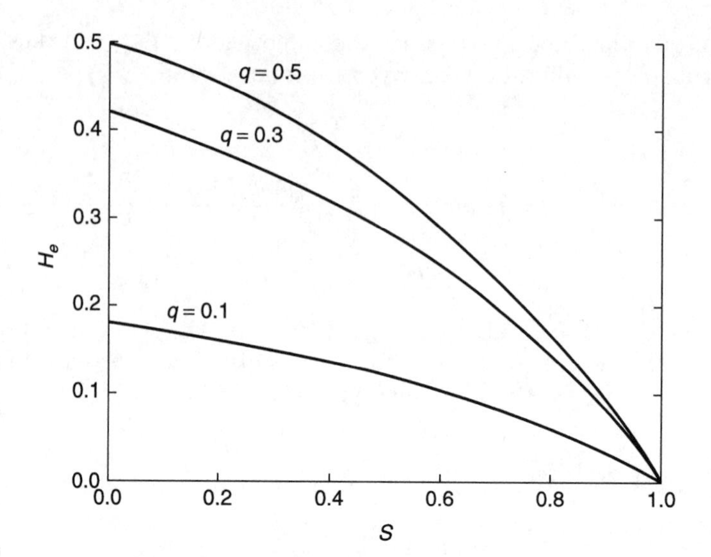

Figure 5.8. The equilibrium heterozygote frequency for different levels of partial selfing and three different allele frequencies.

For the partial-selfing case, the solution is

$$f_t = \frac{S}{2-S}\left[1 - \left(\frac{S}{2}\right)^t\right] + \left(\frac{S}{2}\right)^t f_0 \tag{5.6b}$$

With increasing t, this expression approaches the equilibrium value of

$$f_e = \frac{S}{2-S} \tag{5.6c}$$

or in terms of outcrossing

$$f_e = \frac{1-T}{1+T} \tag{5.6d}$$

c. Estimates of Outcrossing, Self-Fertilization, and Biparental Inbreeding

Two basic approaches have been used to estimate the amount of outcrossing or self-fertilization: an experimental one in which **progeny** arrays from a given mother are examined and individual progeny resulting from outcrosses are identified and an equilibrium one that assumes the observed genotype proportions in the **population** are in inbreeding equilibrium proportions. These approaches should give similar estimates of outcrossing if outcrossing rates have been constant over several generations—that is, if there has been time to reach inbreeding equilibrium proportions and if the effects of selection, gene flow, or genetic drift on genotype frequencies in

TABLE 5.4 The proportion of progeny genotypes expected from (a) a homozygous maternal genotype A_iA_i and (b) a heterozygous maternal genotype A_iA_j when proportions S and T of the matings are the result of self-fertilization and outcrossing, respectively.

Frequency of mating	*Progeny genotypes* A_iA_i	A_iA_j	A_jA_j	*Other (H*)*
(a)				
S	S	—	—	—
T	Tp_i	Tp_j	—	—
(b)				
S	$\frac{1}{4}S$	$\frac{1}{2}S$	$\frac{1}{4}S$	—
T	$\frac{1}{2}Tp_i$	$\frac{1}{2}T(p_i + p_j)$	$\frac{1}{2}Tp_j$	$T(1 - p_i - p_j)$

recent generations are not too great. Because all of these other evolutionary factors can affect genotype frequencies, estimates of the effect of inbreeding in partial-selfing organisms are usually measured using the fixation index, F. However, the statistical expectation of both progeny and population estimates of the fixation index should be equal to the inbreeding coefficient if genotype proportions are solely determined by the mating system.

To estimate the amount of outcrossing in progeny arrays, it is generally necessary to know the genotypes of the maternal parents (the paternal genotype is usually not known) and the progeny. The proportion of progeny types expected from a homozygous maternal parent of genotype A_iA_i are given in Table 5.4a, where p_i and p_j are frequencies of alleles A_i and A_j (or all other alleles besides A_i) in the male gamete (pollen) pool. In this situation, there are two types of progeny from homozygous maternal genotypes. Assuming that the allele frequency in the male gamete pool of A_j is p_j, then the observed proportion of heterozygous progeny is

$$H = Tp_j$$

This expression can be rearranged to give the outcrossing rate as

$$T = \frac{H}{p_j}$$

If there are N_{ij} heterozygous progeny out of N total progeny, the estimate of the outcrossing rate is

$$\hat{T} = \frac{N_{ij}}{p_jN} \tag{5.7a}$$

and $\hat{S} = 1 - \hat{T}$. Using the relationship given in expression 5.6d with the estimate of outcrossing, we can estimate the fixation index F as

$$\hat{F}_{prog} = \frac{1 - \hat{T}}{1 + \hat{T}} \tag{5.7b}$$

where the subscript indicates an estimate from progeny arrays.

When there are highly variable loci, such as microsatellites, then a heterozygous mother becomes more informative, and there are four categories of progeny (Table 5.4b) (Viard *et al.*, 1997a). In this case, a simple estimate of the outcrossing rate when the frequencies of alleles A_i and A_j are low is

$$\hat{T} = \frac{H^*}{1 - p_i - p_j} \tag{5.7c}$$

where H^* includes all heterozygous progeny except A_iA_j. For the complete estimation procedure with multiple alleles and multiple loci, see Ritland (2002) (see program MLTR at http:/www.genetics.forestry.ubc.ca/ritland/programs).

If it is assumed that a population is at inbreeding equilibrium with partial selfing, then a population equilibrium approach can be used to estimate the fixation index. This value can be estimated by using the relationship in expression 5.2c along with the observed (H_O) and expected Hardy–Weinberg (H_E) heterozygosities so that

$$\hat{F}_{pop} = 1 - \frac{\hat{H}_O}{\hat{H}_E} \tag{5.8a}$$

Equating this expression to that in equation 5.6c, then

$$\hat{S} = \frac{2(\hat{H}_E - \hat{H}_O)}{2\hat{H}_E - \hat{H}_O} \tag{5.8b}$$

The progeny and population approaches have been used to estimate the amount of self-fertilization in a number of species. Often for predominately self-fertilizing species, the population equilibrium estimate of F is lower than the estimate from progeny arrays (as in wild oats in Example 5.4). In other words, it seems that inbreeders have more heterozygotes in populations than expected from an outcrossing estimate from progeny arrays (see Brown, 1979, for possible reasons for this effect).

Example 5.4. Wild oats were introduced into California from the Mediterranean in the 1700s but are now known because they give the color to the "golden hills of California." Jain and Marshall (1967) surveyed a number of California populations of two species of wild oats, *Avena fatua* and *A. barbata*, for visible genetic polymorphisms. Virtually all populations

TABLE 5.5 The genotype frequencies and estimates of the frequencies of the recessive allele and the population estimate of the fixation index for two loci at three sites in the wild oat, *A. fatua* (Jain and Marshall, 1967).

	Site 1		*Site 2*		*Site 3*	
	Genotype frequencies		*Genotype frequencies*		*Genotype frequencies*	
BB	0.712	$\hat{q} = 0.219$	0.548	$\hat{q} = 0.417$	0.667	$\hat{q} = 0.303$
Bb	0.138	$\hat{F}_{pop} = 0.597$	0.071	$\hat{F}_{pop} = 0.854$	0.060	$\hat{F}_{pop} = 0.858$
bb	0.150		0.381		0.273	
$LsLs$	0.775	$\hat{q} = 0.162$	0.571	$\hat{q} = 0.395$	0.291	$\hat{q} = 0.677$
$Lsls$	0.125	$\hat{F}_{pop} = 0.539$	0.071	$\hat{F}_{pop} = 0.851$	0.064	$\hat{F}_{pop} = 0.853$
$lsls$	0.100		0.358		0.645	

of *A. barbata* were monomorphic, whereas the *A. fatua* populations were highly polymorphic. An example of their data is given in Table 5.5 for two loci, one that determines black (BB or Bb) or grey (bb) lemma color and the other pubescent ($LsLs$ or $Lsls$) or nonpubescent ($lsls$) leaf sheaf. Heterozygotes were determined by progeny tests. The three sites given in Table 5.5 were chosen because there were predominantly *A. fatua* (site 1), both species mixed (site 2), and predominantly *A. barbata* (site 3). The estimates of the frequencies of the recessive alleles and the fixation index varied among sites but were fairly consistent over loci within a site. However, progeny estimates of outcrossing for *A. fatua* are between 0.02 and 0.05, which would result in an F_{prog} value of 0.9 to 0.96, much higher than the population fixation index observed at site 1. It would appear that other factors besides the mating system are affecting the genotype frequencies in the nearly pure stand of *A. fatua* (see Brown, 1979).

The advent of highly polymorphic markers in many species has made it possible to estimate the mating type for given individuals. For example, Vogl *et al.* (2002), used microsatellite loci to estimate individual inbreeding coefficients in five populations of Monterey pine, *Pinus radiata*, collected in the wild but grown in a common garden (Table 5.6). Although the mean estimated inbreeding coefficient is low in all of the populations, it is somewhat higher in the two island populations. In all of the populations, the estimates of inbreeding are bimodal, with most individual estimates close to zero, as expected from outcrossing, and a few individuals per population with estimates near 0.5, the level expected from self-fertilization.

Within a population, the proportion of self-fertilization may vary because of a number of genetic or environmental factors (Clegg, 1980). For example, the estimated level of outcrossing may vary as the result of genetic differences among families or even among different loci, indicating possible

TABLE 5.6 Estimated individual inbreeding coefficients in five populations of Monterey pine (from Vogl *et al.*, 2002), where the number of individuals is given in parentheses (when there is more than one) and individual inbreeding coefficients over 0.25 are in boldface. At the bottom are given the mean inbreeding coefficients and the estimated outcrossing (proportions of individuals with inbreeding less than 0.25) for the five populations.

	Population				
	Ano Neuvo (34)	*Monterey (30)*	*Cambia (30)*	*Guadalupe (31)*	*Cedros (31)*
Individual	**0.42**	**0.71**	**0.27**	**0.57**	**0.63**
	0.38	0.13	0.18	**0.55**	**0.60**
	0.34	0.08	0.06 (2)	**0.47**	**0.57**
	0.06 (2)	0.04	0.03 (3)	**0.44**	**0.49**
	0.05	0.03	0.02 (3)	**0.27**	**0.36**
	0.02	0.02 (2)	0.01 (11)	0.12	**0.33**
	0.03 (2)	0.01 (8)	0.00 (9)	0.05	**0.26**
	0.01 (8)	0.00 (15)		0.03 (3)	0.24
	0.00 (17)			0.02 (7)	0.11
				0.01 (7)	0.06
				0.00 (7)	0.05
					0.03 (3)
					0.02 (3)
					0.01 (6)
					0.00 (8)
Mean	0.04	0.04	0.03	0.09	0.13
T	0.91	0.97	0.97	0.84	0.77

selective differences, in the same population. In addition, ecological factors such as the density of the population, pollinator activity, and humidity can also affect outcrossing rates. Example 5.5 presents data from a study that examined the extent of such environmentally induced variation in selfing over different times and habitats in a highly selfing snail.

Example 5.5. Most highly selfing organisms are plants. However, a number of snails and slugs have been shown also to be highly selfing. Viard *et al.* (1997b) estimated the amount of selfing in the tropical freshwater snail, *Bulinus truncatus*, from Niger by using four highly polymorphic microsatellite loci at different sampling times and in different habitats. Table 5.7 gives the average observed and expected heterozygosities in five populations with the largest sample sizes from three types of habitats over a period of a year. Because the generation length is quite short, the samples from 2/94 and 1/95 are separated by four to five generations and the samples from 1/95 and 2/95 by about one generation.

Over all the sample times and all the populations, the average observed heterozygosity is quite low, 0.06, and the average expected heterozygosity assuming Hardy–Weinberg proportions is 0.43. Using expression 5.8b to estimate $\hat{S}$, the overall estimate of the selfing rate is 0.93, and does not appear to vary with either sampling time or type of habitat. The similarity of selfing rate over different habitats is apparently related to high movement between habitats.

TABLE 5.7 The observed and expected heterozygosites for four microsatellite loci and the estimated selfing rates for five populations of the selfing snail *Bulinus truncatus* (Viard *et al.*, 1997b). The populations were sampled at three different times and were from permanent (P), semipermanent (SP), or temporary (T) habitats.

		Sampling Time									
		2/94			*1/95*			*2/95*			
Population	*Type of habitat*	H_O	H_E	$\hat{S}$	H_O	H_E	$\hat{S}$	H_O	H_E	$\hat{S}$	$\bar{\hat{S}}$
Boyze I	P	0.08	0.24	0.80	0.03	0.21	0.92	0.04	0.22	0.91	0.88
Tera R	P	0.01	0.40	0.98	0.02	0.55	0.98	0.09	0.64	0.92	0.96
Namanga W	SP	0.01	0.67	0.99	0.09	0.50	0.91	0.01	0.36	0.98	0.96
Mari Sud	SP	0.18	0.65	0.84	0.14	0.66	0.88	0.14	0.74	0.90	0.87
Bala	T	0.00	0.21	1.00	0.01	0.21	0.98	0.01	0.20	0.92	0.97
Mean		0.06	0.43	0.93	0.06	0.43	0.93	0.06	0.43	0.93	0.93

In addition to these genetic and environmental factors, there is also statistical variation in estimates even if there is no true variation of selfing among plants. Ritland and Ganders (1985) examined among-plant (family) variation of outcrossing rates in the *Bidens menziesii*, an endemic plant from Hawaii. Data from two of these populations are given in Figure 5.9, where the histograms indicate the actual estimates and the closed circles give the distribution expected by chance. First, the two populations greatly differ in outcrossing rates, with Puu Kooke having very little outcrossing and a majority of Akumoa plants showing mostly outcrossing. Second, the observed variance over plants is not significantly greater than that expected for Puu Kooke, whereas in the Akumoa sample, there is more observed variation in the estimates than expected. Ritland and Ganders suggest that this additional observed variation over that expected is the result of spatial variation in allele frequencies and selfing rates.

Estimation of the rate of selfing using the above approaches may be influenced by other types of inbreeding that also increase the observed level of homozygosity and make the rate of selfing appear higher than it really is. In one approach to deal with this phenomenon, Ritland (1984) derived estimates of "effective selfing," which included, as a group, different degrees of inbreeding. However, it is often useful to separate uniparental inbreeding (selfing) and **biparental inbreeding** in which an offspring has two different but related parents. One approach to estimate these different effects

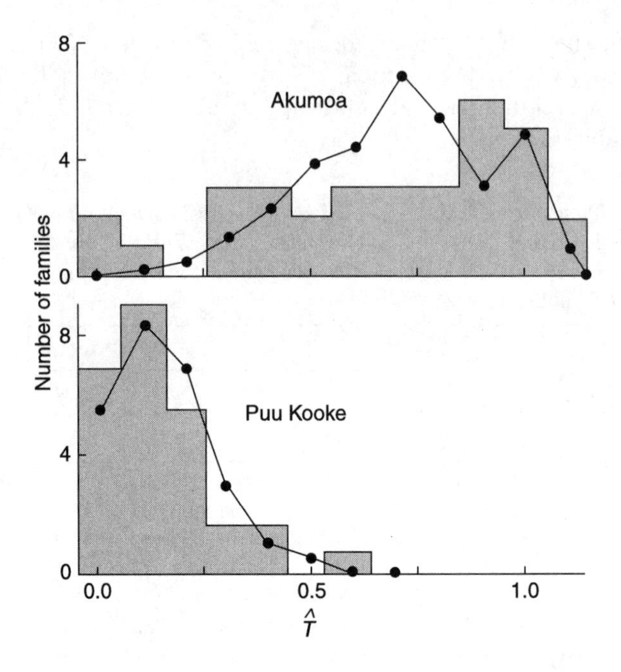

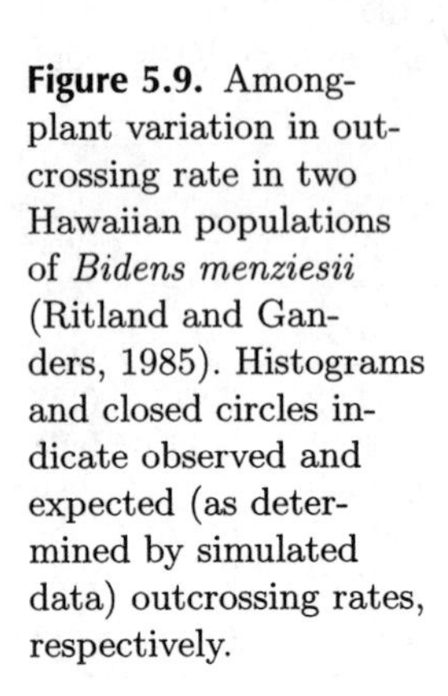
Figure 5.9. Among-plant variation in outcrossing rate in two Hawaiian populations of *Bidens menziesii* (Ritland and Ganders, 1985). Histograms and closed circles indicate observed and expected (as determined by simulated data) outcrossing rates, respectively.

is to compare a multilocus estimate of outcrossing, T_m, with the mean of the single locus estimates, $\overline{T}_s$, as the difference in these values $T_m - \overline{T}_s$. Given that there is biparental inbreeding, this difference should be positive because single-locus estimates include real selfing and all apparent selfing caused by biparental inbreeding, whereas multilocus estimates exclude much of the apparent selfing (Ritland, 2002). This difference occurs because in a given progeny an observed outcross at any locus categorizes the individual the result of outcrossing and overrides apparent selfing at any other loci. If there are a number of highly variable loci, then virtually all nonselfing events could potentially be identified, and $1 - T_m$ should approach the real selfing rate (see Example 5.6 for an application with highly variable loci and that suggests that there is no selfing, but high biparental inbreeding, in an endangered tropical tree species).

Example 5.6. The endangered tree species, *Caryocar brasiliense*, is found in the savanna-like Brazilian cerrado. Using 10 highly polymorphic, microsatellite loci, Collevatti *et al.* (2001a) examined progeny arrays from given trees from four different populations. The mean single-locus estimates of outcrossing were all significantly less than one for each population and averaged 0.80 over the four different populations (Table 5.8). On the other hand, all of the progeny were identified as outcrosses from the multilocus estimator, and the estimate of T_m was 1.0 for all four populations. That is, at least one of the 10 loci identified each progeny as an outcross, demonstrating the high resolution of highly variable loci to identify nonselfed progeny. Because of other evidence of inbreeding depression from selective abortion of seeds with fruits, they suggested that any selfed progeny are selected against.

The large difference $T_m - \overline{T}_s$ suggests that there is extensive biparental inbreeding in this species. *C. brasiliense* is highly genetically structured (Collevatti *et al.*, 2001b), the pollinators are small, territorial bats that have a low flight range (Gribel and Hay, 1993), and there is limited seed dispersal, all factors that are potentially consistent with high biparental inbreeding. To give perspective to this estimate, assume that all of the biparental inbreeding is the result of full-sib mating. In this situation, then 40% of the progeny would be expected to be the result of full-sib mating (Hedrick, 1987b), an extraordinary amount of mating between relatives.

TABLE 5.8 Estimates of multilocus (T_m) and mean single locus ($\overline{T}_s$) outcrossing rates using 10 microsatellite loci in four Brazilian populations of the endangered tree *C. brasiliense* (from Collevatti *et al.*, 2001a).

Population	T_m	$\overline{T}_s$	$T_m - \overline{T}_s$
Caldas Novas	1.00	0.81	0.19
Agua Limpa	1.00	0.87	0.13
Brasilia	1.00	0.77	0.23
Uruacu	1.00	0.77	0.23
Mean	1.00	0.80	0.20

In order for biparental inbreeding to occur, then relatives need to either be nearby or somehow available to reproduce inbred offspring. In populations that show strong genetic substructure, that is, similar genotypes are near each other, then it is assumed that many of these neighboring plants are relatives. To determine the extent of biparental inbreeding in the yellow monkey flowers, *Mimulus guttatus*, Kelly and Willis (2002), transplanted plants between populations and then compared genetic variation in the progeny of transplanted and nontransplanted plants. $T_m - \overline{T}_s$ was not different from zero between these groups, and they suggested that any inbreeding observed in these populations appears to be the result of selfing. Furthermore, data from Sweigart *et al.* (1999) suggested that neighboring plants in one of these populations do not appear to be related, one of the important factors for the occurrence of biparental inbreeding.

d. Regular Systems of Inbreeding

As we have seen, complete self-fertilization reduces heterozygosity very quickly over time. In organisms in which self-fertilization is not possible, the most extreme form of inbreeding is between full-sibs (brother–sister) mating (see Figure 5.10a for a continuous full-sib mating pedigree). To establish

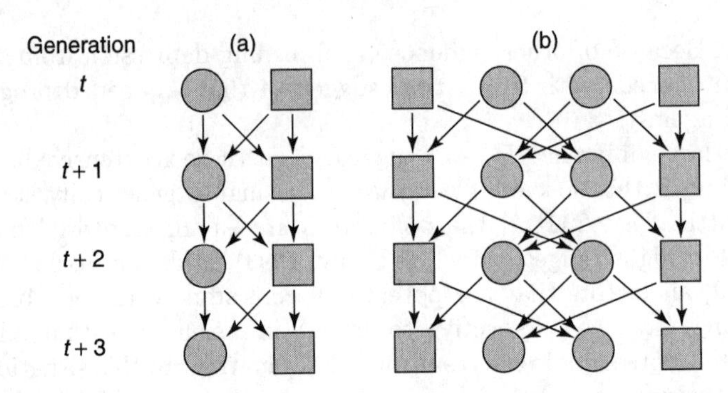

Figure 5.10. Pedigrees showing the regular mating systems of (a) full-sib mating and (b) double first-cousin mating. Circles and squares indicate females and males, respectively, and arrows indicate a parent-to-offspring relationship. The individuals in generation t are assumed to be unrelated.

highly homozygous lines of animals for testing the efficacy of pharmaceuticals and other chemicals, lines of mice and rats have been maintained by full-sib mating for many generations. Some lines of *D. melanogaster* have been full-sib mated for hundreds of generations, making the theoretical likelihood of heterozygosity at any loci quite small.

Continuous **full-sib mating** reduces heterozygosity rather quickly but not as quickly as complete self-fertilization. The effect on heterozygosity from continuous full-sib mating can be calculated by the relationship

$$H_{t+2} = \tfrac{1}{2}H_{t+1} + \tfrac{1}{4}H_t$$

where H_t is the heterozygosity in the tth generation (Crow and Kimura, 1970; Li, 1976). Such continuous full-sib mating is in fact the same as mating within a closed population with two individuals in which self-fertilization is not permitted. Assuming that the original heterozygosity H_0 is 1 (no alleles are IBD), then $H_1 = 1$, $H_2 = 0.75$, $H_3 = 0.625$, $H_4 = 0.5$, $H_5 = 0.406$, and so on. These frequencies are the ratios 1/1, 2/2, 3/4, 5/8, 8/16, 13/32, and so on. In these ratios, the denominator doubles each generation, and the numerator is given by the Fibonacci sequence; each number is the sum of the two preceding numbers (Crow, 1986).

The recursion relationship for the inbreeding coefficient for continuous full-sib mating is

$$f_{t+2} = \tfrac{1}{4} + \tfrac{1}{2}f_{t+1} + \tfrac{1}{4}f_t$$

where f_t is the inbreeding coefficient in the tth generation. The inbreeding coefficients are the complements of the heterozygosities in a given generation; that is, $f_t = 1 - H_t$. Later, we demonstrate how to calculate the inbreeding coefficient for this case from the pedigree.

Another regular system of mating is termed continuous **double first-cousin mating** (see the pedigree in Figure 5.10b). In this scheme, individuals are mated that are first cousins in two ways. In other words, they share in common all four grandparents but are not siblings. The effect on heterozygosity from this mating scheme is given by the relationship

$$H_{t+3} = {}^1\!/_2 H_{t+2} + {}^1\!/_4 H_{t+1} + {}^1\!/_8 H_t$$

This mating scheme is identical to having a population with four individuals with both sib mating and self-fertilization excluded. The recursion relationship for the inbreeding coefficient is

$$f_{t+3} = {}^1\!/_8 + {}^1\!/_2 f_{t+2} + {}^1\!/_4 f_{t+1} + {}^1\!/_8 f_t$$

These and other regular systems of mating are thoroughly discussed by Crow and Kimura (1970) and Li (1976).

A comparison of the rate of decay of heterozygosity with $H_0 = 1$ for complete selfing, full-sib mating, and double first-cousin matings is given in Figure 5.11. Because it is assumed that the initial parents are unrelated, there are one- and two-generation delays for the decline in heterozygosity for full-sib and double first-cousin matings, respectively. For all of these systems of mating, the heterozygosity approaches zero over time, although at quite different rates. Remember that the heterozygosity for the complete-selfing model is halved each generation. The heterozygosity for full-sib mating declines more slowly and that for double first-cousin mating even more slowly.

In order to calculate the asymptotic rate of decline of heterozygosity, let us assume that the relationship between heterozygosity in different generations can be written as

$$\lambda = \frac{H_{t+1}}{H_t} \tag{5.9}$$

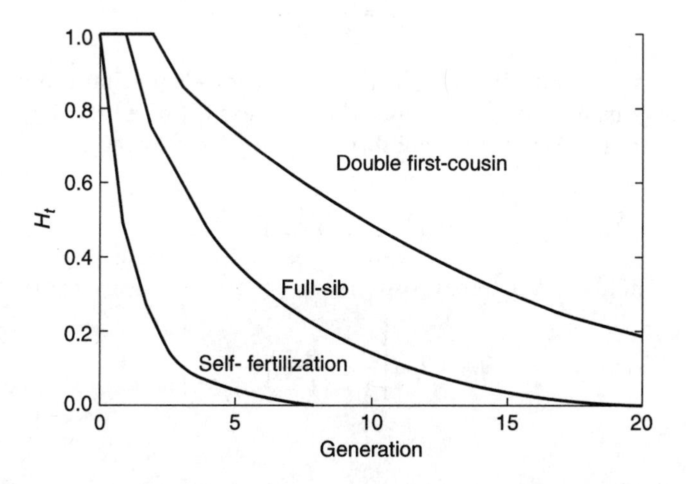

Figure 5.11. The loss in heterozygosity over time for three regular mating systems (after Crow and Kimura, 1970).

where indicates the characteristic rate of decline of heterozygosity or the proportion of heterozygosity remaining after one generation. For self-fertilization, $\lambda = 0.5$ because only 50% of the heterozygosity is retained each generation. For the case of continuous full-sib mating, this relationship becomes

$$\lambda = \frac{1}{2} + \frac{1}{4\lambda}$$

or the quadratic

$$\lambda^2 - \frac{\lambda}{2} - \frac{1}{4} = 0$$

The positive root of this expression is $\lambda = 0.809$, which is the proportion of heterozygosity retained each generation by continuous full-sib mating. For continuous double first-cousin matings, the asymptotic value of λ is 0.920 each generation. These rates of loss of heterozygosity are slower than for populations of sizes 2 and 4, respectively (Chapter 6), because selfing and selfing and sib mating, respectively, are not permitted.

e. Other Levels of Partial Inbreeding

Natural populations of organisms that have individuals of different sexes so that self-fertilization is precluded may have a mixture of matings between close relatives, biparental inbreeding, and unrelated individuals. For example, there may be a certain proportion of matings between full sibs and the remainder between unrelated individuals. By analogy to the partial-selfing situation examined previously, we would expect an increase in homozygosity to an inbreeding equilibrium level determined by the proportion of full-sib mating. In this case, the equilibrium level of inbreeding is

$$f_e = \frac{S_2}{4 - 3S_2}$$

where S_2 is the proportion of full-sib mating (first-degree inbreeding in non-selfing organisms). In general, for other degrees (j) of inbreeding (Hedrick and Cockerham, 1986), the equilibrium level of inbreeding is

$$f_e = \frac{S_j}{2^j - (2^j - 1)S_j} \tag{5.10a}$$

and the equilibrium heterozygosity is

$$H_e = 2pq\left[\frac{1 - S_j}{1 - (1 - 1/2^j)S_j}\right] \tag{5.10b}$$

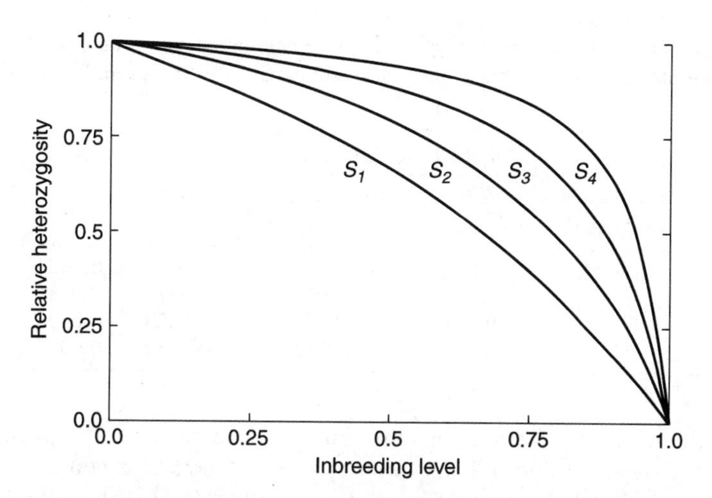

Figure 5.12. The equilibrium heterozygosity relative to that found in a random-mating population for four different levels of partial inbreeding (S_1, partial selfing; S_2, partial full-sib mating; S_3, partial half-sib mating; and S_4, partial first-cousin mating).

Figure 5.12 gives the equilibrium heterozygosity, relative to that in a random mating population, for different levels of partial inbreeding. Obviously, the closer the type of mating and the higher the proportion of inbreeding, the greater the reduction in heterozygosity. For example, 0.25 selfing reduces the heterozygosity over 16%, whereas 0.25 first-cousin mating reduces the heterozygosity only 2%. If a given reduction in heterozygosity is observed, compared with Hardy–Weinberg proportions, then the proportion of different types of mating that may potentially explain the deficiency could be calculated.

The most common types of matings between relatives in humans are uncle-niece ($f = 1/8$), first cousin ($f = 1/16$), first cousin once removed ($f = 1/32$), and second cousin ($f = 1/64$). In a population with different types of between-relative matings, the average level of inbreeding in the population can be calculated by

$$\overline{f} = \sum x_j f_j \tag{5.11a}$$

where x_j is the proportion of the population that is the result of a particular mating type with an inbreeding coefficient of f_j. The average inbreeding equilibrium can be calculated for the same data set as

$$f_e = \frac{\sum x_j/2^j}{1 - \sum x_j(1 - {}^1\!/_{2^j})} \tag{5.11b}$$

where x_1 is the proportion of selfing, x_2 is the proportion of first-degree matings (full-sib and parent-offspring matings), and so on (Hedrick and

TABLE 5.9 The proportion of different matings, the average inbreeding coefficient, and the equilibrium inbreeding in samples from four countries with relatively high inbreeding (Cavalli-Sforza and Bodmer, 1971; Hedrick, 1986a).

		Proportion of marriages					
Country	*Number of marriages*	*Uncle–niece Aunt–nephew*	*First cousin*	*First cousin once removed*	*Second cousin*	$\overline{f}$	f_e
India	6,945	0.0923	0.3330	—	—	0.0324	0.0533
Guinea	739	—	0.1908	0.0054	0.0622	0.0131	0.0173
Japan	152,790	—	0.0615	0.0133	0.0228	0.0046	0.0051
Brazil	212,090	0.0006	0.0263	0.0081	0.0132	0.0022	0.0023

Cockerham, 1986). Data from four samples with relatively high inbreeding are given in Table 5.9. Generally, the most common consanguineous marriage is that between first cousins; in the Indian sample, for example, one-third of the marriages were between first cousins. In this sample, the average inbreeding coefficient was the highest of all of these populations at 0.0324, and the level of inbreeding equilibrium was even higher at 0.0533 (Hedrick, 1986a).

f. Estimation of Inbreeding from Pedigrees

As we discussed above, the inbreeding coefficient (f) is known as the probability of **identity by descent (IBD)**, which implies that homozygosity is the result of the two alleles in an individual descending from the same ancestral allele. The probability of IBD varies depending on the relationship of the parents of the individual being examined. For example, if the parents are unrelated, then it is assumed that there is no possibility that the individual will be homozygous by descent. The other extreme (given sexual reproduction and ignoring intragametophytic selfing) is when the two parents are the same individual and self-fertilization occurs. In this case, the probability of an offspring having alleles IBD is 0.5. To illustrate, assume that the parent is heterozygous and has the genotype A_1A_2. Therefore, half the progeny, A_1A_1 and A_2A_2, will have alleles IBD. We can obtain the inbreeding coefficient f from a pedigree in which there is a mating between relatives by calculating the probability of IBD. Below, an example of this is given in probability terms, and then a general approach, the chain-counting technique, is discussed.

Figure 5.13a gives a pedigree in which two half-first cousins (individuals having one grandparent in common), X and Y, mate to produce an inbred offspring, Z. All unrelated individuals are omitted from the pedigree because they do not contribute to the inbreeding coefficient. The parents of Z have a **common ancestor** (indicated as **CA**—in this case, either a grandmother or a grandfather) that we assume has genotype A_1A_2. In

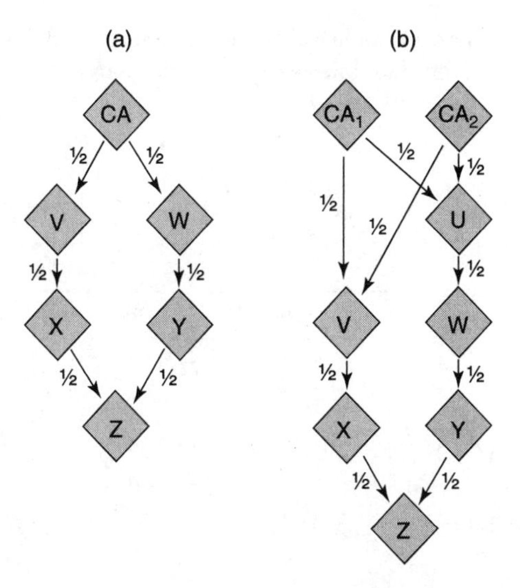

Figure 5.13. Pedigrees for (a) a half-first-cousin mating and (b) a mating between first cousins once removed. Diamonds indicate individuals of either sex, CA the common ancestors, and 1/2 the probability of segregation between generations.

order to calculate the inbreeding coefficient, we must know the probability of Z being either A_1A_1 or A_2A_2—that is, having two alleles IBD. The first alternative, Z being A_1A_1, can happen only if X contributes to Z a gamete containing A_1 and Y also contributes to Z a gamete containing A_1. The probability of this occurring is equal to the probability of A_1 being passed on for three generations from both parents or

$$\Pr(A_1 \text{ from X}) = (\tfrac{1}{2})(\tfrac{1}{2})(\tfrac{1}{2}) = \tfrac{1}{8}$$
$$\Pr(A_1 \text{ from Y}) = (\tfrac{1}{2})(\tfrac{1}{2})(\tfrac{1}{2}) = \tfrac{1}{8}$$

where the 1/2 values are the probabilities of segregation between generations—for example, the probability that A_1 will be passed from CA to V, from V to X, or from X to Z. Because A_1 must come from both parents to Z, the probability of identity of A_1 by descent is the product of the individual probabilities

$$\Pr(A_1A_1 \text{ in Z}) = [\Pr(A_1 \text{ from X})][\Pr(A_1 \text{ from Y})] = \tfrac{1}{64}$$

Likewise, the probability of IBD for A_2 is

$$\Pr(A_2A_2 \text{ in Z}) = [\Pr(A_2 \text{ from X})][\Pr(A_2 \text{ from Y})] = \tfrac{1}{64}$$

The overall probability of IBD (the inbreeding coefficient) in individual Z is then

$$f = \Pr(A_1A_1) + \Pr(A_2A_2)$$
$$= \tfrac{1}{64} + \tfrac{1}{64} = \tfrac{1}{32}$$

TABLE 5.10 The genotypes that are identical by descent generated from two generations of full-sib mating where all alleles in the initial parents (generation 0) are different.

Generation 0		$A_1A_2 \times A_3A_4$			
		↓			
Generation 1		$\frac{1}{4}A_1A_3$, $\frac{1}{4}A_1A_4$, $\frac{1}{4}A_2A_3$, $\frac{1}{4}A_2A_4$			
		↓			
Generation 2		*Males*			
		$\frac{1}{4}A_1A_3$	$\frac{1}{4}A_1A_4$	$\frac{1}{4}A_2A_3$	$\frac{1}{4}A_2A_4$
	$\frac{1}{4}A_1A_3$	$\frac{1}{64}A_1A_1$ $\frac{1}{64}A_3A_3$	$\frac{1}{64}A_1A_1$	$\frac{1}{64}A_3A_3$	—
	$\frac{1}{4}A_1A_4$	$\frac{1}{64}A_1A_1$	$\frac{1}{64}A_1A_1$ $\frac{1}{64}A_4A_4$	—	$\frac{1}{64}A_4A_4$
Females	$\frac{1}{4}A_2A_3$	$\frac{1}{64}A_3A_3$	—	$\frac{1}{64}A_2A_2$ $\frac{1}{64}A_3A_3$	$\frac{1}{64}A_2A_2$
	$\frac{1}{4}A_2A_4$	—	$\frac{1}{64}A_4A_4$	$\frac{1}{64}A_2A_2$	$\frac{1}{64}A_2A_2$ $\frac{1}{64}A_4A_4$

We can use the same approach to calculate the inbreeding coefficient for full-sib mating. Assume that each allele in the initial parents can be identified. Therefore, the parental genotypes can be represented as genotypes A_1A_2 and A_3A_4, as shown in the top line of Table 5.10. Of course, because the initial parents are assumed to be heterozygous (having alleles not IBD), their inbreeding coefficients are 0.0. Likewise, their offspring are all heterozygotes (not IBD), and thus, the inbreeding coefficient in the first generation is also 0.0. However, in the second generation, there is a probability that the alleles in these individuals are IBD. This can be shown in the four-by-four mating scheme in the bottom of Table 5.10. For example, the probability that an individual will be IBD and be A_1A_1 is $1/16$ (summing from the four upper left matings). Likewise, the probabilities of IBD for A_2A_2, A_3A_3, and A_4A_4 are all each $1/16$. Therefore, the overall probability of IBD, the inbreeding coefficient for one generation of full-sib mating, is the sum of these probabilities or $1/4$.

g. The Chain-Counting Technique

Enumerating all of the possible genotypes, as we did for a full-sib mating above, to calculate the inbreeding coefficient becomes very complicated for more complex pedigrees. A straightforward alternative approach often used to calculate the inbreeding coefficient is the chain-counting technique. A **chain** for a given common ancestor ***starts with one parent of the inbred individual, goes up to the common ancestor, and comes***

back down to the other parent (the inbred individual is not in the chain). For the example in Figure 5.13a, the chain would be X-V-**CA**-W-Y. The number of individuals in the chain, N, can be used in the following formula to calculate the expected inbreeding coefficient due to the presence of this common ancestor so that

$$f = (1/2)^N \tag{5.12a}$$

In Figure 5.13a, the chain for the pedigree includes five individuals, and thus, $f = 1/32$, corresponding to our previous result.

If there is more than one common ancestor, then the inbreeding coefficient is the sum of the f values for the different chains or

$$f = \sum_{i=1}^{m} (1/2)^{N_i} \tag{5.12b}$$

where m is the number of common ancestors and N_i is the number of individuals in chain i. This expression can be used to calculate f in the mating between first cousins once removed, X and Y in Figure 5.13b (**once removed** indicates that there is an extra generation in one lineage compared with a normal first-cousin mating). In this case, there are two common ancestors, $\mathbf{CA_1}$ and $\mathbf{CA_2}$, which have the six-link chains X-V-$\mathbf{CA_1}$-U-W-Y and X-V-$\mathbf{CA_2}$-U-W-Y, respectively, so that $f = (\frac{1}{2})^6 + (\frac{1}{2})^6 = \frac{1}{32}$.

When the common ancestor itself is inbred, then the inbreeding coefficient can be calculated from

$$f = \sum_{i=1}^{m} (1/2)^{N_i} \left(1 + f_{CA(i)}\right) \tag{5.12c}$$

where $f_{CA(i)}$ is the inbreeding coefficient of the ith common ancestor (see Example 5.7 for a pedigree with one inbred common ancestor). In calculating inbreeding coefficients from complicated pedigrees, it is important to find all of the chains for a given common ancestor. To help check that all chains are valid, make sure that a given individual appears only once in a chain and that the chain is always descending from the common ancestor to the parents. For complicated pedigrees, the approach outlined by Ballou (1983) is a useful one, and a computer program that systematically determines all relationships is often necessary (Lacy and Ballou, 2001; Population Management 2000 or PM2000 at http://www.vortex9.org/pm2000.html).

Example 5.7. In the breeding of dogs, horses, and other animals, there are often complicated patterns of inbreeding. Captive populations of endangered species also often have complicated pedigrees (Hedrick *et al.*, 1997).

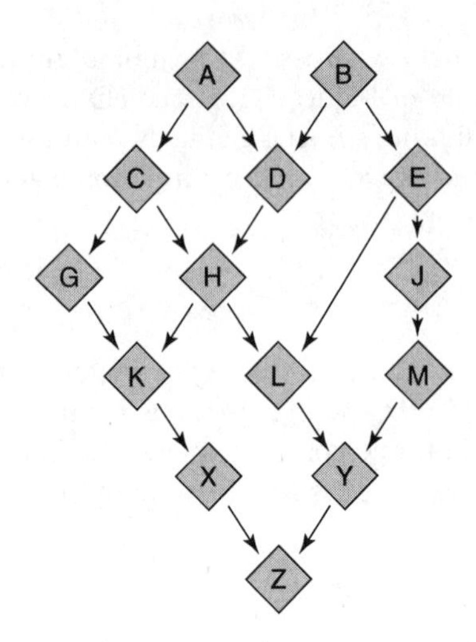

Figure 5.14. A hypothetical pedigree with a complicated history of inbreeding.

For an example of a pedigree from a natural population of great tits, see van Noordwijk and Scharloo (1981). Figure 5.14 gives such a hypothetical pedigree with several common ancestors, one of which is inbred. Table 5.11 lists the common ancestors and the chains that connect the two parents. Two particular complications occur in this pedigree. When B is the common ancestor, there are two ways in which inbreeding can occur. Alleles can be passed from B to parent Y either through E and L or through E, J, and M. As a result, B must be considered a common ancestor twice, as shown in Table 5.11. Second, all the common ancestors are assumed to have inbreeding coefficients of 0.0 except the common ancestor H, which is inbred because its parents C and D are half-sibs. To calculate the inbreeding coefficient of H, the number of individuals in this chain can be counted and is three (C-**A**-D), making the inbreeding coefficient of H $(\frac{1}{2})^3 = 0.125$. The contribution of each of these common ancestors to the inbreeding coefficient

TABLE 5.11 The common ancestors and their chains, inbreeding coefficients, and contribution to the inbreeding coefficient of Z in Figure 5.14.

Common ancestor	*Chain*	N_i	$f_{\text{CA}(i)}$	*Contribution to f*
A	X-K-G-C-**A**-D-H-L-Y	9	0.0	$(\frac{1}{2})^9 = 0.0020$
B	X-K-H-D-**B**-E-J-M-Y	9	0.0	$(\frac{1}{2})^9 = 0.0020$
B	X-K-H-D-**B**-E-L-Y	8	0.0	$(\frac{1}{2})^8 = 0.0039$
C	X-K-G-**C**-H-L-Y	7	0.0	$(\frac{1}{2})^7 = 0.0078$
H	X-K-**H**-L-Y	5	0.125	$(\frac{1}{2})^5(1.125) = 0.0352$
				$f = 0.0509$

is given in the last column of Table 5.11. By summing all of these, we find that the overall inbreeding coefficient for Z is 0.0509.

In managing for genetic conservation of captive populations of endangered species, the main goals include maintenance of genetic diversity, avoidance of inbreeding, and having representation of the founder individuals as nearly equal as possible (Ballou *et al.*, 1995). One approach that generally accomplishes these overlapping goals is to minimize the mean kinship coefficient in the population. The **kinship coefficient** between two individuals is defined as ***the probability that alleles drawn at random from them are IBD*** (Falconer and Mackay, 1996) and is ***equal to the inbreeding coefficient of an offspring from them.*** **Genetically important individuals** in the pedigreed population can be determined by finding what individuals have the lowest mean kinship, that is, the average kinship coefficient between a given individual and all living individuals in the population (Ballou and Lacy, 1995). These individuals can then be mated with other genetically important individuals to which they are not closely related (for a program to carry out these calculations, see Lacy and Ballou, 2001). The Monte Carlo simulation approach called **gene dropping** (MacCluer *et al.*, 1986), in which numbered alleles are randomly passed from parents to offspring, allows versatility in examining these and other related pedigree questions.

There does not appear to be good direct evidence for significant levels of inbreeding in most natural vertebrate populations (however, see the study of white-toothed shrews on p. 276 and a study of a self-fertilizing fish by Turner *et al.*, 1992). Ralls *et al.* (1986) reviewed the extent of inbreeding in birds and mammals and found few cases of mating much different from that expected at random. The highest level of inbreeding that they found was in the splendid fairy wren from Australia, but a follow-up study determining paternity using molecular markers (Rowley *et al.*, 1993) found that there actually may be very few matings between close relatives in this population.

In two isolated island bird populations in which virtually all birds have marked, there appears to be some inbreeding. However, random mating in a finite population includes an expectation of chance mating between close relatives. A long-term population study of approximately 50 breeding pairs of great tits, *Parus major*, on the Dutch island of Vlieland documented an inbreeding level of 0.015 to 0.036, based on the different assumptions used in the calculations (van Noordwijk and Scharloo, 1981). This level appears to be slightly higher than that expected from random mating in a finite population of approximately 100 adults. However, the difference is not inconsistent with factors that may influence the effective population size (see Chapter 6), such as the observed variance in family numbers or the variability in population size over time or space.

Similarly, a population of approximately 90 breeding pairs of song sparrows, *Melospiza melodia*, on Mandarte Island, British Columbia, Canada, has been studied since 1975 (Keller and Arcese, 1998). Although there has been some breeding between close relatives in this population ($f \geq 0.125$), most of these matings occurred after a bottleneck in which the population number was reduced to 6% of its previous size. Detailed examination of the pedigree showed that the mean inbreeding coefficient was very similar to that expected in a finite population; that is, the inbreeding was not higher or lower because of inbreeding preference or avoidance than that expected. In addition, the proportion of different classes of inbreeding was not different that that expected by mating at random.

For **X-linked genes** or **genes in haplo-diploid organisms**, the chain-counting technique, with several modifications, can be used to calculate the inbreeding coefficient. First, the inbreeding coefficient is not meaningful in organisms that have only one allele at a given locus. Therefore, it cannot be calculated for a male haplo-diploid or for an X-linked gene in males (in birds and lizards, the females are the heterogametic sex and males the homogametic sex, and thus, the principles are reversed in the sexes). Second, because males receive all of their X-linked genes or all of their genes in haplo-diploids from their mothers, the probability of passing genes from mother to son is unity. As a result, males can be omitted from the chain, and only females need be counted. Finally, if there is a male-to-male link, because the son receives no genes from his father, the chain of inheritance is broken so that the resultant inbreeding coefficient is zero. The consequences of these effects are that the inbreeding coefficient may be high or zero, depending on the arrangement of the sexes in the pedigree, quite unlike that for autosomal genes.

To incorporate these differences, the chain-counting formula for X-linked genes or haplo-diploids is modified to

$$f = (1/2)^{N_f} \tag{5.13a}$$

where N_f is the number of females in the chain. This formula can be given in general as

$$f = \sum_{i=1}^{m} (1/2)^{N_f} (1 + f_{CA}) \tag{5.13b}$$

where the inbreeding coefficient is summed over all chains and inbreeding in common ancestors is taken into account. Of course, if a common ancestor is a male, then his inbreeding coefficient must be zero.

Three examples of pedigrees to illustrate inbreeding for X-linked or haplo-diploid genes are given in Figure 5.15. In the brother–sister mating in Figure 5.15a, there are two chains (X-$\mathbf{CA_1}$-Y and X-$\mathbf{CA_2}$-Y), one for each common ancestor. However, the chain for common ancestor CA_1 is

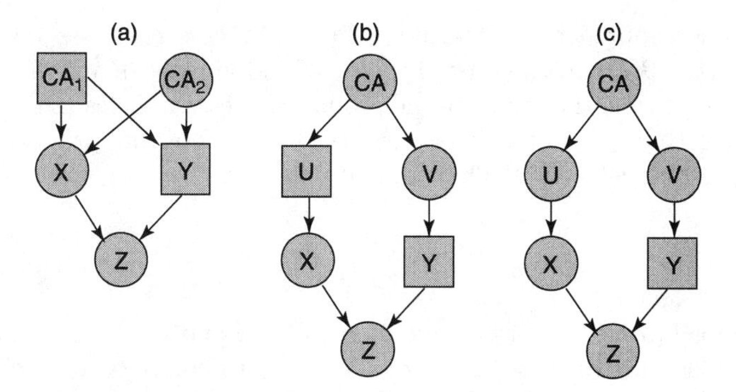

Figure 5.15. Pedigrees to illustrate inbreeding for an X-linked gene or a haplo-diploid organism where squares indicate males and circles females. Pedigree (a) is a brother–sister mating and pedigrees (b) and (c) are two half-first-cousin matings.

broken because of the father–son transmission, CA_1 to Y. The second chain contains two females, and thus, from expression 5.13a, $f = 1/4$, the same value for autosomal genes with a full-sib mating. If Z is a male offspring, it would have an inbreeding coefficient of zero.

Figures 5.15b and 5.15c give examples of half first-cousin matings with the chain X-U-**CA**-V-Y. In Figure 5.15b, there are three females in the chain, and thus, $f = 1/8$; in Figure 5.15c, $f = 1/16$ because there are four females. Of course, there are other half first-cousin pedigrees in which $f = 0$ because of two successive males in the pedigree. There are four different types of first-cousin matings: two with $f = 0$, one with $f = 1/8$, and one with $f = 3/16$. Because of a cultural preference in parts of India for this latter type of mating with $f = 3/16$, where the related parents of the first cousins are female, the inbreeding coefficient for X-linked genes is significantly higher than for autosomal genes (Hedrick and Parker, 1997).

h. Relatedness

It is often useful to know the degree of relatedness of two individuals in a population even though they have not or cannot produce offspring. Wright (1922) suggested that the genetic correlation coefficient or the **coefficient of relationship** (r) be used in such a case; r is now also known as the **relatedness** between two individuals, that is, ***the fraction of alleles that the two individuals share IBD*** (Blouin, 2003). For example, one may want to know the degree of relatedness of a same-sex pair, such as two females or two males. For two diploid individuals that are not inbred, the coefficient of relationship is ***equal to twice the inbreeding coefficient in possible offspring*** (see Li, 1976, for a full discussion)

$$r = 2f \tag{5.14a}$$

For example, the coefficient of relationship for two sisters in a diploid organism that are full sibs from a brother–sister mating would be $1/2$.

Another approach to determine the relatedness is to use **IBD coefficients, k_0, k_1, and k_2**, for the pair of individuals in which k_2 is the proportion of time that the two individuals share two alleles IBD, k_1 is the proportion that they share one allele IBD, and k_0 is the proportion that they share no alleles IBD. The value of r is then

$$r = \frac{k_1}{2} + k_2 \tag{5.14b}$$

because for k_1 only one of two alleles are shared and for k_2 both alleles are shared. The simplest case to illustrate this is to compare identical twins, for example, A_1A_2 and A_1A_2. In this case, $k_0 = 0$, $k_1 = 0$, and $k_2 = 1$, and therefore, $r = 1$. For a parent–offspring pair, where the parent is A_1A_2 and the offspring are A_1- or A_2-, where the dash indicates some other allele, $k_0 = 0$, $k_1 = 1$, and $k_2 = 0$, and $r = 0.5$ (see Table 5.12).

TABLE 5.12 Identity by descent IBD coefficients, k_0, k_1, and k_2, giving the proportion of time that 0, 1, and 2 alleles are shared between the two individuals and the relatedness (r) for several different relationships (after Blouin, 2003).

Relationship	k_0	k_1	k_2	$r = k_1/2 + k_2$
Identical twins	0	0	1	1.0
Parent-offspring	0	1	0	0.5
Full sibs	0.25	0.5	0.25	0.5
Half sibs (or uncle-niece)	0.5	0.5	0	0.25
First cousins	0.75	0.25	0	0.125
Unrelated	1	0	0	0.0

For full sibs, all three IBD coefficients are nonzero. If you look back at Table 5.10, then all of the different possible combinations of full sib parents from initial parents that were A_1A_2 and A_3A_4 are shown. For example, the full-sib pair (A_1A_3, A_1A_3) in the upper left of the table shares two alleles IBD and has a frequency of 1/16 (as do all the 16 combinations). The four diagonal (top left to lower right) combinations of full sibs are the four pairs that share two alleles so that $k_2 = 0.25$. The other four diagonal pairs (top right to lower left) all share no alleles IBD, and thus, $k_0 = 0.25$. The other eight combinations share one allele, and thus, $k_1 = 0.5$, and then $r = 0.5$.

Note that $r = 0.5$ for both parent–offspring and full-sibs but that the distribution of IBD coefficients is different, making it potentially possible to distinguish these two relationships. Molecular data can also be used to distinguish among alternative types of relationships, for example, full-sibs versus half-sibs (Thompson, 1975; Blouin, 2003). However, it is often difficult to distinguish between classes of relationship more distant than siblings and parent–offspring (second and higher degrees of relationship). On p. 622, we discuss how molecular data can be used in paternity analysis.

Interest in the genetic relatedness in the haplo-diploid, **social Hymenoptera** was stimulated by the work of Hamilton (1964, 1972) who pointed out that in a haplo-diploid system, the coefficient of relationship between sisters is 0.75, higher than that of mother–daughter combinations, which is 0.5. From this calculation, it appears that females might more successfully promulgate their alleles by helping their sisters produce offspring rather than by producing female offspring themselves. Hamilton (1964) suggested that the higher relatedness among sisters may be the basis for the development of social behavior in Hymenoptera in which one sister becomes the queen and other sisters become workers (see p. 279).

The basis for these coefficients of relationship is illustrated in Table 5.13, where it is assumed that each parental allele is unique. Because the haploid male parent always contributes the same allele, every female offspring will have allele A_3. In addition, half of the time, female offspring will have the same allele from the female parent and consequently identical genotypes; for instance, both individuals of a sister pair may have genotype A_1A_3, therefore sharing two alleles. The overall coefficient of relationship for sisters is then $r = k_1/2 + k_2 = 0.5/2 + 0.5 = 0.75$, whereas that for the mother–daughter pair is 0.5 because they always share only one allele IBD just as for diploids. These calculations assume that there is only one male parent, a situation that is not true for honeybees and some other hymenoptera. This is significant because as the number of male parents increases, r between sisters approaches 0.5, the r between mother and daughter.

There has been great interest in **estimating the relatedness of pairs of individuals** in natural populations in which there is no pedigree information but there are molecular genotypic data. Of course, in this case, individuals may share alleles that are identical in state but not IBD so

TABLE 5.13 (a) The parental and offspring genotypes, assuming that each parental allele is unique, the number of alleles shared, and the coefficient of relationship for (b) sister–sister and (c) mother–daughter pairs in haplo-diploids.

(a) Parents		A_1A_2 (female) × A_3 (male)		
Female offspring		$\frac{1}{2}A_1A_3$ and $\frac{1}{2}A_2A_3$		
(b) Sister–sister		*Sister*		
		$\frac{1}{2}A_1A_3$	$\frac{1}{2}A_2A_3$	
Sister	$\frac{1}{2}A_1A_3$	2	1	$r = 0.75$
	$\frac{1}{2}A_2A_3$	1	2	
(c) Mother–daughter		*Daughter*		
		$\frac{1}{2}A_1A_3$	$\frac{1}{2}A_2A_3$	
	Mother A_1A_2	1	1	$r = 0.5$

that corrections are necessary. We will give two such estimates, those of Queller and Goodnight (1989) and Li *et al.* (1993), using the terminology in Lynch and Ritland (1999). When loci exhibit codominance, the sampling variance for relatedness estimates is roughly proportional to the total number of alleles at all loci (Ritland, 1996) so that either many diallelic loci or relatively few hypervariable loci are needed for reasonable estimates. For recent discussion and comparisons of different relatedness estimates, see Wang (2002), Blouin (2003), Glaubitz *et al.* (2003) and Milligan (2003). For a list and addresses of software that are available to estimate relatedness see Blouin (2003), for example, the website for the program RELATEDNESS is http:/www.gsofnet.us/GSoft.html.

To calculate the relatedness estimate of Queller and Goodnight (1989), let a and b refer to the two alleles in the first individual x and c and d as the two alleles in the second individual y. To estimate relatedness, we need to know the similarity values that are defined in the following way:

$$\begin{aligned} S_{ab} &= 1 \quad \text{when } x \text{ is a homozygote} \\ S_{ab} &= 0 \quad \text{when } x \text{ is a heterozygote} \\ S_{ac} &= 1 \quad \text{when allele } a \text{ in } x \text{ is the same as allele } c \text{ in } y \\ S_{ac} &= 0 \quad \text{when } a \text{ and } c \text{ are different.} \end{aligned}$$

The other similarity values between individuals, S_{ad}, S_{bc}, and S_{bd}, are defined in the same manner as S_{ac}. In addition, we need to know estimates of the frequencies of alleles a and b in the population (sample); these are designated as p_a and p_b, respectively. Given these values, then the relatedness estimate of Queller and Goodnight (1989) can be written as

$$r_Q = \frac{0.5(S_{ac} + S_{ad} + S_{bc} + S_{bd}) - p_a - p_b}{1 + S_{ab} - p_a - p_b} \tag{5.15a}$$

When necessary (see Example 5.8), this value and the analogous value using individual y (with alleles c and d) as the reference individual should be averaged.

The other estimate of relatedness (Li *et al.*, 1993) uses a different similarity index that gives the average fraction of genes at a locus for which the other individual is identical in state. This similarity index is defined in the following way:

$$\begin{aligned} S_{xy} &= 1 \quad && \text{when } x = ii \text{ and } y = ii \text{ or when } x = ij \text{ and } y = ij \\ S_{xy} &= 0.75 \quad && \text{when } x = ii \text{ and } y = ij \text{ or when } x = ij \text{ and } y = ii \\ S_{xy} &= 0.5 \quad && \text{when } x = ij \text{ and } y = ik \\ S_{xy} &= 0 \quad && \text{when } x = ij \text{ and } y = kl \end{aligned}$$

The measure of relatedness is then

$$r_L = \frac{S_{xy} - S_0}{1 - S_0} \tag{5.15b}$$

where

$$S_0 = \sum_{i=1}^{n} p_i^2(2 - p_i)$$

and p_i is the frequency of allele i (see Example 5.8, which illustrates how to calculate these relatedness values from a set of data).

Example 5.8. To illustrate how the relatedness measures of Queller and Goodnight (1989), r_Q, and Li *et al.* (1991), r_L, are calculated from real data, use genotypes that have been randomly drawn from a population. Table 5.14 gives on the left the eight-locus genotypes for two individuals drawn from a noninbred population in which each of these loci has 10 alleles in equal frequency. By chance, some of the alleles are represented more than once in the two parents at a given locus; for example, at locus A, allele A_{10} is present in both parents and is homozygous in parent 2. These three copies are identical in state; they are not IBD because they were drawn from a non-inbred population.

Two randomly generated full-sib progeny, x and y, are also given for each pair of parents. For example, at locus A, the two progeny are both A_8A_{10} heterozygotes. In this case, all of the S values are 1 except $S_{ab} = 0$, and because $p_a = p_b = p_i = 0.1$, $S_0 = 0.19$. Using these values, then $r_Q = r_L = 1$. For locus E, the two full sib progeny are E_2E_{10} and E_2E_3, and they share the allele E_2, an allele that could be either identical in kind or IBD from parent 2. In this case $S_{ac} = 1$ so that r_Q= 0.375 and $S_{xy} = 0.5$ so that $r_L = 0.383$. In other words, the estimates of the relatedness values are corrected downward from 0.5 for the possibility of identity in kind as a function of the allele frequencies.

For locus B, the two full sibs do not share any alleles, and then $r_Q = -0.25$ and $r_L = -0.235$. In this case, the estimates of relatedness are negative because there is some expected sharing of identity in kind with the given allele frequencies, but in this instance, there was no sharing of alleles. The final possibility is represented by locus F in which one of the full sibs is homozygous and the other is a heterozygote that shares this allele. In this case, $S_{ac} = S_{ad} = 1$ so that $r_Q = 0.444$. When the genotypes of individuals x and y are switched, then $S_{ac} = S_{bc} = S_{ab} = 1$ so that $r_Q = 1$. Taking the average of these two values gives $r_Q = 0.722$. $S_{xy} = 0.75$ so that $r_L = 0.691$. The mean values for these two estimates at the bottom of Table 5.14 are nearly identical and are both close to the 0.5 values expected for these two relationships.

TABLE 5.14 On the left are randomly generated genotypes of two parents and two of their full-sib progeny for eight loci, each with 10 alleles in equal frequency in the population. On the right are the two relatedness estimates given in expressions 5.15*a* and 5.15*b* for each locus and the overall mean between the two full sibs (x and y) and the first parent and the first full sib (1 and x).

	Genotypes of				*Relatedness between*			
	Parents		*Full sib progeny*		x *and* y		*1 and* x	
Locus	*1*	*2*	x	y	r_Q	r_L	r_Q	r_L
A	8 10	10 10	8 10	8 10	1	1	1	1
B	2 7	6 9	7 9	2 6	-0.25	-0.235	0.375	0.383
C	4 10	4 7	4 4	4 4	1	1	0.722	0.691
D	2 5	3 5	2 5	2 3	0.375	0.383	1	1
E	3 10	2 2	2 10	2 3	0.375	0.383	0.375	0.383
F	4 8	1 8	1 8	8 8	0.722	0.691	0.375	0.383
G	2 4	5 10	2 10	4 5	-0.25	-0.235	0.375	0.383
H	4 6	7 10	6 10	6 10	1	1	0.375	0.383
Mean					0.496	0.495	0.575	0.576

First, let us demonstrate that these expressions are correct for unrelated individuals when there is no identity in state. In this case, we can assume that p_a, p_b, and p_i all approach 0. Then, because all the S values are 0, both r_Q and r_L approach 0. Now let us examine the relatedness estimates between a parent and offspring, again assuming that there is no identity in state. Assume that all the p values approach 0, $S_{ac} = 1$, $S_{xy} = 0.5$, and all other S values are 0. Then, both r_Q and r_L approach 0.5, as expected for parent–offspring relatedness with no identity in state.

Relatedness estimate measures are becoming more useful in many different evolutionary and behavior ecology contexts as highly variable microsatellite loci are found in many organisms. SNPs may also be useful for determining relatedness, but many more loci are generally necessary (Glaubitz *et al.*, 2003). Example 5.9 shows how relatedness measures are useful in determining the level of matings between close relatives in white-toothed shrews and the relatedness of same-sex individuals in breeding groups of carrion crows.

Example 5.9. Estimates of relatedness based on molecular markers are being widely used to determine kin relationships of individuals in natural populations. For example, Duarte *et al.* (2003) examined dispersal, mating, and inbreeding in a population of the white-toothed shrew (*Crocidura russala*) using 12 microsatellite loci to determine relationships. In 81 pairs of mated individuals, they observed 14 (17%) that had relatedness estimates averaging 0.5 (Figure 5.16). On further analysis, they found a number of matings between first-degree relatives, including father–daughter, mother–son, and full-sib. The other 67 pairs had an average relatedness not significantly

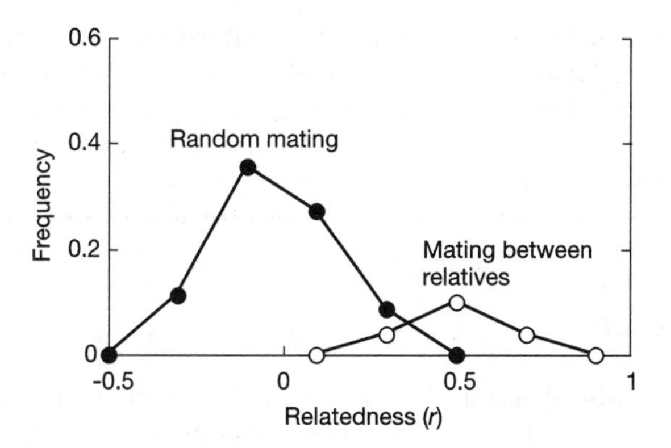

Figure 5.16. Relatedness estimates among 81 mating pairs of white-toothed shrews (from Duarte *et al.*, 2003) where the closed and open circles give the proportions of random mating and mating between first-degree relatives, respectively.

different from 0. They concluded that there is limited and female-biased dispersal then followed by random mating among the remaining individuals with a neighborhood. The high level of breeding between close relatives occurs because some of the individuals within a neighborhood are relatives, and there does appear to be inbreeding avoidance.

Relatedness estimates are fundamental importance in studies of **kin recognition** and kin selection (see p. 278). Baglione *et al.* (2003) reported a case of kin-based cooperative breeding in a population of the carrion crow (*Corvus corone corone*) in northern Spain. Carrion crows form cooperative breeding units that frequently have more than one adult male, and often the extra males are immigrants that, along with the resident male, mate with the resident adult female. Using six microsatellite loci, Baglione *et al.* (2003) determined the relatedness of same-sex immigrant and resident crows cooperating on a territory and compared it with r estimates in the population (Figure 5.17). The mean r for the same-sex immigrant and resident crows within a territory was 0.24 ± 0.07, whereas the mean of pairs

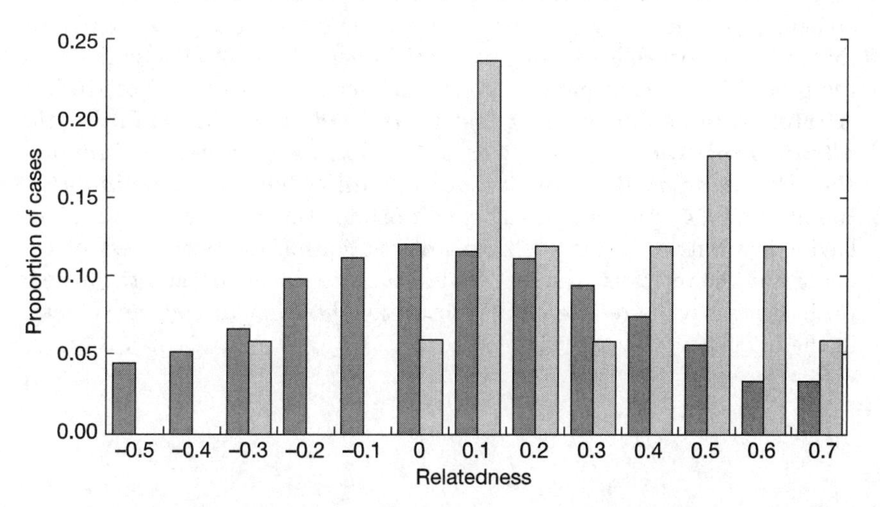

Figure 5.17. Relatedness estimates in a carrion crow population (darkly shaded bars) and in same-sex immigrant and resident crows cooperating on a territory (lightly shaded bars) (from Baglione *et al.*, 2003).

chosen at random in the population had relatedness values not different from zero. The relatedness of same-sex pairs does not appear to be the result of **philopatry** (a tendency of dispersers to settle near their natal territory), but it does appear that resident crows are more tolerant of allowing relatives to settle in their territory.

---

i. Kin Selection

Until now we have been considering models of individual selection—that is, natural selection based on the genotypes or phenotypes of each individual—assuming that these in turn determine the fecundity, survival, mating success, and so on of that individual. It is possible, however, that ***selection may act through the close relatives of individuals***—that is, **kin selection** (Maynard Smith, 1964). In fact, both Fisher (1930) and Haldane were aware of the potential of kin selection, and Haldane is said to have responded to a moralist who asked if he would lay down his life for his brother, "No, but I would consider it for two brothers."

When related individuals interact with one another in a nonrandom way, these actions may affect the fitness of kin in a manner different from that of unrelated individuals. For example, individuals often interact differently with their siblings than with other individuals and, as a result, may increase the fitness of their sibs (see Example 5.9 in which related same-sex carrion crows were found in breeding groups). For kin selection to operate, the population must generally be structured into kin groups, such as families, but these kin groups need not be separate physical entities and may exist as part of a large population.

Of major significance is that kin selection allows the formulation of the conditions for the development of altruistic (social) behavior. The development of altruistic behavior via kin selection does not necessarily require that relatives be able to recognize each other but at least assumes that the probability of encounter is directly related to the closeness of kinship. **Altruistic behavior** assumes that an ***individual or individuals (the altruists or donors) sacrifices some fitness in order to increase the fitness of another individual or individuals (the recipients).*** Hamilton (1964) generalized the conditions for the origin of altruistic behavior in terms of r, the coefficient of relationship or relatedness of the donor and the recipient, the cost (c) in fitness to the donor, and the benefit (b) in fitness to the recipient. His conclusion was that to have an increase in the frequency of an altruistic allele

$$r > \frac{c}{b} \tag{5.16}$$

This theoretical result has been verified using a number of different approaches and appears to be of wide application. For example, r between two sibs is 0.5 so that c would have to be less than $0.5b$, half the benefit to the recipient, for an altruistic allele to increase in frequency.

For a haplodiploid organism, the r value between two sisters is $^3/_4$ (assuming a single fertilization of their mother; see Table 5.13). Therefore, the condition for an altruistic allele to increase is that c is less than $0.75b$, a larger allowable cost than for diploids. Because these conditions are less restrictive than for diploids, Hamilton (1964) suggested that this was the reason that sociality has arisen a number of different times in Hymenoptera, a haplodiploid group, and only once in diploid arthropods (in termites). Although this genetic hypothesis appears to be a reasonable explanation for the multiple origin of social behavior in Hymenoptera (Crozier and Pamilo, 1996), there are other reasonable hypotheses (West Eberhard, 1975). In fact, using molecular markers, examination of the relatedness of nest founders in the social wasp, *Polistes dominulus*, found that 35% of the helper foundresses were unrelated (Queller *et al.*, 2000), in contrast to the expectations of kin selection.

Hamilton's approach has been reformulated in terms of the fitness values of the donor and recipient (West Eberhard, 1975). In this case, the concept used is that of **inclusive fitness**, the ***fitness of the individual plus the effects on the relatives of the individual, each weighted by their coefficient of relationship***. For example, an individual that helps a relative survive is actually increasing its own inclusive fitness because the relative contains some of its alleles by descent. Of course, viewed in this way, such behavior is not altruistic but rather a "selfish" propagation of one's own alleles.

II. INBREEDING AND SELECTION

a. Inbreeding Depression

As we have noted, inbreeding increases the frequency of homozygotes and decreases the frequency of heterozygotes. These deviations from Hardy–Weinberg proportions influence the mean fitness of the population, the rate of allele frequency change, and the conditions for a stable polymorphism. **Inbreeding depression**, ***the decline of fitness due to inbreeding***, seems to be a nearly universal phenomenon, although the extent of inbreeding depression varies for different species (Lynch and Walsh, 1998; Keller and Waller, 2002) and even for different populations of the same species (Kärkkäinen *et al.*, 1996).

In recent years, there has been a great deal of interest in the level of inbreeding depression, particularly because of research on the **evolution of plant mating systems** and on determining fitness in threatened and endangered species. For example, in the absence of inbreeding depression, selfing is favored because of the representation in progeny of an allele that causes selfing is twice as high as an allele that does not cause selfing (Fisher, 1941). Simply, if a maternal plant is homozygous for an allele that causes selfing, for example, $A_S A_S$, then all selfed progeny will also be $A_S A_S$. On the other hand, if a maternal plant is homozygous for an allele that causes outcrossing, for example, $A_T A_T$, then all progeny will be $A_T A$, where A is any other allele. Therefore, the frequency of A_S in the selfed progeny is twice that of A_T in outcrossed progeny. The **twofold advantage of selfing** is reduced when there is inbreeding depression because fewer selfed than outcrossed progeny survive (see the discussion in Uyenoyama *et al.*, 1993, and Example 5.13). Theory suggests that if inbreeding depression is low—that is, the fitness of selfed progeny is greater than 50% that of outcrossed progeny—then the population will evolve to be self-fertilizing (Lande and Schemske, 1985). On the other hand, if the fitness of selfed progeny is less than 50% that of outcrossed progeny, the population will be outcrossing. However, other factors may play an important role in the evolution of selfing; for example, self-fertilization is beneficial when pollinators and/or potential mates are rare because otherwise the production of seeds may be reduced (Herlihy and Eckert, 2002).

Endangered species face extinction from a number of extrinsic ecological threats, most of them human related. However, if the population numbers of a species are low or if there is inbreeding for other reasons, then an endangered species may have a lowered fitness because of inbreeding depression (Hedrick and Kalinowski, 2000). This adds an intrinsic biological factor that may interact with other threats and thereby further increases the probability of extinction (Soulé, 1986). In addition, endangered species may exist only in marginal environments or may be exposed to novel environmental stresses such as introduced species, pollution, or habitat degradation. Recent experiments are broadly consistent in reporting greater inbreeding depression in more stressful environments (Hedrick and Kalinowski, 2000; Joron and Brakefield, 2003) so that endangered species may be particularly vulnerable to reduced fitness from inbreeding and may face a greater probability of extinction (Bijlsma *et al.*, 2000).

The extent of inbreeding depression varies greatly among different organisms. For example, three generations of full-sib mating in Japanese quail resulted in a complete loss of reproductive fitness (Sittman *et al.*, 1966). Some pine species have such a large inbreeding depression that no progeny produced by selfing survive (Lande *et al.*, 1994), and New Zealand birds species that have undergone a bottleneck of less than 150 individuals have high hatching failure (Briskie and Mackintosh, 2004). Many studies have shown the extent of inbreeding depression in *Drosophila*, mice, and humans (see Example 5.10). In some model organisms, such as mice and *D. melanogaster*, lines have been maintained by full-sib mating for hundreds of generations. However, it is not generally known how many lines are lost in an effort to maintain these laboratory lines, and thus, selection between lines may have resulted in the maintenance of the few lines with higher survival. Furthermore, many plant species that have predominately self-fertilization generally show little decline of fitness when inbred, although this is not always true (Husband and Schemske, 1996). It is thought that the lethals or detrimentals that would lower fitness in these species have been selected against, or purged, from these populations (see p. 289 and p. 387).

Example 5.10. Some of the best studies documenting the overall effect of inbreeding are in various *Drosophila* species. A number of species have been examined for different traits, including survival, female and male fertility, weight, and bristle number (see Lynch and Walsh, 1998). In nearly all instances, there is a general trend of greater effect with increased inbreeding, although the amount varies with species and trait.

There is also a large body of data on the effects of inbreeding in humans (Bennett *et al.*, 2002). Some of the best data are summarized in the report of Schull and Neel (1965) on the effects of the atomic bombs in Hiroshima and Nagasaki, Japan. A sample of these data is given in Table 5.15 for infant mortality when the parents have different degrees of consanguinity. Again, there is a trend of increased mortality with an increasing amount of inbreeding.

TABLE 5.15 Percent infant mortality in two Japanese populations where the parental consanguinity was known (Schull and Neel, 1965).

	Parental relationship (f)			
Population	*Unrelated* (0)	*Second cousins* ($\frac{1}{64}$)	*First cousins once removed* ($\frac{1}{32}$)	*First cousins* ($\frac{1}{16}$)
Hiroshima	3.55	4.43	7.18	6.12
Nagasaki	3.42	3.18	4.94	5.25

Another example is the effect of inbreeding on litter size and progeny survival in mice (Connor and Bellucci, 1979). In this study, wild populations of mice were inbred for a number of generations and monitored for the mean value of these characters. As shown in Figure 5.18, there was a nearly linear drop in the litter size as inbreeding increased. Progeny survival, on the other hand, decreased only at high levels of inbreeding. Because these experiments were carried out over a number of generations, there is opportunity for allele frequency change, so inbreeding depression equations that assume no change in frequency may not be applicable.

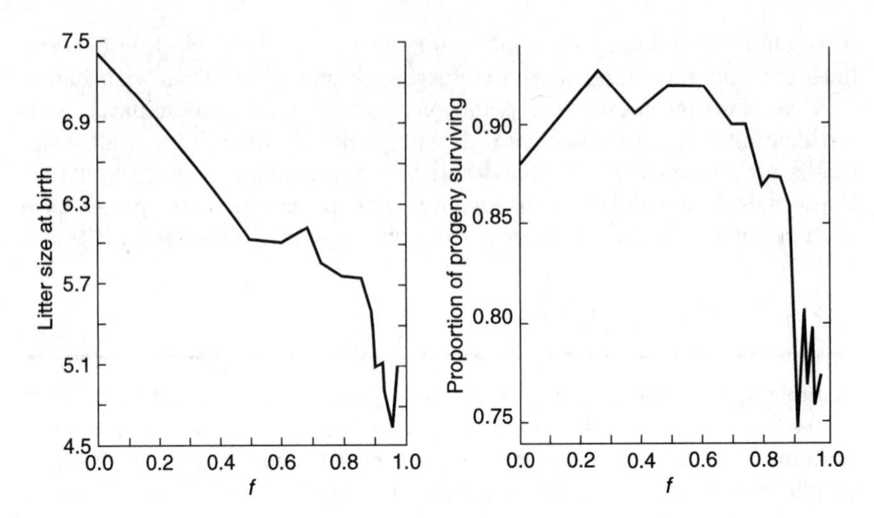

Figure 5.18. The litter size and progeny survival for lines of mice inbred for 20 generations (Connor and Bellucci, 1979).

The level of inbreeding depression can be influenced by a number of other factors. First, various phenotypic traits can be influenced differently by inbreeding. In animals, generally viability is examined, but other fitness components, such as fecundity and mating success, can also be affected by inbreeding (Miller and Hedrick, 1993). Husband and Schemske (1996) found that most self-fertilizing plants had little inbreeding depression early in the life cycle, for example, in germination, but had most of their inbreeding depression late in their life cycle for traits such as growth and development. In addition, other traits that are not components of fitness but are correlated with fitness, such as size or shape, may exhibit inbreeding depression. Second, the extent of inbreeding depression appears to vary with environment. Generally, when the environment is more benign as in a laboratory or greenhouse, there appears to be less inbreeding depression than in more stressful environments, such as more natural conditions or when environmental stressors are present (Bijlsma *et al.*, 1999; Joron and Brakefield, 2003). Third, the genetic basis of inbreeding depression may vary. In *Drosophila* (Charlesworth and Charlesworth, 1987), it appears that approximately half of the reduction in fitness on inbreeding results from lethals with low dominance ($h = 0.02$) and approximately half from detrimentals ($s = 0.01$ to 0.05) with higher dominance ($h = 0.3$) (for some data in plants, see Kärkkäinen *et al.*, 1999; Carr and Dudash, 2003). It is not clear how general these findings are, but it is likely that in some species that the proportion of reduced fitness from lethals is much lower because lethals are predicted to be at lower frequency in small populations (Hedrick, 2002b).

Let us examine how inbreeding at a single locus can affect the mean fitness of a population. If we assume genotype fitness values as given in

Chapter 3 and let the genotype frequencies be as given in expression 5.1, then the mean fitness with inbreeding is

$$\begin{aligned}\overline{w}_f &= w_{11}(p^2 + fpq) + w_{12}\,(2pq - 2fpq) + w_{22}(q^2 + fpq) \\ &= \overline{w} + fpq(w_{11} + w_{22} - 2w_{12})\end{aligned}$$

where the first term is the mean fitness

$$\overline{w} = p^2 w_{11} + 2pqw_{12} + q^2 w_{22}$$

as in expression 3.1, and the second term measures the change in fitness caused by inbreeding. Inbreeding depression can be defined as the difference in fitness between an outbred and an inbred population or

$$\overline{w} - \overline{w}_f = -fpq\,(w_{11} + w_{22} - 2w_{12}) \tag{5.17a}$$

Inbreeding depression is a function of the allele frequencies and the relationship between the fitnesses and is a linear function of the inbreeding coefficient. If there is no inbreeding ($f = 0$) or the heterozygote is exactly intermediate between the homozygotes, additive gene action ($w_{11} + w_{22} - 2w_{12} = 0$), then there is no inbreeding depression. Of course, it is possible, but unlikely, for inbreeding to increase the population fitness if $w_{11} + w_{22} - 2w_{12} > 0$.

Let us assume the widely accepted partial dominance explanation for inbreeding depression (Charlesworth and Charlesworth, 1999; Roff, 2002), and assume that the locus is a detrimental with dominance h so that the fitnesses of A_1A_1, A_1A_2, and A_2A_2 are 1, $1 - hs$, and $1 - s$, respectively, then

$$\overline{w} - \overline{w}_f = sfpq(1 - 2h) \tag{5.17b}$$

Again, if there is additivity ($h = 0.5$), there is no inbreeding depression because $1 - 2h = 0$. Inbreeding obviously has the biggest effect on fitness when the heterozygote is closest to the favorable genotype A_1A_1 in fitness (h lowest). Of course, the observed effect of inbreeding in a population is the summation over all loci. In other words, each of a number of loci may contribute somewhat to the lowering of fitness observed from inbreeding.

To investigate the effects of inbreeding on survival (viability), Morton *et al.* (1956) developed a model that assumes that the loci affecting survival act independently and multiplicatively (see p. 556). In this case, assuming for simplicity that survival is the only trait affecting fitness, then the fitness of individuals with inbreeding f becomes approximately

$$w_f = w_0 e^{-Bf} \tag{5.18a}$$

where w_0 and w_f are the fitnesses when there is no inbreeding or inbreeding to the level f. B is the regression coefficient that measures how fast fitness decreases with inbreeding; it is equal to zero when there is no inbreeding depression. The **number of lethal equivalents** in a zygote is defined as the ***group of genes that, when made homozygous, would on average cause 2B deaths***. For example, if there is one lethal allele present as a heterozygote, there would be one lethal equivalent. Using data documenting the survival proportion for different inbreeding categories, investigators have used this relationship extensively to determine the how much inbreeding depression is present in a number of different species. In particular, this approach has been used to examine the amount of inbreeding depression for survival in a number of captive endangered species (Example 5.11 discusses these general findings and examines inbreeding depression in two endangered species of wolves and an endangered species of salmon).

Example 5.11. Although the effect of inbreeding on fitness has long been appreciated in many organisms, the study of Ralls *et al.* (1979) was the first to document quantitatively the lower survival in inbred than in noninbred captive endangered species. The findings from this initial study had a great impact on population management in zoos and resulted in a reorientation in the breeding programs of captive endangered species to avoid inbreeding as much as possible. A further study by Ralls *et al.* (1988) expanded the number of species examined, and also utilized different inbreeding coefficients within a species, to estimate inbreeding depression for juvenile survival using expression 5.18a. Although the variation in estimated inbreeding depression over species was large, the median number of lethal

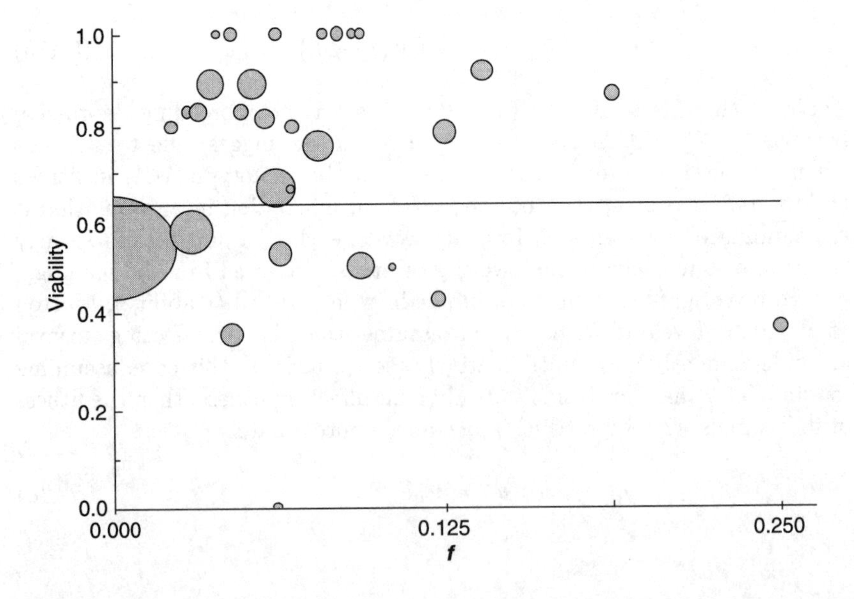

Figure 5.19. The observed average viability of each inbreeding level for red wolves (Kalinowski *et al.*, 1999). The areas of the circles are proportional to the number of individuals born, and the fitted line is based on the model of Morton *et al.* (1956).

equivalents over species, 3.14, was substantial (the similarity to π was a coincidence).

On the other hand, inbreeding in some captive endangered species appears to have no statistically demonstrable effect on some fitness components. For example, in the endangered Mexican and red wolves, the maximum likelihood estimate of $2B$ in both species is 0.00 (see Figure 5.19 for the red wolf data) and the 95% confidence intervals for $2B$ are 0.00 to 1.68 for the Mexican wolf and 0.00 to 0.74 for the red wolf (Kalinowski *et al.*, 1999). These data sets were fairly large and included 251 Mexican wolves over five inbreeding levels and 688 red wolves over 29 levels of inbreeding. However, the concentration of Mexican wolves in two adjacent inbreeding levels ($f = 0.1875$ and $f = 0.25$) and the large number of noninbred red wolves somewhat reduced the statistical power of these data sets (Kalinowski and Hedrick, 1999). On the other hand, a comparison of body size in Mexican wolves with little or no inbreeding to those with higher inbreeding did show evidence of significant inbreeding depression (Fredrickson and Hedrick, 2002).

The effect of inbreeding may only be seen under more stressful conditions than in captive environments (Hedrick and Kalinowski, 2000). For example, Arkush *et al.* (2002) examined disease resistance to three pathogens in the endangered winter-run Chinook salmon families that were the result of random and full-sib ($f = 0.25$) matings from the same mother. For *Myxobolus cerebralis*, for the myxozoan parasite that causes whirling disease, in four of the five families examined (families 1–4 in Table 5.16), significantly fewer inbred fish were resistant than outbred fish with the same mother. Overall, 82% of the outbred fish were resistant, whereas only 59% of the inbred fish were resistant. Expression 5.18*b* can be used to estimate the number of genes that would result in no resistance when homozygous (a measure analogous to the number of lethal equivalents) as $2B = (-2/0.25)\ln(0.594/0.822) = 2.60$. In other words, the equivalent of two to three genes appears to be segregating in winter-run Chinook salmon that if IBD result in susceptibility to whirling disease, a pathogen that is generally fatal to infected fish.

TABLE 5.16 The number of outbred and inbred ($f = 0.25$) winter-run Chinook salmon exposed and resistant to the parasite that causes whirling disease in five families (Arkush *et al.*, 2002).

	Outbred			*Inbred*		
Family	*Exposed*	*Resistance*	*Proportion*	*Exposed*	*Resistance*	*Proportion*
1	60	58	0.97	55	46	0.84
2	58	53	0.91	51	30	0.59
3	58	41	0.71	54	17	0.71
4	58	47	0.81	60	37	0.62
5	30	18	0.60	31	19	0.61
Total	264	217	0.822	251	149	0.594

Expression 5.18a can be rearranged when comparing the fitnesses of noninbred individuals and those with inbreeding f, w_f, to give

$$B = -\frac{1}{f} \ln \left(\frac{w_f}{w_0} \right) \tag{5.18b}$$

an estimate of lethal equivalents. In plants, inbreeding depression is usually measured by comparing the fitness of progeny from self-fertilization to progeny of individuals from outbred or random matings. For this, the following expression is used

$$\delta = 1 - \frac{w_S}{w_O} \tag{5.18c}$$

where w_S is the fitness of self-fertilized progeny and w_O is the fitness of outcrossed progeny (Husband and Schemske, 1996). For example, if $w_S = 0.4$ and $w_O = 0.8$, then $\delta = 0.5$, and because $f = 0.5$, then $B = 2$. Similarly, Lynch and Walsh (1998) extrapolated to the fitness in a completely inbred population and replaced w_S with this value to estimate the total amount of inbreeding depression. The measures in expression 5.18c and Lynch and Walsh (1998) are similar to that in expression 5.18a, except that they are standardized by dividing by the fitness of outcrossed progeny.

It is generally useful to distinguish between **inbreeding depression**, decreased fitness on inbreeding, and **genetic load**, a reduction in population fitness from the optimum possible (Kirkpatrick and Jarne, 2000; Hedrick 2002b). For example, a population that has been small and/or inbred for a long time may have little genetic variation left because of genetic drift and/or inbreeding, including variants that have detrimental effects. However, this population may have become fixed for some detrimental alleles by chance so that it has a lowered fitness. This population may therefore have little or no inbreeding depression (all of the individuals, both noninbred and inbred, will have virtually the same genotypes) but may have substantial genetic load. The extent of genetic load may be determined by comparing the fitness of the population before and after the effects of the small population size or inbreeding or to compare the fitness to that of crosses between individuals from different inbred lines or groups within the population (Barrett and Charlesworth, 1991; Paland and Schmid, 2003).

b. Change in Allele Frequency and Conditions for a Polymorphism

To examine the effect of inbreeding on the progress of directional (adaptive Darwinian) selection and the conditions for a stable polymorphism, let us consider as an example the effect of partial selfing on selection.

Expression 5.5a can be modified to include selection so that

$$
\begin{aligned}
P_1 &= \frac{[Tp^2 + S(P_0 + {}^1\!/\!_4 H_0)]w_{11}}{\overline{w}} \\
H_1 &= \frac{(2Tpq + {}^1\!/\!_2 SH_0)w_{12}}{\overline{w}} \\
Q_1 &= \frac{[Tq^2 + S(Q_0 + {}^1\!/\!_4 H_0)]w_{22}}{\overline{w}}
\end{aligned}
\tag{5.19}
$$

where the mean fitness is

$$\overline{w} = w_{11}[Tp^2+S(P_0+\tfrac{1}{4}H_0)]+w_{12}(2Tpq+\tfrac{1}{2}SH_0)+\ w_{22}[Tq^2+S(Q_0+\tfrac{1}{4}H_0)]$$

As a result of partial selfing, the change in allele frequency due to **directional selection** is generally substantially increased. The cause of this effect can be shown by assuming equilibrium genotype frequencies from complete selfing—that is, no heterozygotes. In this extreme example, the mode of selection reduces to gametic selection where the selection differential is the difference between the two homozygotes, resulting in a doubling of the Δq value as compared with random mating.

Figure 5.20 shows the change in allele frequency and fitness for directional selection for a recessive gene with three values of partial selfing. Even

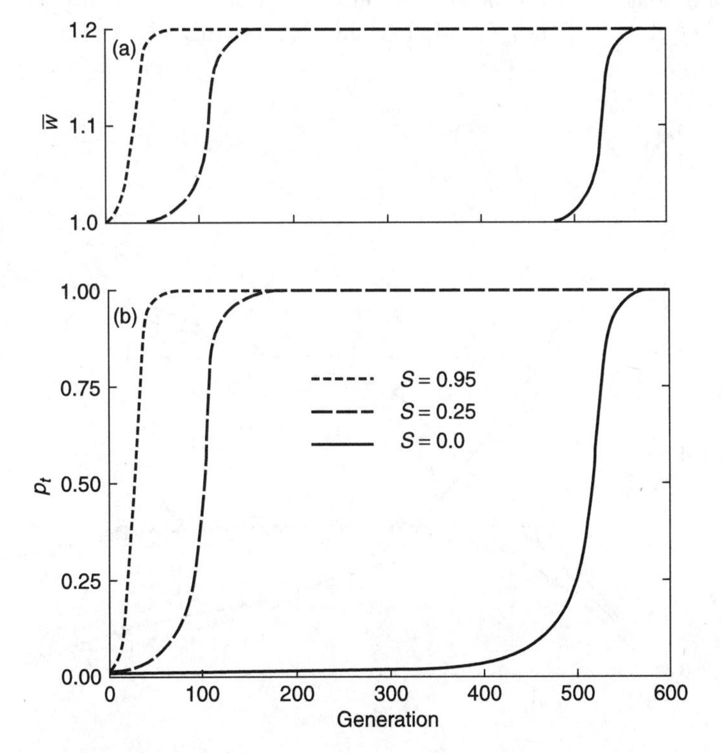

Figure 5.20. The mean fitness (a) and allele frequency (b) over time when selection favors a recessive allele with s = 0.2, an initial frequency of 0.01, and when partial selfing (S) is 0.0, 0.25, or 0.95.

a small amount of partial selfing, such as $S = 0.25$, greatly increases the change in allele frequency. Of course, the most marked effect on genetic change is for a new recessive advantageous mutant, whereas the least effect occurs with a dominant mutant. Some breeding schemes used in artificial selection take advantage of the increased response with inbreeding by alternating between generations of inbreeding and selection or simultaneously carrying out inbreeding and selection. Example 5.12 gives an illustration of change in genotype frequency in the completely selfing plant, *Arabidopsis thaliana*.

Example 5.12. Asmussen *et al.* (1998) and Liu *et al.* (2003) provided an illustration of the impact of selection in the completely selfing model plant *Arabidopsis thaliana,* for variants at the actin gene, *act*2. They initiated a population from a plant heterozygous for the wildtype allele A_1 and a mutant A_2 and followed genotype frequencies over five generations (Figure 5.21). In the first generation, the frequency of the wildtype homozygote was higher than that of the mutant homozygote, and by the second generation, the frequency of A_1A_1 was higher than that of the heterozygote, whereas that for genotype A_2A_2 was lower. The change in genotype frequencies can be predicted with expression 5.19 when $H_0 = 1$ and $S = 1$ with different relative fitnesses (viabilities) for the three genotypes. When Liu *et al.* did this, they found that the change through generation 5 was

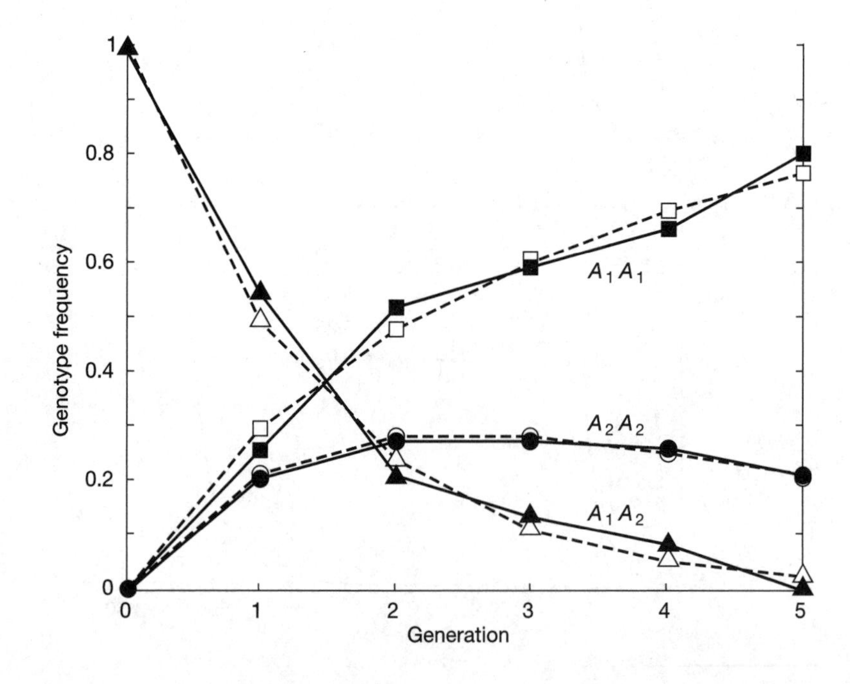

Figure 5.21. The change in genotype frequencies at an actin gene in an experimental population of the completely selfing plant *Arabidopsis thaliana* (Liu *et al.*, 2003). Generation 0 was initiated with a single plant, and the genotypes of the progeny were followed for five generations (closed symbols). The expected change in genotype frequencies, given fitnesses 1, 0.85, and 0.73 for genotypes A_1A_1, A_1A_2, and A_2A_2, are also shown (open symbols).

closely approximated by fitnesses 1, 0.85, and 0.73 for genotypes A_1A_1, A_1A_2, and A_2A_2, respectively. With these fitnesses, the frequency of the mutant allele is less than 0.001 within 20 generations, indicating that the mutant would be lost from the population quickly. Although other combinations of similar fitness values are generally consistent with the genotypic dynamics, it appears obvious that there was fairly strong selection against the homozygote mutant and that the heterozygote had a lower fitness than the wildtype homozygote.

Another consequence of this relationship is that the amount of inbreeding depression may be influenced by the inbreeding and population size history of the population—that is, selection is more effective at **purging** deleterious alleles when inbreeding or genetic drift is present. For example, Hedrick (1994), Wang *et al.* (1999), and Glemin (2003) have shown theoretically that detrimental variants of large effects causing inbreeding depression may be purged by close inbreeding or a bottleneck. However, detrimental variants of smaller effects may be fixed and thereby lower the overall fitness, or increase the genetic load, of the population. A good example is in the study of Barrett and Charlesworth (1991), who selfed a predominantly outcrossing population of water hyacinth for five generations and found that the inbreeding depression was greatly reduced, consistent with a purging explanation. In addition, the effect of inbreeding fixing different detrimental variants in different inbred lines was subsequently uncovered by an increase in fitness in crosses between inbred lines.

Several surveys have evaluated the evidence for purging (Ballou, 1997; Byers and Waller, 1999; Crnokrak and Barrett, 2002) with mixed evidence for its significance. For example, an increase, or rebound, in fitness over time in some studies that is consistent with purging may be the result of adaptation to environmental conditions and needs to be experimentally differentiated. Crnokrak and Barrett (2002) found general support for purging, particularly when changes in inbreeding depression were monitored over generations, but found that different measures of purging in the same study often were not consistent. On the other hand, the often-cited example of the reduction of inbreeding depression in the captive Speke's gazelle population because of past inbreeding (Templeton and Read, 1983) is most easily explained by changes in survival in the population over time (Kalinowski *et al.*, 2000).

If there is **heterozygote advantage**, then inbreeding can drastically reduce the rate of change in allele frequency. For example, assume that there are no heterozygotes because of complete selfing and symmetrical heterozygote advantage. In this case, the lack of heterozygotes caused by the mating system results in no differential selection because the two remaining genotypes, the homozygotes, have the same relative fitness.

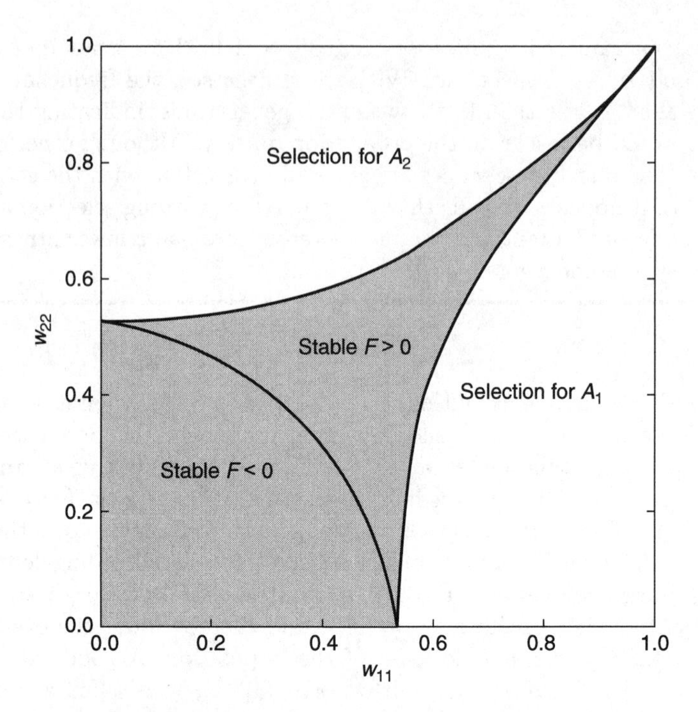

Figure 5.22. The region of heterozygous fitness advantage when partial selfing (S) is 0.95. The unshaded regions indicate fitness values that lead to directional selection, and the shaded regions indicate fitness values that give stable equilibria.

Inbreeding also affects the conditions for a stable equilibrium when there is a heterozygote advantage because the proportions of genotypes are affected. Workman and Jain (1966) and Kimura and Ohta (1971) have derived the conditions for stability with partial selfing as

$$\frac{2w_{11}(1 - w_{11})}{w_{11} + w_{22} - 2w_{11}w_{22}} > S < \frac{2w_{22}(1 - w_{22})}{w_{11} + w_{22} - 2w_{11}w_{22}} \tag{5.20}$$

where w_{11} and $w_{22} < 1$. We can illustrate the region of stability by examining the lower left quadrant of Figure 3.3 with a certain level of selfing (after Hayman, 1953), as in Figure 5.22, where $S = 0.95$. Remember that with random mating this whole fitness region gave a stable polymorphism, but with this partial-selfing example, the region that gives a stable polymorphism is reduced to the amount indicated in the shaded regions. The unshaded area is the region where there is selection favoring either allele A_1 or A_2. The shaded region is divided into two areas depending on whether there are more or fewer heterozygotes than expected from Hardy–Weinberg proportions, $F < 0$ and $F > 0$, respectively.

The general result is that the region of stability is substantially reduced because of inbreeding and that for heterozygous advantage to maintain a polymorphism with partial selfing selection must be quite symmetrical, particularly when selective differences are small (w_{11}, w_{12} approach 1.0). For example, if $S = 0.95$ and $w_{22} = 0.9$, then w_{11} must be between 0.888 and 0.91. One implication of this is that in a highly inbreeding species, slight differences in selection between homozygotes in different populations may lead to large differences in allele frequencies. For example, if again $S = 0.95$ and $w_{22} = 0.9$, then when $w_{11} < 0.888$, A_2 would go to fixation, whereas if $w_{11} > 0.91$, A_1 would go to fixation (Hedrick, 1990b). However, even with high selfing, genetic polymorphism can be maintained if there is spatially different selection (Hedrick, 1998) or if there is frequency-dependent selection (Stahl *et al.*, 1999). Example 5.13 discusses a balanced polymorphism in flower color in the partial-selfing morning glory.

Example 5.13. Flower color polymorphism in the common morning glory, *Ipomoea purpurea*, has been extensively examined at both the molecular and the evolutionary levels (Clegg and Durbin, 2000). In particular, the factors influencing variation at the regulatory locus W, which determines the patterning and degree of flower pigmentation, have been investigated in natural and experimental populations. The three genotypes at this locus, WW (darkly pigmented), Ww (lightly pigmented), and ww (white), can be distinguished, and the frequency of the w allele varies from 0.0 to 0.4, averaging approximately 0.11, in populations from the southeastern United States (Epperson and Clegg, 1986).

Subramaniam and Rausher (2000) set up experimental populations with w allele frequencies (q) at low, intermediate, and high values to determine whether selection would act as expected for a stable polymorphism. Figure 5.23 gives the changes in allele frequency (Δq) of w as a function of the allele frequency. As expected for a stable equilibrium, there was an increase in frequency for the two populations started at low frequencies and a decrease in frequency for the three populations started with higher frequencies. Assuming that $\Delta q = 0$ at both $q = 0$ and 1, then a curve can be

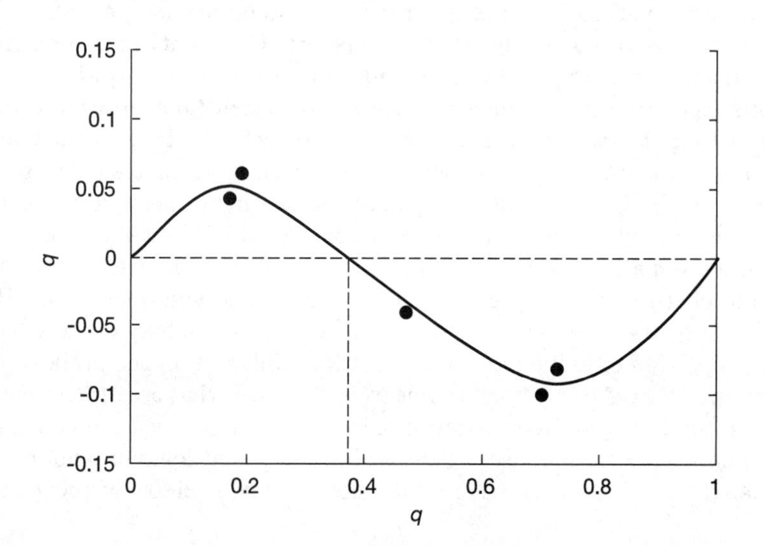

Figure 5.23. The observed change in the frequency of the w allele in experimental populations of the morning glory (after Subramaniam and Rausher, 2000).

drawn through these points, suggesting that there is a stable equilibrium between 0.3 and 0.4 (vertical broken line).

What are the factors that are responsible for this balancing selection? When the white allele is at low frequency, ww plants have a higher selfing rate than the other two genotypes because bumblebees, the main pollinator, visit white plants less frequently. When no other factors are operating, the higher selfing rate for homozygous ww plants will result in a selective advantage for the w allele. This advantage is similar to the twofold disadvantage of sex (p. 583) in that all selfed progeny from these homozygous plants will possess two copies of a maternal allele (and both of them are w) and no copies of the other alleles in the population, most of which are not w.

When the w allele is not at low frequency, pollinators visit all color flowers at the same rate, and selfing rates are similar for the three genotypes. In other words, there does not seem to be an analogous frequency-dependent effect favoring the W allele when dark flowers are rare. Examination of viability and seed number for the three genotypes does not suggest any selective advantage for the W allele or for Ww heterozygotes (Subramaniam and Rausher, 2000). However, the spatial pattern of genotypes at the W locus led Epperson (1990) to suggest that there was a 10% selective disadvantage to ww genotypes. Therefore, although the specific mode of selection keeping the w allele from going to fixation has not been identified, some type of selection appears to be maintaining the w allele at relatively low frequencies.

III. POSITIVE-ASSORTATIVE MATING

Assortative mating is nonrandom mating based on phenotypes rather than genotypes, as is inbreeding. **Positive-assortative mating** or **negative-assortative mating** occur if the ***mated pairs in a population are composed of individuals with the same phenotype more often, or less often, than expected by chance***, respectively. Positive-assortative mating is in some ways analogous to inbreeding in that similar phenotypes, which may have similar genotypes, are more likely to mate than random individuals from the population. As we illustrate below, some assortative-mating models are similar to inbreeding models in that they do not change allele frequencies but do affect genotype frequencies. However, many assortative-mating models do change allele frequencies because the proportion of individuals in the matings differs from the proportion in the population. An important point to remember is that assortative mating affects the genotype frequencies of only those loci involved in determining the phenotypes for mate selection (and genotypes at loci nonrandomly associated with these loci), whereas inbreeding affects all loci in the genome.

There appears to be positive-assortative mating for many traits in humans, such as height, skin color, and intelligence (Vogel and Motulsky, 1997), although the consequent phenotypic correlation is often not very large. In addition, there also appears to be positive correlations among mates that have particular phenotypes, such as deafness, blindness, or small stature. Of course, there are many different genetic (and nongenetic) causes for deafness, blindness, or small stature so that such a phenotypic correlation may not result in a genetic correlation. On the other hand, rather strong positive-assortative mating may occur in plants when a pollinator forages at a given height or is attracted to a given flower color and, as a result, tends to pollinate plants similar to the ones where the pollen was collected. Similar effects may also occur when flowering time is variable, and only plants that flower simultaneously pollinate each other.

In many birds there is strong mate selection, and in some cases, individuals with a phenotype similar to the maternal phenotype may be preferred. For an example of such **sexual imprinting**, see Example 5.14, which discusses a study in hawks with genetically determined variable plumage. Sexual imprinting generally leads to positive-assortative mating if the mother and offspring are alike in phenotype, but it may result in some cases of negative-assortative mating if they are different because of genetic segregation. For example, when both parents are heterozygotes and have a dominant phenotype and the offspring is a homozygote recessive, then sexual imprinting would result in negative-assortative mating.

Example 5.14. The common buzzard (buzzard is a European term for hawk), *Buteo buteo*, which is found over much of central Europe, has three main plumage color types. The Dark morph is the rarest, the Light morph is more common, and the Intermediate is the most common throughout the range. From examination of phenotype frequencies in 162 progeny from known matings, the color morphs are consistent with determination by a single gene with two alleles. The heterozygotes are the Intermediate phenotype, intermediate between the two homozygotes, which are Dark and Light (Krüger *et al.*, 2001).

Krüger *et al.* (2001) monitored the mating types of newly formed pairs in a population from Eastern Westphalia, Germany over a 10-year period from 1989 to 1999 (Table 5.17). The mating types between two similar homozygotes, Dark × Dark and Light × Light, were in substantial excess and the mating between different homozygotes, Dark × Light, was in great deficiency, indicating positive-assortative mating overall. However, the choice of mates does not appear to depend on the phenotypes of the individuals mating but appears to be the result of sexual imprinting; that is, chicks tend to choose mates who are the same morph as their mothers. In addition, there is nearly a much higher survival and lifetime reproductive success for both Intermediate females and males. Therefore, positive-assortative mating among the homozygotes, which acts to increase the frequency of homozygotes, is acting oppositely to the selection for life-history characteristics, which is increasing the frequency of heterozygotes.

TABLE 5.17 The observed numbers of newly formed mating pairs of hawks and the expected numbers assuming random mating (Krüger *et al.*, 2001).

Mating pairs	*Observed*	*Expected*	*Observed/Expected*
Dark × Dark	10	3.6	2.78
Intermediate × Intermediate	136	136.9	0.99
Light × Light	50	37.7	1.33
Dark × Intermediate	52	45.3	1.15
Intermediate × Light	139	144.0	0.97
Dark × Light	4	23.5	0.17

As an example of the potential impact of positive-assortative mating on genotype frequencies, let us examine a situation in which there is complete positive-assortative mating. Assume that allele A_1 is dominant over allele A_2 so that the only matings are between dominants or between recessives (the four mating types listed in Table 5.18). The other two possible mating types, $A_1A_1 \times A_2A_2$ and $A_1A_2 \times A_2A_2$, are between different phenotypes. Allowing for segregation in the production of progeny, the frequencies of the three genotypes in the progeny are

$$
\begin{aligned}
P_1 &= \frac{(P + H/2)^2}{1 - Q} = \frac{p^2}{1 - Q} \\
H_1 &= \frac{(P + H/2)}{1 - Q} = \frac{pH}{1 - Q} \\
Q_1 &= \frac{H^2}{4\,(1 - Q)} + Q = \frac{q^2 + Q\,(p - q)}{1 - Q}
\end{aligned}
\tag{5.21a}
$$

TABLE 5.18 The mating types and frequencies of progeny when there is complete positive-assortative mating at a dominant gene.

		Progeny		
Mating type	*Frequency*	A_1A_1	A_1A_2	A_2A_2
$A_1A_1 \times A_1A_1$	$P^2/(1-Q)$	$P^2/(1-Q)$	—	—
$A_1A_1 \times A_1A_2$	$2PH/(1-Q)$	$PH/(1-Q)$	$PH/(1-Q)$	—
$A_1A_2 \times A_1A_2$	$H^2/(1-Q)$	$H^2/[4(1-Q)]$	$H^2/[2(1-Q)]$	$H^2/[4(1-Q)]$
$A_2A_2 \times A_2A_2$	Q	—	—	Q
	1	$\frac{p^2}{1-Q}$	$\frac{pH}{1-Q}$	$\frac{q^2+Q(p-q)}{1-Q}$

As with inbreeding, there is no change in allele frequency with this model, a situation that occurs because the proportions of individuals participating in matings are the same as in the population. However, the genotype proportions change considerably over time, with the frequency of the homozygotes increasing and that of the heterozygote decreasing. The general recurrence relationship for heterozygosity is

$$H_t = \frac{2pH_0}{2p + tH_0} \qquad (5.21b)$$

which approaches zero as the number of generations gets large, just as for inbreeding.

Of course, it is unlikely that the population has complete assortative mating. The recursion relationships for partial positive-assortative mating of this type, where R is the proportion of positive-assortative mating and T is the proportion of random mating ($R + T = 1$), are

$$\begin{aligned} P_1 &= Tp^2 + \frac{Rp^2}{1 - Q_0} \\ H_1 &= T2pq + \frac{RH_0p}{1 - Q_0} \\ Q_1 &= Tq^2 + \frac{RH_0^2}{4(1 - Q_0)} + RQ_0 \end{aligned} \qquad (5.22)$$

The change in heterozygosity over time is given in Figure 5.24 for three levels of positive assortative mating when there are initially Hardy–Weinberg proportions and $p = 0.5$. With complete positive assortative mating ($R =$ 1.0), the heterozygosity eventually decays to zero but at a rate much slower

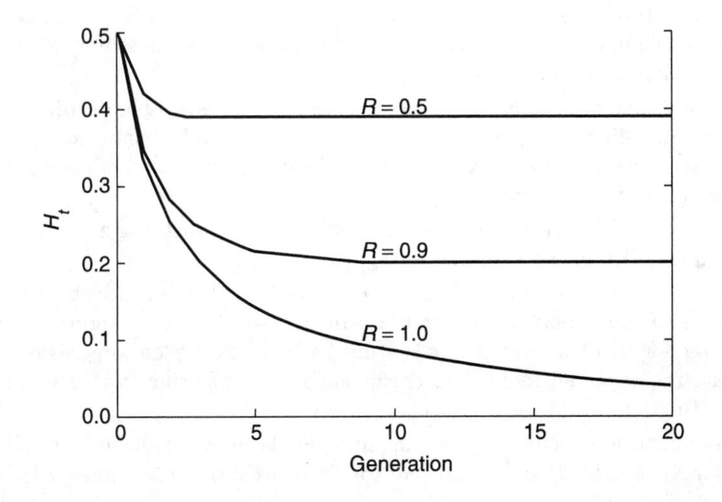

Figure 5.24. The change in heterozygosity over time with three levels of positive-assortative mating.

than for complete self-fertilization. With partial positive-assortative mating, the heterozygosity quickly attains its equilibrium value, which is 0.390 and 0.200 when there is 0.5 and 0.9 positive-assortative mating, respectively. With even high levels of positive-assortative mating, heterozygosity is still maintained at a substantial level.

PROBLEMS

1. In a particular population with a mean inbreeding coefficient of 0.005, a rare recessive disease has an allelic frequency of 0.005. What is the ratio of diseased homozygotes in this population compared to that in a random-mating population? What are the implications of your answer?
2. Assume that the initial heterozygosity is 0.3 and Hardy–Weinberg proportions. What is the expected heterozygosity after two generations with $S = 0$, $S = 0.5$, and $S = 1$? Discuss your results.
3. Assume that in alternating generations the selfing proportion varies between 0.8 and 1.0. If a population begins in Hardy–Weinberg proportions with $q = 0.5$, what are the heterozygosities in the first four generations? Assume that in another population the selfing proportion is 0.9. What is the heterozygosity in the first four generations? Compare these two results graphically.
4. In an experiment to estimate the amount of outcrossing in a population, a group of A_1A_1 individuals was grown interspersed in a population where $p_1 = 0.3$. Of 1000 progeny of the A_1A_1 individuals, 100 were heterozygotes. What is the estimated outcrossing rate? What proportion of the homozygous progeny was produced by outcrossing?
5. The extent of outcrossing appears to vary among loci, families, populations, and related species. Design an experiment that might establish the cause for this variation. What do you think is the evolutionary significance of this variation?
6. Calculate the expected heterozygosity for the first five generations, given $H_0 = 0.5$ in different lines where there is either complete selfing, sib mating, or double-first-cousin mating.
7. In birds and mammals, it appears that the highest level of inbreeding is about 10% full-sib mating. What is the equilibrium level of inbreeding for this level of inbreeding and what is the resulting expected level of heterozygosity compared to the maximum possible?
8. Calculate the expected inbreeding coefficient in an offspring from individuals IVa and IVb in the pedigree in Figure 5.4.
9. Diagram the four types of first-cousin matings that are possible when the sexes of the parents of the first cousins are specified (see Figure 5.15b and 5.15c for two half-first-cousin matings). For haplo-diploid organisms, calculate the expected inbreeding coefficient for an offspring from each of these matings.
10. Assume that a locus is segregating in a population with allelic frequencies 0.9 (for A_1) and 0.1 (for A_2) and that the fitnesses of the genotypes A_1A_1, A_1A_2,

and A_2A_2 are 1, 1, and 0.8, respectively. What is the inbreeding depression from this locus when there is full-sib mating? Assuming that there are 500 identical loci like this one and that the fitness over loci is multiplicative, that is, overall fitness is $\overline{w}_f^{500}$, what is the expected relative fitness when there is full-sib mating?

11. A plant population was selfed, and the relative fitness of the selfed progeny was 0.6 that of outcrossed progeny. What are the estimates of inbreeding depression and the number of lethal equivalents?

12. In Example 5.12, assume that the estimates of the relative fitnesses of the three genotypes are 1.0, 0.87, and 0.7 (slightly different from that given in the text). Using these fitnesses, and assuming $H_O = 1$ and $S = 1$, what are the genotype frequencies and allele frequency of A_2 after one and two generations?

13. Some plant populations may reproduce by outcrossing, selfing, and asexual reproduction. Give an expression for the heterozygosity after one generation when all three occur. What will be the equilibrium frequency of heterozygotes in this situation? How does the level of asexual reproduction affect the equilibrium heterozygosity?

14. Make a graph of the equilibrium heterozygosity for different levels of positive-assortative mating when $p = 0.1$, 0.3, and 0.5. Compare the graph to Figure 5.8 for partial selfing.

15. Using the data in Table 5.14, calculate the level of relatedness using expressions 5.15*a* and 5.15*b* between individuals x and y for loci B, E and F.

6

Genetic Drift and Effective Population Size

> The views of Fisher and Wright contrast strongly on the evolutionary significance of random changes in the population. Whereas, to Fisher, random change is essentially noise in the system that renders the determining processes somewhat less efficient than they would otherwise be, Wright thinks of such random fluctuations as one aspect whereby evolutionary novelty can come about by permitting novel gene combinations. . . . The other aspect of near-neutral genes and a great multiplicity of potential mutations at each locus is the possibility that this may account for polymorphisms, particularly those having no overt effect and detected only by electrophoresis or other chemical trickery.
>
> James Crow and Motoo Kimura (1970)

> Small population size can lead to loss of neutral genetic variation, fixation of mildly deleterious alleles, and thereby reduced population fitness. The rate of this process depends on the effective size of a population, N_e, rather than the actual number of living individuals, N, making the effective size of a population one of the most fundamental parameters in evolutionary and conservation biology.
>
> Kalinowski and Waples (2002)

Since the beginning of population genetics, there has been controversy concerning the importance of chance changes in allele frequencies caused by finite population size. Part of this controversy has resulted from the large numbers of individuals observed in many natural populations, large enough to make chance effects small in comparison to the effects of other factors, such as selection and gene flow. However, if the selective effects or amount of gene flow are small relative to the population size, then long-term genetic change caused by genetic drift may be important. Consideration of this possibility, even when the population size is large, led to the development of the theory of neutrality, in which selectively neutral variants are generated by mutation and changed in frequency by genetic drift (see Chapter 8 for a discussion).

Under certain conditions, a finite population may be so small that genetic drift is significant even for loci with sizable selective effects or when there is gene flow. First, some populations may be continuously small for a relatively long period of time because of limited resources in the populated area, low vagility (tendency or capacity to disperse) between suitable habitats, territoriality among individuals, or other factors. For example, lizard numbers can be limited by perch sites and territoriality, bird populations by nesting sites and territoriality, and the number of colonizing plants by open habitat and vagility between habitats. Isolated populations, whether of land animals or plants on an island, vertebrates or invertebrates in a lake, or other groups living in a circumscribed area, may also have a continuously low population size.

Second, some populations may have intermittent small population sizes. Examples of such episodes are the overwintering loss of population numbers in many invertebrates, periodic crashes of populations in small rodents such as lemmings and voles, epidemics that periodically decimate populations of both plants and animals, and the seasonal desiccation of ponds that affect population numbers of many species. Such population fluctuations generate **bottlenecks**, ***periods during which only a few individuals survive to continue the existence of the population***. A classic example of periodic oscillations is the relative abundance of the lynx and snowshoe hare in Canada, where both species show approximately 9- to 10-year oscillations and the population density fluctuates by an order of magnitude or more. As a result, in periods of low density, individuals of both species often become exceedingly rare.

Small population size is also important when a population grows from a few founder individuals, a phenomenon termed **founder effect**. For example, many island populations appear to have started from a very small number of founders. If a single female who was fertilized by a single male founds a population, then only four genomes, two from the female and two from the male, may start a new population. In plants, an entire population may be initiated from a single seed—only two genomes, if self-fertilization occurs. As a result, ***populations descended from a small founder group may have low genetic variation or by chance have a high or low frequency of particular alleles***. Such initial restrictions in the number of founders also appear to be important in some human populations. For example, some religious isolates in North America, such as the Amish and the Hutterites, were initiated by small numbers of migrants from Europe; some remote sites, such as the island Tristan da Cunha, were settled by a few individuals (see Example 6.1, which discusses a rare type of dwarfism in the Amish and the number of founding mtDNA and Y-chromosome lineages in the Tristan da Cuhna islanders). Furthermore, the high frequency of some rare genetic diseases in the populations of some countries, such as in Finland, is attributed to the founder effect, chance increases in some genetic variants as a result of the relatively small numbers of initial founders (de la Chapelle and Wright, 1998; Norio, 2003).

Example 6.1. The Amish population of Lancaster County, Pennsylvania, has a high incidence of a recessive disorder known as six-fingered dwarfism (Fig. 6.1) or Ellis–van Creveld (EvC) syndrome (McKusick, 1978). From a population of about 13,000, 82 affected individuals in 40 affected sibships were diagnosed as having this disease. If inbreeding is taken into account, the frequency of the recessive allele is estimated to be about 0.066 and the incidence of the disease is about 0.005. Indicative of the restricted number of founders in this population, the 80 parents in these 40 sibships all trace their ancestry to Samuel King and his wife, early members of the community. From this pedigree information, it appears quite certain that the high incidence can be primarily attributed to founder effect. Either Samuel King or his wife carried the recessive allele; and because many individuals in the population are their descendants, the incidence of the disease is now high.

Tristan da Cunha is a small, remote island in the south Atlantic about 2900 km west of South Africa. The British established a garrison on the island in 1816 to prevent the French from rescuing Napoleon who was in exile on St. Helena, 2259 km to the north. From genealogical records, the contributions of only seven women remain: M. L. who came in 1816; M. W., S. W., and M. W. in 1827; S. P. in 1863; and E. S. and A. S. in 1908. Because mtDNA is maternally inherited with no recombination, present-day mtDNA types can be used to trace the ancestry to the founding females. Table 6.1a gives the mtDNA types found in 161 present-day individuals for nine different mtDNA regions first found by SSO probes and then described by sequence differences (Soodyall *et al.*, 1997). S. W. and M. W. were described as sisters from the historical data, but the mtDNA show that they have distinct mtDNA types. M. W. and M. W. were mother and daughter and E. S. and A. S. were sisters, both of which are confirmed by the mtDNA data. In other words, from the genealogical information, four founder mtDNA types were expected, but five were observed. The estimated level of mtDNA diversity using expression 2.18*c* is 0.768.

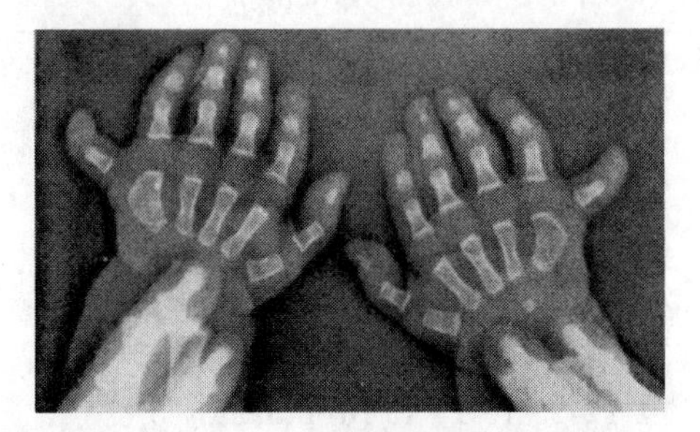

Figure 6.1. An X-ray of the hands of an Amish person with Ellis–van Creveld syndrome, a form of dwarfism in which affected individuals have six fingers on each hand. (McKusick, Victor A., M.D. *Medical Genetic Studies of the Amish: Selected Papers*. p. 96. © 1978. Reprinted with permission of The Johns Hopkins University Press.)

There are seven family names in use in Tristan, corresponding to the number of founding fathers with present-day descendants from public records (Soodyall *et al.*, 2003). Because Y chromosomes are paternally inherited with no recombination, present-day Y-chromosome haplotypes can be used to trace ancestry to the founding males. Within each family, there was a haplotype that could be traced to the known ancestors (Table 6.1b). However, two other haplotypes were also observed, one in family 3 that appears to be a mutation and one in family 4 that appears to be from a migrant. In addition, in families 5, 6, and 7, haplotypes from other families were also found that appear from pedigree examination to be the result of four instances of nonpaternity and the subsequent descendants. Overall, there are nine Y haplotypes, and the estimated level of Y-chromosome diversity using expression 2.18c is 0.847.

TABLE 6.1 (a) The mtDNA sequence differences in Tristan da Cunha islanders showing the number of individuals with the types traced to the founding females (Soodyall *et al.*, 1997) and (b) the seven families and the Y-chromosome haplotypes found in each one (Soodyall *et al.*, 2003). The repeat numbers for the microsatellite alleles that are different from the ancestral family type are given in boldface.

(a)

Founding females	*mtDNA sequence*	*N (proportion)*
S. W.	ACTTGTTTCG	46 (0.29)
M. W. and M. W.	GTTCGCTTCG	34 (0.21)
E. S. and A. S.	GCTTATCTTG	25 (0.16)
M. L.	ATCTGCCCTA	11 (0.07)
S. P.	GTCTGTCCTG	45 (0.28)
Total		161 (1.0)

(b)

Family	*Y-chromosome haplotype*	*N (proportion)*
1	15-12-25-10-14-13	5 (0.066)
2	14-12-24-11-13-13	3 (0.039)
3	14-12-23-11-13-13	9 (0.118)
	14-12-23-**10**-13-13 (mutant)	4 (0.053)
4	14-12-24-10-13-14	8 (0.105)
	16-12-**25**-10-**11**-**13** (migrant)	1 (0.013)
5	14-12-23-10-14-13	16 (0.211)
	14-**14**-**22**-10-**11**-13 (from family 7)	3 (0.039)
6	16-13-24-10-11-13	10 (0.132)
	14-**12**-**23**-10-**14**-13 (from family 5)	1 (0.013)
7	14-14-22-10-11-13	14 (0.184)
	14-**12**-**23**-10-**14**-13 (from family 5)	2 (0.026)
Total		76 (1.0)

Another situation in which small population size may be of great significance is where the population (or species) in question is one of the many threatened or endangered species. For example, only 20 whooping cranes were alive in 1920 because they were hunted and their habitat was destroyed. Thanks to protection and captive rearing, their numbers have now grown to approximately 130 in captivity and to approximately 320 in the wild where two new populations have been established. The northern elephant seal was hunted to near extinction in the 19th century, and as few as 20 individuals are believed to have survived on a remote beach on Isla Guadalupe, Mexico, in the Pacific Ocean. Today it is estimated that there are nearly 200,000 Northern elephant seals, all descended from the small surviving colony (Bonnell and Selander, 1974; Weber *et al.*, 2000; Hoelzel *et al.*, 2002; see also Example 6.12).

Furthermore, some species, such as Przewalski's horses, California condors, black-footed ferrets, Galapagos tortoises from Espanola Island, and Mexican wolves have gone extinct, or were very near extinction, in nature. All of the living individuals of these species are descended from a few individuals that were brought into captivity to establish a protected population: Przewalski's horses, the only extant wild horse species, are descended from 13 animals caught primarily around 1900 (Boyd and Houpt, 1994). California condors, the largest bird in North America, are descended from 13 animals, the last of which were caught in 1987. Black-footed ferrets are descended from six animals caught in Wyoming in 1986 (Seal *et al.*, 1989). The 1200 repatriated Galapagos tortoises from Espanola Island are all descended from the last wild 12 females and three males taken into a captive breeding program in the 1960s (Milinkovitch *et al.*, 2004). All Mexican wolves are descended from seven animals caught primarily in Mexico in the 1970s (Hedrick *et al.*, 1997). For these species, there are reintroduction programs that have used descendents of the captured individuals to establish protected populations in natural habitats: all of these programs have had both their setbacks and successes. The management of these species continues to be of great concern (Ballou *et al.*, 1995), and it remains to be seen whether these species have retained, as they passed through the bottleneck caused by their near extinction, enough genetic variation to adapt to future environmental changes.

The effect of finite population size on genetic variation was investigated in depth in the 1930s and 1940s by Wright (1969, for a summary). In the 1950s, Kimura introduced the diffusion equation approach to understanding the impact of genetic drift (see Kimura and Ohta, 1971). Their elegant work has contributed greatly to our basic knowledge concerning the interplay of genetic drift and other factors such as selection, mutation, and gene flow. Here we concentrate on discrete generation models and illustrate, through some numerical examples, the dynamics of genetic variation in a finite population. As we said at the beginning of Chapter 5, inbreeding and genetic drift appear to have similar overall effects on heterozygosity, but when examining a given locus within a population, the predicted effect is different. Obviously, the fundamental importance of genetic drift in understanding molecular evolution and the very small population sizes in many threatened and endangered species make genetic drift of great significance today in applications of population genetics. Here we discuss the effective population size, an approach that allows the generalization of the effects of genetic drift. We wait to introduce the concept of coalescence—that is, how the effect of genetic drift can be traced backward in the ancestry of a contemporary population—until Chapter 8 on the neutrality theory.

SEWALL WRIGHT (1889–1988)

Sewall Wright, born in Illinois, carried out much of his early research on problems in physiological and developmental genetics and was one of the first scientists to recognize a direct relationship between genes and enzymes (Wright, 1917). Working on the guinea pig, he detailed the complex inheritance patterns for a number of coat-color genes. Although he did not publish his work in book form until the late 1960s and 1970s (his four-volume work is a comprehensive treatment; Wright, 1968, 1969, 1977, 1978), his contributions to inbreeding analysis, the consideration of finite population size, and many other topics are fundamental to population genetics. In fact, genetic drift was sometimes referred to as *the Sewall Wright effect.* Wright used a number of ingenious mathematical approaches to understand the effect of finite population size, and his view of the factors (and their interactions) affecting evolutionary genetic phenomena is central to the thinking and approaches of most modern-day population geneticists. Provine (1986) wrote a biography of Wright, and Hill (1995) wrote a perspective on his early papers. (Photo courtesy of USDA.)

I. THE EFFECT OF GENETIC DRIFT

All the above examples of restricted population size can have the same general genetic consequence: a small population size causes chance alterations in allele frequencies. **Genetic drift** is the ***chance changes in allele frequency that results from the sampling of gametes from generation to generation in a finite population***. Genetic drift has the same expected effect on all loci in the genome. In a large population, on the average, only a small chance change in the allele frequency will occur as the result of genetic drift. On the other hand, if the population size is small, then the allele frequency can undergo large fluctuations in different generations in a seemingly unpredictable pattern and can result in chance fixation or the loss of an allele.

Figure 6.2 illustrates the type of allele frequency change expected in a small diploid population with two alleles. This example uses Monte Carlo simulation with uniform random numbers to imitate the allele changes in four populations (see p. 17). In Figure 6.2, the solid lines are four replicates of a hypothetical diploid population of size $N = 20$ ($2N = 40$), and the broken line is the mean frequency of allele A_2 over the four replicates. All of the replicates were initiated with the frequency of allele A_2 equal to 0.5. One of these simulated replicates went to fixation for the A_2 allele in generation 19, and another lost the A_2 allele in generation 28. The other two replicates were still segregating for both alleles at the end of 30 generations. As shown here, genetic drift may cause large and erratic changes in allele frequency in a rather short time.

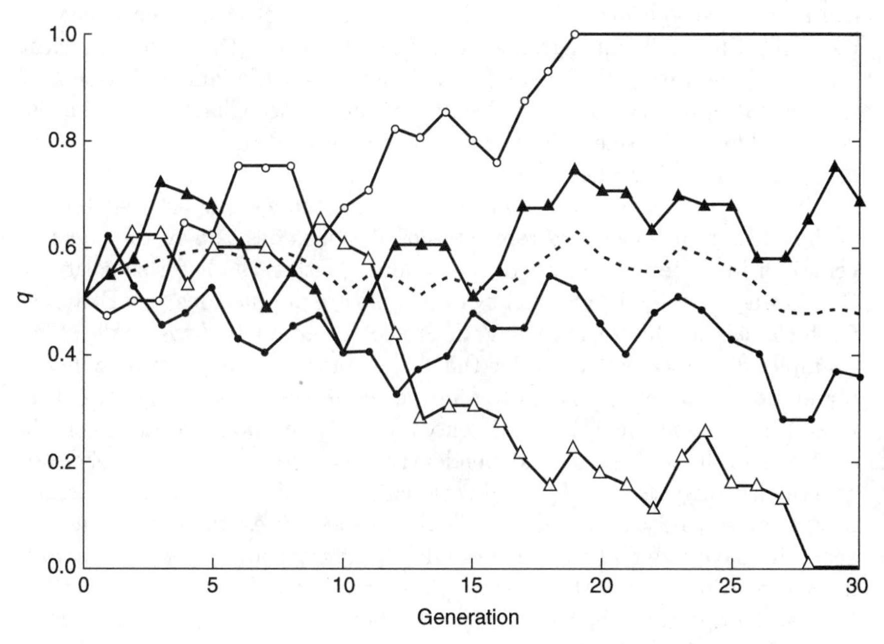

Figure 6.2. Allele frequency over time for four replicates (solid lines) of a population of size 20. The mean frequency of allele A_2 for the four replicates is indicated by the broken line.

On the other hand, the mean of the four replicates varied much less: it ranged from 0.625 in generation 19 to 0.475 in generation 30 but was generally near to the initial frequency of 0.5. If there are enough replicate populations, then there is no expected change in the mean allele frequency, so that

$$\bar{q}_0 = \bar{q}_1 = \bar{q}_2 \cdots \bar{q}_t \cdots \bar{q}_\infty$$

where $\bar{q}_t$ is the mean frequency of A_2 in generation t over all replicates. The constancy of the mean occurs because the increases in allele frequency in some replicates are cancelled by reductions in allele frequency in other replicates.

Individual replicates eventually either go to fixation for A_2 ($q = 1$) or to loss of A_2 ($q = 0$). The ***proportion of populations expected to go to fixation for a given allele is equal to the initial frequency of that allele***. In other words, if the initial frequency of A_2 is q_0, then the **probability of fixation** of that allele, $\boldsymbol{u(q)}$ (proportion of replicate populations fixed for it), is

$$u(q) = q_0 \tag{6.1}$$

For example, if the initial frequency is 0.1, only 10% of the time will a population become fixed for that allele. On the other hand, if the other allele A_1 has an initial frequency of 0.9, 90% of the time it will become fixed. This can be understood intuitively because the amount of change necessary to go from 0.1 to 1.0 is much greater than from 0.9 to 1.0 (there are numerical examples illustrating this below). This finding is a fundamental aspect of the neutrality theory used in molecular evolution (see Chapter 8)—that is, without differential selection, the probability of fixation of a given allele is equal to its initial frequency.

Because the mean allele frequency does not change but the distribution of the allele frequencies over replicate populations does, the overall effect of genetic drift is best understood by examining either the heterozygosity or the variance of the allele frequency over replicate populations (see Example 6.2 for a classic illustration using an eye color variant in *Drosophila*). The examples discussed so far involve the change of a given locus over replicate populations. However, we are often interested in the impact of genetic drift in all the different loci (the total genome) in a given population. If genetic drift affects different genes independently of each other, then the effect of genetic drift can also be observed by looking at multiple genes in the same organism. In reality, this is often difficult because, for example, different loci generally have different numbers of alleles, different allele frequencies, and linkage relationships with other loci that may be influenced by selection. It is still important, however, to remember in our discussion here that although we are talking about the effects of genetic drift at a given locus, genetic drift acts in essentially in the same general manner over all of the loci in a given population.

Example 6.2. A classic illustration of how finite population size affects allele frequency in the manner described was provided by Buri (1956). He looked at the frequency of two alleles at the *brown* locus that affects eye color in *Drosophila melanogaster* in randomly selected populations of size 16.

The alleles, bw^{75} and bw, were chosen because they appeared to be nearly neutral with respect to each other. As a result, the effect of finite population size on allele frequency could be examined almost independently of the effect of selection. Some data from his study are presented here in two ways:

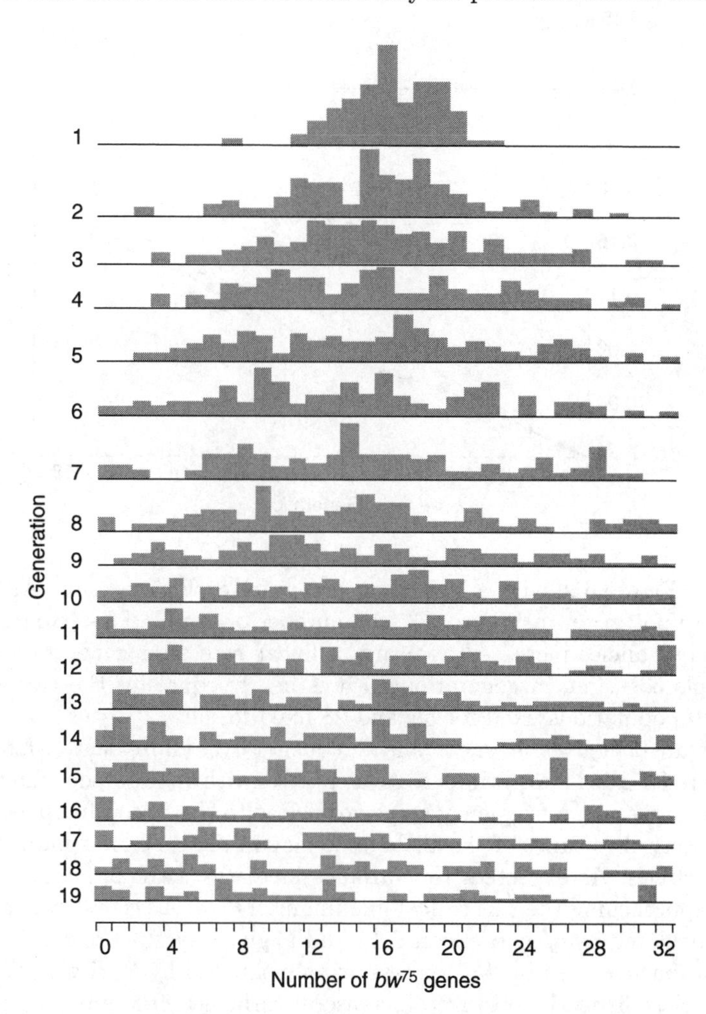

Figure 6.3. The distribution of bw^{75} alleles over time in populations of size 16 for the segregating replicates in an experiment of Buri (after Buri, 1956).

first in Figure 6.3 as the number of the 107 replicate populations that had from 0 to 32 bw^{75} genes in different generations, and then in Figure 6.4 as the mean and variance of the allele frequency over the populations.

The histograms in Figure 6.3 illustrate that the distribution of the allele frequencies over replicates has a greater and greater spread with time. This is a graphical way of presenting the type of data (but for more replicates)

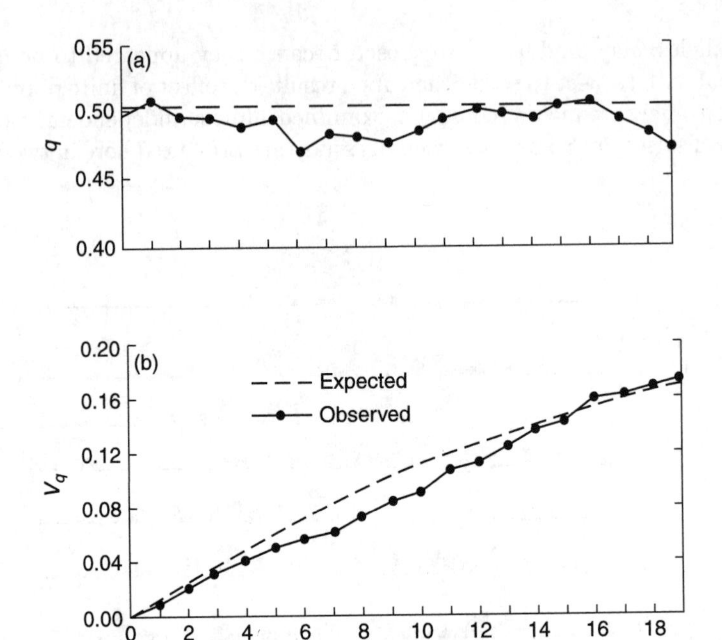

Figure 6.4. The observed and expected mean (a) and variance (b) in allele frequency in the experiment of Buri (1956). The expected variance was generated using expression 6.4a and a population size of nine.

given in Figure 6.2, with a histogram drawn for each generation. Although they are not given in Figure 6.3, the number of populations fixed for one of the two alleles increased at nearly a linear rate after generation 4 (see Example 6.3), and in generation 19 it is nearly equal for the two alleles, with 30 populations fixed for bw and 28 fixed for bw^{75}.

Figure 6.4a gives the mean allele frequency over all replicates (fixed and unfixed) for Buri's experiment. As expected with little or no differential selection, the mean frequency stays very close to the initial frequency, 0.5. The observed variance of the allele frequency in Figure 6.4b is indicated by closed circles. As expected, the variance increases with time and appears to be approaching the theoretical maximum of 0.25. A theoretical variance was calculated using expression 6.4a and is given by the broken line. The population size used in this expression as estimated by Buri was nine and gave a close fit to the observed increase in variance with time.

To derive the expected heterozygosity in a finite population, let us assume that there are N diploid individuals in the population and that each individual contributes two haploid gametes to the next generation (Crow and Kimura, 1970). To generate offspring, let us choose alleles randomly (and with replacement) from these parents. The probability that the same allele is drawn twice is $2N[1/(2N)]^2 = 1/(2N)$. The probability that the two alleles drawn for an offspring are different is $1-1/(2N)$. However, even if the two alleles are different (not from the exact same allele in the parents), it is possible that these alleles are the same because they came from a common ancestor in a previous generation. If we assume, as in Chapter 5, that f_t is the inbreeding coefficient in generation t, the probability that the alleles are identical by descent, then, is

$$f_{t+1} = \frac{1}{2N} + \left(1 - \frac{1}{2N}\right) f_t \tag{6.2}$$

This can be rewritten (add unity to both sides of the equation in the process) as

$$1 - f_{t+1} = \left(1 - \frac{1}{2N}\right)(1 - f_t)$$

Remember from Chapter 5 that $1 - f = H/(2pq)$ so that

$$H_{t+1} = \left(1 - \frac{1}{2N}\right) H_t \tag{6.3a}$$

which indicates that the heterozygosity declines each generation at a rate inversely dependent on the population size.

The relationship between the heterozygosity over several generations can be generated from this expression; it is

$$H_t = \left(1 - \frac{1}{2N}\right)^t H_0 \tag{6.3b}$$

Thus, for a given initial heterozygosity, H_0, the heterozygosity t generations later can be predicted. Expression 6.3b is approximately

$$H_t = H_0 e^{-t/2N}$$

From this expression, we can calculate the approximate time until a given proportion of the heterozygosity is lost. For example, the number of generations until a proportion x ($= H_t/H_0$) of the original heterozygosity is left is

$$t = -2N \ln x$$

We can also determine that the time until 50% of the heterozygosity is lost ($x = 0.5$), the half-life, is 1.39 N.

On p. 479, we show how average observed heterozygosity is affected when a population is subdivided (Wahlund effect), an effect identical to examining the average observed heterozygosity over replicate experimental populations. From that consideration, we can use expression 9.3*a*, which relates the observed heterozygosity to the difference in the expected heterozygosity and the variance in allele frequency (V_q) as

$$H = 2pq - 2V_q$$

If we substitute this in expression 6.3*b* for H_t, assuming that the initial heterozygosity is $2p_0q_0$, and rearrange, the variance in allele frequency in generation t becomes

$$V_{q \cdot t} = p_0q_0\left[1 - \left(1 - \frac{1}{2N}\right)^t\right] \quad (6.4a)$$

As the number of generations increases—that is, as t becomes large—the variance approaches a maximum of p_0q_0. For example, if $p_0 = 0.3$, the variance of the allele frequency will approach a maximum of 0.21 at a rate that is a function of the population size. After one generation ($t = 1$), the variance is

$$V_q = \frac{p_0q_0}{2N} \quad (6.4b)$$

which is the binomial sampling variance. As a general yardstick to measure the effect of genetic drift on allele frequency and compare it with other evolutionary factors, such as selection, gene flow, or mutation, the standard deviation of q for one generation is approximately equal to the average absolute value of the allele frequency change (T. Prout, personal communication). For example, if $p_0 = q_0 = 0.5$ and $N = 50$, then $(V_q)^{1/2} = 0.05$, which is approximately equal to the mean $|\Delta q|$.

a. The Probability Matrix Approach

Although it is impossible to determine precisely how much change in allele frequency in a population is due to genetic drift, we can calculate the probability that the allele frequency will be a certain value. For example, given an allele frequency of 0.4 in a population of size 10, there is an 18% chance that the allele frequency will remain at exactly 0.4 after one generation. Such probabilities can be calculated for different population sizes and allele frequencies, and they give us a general way to examine the effect of genetic drift. These probabilities can be arranged in matrix form and can give the expected change in the distribution of alleles in a population of a given finite size over time.

Such a matrix has as its elements the probabilities of a certain number of alleles of a particular type in the next generation, given a certain number in the previous generation (see p. 26 for an introduction to using matrices). More specifically, the elements of this matrix, called a **probability transition matrix**, are the probability of iA_2 alleles in generation $t + 1$ given jA_2 alleles in generation t. These elements can be calculated from the binomial probability expression as follows:

$$x_{ij} = \frac{(2N)!}{(2N-i)!i!}\left(1 - \frac{j}{2N}\right)^{2N-i}\left(\frac{j}{2N}\right)^{i}$$

where the frequency of A_2 in generation t is $j/2N$ and there are $2N$ alleles in the population.

A simple example of such a matrix is given in Table 6.2 for a population of size two ($2N = 4$). The matrix has five columns corresponding to the possible states in generation t (0, 1, 2, 3, or 4 A_2 alleles) and five rows corresponding to the possible states in generation $t + 1$. The first and last columns have only one nonzero element, in the first and last rows, respectively. This occurs because once a population is homozygous either for A_1 or A_2 it will continue to be homozygous for that allele if the lost allele is not reintroduced back into the population. As a result, these two states, 0 A_2 and 4 A_2 alleles, are termed **absorbing states.** On the other hand, all of the elements in the middle three columns are nonzero, which indicates that there is a probability of a population moving to all other possible states from these states. For example, the probability of 0 A_2 alleles in generation $t + 1$, given 1 A_2 allele in generation t, x_{01}, is $(0.75)^4 = 0.3164$. The probabilities in all columns sum to unity because these values specify all of the possible transitions from a given initial state.

TABLE 6.2 A probability transition matrix for a population of size two ($2N = 4$), where the values indicate the probability of iA_2 alleles in generation $t+1$, given jA_2 alleles in generation t.

	Generation t				
Generation t + 1	0	1	2	3	4
0	1	0.3164	0.0625	0.0039	0
1	0	0.4219	0.25	0.0469	0
2	0	0.2109	0.375	0.2109	0
3	0	0.0469	0.25	0.4219	0
4	0	0.0039	0.0625	0.3164	1

Once we have a transition matrix that gives the probability of change from one state to another, we can evaluate how the distribution of allele frequencies for populations of a given size is expected to change with time. Such a ***distribution of allele frequencies over populations of the same size*** is termed the **allele-frequency distribution** (or often called the **gene-frequency distribution**). We can observe the change of this distribution by assuming some initial distribution of allele frequencies over populations and then calculating distributions in future generations using the transition matrix. More specifically, the proportion of populations that have jA_2 alleles in generation t is $y_{j\cdot t}$, and the sum over all possible population states is unity, or

$$\sum_{j=0}^{2N} y_{j\cdot t} = 1.0$$

If we call the matrix of probability transition values (the x_{ij} values) $\mathbf{X}$ and the vector of population states (the $y_{j\cdot t}$, values) $\mathbf{Y}_t$, then we can specify the distribution of population states from one time to the next by multiplying the transition matrix by the vector of population states (remember, this is the allele-frequency distribution) or

$$\mathbf{Y}_{t+1} = \mathbf{X}\mathbf{Y}_t$$

In other words, the proportion of populations in state i at time $t + 1$ can be obtained by postmultiplication of the matrix $\mathbf{X}$ by the vector $\mathbf{Y}_t$ so that

$$y_{i\cdot t+1} = \sum_{j=0}^{2N} x_{ij}\, y_{j\cdot t}$$

Therefore, the proportion of populations in state i at time $t + 1$ is the sum for all states of the product of the transition to state i from state j and the proportion of populations in state j at time t. If the initial distribution of population states is given as Y_0, then the above recursion relationship can be generalized to

$$\mathbf{Y}_t = \mathbf{X}^t Y_0$$

To illustrate how the population states change over time, let us use the transition matrix given in Table 6.2. Assume that initially all populations had equal numbers of A_1 and A_2 alleles. In other words, two of the four alleles in the zero generation are A_2 so that $y_{2\cdot 0} = 1.0$, and all other initial states are 0.0, making $q_0 = 0.5$. With this initial distribution, the distribution of allele frequencies over populations changes with time; it is given in Table 6.3. One observation is that a high proportion of the populations quickly become homozygous either for A_1 or A_2. In fact, after only three generations, almost 50% of the populations are homozygous either for A_1 or A_2 because of the small population size. Eventually, 50% of the populations become homozygous for A_1 and 50% for A_2.

TABLE 6.3 The distribution of allele frequencies and heterozygosity over generations for populations of size two ($2N = 4$) when $q_0 = 0.5$.

Number of	*Generation*						
A_2 *alleles*	0	1	2	3	4	...	∞
0	0	0.0625	0.1660	0.2490	0.3117	...	0.5
1	0	0.25	0.2109	0.1604	0.1205	...	0.0
2	1	0.375	0.2461	0.1813	0.1356	...	0.0
3	0	0.25	0.2109	0.1604	0.1205	...	0.0
4	0	0.0625	0.1660	0.2490	0.3117	...	0.5
q_t	0.5	0.5	0.5	0.5	0.5	...	0.5
H_t	0.5	0.375	0.2812	0.2109	0.1582	...	0.0

The mean frequency of allele A_2 in generation t can be calculated as

$$q_t = \frac{1}{2N}\sum_{j=0}^{2N} j y_{j \cdot t}$$

The frequency of A_2 in the different generations for this example is given at the bottom of Table 6.3. The frequency of A_2 remains at 0.5 even though the distribution of the allele frequency over replicates continuously spreads until fixation of all of the replicates. Eventually, 0.5 of the replicates become fixed for A_2; this probability of fixation is equal to the initial frequency of A_2.

An important observation is that there is a constant rate of decrease in heterozygosity per generation. The heterozygosity in generation t can be calculated as

$$H_t = 2\sum_{j=0}^{2N}\left(1 - \frac{j}{2N}\right)\left(\frac{j}{2N}\right) y_{j \cdot t}$$

The relationship of heterozygosity between generations is

$$H_{t+1} = \lambda H_t$$

where λ indicates the characteristic rate of decline of heterozygosity (expression 5.9). To illustrate, we can use the heterozygosity calculated for the early generations given in the example in Table 6.3 where H_0, H_1, and H_2 were 0.5, 0.375, and 0.2812, respectively. In this case, $\lambda = 0.75$ for all comparisons between adjacent generations. At first it may seem surprising that this decline in heterozygosity is different than that the expected decline from full-sib mating on p. 260. Although it is not obvious, there is a probability of random self-fertilization here, and thus, the loss of heterozygosity is somewhat faster than when self-fertilization does not occur.

This expression can be rearranged so that

$$\lambda = \frac{H_{t+1}}{H_t}$$

where λ is specific for a particular population size and is equal to

$$\lambda = 1 - \frac{1}{2N}$$

as was shown in expression 6.3*a*.

A second numerical example is given in Table 6.4, where initially all populations had only one A_2 allele so that $y_{1\cdot 0} = 1.0$ ($q_0 = 0.25$) and the other initial states are 0.0. As in the previous example, the distribution quickly spreads, and populations become fixed either for A_1 or A_2. The frequency of A_2 remains at 0.25, and the rate of decline of heterozygosity, λ, is the same as in the previous example even though the initial allele frequency (as well as the heterozygosity) is different. Again, the probability of fixation of A_2 is equal to the initial frequency of A_2, 0.25.

TABLE 6.4 The distribution of allele frequencies and heterozygosity over generations for populations of size two ($2N = 4$) when $q_0 = 0.25$.

Number of A_2 alleles	Generation 0	1	2	3	4	...	∞
0	0	0.3164	0.4633	0.5484	0.6038	...	0.75
1	1	0.4219	0.2329	0.1471	0.1003	...	0.0
2	0	0.2109	0.1780	0.1353	0.1017	...	0.0
3	0	0.0469	0.0923	0.0943	0.0805	...	0.0
4	0	0.0039	0.0336	0.0748	0.1137	...	0.25
q_t	0.25	0.25	0.25	0.25	0.25	...	0.25
H_t	0.375	0.2812	0.2109	0.1582	0.1187	...	0.0

We can use the probability matrix approach to calculate the distribution of allele frequencies over time for finite populations of different sizes. As an example, let us examine a population of size 20 ($2N = 40$) and follow its distribution over time, again initially assuming equal numbers of A_1 and A_2 alleles so that $y_{20\cdot 0} = 1.0$ and all other initial population states are 0.0. Then the distribution of allele frequencies follows the pattern given in Figure 6.5 for generations 1, 5, and 20 (Figure 6.5 is a smoothed representation of the actual distribution). After one generation, a large proportion of the populations is still near the initial frequency of 0.5. After five generations, the spread is much greater, and after 20 generations, more than 16% of the populations are homozygous (half for A_1 and half for A_2). Obviously, the spread of the allele frequency distribution takes place very quickly, and

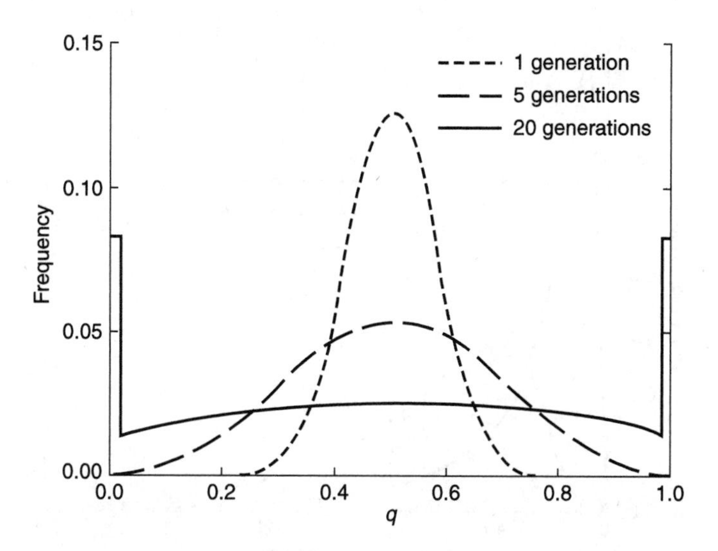

Figure 6.5. The smoothed distribution of allele frequency for a population of size 20 and an initial allele frequency of 0.5 after 1, 5, and 20 generations.

by generation 20, there is a nearly uniform distribution among all population states in which there is still polymorphism. Eventually half the populations become fixed for A_1 and half for A_2 because the initial allele frequency was 0.5.

The **mean time until fixation** of an allele (here A_2) depends on the population size and the initial frequency of the allele. As the population size increases, the effect of genetic drift per generation becomes smaller so that it takes longer for chance changes to accumulate and result in fixation. For a given population size, the further the initial frequency is from unity (the frequency when fixed) the longer it takes for an allele to become fixed. The time can be calculated directly from the iteration of the transition matrix until all populations are fixed as

$$T(q) = \frac{1}{q}\sum_{t=1}^{\infty} t\left(y_{2N \cdot t} - y_{2N \cdot t-1}\right) \tag{6.5a}$$

where the term in parentheses is the proportion of populations that become fixed in generation t. The summation is divided by the allele frequency because only q of the populations will become fixed for A_2.

Kimura and Ohta (1971), using a diffusion approximation for a continuous time model, have given an expression for the mean time until fixation of allele A_2 with an initial frequency of q as

$$T(q) = -\frac{4N(1-q)\ln(1-q)}{q} \tag{6.5b}$$

The mean time until fixation is a linear function of population size and decreases as the initial allele frequency gets higher—that is, becomes closer

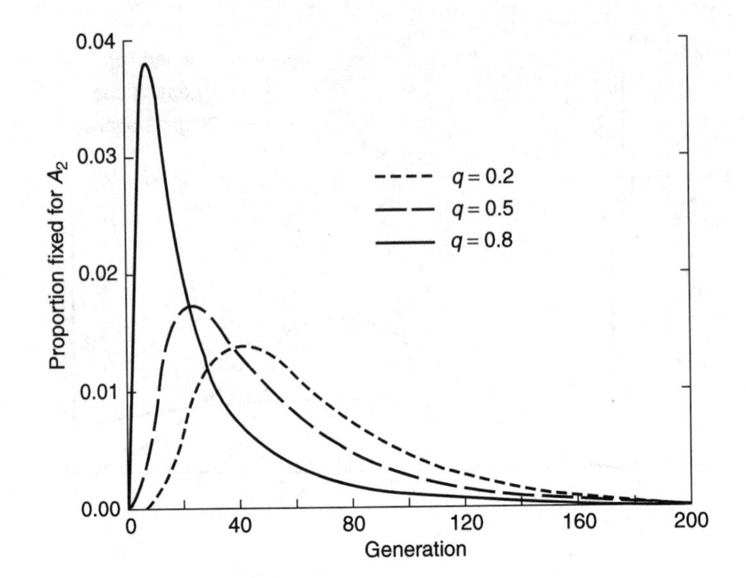

Figure 6.6. The smoothed distribution of populations becoming fixed for A_2 in each generation for three initial allele frequencies when $N = 20$.

to unity. For example, if the initial allele frequencies are $q = 0.2$ and 0.8, then the mean times until fixation become $3.57N$ and $1.61N$, respectively. This approximation and the result obtained by using a transition matrix in expression 6.5a are quite close unless the population size is very small. When q is small, as for a new mutant, expression 6.5b becomes

$$T(q) = 4N \tag{6.5c}$$

We discuss this elegant prediction from neutral theory—that is, the ***expected time to fixation for a neutral mutant is four times the population size***—again in Chapters 7 and 8.

Figure 6.6 gives the proportion of populations fixed for the A_2 allele each generation for three different initial frequencies of A_2 and a population of size 20, using the transition matrix approach. When the initial frequency of A_2 is 0.8, most of the fixation takes place in the first few generations. When the initial frequency is 0.2, the peak fixation period of A_2 is delayed considerably. These differences are due to the total amount of allele frequency change necessary for fixation for these different initial frequencies. The mean times to fixation calculated from expressions 6.5a and 6.5b for $q = 0.2$ are 69.5 and 71.4 generations, respectively, in this example. These are of course somewhat less than $4N$ because the initial allele frequency is significantly greater than $1/(2N)$, the lowest it could be in a polymorphic population. Example 6.3 gives the observed and expected fixation times for the *Drosophila* experiment of Buri (1956) in Example 6.2.

Example 6.3. We can use the data from the classic *Drosophila* experiment of Buri (1956) that we discussed in Example 6.2 to calculate the observed rate that populations become fixed and compare it with the theoretical predictions from expressions 6.5*a* and 6.5*b*. Figure 6.7 gives the observed cumulative proportion of lines fixed for the two alternative alleles, *bw* and bw^{75}. The expected cumulative proportion of fixation was obtained by iterating a transition matrix with $2N = 18$, the effective population size that Buri (1956) found that was a good fit to the observed variance in the experiment. Overall, although there is a lag in fixation for both alleles in the early generations, the observed values are generally a close fit to the expectation.

By using expression 6.5*a*, we can calculate the expected time to fixation over the 19 generations of the experiment and compare it with that observed. The observed average time, 12.8 generations, is slightly longer than that expected, 11.4 generations. Although the total proportions fixed by generation 19 for that observed and expected are very close (0.542 vs. 0.531), the difference in mean times occurs because the fixations observed took place slightly later than those expected. If we use expressions 6.5a and 6.5*b*, the overall expected times to fixation are 23.5 and 25.0 generations. If Buri had wanted to continue his experiment until 95% of the populations were fixed, based on transition matrix results, he would have had to continue the experiment for another 30 generations.

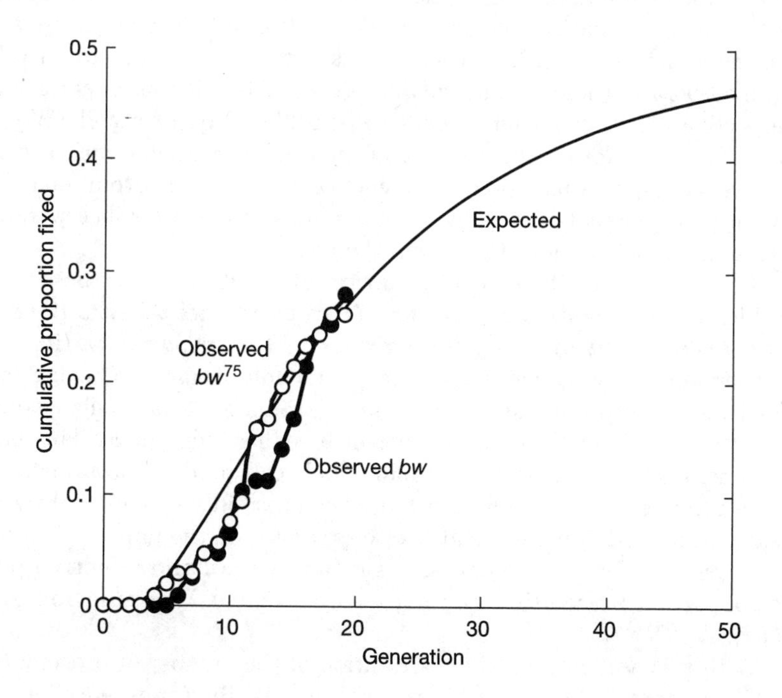

Figure 6.7. The observed cumulative proportion of fixed populations for alleles *bw* (closed circles) and bw^{75} (open circles) (Buri, 1956). Also given is the expected cumulative proportion from iteration of the transition matrix with $2N = 18$.

II. EFFECTIVE POPULATION SIZE

In examining the effects of genetic drift, we have assumed a given population size N. However, the population size that is relevant for evolutionary matters, the number of breeding individuals, may be quite different from the total number of individuals in an area, the **census population size**, that is the appropriate measure for many population ecology studies (Begon *et al.*, 1996; Krebs, 2002). In some cases, the **breeding population size** may be only a small proportion of the total number of individuals—for example, in trees, mammals, or other organisms that mature only after a prolonged juvenile stage much of the population may be prereproductive. In humans and some other vertebrates, there may be postreproductive adults as well.

The size of the breeding population may be estimated with reasonable accuracy by counting indicators of breeding activity such as nests, egg masses, and colonies in animals or by counting the number of flowering individuals in plants. However, even the breeding population number may not be indicative of the population size that is appropriate for evolutionary studies. For example, factors such as variation in the sex ratio of breeding individuals, offspring number per individual, and numbers of breeding individuals in different generations may be evolutionarily important. All of these factors can influence the genetic contribution to the next generation, and a general estimate of the breeding population size does not necessarily take them into account. As a result, the effective population size, a value that incorporates these factors and allows general predictions or statements irrespective of the particular forces responsible, is quite useful (Wright, 1931). In other words, the concept of an ideal population with a given effective size enables us to draw inferences concerning the evolutionary effects of finite population size by providing a mechanism for incorporating factors that result in deviations from the ideal.

The concept of **effective population size, N_e**, makes it possible to consider an ***ideal population of size N in which all parents have an equal expectation of being the parents of any progeny individual***. In other words, the gametes are drawn randomly from all breeding individuals, and the probability of each adult producing a particular gamete equal to $1/N$ where N is the number of breeding individuals. This basic model assumes that these diploid individuals can produce both male and female gametes (monoecy) and includes the possibility of self-fertilization. A straightforward approach that is often used to tell the impact of various factors on the effective population size is the **ratio of the effective population size to breeding (or sometimes census) population size**, that is, N_e/N.

With this assumption, the distribution of the number of progeny (gametes) per parent (k) approaches the Poisson distribution when N is large (see p. 23 about the Poisson distribution). The general terms in the Poisson distribution are

$$P(k) = \frac{e^{-\bar{k}}\bar{k}^k}{k!}$$

where $\bar{k}$ is the mean number of offspring per parent, Np^*. Let us assume that on average each parent has two offspring ($\bar{k} = 2$ because $p^* = 2/N$), as would occur in a population that is not changing in size (remember that each offspring has two parents in a sexual, diploid organism). Then the probability that a given individual has no progeny ($k = 0$) becomes $e^{-2} =$ 0.135, the probability of one offspring is $2e^{-2} = 0.27$, two offspring is 0.27, three offspring is 0.18, and so on (see Table 6.5). One of the most important characteristics of the Poisson distribution is that the mean number of progeny ($\bar{k}$) and the variance in the number of progeny, V_k, are equal. On p. 324, we discuss how differences from the Poisson distribution of progeny can be incorporated into an estimate of effective population size.

TABLE 6.5 The expected number of progeny when the mean number of progeny ($\bar{k}$) is two and there is (a) a Poisson distribution of progeny, (b) all parents have two progeny, and (c) all progeny have the same parent. In the three right-hand columns are the variance of the number of progeny (V_k), the effective population size, and the ratio of the effective population size to the adult number for the three situations.

	Number of progeny									
	0	1	2	3	4	...	$2N$	V_k	N_e	N_e/N
(a) Poisson	0.135	0.270	0.270	0.202	0.090	...	$(e^{-2}2^k)/k!$	2	N	1
(b) Two progeny/parent	0	0	1	0	0	...	0	0	$2N$	2
(c) One parent	0	0	0	0	0	...	1	$2N$	2	$2/N$

Three approaches have been used to calculate the effective population size: inbreeding, variance, and eigenvalue effective population sizes (Crow and Denniston, 1988). The **inbreeding effective population size** relates to increase in inbreeding in a given population to that in the ideal population; the **variance effective population size** relates to the increase in variance in allele frequency in a given population to that in the ideal population, and the **eigenvalue effective population size** relates to the loss of heterozygosity in a given population to that in the ideal population. In other words, the effective population size is the size of an idealized population that would produce the same amount of inbreeding, allele frequency variance, or heterozygosity loss as the population under consideration. We discuss several derivations for expressions to calculate the effective population size using the concept of the inbreeding effective number. The different types of effective population size are generally either identical or quite similar when the population is not changing in size. Under some conditions—for example, when the population is increasing or decreasing in size—Kimura and Crow (1963) have shown that the inbreeding and variance effective population sizes may differ substantially.

The following discussion focuses on how various demographic factors, such as numbers of breeding females and males, variance in reproduction, variance in numbers over generations, and so on, theoretically influence the effective population size. In the next section, we reverse this and discuss how molecular genetic data can be used to estimate the effective population size.

a. Separate Sexes

Before examining how two separate sexes (dioecy) can be incorporated into the concept of effective population size, let us consider a monoecious population of size N. If the probability of an allele coming from each parent is assumed to be $1/N$, then the probability of two alleles coming from the same individual in the parental generation is $(1/N)^2$. Because there are N individuals, the probability of any of the individuals having alleles coming from the same parent is $N(1/N)^2 = 1/N$. In such an idealized population of N individuals, in which each parent has an equal probability of producing each offspring, then $N = N_e$, the effective population size (this is the inbreeding effective population size).

Now consider a dioecious organism and assume that half the gametes must come from individuals of one sex (females) and half from individuals of the other sex (males). For this case, we can use a probability argument similar to that above to derive the inbreeding effective population size. First, the probability that two alleles in different individuals in generation t came from a female in generation $t-1$ is $(1/2)\,(1/2) = 1/4$. If we take this one step further and use the same logic as above, the probability that these two alleles came from the same female is $1/(4N_f)$, where N_f is the number of females in the population. Likewise, the probability that two alleles came from the same male in the previous generation is $1/(4N_m)$, where N_m is the number of males. The combined probability of two alleles coming from the same individual in the previous generation, whether they can from females or males, is then

$$\frac{1}{N_e} = \frac{1}{4N_f} + \frac{1}{4N_m}$$

This expression can be solved for N_e so that the effective population size becomes

$$N_e = \frac{4N_f N_m}{N_f + N_m} \qquad (6.6a)$$

If there are equal numbers of females and males, $N_f = N_m = 1/2N$, and $N_e = N$.

In some species, the number of females and the number of males are often unequal. Frequently, the number of breeding males is smaller than the number of breeding females ($N_m < N_f$) because some males mate more than once. However, the opposite is true in some species, such as honeybees, where the female may mate with and produce offspring from multiple males. Figure 6.8 shows how the effective population size may vary as the proportion of males, N_m/N, varies for different total numbers. N_e is close to N for a substantial range of the proportion of males near 0.5. However, when the proportion is near 0 or 1, then N_e is greatly reduced.

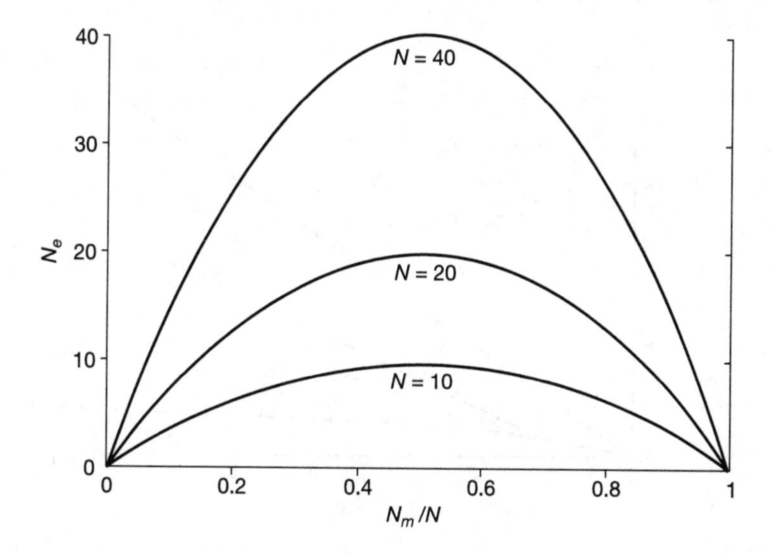

Figure 6.8. The effective population size as a function of the proportions of males, N_m/N, for three different total numbers of individuals.

To evaluate the impact of different numbers of the two sexes on the ratio N_e/N, assume that $x_f = N_f/N$ and $x_m = N_m/N$ are the proportions of females and males, respectively, and $x_m = 1 - x_f$, then

$$\frac{N_e}{N} = \frac{4x_f N x_m N}{N^2}$$

$$= 4(x_f - x_f^2) \tag{6.6b}$$

Taking the derivative of this expression, setting it to 0, and solving for x_f, the maximum N_e/N ratio of unity occurs when $x_f = 0.5$. As examples of values, if 10% or 90% of the breeding animals are females ($x_f = 0.1$ or 0.9), then $N_e/N = 0.36$ from these unequal sex ratios.

Let us assume the most extreme situation possible: one male mates with all of the females in a colony or harem, as is thought to occur in some vertebrate populations in which males control female harems, such as bighorn sheep and elephant seals (see Examples 6.10 and 6.11). In this case, expression 6.6a becomes

$$N_e = \frac{4N_f}{N_f + 1} \tag{6.6c}$$

The maximum value of this expression, when N_f becomes large, is 4.0. In other words, because each sex must contribute half the genes to the progeny, restricting the number of breeding individuals of one sex can greatly reduce the effective population size. Figure 6.9 gives the effective population size for different numbers of females when $N_m = 1$, $N_f/2$, or N_f to illustrate the impact of unequal number of the two sexes on N_e.

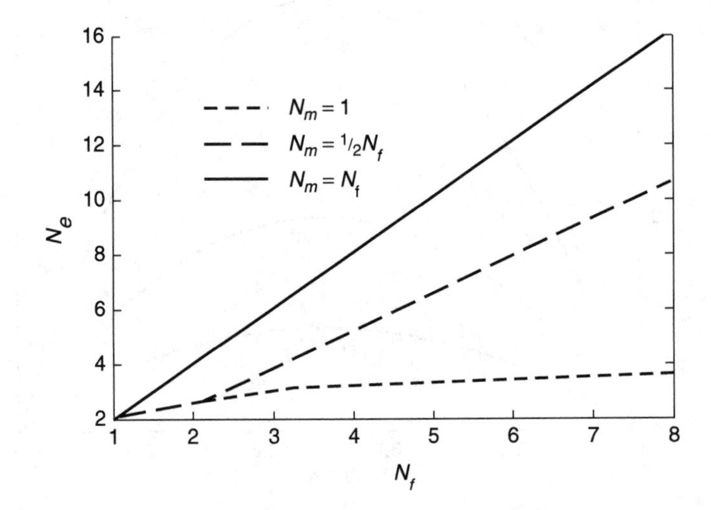

Figure 6.9. The effective population size when $N_m = N_f$, $N_m = \frac{1}{2}N_f$, and $N_m = 1$.

In the past, the numbers of breeding females and males generally have been estimated from behavioral observations. However, genetic examinations in a number of cases have found that behavioral data are not consistent with actual paternity or other genetic data. For example, an examination of the mtDNA and nuclear DNA variation in the southern elephant seal suggests that the estimated sex ratio based on behavioral observations of approximately 40 females per male may greatly overestimate the sex ratio (Slade *et al.*, 1998). The effective sex ratio estimated indirectly from the genetic data is only approximately four or five females per male. The difference in these estimates may be partly due to an overestimate of copulatory success in the behavioral estimate and the short time that a dominant male is a "beachmaster" (1 or 2 years), but other factors may also influence the indirect genetic estimates.

For alleles at an **X-linked gene** or alleles in a **haplo-diploid organism**, the effective population size is somewhat different than that for autosomal genes because females contain two-thirds and males one-third of the alleles. In this case, the effective population size is (Wright, 1931)

$$N_e = \frac{9N_f N_m}{2N_f + 4N_m} \tag{6.7a}$$

If there are equal numbers of females and males ($N_f = N_m = {}^1\!/_2 N$), then $N_e = {}^3\!/_4 N$ as expected because the males are haploid. In some social Hymenoptera, there may be only one breeding female or queen. When this is so ($N_f = 1$), this expression becomes

$$N_e = \frac{9N_m}{2 + 4N_m} \tag{6.7b}$$

The maximum of this expression when N_m becomes large is 2.25. This equation is plotted in Figure 6.10 along with the effective population size when $N_f = N_m$ and when $N_f = {}^1\!/_2 N_m$. Effective population size is important when we are considering a honeybee colony with 20,000 or more bees (most of which are nonreproductive, worker females), of which only one is a breeding female who mates with perhaps a dozen males to produce all of the progeny.

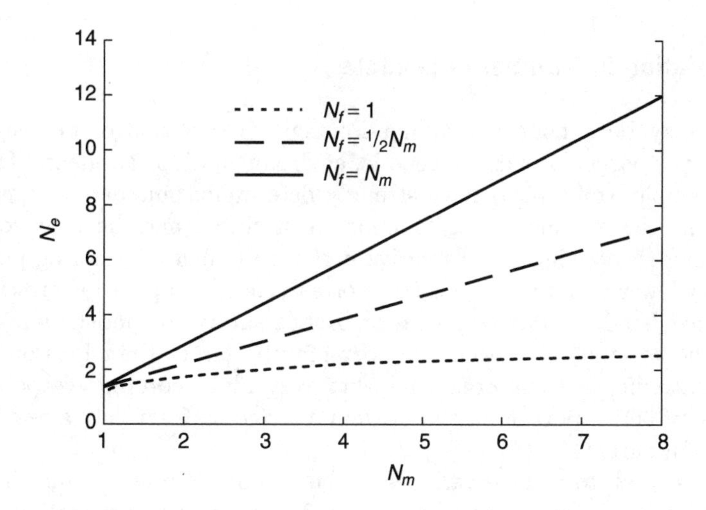

Figure 6.10. The effective population size for an X-linked or haplodiploid gene when $N_f = N_m$, $N_f = \frac{1}{2}N_m$, and $N_f = 1$.

If we estimate the ratio N_e/N as above, then

$$\frac{N_e}{N} = \frac{9(x_f - x_f^2)}{4 - 2x_f} \tag{6.7c}$$

If the proportions of females are 0.1 and 0.9, then the N_e/N ratios are 0.21 and 0.37, respectively. When the proportion of haploid males is high, then this ratio is much lower than it was above for an autosomal locus with the same proportion of males.

To find the maximum N_e/N ratio, we can again take the derivative of this expression, set it to 0, and solve. In this case, using the rule for the derivative of the quotient of two functions gives

$$\frac{dy}{dx_f} = \frac{(4-2x_f)(9-18x_f) - 9(x_f - x_f^2)(-2)}{(4-2x_f)^2} = 0$$

After simplification,

$$x_f^2 - 4x_f + 2 = 0$$

and $x_f = 2-\sqrt{2} = 0.586$. In other words, the maximum N_e/N ratio of 0.773 occurs when there are 58.6% females. The maximum ratio when $x_f > 0.5$ occurs because females have two copies of the gene, making a higher ratio when there are more females.

b. Variation in Number of Gametes

There may be a nonrandom (non-Poisson) distribution of progeny (gametes) per parent because of genetic, environmental, or accidental factors. For example, some birds have strongly determined numbers of eggs in a clutch so the variance of egg number in a clutch may be near zero. In some human populations, a relative uniform number of offspring per parent may lower variation because of efforts to control population growth. On the other hand, if whole clutches or broods survive or perish as a group, then the variance of progeny number may be larger than Poisson. Even more extreme, in some organisms with very high reproductive potential, a substantial proportion of the progeny may come from only a few highly successful parents.

In general, to include variance in the number of progeny and the population is changing in size so that $\bar{k} \neq 2$, the effective population size is approximately

$$N_e = \frac{N\bar{k} - 1}{\bar{k} - 1 + \frac{V_k}{\bar{k}}} \tag{6.8a}$$

where V_k is the variance in the number of progeny (Kimura and Crow, 1963; Crow and Denniston, 1988). The ratio of the effective size to the census size in this case is approximately (Crow, 1954)

$$\frac{N_e}{N} = \frac{2}{1 + \frac{V_k}{\bar{k}}} \tag{6.8b}$$

If $V_k = \bar{k}$, then this ratio is equal to unity, as expected.

When the population is constant in size, $\bar{k} = 2$, the effective population size is

$$N_e = \frac{4N - 2}{V_k + 2} \tag{6.8c}$$

If $V_k = \bar{k}$, a Poisson distribution of progeny, then for both of expressions 6.8*a* and 6.8*c*, $N_e \approx N$ (see Table 6.5).

If $V_k = 0$, as it may in an artificial population where exactly two progeny from each individual are allowed to survive and reproduce, then $N_e = 2N$ or $N_e/N = 2$. Therefore, if V_k is kept low, the effects of finite population size can be avoided to some extent, and the effective population size may actually be larger than the breeding or census number. In some human populations, such as in Japan and Sweden, the family size has become fairly uniform through birth control, and this trend may actually make N_e greater than the breeding population size (Example 6.4 discusses a change in the N_e/N ratio over time in the Japanese).

Example 6.4. Birth records in human populations are often useful sources of demographic information. Imaizumi *et al.* (1970) examined in detail the records for approximately 1000 families over several generations in a rural community in Japan. One measure of fertility that they used was the total number of children surviving beyond the age of 18. The families were divided in five cohorts according to birthdate of the female parent so that any temporal trends in fertility could be recognized. A summary of these data is given in Table 6.6, where $\bar{k}$ and V_k are the mean and variance of family size. The ratio of the effective population size to the actual size can be calculated from expression 6.8*b*.

It is obvious from Table 6.6 that both the mean and the variance of family size decreased substantially over time. The mean and variance were nearly equal in the early cohorts, but in the last cohort (1921–1930) the variance was much smaller. As a result, the N_e/N ratio is 1.43; that is, the effective size is substantially larger than the actual size. Imaizumi *et al.* attributed this to the widespread use of birth control and the desire in most families to have only a few—generally two—children.

TABLE 6.6 The mean and variance of total births for five cohorts in a rural Japanese population and the ratio of effective to actual population size using expression 6.8*b* (after Imaizumi *et al.*, 1970).

	Birthdate of mother				
	1881–1890	1891–1900	1901–1910	1911–1920	1921–1930
$\bar{k}$	4.60	4.80	4.28	3.28	2.74
V_k	4.58	5.12	4.79	2.75	1.09
N_e/N	1.00	0.97	0.94	1.09	1.43

Often, however, the variance in progeny number may also be larger than the mean, and as a result, N_e/N is lower than unity. Evaluating data from *D. melanogaster* and taking into account the variance in the number of gametes, Crow and Morton (1955) found that the ratio N_e/N was between 0.74 and 0.90 in several populations and that the ratio of N_e/N in human populations was between 0.69 and 0.95. Nunney (1993, 1996) has shown that, theoretically, the N_e/N ratio within a generation for many organisms should usually be within the range of 0.25 to 0.75.

However, in some organisms, such as shellfish or fishes (Hedgecock *et al.*, 1992; Li and Hedgecock, 1998), there may be both very high fecundities and very high mortalities of the early life stages (type III survivorship curves; Krebs, 2002). In addition, in a given year, most of small number of recruited young, relative to the very large number of offspring produced, may be from a few parents. This combination of high fecundity and chance success of the progeny of broods of a few parents may result in a quite high variance in progeny number, and consequently, the N_e/N ratio may be quite small.

To illustrate this effect, assume as an extreme that one individual produces all of the progeny. The value of the variance is

$$V_k = \sum_i x_i (k_i - \bar{k})^2$$

where x_i is the proportion of progeny produced by parents with k_i progeny. If all x_i =0 except that $x_{2N} = 1$ (see Table 6.5), then

$$\begin{aligned} V_k &= \frac{2N-1}{2N} \sum_i (0-2)^2 + \frac{1}{2N}(2N-2)^2 \\ &= \frac{8N - 4 + 4(N^2 - 8N + 4)}{2N} \\ &= 2N \end{aligned} \tag{6.9a}$$

In other words, the variance in progeny number is equal to twice the number of progeny produced by the successful parent. Using expression 6.8*b* and

$V_k = 2N$, then

$$\frac{N_e}{N} = \frac{2}{N+1} \tag{6.9b}$$

As the number of progeny produced by the successful individual increases, then this ratio approaches 0.

Sometimes it is useful to calculate the effective population size for the two sexes separately because the variance in progeny number may be significantly different between the two sexes. The **female effective population size** and **male effective population size** are

$$N_{ef} = \frac{N_f\bar{k}_f - 1}{\bar{k}_f - 1 + \dfrac{V_{kf}}{\bar{k}_f}} \tag{6.10a}$$

$$N_{em} = \frac{N_m\bar{k}_m - 1}{\bar{k}_m - 1 + \dfrac{V_{km}}{\bar{k}_m}} \tag{6.10b}$$

respectively (Lande and Barrowclough, 1987, see Example 6.6 later). These values can then be combined to calculate the overall effective population size as

$$N_e = \frac{4N_{ef}N_{em}}{N_{ef} + N_{em}} \tag{6.10c}$$

We can use this approach to calculate the number of effective female and male founders in the Tristan da Cuhna population discussed in Example 6.1 (here we are estimating the effective number of founders for nuclear genes using mtDNA and Y-chromosome data, not estimating the effective size for mtDNA or Y-chromosome genes as discussed below). First, the proportions given in Table 6.1 can be standardized so that the mean number of progeny per female or male is 2 (multiply the female proportions by 10 and the male proportions by 18). Then $V_{kf} = 0.66$ and $V_{km} = 2.01$ so that $N_{ef} = 6.77$ and $N_{em} = 8.21$. In other words, there are slightly fewer effective female than male founders, but $N_{ef}/N_f = 1.35$ and $N_{em}/N_m = 0.91$; that is, the ratio of effective founders to the number of founders is higher for females than males. Overall, the effective number of founders of the population is $N_e = 14.84$. This approach has been used to estimate the effective population size in a population of pumas from Yellowstone Park (Culver *et al.*, 2004, see Example 6.5).

Example 6.5. Only limited published data exist on the number of lifetime offspring produced by individual females and males in a population. One such data set is that of the northern Yellowstone pumas collected from

1987 to 1995 (Murphy, 1998) to estimate effective population size using a demographic approach. In this case, using microsatellite markers, parentage of 70% of the litters was determined over a nine-year period, a nearly complete reproductive history of a single generational cohort (Table 6.7). Two males fathered 23 and 15 offspring, 72% of all of the genotyped kittens. Also, 15 males that were present on the study area did not have any offspring during this period.

From these data, the mean number and variance in number of offspring for females and males can be calculated. The variance in offspring in males is 11.8 times its mean, reflecting the very unequal reproduction noted above. Using these values in expressions 6.10a and 6.10b, then $N_{ef} = 9.14$ and $N_{em} = 4.45$. The ratio of the effective population size to the census number of potentially breeding adults for females is $N_{ef}/N_f = 9.14/14 = 0.653$ and for males is $N_{em}/N_m = 4.45/24 = 0.185$ (the census number used here is a low estimate because a number of other animals were present on the study area at some time between 1987 and 1995). If there were random reproduction, these values should approach unity, but here, particularly for males, it is much below unity. The effective population sizes for each sex can then be used in expression 6.10c to obtain an estimate of the overall effective population size $N_e = 11.97$. The ratio of this to overall census number of potentially breeding adults is $N_e/N = 11.97/38 = 0.315$.

TABLE 6.7 The number of kittens for different females and males in northern Yellowstone Park for cougar litters born from 1987 to 1995 (Murphy, 1998; Culver *et al.*, 2004).

Females		*Males*	
Number of kittens	*Number of females*	*Number of kittens*	*Number of males*
0	2	0	15
2	1	2	1
3	3	3	4
4	1	4	2
5	2	15	1
7	2	23	1
8	1		
9	1		
17	1		
$\bar{k}_f = 5.21$	$V_{k.f} = 19.10$	$\bar{k}_m = 2.50$	$V_{k.m} = 29.39$

For genes that are inherited only through one sex such as mtDNA, cpDNA, and the Y chromosome, the effective population size for the appropriate sex determines the effect of genetic drift on those genes. In all three of these cases, if there is an equal sex ratio, the expected effective population size is $N_e/4$ because these genes are transmitted in only one sex, and in this sex, they are of only one type—that is, they are haploid. Because mtDNA is generally maternally inherited and cpDNA appears always maternally inherited, the **mtDNA effective population size** and the **cpDNA effective population size** are

$$N_e = \frac{N_{ef}}{2} \tag{6.11a}$$

or half the female effective population size. If the number of males breeding or the male effective population size is small, then the effective size for such a gene may actually be greater than for a nuclear gene in the same organism. For example, if N_m is 1, as discussed above, then the maximum value of N_e for a nuclear gene is 4. Because N_{ef} can be much larger than 8, obviously N_e for an organellar gene can be larger than for a nuclear gene. As an example, consider elephant seals (see also Example 6.11) in which one male may have a harem of many females so that in a population $N_{em} = 4$ and $N_{ef} = 200$. In this case, N_e for mtDNA is 50, and N_e for nuclear genes is 15.7, approximately 31% of that for mtDNA genes.

The **Y chromosome effective population size**, and for mtDNA when it is inherited paternally as in conifers, is

$$N_e = \frac{N_{em}}{2} \tag{6.11b}$$

or half the male effective population size. In organisms with a low male effective size, then the effective size for such a gene could be much smaller than that of a nuclear gene in the same organism. Again, for the elephant seal example used above ($N_{em} = 4$ and $N_{ef} = 200$), for Y chromosome genes, $N_e = 2$, approximately one-eighth of the 15.7 estimated N_e for a nuclear gene.

c. Variation in Time

When the population size varies greatly in size in different generations, it can have a large impact on the overall effective population size. The variation in population size could result from regular cyclic variation in population numbers, periodic decimation of the population because of disease or other factors, or seasonal variation in population numbers. When this occurs, the lowest population numbers determine, to a large extent, the overall effective population size because after these bottlenecks, all remaining individuals are descendents of the bottleneck survivors.

The effect of variation in population size can be shown by examining the heterozygosity over time (Crow and Kimura, 1970). Previously, we saw that

$$H_t = \left(1 - \frac{1}{2N}\right)^t H_0$$

If N varies from generation to generation, then

$$\begin{aligned}\frac{H_t}{H_0} &= \left(1 - \frac{1}{2N_0}\right)\left(1 - \frac{1}{2N_1}\right)\left(1 - \frac{1}{2N_2}\right)\cdots\left(1 - \frac{1}{2N_{t-1}}\right) \\ &= \prod_{i=0}^{t-1}\left(1 - \frac{1}{2N_{e.i}}\right)\end{aligned}$$

where $N_{e.i}$ is the effective population size in generation i. The overall effective population size is the one that causes the same reduction in heterozygosity as the varying $N_{e.i}$ values, and thus,

$$\prod_{i=0}^{t-1}\left(1 - \frac{1}{2N_{e.i}}\right) = \left(1 - \frac{1}{2N_e}\right)^t$$

Solving this, then the overall effective population size is

$$N_e = \frac{1}{2\left\{1 - \left[\prod_{i=0}^{t-1}\left(1 - \frac{1}{2N_{e.i}}\right)\right]^{1/t}\right\}} \tag{6.12a}$$

If the $N_{e.i}$ values are not too small, then the effective population size becomes approximately

$$N_e = \frac{t}{\sum \frac{1}{N_i}} \tag{6.12b}$$

Therefore, the effective population size is approximately the harmonic mean of the effective population sizes in individual generations.

For example, assume that the population in subsequent generations has effective population sizes of 10, 100, and 1000. Given that $H_0 = 0.5$, we expect that $H_1 = 0.475$, $H_2 = 0.473$, and $H_3 = 0.472$ because of these finite population sizes. Applying expression 6.12b gives the effective population size as 27.0. Therefore, the heterozygosity declines so that $H_1 = 0.491$, $H_2 = 0.482$, and $H_3 = 0.473$, reaching essentially, in generation 3, the same heterozygosity as when the population size was variable.

To illustrate the importance of a bottleneck in determining effective population size, assume that a population of insects increases 10-fold each of two generations in the summer and then returns to its original low level because of winter mortality; for example, the population sizes are N, $10N$, and $100N$. The mean census number (arithmetic mean) over the three different generations is $36.7N$. However, the effective population size as calculated from expression 6.12*b* is only $2.7N$, more than an order of magnitude less. In this case, the N_e/N ratio is only 0.074—that is, the effective population size is only 7.4% of the average census number (see the discussion below concerning the potential problems comparing harmonic means of N_e and arithmetic means of N). Example 6.6 shows how variation in progeny number in the two sexes and variation over generations can be included in an overall estimate of effective population size.

Example 6.6. As we have seen, the effective population size may be influenced by several factors, including unequal sex ratio, nonrandom variance in progeny number in the two sexes, and variation in effective size over generations. Lande and Barrowclough (1987) gave a useful example to illustrate how all these factors may be incorporated into one estimate of effective population size. Table 6.8 gives the number of progeny produced by individual females and males over three generations in the growing population in their example. Note that there are more females than males contributing each generation, so the mean number of progeny produced per female is less than that per male (Table 6.9). In addition, the variance in progeny reproduction per female is lower than that per male. Using expression 6.10*a*

TABLE 6.8 The number of progeny for females and males in an example of a growing population over three generations (Lande and Barrowclough, 1987).

Generation	*Females*	*Number of progeny*	*Males*	*Number of progeny*
1	A,B	4	M	9
	C	3	N	3
	D	1		
2	A	5	M	9
	B,C	4	N	5
	D	3	O,P	3
	E,F	2		
	G,H	0		
3	A,B	5	M	12
	C,D	4	N	9
	E,F,G	3	O	7
	H,I,J,K	2	P	5
	L	1	Q	3
			R,S,T	0

TABLE 6.9 The mean and variance in the number of progeny for females and males in the three generations of data given in Table 6.7.

Generation	*Number of females*	*Number of males*	$\bar{k}_f$	$\bar{k}_m$	$v_{k \cdot f}$	$v_{k \cdot m}$
1	4	2	3.0	6.0	2.00	18.00
2	8	4	2.5	5.0	3.43	8.00
3	12	8	3.0	4.5	1.64	20.86

and these data, we find that the effective population size for females in the first generation is

$$N_{ef} = \frac{(4)(3) - 1}{3 - 1 + \frac{2}{3}} = 4.12$$

Likewise, $N_{em} = 1.38$. Note that the effective population size for females is slightly greater than the census number, because the female variance in progeny number is lower than the mean progeny number, and that the effective number of females is over three times larger than the effective number of males. The overall effective size for this generation is then

$$N_e = \frac{4(4.12)(1.38)}{4.12 + 1.38} = 4.13$$

For generations 2 and 3, the effective sizes for males are again lower than those for females because of the higher variance in progeny number and number of individuals. Using these values, the overall effective sizes for generations 2 and 3 are 8.97 and 13.10, respectively.

Substituting these effective population sizes for each generation in expression 6.12a yields the overall effective population size:

$$\begin{aligned} N_e &= \frac{1}{2\left\{1 - \left[\left(1 - \frac{1}{(2)(4.3)}\right)\left(1 - \frac{1}{(2)(8.97)}\right)\left(1 - \frac{1}{(2)(13.10)}\right)\right]^{1/3}\right\}} \\ &= 6.91 \end{aligned}$$

In other words, the loss of genetic diversity in the population is the same as though there were a constant effective population size of 6.91. In this example, the arithmetic mean of the effective sizes over generations is 8.73, somewhat greater than the estimated effective size of 6.91. This indicates that the low effective size in the first generation served to lower the overall effective size. The effective sizes for the generations can also be substituted in expression 6.12b to give

$$N_e = \frac{3}{(1/4.13) + (1/8.97) + (1/13.10)} = 6.98$$

In this case, with only a few generations and not much variation in effective size over generations, the harmonic mean approximation is very close to the exact formula.

Frankham (1995) examined 102 published estimates of effective population size in vertebrate, invertebrate, and plant species. The overall low observed ratio for N_e/N of 0.11 that he calculated for these species appeared to be primarily the result of variable effective population size over time. Subsequently, Vucetich *et al.* (1997) demonstrated analytically that variable population size over generations can greater lower the N_e/N ratio. However, these conclusions are potentially confounded by a statistical artifact in the computation of the N_e/N ratio (Waples, 2002). Specifically, in these studies, N_e/N was calculated as the ratio of the harmonic mean of N_e divided by the arithmetic mean of N.

To illustrate this effect, Table 6.10 gives a numerical example in which the within-generation ratios are all 0.4, similar to what Frankham (1995) observed using adjustments for sex ratio differences and variance in reproductive success. When the overall ratio is calculated as the harmonic mean of N_e divided by the arithmetic mean of N, then it is $193/1730 = 0.11$, the same as the value that Frankham (1995) calculated overall. When either the arithmetic means or the harmonic means are used for both N_e and N, then the ratio is 0.4, as it is within each generation. This suggests that calculating these ratios over generations must be done with full knowledge of these potential biases (Kalinowski and Waples, 2002). However, it is important to emphasize again, independent of these measurement concerns, that variation in the effective population size over generations may greatly reduce the overall effective population size because the harmonic mean, which gives heavy weighting to low numbers, is the appropriate mean to use.

TABLE 6.10 An example illustrating the potential statistical artifact introduced when the ratio N_e/N is calculated as the harmonic mean of N_e over generations divided by the arithmetic mean of N over generation (Waples, 2002). The within-generation N_e/N ratio is 0.4 for all generations.

Generation	N_e	N	N_e/N
1	400	1000	0.4
2	200	500	0.4
3	60	150	0.4
4	800	2000	0.4
5	2000	5000	0.4
Arithmetic mean	692	1730	0.4
Harmonic mean	193	482	0.4
N_e (harmonic)/N (arithmetic)			0.11

d. Other Factors That May Influence Effective Population Size

First, the amount of **inbreeding** in a population can reduce the effective population size. In a general way, the effective population size, as a function of the amount of inbreeding, is

$$N_e = \frac{N}{1+f} \tag{6.13}$$

where f is the inbreeding coefficient as defined in Chapter 5. In other words, inbreeding decreases N_e only slightly when it is on the order found in human populations (mean f ranges from 0.0 to approximately 0.05). However, in highly selfed plants or animals, f may approach 1 and result in N_e approaching $1/2N$. This can be understood intuitively because with nearly complete inbreeding the loss of variation is similar to that in a haploid population of size N. However, in a haploid population of size N there are only N gametes, which makes it equivalent to a diploid population of size $1/2N$ (Cabarello, 1994).

Charlesworth (2003) summarized the studies comparing the sequence diversity in taxa with either populations of congeneric species or different populations within the same species that have different levels of inbreeding. Overall, the level of variation in the highly inbreeding populations was much lower than the outcrossing populations, generally even more than predicted by the maximum 50% reduction predicted when the inbreeding coefficient approaches unity. She suggested that the reduced effective recombination in inbreeders (see p. 548) and different life history characteristics in inbreeders and outbreeders may also contribute to differences in variation between the types. Example 6.7 gives the nucleotide diversity of a mustard plant for low, intermediate, and high selfing populations where there is an inverse relationship between the levels of self-fertilization and genetic variation.

Example 6.7. Comparisons of the effect of the mating system on genetic variation are most appropriate when different populations of the same species vary in level of inbreeding. Charlesworth (2003) recently documented the amount of nucleotide diversity at six genes in the mustard plant, *Leavenworthia crassa*. She examined three different types of populations, those that were self-incompatible (S near 0), those with some self-compatibility (intermediate S values), and those with self-compatibility (S near 1). Table 6.11 gives the mean nucleotide diversity within the three types of populations. Although there is significant variation over loci, the observed amount of variation for each population is consistent with its level of selfing: highest variation for low selfing and lowest variation for high selfing.

TABLE 6.11 DNA sequence diversity for six loci in *L. crassa* populations with different levels of selfing (Charlesworth, 2003).

	Level of selfing		
Locus	*Low*	*Intermediate*	*High*
Adh1	0.036	0.000	0.000
Adh2	0.008	0.006	0.014
Adh3	0.017	0.007	0.000
Gapc	0.028	0.017	0.014
Nir1	0.023	0.022	0.007
PgiC	0.000	0.013	0.011
Mean	0.019	0.013	0.008

Second, many organisms have populations that contain prereproductive (and some postreproductive) individuals as well as reproductive individuals. In addition, individuals of different ages may vary in both their birth and death rates because of environmental and genetic factors. As a result, individuals of different ages can potentially make very different contributions to the genetic continuity of a population. For example, a postreproductive individual will make no additional contribution to the population even if it remains alive for a long period. On the other hand, an individual that is just reaching reproduction can potentially make a large genetic contribution to the population.

Incorporating detailed **age structure** into an estimation of the effective population size is complicated, particularly when generations overlap and there are age differences in survival and fecundity, and several different approaches have been developed (see Cabarello, 1994). For example, Nei and Imaizumi (1966) suggested that the effective population size is approximately

$$N_e = TN_a \tag{6.14}$$

where T is the mean age of reproduction in years (generation length) and N_a is the number of individuals born per year who survive to reproductive age. For example, if we assume that $T = 5$, similar to that in many larger mammals, and $N_a = 50$, then the approximate effective population size is 250. In applying such an approach to demographic information on white females in the United States, Felsenstein (1971) found that the ratio of effective population size to census number was approximately 0.34. Nunney and Elam (1994) have suggested an approach in which summary demographic data can be incorporated in an estimate of the ratio of the effective population size to the number of adults. They evaluated estimates in a number of species, including spotted owls, grizzly bears, and the snail *Cepea nemoralis*, and found that their approach was generally robust.

Finally, in many natural situations, no sharp boundaries separate populations, and thus, it is impossible to estimate the effective size of a distinct population. To evaluate the effect of finite population size in species where there is a continuous distribution over space, Wright (1943b) introduced the concept of **neighborhood size**. The size of a neighborhood is the number of individuals in a circle with a radius twice the standard deviation of the per generation gene flow ($2V^{1/2}$) in one direction. Therefore, if we know the density of individuals (d) and area of the neighborhood circle ($4\pi V$), we can estimate the effective population size in the neighborhood as

$$N_e = 4\pi V d \tag{6.15}$$

For example, Beattie and Culver (1979) estimated that in a violet (*Viola pedata*) population, $V = 4.54$ m^2 and $d = 9.6/\text{m}^2$ so that the estimate of effective size is 547. In animals, such estimates are complicated because gene flow is a function of a number of factors, including density, food sources, and genetic variation. In plants, estimates of neighborhood size may need to account for gene flow in different life stages. For instance, in the violet example, dispersal was in gametes mediated by pollinators and in seeds both by ants and ballistic ejection from the capsule.

e. Genetic Techniques for Estimating Effective Population Size

Estimating the effective population size in many situations from demographic data is dependent on information that is unavailable or difficult to obtain from natural populations, such as the variance in the number of progeny. As a result, approaches to estimate effective population size using genetic data have a great appeal. One approach that uses genetic information is to determine the paternity and maternity of a progeny cohort and then use the demographic approaches given above to estimate the consequent effective population size. For example, the puma data given in Example 6.5 determined maternity and paternity from a known number of parents using microsatellite loci, and the winter run Chinook data set given in Example 6.8 uses microsatellite loci to determine maternity and paternity from known spawners (and matings) used to produce progeny.

Example 6.8. Winter run Chinook salmon (*Oncorhynchus tshawytscha*) from the Sacramento River (California) are federally listed as an endangered species. A program to mitigate the factors causing endangerment has included the annual supplementation of young raised at a fish hatchery. In the 1994 brood year, 43,346 progeny from 16 female and 10 male wild-caught spawners were released. After spending more than 2 years in the Pacific Ocean, 93 returning spawners from this cohort were identified,

TABLE 6.12 The number of winter run Chinook salmon progeny released and the number of returning spawners from the different females and males in the 1994 brood year and the ratio of the proportion of returns to release for each parent (Hedrick et al., 2000).

Female Parents				*Male Parents*			
ID	*Releases*	*Returns*	*Ratio*	*ID*	*Releases*	*Returns*	*Ratio*
3	3444	10	1.35	B	4433	9	0.95
4	3055	5	0.77	C	3152	9	0.95
5	2499	7	1.29	D	4360	16	1.61
6	2361	6	1.12	E	6013	8	0.62
7	2421	3	0.57	F	5223	15	1.34
8	2292	2	0.42	G	5098	6	0.51
9	2338	5	1.00	H	4432	10	1.06
11	2230	7	1.39	I	6353	16	1.17
12	2701	3	0.52	J	3012	3	0.46
13	3946	8	0.93	K	1270	1	0.38
14	1364	2	0.69				
15	3426	10	1.37				
16	2855	10	1.64				
17	2766	7	1.17				
18	3088	4	0.61				
19	2270	4	0.75				
Total	43,346	93			43,346	93	

and their maternity and paternity were determined by microsatellite loci (Table 6.12).

The numbers released were fairly even across both females and males, with percentage parentage ranging from 3.2% (ID 14) to 9.2% (ID 13) for females and from 2.9% (ID K) to 114.7% (ID I) for males. The ratio of the variance to mean reproduction was 0.43 and 0.45 for females and males, respectively, much less than unity. This low variance resulted primarily from a breeding protocol at the hatchery instituted to equalize the production over different females and males (Hedrick *et al.*, 1995). However, it was unknown whether the variance over parents would greatly increase in the returns collected 3 years later.

Every female and male parent was represented in the 93 returning spawners, and the numbers were spread fairly evenly over the 16 females and 10 males (Table 6.12). The ratio of the proportion released to the proportion returning for a given individual gives a perspective on the relative return rate; these ratios ranged from 0.38 (ID K) to 1.64 (ID 16). The ratio of the variance to mean reproduction for the returning spawners was 0.51 and 0.55 for the females and males, only slightly higher than that in the released individuals. If we use expression 6.10*c*, the observed effective population size calculated for the returning spawners was 30.2, not significantly different from that predicted if the salmon returning were a random sample of those released. In this case, it appears that the breeding protocol resulted in a rather even distribution of releases, and there was only a small increase in the variance of reproduction across parents in the returning spawners.

In most situations, such detailed information on parents and offspring is not available. As a result, several different techniques have been developed to estimate the relatively short-term effective population by determining the effect of genetic drift on various population genetic measures. In all approaches, a very small effective population size has the largest effect, and in larger populations, there may be very little signal from genetic drift. One of the general problems with these measures is that they have poor precision, although this could potentially be overcome with larger numbers of marker loci and improved statistical techniques.

The most widely used genetic technique to estimate short-term effective population size depends on the fact that, given that no other evolutionary factors are important, genetic drift results in an expected change in allele frequency over generations. As the population size becomes smaller and the number of generations is longer, a greater change in allele frequency from genetic drift is expected. The estimation method using **temporal change of allele frequencies** is based on the expected variance in allele frequency that we discussed in expression 6.4a. However, the variance is a function of the initial allele frequencies so that estimates need to contain a standardization to account for different initial allele frequency values (Waples, 1989).

Let us assume that the initial frequency of allele A_i is $p_{i.0}$ and that after t generations of genetic drift it is $p_{i.t}$. The theoretical value of the standardized variance (F) (Wright, 1931) is

$$F = \frac{(p_{i.0} - p_{i.t})^2}{p_{i.0}(1 - p_{i.0})} \tag{6.16a}$$

The expectation E of the numerator in this expression is

$$E\left[(p_{i.0} - p_{i.t})^2\right] = p_{i.0}(1 - p_{i.0})\left[1 - \left(1 - \frac{1}{2N}\right)^t\right]$$

so that the expectation of F becomes

$$E(F) = \left[1 - \left(1 - \frac{1}{2N}\right)^t\right]$$

If t is not too large, then an estimate of the variance effective population size is

$$\hat{N}_e = \frac{t}{2F} \tag{6.16b}$$

To estimate the effective population size, we need an estimate of the standardized variance. Nei and Tajima (1981) suggested the expression

$$\hat{F} = \frac{1}{n}\sum_{i=1} \frac{(\hat{p}_{i.0} - \hat{p}_{i.t})^2}{(\hat{p}_{i.0} + \hat{p}_{i.t})/2 + \hat{p}_{i.0}\hat{p}_{i.t}} \qquad (6.17a)$$

where n is the number of alleles at a locus. In addition, F is influenced by the size of the two samples such that it is increased by the sampling effects at both sampling times. To adjust for this effect, the estimate of F can be reduced by the reciprocals of the two sample sizes, $2N_0$ and $2N_t$, as

$$\hat{F}' = \hat{F} - \frac{1}{2N_0} - \frac{1}{2N_t} \qquad (6.17b)$$

Therefore, an estimate of the effective population size is

$$\hat{N}_e = \frac{t}{2\hat{F}'} \qquad (6.17c)$$

For an application of this approach to a Swedish population of brown trout monitored for many years, see Example 6.9.

Example 6.9. A natural population of brown trout (*Salmo trutta*) in central Sweden has been studied since the 1970s by Nils Ryman and his colleagues. They have monitored changes in both allozyme frequencies and mtDNA haplotypes. Table 6.13 gives the frequency of the most common mtDNA haplotype (the frequency of the other haplotype is the complement of this) for 14 annual cohorts (Laikre *et al.*, 1998). Using expression 6.16*a*, the estimate of the standardized variance between adjacent cohorts was then estimated, and it ranged from a high of 0.286 for the comparison between 1975 and 1976, because of the large increase in the common haplotype frequency between these two cohorts, to near zero in several comparisons.

Because the sample size for mtDNA haplotypes is half that of a nuclear gene, expression 6.17*b* needs to be modified to

$$\hat{F}'_{mt} = \hat{F} - \frac{1}{N_t} - \frac{1}{N_{t+1}}$$

where N_t is the sample size in year t (Laikre *et al.*, 1998). The sample size–corrected F estimates are somewhat lower than the uncorrected F estimates (Table 6.13), and some of them become negative because the expected effect of sampling is larger than the uncorrected F estimate.

Because brown trout have a generation length much longer than one year, they have overlapping generations, and mtDNA gives only an estimate

TABLE 6.13 The estimated frequency of the most common mtDNA haplotype p_i, the sample size N, and the estimated $\hat{F}$ and sample size corrected $\hat{F}'_{mt}$ between adjacent years in 14 annual cohorts from a population of brown trout in central Sweden (from Laikre *et al.*, 1998).

Cohort	p_i	N	*Cohorts for estimate*	$\hat{F}$	$\hat{F}'_{mt}$
1975	0.640	50			
1976	0.880	50	1975 and 1976	0.286	0.246
1977	0.694	49	1976 and 1977	0.199	0.159
1978	0.800	50	1977 and 1978	0.057	0.017
1979	0.596	52	1978 and 1979	0.190	0.151
1980	0.621	66	1979 and 1980	0.002	–0.032
1981	0.700	50	1980 and 1981	0.028	–0.007
1982	0.675	40	1981 and 1982	0.003	–0.042
1983	0.620	50	1982 and 1983	0.013	–0.032
1984	0.760	50	1983 and 1984	0.091	0.051
1985	0.735	49	1984 and 1985	0.003	–0.037
1986	0.680	50	1985 and 1986	0.014	–0.026
1987	0.660	50	1986 and 1987	0.002	–0.038
1988	0.521	48	1987 and 1988	0.077	0.036
Mean/total	0.680	704		0.074	0.032

of the female effective population size, expression 6.16*c* needs to be modified to

$$\hat{N}_{ef} = \frac{C}{G\hat{F}'_{mt}}$$

where C is a correction because of the overlapping generations and G is the estimated female generation length. From demographic data, Laikre *et al.* (1998) estimated that $C = 18.1$ and $G = 8.3$. Using the mean sample size–corrected F of 0.032 over all adjacent cohort pairs at the bottom of Table 6.13, the estimated female effective size is 68.5 (this is slightly different from the estimate in Laikre *et al.* because they used a different $\hat{F}$ estimator). Although a good estimate of the actual census numbers in the lake is not available, the numbers appear to be much larger than this estimate of effective population size, suggesting that N_e/N is much less than unity.

Recently, new statistical approaches have been developed to estimate more efficiently the effective population size from variation in allele frequency (Anderson *et al.*, 2000; Wang, 2001; Berthier *et al.*, 2002; Beaumont, 2003), and a number of highly polymorphic loci have been examined (Turner *et al.*, 2001, 2002). However, the specific assumptions that are intrinsic to these models (Waples, 1989), including those related to evolutionary factors, need to still be considered. In addition, for larger effective population sizes, the effects of genetic drift may be overwhelmed by the effects of sampling. Overall, Waples (1989) suggested that large $\hat{N}_e$ values may be ambiguous (sometimes small populations may give large estimates) but that small $\hat{N}_e$ values appear to be a good indicator that the effective population size is in fact small.

Second, the effective population size (actually the effective number of parents or breeders) can be estimated from the **heterozygote excess** found in their progeny. When there are small numbers of parents, then the allele frequencies in the female and male parents will differ because of binomial sampling error (Pudovkin *et al.*, 1996). On p. 72, we discussed how differences in parental female (p_f) and male (p_m) frequencies for allele A_1 resulted in an excess of heterozygotes in their progeny but did not specify the cause of differences. The observed heterozygosity from expression 2.6*b* is

$$H = 2\overline{pq} + {}^1\!/\!_2(p_f - p_m)^2 \tag{6.18a}$$

where $\bar{p} = {}^1\!/\!_2(p_f + p_m)$ and $\bar{q} = {}^1\!/\!_2(q_f + q_m)$. When N parents of each sex are randomly drawn from an infinite population with allele frequencies p and q, the right-hand term in the above expression is half of the variance of the difference in allele frequencies between the sexes and can be given as the variance in the difference between two binomial samples, or $2pq/N$ (Pudovkin *et al.*, 1996), so that

$$H = 2\bar{p}\bar{q} + \frac{pq}{N_e} \tag{6.18b}$$

Assuming that $\bar{p} = p$ and $\bar{q} = q$, solving for the effective number of parents gives the estimate

$$\hat{N}_e = \frac{pq}{H - 2pq} \tag{6.18c}$$

Pudovkin *et al.* (1996) also gave a slightly different, more exact derivation.

Although this estimation approach has some advantages (e.g., only one cohort needs to be sampled), theoretical evaluation by Luikart and Cornuet (1999) showed that the confidence intervals were quite large except when the effective number of parents was less than 10 and there were large numbers of both offspring and loci. Furthermore, when they used this approach on 10 data sets where there were known to be only a few parents, half of the estimates were very large or negative, and in most of the rest, the upper confidence interval was infinity.

Genetic drift also generates a statistical association between alleles at different loci, or **linkage disequilibrium** (see Chapter 10). Theory predicts that there is an expected amount of disequilibrium between loci generated in a finite population with a given amount of recombination between the loci. Such theoretical equations can be solved to estimate N_e when the disequilibrium and recombination between loci is estimated (see p. 342). In addition, the neutral theory can also be used to estimate effective population size. In this case, the combination of genetic drift and mutation results in an expected amount of DNA sequence variation and phylogenetic pattern of sequences from which an estimate of the long-term effective population size is possible (see p. 462). Of course, in this case, the cumulative effect of genetic drift is measured over many generations in the past, and the effect of a recent generation with a small population size may not be apparent.

Estimates of effective population size based on genetic approaches and demographic techniques should give similar estimates. A survey of published data by Frankham (1995) suggested that estimates using the two types of estimates are fairly consistent, although detailed evaluation of more recent comparisons would be useful. Different approaches, either the various genetic or demographic approaches, may give complementary information and potentially give a fuller picture of the effective population size over a greater time period.

f. The Founder Effect and Bottlenecks

A population may descend from only a small number of individuals either because the population is initiated from a small number of individuals, causing a **founder effect**, or because a small number of individuals survived in a particular generation or consecutive generations, resulting in a population **bottleneck**. These situations can lead to chance changes in genetic variation so that allele frequencies are different from those in the ancestral population, resulting in lower heterozygosity and fewer alleles.

First let us examine the potential importance of the founder effect or a bottleneck on the amount of genetic variation. One simple way to understand this effect is to compare the heterozygosity in the population in which the founder group came from, or that before the bottleneck, and that in the population after the event. For example, expression 6.3*a* can be solved for the effective population size as

$$N_e = \frac{H_t}{2(H_t - H_{t+1})} \tag{6.19a}$$

where H_t and H_{t+1} are the heterozygosities in the original population and the founding group. Assuming that there are t generations of small numbers as in a bottleneck with size N_e, then expression 6.3*b* can be solved as

$$N_e = \frac{1}{2\left(1 - e^{[\ln(H_t/H_0)]/t}\right)} \tag{6.19b}$$

where H_0 is the heterozygosity before the bottleneck. For example, if $H_0 = 0.7$, $H_t = 0.6$, and $t = 5$, then $N_e = 16.4$. Obviously the greater the

reduction in heterozygosity from the founding event or the bottleneck, the lower the estimate of effective population size in the founder group or the bottleneck generations.

In an effect related to the reduction in heterozygosity, a founder event (or a bottleneck) can also quickly **generate genetic distance** between the ancestral population and the newly founded or bottlenecked population. This can be understood intuitively if we assume that there are 10 alleles in the ancestral population, and the founder (or bottleneck) generation consists of one fertilized female ($2N = 4$). As a result, at least six alleles must be lost in the bottleneck, resulting in significant changes in allele frequencies in the descendant population and genetic distance from the ancestral population in one generation. The expected standard genetic distance (Nei, 1987) after a founder event or a bottleneck is (Chakraborty and Nei, 1977; Hedrick, 1999b)

$$D_t = -\tfrac{1}{2}\ln\left(\frac{1 - H_0}{1 - H_t}\right) \tag{6.20a}$$

Assuming that there is a founding event or bottleneck of t generations as

$$D_t = -\tfrac{1}{2}\ln\left[\frac{1 - H_0}{1 - H_0(1 - \frac{1}{2N_e})^t}\right] \tag{6.20b}$$

where N_e is the effective population size in the generations during the founder event or bottleneck.

Figure 6.11 gives the expected genetic distance generated by a bottleneck that results in heterozygosity that is $(1 - (1/2N_e)^t$ of that before

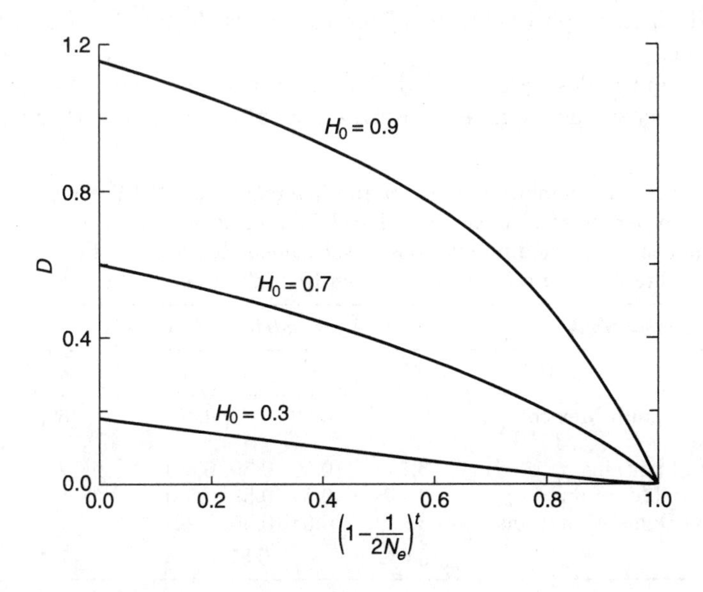

Figure 6.11. The expected genetic distance generated by a bottleneck in one population that results in a reduction of heterozygosity for different initial heterozygosities.

the bottleneck for different initial heterozygosities. When there is high initial heterozygosity, such as found for microsatellite loci, the effect can be quite large. For example, say there is a one-generation bottleneck of two individuals (0.75 on the horizontal axis). Then for H_0 of 0.7 and 0.9, the genetic distances are 0.230 and 0.589, respectively (see Example 6.10 for an application of these approaches to bighorn sheep from Tiburon Island in Mexico).

Example 6.10. Bighorn sheep (*Ovis canadensis*) have greatly declined in numbers and distribution in the past century because of disease, hunting, and other factors. As a result, there have a number of introductions throughout western North America in an effort to establish more viable populations. For example, in early 1975, 20 desert bighorn sheep (4 males and 16 females) were captured in mainland Sonora, Mexico, and translocated to nearby Tiburon Island in the Sea of Cortez (Montoya and Gates, 1975). The translocated population grew quickly, and by 1999, an estimated 650 sheep were on the island, all descended from this small founder group.

In 1998, 14 wild sheep were captured and analyzed for genetic variation at 10 microsatellite loci and one MHC locus (Hedrick *et al.*, 2001a). As shown in Table 6.14, these data were compared with the variation from three populations of the same subspecies from neighboring Arizona (samples from the original founders and the population in Sonora from which they were obtained were not available). The heterozygosity for both the microsatellite loci and the MHC locus was much lower in the Tiburon Island population than in the Arizona populations. If we use the mean heterozygosity for the microsatellite loci for Arizona and Tiburon populations as H_t and H_{t+1}, respectively, then from expression 6.19*a*, $N_e = 1.9$, a very low value.

The four males were ages 1, 1, 2, and 7 years, suggesting that the older ram may have made a greater initial contribution than the other males. If

TABLE 6.14 The number of alleles n and heterozygosity H (95% confidence intervals in parentheses) in the introduced Tiburon Island population of bighorn sheep and that for three populations of the same subspecies for ten microsatellite loci and a MHC locus (from Hedrick *et al.*, 2001a).

Population	*Microsatellite*		*MHC*	
	N	H	n	H
Tiburon Island, Mexico	2.5	0.42 (0.38, 0.46)	5	0.74
Arizona				
Kofa Mountains	3.7	0.60 (0.55, 0.64)	7	0.96
Stewart Mountains	3.1	0.54 (0.50, 0.58)	8	0.87
Castle Dome Mountains	3.9	0.58 (0.55, 0.62)	7	0.85
Mean	3.6	0.57	7.3	0.89

it is assumed that older ram was the only breeding male in the first cohort of males, then the effective size of the founder generation could have been, using expression 6.6*a*, $N_e = 4(16)(1)/(16+1) = 3.8$. Although this does not explain all of the loss of heterozygosity, it may explain much of it. In addition, in the subsequent early generations, there may have been more loss of variation due to small numbers, particularly because of few effective breeding rams.

In addition, the average observed genetic distance between the Tiburon Island population and the Arizona populations was 0.312. Using the observed heterozygosities in these populations in expression 6.20*a*, the genetic distance expected from genetic drift was 0.154, 49.5% of the total observed. In other words, approximately half of the genetic distance between these populations appears to be the result of the recent loss of genetic variation and approximately half because of prior genetic differentiation.

The Tiburon Island population is being used as the source stock for translocations throughout much of northern Mexico. The low amount of genetic variation observed in this population suggests that there has been substantial genetic drift in this population, probably resulting in fixation of some detrimental alleles and reduced future adaptive potential. As a result, it seems advisable to supplement the Tiburon Island population with additional unrelated animals to counter these detrimental genetic effects and to supplement the translocated populations with unrelated animals.

Nei (1987) suggested that genetic distance can be corrected for this effect by assuming that the genetic identity (see p. 108) is

$$I = \frac{J_{xy}}{J_x} \tag{6.20c}$$

where J_{xy} is the product of the allele frequencies in the two populations and J_x is the homozygosity in the population that has not gone through a founder event or a bottleneck. For example, Paetkau *et al.* (1997) found that brown bears on Kodiak Island and black bears on Newfoundland Island had much lower heterozygosities for microsatellite loci than found in other samples and also a higher genetic distance between these island populations and mainland populations (1.50) than predicted by geographic distance. The corrected value using expression 6.20*b* was 1.06, illustrating that both loss of genetic variation and genetic differentiation had both contributed to this high genetic distance value.

Another way to illustrate the impact of a founder effect on genetic variation is to calculate the probability of polymorphism in the founders.

When there are Hardy–Weinberg proportions in the parental population, the probability of polymorphism in a founder group of size N is

$$R = 1 - \left[\left(p_1^2\right)^N + \left(p_2^2\right)^N + ... \left(p_i^2\right)^N \right]$$
$$= 1 - \sum p_i^{2N} \tag{6.21}$$

which is one minus the probability of monomorphism. Figure 6.12 gives the value of R as a function of founder size for several allele frequencies when there are only two alleles. If the alleles are equal in frequency, then the founder size does not need to be very large for a high probability of inclusion of both alleles. For example, the founder population need be only three individuals ($2N = 6$) or larger for there to be a greater than 95% chance of including both alleles when they are equal in frequency. If the frequencies in the parental population of the two alleles are far from equal (e.g., 0.95 and 0.05), then the founder number needs to be 30 or larger for there to be a 95% chance of including both alleles (in reality, of including the rarer allele).

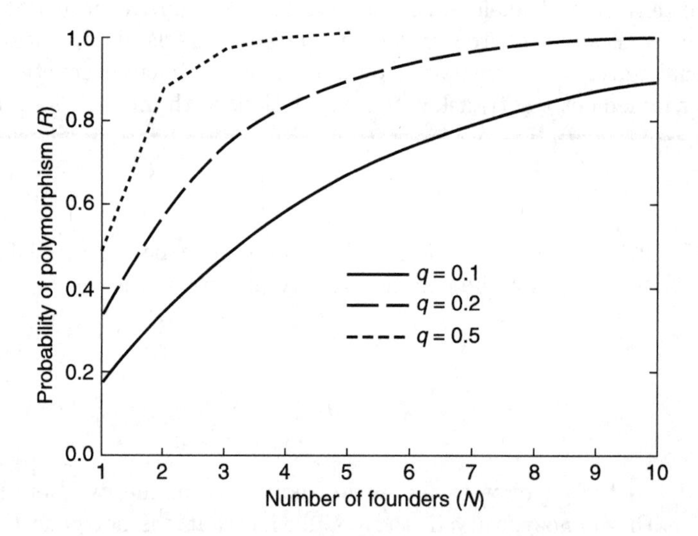

Figure 6.12. The probability of polymorphism, given different numbers of founders in a sample for three different allele frequencies.

For highly variable loci in particular, the loss of alleles occurs more rapidly than the loss of heterozygosity. For example, a microsatellite locus with 20 alleles in a bottleneck of five ($2N = 10$) would lose at least half of its alleles, but only 10% of its heterozygosity would be expected to be lost. One way to quantify this effect is to calculate the expected number of alleles remaining after a founder event

$$n_{t+1} = n_t - \lfloor (1 - p_1)^{2N} + (1 - p_1)^{2N} + \ldots (1 - p_i)^{2N} \rfloor$$
$$= n_t - \sum (1 - p_i)^{2N} \tag{6.22a}$$

where $(1-p_i)^{2N}$ is the probability that allele A_i is lost from the population as the result of the founder event (Denniston, 1978).

To illustrate this effect, assume the "triangular" distribution of allele frequencies used by Pudovkin *et al.* (1996). In this case, the frequency of allele i when there are n alleles is

$$p_i = \frac{i}{n(n+1)/2} \tag{6.22b}$$

For example, when there are two alleles, $p_1 = 0.333$ and $p_2 = 0.667$, and when there are three alleles, $p_1 = 0.167$, $p_2 = 0.333$, $p_3 = 0.5$. This distribution is intermediate between assuming that all allele frequencies are equal to $1/n$ and the neutrality distribution (see Chapter 8), in which there are generally one or a few common alleles and a number of rare alleles. Also use the standardized allele number measure suggested by Allendorf (1986)

$$A' = \frac{n_{t+1} - 1}{n_t - 1} \tag{6.22c}$$

that scales the loss of alleles between 0 and 1 so that it can be compared to the loss of heterozygosity.

Figure 6.13 shows the loss of alleles for three different founder sizes when there are different numbers of alleles with a triangular distribution. For example, when $N = 5$ and $n_t = 5$, then $n_{t+1} = 4.09$ and $A' = 0.77$. This 23% reduction in the number of alleles is much greater than the 10% reduction in heterozygosity for $N = 5$. In fact, when $N = 5$ and there are four alleles or more, the reduction in the standardized allele number is greater than the loss of heterozygosity. Of course, when there are many alleles, even when the founder size is larger, the loss of allele number is

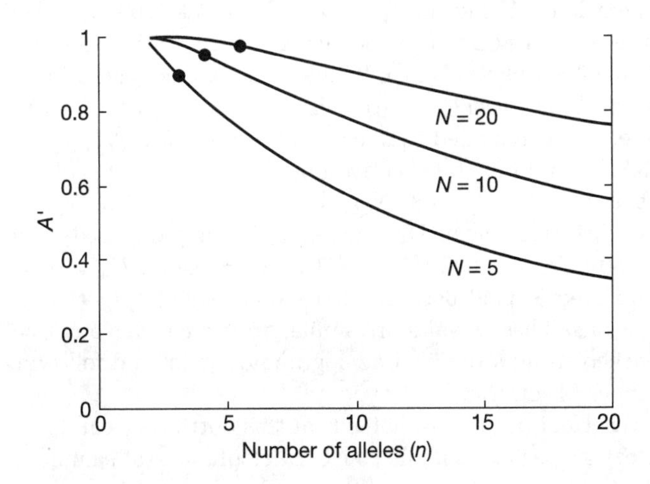

Figure 6.13. The standardized number of alleles A' after a one-generation founder event of size N given the number of alleles in the ancestral population before the founder event in a triangular distribution. The solid circles indicate the expected proportion of heterozygosity retained for the three founder sizes.

much greater than the loss of heterozygosity. On the basis of this differential rate of loss and using the neutrality distribution, Cornuet and Luikart (1996) have devised tests to detect bottlenecks, and Luikart and Cornuet (1998) gave a number of examples of populations that have gone through genetic bottlenecks (see p. 383) (see Example 6.11 for an application of genetic bottleneck theory to explain the low genetic variation in the northern elephant seals).

Example 6.11. The northern elephant seal was thought to have been hunted to extinction at the end of the 19th century when the "last" 153 animals were killed by collectors in 1884. Fortunately, some animals apparently survived on a remote beach on Isla Guadalupe, Mexico, and their descendants were rediscovered in 1892. However, the ancestors of the present-day population of approximately 200,000 stretching up to central California may have numbered as few as 20 (Bonnell and Selander, 1974).

Hoelzel *et al.* (2002) found two mtDNA haplotypes with estimated frequencies of 0.27 and 0.73 in the contemporary northern elephant seal population, giving a haplotype diversity estimate of 0.40. Hoelzel *et al.* (2002) also determined mtDNA haplotypes in prebottleneck museum and midden samples and estimated the mtDNA diversity in these samples as 0.80.

To determine what bottleneck could result in this loss of variation, assume that the loss of mtDNA diversity can be described by the expression

$$H_t = H_0 \prod_{i=1}^{t} \left(1 - \frac{1}{N_{ef.i}}\right)$$

where the original mtDNA diversity is $H_0 = 0.80$, the observed contemporary diversity is $H_t = 0.40$, and $N_{ef.i}$ is the effective female population size in generation i. Using the approach of Hedrick (1995b) and examining various sizes and duration of the bottleneck, but allowing the population to grow to known census levels in 1922 and 1960, a one-generation bottleneck of census size 15 is consistent with this observed loss of mtDNA variation. In this case, it was assumed that $R = 1.7$ (see p. 15), $N_{ef}/N_f = 0.25$, and $N_{ef.0} = 3.8$. Therefore, the effective population size of females grew from the bottleneck as $N_{ef.0} = 3.8$, $N_{ef.1} = 6.4$, $N_{ef.2} = 11.0$, $N_{ef.3} = 18.7 \ldots$ From this expression, the mtDNA diversity initially declined rather quickly as $H_0 = 0.804$, $H_1 = 0.592$, $H_2 = 0.501$, $H_3 = 0.455$, $H_4 = 0.431$ ... and then asymptoted at 0.40 because of the very large population size after a few generations. These results are similar to the conclusions Hoelzel *et al.* (2002) reached using a detailed demographic model and following the loss in the number of mtDNA haplotypes.

However, Hedrick (1995b) noted that this bottleneck cannot explain the lack of allozyme variation if the southern elephant seal sample is taken to represent the ancestral diversity. Furthermore, the microsatellite variation estimated in the 19th century northern elephant seal samples by Hoezel *et al.* (2002) was 0.565, and in the modern samples, it was 0.432. This is the opposite situation from the allozymes because the bottleneck size that explains the loss of mtDNA diversity results in a much greater loss of microsatellite variation than observed. In other words, it is hard to reconcile the prebottleneck and postbottleneck diversity at the nuclear genes with a bottleneck that is consistent with the observed loss of mtDNA variation.

III. SELECTION IN FINITE POPULATIONS

Remember that when there is no differential selection at a locus, an allele may become fixed or lost as a result of genetic drift. The probability of fixation is equal to the initial frequency of the allele so that when the allele is rare, the probability of fixation is quite low. In contrast, in an infinite population, which by definition has no genetic drift, a **favorable allele** always increases and asymptotically approaches fixation. In a finite population, however, a favorable allele may not always be fixed because it may be lost because of the chance effects of genetic drift.

The **probability of fixation** of a favorable allele in a finite population, $u(p)$, is a function of the initial frequency of the allele, the amount of selection favoring the allele, and the finite population size. For a model in which it is assumed that time is continuous, Kimura (1962) developed a general diffusion equation to incorporate these factors and to calculate the probability of fixation of allele A_1. For a relatively easy-to-follow exposition of the derivation of this general equation, see Kimura and Ohta (1971).

a. Directional Selection

If it is assumed that the relative fitnesses of the three genotypes A_1A_1, A_1A_2, and A_2A_2 are $1 + s$, $1 + hs$, and 1, respectively, the values given for directional or adaptive Darwinian selection in row c2 of Table 3.4, then the general diffusion equation becomes

$$u(p) = \frac{\int_0^p e^{-2Ns[(2h-1)x(1-x)+x]}\,dx}{\int_0^1 e^{-2Ns[(2h-1)x(1-x)+x]}\,dx} \tag{6.23a}$$

where N is the effective population size. This general expression for the probability of fixation depends on the initial frequency of allele A_1 (p), the level of dominance (h), the population size (N), and the selective advantage (s). There are four parameters involved, but the system reduces to three parameters because N and s always appear as a product. Even though the deviation of this expression contains several basic assumptions, the expression appears to be generally accurate even for discontinuous-time models unless the population size is fairly small. When there is additivity ($h = 0.5$), this expression reduces to the much simpler form

$$u(p) = \frac{1 - e^{-2Nsp}}{1 - e^{-2Ns}} \tag{6.23b}$$

The relationship between the different parameters and their effect on the probability of fixation is illustrated in Figures 6.14 and 6.15. In Figure 6.14, the probability of fixation is calculated for three levels of dominance and different initial allele frequencies for $Ns = 2.0$. An Ns value of 2.0 can result, for example, from a combination of $N = 1000$ and $s = 0.002$ or $N = 100$ and $s = 0.02$. The initial allele frequency has a large effect on the probability of fixation, with $u(p)$ increasing quickly as p increases from a low value. The difference in $u(p)$ for different levels of dominance is also substantial at low allele frequences. For example, if $p = 0.1$, the probabilities of fixation for $h = 0.0$, 0.5, and 1.0 are 0.223, 0.335, and 0.461, respectively. In general, an increasing level of dominance significantly increases $u(p)$ unless $u(p)$ is already near 1.0.

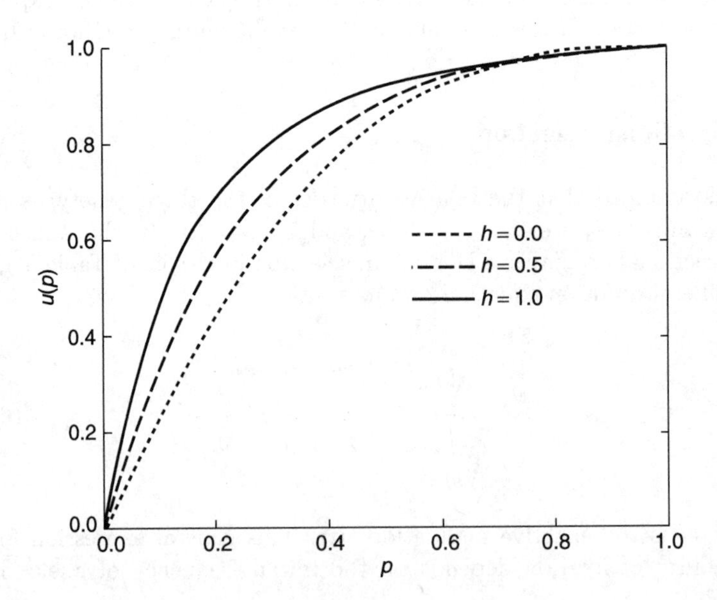

Figure 6.14. The probability of fixation for different initial allele frequencies and three levels of dominance when $Ns = 2.0$.

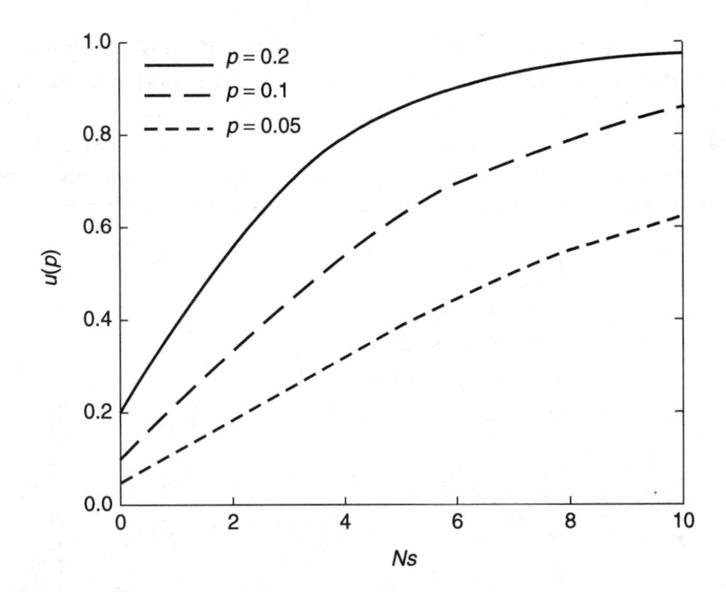

Figure 6.15. The probability of fixation for different Ns values and three initial allele frequencies when there is additive gene action ($h = 0.5$).

When $Ns \ll 1$—that is, $s \ll 1/N$ —the change in allele frequency is primarily determined by genetic drift: $u(p) \approx p$. On the other hand, if $Ns \gg 1$, then the change in allele frequency is primarily determined by selection and $u(p) \gg p$. To illustrate the effect of the size of Ns on $u(p)$, Figure 6.15 gives the probability of fixation for several initial allele frequencies for various levels of Ns when there is additivity ($h = 0.5$). As Ns increases, the probability of fixation also increases quite dramatically so that even if p is only 0.1, the probability of fixation when $Ns = 5$ is already 0.631.

When time is discontinuous, the transition matrix approach can be used to calculate the probability of fixation of a favorable allele and to follow the change in allele frequency distribution over time. In such a situation, the elements in the matrix must be modified to reflect selection as well as genetic drift. This can be done by assuming that selection changes allele frequency before sampling so that

$$x_{ij} = \frac{(2N)!}{(2N-i)\,!i\,!}\,(p')^{2N-i}\,(q')^{i} \tag{6.24a}$$

where

$$q' = \frac{(1+hs)\,pq + q^2}{1 + 2hspq + sp^2} \tag{6.24b}$$

and where $q = j/2N$ and $p = 1 - j/2N$. An example is given in Figure 6.16, where $N = 20$, $s = 0.1$, $h = 0.5$, and the initial frequency of A_1 in all the populations was 0.5. The probability of fixation for these parameter values, from expression 6.23*b*, is 0.78. The distribution of the frequency of the favorable allele reflects the effect of directional selection even after five generations. After 20 generations, 20.6% of the populations are fixed for the favorable allele and only 3.1% for the unfavorable allele. To appreciate the effect of selection, compare Figure 6.5 and Figure 6.16, which have identical parameters except for the presence of selection in the latter figure.

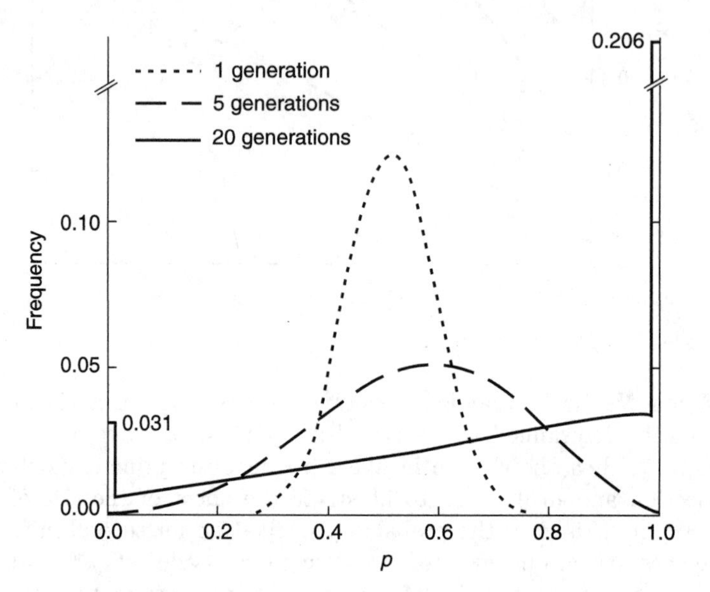

Figure 6.16. The smoothed distribution of allele frequency for a population of size 20 and an initial allele frequency of 0.5 with selection such that the fitnesses of A_1A_1, A_1A_2, and A_2A_2 are 1.1, 1.05, and 1.0, respectively, after 1, 5, and 20 generations.

In an infinite population, a **detrimental allele** always decreases in frequency and asymptotically approaches loss. In contrast, in a finite population, an unfavorable allele, particularly if its detrimental effects are not large, may increase in frequency by chance and may potentially become fixed. This effect, in which a detrimental allele behaves much like a neutral allele in a small population was pointed out by Wright (1931). Ohta (1973) discussed this phenomenon in terms of molecular evolution and described it as the **nearly neutral model** (see p. 418). She suggested that the relative impact of genetic drift and selection varies with the population size so that detrimental variants may be effectively neutral in a small population, whereas in large populations, they become selected against.

Important concerns for many endangered species are that the existing population generally is small, the species has gone through a bottleneck in recent generations, or the captive or extant population descends from only a few founder individuals. All of these factors may cause extensive genetic drift with a potential loss of genetic variation for future adaptive selective change. In addition, small effective population sizes may result in chance increases in the frequency of detrimental alleles because the absolute value of Ns is so low. For example, the captive population of Scandinavian wolves, initiated from four founders, had a high frequency of hereditary blindness

(Laikre *et al.*, 1993), and the captive California condor population, initiated from 14 founders, had a high frequency of a lethal form of dwarfism (Ralls *et al.*, 2000). Although a management strategy of selecting against carriers may reduce the frequency of these detrimental alleles, it may also result in a reduction in variation at other genes.

b. Balancing Selection

A good approach to understand the impact of balancing selection in a finite population is to compare the effect to the situation when there is no selection, or neutrality. For neutrality, we know that the expected rate of loss of heterozygosity per generation is $1-1/(2N)$. For any given balancing selection regime, we can define the asymptotic decline in heterozygosity per generation as d so that the heterozygosity decays as

$$H_{t+1} = (1-d)\,H_t$$

where d indicates the asymptotic—that is, independent of the starting allele frequency distribution, amount of loss from unfixed allele frequency states, and the amount of gain for the absorbing states. When there is no selection, $d = 1/(2N)$, and this expression reduces to the neutrality expression 6.3*a*.

The rate of decay may also be thought of as the asymptotic probability that an unfixed population will become fixed at either $q = 0.0$ or 1.0 in the next generation. The ratio of this quantity for a neutral locus over that for a locus undergoing selection is called the **retardation factor** (Robertson, 1962) and is

$$rf = \frac{1}{2Nd} \tag{6.25}$$

The retardation factor is unity when there is no selection—that is, when $d = 1/(2N)$. Selection can slow the rate of fixation compared with neutrality, $d < 1/(2N)$, and can make rf greater than one. Selection can also increase the rate of fixation, $d > 1/(2N)$, making the retardation factor smaller than one. The effect of virtually any selection model can be examined by calculating the effect of selection on allele frequency and then using these values to calculate the elements in the transition matrix as shown in expression 6.24*a*.

The retardation factor is particularly useful in assessing the impact of various balancing selection models on retaining genetic variation in finite populations. For example, Robertson (1962) investigated in detail the effect of finite population size when selection favored the heterozygote. He assumed that the relative fitnesses of genotypes A_1A_1, A_1A_2, and A_2A_2 were $1-s_1$, 1, and $1-s_2$, respectively, and then calculated the retardation factor for different values of $N(s_1+s_2)$. These are plotted in Figure 6.17 for different equilibrium values in an infinite population.

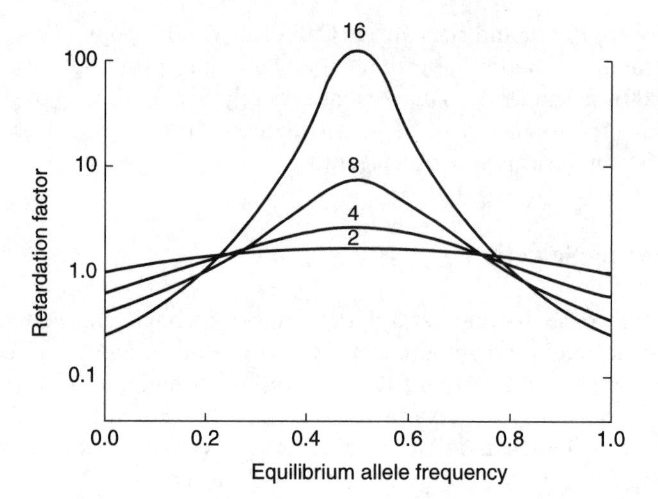

Figure 6.17. The retardation factor for the heterozygote advantage model for different values of $N(s_1 + s_2)$ (after Robertson, 1962).

First, the vertical axis is plotted on a logarithmic scale. This is done because when $N(s_1 + s_2)$ is 16 and the equilibrium frequency is near 0.5, the rate of loss of heterozygosity is quite low and the retardation factor becomes large. One of the most revealing findings of this analysis is that even though selection may result in a balanced polymorphism in an infinite population, in a finite population, less genetic variation may be retained than in a population with no selection. In other words, in some situations, balancing selection in a finite population actually increases the rate of decline of heterozygosity compared with neutrality. When q_e is below 0.2 or above 0.8, the retardation factor is generally smaller than unity, the value attained when there is no selection. In other words, in populations with heterozygote advantage and relatively unequal homozygote fitness values, genetic variation is actually eliminated faster than in populations with no selection.

It is possible to understand the basis for the differing rates of loss of genetic variation in a finite population for various selection models by examining the allele frequency change in an infinite population (Hedrick, 1972). In general, the most important factors increasing the retardation factor are the position of the allele frequency equilibrium, as shown by the heterozygous advantage model, and the magnitude of Δq at allele frequencies near 0 and 1. Having a high positive Δq value near 0 and a high negative Δq near 1 protects the population from becoming homozygous because of the chance effects in a finite population by pushing the allele frequency away from these absorbing states.

PROBLEMS

1. In how many generations will the expected heterozygosity be 5% of the initial value in populations of size 10 and in populations of size 100?

2. Calculate the probability matrix for $N = 2$ with no selection. Assuming that the initial gene-frequency distribution is (0.2, 0.2, 0.2, 0.2, 0.2), what are the gene-frequency distribution and heterozygosity after one and two generations?
3. What is the expected time to fixation for an allele with an initial frequency of 0.1 when the population size is 10, 100, and 1000? Why does the time to fixation depend on population size?
4. What is the probability of polymorphism in a founder group of size 4 when there are two alleles of equal frequency? What is the probability of polymorphism in a founder group of size 4 when there are ten alleles of equal frequency?
5. What is the expected increase in genetic distance between two populations when there is a one-generation bottleneck of $N = 3$ in one of the populations when the initial heterozygosity is 0.6? What is the expected increase in genetic distance between two populations when there is a ten-generation bottleneck of $N = 10$ in one of the populations when the initial heterozygosity is 0.8?
6. What is N_e when $N_f = 5$ and $N_m = 1$ for a diploid organism? What is N_e when $N_f = 1$ and $N_m = 10$ for a haplo-diploid organism?
7. What is N_e if in four consecutive generations the population sizes are 5, 50, 10, and 100? How different are your answers using expressions 6.12*a* and 6.12*b* ?
8. From demographic data in a bighorn sheep population, the estimated effective population size for females is $N_{ef} = 100$ and for males is $N_{em} = 10$. What are the expected effective population sizes for an mtDNA gene, a Y-chromosome gene, and an autosomal gene?
9. What is the probability of fixation for an additive favorable allele when its initial frequency is 0.1, $N = 10$, and its selective advantage is 0.01, 0.1, and 0.25?
10. Calculate the probability matrix for $N = 2$ when $\Delta q = {}^1\!/_2 sq(1 - q)$ and $s = 0.1$. Given the same initial gene-frequency distribution as in question 2 (0.2, 0.2, 0.2, 0.2, 0.2), find the gene-frequency distribution after one and two generations. Compare these distributions with those for no selection.
11. How would you determine empirically and experimentally the importance of genetic drift affecting variation at a given locus?
12. Discuss other factors besides a bottleneck that may account for the low heterozygosity in allozymes and the relatively high heterozygosity in microsatellite loci in the northern elephant seals discussed in Example 6.11.
13. Calculate the top three estimates of F and the sample-size corrected F in Table 6.13.
14. Calculate the arithmetic and harmonic means for N_e and N given in Table 6.10. What ratio, and why, do you think should be used to evaluate the relative differences in N_e and N?
15. If you were going to design a laboratory experiment, and assuming that resources and labor are not limiting, to follow the effect of genetic drift, how would you improve upon the experiment of Buri discussed in Examples 6.2 and 6.3?

7

Mutation

> Such expressions have given your opponents the advantage of assuming that favorable variations are rare accidents, or may even for long periods never occur at all and thus [the] argument would appear to many to have great force. I think it would be better to do away with all such qualifying expressions, and constantly maintain (what I certainly believe to be the fact) that variations of every kind are always occurring in every part of every species, and therefore that favorable variations are always ready when wanted.
>
> Alfred Wallace in a letter to Charles Darwin

> Among the mutations that affect a typical gene, different kinds produce different impacts. A very few are at least momentarily adaptive on an evolutionary scale. Many are deleterious. Some are neutral, that is, they produce no effect strong enough to permit selection for or against; a mutation that is deleterious or advantageous in a large population may be neutral in a small population where random drift outweighs selection coefficients. The impact of mutation is quite different in different DNA sequences. It is maximal in a conventional gene or exon, and at least transitorily less in a gene whose function is required rarely or is redundant.
>
> John Drake, Brian Charlesworth, Deborah Charlesworth, and James Crow (1998)

Mutation is a particularly important process in population genetics and evolution because it is the original source of genetic variation in a population. The process of mutation is multilevel and may involve change in a single nucleotide, several nucleotides, part of a gene, part of a chromosome, a whole chromosome, or sets of chromosomes. In general, we focus here on mutations within a gene that produce either a new allele or a new nucleotide sequence and not on chromosomal mutations. The immediate cause of a mutation may be, for example, a mistake in DNA replication, an insertion of a transposable element, a physical breakage of the chromosome, or a failure in disjunction of meiosis. Specific **mutagens**, such as ***chemicals or radiation that induce mutation***, generally cause certain types of mutations. For example, the chemical ethylmethane sulfonate (EMS) causes the replacement of a cytosine with a thymine, ultraviolet radiation causes the formation of thymine dimers and the subsequent insertion of the wrong nucleotides during replication, and γ or X-rays often causes severe damage, such as either single- or double-stranded breaks in the DNA molecule. Here we center on **spontaneous mutations**, ***mutations for which the immediate cause is not known***, but remember that environmental mutagens may greatly increase the mutation rate in a given population. Such factors as ultraviolet light, background radiation, and chemical pollutants may have a significant effect on the mutation rate and in turn influence the basic amount and type of genetic variation present in a population.

Mutation at a single gene may involve different degrees of change in DNA and protein sequences. For example, mutation may be just a substitution in a single-base pair in the DNA molecule, such as a replacement of one nucleotide for another (we discuss single-nucleotide changes in more detail in Chapter 8). However, a single-nucleotide change can effect a long stretch of DNA if it results in an insertion or a deletion of a nucleotide. For short, ***mutations that are either insertions or deletions*** are called **indels**. If an indel ***changes the reading frame in all of the subsequent codons in a transcript***, it is called a **frameshift mutant**. The result of such a mutant could be the synthesis of only a small part of a polypeptide because of the introduction of a stop codon, which would result in a nonfunctional protein. For microsatellite loci that are composed of repeats of a module, mutation is generally an increase or decrease of the number of repeats by one, but sometimes the change is greater, as we discuss below.

It is often difficult to determine the fitness of newly produced mutants. In general, visible mutations induced by EMS, γ radiation, or other mutagens cause a reduction in fitness. Spontaneous mutations that change the phenotype also usually result in a reduction in fitness. Interestingly, high percentages of the spontaneous visible mutants described in *D. melanogaster* are the result of insertions of transposable elements, and some disease mutations in humans are also the result of insertions of transposable elements (Ostertag *et al.*, 2003) (see Example 7.1 for data documenting the increase in two transposable elements in *D. melanogaster* in recent decades). Many spontaneous mutations may have detrimental effects if they influence the normal functioning of a gene. On the other hand, many mutants may be neutral, or almost neutral, with respect to other variants if they result in a nucleotide change that does not translate into an amino acid change or if they change DNA in a minor way in noncoding regions.

Example 7.1. Transposable elements are pieces of DNA that are capable of moving and replicating themselves within the genome of an organism. They have been found in most eukaryotes and are estimated to make up more than 50% of the maize genome, 15% of the *D. melanogaster* genome and 45% of the human genome (Kidwell and Lisch, 2001; Biémont *et al.*,

2003; Deininger *et al.*, 2003). The nature of the role of transposable elements has been characterized variously as selfish, junk, or neutral DNA or as agents of mutation or adaptation since they have been discovered. In an encompassing viewpoint, Kidwell and Lisch (2001) suggested that transposable elements may act from an extreme parasite to a mutualist, depending on the element, the organism, and the place in the genome. Overall, it appears that they have had an important impact on a number of aspects of genome evolution in many organisms (Kidwell and Lisch, 2001; Deininger *et al.*, 2003; Kazazian, 2004), and there is evidence of positive selection associated with some transposable elements (Maside *et al.*, 2002; Schlenke and Begun, 2004).

The phenomenon of **hybrid dysgenesis**, in which crosses between certain *D. melanogaster* strains result in male recombination (ordinarily there is no recombination in *Drosophila* males), elevated mutation rates, and sterility, was found to be the result of the presence of transposable elements in one of the parent strains and absence in the other. In the late 1970s, it was noticed that older laboratory strains did not appear to have the transposable elements I and P, whereas recent strains did. For example, Kidwell (1983) measured the frequency of I and P in *D. melanogaster* strains collected by decade from the 1920s to the 1970s (Figure 7.1). Although early strains did not contain either element, strains with the I element increased to around 50% by the 1940s and to nearly 100% in the 1970s; strains with the P element began increasing in the 1950s and reached 87.5% in the 1970s.

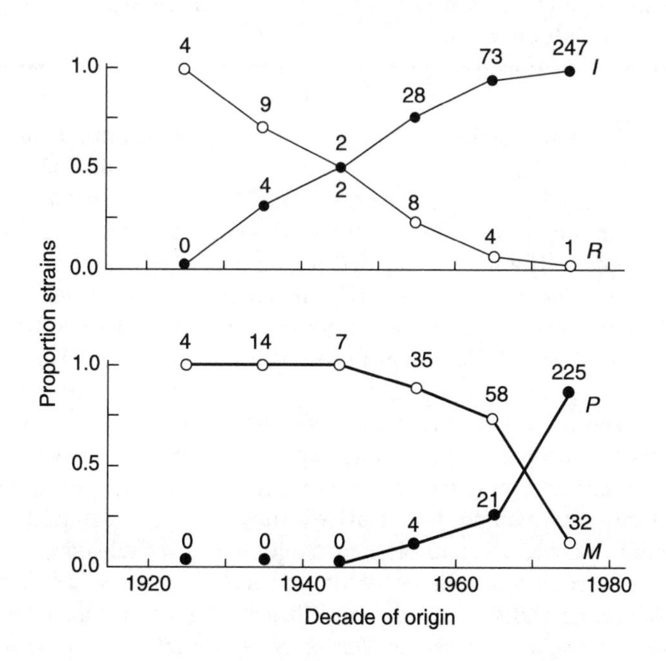

Figure 7.1. The estimated proportion of strains with I (top) and P (bottom) elements as detected by hybrid dysgenesis in strains collected in the decades from the 1920s to the 1970s (after Kidwell, 1983). The category R indicates strains without the I element; the category M indicates strains without the P element; and the number of strains in each category are indicated.

Two hypotheses could explain these results: the I and P elements invaded *D. melanogaster* from another species and then spread worldwide, or the older laboratory stocks had lost the elements over time in the laboratory. P-like elements have been found in other *Drosophila* species, and Daniels *et al.* (1990) found a P element in *D. willistoni* that was only one nucleotide different in 2907 base pairs from the complete P element in *D. melanogaster*. Because the two species are thought to have diverged 60 million years ago, the great similarity between the elements in the two species could be explained only by a recent horizontal transfer between them. Although *D. melanogaster* is a cosmopolitan species today, this ubiquity is thought to be rather recent. *D. melanogaster* is thought to have originated in western Africa and then to have spread as the result of human activity. On the other hand, *D. willistoni* is a new-world species found primarily in Central and South America, so the coexistence of the two species probably started only in the twentieth century. The mechanism of horizontal transfer is unknown, but both parasitic mites and insect viruses have been suggested.

To examine the spread of P elements experimentally in a population, Kidwell *et al.* (1988) and Good *et al.* (1989) introduced P elements into strains without P elements and monitored their increase. For example, in two lines initially with 5% P element chromosomes, Good *et al.* observed that nearly 100% of the chromosomes had P elements by generation 10. In other words, P elements can spread very quickly in a population once they are introduced. This spread is by transposition, in which a copy of the P element jumps to a new position in the genome while the original P element remains functional.

A hypothetical distribution of the fitness of new spontaneous mutants is given in Figure 7.2. (For heuristic purposes, let us assume that each allele has a mean or a marginal fitness value.) The distribution is basically bimodal, with some mutations being so **detrimental** that they result in lethality or near lethality (the shaded region on the left) and others having a less extreme detrimental effect (the unshaded region on left). Another group of mutations, presumably including many single-base substitutions, results in only a small change in fitness if any (the shaded region on the right). This category can be thought of as **neutral** or **nearly neutral** mutations (see further discussion of these terms on p. 411 and p. 418). A third category, probably quite small but important (unshaded region to far right), includes the **advantageous** mutations that confer an increased relative fitness. Of course, the relative fitness of a given mutant depends on the environment; for example, a pesticide-resistant allele would be beneficial in the presence of the pesticide and may be neutral or detrimental in its absence. In addition, the relative fitness of a mutant also depends on the alleles present at other genes; for example, a mutant may be beneficial in one genetic background and neutral or detrimental in another.

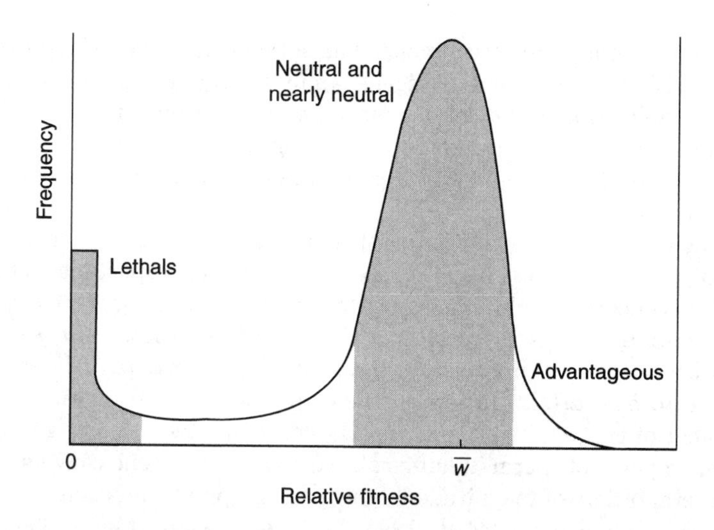

Figure 7.2. A hypothetical distribution of relative-fitness values for new mutants, where $\overline{w}$ is the mean population fitness.

Estimates of the shape of this distribution and the average detrimental mutational effect of 1% to 2% in the past were mainly from *Drosophila* (Simmons and Crow, 1977) (although see discussion and references on p. 398). The estimation of lethal mutation rate in *Drosophila* from a number of studies is approximately 0.026 per generation (Table 7.6). The relative sizes of these categories of mutation are open to speculation, but Mukai *et al.* (1972) estimated that approximately 10 to 20 times as many mutations have a slight effect on viability in *D. melanogaster* as those resulting in lethality. Kidwell and Lisch (2001) suggested that transposable elements may generate a variety of potentially adaptive mutants, and Nei (1987) hypothesized that the availability of advantageous mutants may often limit adaptive change. Experimental evolutionary research in *E. coli* has demonstrated that adaptive change is dependent on the generation of rare beneficial mutations (Elena *et al.*, 1996). The relatively few advantageous mutants and their generally small fitness advantage have been confirmed in *E. coli* by Imhof and Schlotterer (2001) who identified 66 adaptive mutants and found that most of them had a small effect, whereas a few had a large effect (see comments of Orr, 2003). Using site-directed mutagenesis in an RNA virus, Sanjuan *et al.* (2004) found that about 40% of random mutations were lethal and that about 4% of the mutants were beneficial.

To understand genome evolution (see Li, 1997, for an extensive discussion and Lynch and Conery, 2003, for recent insights), mutation by gene duplication (and deletion) and gene conversion are considered important. A substantial proportion number of human genes, such as histone, globin, rRNA, tRNA, and MHC genes, exist in **multigene families**, that is, a ***group of homologous genes with related functions that are often closely linked on a chromosome***. From the available complete genome sequence data, it is estimated that in bacteria 17% to 44% of the genes are duplicate, and in eukaryotes, 30% to 65% are duplicate (Zhang, 2003). Generally, the members of a multigene family are thought to have arisen by duplication from an ancestral gene. The subsequent increase and decrease of the number of genes in a multigene family are thought to be the result of further duplication and deletion, generally through unequal crossing over. Ota and Nei (1994a) suggested that multigene families may evolve by a birth and death process in which genes are continuously added or lost to the multigene family.

These duplicated genes may either diverge from each other or be homogenized by a process called concerted evolution. Important factors involved in concerted evolution are gene conversion and unequal crossing over. **Gene conversion** is a process by which ***part of nucleotide sequence from one allele is replaced by the homologous nucleotide sequence from another allele***. Rates of gene conversion have been estimated for a number of genes in yeast and average approximately 5% per generation. Screening pools of sperm in humans gave estimates of gene conversion that are several orders of magnitude larger than the point mutation rate for an *HLA* gene (Zangenberg *et al.*, 1995), and Jeffreys and May (2004) found that regions of high crossing over are also regions of a high rate of gene conversion. Ohta (1980) presented a model to examine the evolution of a multigene family that contains a tandem array of genes that can undergo both gene conversion and unequal crossing over.

I. ALLELE FREQUENCY CHANGE CAUSED BY MUTATION

To evaluate the impact of mutation on genetic variation, let us initially assume that at each locus there are two principal types of alleles: wild-type alleles and detrimental alleles. Both of these classes appear to be very heterogeneous. For example, the neutrality theory that is discussed further in Chapter 8 assumes that most wild-type variants in nucleotide sequences at a gene are neutral with respect to each other. Detrimental alleles may also be quite heterogeneous, as suggested by the variants at genes that cause human genetic diseases (McKusick, 1998; Online Mendelian Inheritance in Man, OMIM) or morphological variation in *D. melanogaster* (Lindsley and Zimm, 1992).

To begin our discussion of mutation, let us assume that ***mutation is reversible and can occur from the wild-type to the detrimental category or from the detrimental to the wild-type category***. These types of mutations have conventionally been called **forward mutations** and **backward mutations**, respectively. Generally, it is assumed that forward mutations occur more frequently because they are mutations that result in a malfunction of the gene (see Example 7.2 for estimates of forward and backward mutation rates in a large study in mice and a discussion of mutation later in the chapter). Such a malfunction in a normal allele can presumably occur in a multitude of ways, whereas the repair of such a problem, a backward mutation, probably occurs much less frequently because only a limited number of mutations can compensate for the original mutation other than a specific reversion. However, when the original mutant is a frameshift mutant, then a second frameshift mutant at a number of sites may compensate for the first mutant. If a mutation is caused by the insertion of a transposable element, reversion may occur at a substantial rate when the element is excised, although some sequence from the element may be retained.

Example 7.2. Because mutations at individual loci are generally rare events, large-scale experiments are necessary to estimate mutation rates adequately (see p. 390 for a discussion of estimation techniques). One such massive study was undertaken to estimate both the forward and the backward spontaneous mutation rates at five coat-color loci in mice (Schlager and Dickie, 1971). Over a period of six years, more than seven million mice at the Jackson Laboratory in Bar Harbor, Maine, were examined for spontaneous mutations at the *nonagouti, brown, albino, dilute,* and *leaden* loci. The estimated mutation rate for each locus and the overall values are given in Table 7.1. The overall forward mutation rate was 11.2×10^{-6} and the overall backward mutation rate was 2.5×10^{-6}, a result that illustrates the general finding that the rate of forward mutation is higher than the rate of backward mutation.

TABLE 7.1 Spontaneous mutation rates at five specific coat-color loci in mice (Schlager and Dickie, 1971).

Locus	*Number of gametes tested*	*Number of mutations*	*Mutation rate* ($\times 10^{-6}$)	*95% confidence limits* ($\times 10^{-6}$)
	Mutations from wild type (forward)			
Nonagouti	67,395	3	44.5	9.2–130.1
Brown	919,699	3	3.3	0.7–9.5
Albino	150,391	5	33.2	10.8–77.6
Dilute	839,447	10	11.9	5.2–21.9
Leaden	243,444	4	16.4	4.5–42.1
Total	2,220,376	25	11.2	7.3–16.6
	Dominant mutations (backward)			
Nonagouti	8,167,854	34	4.2	2.9–5.8
Brown	3,092,806	0	0	0–1.2
Albino	3,423,724	0	0	0–1.1
Dilute	2,307,692	9	3.9	1.8–11.1
Leaden	266,122	0	0	0–13.9
Total	17,236,978	43	2.5	1.8–3.4

More recently, control data for several large radiation experiments in mice have been summarized (Russell and Russell, 1996). Overall, for seven recessive, visible markers, 69 forward mutations were observed in 1,485,036 progeny, and the estimated per-locus mutation rate was 6.6×10^{-6}. Although this is somewhat lower than the estimate in Table 7.1, if mosaics (partial-body mutants) are included in the Russell and Russell summary, then the per-locus estimate becomes 11.0×10^{-6}, very similar to the estimate by Schlager and Dickie.

a. Forward and Backward Mutation

To examine the effect of mutation on genetic variation in a population, let us assume that ***the rate of mutation from wild-type alleles (A_1) to detrimental alleles (A_2) is u***, the forward mutation rate, and ***the rate of mutation from detrimental to wild-type alleles***, the backward mutation rate, is $\boldsymbol{v}$. The change in the frequency of A_2 due to mutation alone is then

$$\begin{aligned}\Delta q &= up - vq \\ &= u - q(u+v)\end{aligned} \tag{7.1a}$$

In other words, p is the proportion of alleles that can mutate from A_1 to A_2 with a rate u, and q is the proportion that can mutate from A_2 to A_1 with a rate v. This expression is linearly related to the allele frequency and reaches its maximum positive value when $q = 0$ ($\Delta q = u$) and its maximum negative value when $q = 1$ ($\Delta q = -v$). Because u and v are generally quite small, the expected change caused by mutation is also quite small. Figure 7.3 illustrates this relationship when the forward and backward mutation rates are the same, 10^{-5}, and when the backward mutation rate is 10^{-6} (an order of magnitude lower) and the forward mutation rate is still 10^{-5}.

It is obvious from Figure 7.3 that a stable equilibrium can result from a balance between forward and backward mutation rates. The equilibrium frequency can be obtained by setting expression 7.1a equal to zero and solving for q; thus, the equilibrium frequency is

$$q_e = \frac{u}{u+v} \tag{7.1b}$$

If the forward mutation rate is higher than the backward mutation rate ($u > v$), then it would be expected that the frequency of detrimental alleles would be higher than that of wild-type alleles. This does not occur, of course, because purifying selection reduces the frequency of detrimental alleles, a topic we discuss on p. 368.

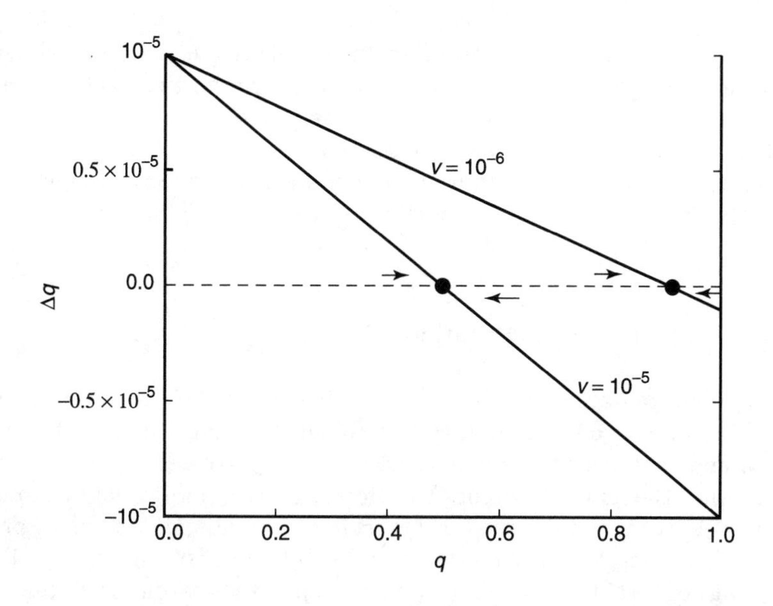

Figure 7.3. The change in allele frequency due to mutation when $u = 10^{-5}$ and v is either 10^{-5} or 10^{-6}. The closed circles indicate allele frequency equilibria.

How effective of a force is mutation in changing the allele frequency in a population over a number of generations? If we assume that v is small compared with u, then the frequency of A_1 after one generation with mutation is

$$\begin{aligned} p_1 &= p_0 - up_0 \\ &= (1 - u)p_0 \end{aligned}$$

Generalizing this relationship to t generations yields

$$p_t = (1 - u)^t p_0 \tag{7.2a}$$

From this we can see, for example, that it takes nearly 70,000 generations (69,314 to be exact) to halve the frequency of the wild-type allele ($p_t = \frac{1}{2}p_0$) when $u = 10^{-5}$. In other words, the change in allele frequencies effected by mutation alone is extremely slow, even when the forward mutation rate is fairly large and there is no backward mutation.

If we do not make the assumption that v is small, then the frequency of A_1 after one generation is

$$p_1 = (1 - u)p_0 + v(1 - p_0)$$

This can be generalized to

$$p_t = \frac{v}{u + v} + \left(p_0 - \frac{v}{u + v}\right)(1 - u - v)^t \tag{7.2b}$$

where $v/(u+v)$ is the equilibrium frequency of A_1 (Neuhauser, 2001). As t increases, the contribution from the second term on the right decreases because $(1-u-v)^t$ approaches zero. If we assume that $u = 10^{-5}$, $v = 10^{-6}$, and $p_0 = 0.5$, then after 69,314 generations, $p_t = 0.282$. In other words, the presence of backward mutation only reduces the expected reduction in the wild-type allele slightly, from 50% to 44%, over this period.

b. The Fate of a Single Mutation

When a new mutation occurs, it is the only copy in the entire population. In other words, the population consists of all A_1A_1 individuals before mutation and contains only one heterozygous A_1A_2 individual, with the rest A_1A_1, after the mutation event. Therefore, the A_1A_2 individual must mate with an A_1A_1 individual, and depending on the number or type of offspring from this mating, the new mutant may be lost (no offspring or only A_1A_1 offspring) or retained (some A_1A_2 offspring). If the mating produces only one offspring, there is a 50% chance that it will be A_1A_1, which means that the mutant A_2 will be lost in the first generation. Even if the single offspring is A_1A_2, there will still be only one copy of A_2 in the next generation. If the mating results in two progeny, the probability of losing A_2 —that is, of having two A_1A_1 progeny—is $(1/2)(1/2) = 1/4$. The probability of one copy of A_2 still remaining is $1/2$, and the probability of two copies, two A_1A_2 progeny, is $1/4$. In other words, the frequency may decrease, remain unchanged, or increase depending on the type and number of progeny produced.

Assume, as on p. 318, that there are k progeny produced and that the distribution of family size is Poisson. Let us also assume that the population is quite large and is constant in size ($\bar{k} = 2$). Table 7.2 gives the frequencies of different family sizes under these assumptions as well as the probability of the loss of the new mutant A_2, given a particular family size. In general, the probability of having all A_1A_1 progeny in a family of size k is $(1/2)^k$.

By summing the product of the frequency of a particular family size and the probability of the loss of the mutant given that family size, we can

TABLE 7.2 The Poisson distribution of family size with $\bar{k} = 2$ and the probability of loss of mutant A_2 associated with different family sizes.

	Number of progeny					
	0	1	2	3	...	k
Frequency	e^{-2}	$2e^{-2}$	$\frac{2^2}{2!}e^{-2}$	$\frac{2^3}{3!}e^{-2}$	...	$\frac{2^k}{k!}e^{-2}$
Probability of loss of A_2	1	$\frac{1}{2}$	$(\frac{1}{2})^2$	$(\frac{1}{2})^3$	...	$(\frac{1}{2})^k$

calculate the overall probability of loss. In other words, the total probability that the new mutant will be lost in one generation is

$$
\begin{aligned}
x_1 &= \sum_{k=0}^{\infty} \left(\frac{2^k}{k!}\right) e^{-2} \left({}^1\!/_2\right)^k \\
&= e^{-2} \sum_{k=0}^{\infty} \frac{1}{k!} \\
&= e^{-2} \left(1 + 1 + \frac{1}{2!} + \frac{1}{3!} \cdots\right) \\
&= e^{-2} e \\
&= e^{-1}
\end{aligned}
$$

In other words, approximately 0.368 of the time, a new mutant will be lost in one generation. However, a retained mutant is still at a low frequency and may be lost in the next generation. In fact, by the second generation, a new mutant is expected to be lost over half the time (Table 7.3). For comparison, the probability of loss when the mutant has a small selective advantage is also given in Table 7.3 (we discuss advantageous mutants later on p. 385). There is virtually no difference between the probability of loss of this slightly selectively advantageous allele and the neutral mutant in the early generations. The probability of fixation of the neutral mutant is essentially zero because of the large population size; however, the probability of fixation of the selectively advantageous mutant is 0.02 (approximately twice its selective advantage).

TABLE 7.3 The probability of loss and of survival of a new mutant when there is neutrality or a 1% selective advantage (after Fisher, 1930).

	Neutral		$s = 0.01$	
Generation	*Loss*	*Survival*	*Loss*	*Survival*
1	0.368	0.632	0.364	0.636
2	0.532	0.468	0.526	0.474
3	0.626	0.374	0.620	0.380
4	0.688	0.312	0.681	0.319
5	0.732	0.268	0.725	0.275
⋮	⋮	⋮	⋮	⋮
15	0.887	0.113	0.878	0.122
⋮	⋮	⋮	⋮	⋮
127	0.985	0.015	0.973	0.027
⋮	⋮	⋮	⋮	⋮
∞	1.0	0.0	0.98	0.02

II. MUTATION–SELECTION BALANCE

Selection is the major force that keeps detrimental alleles from increasing in frequency. For example, many individuals with recessive diseases in humans would not survive to adulthood without intensive medical care. *Drosophila* males with eye mutants are at a great disadvantage in mating, and plants homozygous for recessive chlorophyll mutants would not survive to reproduce. This purifying selection keeps the frequency of detrimental and lethal alleles at a given locus quite low. However, because thousands of genes may be influenced by the opposite effects of mutation and selection, detrimental and lethal alleles may be quite important when considering the entire genome. In the following discussion, which illustrates the joint effects of selection and mutation and the balance between them, we assume that there are only two alleles, and one of those, the mutant (A_2), causes a reduction in viability.

a. Recessive Mutants

If the detrimental allele is recessive, then the change in allele frequency caused by selection is

$$\Delta q_s = -\frac{sq^2p}{1 - sq^2}$$

as given on p. 127. The increase in allele frequency due to mutation is approximately

$$\Delta q_{mu} = up$$

if we assume that the backward mutation rate is low compared with the forward mutation rate, u. Because the two forces have opposite effects on allele frequency, they balance each other, and thus, at some point

$$\Delta q_{mu} + \Delta q_s = 0$$

or

$$up = \frac{sq^2p}{1 - sq^2}$$

If q^2 is small, then the denominator on the right-hand side of the last expression can be assumed to be approximately unity. The expression then can be solved for the equilibrium frequency of the recessive genotype as

$$q_e^2 = \frac{u}{s} \tag{7.3a}$$

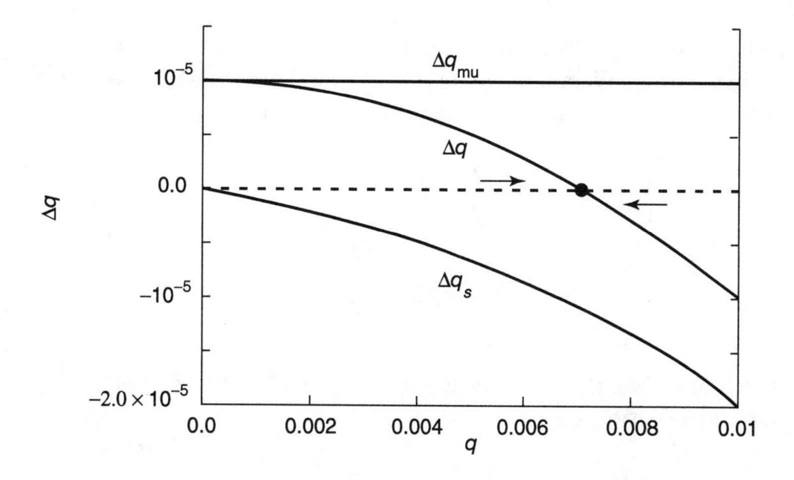

Figure 7.4. The change in allele frequency due to both mutation to and selection against recessive alleles. The line labeled Δq is the summation of Δq_{mu} and Δq_s, and the closed circle indicates the equilibrium.

and for the equilibrium allele frequency as

$$q_e = \left(\frac{u}{s}\right)^{1/2} \tag{7.3b}$$

The balance between selection and mutation can be understood by examining Δq_{mu} and Δq_s at low frequencies of A_2. Figure 7.4 gives these values when $u = 10^{-5}$ and $s = 0.2$; Δq_{mu} is approximately 10^{-5} throughout this low range in allele frequencies, whereas Δq_s is 0.0 when $q = 0.0$ and becomes increasingly negative as q increases. The equilibrium for this example is

$$q_e = \left(\frac{10^{-5}}{0.2}\right)^{1/2} = 0.00707$$

In other words, this is the point at which $\Delta q_{mu} + \Delta q_s = 0$, as shown in the resultant Δq curve and the solid circle indicating the equilibrium.

If the allele is a recessive lethal, then $s = 1.0$, and the equilibrium genotype and allele frequencies are the mutation rate u, and $u^{1/2}$, respectively. The equilibrium allele frequency is obviously increased as the result of either a higher mutation rate or a lower selective disadvantage. For example, if the rate of mutation caused by some mutagenic factor(s) is increased 10-fold, then the equilibrium genotype frequency is also increased 10-fold. Similarly, if a disease that was formerly lethal before the age of reproduction ($s = 1$) and now is only slightly disadvantageous ($s = 0.1$) because of better medical care, such as for phenylketonuria, then again the equilibrium genotype frequency is increased 10-fold.

The mutation–selection equilibrium also can be obtained without assuming Hardy–Weinberg genotype proportions. With a recessive, the change in allele frequency resulting from selection is

$$\begin{aligned}\Delta q &= q_1 - q_0 \\ &= \frac{H/2 + Q\,(1-s)}{1 - sQ} - q_0 \\ &= -\frac{sQ\,(1-q)}{1 - sQ}\end{aligned}$$

If we assume that the effect of mutation is as above, then combining mutation and selection yields

$$\begin{aligned}\Delta q &= u(1-q) - \frac{sQ(1-q)}{1-sQ} \\ &= \frac{(1-q)\,[u - sQ(u+1)]}{1-sQ}\end{aligned}$$

There is an equilibrium when the term in the brackets is zero or

$$u - sQ(u+1) = 0$$

When we solve for the equilibrium genotype frequency of A_2A_2, this becomes

$$Q_e = \frac{u}{s(u+1)}$$

which is approximately

$$Q_e \approx \frac{u}{s} \tag{7.3c}$$

if $u \ll s$. With Hardy–Weinberg proportions, this equilibrium value is equal to expression 7.3*a*.

This approach can also be used to investigate the effect of inbreeding on the equilibrium frequency caused by a mutation–selection balance. Let us assume that the genotype frequency of A_2A_2 is modified by inbreeding and that the genotype frequency can be expressed, as in Chapter 5, as

$$\begin{aligned}Q &= q^2 + fq(1-q) \\ &= (1-f)q^2 + fq\end{aligned}$$

When we substitute for Q into expression 7.3*c*, this becomes

$$(1-f)q^2 + fq = \frac{u}{s}$$

Remember that the genotype frequency at equilibrium is only a function of the mutation rate and selection against the recessive. However, the equilibrium allele frequency is a function of the inbreeding coefficient as well as selection and mutation. The equilibrium allele frequency can be found by solving the above quadratic equation so that the equilibrium allele frequency is given in terms of f, u, and s as

$$q_e = \frac{-f + \left[f^2 + 4(1-f)\frac{u}{s}\right]^{1/2}}{2(1-f)} \tag{7.4a}$$

(Haldane, 1940). Morton (1971) gives the approximate equilibrium value for this situation as

$$q_e = \frac{u}{fs} \tag{7.4b}$$

when q_e is small compared with f. This equilibrium value is obtained by assuming that the q^2 term in the above quadratic equation is small relative to the other terms.

An important prediction from this formulation is that the equilibrium allele frequency is decreased as the amount of inbreeding is increased. An example of this effect is given in Figure 7.5 for several values of inbreeding when $u = 10^{-5}$. One can reach an intuitive understanding of this phenomenon by realizing that inbreeding exposes more detrimental alleles to selection by increasing the frequency of A_2A_2 homozygotes and that this in turn reduces the frequency of the allele A_2 at equilibrium. For example, if $s = 0.5$, the equilibrium allele frequency for $f = 0.0$ is more than 10 times that for $f = 0.05$, whereas the frequency of the homozygote (the incidence of a recessive disease) is the same in both cases. One implication of

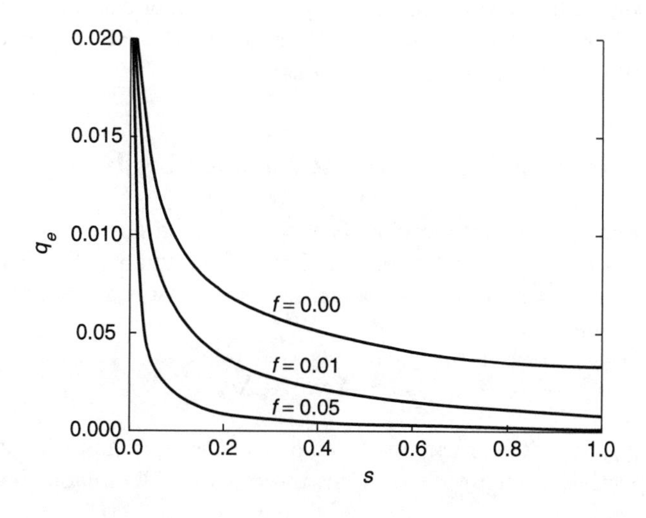

Figure 7.5. The mutation–selection equilibrium allele frequency for different levels of selection and three values of inbreeding ($u = 10^{-5}$).

this effect is that if an equilibrium population with substantial inbreeding had a large decrease in inbreeding, then there would be a temporary reduction (probably lasting hundreds of generations) in the incidence of recessive diseases until mutation restored the population to the equilibrium value.

Lande and Schemske (1985) showed that if inbreeding results from a level of selfing S, when the level of selfing is greater than a few percent, the equilibrium heterozygosity for a recessive is approximately

$$H_e \approx 4u\left(\frac{Ss+1-S}{Ss}\right) \tag{7.4c}$$

For recessive lethals ($s = 1$), this reduces to

$$H_e \approx \frac{4u}{S}$$

From these expressions, it is obvious that the equilibrium heterozygosity is inversely proportional to the level of selfing (see also Example 7.3 below). For all levels of selection, when $S = 1$, $H_e \approx 4u$. When there is more random mating, then the equilibrium heterozygosity is much higher. For example, when $S = 0.1$ and if $s = 0.1$ or $s = 1$, then $H_e \approx 364u$ or $H_e \approx 40u$, respectively.

The equilibrium frequency of allele A_2 is approximately

$$\begin{aligned} q_e &\approx 2u\left(\frac{Ss+1-S}{Ss}\right)+\frac{u}{s} \\ &\approx \frac{u(2Ss+2-S)}{Ss} \end{aligned} \tag{7.4d}$$

A small amount of selfing can greatly decrease the equilibrium allele frequency from that in a random-mating population. For example, if $u = 10^{-6}$ and $s = 0.1$, when $S = 0$, $q_e = 0.00316$, and when $S = 0.1$, $q_e = 0.000192$, more than a 16-fold decrease in frequency.

b. Other Levels of Dominance and Mutation Load

When the mutant is not completely recessive, then the equilibrium allele frequency is highly dependent on the level of dominance. To examine the effect of different levels of dominance, let us assume that the total change in allele frequency is

$$\Delta q = \Delta q_{mu} + \Delta q_s$$

where Δq_{mu} and Δq_s are again the changes caused by mutation and selection, respectively. Therefore, with an intermediate dominance level where the relative fitness of the heterozygote is 1 - hs, this expression becomes approximately

$$\Delta q = up - spq[h - (2h - 1)q]$$

Assuming $\Delta q = 0$, this expression becomes the quadratic equation

$$s(1 - 2h)q^2 + shq - u = 0$$

with the solution

$$q_e = \frac{-sh + \left[s^2h^2 + 4us\,(1 - 2h)\right]^{1/2}}{2s\,(1 - 2h)} \qquad (7.5a)$$

If $h \gg 0$ and q_e is small, the equilibrium allele frequency is approximately

$$q_e \approx \frac{u}{hs} \qquad (7.5b)$$

Morton(1971).

If there is complete dominance ($h = 1$), then the equilibrium frequency of A_2 is approximately

$$q_e \approx \frac{u}{s} \qquad (7.5c)$$

The frequency of the mutant phenotype (only the heterozygote is considered because the dominant homozygote, A_2A_2, is very rare and probably lethal) at equilibrium is

$$\begin{aligned} H_e &= 2p_eq_e \\ &\approx \frac{2u}{s} \end{aligned} \qquad (7.5d)$$

assuming that p_e is approximately unity.

The equilibrium allele frequency from expression 7.5a is given in Figure 7.6 for three levels of selection when $u = 10^{-5}$. As h increases, the equilibrium is reduced very quickly. For example, with $s = 1.0$ for a recessive ($h = 0$) and a dominant ($h = 1$) allele, the equilibrium values are 0.00316 and 0.00001, respectively. With $h = 0.01$, a virtually undetectable disadvantage in the heterozygote, the equilibrium allele frequency becomes 0.00092, a reduction of over 70% from a recessive. Example 7.3 gives an application of the potential impacts of dominance, inbreeding, and mutation on the basis of the difference in inbreeding depression in Scots pine from the north and south of Finland.

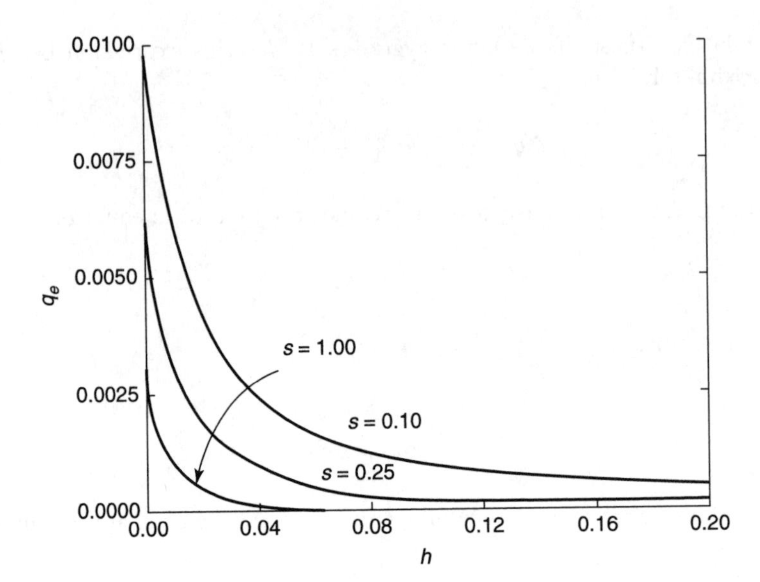

Figure 7.6. The mutation–selection equilibrium frequency for different levels of dominance and three levels of selection ($u = 10^{-5}$).

Example 7.3. Kärkkäinen *et al.* (1996) showed that populations of Scots pine, *Pinus sylvestris*, from northern Finland have a statistically significant lower number of lethals than populations from southern Finland. They suggested that this 34% reduction in inbreeding depression could be the result of more self-fertilization in northern than in southern populations or stronger selection against detrimental alleles in the northern than in the southern populations. Scots pine, the dominant tree in most of Finland, reached the northern areas of Finland only within the last 8000 years, approximately around 100 generations ago. As a result, the evolutionary changes would have to have taken place within this rather short time.

To determine whether differences in selfing or selection could have resulted in the reduced inbreeding depression observed in northern Finland, Hedrick *et al.* (1999a) constructed a model that included selfing, selection, and mutation. They found that a relatively small increase in selfing from 0.083 in the south to 0.146 in the north or a small increase in selection against heterozygotes from $h = 0.02$ in the south to 0.054 in the north could result in the 34% decrease in inbreeding depression. In addition, changes in both of these parameters could have reduced the frequency of heterozygotes, which is directly proportional to the number of lethals, within 100 generations. Figure 7.7 gives the expected reduction of inbreeding depression over 100 generations with the inbreeding depression in the south as the starting point. Obviously, an increase in selfing or an increase in dominance can theoretically reduce the inbreeding depression to the level in the north (indicated by the broken line) within 100 generations. Also given is

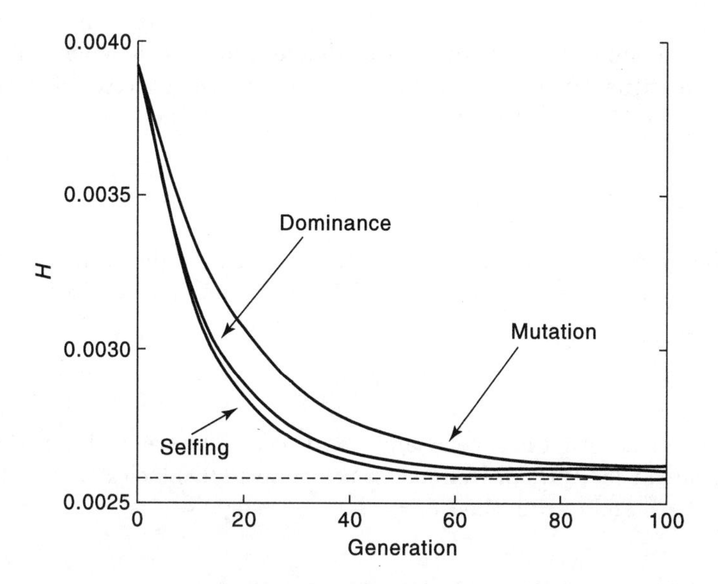

Figure 7.7. The change in heterozygosity starting with equilibrium proportions in the southern Finland population and the changes expected as a result of changes in selection against heterozygotes (dominance), selfing, and mutation (Hedrick *et al.*, 1999a).

the effect of a 34% decrease in mutation rate, and it also could result in a reduction of inbreeding depression over this time span. However, there is evidence that the rate of selfing and the intensity of selection are higher in the north, whereas there is no such evidence for differential mutation rates, so the explanations proposed by Kärkkäinen *et al.* appear most likely.

As we saw on p. 143, the genetic load is generally defined as the reduction in fitness in a population compared with the fitness if the population were composed only of the optimum genotype. In this case, the optimum genotype is A_1A_1 and the load becomes

$$\begin{aligned} L &= 1 - \overline{w} \\ &= sq(2ph + q) \end{aligned} \tag{7.6a}$$

Let us examine how mutation may reduce the fitness—that is, determine the extent of the **mutation load**.

For a recessive mutant ($h = 0$), $L = sq^2$. From expression 7.2a, at equilibrium $u = sq^2$ so that the load is approximately equal to the mutation rate, $L \approx u$. If there is complete dominance, $h = 1$, then there will be very few A_2A_2 genotypes, and thus, the load becomes $L = 2pqhs$. Using the equilibrium frequency of the heterozygote from expression 7.5d, $2u/s$, the load then becomes $2u$. In other words, the **mutation load** is ***between the mutation rate and twice the mutation rate, depending on the level of dominance***. The surprising fact is that the expected influence of mutation on fitness is related to the mutation rate and not to the extent of selection (Haldane, 1937).

If we assume that the fitnesses at different loci are independent, then the overall fitness can be obtained by multiplying the fitnesses of the genotypes (see p. 556). In other words, if the fitness at locus i is $\overline{w}_i$, the overall fitness is

$$\overline{w} = \overline{w}_i^n$$

and the overall genetic load is again $L = 1 - \overline{w}$. As shown by Crow and Kimura (1970), the total load caused by mutation can be given approximately as

$$L \approx C \sum u_i \tag{7.6b}$$

where C is a constant between 1 and 2 as discussed above, and u_i is the mutation rate at locus i. For example, if there are 10,000 loci with a mutation rate of 10^{-5} and $C = 2$, then $L = 0.2$.

III. MUTATION IN A FINITE POPULATION

As discussed earlier, a new mutant can be lost because it is not included in the progeny or descendents of the individual in which it occurred in an infinite population. However, another factor that affects the probability of incorporation of a new mutant is the size of the population. In fact, the probability of a new mutation becoming incorporated into a finite population is quite low, even if the mutation has a selective advantage.

a. Neutrality

As we see in the discussion of the maintenance of molecular variation in Chapter 8, the ***joint consideration of mutation and genetic drift (with no differential selection)*** forms the basis of the **neutral theory**. To develop some of the fundamentals of this theory, let us first assume that a new mutant, A_1, occurs in a population that otherwise consists only of A_2 alleles. (The new mutation is allele A_1, not A_2, as earlier.) The initial frequency of the mutant is then

$$p_0 = \frac{1}{2N}$$

Assume that the two alleles are neutral with respect to each other. From the discussion of genetic drift in a finite population in Chapter 6, we then know that the probability of fixation of the new mutant is equal to its initial frequency or

$$u(p) = \frac{1}{2N} \tag{7.7a}$$

and that the probability of loss of the new mutant (equal to the fixation of the original allele) is

$$u(q) = 1 - \frac{1}{2N}$$

where $u(p)$ and $u(q)$ indicate the probability of eventual fixation and loss, respectively, of the new allele, A_1. In other words, unless the population is quite small, a new neutral mutant will nearly always be lost from the population.

The two alternatives, fixation or loss of the new mutant, involve quite different amounts of time. Because the change of frequency necessary when loss occurs is small, from $1/(2N)$ to zero, the average amount of time for loss to occur is short. On the other hand, in the small proportion of situations when fixation of the new mutant does occur, it requires a substantial change in allele frequency, from $1/(2N)$ to unity, and the time required is much larger. Figure 7.8 illustrates the loss of four mutants and the eventual fixation of a mutant in a small population.

Kimura and Ohta (1971) gave a formulation for the average time to fixation and average time to loss of a new mutant, assuming neutrality of the alleles. When the frequency of the allele is low ($p \to 0$) as for a new mutant, the expected time to fixation is

$$T_1(p) = 4N_e \tag{7.7b}$$

The time to fixation has a very broad distribution with a long tail, as was shown in Figure 6.6, for the time to fixation for an allele with an initial frequency of 0.2. Nei (1987) gave the standard deviation of the time to fixation as $2.14N$, illustrating the very wide distribution around the expected fixation time of $4N_e$.

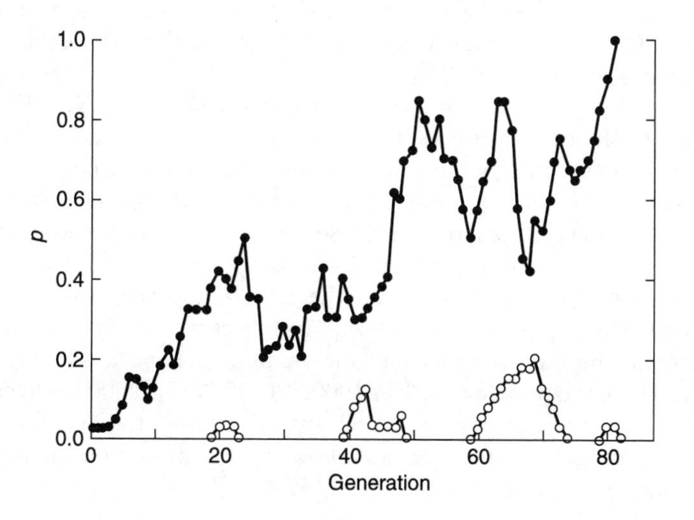

Figure 7.8. The change in allele frequency over time due to genetic drift for several mutants. One mutant (closed circles) eventually becomes fixed, whereas the others (open circles) are lost from the population. Note that the frequency of the ancestral allele is not given, and it is one minus the sum of the allele frequencies given.

The expected time to loss of a new mutant is

$$T_0(p) = 2\left(\frac{N_e}{N}\right) \ln(2N) \tag{7.7c}$$

If it is assumed that $N_e = N$, then the ratio of time to fixation over the time to loss is $2N/[ln(2N)]$. For example, if $N = 500$, then fixation takes an average of 145 times longer than loss. Because the process of fixation takes so long, alleles observed during this process would be described as polymorphic, but this polymorphism is transient rather than permanent.

If there are a large number of potential alleles at a locus, then mutation will increase the number of alleles, and genetic drift will reduce the number of alleles. The properties of the equilibrium resulting from the balance of these two factors for this model, called the **infinite-allele model (IAM)** because each mutation is assumed to be to a new, unique allele, have been given by Kimura and Crow (1964) using inbreeding coefficients. To illustrate this derivation, assume the expected homozygosity in generation t is

$$f_t = \frac{1}{2N_e} + \left(1 - \frac{1}{2N_e}\right) f_{t-1}$$

Let u be the mutation rate to new alleles at the locus. The probability of identity is now modified by the probability that both alleles do not mutate, or $(1 - u)^2$, thus

$$f_t = \left[\frac{1}{2N_e} + \left(1 - \frac{1}{2N_e}\right) f_{t-1}\right] (1 - u)^2 \tag{7.8a}$$

To understand the pattern of change in genetic variation over time because of the joint effects of mutation and genetic drift, let us examine the change of heterozygosity (remember $H = 1 - f$) for three different effective population sizes. If $H_0 = 0$ ($f_0 = 1$) and $u = 10^{-5}$, then the expected pattern of change from expression 7.8a is given in Figure 7.9. First, the approach to the asymptotic value only takes place over many generations, as would be expected for the combination of mutation, which only slowly changes genetic variation, and relatively large population sizes so that the impact of genetic drift is fairly small. Second, the asymptotic level of variation is higher when the population size is larger (we derive the equilibrium value of heterozygosity below), and the rate of approach to this value is faster when the population size is lower. In this case, the heterozygosity has gone 90% of the way to its asymptotic value by generations 4500, 38,000, and 95,000 for the population sizes 1000, 10,000, and 100,000, respectively. This faster rate occurs primarily because the lower the $N_e u$ value, the greater impact genetic drift has compared with mutation on the dynamics of the genetic variation change.

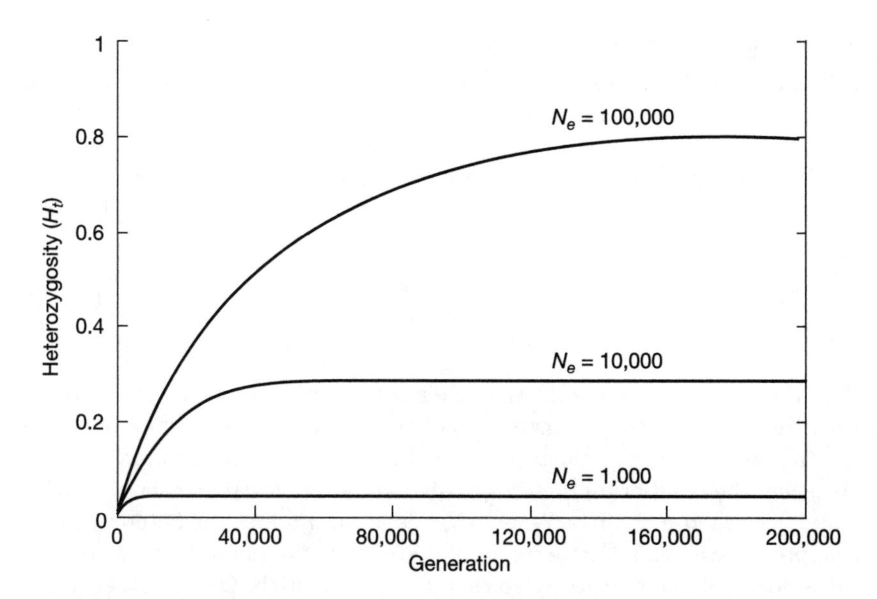

Figure 7.9. The expected change in heterozygosity from mutation and genetic drift when $H_0 = 0$ and $u = 10^{-5}$ for three different effective population sizes.

If it is assumed that there is an equilibrium between mutation producing new alleles and finite population size eliminating them, then $f_t = f_{t-1} = f_e$. Therefore, expression 7.8a becomes, if the terms with u^2 are ignored,

$$f_e \approx \frac{1 - 2u}{4N_e u + 1 - 2u}$$

and if we ignore the small term $2u$ in both the numerator and denominator, then

$$f_e \approx \frac{1}{4N_e u + 1} \qquad (7.8b)$$

Because the proportion of heterozygotes is $H = 1 - f$, then

$$H_e = \frac{4N_e u}{4N_e u + 1} \qquad (7.8c)$$

the equilibrium heterozygosity for the infinite allele, neutral model. Often, because N_e and u appear as a product, the notation $\theta = 4N_e u$ is used so that

$$H_e = \frac{\theta}{\theta + 1} \qquad (7.8d)$$

This equilibrium is different from the equilibria that we have discussed before because the allele frequencies, and even the identity of the alleles, are constantly changing, and it is only the distribution of alleles that remains more or less constant.

We can also calculate the expected number of effective alleles

$$n_e = \frac{1}{1 - H_e}$$
$$= 4N_e u + 1 \quad (7.8e)$$

For illustration, let us assume that the mutation rate is either 10^{-5} or 10^{-7}; then the expected range of heterozygosity at equilibrium can be calculated for different effective population sizes (Figure 7.10). For example, if N_e is 10^4, then the equilibrium heterozygosity would be nearly 0 when $u = 10^{-7}$ and 0.286 when $u = 10^{-5}$. When $4N_e u$ is around unity, the heterozygosity is approximately 0.5. Furthermore, if $4N_e u \gg 1$, then mutation determines the amount of heterozygosity so that H_e is quite high. On the other hand, if $4N_e u \ll 1$, then genetic drift becomes the major factor and H_e is low.

Another mutation model that was developed for allozyme variation and has also been applied to mutation for microsatellite loci is called the **stepwise-mutation model (SMM)**, in which it is assumed that mutation only occurs to adjacent states (Ohta and Kimura, 1973; Ellegren, 2004). For microsatellite loci where different alleles have different numbers of repeats of a motif, if mutation occurred only by adding or deleting one repeat, it

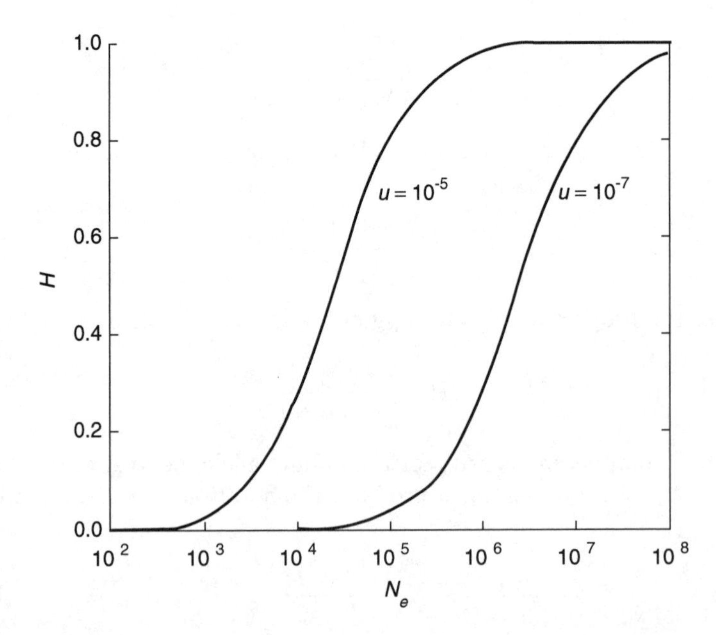

Figure 7.10. The equilibrium heterozygosity for the infinite-allele model (IAM) for different effective population sizes when $u = 10^{-5}$ and when $u = 10^{-7}$.

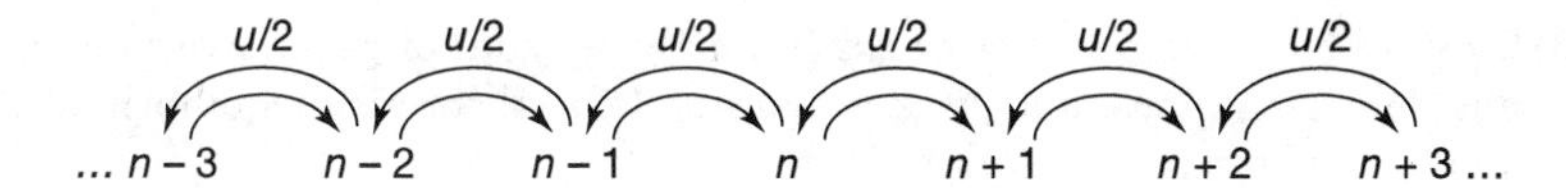

Figure 7.11. The stepwise mutation model (SMM) as applied to a microsatellite locus where n is the number of repeats and u is the total mutation rate to alleles with either one more or one less repeat.

would be consistent with the stepwise model (Figure 7.11). With the SSM, unlike the IAM, mutation may produce alleles that are already present in the population. For example, if a population has alleles with n, $n+1$, and $n+2$ repeats, then only mutants to $n-1$ repeats from alleles with n repeats and mutants to $n+3$ repeats from alleles with $n+2$ alleles produce new mutant alleles. All of the other mutations, n to $n+1$ repeats, $n+1$ to n or $n+2$ repeats, and $n+2$ to $n+1$ repeats, do not produce new alleles. Because mutation does not always produce new mutations under the SMM, one would expect that both the generation of variation and the equilibrium level of heterozygosity to be less than for the IAM.

For the SMM, Ohta and Kimura (1973) found the equilibrium homozygosity (f) to be

$$f_e = \frac{1}{(8N_e u + 1)^{1/2}} \tag{7.9a}$$

so that the equilibrium heterozygosity is

$$H_e = 1 - \frac{1}{(8N_e u + 1)^{1/2}} \tag{7.9b}$$

and the effective number of alleles is

$$n_e = (8N_e u + 1)^{1/2} \tag{7.9c}$$

For $4N_e u = 1$, the equilibrium heterozygosity for the SMM is 0.423—85% that of the IAM. Unless $4N_e u$ is very small where the two values become very similar, this is close to the proportional difference for the two heterozygosity values. For example, if $4N_e u = 20$, then for the SSM, H_e is 0.782—82% that of the IAM. Example 7.4 shows how these expressions may be useful to understand the variation for molecular markers in the cheetah.

Example 7.4. In a series of studies, the variation at different genetic markers in cheetahs has been compared with that in other large cats (summarized in Driscoll *et al.*, 2002). For allozymes, cheetahs had a heterozygosity of only 0.0072 over 47 loci (Table 7.4), and they also have low variation at MHC

TABLE 7.4 The observed heterozygosity for three types of genetic variants in cheetahs, mountain lions, and African lions and the theoretical equilibrium values for two effective population sizes.

<table>
<tr><th></th><th>Allozymes</th><th>Minisatellites</th><th>Microsatellites</th></tr>
<tr><td>Observed</td><td></td><td></td><td></td></tr>
<tr><td>Cheetahs</td><td>0.0072</td><td>0.435</td><td>0.47</td></tr>
<tr><td>Mountain lions</td><td>0.018–0.067</td><td>0.579</td><td>0.35</td></tr>
<tr><td>African lions</td><td>0.037</td><td>0.481</td><td>0.47</td></tr>
<tr><td>Theoretical</td><td></td><td></td><td></td></tr>
<tr><td>$N_e = 200$</td><td>0.0008</td><td colspan="2">0.380</td></tr>
<tr><td>$N_e = 2000$</td><td>0.0079</td><td colspan="2">0.757</td></tr>
</table>

genes (Yuhki and O'Brien, 1990; Drake *et al.*, 2004). However, for mtDNA, minisatellites, and microsatellites (Menotti-Raymond and O'Brien, 1993; Driscoll *et al.*, 2002), levels of genetic variation have been documented that are as high as in other large cats (Table 7.4).

The different levels of variation can be explained by different mutation rates at the two types of loci (Hedrick, 1996). For example, if $u = 10^{-6}$ for allozymes and $u = 10^{-3}$ for minisatellite and microsatellite loci, then the equilibrium heterozygosity is given for two different effective population sizes in Table 7.4 (calculations for the allozyme equilibrium heterozygosity used the infinite-allele model and expression 7.8*c*, and that for minisatellites and microsatellites used the stepwise-mutation model and expression 7.9*b*). With these parameters, the heterozygosity for allozymes is low and similar to that observed, and that for minisatellites and microsatellites is much higher and similar to that observed.

If $4N_eu = 1$, then n_e for the IAM and SSM are quite similar, 2 and 1.73, respectively. However, if $4N_eu$ is large, for example 20, then n_e is 21 and 6.4 for the IAM and SSM models, respectively, a substantial difference (because the estimated effective number of alleles is very sensitive to small changes in heterozygosity when H_e is large, the confidence interval on these estimates is broad). Another way to compare the IAM and SSM models is to determine the mutation rate necessary for the SSM to result in the same heterozygosity as the IAM. Using expressions 7.8*c* and 7.9*a*, the necessary mutation rate, u_s, for the SMM is

$$u_s = u(2N_eu + 1) \qquad (7.9d)$$

where u is the mutation rate for the IAM. For example, if $4N_eu = 1$ and $u = 0.001$, then u_s needs to be 0.0015, 50% larger than the mutation rate in the IAM model to result in the same equilibrium heterozygosity. This expression can be solved for u and used in expression 7.8*a* to examine the generation of genetic variation from the SSM. Because the effective u in the SMM is lower than for the IAM, the generation of genetic variation is always somewhat slower and depends on the values of N_e and u_s.

Finally, we can compare the probability of **identity by descent** and **identity in state** for the IAM and SMM (see Estoup *et al.*, 2002) for microsatellite loci, for which identity in state for two alleles is defined as having the same number of repeat units. Under the IAM, the probability of identity by descent and identity in state are the same because any mutation in the historical lineage makes changes in both. However, mutation under the SMM may result in a mutation that is identical in state to one that is already present. Therefore, the probability of drawing two alleles identical by descent, given that they have the same state, is the ratio of homozygosity for the IAM to the homozygosity for the SMM or

$$\frac{1 - H_{IAM}}{1 - H_{SMM}} = \frac{(8N_e u + 1)^{1/2}}{4N_e u + 1}$$

For example, if $\theta = 0.1$, 1, and 10, then this ratio is 0.996, 0.6, and 0.416, respectively. In other words, if θ is small, then nearly all of the identity in state for the SMM reflects identity by descent, whereas if θ is large, then a high proportion of the identity in state does not reflect identity by descent and is the result of mutation to pre-existing repeat classes.

Mutations for microsatellite loci generally do not appear to be consistent with either the IAM or the SSM. As a result, Di Rienzo *et al.* (1994) suggested a **two-phase model (TPM)** that consists of a proportion p of single-step mutations and the remaining proportion $1 - p$ of larger step mutations. They proposed that the larger step mutations have a symmetrical geometric distribution with a variance V_g. The overall variance in repeat number V is then

$$V = p + (1 - p)V_g$$

If $p = 1$, then $V = 1$ as for the SSM. For example, if $V = 5$ and $p = 0.7$, then $V_g = 14.3$. Because the typical step size of a multistep repeat is $(V_g)^{1/2}$, then the typical multiple step size in this example is 3.8.

Earlier we examined the rate of generation of genetic variation when a population has no variation. In a biological context, what is the effect of a genetic **bottleneck** in population size on the amount of genetic variation over time? To examine this, let us assume that a population starts at mutation–genetic drift equilibrium for the IAM, goes through a bottleneck (or a founder event), and then grows to a large size so that mutation can then regenerate the genetic variation that was lost in the bottleneck and subsequent generations (Nei *et al.*, 1975). The expected genetic variation after the bottleneck is determined by heterozygosity before the bottleneck, the bottleneck size, and the intrinsic rate of increase after the initiation of the population. The bottleneck size has a large effect on the number of alleles in the population, as we discussed on p. 346, because the low population number often eliminates alleles of low frequency. The average heterozygosity, on the other hand, is influenced to some extent by the bottleneck size but primarily by the rate of population growth after the founding event. The heterozygosity is reduced by finite population size, as indicated by equation 6.3a, so that if $N = 2$, the smallest population size for an outbreeding organism, then $H_{t+1} = 0.75H_t$; the heterozygosity is reduced to 75% of its previous level each generation.

When population growth is slow from a small bottleneck number, then heterozygosity is lost each generation until the population grows to a substantial size, as illustrated in Figure 7.12 (Nei *et al.*, 1975). The equilibrium heterozygosity before the bottleneck in this example was 0.138, the average amount observed in a survey of allozyme variation in *Drosophila*, and then a founder population of size 2 (solid lines) or 10 (broken lines) was initiated in which the intrinsic rate of population growth per generation, r, was 0.1 or 1.0 ($r = 0$ indicates no change in population size and when $r = 1.0$, there is fast population growth, see p. 16). Mutation with $u = 10^{-8}$ under the IAM was allowed to regenerate the lost heterozygosity.

The heterozygosity when there is a founder number of two and $r = 0.1$ is reduced to less than 0.01 (less than 7% of the initial value). Most of the loss of heterozygosity takes place over the first 10 generations. However, if the intrinsic rate of increase is 10 times as high ($r = 1.0$), then the heterozygosity is reduced to only approximately 0.09 (about 65% of the initial value) because the population size rebounds quickly and loss only occurs over the first several generations. On the other hand, the restoration of genetic variation by mutation to the equilibrium level takes many generations

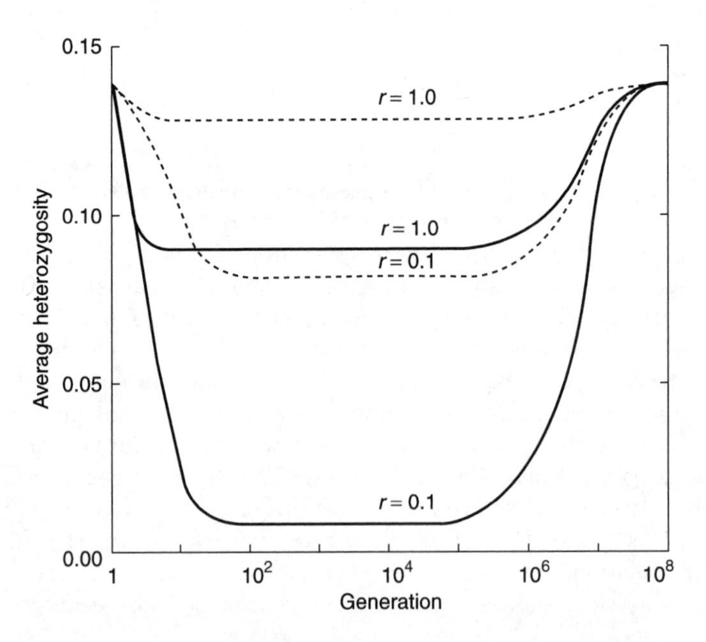

Figure 7.12. The heterozygosity over time after a founding event (bottleneck) of two (solid lines) or ten (broken lines) individuals. The r values indicate the intrinsic rate of increase after the founding event (from Nei *et al.*, 1975).

and is on the order of the reciprocal of the mutation rate, approximately 10^8 generations in this case. In other words, this asymmetry of effects of extreme genetic drift and mutation implies that the signature of an extreme bottleneck may be present in a population for many generations after the event, depending on the mutation rate.

b. Selection

Now let us examine the fate of a mutant when it is not neutral but has a selective effect. When a single **advantageous mutant** exists in a population and there is selection for different genotypes at the locus, then the probability of fixation can be used to determine how often the new mutant will be incorporated (see p. 349). When the new mutant is additive so that the fitnesses of A_1A_1, A_1A_2, and A_2A_2 are $1 + s$, $1 + \frac{1}{2}s$, and 1, respectively, then the probability of fixation is (expression 6.23*b*)

$$u(p) = \frac{1 - e^{-2N_e sp}}{1 - e^{-2N_e s}}$$

If there is only one mutant, then $p = 1/(2N)$ and

$$u\left(\frac{1}{2N}\right) = \frac{1 - e^{-(N_e/N)s}}{1 - e^{-2N_e s}}$$

Assuming that the mutant is advantageous ($s > 0$) and that the selective advantage is not too large, then

$$u\left(\frac{1}{2N}\right) \approx s\left(\frac{N_e}{N}\right)$$

This becomes the result of Fisher (1930) (Table 7.3) if one assumes that $N_e = N$ and realizes that s here is half that used by Fisher. In other words,

$$u\left(\frac{1}{2N}\right) \approx 2s^* \qquad (7.10a)$$

where s^* is the selective difference between the mutant and the original allele.

The foregoing calculation assumes that the population size is constant, but if the population size is increasing or decreasing, then this probability is somewhat altered (Otto and Whitlock, 1997). For example, the probability of fixation of a mutant in a population with intrinsic rate of increase of r is approximately

$$u\left(\frac{1}{2N}\right) \approx 2(s^* + r) \qquad (7.10b)$$

When there is an increasing population size, the likelihood of the inclusion of a lineage with an advantageous mutation is increased in the first few generations so that there is an increase in the probability of fixation.

The joint effect of mutation and genetic drift is also important for mutants that have other types of selection. For example, **lethals** or **detrimentals** have a lower expected frequency in a small, finite population than in an infinite population because ***selection will push them to a low frequency and then they will be lost from the population by genetic drift***. As we showed in expression 7.2*b*, the expected equilibrium frequency of a recessive detrimental allele A_2 in an infinite population is $q_\infty = u/s$, an expectation that also holds when $2N_eu$ is greater than unity.

Wright (1937) showed that when $2N_eu$ is less than unity that the expected frequency of A_2 is approximately

$$q_{N_e} \approx u\left(\frac{2\pi N_e}{s}\right)^{1/2} \tag{7.11a}$$

In other words, when the population size is relatively small, the frequency of a recessive detrimental may be greatly reduced. One way to understand this effect is to examine the ratio of these two expressions

$$\frac{q_{N_e}}{q_\infty} \approx (2\pi N_e u)^{1/2} \tag{7.11b}$$

(Hedrick, 2002b). For example, if $N_eu = 0.0001$ or 0.01, then this ratio is approximately 2.5% or 25%, respectively, illustrating that the frequency of a completely recessive lethal or detrimental may be only a small proportion of that in an infinite, or very large, population.

Nei (1968, 1969) using an approximation of the general formula of Wright (1937) showed that when $h > 0$, the equilibrium is approximately

$$q \approx \frac{u}{hs} \tag{7.11c}$$

which is equivalent to expression 7.4*b*, which we gave for an infinite population. Unlike expression 7.11*b* for $h = 0$, this expression is independent of the effective population size. In other words, we have two different approximate theoretical findings; that is, when $h = 0$, the equilibrium is dependent on N_e, but when $h > 0$, it is not.

To examine numerically how sharp the transition is from $h = 0$ to $h > 0$, Hedrick (2002b) used the transition matrix approach outlined on p. 351. Those results demonstrated that even if $h > 0$, the expected frequency of a lethal ($s = 1$) is highly dependent on the population size. To illustrate, Figure 7.13 gives the ratio of observed frequency in a population of size N_e to that expected in an infinite population or q_{N_e}/q_∞. When N_e is low, even when $h = 0.05$, this ratio is significantly below one. For example, when

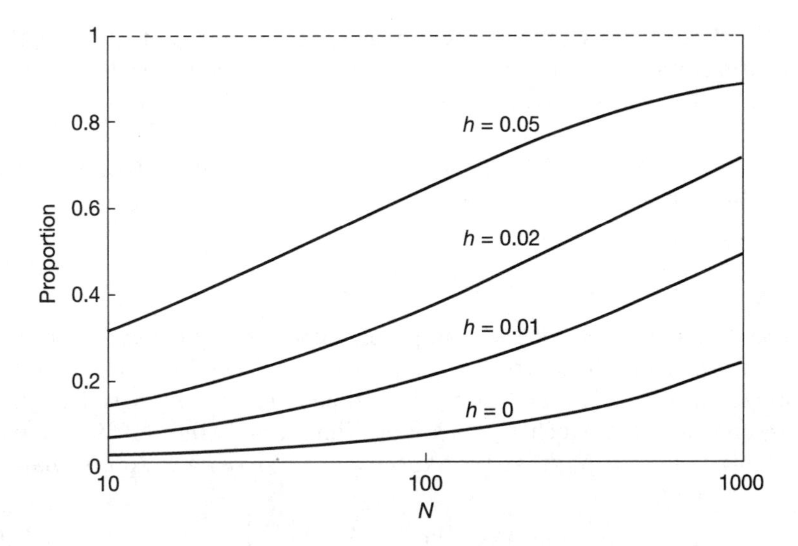

Figure 7.13. The ratio of observed equilibrium allele frequency in a finite population to that expected in an infinite population for four different levels of dominance h when $u = 3 \times 10^{-6}$. The broken line indicates the expectation when the equilibrium is independent of the population size (from Hedrick, 2002b).

$N_e = 100$ and $h = 0.01$ and 0.02, the equilibria are 22% and 38%, respectively, of that expected in an infinite population. Even for a population size of 1000 and $h = 0.01$, the equilibrium is only 50% of that in an infinite population.

From these results, it is apparent that for the average level of dominance estimated for lethals in *Drosophila* ($h = 0.02$) (Simmons and Crow, 1977) a small population size may greatly decrease the equilibrium lethal allele frequency. Because the expected number of lethal equivalents from lethals is a direct function of the allele frequency (Hedrick *et al.*, 1999a), the expected impact of viability from inbreeding is a function of the past population size. For example, assume that the expected number of lethal equivalents from lethals in a finite population is 20% that in an infinite population. Therefore, if it is assumed that half of the genetic variation causing inbreeding depression is from lethals (and the detrimental half is unaffected by population size because h is higher), then overall, the number of lethal equivalents if approximately 60% that expected in an infinite population.

In other words, **purging** of mutants of large detrimental effects and even those that are not completely recessive will occur in small finite populations. For conservation genetics, this implies that species chronically low population numbers would be expected to have less variation influencing inbreeding depression. In this case, schemes to purge variation that influences inbreeding depression would likely have little positive impacts because most of the easily purged variation would have been already removed. On the other hand, if a population had been recently reduced from large numbers to small numbers, as for example, the four endangered big fishes of the lower Colorado River (Garrigan *et al.*, 2002), then they may have substantial detrimental genetic variation. An attempt to purge such a species of detrimental genetic variation may result in fixation of detrimental alleles and lowering of the fitness of the total population (Hedrick, 1994).

In addition, it is useful to know the **persistence** or the number of generations until loss of a deleterious mutation. First, let us assume that in an infinite population there is a new mutation with fitness $1 - hs$ in heterozygotes and that the deleterious effect (hs) is large enough that selection against homozygotes can be ignored. In this case, the expected number of generations until loss is

$$T_0(p) = \frac{1}{hs} \tag{7.12a}$$

where this is the reciprocal of the probability of being eliminated in any particular generation (Morton *et al.*, 1956; Garcia-Dorado *et al.*, 2003). For example, if $hs = 0.05$, then the expected time to loss is 20 generations. In a finite population in which $N_e \gg 1/4hs$, Kimura and Ohta (1969) showed, assuming that $N = N_e$, that the expected time to loss was approximately

$$T_0(p) = 2\left[\ln\left(\frac{1}{hs}\right) + 1 - \gamma\right] \tag{7.12b}$$

where $\gamma = 0.577$ (Euler's constant). Again assuming that $hs = 0.05$, then the expected time to loss is 6.84 generations, considerably less than in an infinite population. Remember that for a neutral mutant the expected time to loss is $2\ln(2N)$ (expression 7.7c) so that, for example, if $2N = 200$, then the expected time to loss is 10.6 generations.

If $2Ns < 1$—that is, if $s < 1/(2N)$—then genetic drift becomes a stronger factor influencing allele frequency than selection, and the mutant becomes effectively neutral (Kimura, 1983). Lynch and Gabriel (1990) and Lande (1994) have shown that in small populations, detrimental mutations with a selective disadvantage less than $1/(2N)$ become fixed much as if they were neutral. As a result, ***over time the fitness declines and the population decreases in size so that detrimental mutants of larger effect become effectively neutral and subsequently are more likely to be incorporated***. This feedback process has been named the **mutation meltdown** and in theory may result in the extinction of small populations (e.g., Lynch *et al.*, 1995). However, it is not clear how significant a factor it may be in actual situations (see Example 7.5 for an experimental test in yeast) because extinction probability caused by other factors may be high in such small populations (Lande, 1995). In addition, it has been shown theoretically that back mutations (Lande, 1998) and experimentally that compensatory mutations (Burch and Chao, 1999; Estes and Lynch, 2003; see Example 7.5) may ameliorate the mutation meltdown effect.

Example 7.5. Although inbreeding depression and loss of potentially adaptive variation are widely accepted as genetic concerns for small populations of endangered species, Lynch *et al.* (1995) and Lande (1995) also suggested that mutation meltdown is an additional genetic problem for small populations. To examine experimentally the potential of mutation meltdown, Zehl *et al.* (2001) established 12 replicate populations of the yeast *Saccharomyces cerevisiae* from two isogenic strains that had mutation rates differing by approximately 200-fold. They maintained the replicates at an effective population size of approximately 250 for approximately 2900 generations. The probability of extinction under the mutation meltdown scenario is a function of the mutation rate. Consistent with this prediction, 0 of the 12 replicates of the wild-type strain with the lower mutation rate went extinct. However, 2 of the 12 replicates of the strain with the higher mutation rate went extinct after approximately 1975 and 2575 generations. Detailed examination of one of these strains from frozen samples demonstrated that it went extinct shortly after it accumulated a genetic factor resulting in a major fitness decline.

Assuming that mutation meltdown is a threat to small populations, is it possible that the impact of accumulated detrimental mutations can be reversed? To evaluate this possibility, Estes and Lynch (2003) examined the potential for recovery of fitness in 74 lines of the nematode *Caenorhabditis elegans*, which had been kept as single individuals for 280 generations to determine the rate of accumulation of detrimental mutations (see p. 398). These lines, indicated as MA (mutation accumulation) lines in Figure 7.14, had much lower mean productivity and survival that simultaneously tested C (control) lines that had been kept as frozen samples. The MA lines were expanded to large population sizes and maintained for 80 generations. At the end of this time period, these lines (MA-LP in Figure 7.14) had increased their reduced productivity and survival and achieved values not significantly different from the initial control. The rate of fitness recovery in the MA-LP lines was at least three times as fast as the earlier rate of fitness loss in the MA lines. Although back mutations, accumulation of beneficial

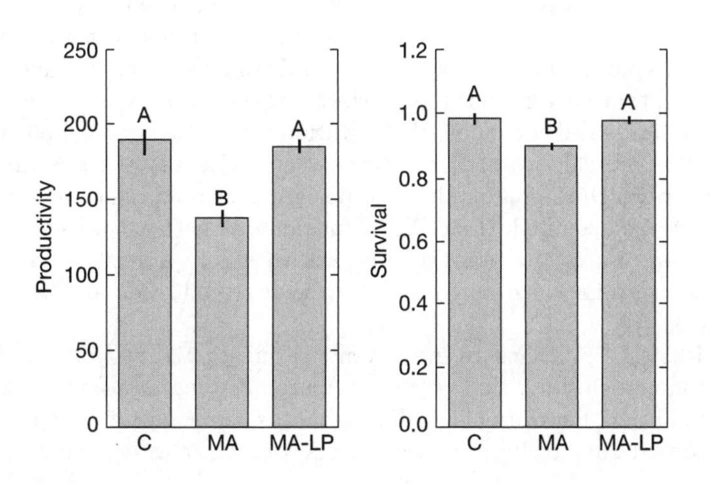

Figure 7.14. The mean progeny production and survival for ancestral control (C) lines, mutation accumulation (MA) lines, and mutation accumulation lines after 80 generations at a large population size (MA-LP) (from Estes and Lynch, 2003). Error bars indicate one standard error, and means labeled with the same letter are not significantly different from each other.

mutations, and compensatory mutations could all theoretically contribute to this reversal, Estes and Lynch (2003) provided evidence supporting the significance of compensatory mutations for this fast fitness recovery.

Finally, many chromosomal mutants affect meiosis in heterokaryotypes, resulting in lowered fertility of these individuals. As a result, the net fitness of the heterokaryotype may be substantially smaller than that of the original homokaryotype or of the homokaryotype for the new chromosomal mutant. Because a mutant chromosomal type would initially be in low frequency, well below the unstable equilibrium generated by heterozygote disadvantage, selection should always reduce the frequency of the mutant. However, many closely related species differ by some chromosomal type that potentially may produce a heterokaryotypic disadvantage. White (1978) suggested four factors that separately or in combination might lead to the fixation of a chromosomal variant: genetic drift, meiotic drive, selective advantage of the new homokaryotype, and inbreeding. These and other factors have been examined theoretically by Bengtsson and Bodmer (1976), Lande (1979), Hedrick (1980b), Barton and Rouhani (1991), Pialek and Barton (1997), Huai and Woodruff (1998), and Navarro and Barton (2003), but it is not clear which are important in natural situations.

IV. ESTIMATION OF MUTATION RATES

Estimating mutation rate is difficult, primarily because mutation is such a rare event. As a result, in order to obtain fairly accurate estimates of the mutation rate, an extremely large number of observations must be made, as exemplified by the studies of Schlager and Dickie (1971) and Russell and Russell (1996) that involved millions of mice (see Table 7.1). Even when a new mutant is observed, it must be checked to make sure that it is not a contaminant from another source. In addition, mutants that are identified by phenotypic change may be difficult to distinguish from mutants at different loci that have similar phenotypes. In these cases, unless the allelism of the mutants is determined, the mutation rate is to a general phenotypic class. The general approach to estimate mutation rates is the direct approach, in which the parental genotypes are known and new mutants that differ are then counted. In addition, for mutants with a negative effect on fitness, we discuss the indirect approach to obtain a general estimate of mutation rate in which an equilibrium between mutation and selection is assumed.

Although mutations are rare events, a substantial proportion of them may occur in clusters; that is, two or more offspring in a family have the same mutation. For example, in *D. melanogaster*, Woodruff *et al.* (1996) estimated that 20% to 50% of mutations may occur in clusters and in pipefish,

Jones *et al.* (1999) found that 40% of the mutations at a microsatellite locus were in clusters. Because of such mutation clusters, different researchers have proposed various ways of counting mutations, including counting every mutant individual (offspring), counting each independent mutation once, or even disregarding all clustered mutations. Fu and Huai (2003) suggested that counting all mutant individuals provides the correct approach for estimating the mutation rate. In addition, theory that includes mutation clusters may result in evolutionary predictions different from traditional predictions (Woodruff *et al.*, 1996).

Perhaps more is known about mutations in humans than in any other organism because of the detrimental effects of many mutants on health. For a number of years, McKusick (1998) maintained a catalogue of variants in humans, many of which causes genetically based diseases. Now this catalogue is online, the Online Mendelian Inheritance in Man (OMIM) (http://ncbi.nlm.nih.gov/OMIM), and contains information on over 10,000 human genes and is an excellent background source for specific genetic diseases and provides an initial entry into the literature. In addition, there are many databases or sites dedicated to the documentation of information on particular genetic diseases, such as albinism, cystic fibrosis, Tay-Sachs, etc. For many of these diseases, a number of genetic mutations are responsible for the disease. For example, approximately 1000 mutations have been reported to cause cystic fibrosis (Mateu *et al.*, 2002) and over 1500 to cause hemophilia B (Giannelli *et al.*, 1997). In fact, a number of different genes may result in a given disease. For example, at least 11 different genes have mutations that produce oculocutaneous albinism (OCA, albinism that affects both the skin and eyes) (King *et al.*, 2003). In some situations, a high proportion of mutant alleles in a population can be traced back to a single or a few mutations (Risch *et al.*, 2003; Yi *et al.*, 2003), a topic that we return to when we discuss the age of specific alleles (see p. 580).

a. Dominant or Codominant Mutants

The **direct, single-generation approach** has been widely used to estimate the mutation rate for dominant or codominant mutants. In this case, the genotypes of both parents are known, and neither of them can carry a dominant or codominant mutant. If an offspring carries a new dominant or codominant allele, it must be the result of a new mutation. In this case, because each offspring receives two gametes that can potentially carry a mutant, the mutation rate is

$$\hat{u} = \frac{x}{2N} \tag{7.13a}$$

where x is the number of mutant offspring and N is the total number of offspring examined. This is the procedure used by Schlager and Dickie

(1971)(Example 7.1) to estimate backward mutation rates for several loci simultaneously, and it has also been used to estimate the mutation rate in a number of dominant human disorders. For example, hospital records of 94,075 children in Lying-in Hospital in Copenhagen were examined for cases of achondroplasia, a dominant form of dwarfism (Stern, 1973). Ten of their newborns were achondroplasiac dwarfs, but two had an affected parent. Therefore, eight offspring appeared to be new mutants, and the estimate of the mutation rate is $\hat{u} = 8/[(2)(94,075)] = 4.2 \times 10^{-5}$. Example 7.6 describes the application of this general approach to the relative minisatellite mutation rates in humans exposed to radiation at Chernobyl and a control group. The same procedure is used in *E. coli* and other haploid organisms where each new cell is a potential mutant. The above expression for haploid organisms is

$$\hat{u} = \frac{x}{N} \tag{7.13b}$$

where N is the number of cells examined.

Example 7.6. It has long been known that high levels of radiation cause mutations—often chromosomal breaks—in experimental systems. However, there is little evidence that ionizing radiation increases the general germline mutation rates for all genes in humans. Examination of mutations at minisatellite loci, which have a high base mutation rate, provides a more sensitive assay of mutation rate and makes it possible to detect a statistically significant increase in mutation rate in a relative small sample size.

The accident at the Chernobyl nuclear power station in Ukraine on April 26, 1986, resulted in the largest reported accidental release of radioactivity. To determine the effect of this release on the germline mutation rate, Dubrova *et al.* (2002) estimated the mutation rates for eight minisatellite probes in children of parents exposed to the Chernobyl radiation and compared them with rates in matched control subjects of children from unexposed parents from the same area born before the accident. Overall, the control group had 39 mutations in 1407 scored band, whereas the exposed group had 137 mutations in 3475 scored bands (Table 7.5) with an overall ratio of mutation rates in exposed to control of 1.42. The mutations in the control and exposed groups involved similar gains and losses of repeat units, suggesting that radiation increased the rate of normal mutation rather than inducing different types of mutational events.

Further examination showed that all of the increased rate of mutation occurred in paternal mutations, that is, in children of fathers exposed to the radiation. Table 7.5 gives the rates observed separately into paternal and maternal mutations, and the ratios of mutation rates in exposed to controls are 1.56 for paternal mutations and 1.01 in maternal mutations. A comparison of these estimates to earlier data (Dubrova *et al.*, 1996) found

TABLE 7.5 The estimated mutation rate for eight minisatellite loci (the four with the highest rates are given separately and the other four as a group) in humans exposed (E) to radiation in Chernobyl compared to a control (C) group divided into (a) paternal mutations and (b) paternal mutations (Dubrova *et al.*, 2002).

	Control			*Exposed*			
	Bands	*Mutations*	*Mutation rate*	*Bands*	*Mutations*	*Mutation rate*	*Ratio (E/C)*
(a) Paternal							
B6.7	82	3	0.037	196	18	0.092	2.51
CEB1	90	16	0.178	219	38	0.174	0.98
CEB15	87	2	0.023	218	18	0.083	3.59
MS1	94	2	0.021	223	13	0.058	2.74
Other	353	6	0.017	892	25	0.028	1.65
Total	706	29	0.041	1748	112	0.064	1.56
(b) Maternal							
B6.7	80	0	0	196	4	0.020	—
CEB1	86	1	0.012	208	2	0.010	0.83
CEB15	91	0	0	211	3	0.014	—
MS1	88	7	0.080	226	12	0.053	0.67
Other	356	2	0.006	886	4	0.005	1.24
Total	701	10	0.014	1727	25	0.014	1.01

that the paternal rates were significantly elevated and indistinguishable in exposed groups from both Ukraine and Belarus and that maternal rates were lower in the two exposed cohorts and did not differ from those in the control groups. Dubrova *et al.* (2002) discussed the basis for this sex difference in mutation rates and suggested that because all mothers in this study were at least 8 years old at the time of the accident, the female germlines were already past the mutation-susceptible stages.

Kodaira *et al.* (1995) carried out a similar study in children from families exposed to atomic bomb radiation in Japan. In contrast to the findings of Dubrova *et al.* (1996, 2002), they found very similar mutation rates in exposed and control samples (12 mutations in 1111 bands for a mutation rate of 0.011 in the exposed group, and 13 mutations in 1041 bands for a mutation rate of 0.012 in the control group) (Satoh and Kodaira, 1996).

Trinucleotide repeats are involved in a number of human diseases (Cummings and Zoghbi, 2000), including fragile X and Huntington disease. For dominant Huntington disease, individuals with a higher repeat number have greater disease severity and an earlier age of disease onset (The U.S.-Venezuela Collaborative Research Project, 2004). To estimate the mutation rate in repeat number in the alleles causing Huntington disease, Leeflang *et al.* (1999) examined single sperm in 26 different individuals with Huntington disease, that is, individuals having an allele with

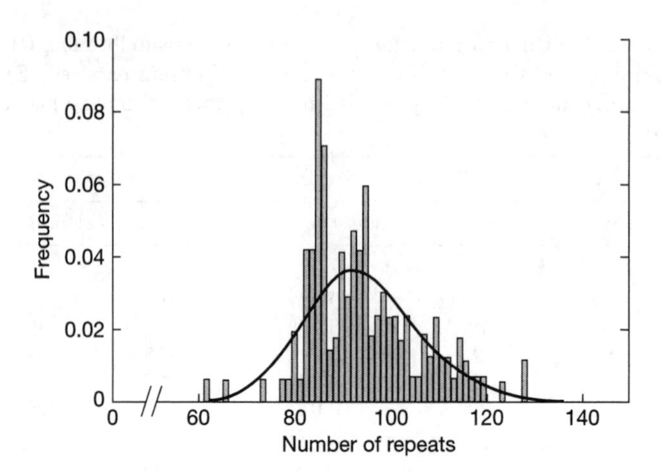

Figure 7.15. The observed frequencies of repeat mutations of different sizes in 168 sperm from an individual with a Huntington disease allele with 62 repeats (the smoothed line gives a theoretical distribution). (Courtesy of Leeflang, E.P., S. Tavare, P. Marjoram, *et al.* 1999. Analysis of germline mutation spectra at the Huntington's disease locus supports a mitotic mutation mechanism. *Hum. Molec. Genet.* 8:173–183.)

37 to 62 CTG repeats (normal alleles have approximately 15 repeats). Overall, the average mutation rate in sperm was 82%, and for individuals with an allele with over 50 repeats, it averaged 98%. Figure 7.15 gives the distribution of the number of repeats observed in sperm from an individual with an allele with 62 repeats. In this case, 167 out of 168 sperm had an increase in the number of repeats, and the average increase in the number of repeats was 32.5. Several mechanisms have been proposed to explain these extraordinary expansions of trinucleotide repeat number, and Yoon *et al.* (2003) have found that they can occur both during meiosis and after meiosis.

If the genotypes of the parents are unknown, then an **indirect method** can be used to estimate the mutation rate for a dominant mutant with a detrimental effect. This approach assumes that the allele frequency is near the equilibrium resulting from the mutation-selection balance so that the frequency of heterozygotes (H) is as given in expression 7.4b. This expression can be rearranged so that

$$\hat{u} = \frac{sH}{2} \tag{7.13c}$$

where $\hat{u}$ is an estimate of the mutation rate. Estimations of mutation rates for dominant alleles using this approach should be used cautiously because the population may not be at the equilibrium frequency and estimation of s (assuming $s < 1$) may be problematic. In addition, the level of selection may vary among individuals because the extent of mutant expression may vary because of genetic background or environmental effects.

b. Recessive Mutants

Estimating mutation rates for recessive mutants is a somewhat more difficult problem than for dominant mutants. A **direct, single-generation method** is to cross homozygous dominant and homozygous recessive individuals and examine the offspring for recessive individuals (Russell and Russell, 1996). Because the offspring should be heterozygous unless there is a mutation, the estimate of the forward mutation rate in this case is

$$\hat{u} = \frac{x}{N} \qquad (7.14a)$$

where x is the number of recessive progeny and the denominator is N because only one of the gametes can mutate from the dominant to the recessive type. However, mutants may not be observed immediately because the mutation could have occurred in a previous generation in the homozygous dominant strain. A correction can be made for this possibility depending on the breeding system employed in maintaining the inbred lines (Falconer, 1949).

Because X-linked recessives are expressed in males, there is a nearly direct method to analyze for X-linked mutants in humans (and potentially X-linked recessives in other mammalian species and all genes in haploid-diploid species). Given family information, a probability statement can be made about whether the new mutation occurred in the formation in the egg of the mother of the male in which it was identified or in a previous generation. With this technique, mutation rates have been estimated for a number of X-linked diseases in humans (Vogel and Motulsky, 1997).

As shown in Chapter 1, in *Drosophila*, single chromosomes can be tested for the presence or absence of a lethal. Table 7.6 gives a summary of some of this research, which estimates the spontaneous mutation rate to lethals. for chromosomes I (the X chromosome) and II in *D. melanogaster* (Crow

TABLE 7.6 Spontaneous lethal mutation rates for chromosome II and the X chromosome for chromosomes extracted from wild populations and from laboratory stocks of *D. melanogaster* (Crow and Temin, 1964).

	Number of chromosomes examined	*Number of lethals obtained*	*Lethal mutation rate*
Chromosome II			
Nature (3 studies)	19,997	115	0.0058
Laboratory (7 studies)	38,499	179	0.0046
Total	58,496	294	0.0050
X chromosome			
Nature (6 studies)	271,050	669	0.0025
Laboratory (14 studies)	346,719	898	0.0026
Total	617,769	1567	0.0025

and Temin, 1964). Although the mutation rate is nearly twice as high for chromosome II as for the X, this agrees well with the fact that the estimated number of genes on chromosome II is approximately twice that on the X chromosome. Because chromosome II constitutes approximately 40% of the *D. melanogaster* genome, the estimated rate of lethal mutations per generation per haploid genome is approximately 0.013.

The indirect method for detrimental recessives uses expression 7.3*c* and gives an estimate of the forward mutation rate as

$$\hat{u} = sQ \tag{7.14b}$$

where Q is the frequency of the affected, recessive homozygote individuals. In addition to the problems suggested for the indirect approach for dominant, detrimental mutants, in this case the relative fitness of the heterozygote greatly influences the equilibrium genotype frequency, as discussed previously. If, for example, there is even a slight disadvantage in the heterozygotes, say 1%, then the above expression cannot be used with any accuracy.

For example, Klekowski and Godfrey (1989) examined the rate of mutation to recessive chlorophyll-deficient variants in a highly selfing mangrove species that retains its seedlings on the maternal plants. By examining the progeny groups, they could estimate the frequency of heterozygous maternal plants. From these frequencies, they estimated the genomic rate of mutation (over perhaps 300 genes) to chlorophyll-deficient mutants in two populations to be 6.6×10^{-3}.

c. Mutation Accumulation Estimates

In a few model organisms, cumulative mutation rates have been estimated by maintaining a number of independent lines, founded from an identical initial chromosome or a single individual, over a number of generations. Generally, such mutation accumulation lines are examined only at the end of the experiment, and the mutation rate for particular loci is estimated as the number of observed mutations divided by the number of allele generations (the product of the number of generations, the number of lines, and the number of alleles or loci examined). In *D. melanogaster*, mutation accumulation lines are generally constructed using techniques that are modifications of the method (discussed on p. 47) used to identify lethals. For example, examining such lines for allozyme mutants, Voelker *et al.* (1980) found four mutants in over 3-million mutation possibilities for an estimated mutation rate of 1.28×10^{-6}. Example 7.7 gives estimates at a large number of microsatellite (dinucleotide repeat) loci in *D. melanogaster*.

Example 7.7. Estimates of mutation rates at microsatellite loci in mammals have generally been quite high. For example, in a large study in humans, the mutation rate was 2×10^{-4} (Huang *et al.*, 2002) (see Example 7.9). However, three studies of mutation accumulation lines suggest that the mutation rate for microsatellite loci in *D. melanogaster* is much lower (Table 7.7). Schug *et al.* (1998) examined 49 dinucleotide repeat loci for a total of 321,930 allele generations and observed three mutants, all of which differed from the original allele by only one repeat. Schlötterer *et al.* (1998) observed 9 mutations at 24 loci (22 were dinucleotide repeats and 2 were trinucleotide repeats) for a total of 1,428,000 allele generations. All 9 mutations were on chromosomes originally containing the 28-dinucleotide-repeat allele at the *DROYANETSB* locus, and the 9 mutants differed from the original allele by −12, −6, −5, −1 (2 mutants), +1 (2 mutants), +3, and +4 repeats. Vazquez *et al.* (2000) observed two mutations at 28 loci, one of which was at the same locus for which Schlötterer *et al.* (1998) observed nine mutations.

The average mutation rate over these studies was 0.66×10^{-5}, lower than the estimates for mammalian microsatellite loci. The exception is the 28-repeat allele at *DROYANETSB*, one of the longest microsatellite alleles seen in *D. melanogaster*, which had a mutation rate of 3.0×10^{-4}. It appears that the low overall mutation rate (and low variation) for microsatellite loci in *D. melanogaster* is related to their low average number of repeats

TABLE 7.7 Estimates of mutation rates for microsatellite loci in *D. melanogaster* in two mutation accumulation studies. The estimates in Schlötterer *et al.* (1998) are given separately for the two alleles at the *DROYANETSB* locus, which have 28 and 15 dinucleotide repeats.

Study	*Locus*	*Number of mutations*	*Allele generations*	$\hat{u}$ ($\times10^5$)
Schug *et al.*	*DMU1951*	1	6,570	15.2
	DMZ3K25Z	1	6,570	15.2
	DM86	1	6,570	15.2
	46 other loci	0	302,220	0
Schlötterer *et al.*	*DROYANETSB-28*	9	29,750	30.3
	DROYANETSB-15	0	29,750	0
	23 other loci	0	1,368,500	0
Vazquez *et al.*	*DROYANETSB*	1	14,000	7.1
	DMSGG3	1	14,000	7.1
	26 other loci	0	364,000	0
All studies		14	2,113,930	0.66

compared to mammalian loci. Further, the directional change to smaller repeat number in long 28-repeat alleles (average reduction of 1.8 repeats) suggests that there is an upper constraint on repeat number.

Mutation accumulation lines have also been used to estimate the rates of deleterious mutation. Estimation of mutation rates for genes that influence fitness-related traits has been the subject of extensive recent research (Drake *et al.*, 1998). In general, it is assumed that these mutants reduce fitness and are eliminated by selection. The larger the detrimental effect, both as a homozygote and as a heterozygote, the lower the expected allele frequency for detrimental mutants. Estimation of the mutation rate and the effect of the mutants on viability have been determined in several large mutation accumulation experiments in *D. melanogaster* (see data from the experiment of Mukai, 1964, in Example 7.8). From these experiments, which include a number of replicate lines, the decline in mean viability per generation, ΔM, and the increase in among-line variance in viability per generation, ΔV, can be calculated. Using these values and the expressions

$$U \geq \frac{(\Delta M)^2}{\Delta V} \tag{7.15a}$$

$$\overline{s} \leq \frac{\Delta V}{\Delta M} \tag{7.15b}$$

a minimum estimate of U, the rate of mutation affecting viability per haploid chromosome per generation and a maximum estimate of s, the average effect of the mutations can be calculated (Drake *et al.*, 1998).

Table 7.8 summarizes four major mutation accumulation experiments, including the one discussed in Example 7.8, that have estimated these values for viability of the second chromosome in *D. melanogaster*. Extrapolating to a diploid genome (chromosome II is approximately 40% of the

TABLE 7.8 The estimated values of decline in mean viability per generation (ΔM) and the increase in variance among lines (ΔV) are given for four different studies. By using these values and expressions 7.15*a* and 7.15*b*, the rate of mutation affecting viability per haploid second chromosome per generation (U) and the average effect of the mutations ($\overline{s}$) are calculated. Also given is the estimated per-zygote mutation rate U_Z, which is obtained by multiplying U by 5.

Study	ΔM	ΔV	U	$\overline{s}$	U_Z
Mukai (1964)	0.0038	0.00010	0.14	0.027	0.70
Mukai *et al.* (1972)	0.0040	0.000094	0.17	0.023	0.85
Onishi (1977)	0.0017	0.000051	0.058	0.030	0.29
Fry *et al.* (1999)	0.0024	0.00027	0.021	0.113	0.10

genome), the mutation rate is nearly one per generation from the two Mukai experiments, lower from the Onishi, and only 0.1 per generation from Fry *et al.* (1999). Furthermore, Fry *et al.* (1999) found that the estimate of the effect of the mutations was three to five times as large as in the earlier experiments (see Fry, 2001, 2004, for other analyses).

Example 7.8. Inbred lines of *Drosophila* have been utilized to observe the effects of mutation on viability, the frequency of lethal genes in a population, and other traits. For example, Mukai (1964) estimated the effect of spontaneous mutation on viability in *D. melanogaster* using an isogenic stock for the second chromosome in which the relative viability of these chromosomes was periodically estimated. The mean relative viability (for 72 such stocks) and the phenotypic variance among these lines are given in Figure 7.16. The accumulation of mutants over time both reduced the viability of the lines and increased the variance among lines. By making

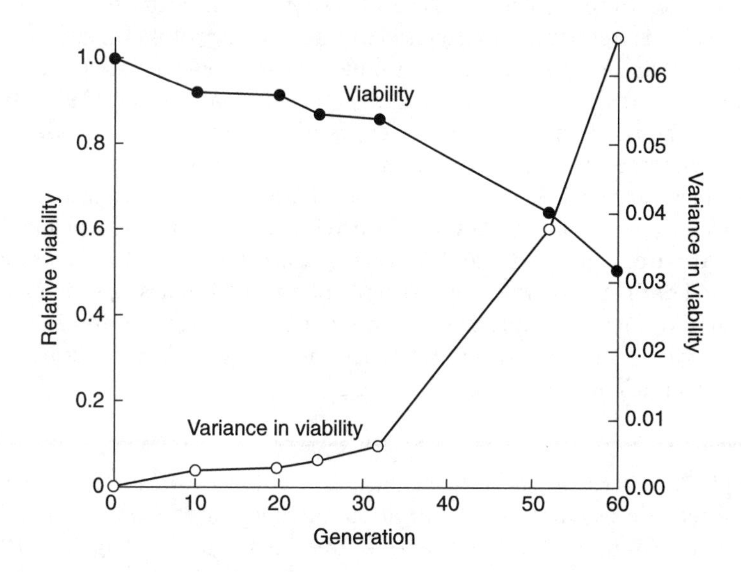

Figure 7.16. The relative viability and the variance in viability over time for 72 lines that were initially isogenic for the second chromosome in *D. melanogaster* (after Mukai, 1964).

various assumptions, Mukai estimated the number of polygenic mutations affecting viability per generation as 0.141. Of particular importance is that the amount of variation in viability by the end of the experiment is similar in magnitude to that in a natural population, which indicates that populations genetically depauperate for variation in a quantitative trait may regenerate variation in a relatively short time.

Some estimates of the mutation rate in other organisms are also much lower than the earlier *D. melanogaster* estimates (Bataillon, 2000; Zehl and de Visser, 2001; Shaw *et al.*, 2002). On the other hand, indirect estimates (Drake *et al.*, 1998) of the genomic mutation rate for fitness-related traits in *Daphnia* (Deng and Lynch, 1997), selfing plants (Johnston and Schoen, 1995), and *D. melanogaster* (Hughes, 1995) have given estimates not greatly different from the earlier *D. melanogaster* estimates. For discussion about possible explanations of differences in the genomic mutation rate estimates, see Keightley and Caballero (1997), Drake *et al.* (1998), Lynch *et al.* (1999), and Estes *et al.* (2004).

d. Factors Influencing the Mutation Rate

There is evidence that the mutation rate is influenced by a number of factors other than the mutagens in the environment (such as radiation discussed in Example 7.6 and in Forster *et al.*, 2002). For example, the mutation rate is influenced by the GC content of genes (Nei and Kumar, 2000) and CG dinucleotides (often called CpG) are hotspots of mutation to TG because the cytosine is frequently methylated in higher eukaryotes (Li, 1997). Other factors that influences mutation rate include sex, age, the ancestral allele, number of cell divisions, and mutator genes (Drake *et al.*, 1998; Crow, 2000). Although a maternal age effect is often found for traits associated with chromosomal variants in humans (e.g., the incidence of Down's syndrome increases with mother's age), for a number of gene mutations, there is a strong bias toward **paternal mutation** and an increase in incidence with **paternal age** (Vogel and Motulsky, 1997; Crow, 2000). Example 7.6 presented data supporting a higher mutation rate in males than females for minisatellite loci. Example 7.9 presents data for microsatellite data in humans in which there is no sex difference for dinucleotide repeats but there is for tetranucleotide repeats.

Example 7.9. Because the rate of mutation is low, even for microsatellite loci, direct determination of mutation rates requires examining a large number of potential mutational events. In such a study, Huang *et al.* (2002) examined 362 dinucleotide loci, evenly spaced over the human genome, for 1380 parent–offspring transitions in subjects of European ancestry. They documented 97 mutations in 499,560 transitions for an average mutation rate of 1.94×10^{-4}. These mutations were distributed over 19 of the 22 autosomes, with one mutation at 45 loci, two mutations at 18 loci, three mutations at 4 loci, and four mutations at 1 locus.

The distribution of the number of repeats is given in Table 7.9 with significantly more contraction than expansion mutations and no significant difference in mutation rate for the two sexes. Most different from earlier studies is that 61 (63%) of the 97 mutations changed by more than one repeat, quite unlike the expectations of the SMM in which all mutants are

TABLE 7.9 Distribution of the number of repeats in the 97 mutations observed by Huang *et al.* (2002).

	Number of changes in repeat number									
Category	*1*	*2*	*3*	*4*	*5*	*6*	*7*	*8*	*10*	*Total*
Expansion	12	10	5	4	2	3	0	1	0	37
Contraction	24	15	8	3	2	4	2	0	2	60
Paternal	16	12	2	4	0	3	1	0	2	40
Maternal	13	12	11	3	3	4	1	0	0	47
Total	36	25	13	7	4	7	2	1	2	97

assumed to change by only one repeat. The mean number of changes in repeat number was 2.67 for the 37 expansion mutants and 2.60 for the 60 contraction mutants. The number of changes in repeat number was negatively correlated with the standardized allele size (a measure in which the shortest alleles are close to 0 and the longest close to 1) (Figure 7.17). In other words, long alleles are more likely to decease in size from mutation, and short alleles are more likely to increase in size.

In another study of mutation rates for 122 human tetranucleotide microsatellite loci, Xu *et al.* (2000) verified 236 mutations in 287,786 parent–offspring transitions for an estimated mutation rate of 8.2×10^{-4}. However, 179 (76%) of these mutants were paternal, and only 57 were maternal, giving a more than threefold higher mutation rate for males than for females. In addition, Xu *et al.* (2000) found a bias toward contraction mutants with a greater reduction in repeat number when the parental allele was large but did not find a similar bias for expansion mutants. These two studies suggest that a negative size-dependent bias in mutation may be the basis for the observed constraints on microsatellite allele sizes.

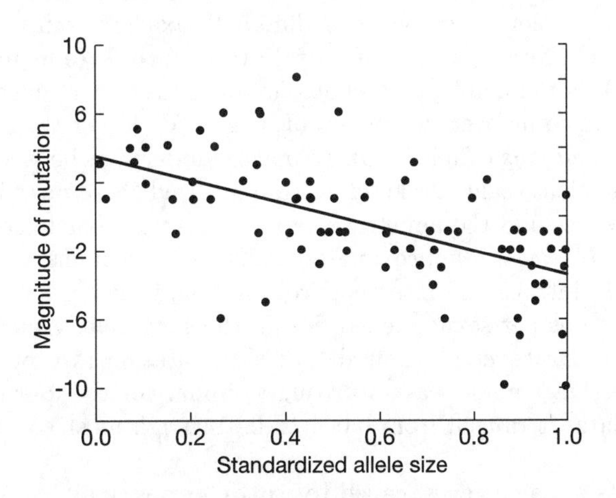

Figure 7.17. The changes in the number of repeats of 97 microsatellite mutation as a function of the standardized allele size (the line is a fit to these points). (Courtesy of Huang, Q.-Y., F.-H. Xu, H. Shen, *et al.* 2002. Mutation patterns at dinucleotide microsatellite loci in humans. *Amer. J. Hum. Genet.* 70:625–634.)

It is possible, using molecular markers, to determine the paternal origin of a new disease mutant, and in a study of Apert syndrome, a disease with skull and limb deformities, all 57 independent, new mutations were of paternal origin (Moloney *et al.*, 1996). Achondroplasia, a form of dwarfism, is one of a number of diseases that shows a large increase in frequency with increasing paternal age, relative to the population average (Risch *et al.*, 1987). Both diseases appear to originate as simple point mutations at related genes; all 57 Apert syndrome cases examined by Moloney *et al.* (1996) involved C-to-G tranversions at two adjacent sites, and all 16 independent cases of achondroplasia examined in Shiang *et al.* (1994) were glycine to arginine replacement at a single site (GGG to either AGG or CGG). As a result, Crow (2000) suggested that point mutations may often occur in males, particular older males, because of the higher number of cell divisions before gamete production than is found in females. However, examination of mutation rates in sperm for the gene causing achondroplasia from males of different ages showed only a small increase in mutation rate with age, not nearly enough to explain the observed increase in the disease with paternal age (Tiemann-Boege *et al.*, 2002).

Ellegren and Fridolfsson (1997) presented interesting confirmation of the higher mutation rate in males from a nearly fourfold higher estimate of the synonymous and intron substitution rates (see p. 421 for discussion of substitution rates) for the *CHD* gene on the Z chromosome in birds than for a homologous gene on the W chromosome. Because male and female birds are ZZ and ZW, respectively, two-thirds of the time Z chromosomes are in males and W chromosomes are found in only females. Therefore, a higher mutation rate in males would be expected to result in a higher substitution (mutation) rate for the *CHD* gene on the Z chromosome than for the *CHD* gene on W chromosome, as was observed (for discussion of estimation of sex differences in mutation rates, see Li *et al.*, 2002).

When mutation rates between different taxa are compared, there is often a large difference in mutation rate **per sexual generation** (Drake *et al.*, 1998). For example, the genome mutation rate per generation in humans appears to be over two orders of magnitude larger than that for the nematode *C. elegans* (Table 7.10). To make comparisons between species on a more equivalent scale, Drake *et al.* standardized these rates by the effective genome size and the number of replications per sexual generation. For example, in *C. elegans*, approximately 0.225 of the genome is in functional genes (not in introns and intergenic regions), and there are approximately 9.1 cell divisions per sexual generation in the germ line. Using these standardizations, Drake *et al.* showed that the mutation rate per replication for the effective genome was surprisingly similar for the four species with adequate data; it ranged from 0.004 to 0.014 (rightmost column of Table 7.10).

Variants at some genes, called **mutator genes**, may increase the general mutation rate for many loci. Often mutator genes code for, or are involved with, the enzymes that are important in DNA replication or DNA

TABLE 7.10 The mutation rate per genome standardized by the effective size of the genome (G_e) compared to the size of the total genome (G) and the number of cell replications per sexual generation (Rep/Gen) (Drake *et al.*, 1998).

			Mutation rate			
			Sexual generation		*Replication*	
Species	G_e/G	*Rep/Gen*	G	G_e	G	G_e
C. elegans	0.225	9.1	0.16	0.036	0.018	0.004
D. melanogaster	0.094	25	1.5	0.14	0.058	0.005
Mouse	0.030	62	30	0.9	0.49	0.014
Human	0.025	400	64	1.6	0.16	0.004

repair. Obviously, such an incompletely functional enzyme may result in the production of a number of mutants (see Example 7.10 for data showing that null homozygotes for an enzyme involved in mismatch repair have a greatly increased mutation rate for microsatellite loci in *D. melanogaster*).

Example 7.10. The MutS protein in *Escherichia coli* recognizes base-pair mismatches and is involved in initiating repair of these errors. Mutants at the *mutS* and *mutL* genes result in an increase in point mutation and instability of simple DNA repeats (Strand *et al.*, 1993). This repair system appears to be evolutionarily conserved and has been found in eukaryotes from fungi to humans. Mutations in a closely related protein in yeast cause instability in simple repeats, and several forms of cancer characterized by repeat instability have been induced in humans by similar mutants.

Flores and Engels (1999) cloned a gene homologous to *mutS* in *D. melanogaster* called *spellchecker 1* (*spel1*) and constructed lines that had deletions of this region. They established a number of such lines and compared the number of mutations for eight microsatellite loci in lines homozygous for a deletion in the *spel1* region and in a control that was heterozygous for the deletion. Table 7.11 gives the observed number of mutations after 10 to 12 generations and the approximate mutation rate for the *spel1* null homozygotes and the control heterozygotes. Only three mutations were observed in the control heterozygotes, whereas 189 were observed in the null *spel1* lines, and the mutation rate was nearly two orders of magnitude higher in the null homozygotes. Approximately 90% of the mutant alleles differed from a parental allele by only a single repeat. The rate of mutation in the control heterozygotes appears higher than that observed in the studies in Example 7.7, which suggests that two copies of *spel1* may be necessary for normal repair.

An increase in mutation rate does not necessarily result in an increase in fitness. A direct demonstration that it does provide an advantage was given by Cox and Gibson (1974) in *E. coli* populations grown in chemostats.

TABLE 7.11 Mutations at eight microsatellite loci for null homozygotes and control heterozygotes for the mutant *spellchecker 1*, where N is the sample size.

	spel1 null homozygotes			*Control heterozygotes*		
Locus	*Number of mutations*	N	*Mutation rate**	*Number of mutations*	N	*Mutation rate**
elf1 $(CAG)_7$	0	378	0	0	192	0
mam $(CAG)_7$	1	188	0.005	0	190	0
sev $(AC)_{14}$	3	97	0.031	2	98	0.020
35 F $(AT)_{17}$	18	192	0.094	0	190	0
w$(AT)_{13}$	9	92	0.098	1	96	0.010
Ula1 $(AT)_{15}$	44	380	0.116	0	192	0
tenA $(AT)_{14}$	14	94	0.149	0	95	0
AbdB $(CA)_{26}$	100	378	0.265	0	192	0
Overall	189	1629	0.116	3	1245	0.002

*The lines were not independent in the first generations, so this is only an approximate rate.

They competed strains with equivalent growth rates that differed only in a mutator allele. The results are given in Figure 7.18 as the ratio, over time, of the proportion of the mutator strain to the proportion of the wildtype strain. As shown in Figure 7.18b, sometimes the wildtype strain initially increased. However, in 10 of 11 experiments, the mutator strain outcompeted the wild type. The mutator strains accumulated "sticky" mutants that allowed the bacteria to adhere to the glass walls of the chemostat more

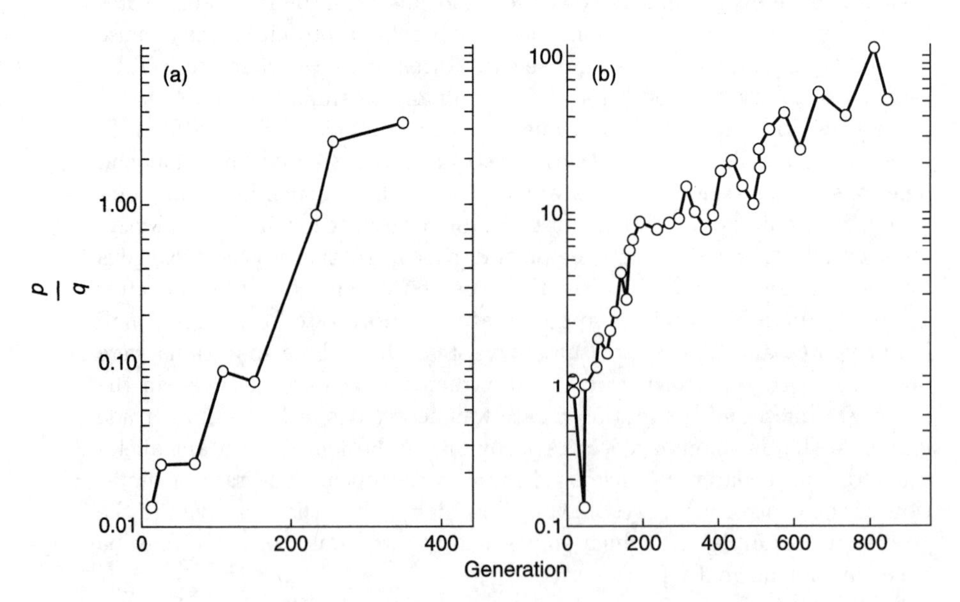

Figure 7.18. The change in the ratio of the proportion of a mutator strain to the wild type in *E. coli* in two experiments (after Cox and Gibson, 1974).

effectively. However, the benefit of a mutator gene on adaptive change in asexual organisms is complicated and depends on the rate of beneficial mutation without the mutator gene and other factors (de Visser *et al.*, 1999).

A high mutation rate is generally detrimental because it lowers the fitness of the population through the production of disadvantageous mutants. When the environment is novel or variable or the population is depauperate of genetic variation, an increase in genetic variation may be beneficial; that is, a high mutation rates produced by mutator genes can potentially provide short-term advantages in adapting to new environments (Giraud *et al.*, 2001; Tanaka *et al.*, 2003) (see Example 7.10 for data showing rapid increases in an *E. coli* strain with a mutator allele in a novel environment). Without new mutants being generated, the population fitness may be slowly reduced because this lack of new variability limits adaptation to new environments. A number of theoretical studies have examined these factors and have considered the evolution of an optimum mutation rate (Kimura, 1967; Leigh, 1970; Drake *et al.*, 1998). However, the change in mutation rate mediated by an overall change in fitness must be quite slow, and thus, such an optimum itself must be reached only slowly. Furthermore, it may be that the optimum mutation rate is the minimum possible without large expenditures of energy to reduce the mutation rate further—for example, to improve the repair systems.

PROBLEMS

1. What further experiments would you propose to examine whether the I and P elements have recently invaded *D. melanogaster*?
2. What is the equilibrium frequency if $u = 10^{-6}$ and $v = 10^{-7}$? Draw a Δq curve for these mutation rates. If $u = 10^{-6}$ and v is assumed to be zero, how many generations will it take for the frequency of A_2 to increase from 0.01 to 0.05?
3. What is the probability of loss of a new mutation in the first generation if all families are of size 2? What is the probability of loss of a new mutation in the first generation if half of the families have no offspring and half have 4? Compare these values to those obtained when there is a Poisson distribution of offspring and interpret the differences.
4. Assume that $u = 10^{-6}$ and $s = 0.4$. What is the expected equilibrium frequency of A_2 ? Draw the Δq curve between 0.0 and 0.01.
5. It has been suggested that both the increase in mutation rate from environmental mutagens and the better medical care of people with genetic diseases will increase the frequency of individuals with recessive genetic diseases. Which factor do you think will have a bigger impact and why?

6. The equilibrium frequency of A_2 when inbreeding, mutation, and selection occur is given by expressions 7.4*a* and 7.4*b*. Assuming $f = 0.02$ and $u = 10^{-5}$, what is q_e when $s = 0.1$, 0.2, and 0.5 for both expressions?

7. What is the equilibrium heterozygosity and allele frequency when $S = 0.25$, $s = 1$, and $u = 10^{-5}$? What is the equilibrium heterozygosity in a random-mating population and why are the two values different?

8. The equilibrium frequency of A_2 when there is selection with intermediate dominance and mutation is given by expressions 7.5*a* and 7.5*b*. Assuming $s = 0.1$ or 0.5, $h = 0.8$, and $u = 10^{-6}$, what is q_e for both expressions? What is the mutation load for these two parameter combinations?

9. How long does it take for an average neutral mutant allele to become fixed or to become lost if $N = 2000$ and if $N = 100$? Discuss the difference between these answers.

10. Assuming that $4N_eu = 0.1$, 1, and 10, calculate the equilibrium heterozygosity for the infinite-allele and stepwise-mutation models. Interpret your results.

11. Do you think that the data on genetic variation in cheetahs given in Example 7.4 can be used either to predict the probability of extinction of cheetahs or to support the hypothesis that they have gone through a bottleneck? Suggest further data or analysis that could support or falsify your answer.

12. What is the probability of fixation of a favorable mutant with $s = 0.1$ when $N = 100$ and $N_e = 20$, 100, and 200? Interpret your results.

13. Do you think the data in Example 7.6 are good evidence that the accident at Chernobyl caused an increase in mutation rate? What further data or analysis would be useful to determine more precisely the effect of the Chernobyl accident on mutation rate?

14. For a lethal dominant disease, 20 new mutations were observed in 200,000 births. What is the estimated mutation rate? In a survey of the population, the incidence of the disease is 0.00001. What is the estimated mutation rate using this information? Why might the two estimates differ?

15. For a recessive mutant, such as the coat-color loci discussed in mice, 10 mutants were observed in a million progeny from homozygous dominant by heterozygote matings. What is the estimated mutation rate? No recessive mutants have been detected in natural populations. How can this be explained?

16. Calculate the minimum estimate of U and the maximum estimate of s for the four experiments in Table 7.8 from the given estimates of ΔM and ΔV.

8

Neutral Theory and Coalescence

> Variation neither useful nor injurious would not be affected by natural selection, and would be left either a fluctuating element, as perhaps we see in certain polymorphic species or would ultimately become fixed, owing to the nature of the organism and the nature of the conditions.
>
> Charles Darwin (1859)

> Life can only be lived forwards, but must be understood backwards.
>
> Soren Kierkegaard

> The major theories of molecular evolution—Kimura'a neutral model, Ohta's slightly deleterious model, and Gillespies's balancing and episodic selection models—are each consistent with at least some aspects of allozyme and DNA data. None of these models, however, can account for all available empirical observations. An understanding of the evidence, we believe, will require a comprehensive theory that emphasizes strong and weak forces acting simultaneously under constraints of genetic linkage and population size.
>
> Martin Kreitman and Hiroshi Akashi (1995)

Molecular genetic variation, both as allele variation and as nucleotide sequence variation, provide the fundamental description of genetic variation in an organism. The forces that influence the amount and pattern of molecular variation—selection, inbreeding, genetic drift, gene flow, mutation, and recombination—are the same ones that influence other genetic variation, although their relative effects may be different. As a result, throughout this book we have used examples of molecular variants to illustrate the impact of these forces on genetic variation.

The first real measure of the extent of molecular variation within populations was provided by surveys of allozymes by Harris (1966) in humans and Lewontin and Hubby (1966) in *D. pseudoobscura*, both of whom found extensive genetic variation. With this stimulus, hundreds of studies followed documenting the amount of allozyme variation in various organisms in the 1970s and 1980s. The high amount of genetic variation found was not consistent with the classical view of genetic variation (see p. 28), which predicted that only a small proportion of genes would be variable. Partly to explain this large amount of genetic variation, Kimura (1968) and King and Jukes (1969) suggested that most polymorphisms are selectively neutral, and Kimura (1968) developed the **neutral theory**, or **neutrality**, in which ***genetic variation is primarily influenced by mutation generating variation and genetic drift eliminating it***. This elegant theory has become the centerpiece for understanding molecular evolution and has provided the null hypothesis for examining the amount and pattern of molecular genetic variation. In this chapter, we consider the predictions of the neutral theory, statistical tests related to the neutral theory, and an introduction to coalescent approaches to understanding molecular variation.

I. NEUTRAL THEORY AND PREDICTIONS ABOUT MOLECULAR VARIATION

The finding of large amounts of genetic variation for loci that determine proteins (allozyme variation) resulted in a controversy concerning the factors responsible for this variation. The initial opinions can be simply stated as two dichotomous views: one that held that the large majority of variants were maintained by some form of balancing selection and the second being the neutral theory, which held that the large majority of molecular variants were neutral with respect to each other, that is, the variation had no associated fitness differential. In the next two decades, it became apparent that many of these observations were generally consistent with predictions from neutral theory (see summaries in Kimura, 1983, and Nei, 1987), suggesting that widespread balancing selection was unlikely. Before we discuss the neutral theory, however, let us summarize some of the discussion surrounding balancing selection and maintenance of molecular variation.

a. Balancing Selection

When Lewontin and Hubby (1966) discovered the extensive amount of genetic variation within populations, they suggested that this variation was probably maintained by balancing selection, a view that reflected that of Dobzhansky (1955). Balancing selection can take many diverse forms, and since the discovery of molecular variation, virtually all of them have been considered by some researchers as contributing to the selective maintenance of molecular variants. We discussed in Chapters 3 and 4 a number of selection models that can result in the maintenance of genetic variation and may influence the amount of molecular variation, all of which may be termed forms of balancing selection. The simplest explanation for the presence of large numbers of molecular variants is that the variants are maintained by an advantage of the heterozygote at each of the polymorphic loci. However, Lewontin and Hubby (1966) demonstrated that a **heterozygous advantage** model with multiplicative fitness over loci leads to an intolerably low fitness for most populations, suggesting that such a simple model is not a feasible hypothesis.

To examine this argument, let us assume that the mean fitness at equilibrium for the ith biallelic locus with a heterozygous advantage (see p. 140) is

$$\overline{w}_i = 1 - s_1p^2 - s_2q^2$$

where s_1 and s_2 are the selective disadvantages of the two homozygotes. If there is independence on a multiplicative scale, and if there are m such loci, the mean fitness over all such loci becomes

$$\overline{w} = \overline{w}_i^m \tag{8.1}$$

Because each $\overline{w}_i$ is less than unity and if the number of such loci is large, the product of the fitnesses of the individual loci quickly approaches zero. For example, if $s_1 = s_2 = s$, resulting in an equilibrium so that $p_e = q_e = 0.5$, then $\overline{w}_i = 1 - {}^1\!/_2 s$ and expression 8.1 becomes

$$\overline{w} = (1 - {}^1\!/_2 s)^m$$

If there are 1,000 such loci and $s = 0.02$, then $\overline{w}$ would be approximately 0.00004, a very low fitness indeed. Assuming that selection occurs in preadult viability, then this fitness value implies that only 1 of 25,000 zygotes would survive to adulthood.

After Lewontin and Hubby published these conclusions, King (1967), Sved *et al.* (1967), and Milkman (1967) simultaneously suggested that alternatives with epistasis (and forms of truncation selection) might be important. In other words, the fitnesses at different loci may not combine multiplicatively, a situation that is defined as epistasis when considering viability selection (see p. 556). With such models, the very low population fitness for multiple homozygous individuals could be avoided.

Of course, these considerations were for biallelic loci, and many loci have more than two alleles. As we saw on p. 158, the conditions for the maintenance of multiple alleles via heterozygous advantage require that the relative fitness values be quite well balanced. In fact, Lewontin *et al.* (1978) showed that only a small proportion of randomly generated fitness values with multiple alleles satisfied the conditions for maintenance of all of the alleles. For example, with "complete heterosis" arrays, where all heterozygotes had to be more fit than all homozygotes, only 10% of the fitness arrays resulted in maintenance when there were five alleles. Spencer and Marks (1988, 1991) used an approach that sequentially introduced mutant alleles to a stable array of alleles and found that the number of alleles that could be maintained was higher than that found by Lewontin *et al.* (1978) but the distribution of these alleles in most cases was not different from that expected under the neutral theory (see p. 429).

Another assumption in the calculation of the fitness values by Lewontin and Hubby (1966) is that allele frequencies at different loci are independent of each other—that is, there is no linkage disequilibrium (see p. 526). However, as Franklin and Lewontin (1970) showed, when linked loci have a heterozygous advantage, then large statistical associations of allele frequencies at different loci, or strong **linkage disequilibrium,** may be generated even without epistasis (see the discussion on p. 562). Such associations in the extreme can lead to the situation in which only two complementary gametic types are in high frequency. As a result, if there are Hardy–Weinberg proportions of two such multilocus gametes, then half of the population would be heterozygous at all loci and have a high relative fitness. Even with some breakdown into several nearly complementary multilocus gametes, such associations could still maintain a high mean fitness.

A number of other types of balancing selection besides heterozygous advantage were proposed as having potential importance in the maintenance of molecular variation. For example, frequency-dependent selection was advocated as a mechanism responsible for maintaining allozyme variation (Kojima, 1971). One appeal of this model is that all genotypes may have the same relative fitness at equilibrium—that is, there is no genetic load (see p. 221). Perhaps the most strongly advocated balancing-selection hypothesis is that selection varying in time and/or space is responsible for maintaining allozyme variants (Gillespie, 1978, 1991). In this situation, the theoretical conditions for polymorphism are broader than for heterozygous advantage (see p. 204), and when there is limited gene flow or habitat selection, they become even less restrictive. However, under this model as in others, when the selective effects per locus are small, the conditions for maintenance of polymorphism become more restrictive.

A number of different experimental approaches have been used to examine the extent of selection in maintaining genetic variation, particularly allozyme variation (for reviews and differing opinions see Gillespie, 1991; Lewontin, 1991; Watt, 1994; Mitton, 1998; Eanes, 1999). For example, a promising approach was the examination of the allozymes for particular biochemical properties that might be consistent with balancing selection. Generally, in these studies, extracts of the allozymes were measured for their *in vitro* activity under different environmental conditions such as temperature, salinity, and concentration of the substrate (Eanes, 1999). In fact, many allozymes differ in some *in vitro* biochemical properties, but the selective importance of these differences in natural populations often remains unclear. However, as we will discuss more below, advances in nucleotide sequence analyses have provided more resolution in studying the maintenance of allozymes and molecular variation in general. For example, analyses of the nucleotide variants underlying many of the previously characterized allozyme latitudinal clines in *D. melanogaster* have revealed the potential bases of selection not seen with the allozymes alone (Duvernell and Eanes 2000; Verrelli and Eanes 2001).

One suggestion was that it would be difficult to detect selective differences at a single locus and that selection would be apparent only when a number of loci are examined simultaneously. Lerner (1954) had suggested that natural selection favors individuals with increased heterozygosity: the more loci heterozygous, the higher the relative fitness. A large number of studies have looked at such heterozygosity–fitness correlations (see Example 10.8), and David (1998) concluded that "the correlations appear intrinsically weak and often inconsistent across samples." To examine this association directly, Mukai (1977) carried out a large experiment that measured the relative viability and fecundity of females in a population of *D. melanogaster* in which flies were heterozygous for different numbers of allozyme loci. They did not find any significant differences in either viability or fecundity for second chromosomes that were heterozygous for different numbers of allozyme loci, even with the statistical power to detect differences of only a few percent (Table 8.1).

TABLE 8.1 The average viability and fecundity for individuals heterozygous for different numbers of allozyme loci on the second chromosome of *D. melanogaster* (Mukai, 1977).

Number of heterozygous loci	*Average viability*	*Average fecundity*
0	1.000	1.000
1	0.999 ± 0.008	0.989 ± 0.011
2	1.001 ± 0.010	1.009 ± 0.013
2 or more	1.000 ± 0.009	1.009 ± 0.013
3 or more	0.984 ± 0.019	1.006 ± 0.021

Although the finding of substantial molecular variation stimulated extensive research into theoretical analyses of balancing selection and experimental examination of selection, overall, it appears that factors other than balancing selection are most important in influencing and maintaining molecular variation. As a result, in developing concepts in the rest of this chapter, we generally assume that factors basic to neutral theory, genetic drift and mutation, play the major roles in influencing the amount and pattern of molecular genetic variation. However, later in the chapter we discuss some statistical and theoretical approaches that have been used to examine the extent of the role of selection influencing molecular variation.

b. The Neutral Theory

The neutral theory that Kimura (1968) introduced to explain molecular genetic variation, based on theory he developed in the proceeding decade, in many ways revolutionized population genetics. In contrast to much of the thinking of the day, Kimura assumed that selection played a minor role in determining the maintenance of molecular variants and proposed that different molecular genotypes have almost **identical relative fitnesses**; that is, they are ***neutral with respect to each other***. Of course, this does not mean that the different allele forms have no effect on fitness but, rather, that their effects are equivalent. Kimura assumed that the majority of molecular variants that become polymorphic are affected primarily by the interplay of mutation, generating new variation, and genetic drift leading to the eventual fixation of these variants.

The actual definition of selective neutrality depends on whether changes in allele frequency are primarily determined by genetic drift. In a simple example, if s is the selective difference between two alleles at a locus, and **if $s < 1/(2N)$**, the ***alleles are said to be neutral with respect to each other***. This definition implies that allele variants may be effectively neutral in a small population and not in a large population. Neutral theory does not claim that allele substitutions responsible for evolutionarily adaptive traits are neutral. It does suggest that most allele substitutions have no selective advantage over those that they replace, and hence, the term **non-Darwinian evolution** is sometimes used to describe this process. However, this term is somewhat a misnomer because Darwin (1859) in the quote at the beginning of the chapter, even before modern genetics was known, realized that not all traits were under selection and that chance determined the fate of neutral ("neither useful nor injurious") variation. It should be noted that many new mutations never become polymorphic because they are quickly eliminated from the population because of their deleterious effects in combination with genetic drift.

As we discussed on p. 379, when mutation and genetic drift are the only factors affecting the genetic variation (there are no selective differences between the genotypes), the equilibrium heterozygosity in the population for the infinite allele model is expected to be

$$H_e = \frac{4N_e u}{4N_e u + 1} = \frac{\theta}{\theta + 1}$$

where $\theta = 4N_e u$ (Kimura and Crow, 1964). Because the production of new mutants is generally quite slow and the population sizes in most species are relatively large, making the effect of genetic drift small, a long time is needed for this equilibrium heterozygosity to be reached (see p. 378). Of course, low population sizes at some time in the past would reduce the heterozygosity to below the equilibrium expected from a contemporary large population size.

This theory predicts that, as $4N_e u$ increases, the amount of heterozygosity would also increase, as shown in Figure 7.10. Therefore, neutrality predicts that an increase in either effective population size or mutation rate would result in an increase in heterozygosity. Surveys of microsatellite loci,

MOTOO KIMURA (1924–1994)

Perhaps the leading successor to the theoretical heritage of Haldane, Fisher, and Wright was Motoo Kimura (left in photo) (Crow, 1995), widely known for his neutral theory of molecular evolution. In fact, one of his first contributions (Kimura, 1954) was to examine the relative effects of genetic drift and selection, the subject of intense debate between Fisher and Wright. Kimura had a long collaboration with James Crow, culminating in their influential book on population genetics theory (Crow and Kimura, 1970). When Kimura proposed the neutral theory there were many detractors, and he spent a great deal of time and energy defending the then-novel perspective. Presently this theory has become so widely accepted that its origin with Kimura often is not cited. In addition to his contribution in founding and elaborating the neutral theory, Kimura authored papers on a diverse array of theoretical topics in population genetics including linkage, quantitative traits, sexual reproduction, and population structure. Many of Kimura's papers have been reprinted (Kimura, 1994) with a commentary by Naoyuki Takahata. Kimura became probably the most noted evolutionary biologist in Japan ever (Nei, 1995) and made the National Genetics Institute at Mishima a world center for theoretical population genetics. A number of his colleagues at Mishima—Takeo Maruyama, Tomoko Ohta (pictured here on Kimura's left), Fumio Tajima, and Naoyuki Takahata—have also made many fundamental contributions to population genetics theory. (Photograph of Motoo Kimura and Tomoko Ohta at the home of Motoo and Hiroko Kimura in Mishima, Japan, in May of 1994, courtesy of Will Provine.)

which have a higher mutation rate that allozymes, show that they have a higher heterozygosity, consistent with predictions from the neutral theory. In another outcome consistent with this prediction, Nei and Graur (1984) found a positive association of allozyme heterozygosity with a general estimate of population size over different species. However, in this case, the maximum level of observed heterozygosity was less than 0.25, far below that predicted, because the effective population size of the different species ranged over more than five orders of magnitude.

A more specific test of the effect of population size on heterozygosity is provided by comparing the genetic variation within a species for autosomal, Y-linked, and X-linked loci. Assuming that the mutation rate is the same for the different loci, the lower effective population size for Y-linked and X-linked loci (see p. 322 and p. 329) would predict that Y-linked and X-linked loci would have a lower heterozygosity than autosomal loci. Nachman *et al.* (1998) summarized the available human data and found that after standardizing them the values over the different types of loci were in general agreement (Table 8.2). However, there was substantial variation within their data on seven X-linked introns that was correlated with the

TABLE 8.2 Summary of the nucleotide diversity from nine different studies (from Nachman *et al.*, 1998). To standardize the values relative to autosomal estimates, the Y-chromosome estimates were multiplied by 4 and the X-chromosome estimates were multiplied by 4/3.

Locus	*Chromosome*	*Sample size*	*Length (bp)*	π	$\pi_{\text{standardized}}$
49 loci	Autosomes	2	8,537	0.00110	0.00110
Zfy intron	Y	38	729	0.00000	0.00000
YAP region	Y	16	2,638	0.00037	0.00148
Sry region	Y	5	18,300	0.00008	0.00031
β-globin	11	349	2,670	0.00180	0.00180
Lpl	8	142	9,700	0.00200	0.00200
Pdha1 intron	X	8	1,769	0.00113	0.00151
Zfx intron	X	29	1,151	0.00040	0.00053
7-loci introns	X	10	11,365	0.00063	0.00084

amount of recombination in the region (see p. 579). Example 8.1 examines heterozygosity for a large number of X-linked and autosomal microsatellite loci in humans and shows that the variation is consistent with neutral theory predictions.

Example 8.1. For neutrally evolving genes, the amount of genetic variation at equilibrium is expected to be a function of the mutation rate and the effective population size. As we discussed on p. 322, the effective population size for an X-linked gene or a gene in a haplo-diploid organism is expected to be $\frac{3}{4}$ that for an equivalent gene in a diploid organism or at an autosomal locus. Comparisons between diploid and haplo-diploid organisms also may include other differences between the species (Hedrick and Parker, 1997), but comparisons between autosomal and X-linked genes in the same organism should avoid confounding with other unknown factors.

Dib *et al.* (1996) gave the heterozygosity for 5264 microsatellite loci in humans that could be used as markers throughout the human genome. Figure 8.1 gives the heterozygosity for the 216 X-chromosome loci by location that they reported (Hedrick and Parker, 1997). Although there does appear to be some clustering of heterozygosity levels along the chromosome, the loci near the ends do not appear to have significantly lower heterozygosity, as they do on the *D. melanogaster* X chromosome. This is not unexpected, because recombination at the ends of the human X chromosome is not reduced (see the discussion on p. 579). The range of heterozygosity over the 22 autosomes is from 0.69 to 0.73, and the mean of the 5048 autosomal loci is 0.70, as indicated by the broken line in the figure. The mean for the 216 X-chromosome loci, 0.65, is lower than for any of the autosomes.

Assuming that the mutation rates for X-linked and autosomal genes are the same, then on the basis of the difference in the effective population size for the X and autosomal (A) loci, the ratio θ_X/θ_A should be approximately

Figure 8.1. Heterozygosity for 216 microsatellite loci on the human X chromosome shown by map location (Hedrick and Parker, 1997). The horizontal solid and broken lines indicate the mean heterozygosity for the 216 X-linked genes and the 5048 autosomal genes, respectively.

0.75. For the infinite-allele model, rearranging expression 7.8*d* yields

$$\theta = \frac{H_e}{1 - H_e}$$

Using the average heterozygosities, the estimates of θ for the X chromosome and the autosomes are 1.857 and 2.333 with a ratio of 0.796. For the stepwise-mutation model, rearranging expression 7.9*b* yields

$$\theta = \frac{1}{2}\left[\frac{1}{(1 - H_e)^2} - 1\right]$$

In this case, estimates of θ for the X chromosome and the autosomes are 3.58 and 5.06 with a ratio of 0.708. In other words, the estimates of the ratio from two mutation models are fairly close to each other and exactly bracket the expected ratio of 0.75 based on the effective population. This supports the hypothesis that variation at these microsatellite loci is primarily maintained by a balance of mutation and genetic drift and suggests that mutation at these loci may be a mixture of the infinite-allele and stepwise-mutation models.

As we noted in Chapter 7, neutral theory allows a number of simple predictions to be made about genetic variation. For example, the expected probability of fixation and time to fixation of a new neutral mutant are $1/(2N)$ and $4N$, respectively (expressions 7.6a and 7.7b). Kimura (1968) also showed that the neutral theory was consistent with a **molecular clock**; that is, there is a constant rate of substitution for molecular variants. These simple but elegant mathematical predictions, and others, discussed below, provide the basis for the most important developments in evolutionary biology in the past half-century.

c. The Molecular Clock

To illustrate the mathematical basis of the molecular clock, let us assume that genetic drift and mutation are the determinants of changes in frequencies of molecular variants so that the rate of allele substitution can be understood in the following manner. Let the mutation rate to a new allele be u so that in a population of size $2N$ there are $2Nu$ new mutants per generation. Because the probability of fixation of a new mutant is $1/(2N)$ and assuming an equilibrium situation, the rate of allele substitution is the product of the number of new mutants per generation and their probability of fixation or

$$k = 2Nu\left(\frac{1}{2N}\right)$$
$$= u \qquad (8.2a)$$

In other words, the elegant prediction from the neutral theory is that the **rate of substitution** is ***equal to the mutation rate at the locus*** and is constant over time. The substitution rate is independent of the effective population size, a fact that may initially be counterintuitive. This independence occurs because in a smaller population there are fewer mutants; that is, $2Nu$ is smaller, but the initial frequency of these mutants is higher, which increases the probability of fixation, $1/2N$, by the same magnitude by which the number of mutants is reduced.

For any type of mutation model, if the rate of substitution is assumed to be equal to the mutation rate u, then the **expected time between neutral substitutions** is $\mathbf{1/u}$. This can be seen intuitively by flipping a coin to determine the average number of coin flips until a head is first seen: the answer is 1 divided by the probability of a head $1/2$, and thus, the average number of coin flips is 2. In other words, the time between the substitutions is a function only of the mutation rate and is independent of the population size, just like the substitution rate. For example, if the mutation rate is $u = 10^{-9}$ per nucleotide per generation, then the expected time between new substitutions is 10^9 generations.

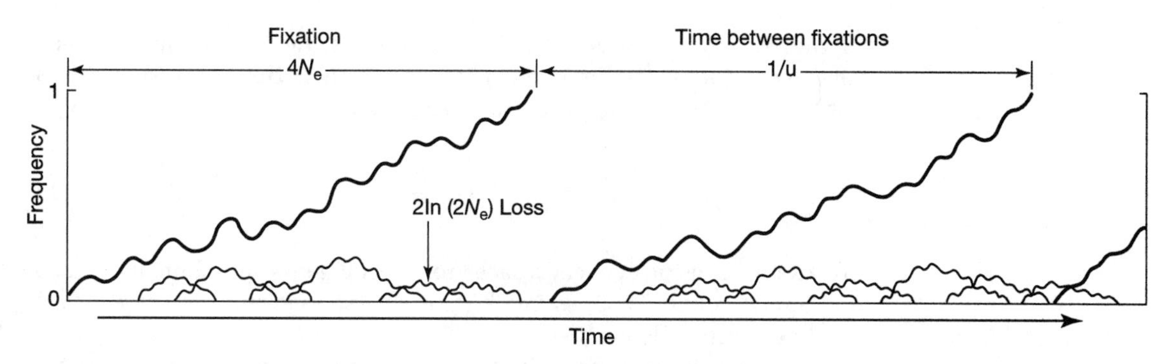

Figure 8.2. The change in frequency of new neutral mutants over time (after Kimura, 1983). The two mutations that eventually become fixed are represented by thick lines, whereas the many mutations that are lost are represented by thin lines. In general, the time is smallest for loss of mutants, larger for fixation of mutants, and longest for the time between fixation of mutants, with the expected values a function of N_e and u.

Figure 8.2 illustrates some of the properties of the change in frequency of neutral mutants after their appearance in the population (after Kimura, 1983, p. 34). First, most of the mutants are lost after a few generations (actually many more are lost than are shown), and only two are eventually fixed, reflecting the probabilities of fixation and loss of new mutants of $1/(2N)$ and $(2N - 1)/(2N)$, respectively. Second, the time to fixation is much longer than the time to loss, reflecting the expected times to fixation and loss of $4N_e$ and $2\ln(2N_e)$, respectively (assuming $N = N_e$). Finally, as we discussed above, the expected time between the fixation of mutations is $1/u$.

Using the neutral rate of substitution from expression 8.2*a*, we can also predict the **amount of divergence *K*** between two nucleotide sequences, or the number of different nucleotide sites, over t generations. The expected number of differences per site is then

$$K = 2ut \tag{8.2b}$$

where the 2 is necessary because substitution can occur in the sequences for both lineages (below time will often be in units of years, not generations).

This expression can be solved for t, the time since the two sequences diverged from a common ancestor as

$$t = \frac{K}{2u} \tag{8.2c}$$

K can be measured from sequence data, and if u can be estimated in some manner, then an estimate of t is possible. For example, if $K = 0.02$ (2% divergence) and u is assumed to be 10^{-8}, then t, the time since divergence of the two sequences is 1,000,000 generations. Or, given the time since divergence (here it is in years, not generations), this expression can be used to estimate the mutation rate per year as

$$u = \frac{K}{2t} \tag{8.2d}$$

For example, using sequence data from 5,669 genes in mammals, Kumar and Subramanian (2002) estimated that the average mutation rate is 2.2×10^{-9} per nucleotide per year.

The probability of fixation of a new selectively advantageous mutant is approximately $2s$ (expression 7.10a). If we use the same approach as above, the rate of allele substitution becomes

$$\begin{aligned} k &= 2Nu(2s) \\ &= 4Nus \end{aligned} \tag{8.2e}$$

Obviously, if $4Ns \gg 1$, then the rate of substitution for selectively advantageous mutants will be much higher than for neutral mutants, given that the mutation rate u is the same.

On the other hand, the **nearly neutral model** of Ohta (1973), in which there is slightly deleterious selection against alleles, predicts that the rate of substitution decreases with increasing population size. This dependence occurs because slightly detrimental variants may be effectively neutral in a smaller population; that is, $s < 1/(2N)$, whereas in a larger population, they are selected against because $s > 1/(2N)$. In fact, the efficacy of selection for codon bias (see p. 451) appears to depend in part on population size (Akashi, 1995, 1999) consistent with the nearly neutral model. One of the initial motivations for the nearly neutral model was to provide a way of accounting for the observation that rates of amino acid substitution appear to be roughly constant per year rather than per generation.

In the late 1960s, data from amino acid sequences of proteins seemed to indicate that mutations in protein evolution accumulate at a constant rate (King and Jukes, 1969). Assuming that such a regular replacement occurs, then the differences in amino acid or nucleotide sequence between two species may serve as a **molecular clock**, indicating the time since two species diverged from a common ancestor (Bromham and Penny, 2003; Hedges and Kumar, 2003). Kimura (1968) suggested that the idea of a molecular clock is consistent with the hypothesis that molecular variants may be neutral with respect to each other, as we discussed above. In other words, if the replacement of molecular variants over time is a function only of genetic drift and mutation, it should result in a relatively regular turnover (substitution or replacement) of molecular variants over time.

Before modern techniques of DNA sequencing were developed, the amino acid sequences of proteins were the molecular data used to determine the rate of substitution and the relationship between organisms. Because DNA sequencing is now much easier than determining amino acid sequences, generally amino acid sequences are inferred from DNA sequences. Here we start by introducing the more simple approach used to determine the **rate of amino acid substitution** for proteins and then discuss estimation of substitution rate for DNA sequences.

Assume that the number of differences in the homologous amino acid sequence of a particular protein for two species is known. Let N be the number of amino acids that can be compared in the organisms, a number that excludes any insertions or deletions (indels) because we are concerned only with amino acid substitution, and let d_{aa} be the proportion of different amino acid sites between the two sequences. When the amino acids at a particular position are different, this indicates that at least one amino acid replacement has occurred in the course of evolution. We can estimate K_{aa}, the mean number of amino acid substitutions per site, by assuming that there is independence at different sites. Therefore, the probability of no substitution at any site from the Poisson distribution is

$$\Pr(0 \text{ substitution}) = e^{-K_{aa}}$$

and the probability of one or more substitutions is

$$\Pr(\geq 1 \text{ substitution}) = 1 - e^{-K_{aa}}$$

or one minus the probability of no substitutions.

Assuming that the proportion of sites at which the sequences differ is

$$d_{aa} = 1 - e^{-K_{aa}}$$

the estimate of the number of amino acid substitutions per site becomes

$$K_{aa} = -\ln(1 - d_{aa}) \tag{8.3a}$$

(Zuckerlandl and Pauling, 1965) with variance

$$V(K_{aa}) = \frac{d_{aa}}{(1 - d_{aa})N} \tag{8.3b}$$

(Kimura, 1969). The standard error is the square root of the variance here. The proportion of sites that differ, d_{aa}, is always less than K_{aa}, because K_{aa} is corrected for multiple changes (hits) at a site.

The rate of substitution per amino acid site per year is

$$k_{aa} = \frac{K_{aa}}{2T} \tag{8.3c}$$

where T is the number of years since the divergence of the two species from their common ancestor. Example 8.2 calculates the amino acid substitution rate for some hemoglobin data and presents some classic data that was one of first illustrations of the molecular clock.

Example 8.2. The amino acid sequences of protein molecules in a number of organisms were compared soon after the concept of the molecular clock was proposed. For example, the α chain of the hemoglobin molecule was sequenced, and the sequence in humans and carp differed at 68 out of 140 sites, $d = 68/140 = 0.486$. Therefore, $K_{aa} = 0.665 \pm 0.082$, and because the time since divergence of humans and carp is about 450 million years (Kumar and Hedges, 1998), $k_{aa} = 7.4 \times 10^{-9}$.

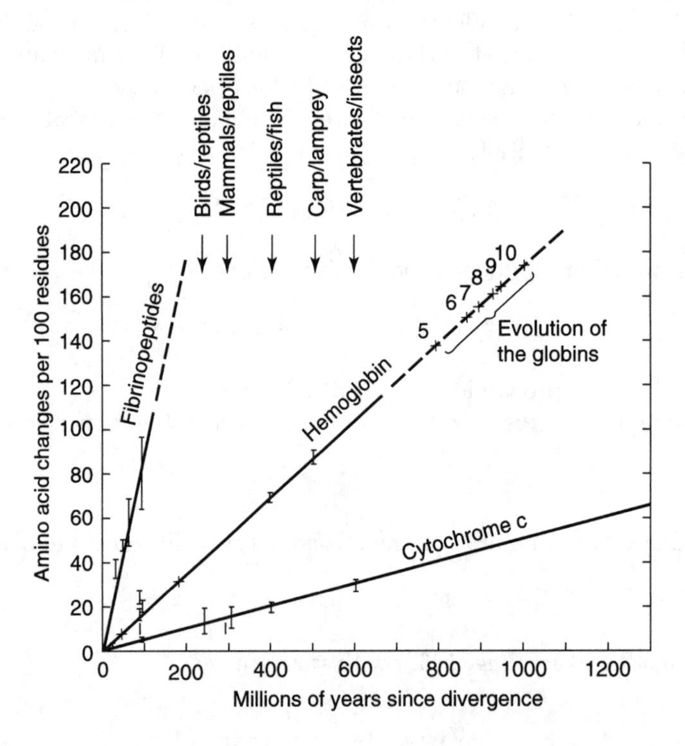

Figure 8.3. The amount of amino acid substitution for three proteins having different rates of substitution. The horizontal axis gives the times since divergence of various organisms in millions of years (after Dickerson, 1971).

Dickerson (1971) compared the fibrinopeptides, hemoglobin, and cytochrome c in organisms that had diverged at different times (Figure 8.3). From these data for molecules from the same organisms, it appears that the rate of amino acid substitution—molecular clock calibration—is high for fibrinopeptides, intermediate for hemoglobin, and low for cytochrome c. Kimura (1983) estimated the rates of amino acid substitution, k_{aa}, to be 8.3×10^{-9}, 1.2×10^{-9}, and 0.3×10^{-9} for fibrinopeptides, hemoglobin, and cytochrome c, respectively. Although the rates differ greatly, within each molecule there appear to be fairly uniform rates of substitution among lineages. Dickerson (1971) attributed the differences between molecules to the relative importance of the three-dimensional structure of these molecules. For example, histone IV, which has a very low k_{aa} value of 0.006×10^{-9}, fits in tightly with the DNA molecule, so nearly the whole amino acid sequence is critical for its three-dimensional structure. With fibrinopeptides, on the other hand, only a small part of the molecule appears to be important in binding.

A number of refinements in the theory used in estimating the mean number of substitutions per site have been proposed (see Nei and Kumar, 2000, for comparisons of these approaches). In addition, Dayhoff *et al.* (1978) empirically determined, from a survey of comparisons between a number of proteins, the probability of any given amino acid being replaced by any of the other 19. From this information, they constructed a 20 × 20 matrix that gives the probabilities for all possible transitions, which can then be used to predict changes in amino acid sequence (Nei and Kumar, 2000). These probabilities reflect the observation that changes between amino acids similar in biochemical properties are much more likely than are changes between greatly different amino acids.

Because most of the data generated today are nucleotide sequence data, having a measure of the substitution rate for DNA data, or the **rate of nucleotide substitution**, is essential. However, because there are only four nucleotides, the problem of back mutation, say C → T → C, becomes very important. The simplest approach to determining the mean number of nucleotide substitutions per site is to assume that substitution occurs randomly among the four types of nucleotides with a rate α between any pair of nucleotides (Jukes and Cantor, 1969). Therefore, given a particular initial nucleotide, the probability of still having this nucleotide at time t is

$$\mathrm{Pr}_t = 1 - 3\alpha$$

because there are three types of changes that produce different nucleotides (3α is the total mutation rate). The probability of still having this nucleotide at time $t + 1$ is

$$\mathrm{Pr}_{t+1} = (1 - 3\alpha)\mathrm{Pr}_t + \alpha(1 - \mathrm{Pr}_t).$$

The first term here is the probability that the nucleotide has continued to remain unchanged from time t to time $t + 1$, and the second term allows for the probability for back mutation. In this latter case, at time t, the nucleotide was different from the original one $(1 - \mathrm{Pr}_t)$, but with a probability α, it changed back to the original nucleotide. This expression can be written as the amount of change as

$$\Pr_{t+1} - \Pr_t = -4\alpha \Pr_t + \alpha$$

For convenience, let us assume that time is continuous so that the following differential equation is approximately equal to the above discrete time difference equation or

$$\frac{d\Pr}{dt} = -4\alpha \Pr_t + \alpha$$

This expression can then be solved to obtain

$$\Pr_t = \frac{1}{4} + \frac{3}{4} e^{-4\alpha t}$$

The probability that two sequences, separated for time t, continue to retain the same nucleotide is

$$\Pr_t = \frac{1}{4} + \frac{3}{4} e^{-8\alpha t}$$

because the total time along the lineages during which changes could occur is $2t$. If the probability that the nucleotide sites are different is $d = 1 - \Pr_t$, then

$$d = \frac{3}{4}\left(1 - e^{-8\alpha t}\right)$$

and taking logarithms of both sides yields

$$8\alpha t = -\ln\left(1 - \frac{4d}{3}\right)$$

The expected number of substitutions per site in a lineage is $3\alpha t$ (remember 3α is the mutation rate), and because there are two lineages being compared, then $K = 6\alpha t$. Thus, by substitution, an estimate of the number of substitutions is

$$K = -\frac{3}{4} \ln\left(1 - \frac{4d}{3}\right) \tag{8.4a}$$

(Jukes and Cantor, 1969), where d is the proportion of nucleotide sites that differ. The variance of this estimate is approximately

$$V(K) = \frac{d(1-d)}{N(1-4d/3)^2} \tag{8.4b}$$

(Kimura and Ohta, 1972). An estimate of the rate of substitution per nucleotide site per year is then

$$k = \frac{K}{2T} \tag{8.4c}$$

where T is as above.

The assumption that all types of nucleotide mutations (substitutions) are equally likely is known empirically not to be true. From the fundamental chemical perspective, there are two types of changes in nucleotides: **transitions**, in which a pyrimidine replaces a pyrimidine (C $\leftrightarrow$ T) or a purine replaces a purine (A $\leftrightarrow$ G), and **transversions**, in which a purine replaces a pyrimidine or vice versa. In particular, the rate of transitions is usually higher than the rate of transversions; sometimes it is an order of magnitude higher in mtDNA. It has generally been assumed that much of this difference in observed substitution rate was the result of differential mutation rate. In fact, Denver *et al.* (2000) observed 13 transitions and 3 transversions in complete mtDNA sequences of 74 *C. elegans* mutation accumulation lines, consistent with the transition excess seen in phylogenetic comparisons.

To accommodate different mutation rates for transitions and transversions, Kimura (1980) developed the two-parameter model in which the rate of transition mutation is α and the rate of transversion mutation is β (Figure 8.4). To circumvent the estimate of two parameters, Felsenstein (1995) assumed that the transition/transversion ratio α/β is known from other data and suggested using the estimate of this ratio. A number of more sophisticated approaches have been proposed, including using estimates of all substitution rates in the 4×4 nucleotide matrix. However, when the expected number of substitutions per site is less than 0.5, these different approaches give similar estimates of substitutions (Nei and Kumar, 2000).

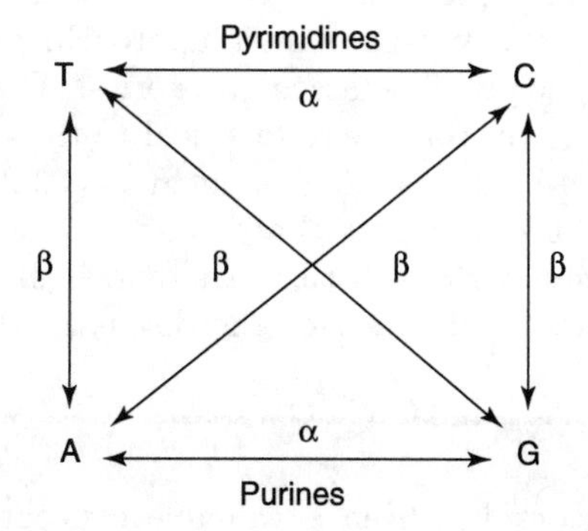

Figure 8.4. The rates of nucleotide substitutions for transitions (α) and transversions (β) under the two-parameter model of Kimura (1980).

Example 8.3 gives a calculation of the substitution rate for globin genes in the rabbit and mouse.

Example 8.3. Large amounts of DNA sequence data are available, but to illustrate calculation of the number of nucleotide substitutions, a small data set is useful. Nei (1987) summarized the observed numbers of different pairs of nucleotides for the 139 homologous codons for the rabbit α and the mouse β^{maj} globin genes (Table 8.3). Here the pairs of nucleotide sites for the two sequences are categorized as identical (the same nucleotide in both sequences), transitional (the sequences have different purines or different pyrimidines), or transversional (there is a purine in one sequence and a pyrimidine in the other) and are also identified as in the first, second, or third position of the codons.

TABLE 8.3 The observed numbers of the ten possible pairs of nucleotides between the DNA sequences of the rabbit α and the mouse β^{maj} globin genes separated by codon position (Nei, 1987).

	Type of pair										
	Identical				*Transitional*		*Transversional*				
Position	*AA*	*TT*	*CC*	*GG*	*AG*	*TC*	*AT*	*AC*	*TG*	*CG*	*Total*
First	18	8	19	34	15	9	8	10	8	10	139
Second	32	35	20	7	11	5	4	11	2	12	139
Third	1	4	35	27	5	30	2	3	12	20	139

For the first codon position, the number of nucleotide differences is 60, so the proportion of sites that are different is $d = 60/139 = 0.432$. Using expression 8.4a, the estimate of $K = 0.64 \pm 0.10$. The estimated number of substitutions is lower for the second position, $K = 0.42 \pm 0.07$, and higher for the third position, $K = 0.88 \pm 0.14$. Nei also calculated estimates using Kimura's two-parameter approach, in which transitions and tranversions are kept in separate categories. The estimates for the first and second codon positions were identical to those above, but for the third position, the estimate was higher at $K = 0.92 \pm 0.16$. When the number of substitutions is greater than 0.5, the Jukes–Cantor approach underestimates the number of substitutions, and Kimura's two-parameter approach gives a better estimate (see the discussion in Nei and Kumar, 2000).

The molecular clock has been a valuable concept and has generally been useful in understanding the relationships between genes and species. As we mentioned in Example 8.2, the molecular clock runs at different times for different genes, apparently because of differing constraints for different molecules. As a result, genes that have a rapidly running clock are useful in comparing closely related taxa, whereas genes with a slower running clock can be useful for distantly related organisms.

In addition, different organisms may have different rates of neutral mutation, which in turn would result in different rates of substitution. Most extreme is the rate of substitution in some viruses, which may change at rates six orders of magnitude faster than for eukaryotes. For example, Buonagurio *et al.* (1986) found that the *NS* genes in influenza virus strains showed a constant rate of substitution of 0.0019 nucleotides per year over a period of 50 years. Example 8.4 discusses the use of the molecular clock to date the origin of the human immunodeficiency virus (HIV) to approximately 1931, another virus which has a very high rate of substitution.

Example 8.4. The most serious viral epidemic caused by the spread of an animal virus into humans is that of AIDS caused by spread of the HIV. It is estimated that at least 42 million people worldwide are presently infected with HIV and that the mortality rate is nearly 100% (Rambaut *et al.*, 2004). The two different HIV strains, HIV-1 and HIV-2, appear to descend from the simian immunodeficiency virus (SIV), originally transmitted to humans from chimpanzees (HIV-1) and sooty mangaby monkeys (HIV-2) (Rambaut *et al.*, 2004). Within HIV-1, group M has a worldwide distribution and represents a single phylogenetic lineage.

Assuming a molecular clock and known sampling dates of HIV samples, Korber *et al.* (2000) dated the origin (common ancestor) of the HIV-1 group M using data from the *env* (envelope) gene that encodes for the proteins on the outer envelope of the virus. Figure 8.5 gives the estimated

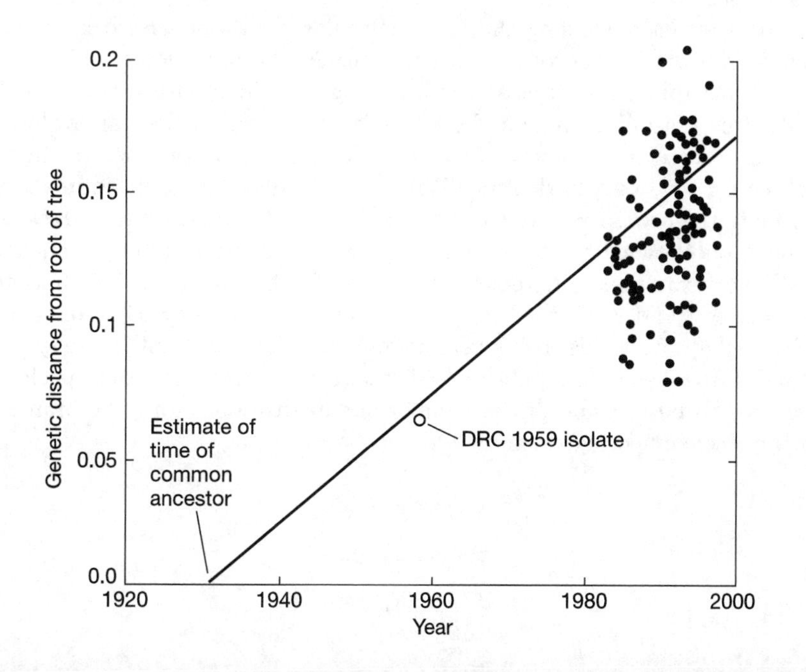

Figure 8.5. The estimated genetic distance from the root of the phylogenetic tree for different group M strains of HIV-1 collected at different times. The line indicates the best fit to the data and intersects the horizontal axis at 1931. The open circle at 1959 indicates the oldest known HIV-1 sample. (Reprinted with permission from Korber, B., M. Muldoon, J. Theiler, *et al.* 2000. Timing the ancestor of the HIV-1 pandemic strains. *Science* 288:1789–1796. © 2000 AAAS.)

distance from the root of the phylogenetic tree for these different strains plotted against their date of collection. A linear fit to these data intersects the horizontal axis at 1931, suggesting that these strains began diverging only approximately 50 years before the worldwide AIDS epidemic began in the early 1980s. In addition, the oldest HIV sequence from 1959 in the Democratic Republic of Congo (DRC) closely fits to the line predicted from the other data points. From this analysis, the estimated rate of substitution per year was 0.0024 for the *env* gene and 0.0019 for the *gag* (group-specific antigen) gene (Korber *et al.*, 2000).

If the clock operates randomly, then it should be described by Poisson processes in which the mean number of substitutions and the variance of the number of substitutions should be equal. However, shortly after the development of the neutral theory, Ohta and Kimura (1971) estimated that for three proteins the variance was on average 1.75 times the mean. Gillespie (1989) found that for some molecules the ratio of the variance to the mean may be very large and, with more conservative weighting, estimated the ratio over 20 proteins from mammals as 6.95 and 4.54 for nonsynonymous and synonymous substitutions, respectively. In contrast, Zeng *et al.* (1998) estimated the ratios as 1.64 (not significantly greater than unity) and 4.36 for nonsynonymous and synonymous substitutions, respectively, for 24 genes in three *Drosophila* species. Gillespie (1991) suggested that the overdispersion in the molecular clock he observed was the result of varying episodes of no change and periods of rapid substitution, whereas Takahata (1991), Ohta (1995), and Zeng *et al.* (1998) have suggested alternative explanations.

It is rather surprising that the molecular clock appears to keep time based on years instead of generations because the mutation rate in many eukaryotes appears most closely related to generations (Drake *et al.*, 1998). With this reasoning, unless there is a counterforce, molecular evolution should proceed more slowly per year for longer generation lineages. In the **relative-rate test,** Sarich and Wilson (1973) developed an approach to estimate the overall rate of substitution in different lineages that does not depend on knowing the times of divergence (see Tajima, 1993, for another test). For example, to determine the rate of substitutions in the lineages for species 1 and 2, we need to have a less related species 3 as an outgroup (Figure 8.6). With this tree, where node A indicates the point where species 1 and 2 diverged, the number of substitutions between any two species is assumed to be the sum of the number of substitutions along the branches of the tree connecting them or

$$\begin{aligned} d_{13} &= d_{A1} + d_{A3} \\ d_{23} &= d_{A2} + d_{A3} \\ d_{12} &= d_{A1} + d_{A2} \end{aligned} \tag{8.5a}$$

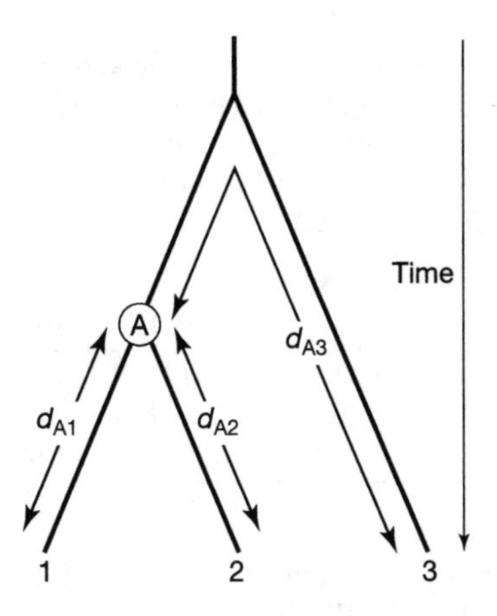

Figure 8.6. A phylogenetic tree to illustrate the relative-rate test that examines the rate of evolution in the lineage from node A to species 1 and that from node A to species 2. Species 3 is an outgroup and the d values indicate substitutions that have occured between two points.

These expressions can be solved to give estimates

$$d_{A1} = \frac{d_{12} + d_{13} - d_{23}}{2}$$

$$d_{A2} = \frac{d_{12} + d_{23} - d_{13}}{2}$$

$$d_{A3} = \frac{d_{13} + d_{23} - d_{12}}{2}$$

The time since species 1 and 2 have a common ancestor at node A is by definition the same, and thus, we can compare the rate of substitutions in these two lineages by comparing d_{A1} and d_{A2}. If the rates are the same, then d_{A1} and d_{A2} should be the same; that is, $d_{A1} - d_{A2} = 0$. If we subtract the expression for d_{23} from that for d_{13}, then

$$d_{A1} - d_{A2} = d_{13} - d_{23} \qquad (8.5b)$$

so we can compare the rates of substitutions to 1 and 2 by comparing the estimates between these species and species 3.

Controversy has existed over whether there has been a slowdown in molecular substitution in the hominid lineage (humans and apes) since their separation from the Old World monkeys. To examine this, Li (1997) and Yi *et al.* (2002) calculated the number of nucleotide substitutions for noncoding regions of Old World monkeys, humans, and the outgroup New World monkeys and carried out the relative rate test (Table 8.4). For all

TABLE 8.4 The estimated number of nucleotide substitutions per 100 sites, d_{ij}, between an Old World monkey (species 1), humans (2), and the outgroup New World monkey (3). The rates in the human lineage, d_{23}, are significantly lower at the 1% level than those for the Old World monkey lineage, d_{13}, for all three sets of data (Li, 1997; Yi *et al.*, 2002).

	Nucleotides compared	d_{12}	d_{13}	d_{23}	$d_{13}-d_{23}$
Pseudogene (Li)	8,871	6.7	11.8	10.7	1.1**
Introns (Yi *et al.*)	15,304	6.8	14.0	13.0	1.0**
Flanking regions (Li)	936	7.9	14.9	11.7	3.2**

three types of sequence, the estimated rate in the human lineage was significantly slower than in the Old World monkey lineage. Although the basis of these differences is unknown, they are consistent with a longer generation length in the human lineage than the Old World monkeys.

II. TESTS OF THE NEUTRAL THEORY AND EVIDENCE OF SELECTION

One of the appealing aspects of the neutral theory is that if it is used as a null hypothesis, then predictions about the magnitude and pattern of genetic variation are possible. Because of the controversy about the basis of allozyme variation, examination of allozyme variation and comparison to that predicted from the neutral theory was one of the first large tests of the neutral theory. In particular, Nei and his coworkers carried out extensive data analysis to determine whether the patterns of variation found for allozymes were consistent with the neutral theory (for a summary, see Nei, 1987). Overall, they found that "the general pattern of protein polymorphism can be explained by the neutral theory, particularly when the bottleneck effect and slightly deleterious selection were taken into account" (Nei, 1987). Of course, these comparisons of large numbers of loci generally did not explicitly examine particular loci or alleles that could be under balancing selection.

In recent years, the focus has been more on determining the presence and extent of selection on variation in DNA sequences. Analysis of sequence variation has allowed detection of very low levels of selection ushering in the "age of weak selection" (Wayne and Simonsen, 1998). In fact, the cumulative effect of many generations of selection seen in sequence data appears to allow documentation of selective effects that are on the order of the reciprocal of the effective population size (Akashi, 1999). Although there is still emphasis on testing for differences from neutrality, there is increasing interest in detecting evidence for selection and the loci involved. Below we introduce some of the more commonly used tests and approaches to determine whether molecular variation is consistent with neutral theory and potentially detecting purifying and positive selection. For reviews and perspectives on these tests and related material, see Kreitman (2000), Gerber *et al.* (2001), Nielsen (2001), Fay and Wu (2003), and Bamshad and Wooding (2003). Software to carry out Tajima's test, HKA test, MK test, and other statistical tests of neutrality are provided at DNAsp (http://www.ub.es/dnasp/).

a. Ewens–Watterson Test

One of the first, and quite elegant, tests of the neutral theory was developed from the infinite allele model. We earlier showed (expression 7.8*b*) that the neutral theory equilibrium homozygosity is

$$f_{eq} = \frac{1}{4N_e u + 1} \tag{8.6a}$$

Ewens (1972) showed that under neutrality equilibrium, the expected number of different alleles n in a sample of size $2N$ is

$$E(n) = \sum_{i=0}^{2N-1} \frac{\theta}{\theta + i} \tag{8.6b}$$

where $\theta = 4N_e u$.

If θ is very small, then the expected number of different alleles is only slightly larger than one, even in very large samples. On the other hand, if θ is large, as for high mutation rate microsatellite loci, then $E(n)$ can potentially be very large, and many of the alleles sampled will be different from the alleles sampled previously. For example, if $\theta = 10$ and $2N = 200$, then the expected number of alleles is 30.9. The expected number of alleles increases in a nearly linear way with increasing θ, with the slope of the increase slightly greater for larger sample sizes. We can see how, for a given θ value, $E(n)$ changes with increased sample size. For example, if $\theta = 2$ and $2N = 2$, 10, 100, and 1000, then $E(n) = 1.67$, 4.04, 8.39, and 12.97, respectively. Obviously, as $2N$ increases, then $E(n)$ increases but at a decreasing rate.

From these relationships, Watterson (1978) developed a test to determine whether the Hardy–Weinberg homozygosity in a sample of size $2N$ with n different alleles is consistent with the homozygosity when there is a mutation-genetic drift equilibrium under neutral theory. This test circumvents the need to estimate $4N_e u$ because the equilibrium value depends only on two known values, the sample size $2N$ and the number of different alleles n. To illustrate the basis of this test, Table 8.5 shows the hypothetical distribution of four alleles observed in a sample of $2N = 200$. In

TABLE 8.5 Allele frequencies and the observed Hardy–Weinberg homozygosity, f_e, from three different hypothetical samples with four alleles and a sample size of 200.

Sample	*Allele frequencies* A_1	A_2	A_3	A_4	f_e	*Comments*
1	0.35	0.30	0.20	0.15	0.275	Too even
2	0.76	0.17	0.06	0.01	0.610	Neutrality
3	0.965	0.025	0.005	0.005	0.932	Too uneven

sample 1, the frequencies of the alleles are more uniform than expected under neutral theory, in sample 2, they are close to neutrality, and in sample 3, the frequency of the most common allele is higher than expected from neutrality.

To carry out the test proposed by Watterson (1978), the expected Hardy–Weinberg homozygosity

$$f_e = \sum \hat{p}_i^2 \tag{8.6c}$$

for a sample is compared with the equilibrium homozygosity f_{eq} under neutral theory. For example, as in Table 8.5, the sample size is $2N$ =200, and the observed number of alleles n is 4; then the equilibrium homozygosity f_{eq} is 0.608. The equilibrium homozygosity values of f_{eq} is given in Appendix C of Ewens (1979) and a large table of virtually all combinations of $2N$ and n is given on the website of G. Thomson (http://allele5.biol.berkeley.edu under Software, the Watterson homozygosity test for neutrality) (see also Slatkin, 1994a, 1996). This test compares the expected homozygosity assuming Hardy–Weinberg proportions to the equilibrium homozygosity under the neutral theory (it is not comparing the observed homozygote frequencies to that expected under Hardy–Weinberg as we did on p. 63). As a result, for the data in Table 8.5, the expected homozygosity f_e is much lower than the equilibrium homozygosity f_{eq} for sample 1, much higher for sample 3, and close to the equilibrium value for sample 2.

To illustrate how the equilibrium homozygosity f_{eq} under neutral theory varies depending on sample size and number of alleles, Figure 8.7 gives it for $2N$ = 100, 200, and 400 and for n from 2 to 20. The equilibrium homozygosity only gradually increases as the sample size increases for a given number of alleles, whereas it greatly decreases as the number of alleles increases. As the number of alleles increases, the equilibrium heterozygosity, $H_{eq} = 1 - f_{eq}$, increases.

Keith *et al.* (1985) examined the frequencies of the *Xanthine dehydrogenase* alleles in 89 lines of *D. pseudoobscura* collected from a winery in California. They found 15 different alleles, with frequencies of 0.584, 0.101, 0.090, 0.045 (2 different alleles), 0.022 (2 different alleles), and 0.011

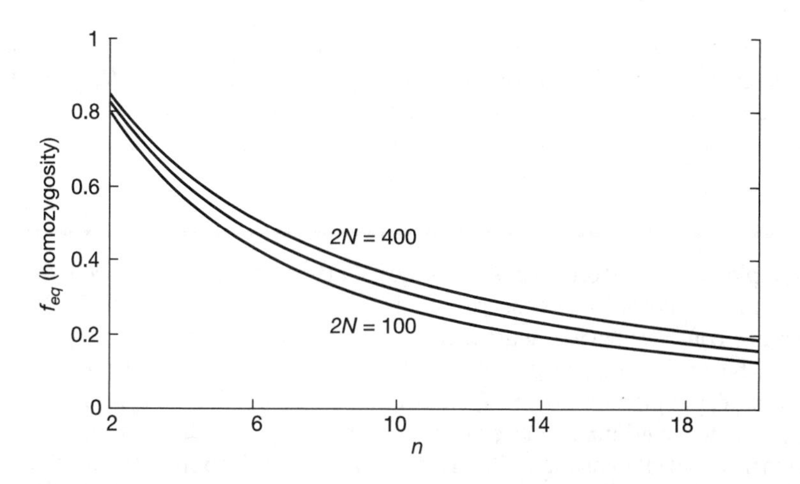

Figure 8.7. The equilibrium homozygosity, f_{eq}, under the neutral theory for n (2 to 20) alleles in samples of 100, 200, and 400.

(8 different alleles). In this case, f_e was 0.366, whereas the neutral theory equilibrium homozygosity f_{eq} is only 0.169. In this case, they rejected neutrality because of a higher expected homozygosity (lower heterozygosity) for the given number of alleles and sample size, similar to the deviation in sample 3 on Table 8.5. In other words, the most common allele was more common, and there were more single copies of alleles (singletons) than expected by the neutral theory. Such a distribution is consistent with purifying selection eliminating detrimental alleles that are continually being generated by mutation.

We should note that this test compares a population against neutral equilibrium expectations, not just that there is no differential selection against the variants (neutrality). For example, as we discussed on p. 346, a bottleneck generally reduces the number of alleles, particularly rare ones, faster than the amount of heterozygosity. As a result, testing of a recently bottlenecked population gives a pattern similar to a population with balancing selection, that is, given the number of alleles and the sample size, we see a more even distribution of allele frequencies than expected by the neutral theory.

In fact, the bottleneck test developed by Cornuet and Luikart (1996) detects the deviation in allele frequency distribution from neutrality caused by a bottleneck. On the other hand, a recent selective sweep can give a very uneven allele frequency distribution because of the replacement of a large proportion of the population variation by the positively selected mutant, similar to that expected for purifying selection. To distinguish between patterns of genetic variation consistent with demographic or neutral events and ones that are the result of natural selection, these processes can be modeled together. Although this is a complex venture with many assumptions, algorithms like those developed by Hudson (2002) enable one to simulate historical population demographic events, that is, bottlenecks or expansions, with estimates of θ and N_e in order to test a nucleotide sequence dataset for deviations from neutrality. Example 8.5 gives some data showing the application of the Ewens–Watterson test to *HLA* data in humans where there is a deviation towards more evenness than neutrality equilibrium, similar to the deviation in sample 1 of Table 8.5.

Example 8.5. The homozygosity test of Watterson (1978) has been used to determine whether the observed allele frequency distributions of different loci are consistent with neutrality predictions. The first example, showing that the allele frequency distribution was more even than expected and the expected homozygosity f_e was lower than equilibrium homozygosity f_{eg}, was for the human major histocompatibility loci, *HLA-A* and *HLA-B* (Hedrick and Thomson, 1983). In this case, over 22 populations, 25 of 44 data sets exhibited statistically significant deviations from neutrality in the direction of more even allele frequencies and they concluded that balancing selection was the most likely cause of this effect.

Markow *et al.* (1993) examined the distribution of these two loci in the Havasupai, a small tribe (< 600) that inhabits an isolated side canyon of the Grand Canyon in Arizona. In the sample of 122 individuals ($2N = 244$), for *HLA-A* and *HLA-B* three (see Parham *et al.*, 1997) and eight different alleles, respectively, were observed (Table 8.6). For both loci, the distribution of alleles was more even than expected from neutrality, and the average homozygosity f_e was 48% that expected under neutrality. The deviation from neutrality in this case may also be attributable to balancing selection, but because this is a small, isolated population, recent genetic drift that would cause the loss of rare alleles may have mimicked the effect of balancing selection. In an unpublished study of a large number of microsatellite

TABLE 8.6 Allele frequencies, the expected Hardy–Weinberg and equilibrium homozygosity, and the statistical significance level for two *HLA* loci in the Havasupai (Markow *et al.*, 1993; Parham *et al.*, 1997).

HLA-A		*HLA-B*	
Allele	*Frequency*	*Allele*	*Frequency*
A2	0.545	B5v	0.119
A24	0.184	B27	0.037
A31	0.270	B35	0.164
		B39	0.061
		B48	0.422
		B51	0.086
		B60	0.102
		B61	0.008
	$f_e = 0.404$		$f_e = 0.242$
	$f_{eq} = 0.714$		$f_{eq} = 0.611$
	$P = 0.04$		$P = 0.11$

loci in the same individuals, the H_e was only 3% larger than H_{eq}, suggesting that the excess of heterozygosity seen at the *HLA* loci was not due to a bottleneck or to genetic drift. Salamon *et al.* (1999) examined the frequency distribution of amino acids at specific sites for class II *HLA* loci in 26 different populations and found numerous examples of statistically significant deviations from neutrality.

---

b. Tajima Test

Tajima (1989) developed a test that is an analog to the Ewens–Watterson test but explicitly accounts for mutational events. In other words, Tajima's test takes into account nucleotide sequence information as both the frequencies of different sequences and the differences between them, whereas the Ewens–Watterson test only considers their frequencies and not the differences between the observed alleles. As we have discussed, one way in which nucleotide variation is commonly quantified is with the measure of nucleotide diversity π as estimated in expression 2.23*b*. Under the **infinite-sites model**, which assumes that the number of nucleotide sites is large enough that each new mutation occurs at a site that has not mutated before (Kimura, 1969), the expectation of π found under neutrality is

$$E(\pi) = \theta_\pi \tag{8.7a}$$

where $\theta_\pi = 4Nu$ (the subscript π indicates that this measure of θ is based on nucleotide differences).

Another measure of nucleotide variation under the infinite-sites model is the expected number of sites segregating S for different nucleotides

$$E(S) = a_1\theta_S$$

where

$$a_1 = \sum_{i=1}^{n-1} \frac{1}{i}$$

in a sample of n alleles (the subscript S indicates that this measure of θ is based on the number of segregating sites). This expression can be rearranged to give an estimate of θ_S as

$$\hat{\theta}_S = \frac{S}{a_1} \tag{8.7b}$$

Although these two estimates reflect different types of information, they are expected to be of equal magnitude under the neutral theory. The estimate related to the number of differences θ_π given by 8.7a is similar to measures of heterozygosity and is not influenced much by rare alleles, whereas the estimate from segregating sites θ_S counts all variable sites equally and may be strongly influenced by rare alleles.

Using these estimates, Tajima (1989) developed a test to compare the difference between these two estimates, $d = \hat{\theta}_\pi - \hat{\theta}_S$. Tajima (1989) used the expected total numbers rather than mean value per site, and we follow this notation here; that is, he used k, the average number of nucleotide sites that are different rather than the mean per site π. The statistic D that he introduced (generally now known as Tajima's D), or d divided by the standard deviation of d, is

$$D = \frac{d}{V(d)^{1/2}} = \frac{k - S/a_1}{\left[e_1 S + e_2 S(S-1)\right]^{1/2}} \tag{8.7c}$$

where

$$e_1 = \frac{1}{a_1}\left(\frac{n+1}{3(n-1)} - \frac{1}{a_1}\right) \quad \text{and} \quad e_2 = \frac{c}{a_1^2 + a_2}$$

where

$$a_2 = \sum_{i=1}^{n-1} \frac{1}{i^2} \quad \text{and} \quad c = \frac{2(n^2+n+3)}{9n(n-1)} - \frac{n+2}{a_1 n} + \frac{a_2}{a_1^2}$$

To illustrate the computation of D, Tajima (1989) used a 900-nucleotide sequence of human mtDNA from Aquadro and Greenberg (1983). In this sample of seven sequences ($n = 7$), the estimated average number of nucleotide differences k was 15.38, and the number of segregating sites S was 45. Therefore, $d = -2.99$, $a_1 = 2.45$, $a_2 = 1.49$, $e_1 = 0.0148$, $c = 0.0358$, and $e_2 = 0.00478$, making $D = -0.938$. From Tajima (1989), the 95% confidence interval for D when $n = 7$ is -1.498 to 1.728, making D for these data not statistically significant different from zero. As an illustration of the basic data used for these calculations, Table 8.7 gives the nine polymorphic sites in a 26.5-kb region of the Y chromosome in 24 individuals from Papua New Guinea (Hammer *et al.*, 2003). There are four singletons, sites 2–4. The estimates are $\theta_\pi = 0.000075$ and $\theta_S = 0.000091$ for these data; Tajima's $D = -0.592$, a nonsignificant value.

Garrigan and Hedrick (2003) calculated both the Ewens–Watterson and Tajima's tests for *HLA-A* data from 12 different human populations. Three of the populations were significantly different from neutral expectations for the Ewens–Watterson test, whereas 11 of 12 were significant

TABLE 8.7 The nine polymorphic sites on a 26.5-kb region of noncoding DNA on the Y chromosome from 24 individuals from Papua New Guinea (Hammer *et al.*, 2003).

		Polymorphic sites								
		1	2	3	4	5	6	7	8	9
	Consensus	T	A	C	T	G	T	G	C	T
Individual	1	-	-	-	-	-	-	-	-	-
	2	-	G	-	-	-	-	-	-	-
	3	-	-	A	-	-	-	-	-	-
	4	-	-	-	-	-	-	-	T	-
	5	-	-	-	-	-	-	-	T	-
	6	-	-	-	-	-	-	-	T	-
	7	-	-	-	-	-	-	-	T	-
	8	-	-	-	-	-	-	-	T	-
	9	-	-	-	-	-	-	-	T	-
	10	-	-	-	-	-	-	-	T	-
	11	-	-	-	-	-	-	-	T	-
	12	-	-	-	-	-	-	-	T	-
	13	-	-	-	-	-	-	-	T	-
	14	-	-	-	-	-	-	-	T	-
	15	-	-	-	-	-	-	-	T	-
	16	-	-	-	G	-	-	-	T	-
	17	-	-	-	-	A	-	-	T	-
	18	-	-	-	-	-	C	-	T	C
	19	-	-	-	-	-	C	A	T	C
	20	-	-	-	-	-	C	A	T	C
	21	-	-	-	-	-	C	A	T	C
	22	-	-	-	-	-	C	A	T	C
	23	C	-	-	-	-	C	A	T	C
	24	C	-	-	-	-	C	A	T	C

for the Tajima's test. The difference in proportion of significance occurs because the Ewens–Watterson test does not take into account that the *HLA-A* alleles are highly divergent, whereas Tajima's D does (the average number of pair-wise base differences is approximately 18).

If the population is at neutral equilibrium, then Tajima's D statistic should be zero because $d = 0$. However, if some of the variants are slightly deleterious, then this should not influence π very much, but it would increase S. As a result, the estimate of θ_π, would be smaller than that of θ_S, and the difference would be negative. Heterozygote advantage or a recent bottleneck, on the other hand, would reduce S; therefore, the estimate of θ_π would be larger than that of θ_S, and the difference would be positive. If the population is growing from an equilibrium situation, then S will grow faster than π, and the difference in the estimates should be negative. However, distinguishing between the contributions of selection and demographic events may be quite difficult, and the power of the test may be limited to detecting particularly strong or relatively recent bouts of positive selection

(Braverman *et al.* 1995; Simonsen *et al.*, 1995; Fu, 1997). This test was first applied to the *Drosophila* datasets in the 1990s that may have been more appropriate because their population demographic histories are more consistent with equilibrium assumptions (the exception being samples from some African populations).

However, if the assumption of population equilibrium is strongly violated, then low D values may result from selective sweeps, population growth, weak purifying selection, or even sampling design (Ptak and Przeworski, 2002; Hammer *et al.*, 2003), whereas high D values may result from balancing selection or from population bottlenecks. For example, the majority of nucleotide datasets from human populations shows significant or near significant negative D values. Because this is largely a genome-wide observation, this pattern is more consistent with the recent expansion of the human population in the last 10,000 years than with a model of positive selection for all of these genes (see review by Tishkoff and Verrelli, 2003).

Although they are somewhat similar, other recent statistical tests that examine the polymorphism frequency distributions have been designed that are slightly more advantagous than the statistic proposed by Tajima (1989). The statistical tests of neutrality of Fu and Li (1993) and Fu (1997), known as Fu and Li's D and Fu's F statistics, are based on the number of mutations in the external and internal branches of a phylogenetic tree (see p. 599), or recent versus ancestral mutations. Deviations from neutrality may be consistent with balancing selection, that is, increased number of internal branch mutants, or purifying selection, that is, increased number of external branch mutants or singletons. On the other hand, Fay and Wu (2003) introduced their H statistic that is apparently well suited for distinguishing patterns of variation due to recent population expansion from that due to directional selection, that is, a recent increase in frequency of a positively selected variant. Because directional selection is often acting on recent mutations and not older pre-exisiting ones, Fay and Wu's H statistic looks for derived alleles that are significantly greater in frequency than expected by the neutral theory.

c. Hudson–Kreitman–Aguade (HKA) Test

The neutral theory predicts both the rates of substitution over time and the amount of polymorphism within species. In fact, Kimura (1983) stated, "Eventually it will be found, if the neutral theory is valid, that molecules or parts of one molecule which are more important in function, and which therefore evolve more slowly, will show a lower level of heterozygosity." Hudson *et al.* (1987) used this positive association of within-population (or species) variation and divergence between species that is expected by the neutral theory as the basis of their **HKA test**. Both the amount of polymorphism within species and the evolutionary rate of change, or the amount of divergence between species, are increasing functions of the mutation rate, and thus, loci with a high mutation rate should have both a high heterozygosity within species and a high divergence between species.

To use the HKA test, which is a rather complicated goodness-of-fit test, there needs to be data for at least two different genetic regions in at least two different species so that both the within-species variation and the between-species divergence can be compared for the different regions. The expected values of the within- and between-species measures can be shown, using the infinite sites model (see p. 433), to be a function of the time t since the two species diverged, the relative effective population sizes of the two species, and the θ value for each gene (Hudson *et al.*, 1987; Kreitman and Hudson, 1991). Although these factors may not be purely identical across all genomic regions (there may be different mutation rates), we still expect that these factors will influence estimates of within- and between-species variation in a similar manner across different genes, and, therefore, under neutrality, the ratios of these two estimates should be similar (however, see Ingvarsson, 2004).

To illustrate the general form of the overall data, Table 8.8 gives the estimated variation within species and divergence between species for the third position codon sites at the *Adh* gene and the noncoding 5′ flanking region (Hudson *et al.*, 1987). In this case, the ratio of within-species to between-species variation is much higher for the *Adh* gene than the flanking region, and Hudson *et al.* (1987) were able to reject neutrality in this case. Notice that the level of divergence between species is very similar for the two genes but that the amount of variation within species is 4.59 higher for the *Adh* gene than for the flanking region. As a result, Hudson *et al.* (1987) concluded that the deviation from neutrality occurred because of the high within-species variation at the *Adh* gene.

One assumption of the HKA test is that one has identified a putative neutral gene region with which to compare the selected gene locus. In the original dataset (Table 8.8), the 5′ *Adh* region was chosen under the assumption that non-coding regions were commonly accepted as neutral

TABLE 8.8 Estimates of the amount variation within species and the amount of divergence between species for *D. melanogaster* and *D. sechellia* at the *Adh* gene and the 5′ flanking region (Hudson *et al.*, 1987).

	Adh	*Flanking region*	*Ratio (Adh/flanking)*
Within species	0.101	0.022	4.59
Between species	0.056	0.052	1.08
Ratio (within/between)	1.80	0.42	

regions. However, based on much data to date, we now know that promoter regions may actually be under strong purifying selection (Wray *et al.* 2003). Nachman and Crowell (2000) also found that two introns of the human Duchenne muscular dystrophy (DMD) gene had contrasting levels of polymorphism and divergence when subjected to the HKA test, thus, suggesting that introns could not simply be regarded as "neutral" for statistical tests. To examine the ratio of polymorphism to divergence when a gene cannot be divided *a priori* into discrete regions, McDonald (1996, 1998) developed statistical tests to evaluate sliding windows of the polymorphism/divergence ratio. Example 8.6 gives more data from the *Adh* region in *Drosophila* and data from the gene involved with branching pattern in maize and teosinte, both of which are rejected by the HKA test, although in the opposite direction.

Example 8.6. Let us examine the pattern of nucleotide variation over adjacent genetic regions in two cases in which the HKA test has rejected neutrality. First, the classic case of the *Adh* gene in *Drosophila* shows higher genetic variation around the site of amino acid polymorphism in *D. melanogaster* (Kreitman, 1983), whereas in the two adjacent regions, the 5′ flanking region and the 3′ *Adh-dup* region, the amount of variation is less. Figure 8.8 gives a sliding window of the observed and expected average pairwise difference between the *Adh* F and S alleles over the more than 4500 nucleotides compared over these regions (Kreitman and Hudson, 1991). To calculate the sliding-window values, the number of differences over 100 silent sites is plotted at each nucleotide position. The largest difference between the observed and expected variations is centered on the region around the amino acid replacement at position 1490. Pairwise HKA tests for the three regions demonstrated that there is significance at the 5% level between the 5′ flanking region and the *Adh* locus and significance at the 2% level between the *Adh* and *Adh-dup* loci. The statistical departure from neutrality results from the higher-than-expected within-species variation for the *Adh* locus and is consistent with balancing selection acting on the amino acid difference (Kreitman and Hudson, 1991; however, see Begun *et al.*, 1999).

Another example of the pattern of genetic variation that is rejected by the HKA test is for the *teosinte branched 1* (*tb 1*) locus in maize (Wang *et al.*, 1999). In this case, the variation was compared between maize and the ancestor of maize, teosinte, and the 5′ regulatory, nontranscribed region (NTR) and the coding region of *tb 1*. Figure 8.9 gives the amount of diversity for a 300-base-pair sliding window over the region for both species. The largest difference between the species is the 97% reduction in variation for maize in the NTR region. The HKA test showed that this reduction was highly significant (at the 0.4% level) for the comparison between the NTR

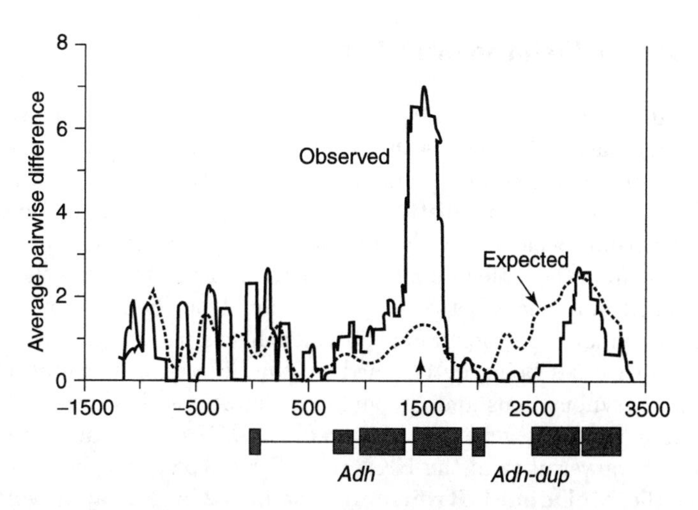

Figure 8.8. A sliding window of the observed and expected genetic variation over the 5′ flanking region, the *Adh* gene, and the *Adh-dup* gene in *Drosophila* (after Kreitman and Hudson, 1991). The expected values are based on a no-selection model calculated from between-species divergence. The arrow indicates the position of the amino acid polymorphism.

and the coding region. Domestication—selection for traits of interest to humans—should reduce the extent of genetic variation at genes influencing these traits. The NTR variant found in maize causes the side branches to shorten and the plant to carry ears instead of tassels. Consistent with the prediction of directional selection and reduced variation from domestication, the NTR region in maize is nearly monomorphic, whereas the NTR region in teosinte is highly variable, as are both species in the adjacent coding region of *tb 1*.

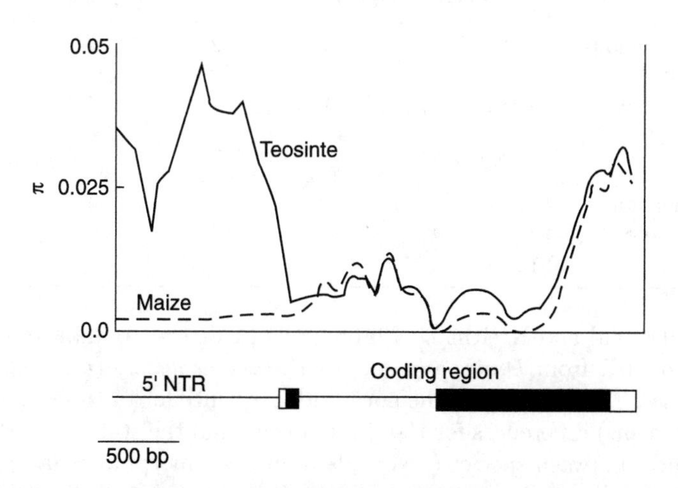

Figure 8.9. A sliding window of the polymorphism of the *teosinte branched 1* gene for maize and teosinte (after Wang *et al.*, 1999).

d. McDonald–Kreitman (MK) Test

McDonald and Kreitman (1991) proposed a simpler statistical test based on the same basic principles as the HKA test for synonymous and nonsynonymous sites in the coding region of a gene. If the observed variation is neutral, then the rate of substitution between species and the amount of variation within species are both a function of the mutation rate. Therefore, under neutrality, the ratio of nonsynonymous (replacement) to synonymous (silent) fixed differences between species should be the same as the ratio of nonsynonymous to synonymous polymorphisms within species, with the ratio in both cases being determined by the difference in mutation rates between nonsynonymous and synonymous mutations. Because patterns of variation at the same gene are compared, the MK test is considered more statistically powerful than the HKA test. Table 8.9a gives the form of the data for the **McDonald–Kreitman test** in a 2 × 2 array in which the cells contain the numbers of nonsynonymous fixed (N_F), synonymous fixed (S_F), nonsynonymous polymorphic (N_P), and synonymous polymorphic (S_P) substitutions. Under neutrality, the ratio N_F/S_F should be equal to the ratio N_P/S_P.

TABLE 8.9 The number of nonsynonymous (replacement) and synonymous substitutions for fixed differences between species and polymorphism within species (a) in general, (b) for *Adh* in three *Drosophila* species (McDonald and Kreitman, 1991), and (c) for *G6pd* in *D. melanogaster and D. simulans* (Eanes *et al.*, 1993). Below, data from humans and chimpanzees are given for (d) mtDNA gene *ND3* (Nachman *et al.*, 1996), (e) *G6pd* (Verrelli *et al.*, 2002), and (f) *HLA-B* (Garrigan and Hedrick, 2003).

	(a) *General*		(b) *Adh*		(c) *G6pd*	
	Fixed	*Polymorphic*	*Fixed*	*Polymorphic*	*Fixed*	*Polymorphic*
Nonsynonymous	N_F	N_P	7	2	21	2
Synonymous	S_F	S_P	17	42	26	36
Ratio	$N_F/S_F = N_P/S_P$		0.41	0.05	0.81	0.06
	(c) *ND3*		(d) *G6pd*		(e) *HLA-B*	
Nonsynonymous	4	8	0	5	0	76
Synonymous	31	10	44	23	0	49
Ratio	0.13	0.8	0.0	0.28	—	1.61

McDonald and Kreitman (1991) applied this test to data from coding region of *Adh* from *Drosophila* species *D. melanogaster, D. simulans*, and *D. yakuba*. Table 8.10 gives the nucleotide sequence for the nonsynonymous (replacement) differences for the three species and the status as far as fixed differences between species (seven sites) and polymorphic variation within species (two sites). For example, position 781 is considered a fixed difference

TABLE 8.10 Nucleotide sequence data for the nine nonsynonymous (replacement) differences for the *Adh* gene in three species of *Drosophila* (McDonald and Kreitman, 1991). In the right column, the status, fixed for differences between the species or polymorphic in one of the species, is indicated. A polymorphism at position 1490, indicated by an asterisk, results in the difference in the *F* and *S* allozyme alleles.

		Species			
Position	*Consensus*	*D. melanogaster*	*D. simulans*	*D. yakuba*	*Status*
781	C	T T T T T T T T T T T T	- - - - - -	- - - - - - - - - - - -	Fixed
808	A	- - - - - - - - - - - -	- - - - - -	C C C C C C C C C C C C	Fixed
859	C	- - - - - - - - - - - -	- - - - - -	G G G G G G G G G G G G	Fixed
1089	C	- - - - - - - - - - - -	A A A A A A	- - - - - - - - - - - -	Fixed
1101	C	- - - - - - - - - - - -	- - - - - -	A A A A A A A A A A A A	Fixed
1490*	A	- - - - - - C C C C C C	- - - - - -	- - - - - - - - - - - -	Polymorphic
1555	C	- - - - - - - - - - - T	- - - - - -	- - - - - - - - - - - -	Polymorphic
1573	C	- - - - - - - - - - - -	- - - - - -	C C C C C C C C C C C C	Fixed
1657	A	- - - - - - - - - - - -	- - - - - -	T T T T T T T T T T T T	Fixed

because all 12 *D. melanogaster* are T and all 18 *D. simulans* and *D. yakuba* samples are C. Position 1490. The one that results in the F and S allozyme variants in *D. melanogaster* is polymorphic in *D. melanogaster* for A and C but fixed for C in the other two species.

The summary of these data and that for the 59 synonymous sites is given in Table 8.9b. Here the ratio of nonsynonymous to synonymous fixed differences is $7/17 = 0.41$, whereas the ratio of nonsynonymous to synonymous polymorphisms is only $2/42 = 0.05$. This pattern is statistically significant from the null hypothesis using a goodness of fit test. McDonald and Kreitman (1991) suggested that the excess of nonsynonymous substitutions results from fixation of selectively advantageous mutations. Sawyer and Hartl (1992) estimated that a selective advantage of only 1.5×10^{-6} is sufficient to explain these observed data. Eanes *et al.* (1993) described a similar but more extreme situation (Table 8.9c) for the *Glucose-6-phosphate dehydrogenase* (*G6pd*) gene in *Drosophila*.

Similar comparisons can be made between human and chimpanzee data for different genes. For example, Nachman *et al.* (1996) found that for sequences of the NADH dehydrogenase subunit 3 (*ND3*) from a sample of 61 humans and 5 chimpanzees, the ratio of nonsynonymous to synonymous differences was lower between than within species (Table 8.9d). In this case the ratio between (N_F/S_F) was 0.13, whereas that within (N_P/S_P) was 0.8. A summary of the available data for other mtDNA genes was examined, and a similar excess of within-species nonsynonymous polymorphisms was observed. This pattern is consistent with the hypothesis that most mtDNA protein polymorphisms are slightly deleterious and are eliminated before they become fixed in different species. Nielsen and Weinrich (1999) and Piganeau and Eyre-Walker (2003) examined 25 and 18 mtDNA data sets, respectively, and most of them showed this same general pattern.

In Table 8.9e and 8.9f, data are given for two other genes, *G6pd* and *HLA-B*, known, from other information, to be under positive selection in humans. Although the pattern for *G6pd* superficially is similar to that for *ND3*—that is, $N_F/S_F < N_P/S_P$, suggesting strong historical purifying selection—polymorphic nonsynonymous variants are in high frequency in contrast to the prediction of low frequency under purifying selection (Verrelli *et al.*, 2002). In Table 8.9f for *HLA-B* (and three other *HLA* genes), Garrigan and Hedrick (2003) found an extreme pattern, reflective of very strong balancing selection. First, there were 1.61 times as many nonsynonymous polymorphic sites as polymorphic synonymous sites, a pattern not seen in other data. In addition, the polymorphic nonsynonymous variants are often in intermediate frequency, consistent with the hypotheses that there are maintained by balancing selection. Second, for these genes, there were no fixed differences between humans and chimpanzees, either for nonsynonymous of synonymous variants. This suggests that all variants are strongly influenced by close linkage to sites that persist at intermediate frequency in the population as a consequence of balancing selection.

Fay *et al.* (2001, 2002) suggested that the total effect of positive selection that results in a higher nonsynonymous to synonymous ratio of fixed differences than nonsynonymous to synonymous ratio for polymorphic sites (as for the *Adh* and *G6pd* in Table 8.9b and 8.9c) may be somewhat obscured because slightly deleterious nonsynonymous mutations may increase the N_P/S_P ratio but not the N_F/S_F ratio. This effect can be reduced by comparing common polymorphisms to fixed difference data because deleterious mutations are kept in low frequencies. Table 8.11 presents their data from *Drosophila* and humans, illustrating that the nonsynonymous to synonymous ratio for rare polymorphisms is higher than for common polymorphisms. From these data, because the nonsynonymous/synonymous ratio for fixed differences is even higher than the nonsynonymous/synonymous ratio for common polymorphisms (and other considerations), Fay *et al.* (2002) concluded that the rate of adaptive evolution is higher than expected by neutral theory (see also Smith and Eyre-Walker, 2002).

TABLE 8.11 The number of fixed differences and polymorphic sites that were considered either rare (< 0.125 or < 0.05) or common for (a) 35 autosomal genes in *Drosophila* (Fay *et al.*, 2002) and (b) SNP data from 181 genes in humans (Fay *et al.*, 2001).

	(a) *Drosophila*			(b) *Human*		
		Polymorphic			*Polymorphic*	
	Fixed	*< 0.125*	*Common*	*Fixed*	*< 0.05*	*Common*
Nonsynonymous	421	79	44	3660	178	70
Synonymous	521	126	118	4151	147	122
Ratio	0.81	0.63	0.37	0.88	1.2	0.57

e. Synonymous and Nonsynonymous Nucleotide Substitutions

As we have seen, the expected rate of substitution for neutral nucleotide sites is equal to the mutation rate. Differences at sites in noncoding regions, such as introns, those in the third position of coding regions, synonymous changes, have generally been thought to be neutral or under low selection. If there is **purifying selection** so that detrimental mutants are eliminated, then this should reduce the rate of substitution below that for neutral sites. On the other hand, if there is advantageous selection for new mutants, then the rate of substitution may be higher than for neutral variants, an observation called **adaptive** or **positive Darwinian selection**. Quantifying the relative rates of synonymous substitutions and nonsynonymous substitutions per site can give insight into the relative importance of neutrality versus these different types of selection.

We first present the approach of Nei and Gojobori (1986) to calculate the **rate of synonymous substitutions (d_S)** and **rate of nonsynonymous substitutions (d_N)**. Then below we introduce the approach of Li *et al.* (1985), who calculate the **number of synonymous substitutions per synonymous site (K_S)** and **number of nonsynonymous substitutions per nonsynonymous site** (known either as K_N or K_A). Both of these approaches are in wide use, making it important to understand the basis of the calculations involved. Nei and Kumar (2000) compared these and different approaches to estimate the relative values of synonymous and nonsynonymous substitution. To calculate these values for sequences and the related statistics, software packages such as MEGA and PAML are recommended. To reiterate what we said above, if there is neutrality, then $dN/dS = 1$, if there is purifying selection, then $dN/dS < 1$, and if there is positive selection, then $dN/dS > 1$ (note that Yang and Bielawski, 2000, assumed that $\omega = dN/dS$; see also p. 449).

To calculate the values for the Nei and Gojobori (1986) approach, let f_{ij} be the **proportion of synonymous changes of all changes at a site for codon j**, excluding nonsense mutations, where the index i indicates the position of the nucleotide within the codon. The **number of potentially synonymous changes for codon j (s_j)** and the **number of potentially nonsynonymous changes for codon j (n_j)** are then

$$s_j = \sum_{i=1}^{3} f_{ij} \qquad \text{and} \qquad n_j = 3 - s_j$$

For this calculation, it is useful to have a DNA version of the genetic code that inserts T instead of U (Table 8.12). For example, for the leucine codon TTA, $s_j = 0.333 + 0 + 0.333 = 0.667$ because all nucleotide changes at the second position result in nonsynonyomous changes, and only one of the three changes at both the first and the third positions results in a synonymous change. When a nucleotide change results in a stop codon,

TABLE 8.12 The number of potentially synonymous sites (s_j) in parentheses for the DNA genetic code. The amino acid for each codon is given by the one-letter symbol as in Table 1.1.

Second position							
T		*C*		*A*		*G*	
TTT (0.333)	F	TCT (1.0)	S	TAT (1.0)	Y	TGT (0.5)	C
TTC (0.333)	F	TCC (1.0)	S	TAC (1.0)	Y	TGC (0.5)	C
TTA (0.667)	L	TCA (1.0)	S	TAA (Stop)		TGA (Stop)	
TTG (0.667)	L	TCG (1.0)	S	TAG (Stop)		TGG (0.0)	W
CTT (1.0)	L	CCT (1.0)	P	CAT (0.333)	H	CGT (1.0)	R
CTC (1.0)	L	CCC (1.0)	P	CAC (0.333)	H	CGC (1.0)	R
CTA (1.333)	L	CCA (1.0)	P	CAA (0.333)	Q	CGA (1.5)	R
CTG (1.333)	L	CCG (1.0)	P	CAG (0.333)	Q	CGG (1.333)	R
ATT (0.667)	I	ACT (1.0)	T	AAT (0.333)	N	AGT (0.333)	S
ATC (0.667)	I	ACC (1.0)	T	AAC (0.333)	N	AGC (0.333)	S
ATA (0.667)	I	ACA (1.0)	T	AAA (0.333)	K	AGA (0.833)	R
ATG (0.0)	M	ACG (1.0)	T	AAG (0.333)	K	AGG (0.667)	R
GTT (1.0)	V	GCT (1.0)	A	GAT (0.333)	D	GGT (1.0)	G
GTC (1.0)	V	GCC (1.0)	A	GAC (0.333)	D	GAC (1.0)	G
GTA (1.0)	V	GCA (1.0)	A	GAA (0.333)	E	GGA (1.0)	G
GTG (1.0)	V	GCG (1.0)	A	GAG (0.333)	E	GGG (1.0)	G

this change is disregarded. Consider the arginine codon CGA, which is fourfold degenerate for the third position. For the first position, the change C $\rightarrow$ A is synonymous, C $\rightarrow$ G is nonsynonymous, and C $\rightarrow$ T results in a stop codon so that $f_{1j} = 0.5$. Therefore, for this codon, $s_j = 0.5 + 0 + 1.0 = 1.5$. Table 8.12 gives the number of potentially synonymous changes for the 61 different amino acid codons (the three stop codons are omitted).

The **total potential number of synonymous changes (S)** and the **total potential number of nonsynonymous changes (N)** can be calculated as

$$S = \sum_{j=1}^{C} s_j \qquad \text{and} \qquad N = 3C - S \tag{8.8a}$$

where s_j is the number of potentially synonymous sites for the jth codon and C is the number of codons. Generally, two sequences are compared, and the average values of S and N for the two sequences are used.

Now we need to calculate the **observed numbers of synonymous and nonsynonymous nucleotide substitutions** between two sequences. The sequences are compared codon by codon, and there are three types of comparisons: differences at one, two, or three nucleotide sites. Obviously, for most genes, a large proportion of codons that differ will differ at only one site. For some genes, however, such as variable regions in MHC or comparisons of sequences between distantly related organisms, some codons may differ at two or perhaps three nucleotide sites. Although to calculate all the possibility changes, all elements of a 64 × 64 matrix are necessary, we show how differences in one and two nucleotides can be calculated.

First, let us examine the situation in which there is only one nucleotide difference and determine whether the difference is synonymous or nonsynonymous. For example, if the codon sequences are TTT (Phe) and TTC (Phe), there is one synonymous difference. Assuming that s_{dj} and n_{dj} are the observed number of synonymous and nonsynonymous differences for codon j, then $s_{dj} = 1$ and $n_{dj} = 0$.

When there are two nucleotide differences between the codons in the different sequences, then there are two pathways that can result in this difference. For example, if the codons are CGT (Arg) and AGA (Arg), the two pathways are

$$\text{CGT(Arg)} \leftrightarrow \text{AGT(Ser)} \leftrightarrow \text{AGA(Arg)}$$

and

$$\text{CGT(Arg)} \leftrightarrow \text{CGA(Arg)} \leftrightarrow \text{AGA(Arg)}$$

The first pathway has two nonsynonymous substitutions, and the second has two synonymous substitutions. If we assume that the two pathways are equally likely, then the average number of observed synonymous differences, s_{dj}, is 1.0, and the average number of nonsynonymous differences, n_{dj}, is also 1.0. If a pathway includes a stop codon, it is not considered. When there are three differences between codons, then there are six different pathways (see Nei and Kumar, 2000). The total number of observed synonymous and nonsynonymous differences are the sum over all the codons or

$$S_d = \sum_{j=1}^{C} s_{dj} \quad \text{and} \quad N_d = \sum_{j=1}^{C} n_{dj} \tag{8.8b}$$

The proportion of synonymous and nonsynonymous differences can then be estimated by

$$p_S = \frac{S_d}{S} \quad \text{and} \quad p_N = \frac{N_d}{N} \tag{8.8c}$$

To estimate the numbers of synonymous (d_S) and nonsynonymous (d_N) substitutions per site, expression 8.4a using the Jukes–Cantor method can be used by inserting p_S and p_N for d, respectively, as

$$d_S = -\frac{3}{4}\ln(1 - \frac{4p_S}{3}) \quad \text{and} \quad d_N = -\frac{3}{4}\ln(1 - \frac{4p_N}{3}) \tag{8.8d}$$

Although this approximation may not hold in all instances, Ota and Nei (1994b) have shown that it provides good estimates in most situations.

Examination of sequences for most genes gives estimates of synonymous and nonsynonymous rates consistent with purifying selection; that is, $d_N < d_S$. For example, Li (1997) found that for 47 genes in humans and rodents, the ratio of nonsynonymous to synonymous substitutions was 0.21, and for 32 genes in *D. melanogaster* and *D. pseudoobscura*, the ratio was 0.12. On the other hand, many MHC genes generally have a higher rate of nonsynonymous than synonymous, particularly for antigen-binding codons. Example 8.7 shows how to calculate substitution rates using the above approach and provides a d_N/d_S ratio estimate greater than unity for MHC sequences from the endangered Przewalski's horse.

Example 8.7. Przewalski's horse, *Equus przewalskii*, is the only extant species of horse other than the domestic horse (Boyd and Houpt, 1994). It was once widespread in Eurasia but now is assumed extinct in nature, the last confirmed observation of a wild animal having occurred in the 1960s. The captive population, which now numbers over 1200 in facilities around the world, is descended from 13 animals. In recent years, Przewalski's horses have been released to reserves in Mongolia and China in an effort to reestablish wild populations.

Hedrick *et al.* (1999b) examined variation for exon 2 of a DRB MHC gene in 14 Przewalski's horses that were broadly representative of the 13 founders. They concluded, on the basis of several lines of evidence, that there were two DRB genes, one with four alleles and one with two alleles. This low number of alleles is consistent with the small number of founders and is less than the number in domestic horses (Hedrick *et al.* 1999b) and cows (Mikko and Andersson, 1995).

As an illustration of how rates of substitution can be calculated, Table 8.13 gives the sequence for 10 codons, out of 78 total codons examined, for two alleles at the first locus, *Eqpr* DRB*1 and *Eqpr* DRB*5. The three codon differences between these alleles (total of five nucleotide differences), all of them nonsynonymous, are at codon positions 52, 56,

TABLE 8.13 The nucleotide sequence for the 10 codons in which there are differences in exon 2 (underlined in both sequences) between the two Przewalski's horse *MHC* alleles, *Eqpr* DRB*1 and *Eqpr* DRB*5 (Hedrick *et al.*, 1999b) and the values used in estimating the synonymous and nonsynonymous substitution rates.

	Codon									
	48	49	50	51	52	53	54	55	56	57
Eqpr DRB*1	CGG	GCG	GTG	ACC	AAG	CTG	GGG	CGG	ACG	GAC
Eqpr DRB*5	CGG	GCG	GTG	ACC	GAG	CTG	GGG	CGG	CGG	AGC
s_j	1.33	1.0	1.0	1.0	0.33	1.33	1.0	1.33	1.0	0.33
s_d	0	0	0	0	0	0	0	0	0.5	0
n_d	0	0	0	0	1	0	0	0	1.5	2

and 57 and are all in this 10-codon sequence. From expression 8.9*a*, and the values for s_j calculated in Table 8.13, $S = 9.67$ and $N = 20.33$. From expression 8.9*b* and the values in Table 8.13, $S_d = 0.5$ and $N_d = 4.5$. Then, from expression 8.9*c*, $p_S = 0.052$ and $p_N = 0.221$, and then from expression 8.9*d*, $d_S = 0.054$ and $d_N = 0.262$, so the ratio $d_N/d_S = 4.85$. Hughes and Nei (1988) first found a d_N/d_S ratio significantly greater than unity in their examination of codons from antigen-binding sites in the human MHC, and since then similarly high ratios have commonly been found for MHC genes in other organisms.

Li *et al.* (1985) used another approach to estimate the number of synonymous and nonsynonymous substitutions based on the classification of sites as **nondegenerate** (all possible changes are nonsynonymous), **twofold degenerate** (one of the three possible changes is synonymous), and **fourfold degenerate** (all possible changes are synonymous). For example, for phenylalanine codon TTT, the first two positions are nondegenerate (any change result in a different amino acid), and the third position is twofold degenerate (change of T to C still results in phenylalanine) so that this codon can be indicated as 002 for the three sites in the codon. Or for leucine codon CTA, the first position is twofold degenerate, the second in nondegenerate, and the third is fourfold degenerate so that it can be indicated as 204. For simplicity, the threefold degenerate sites in the third positions of the three isoleucine codons are considered as twofold degenerate sites.

With this categorization of the sites on each of the two sequences to be compared, we can count the total number of nondegenerate (L_0), twofold degenerate (L_2), and fourfold degenerate (L_4) sites for each sequence and compute their averages. All substitutions at fourfold sites are synonymous, and all substitutions at nondegenerate sites are nonsynonymous. Li *et al.* (1985) suggested that about one-third of the twofold degenerate sites are potentially synonymous sites and that two-thirds are potentially nonsynonymous sites. Therefore, the **total numbers of potentially synonymous** and **nonsynonymous sites** are

$$S = \frac{L_2}{3} + L_4 \qquad \text{and} \qquad N = L_0 + \frac{2L_2}{3} \tag{8.9a}$$

Now we compare the two sequences, codon by codon, to determine how many observed differences are synonymous (S_d) or nonsynonymous (N_d). Li *et al.* (1985) actually estimated the number of observed differences by using all possible evolutionary pathways for codons that differ by more than one nucleotide, as we showed above for the Nei–Gojobori method. They also estimated the number of transitional and transversional substitutions per site for each to the three classes of nucleotide sites and provided estimates of observed changes based on these relative values. However, to

TABLE 8.14 Two examples of 10-codon sequences for a gene where the nucleotides that are different are underlined and indicated at the bottom as synonymous (S) or nonsynonymous (N) differences. Below each sequence, each site is indicated as nondegenerate (0), twofold (2), or fourfold (4) degenerate.

	Codon									
Sequence	1	2	3	4	5	6	7	8	9	10
1	GGG	TCT	CGG	TAT	TTA	GGT	TTC	ACC	GTT	AAA
	004	004	004	002	202	004	202	004	004	002
2	GGG	TCC	CGG	TAC	TTA	GGT	TCC	ACC	GTC	AAA
	004	004	004	002	202	004	004	004	004	002
Difference		S		S			N		S	

illustrate this approach here, we can compute the observed number of synonymous substitutions per synonymous site (K_S) and the observed number of nonsynonymous substitutions per nonsynonymous site ($K_N = K_A$) as

$$K_S = \frac{S_d}{S} \quad \text{and} \quad K_N = K_A = \frac{N_d}{N} \tag{8.9b}$$

To illustrate the calculation of these values, Table 8.14 gives an example of two sequences of a gene 10 codons in length. For the first and second sequences, L_0, L_2, and L_4 are 18, 6, and 6, and 19, 4, and 7, respectively, so that the averages of their L_0, L_2, and L_4 values are 18.5, 5, and 6.5. Therefore, using expression 8.8*a*, $S = 5/3 + 6.5 = 8.17$ and $N = 18.5 + 3.33 = 21.83$. Because there were three synonymous and one nonsynonymous differences between the sequences, then using expression 8.8*b*, $K_S = 3/8.17 = 0.367$ and $K_N = K_A = 1/21.83 = 0.046$. In this case, K_S is much larger than K_N, and the ratio $K_N/K_S = 0.125$, consistent with purifying selection.

Another approach that has been taken to analyze coding region sequence data is the maximum likelihood method based on explicit codon substitution models (Goldman and Yang, 1994; Yang and Bielawski, 2000; PAML, http://abacus.gene.ucl.ac.uk/software/paml.html). In this approach, the codon is the unit of evolution and the rate of substitution from codon i to codon j is

$$q_{ij} = \begin{cases} 0 & \text{if codons } i \text{ and } j \text{ differ at more than one position} \\ \pi_j & \text{for synonymous transversions} \\ \kappa\pi_j & \text{for synonymous transitions} \\ \omega\pi_j & \text{for nonsynonymous transversions} \\ \omega\kappa\pi_j & \text{for nonsynonymous transitions} \end{cases}$$

where π_j is the equilibrium frequency of codon j, κ is the ratio of the rate of transition substitution to the rate of tranversion substitution, and

$\omega = dN/dS$. The matrix of transitions for all 61 sense codons is used to determine the probability that codon i becomes j after time t. The parameters in the model are estimated from the data by maximum likelihood and are used to calculate dN and dS. This approach has been used to identify sites under positive selection in reproductive proteins (Swanson *et al.*, 2000), plant chitinases (Bishop *et al.*, 2000), and other genes including β-globins (Yang *et al.*, 2000).

An amino acid substitution may be either **conservative**, that is, little difference in the physicochemical properties between the two amino acids, or **radical**, in which there is a large differences in the physicochemical properties. Zhang (2000) used categorizations of the 20 amino acids by charge, by polarity, or by polarity and volume and defined conservative substitutions as changes between amino acids within groups and radical substitutions and changes between groups. Smith (2003) suggested that the ratio of radical to conservative substitutions has properties somewhat similar to dN/dS; that is, the numerator of both ratios is the mutation class under stronger purifying or positive selection. For example, under purifying selection, both the substitution rate for radical and nonsynonymous substitutions would be reduced compared with conservative and synonymous substitutions. Examining variation in *Drosophila*, Smith (2003) found support for purifying selection against radical substitutions, but the proportions of radical and conservative substitutions in adaptive change were similar.

f. Codon Usage

King and Jukes (1969) suggested that synonymous mutations would be expected to be neutral with respect to fitness because there is no resultant change in amino acid. Assuming random mutation and neutrality of synonymous substitutions, then all of the codons for the same amino acid should be in equal proportions.

However, there is extensive evidence that there is unequal usage of codons that code for the same amino acid, in contrast with the predictions of neutrality theory. For example, Figure 8.10a gives the frequency of codon usage for the different leucine codons for highly expressed genes in *E. coli* and in yeast, *Sacharomyces cerevisiae* (two pairs of codons in each species are recognized by a single transfer RNA species). Obviously, one codon is highly preferred for each species, CUG for *E. coli* and UUG for yeast. The bias in codon usage can be measured as the observed frequency of a codon relative to that expected under the assumption of equal codon usage (Sharp *et al.*, 1986) or the Relative Synonymous Codon Usage

$$RSCU = \frac{X_i}{\overline{X}} \tag{8.10}$$

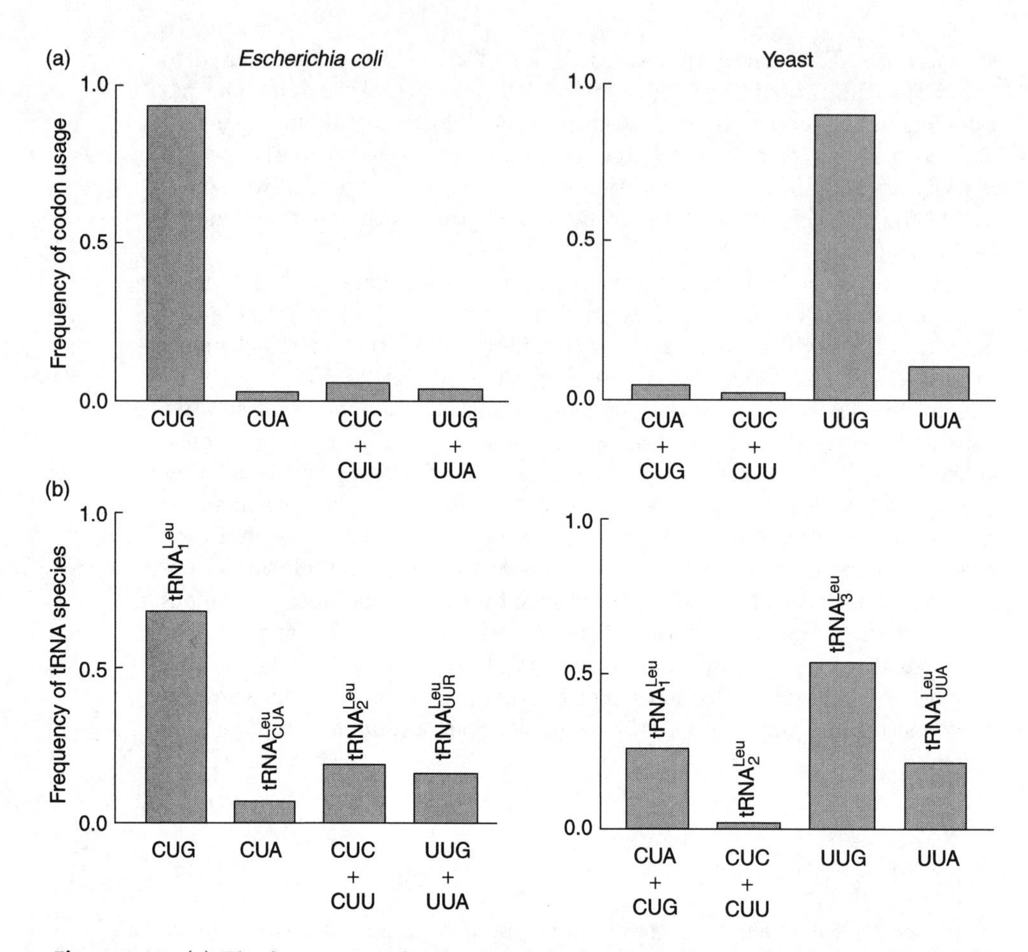

Figure 8.10. (a) The frequencies of codon usage for leucine codons in highly expressed genes and (b) the frequencies of the different leucine tRNA species in *E. coli* (left) and yeast (right) (Li and Graur, 1991).

where X_i is the observed number of the ith type codon for the amino acid and $\overline{X}$ is the average X_i over all codons for the amino acid. When all codons for an amino acid are equally used, then $RSCU = 1$ for all codons. To illustrate the calculation of $RSCU$, in *E. coli* for the RNA polymerase genes *rpo B* and *rpo D*, the number of times that leucine codons CUG, CUA, CUC, CUU, UUG, and UUA are used is 141, 1, 18, 11, 8, and 2, respectively (Ikemura, 1985). Therefore, $\overline{X} = 181/6 = 30.17$, and for codons CUG and CUA, $RSCU = 141/30.17 = 4.67$ and $1/30.17 = 0.03$, respectively, over two orders of magnitude different between the most used and least used codons. For genes that are not expressed as highly, there is generally much more equal use of the various codons and the $RSCU$ values are closer to unity (Sharp *et al.*, 1988), as would be expected if the level of selection if lower and genetic drift has a greater effect.

Ikemura (1982) showed that codon usage is positively correlated with the relative abundance of the appropriate tRNA species. As an illustration, Figure 8.10b gives the relative abundance of the four tRNA species for leucine codons. For example, the tRNA species for CUG in *E. coli* and the tRNA species for UUG are the most abundant, and these are the most common leucine codons used for these two species, as shown in Figure 8.10a.

Another measure of the extent of codon bias for a total sequence is the effective number of codons (Wright, 1990), which ranges from 20 if only one codon is used for each amino acid to 61 if there is equal usage among codons for a particular amino acid. Figure 8.11 gives the distribution of this measure in a large number of genes in *D. melanogaster* and *E. coli* (from Powell and Moriyama, 1997). In both organisms, there was substantial codon bias, with mean values of the measure or 46.2 for *D. melanogaster* and 45.0 for *E. coli*, and a wide distribution over genes.

What is the evolutionary genetic explanation for the observed codon bias? In theory, selection either for or against particular codons, differential mutation rates for particular codons, and genetic drift appear to be the main potential factors that could influence codon usage. For example, Akashi (1999) suggested that $N_e s$ values of approximately 1 to 2, where s is the selection coefficient against an unpreferred codon, are responsible for the codon bias patterns in *Drosophila*. Given that estimates of *Drosophila* effective population size are of the order 10^6, then s would be about 10^{-6}. By directly introducing unpreferred codons into the *D. melanogaster Adh* gene, Carlini and Stephan (2003) confirmed that the level of expression was reduced and estimated that $s > 10^{-5}$. For the *gnd* gene in *E. coli*, Hartl *et al.* (1994) estimated that the observed biased codon usage could result from $s = 7 \times 10^{-9}$. However, McVean and Charlesworth (1999) showed that

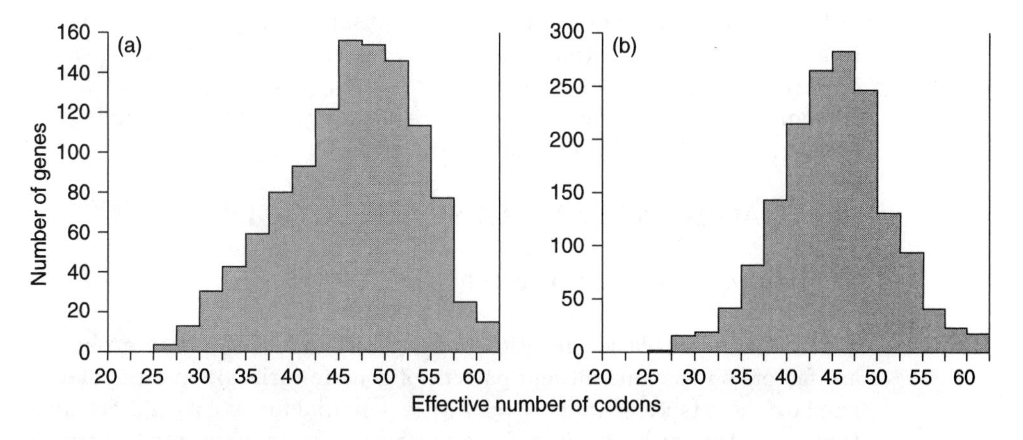

Figure 8.11. The distribution of codon usage bias as measured by the effective number of codons over (a) 1133 genes in *D. melanogaster* and (b) 1586 genes in *E. coli* (Powell and Moriyama, 1997).

if $N_e s \geq 1$, it results in fixation of the preferred codon, even more extreme codon bias than observed in the surveys of codon usage.

Recently, Maside *et al.* (2004) have estimated that $N_e s = 0.65$ in three other species of *Drosophila*, consistent with the bias in codon usage observed. However, when selection is on the order of 10^{-5} to 10^{-6}, then selection at other linked genes may overwhelm these small selective effects. Betancourt and Presgraves (2002) showed that in *Drosophila* there was a negative correlation with usage of the preferred codon and the rate of nonsynonymous substitutions (d_N). In other words, positive selection as indicated by d_N interferes with selection favoring optimal codon usage, a general phenomenon known as the Hill–Robertson effect (Hill and Robertson, 1966) in which selection at one locus interferes with selection at other loci (see p. 580). Kim (2004) has shown theoretically that this genetic hitchhiking effect (see p. 571) may be important in overwhelming selection for codon usage because the codon usage selective values are thought to be so small.

In humans, there is codon bias for some codons in some genes and not for others. For example, $RSCU = 3.88$ and 0.15 for leucine codons CUG and CUA, respectively, in genomic regions in which there is high GC content (Sharp *et al.*, 1988). On the other hand, $RSCU = 1.38$ and 0.57 for the same leucine codons, not very different from equal codon usage ($RSCU = 1$), when the codons are in regions of high AT content. Consistent with importance of the level of GC content and other factors influencing the mutation rate, Chen *et al.* (2004) provided support that mutation is the primary determinant of genome-wide codon bias (see the hypothesis of Grantham *et al.*, 1980) and that selection is of secondary significance. In addition, the effective population size in humans is thought to have been approximately 10,000 (see p. 463), suggesting that either selection or mutation (or both) would have to be higher to counter the greater effect of genetic drift in human populations. The factors causing nonneutrality patterns of codon usage are the continuing subject of research, and because the values of purported selection and mutation are small, the effects of genetic drift low, and genome specific factors may be very significant, determining the precise roles of these factors may be difficult.

III. COALESCENCE AND GENE GENEALOGIES

a. An Introduction to Coalescence

Traditionally, population genetics examines the impact of various evolutionary factors on the amount and pattern of genetic variation in a population and how these factors influence the future potential for evolutionary change. Generally, the evolution is a forward process, examining and predicting the future characteristics of a population. However, rapid accumulation of DNA sequence data over the past two decades has changed the orientation of much of population genetics from a prospective one investigating the factors involved in observed evolutionary change to a retrospective one inferring evolutionary events that have occurred in the past (Fu and Li, 1999). That is, understanding the evolutionary causes that have influenced the DNA sequence variation in a sample of individuals, such as the demographic and mutational history of the ancestors of the sample, has become the focus of much population genetics research.

When determining DNA variation in a population, a sample of alleles is examined. Each one of these alleles may have a different history, ranging from descending from the same ancestral allele, that is, identical by descent, in the previous generation to descending from the same ancestral allele many generations before. The point at which this common ancestry for two alleles occurs is called **coalescence**. If one goes back far enough in time in the population, then all alleles in the sample will coalesce into a single common ancestral allele. The coalescent approach, generally credited to Kingman (1982), is the most dynamic area of theoretical population genetics because it is widely used to analyze DNA sequence data in a population. Much of the research in this area is quite mathematical; reviews and introductions have been written by Hudson (1990), Fu and Li (1999), and Nordborg (2001), and the edited volume by Slatkin and Veuille (2002) contains a number of articles using coalescent approaches. Here we introduce some of the basic theory and graphically illustrate the concepts central to the coalescent theory.

Basic coalescent theory depends on the **Wright–Fisher model** (Fisher, 1930; Wright, 1931), that is, the simple representation of a population that Sewall Wright and Ronald Fisher used in developing the principles of population genetics. We have used this idealized depiction of a population throughout the book. More specifically, the Wright–Fisher model assumes nonoverlapping generations of individuals, random mating among individuals, a constant population size of N diploid individuals, and random reproduction over individuals resulting in a Poisson distribution of progeny (Hey and Machado, 2003). Much of our discussion in previous chapters has focused on situations in which some of the assumptions of the Wright–Fisher model are violated, such as the presence of non-random mating or selection, and allowing for these and other factors has been the focus of extensions of the basic coalescent theory in recent years.

To illustrate the coalescent process, Figure 8.12 gives a hypothetical example of the ancestry of five alleles sampled in the present generation, generation 5. Generally in the treatment of population genetics, we predict genetic changes in future generations. If we go down in Figure 8.12, forward in time, we can see the effect of genetic drift, some alleles being lost (such as the middle allele in the first generation because it has no descendants) and some alleles increasing in frequency, such as the right-hand allele (it has two descendants in the second generation). After five generations, only the right-hand allele remains; the other four original alleles have been lost.

Rather than following all of the individuals in the population each generation, as in traditional population genetics, the theory of coalescence

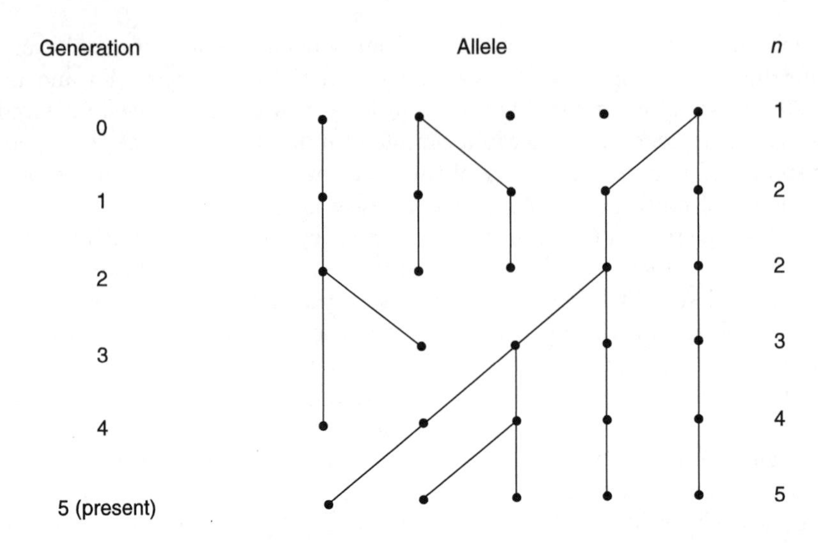

Figure 8.12. A hypothetical example illustrating the ancestry of five alleles sampled in the present generation. If we go forward in time, top to bottom, then we see the effects of genetic drift resulting in the fixation of a single allele in generation 5. If we go back in time, bottom to top, then we see the coalescence of the five sampled alleles in the right-hand ancestral allele in generation 0. Here n is the number of ancestral alleles from which the five alleles sampled in the present generation are descended.

allows us to only examine the alleles ancestral to those sampled in the present generation. If we go up Figure 8.12, back in time, we see that the five alleles sampled in the present generation 5 are descended from four alleles in generation 4. In other words, there is a coalescence because two alleles in generation 5 are descended from the same ancestral allele in generation 4. If we continue on up the figure, the number of ancestral alleles either declines because of additional coalescence events, or stays the same in each generation, until only one ancestral allele remains in generation 0.

In this hypothetical example, **only genetic drift** is assumed to affect the alleles; we have not included any mutation, and thus, all of the descendant alleles in the sample are actually assumed to be identical. However, if there is mutation, the observed alleles that have a common ancestor may have different sequences (we discuss the incorporation of mutation below).

In the incorporation of DNA sequence data into **gene trees** or **genealogies**, which shows the evolutionary relationship between the sequences, the **most recent common ancestor (MRCA)** of the different sequences is determined. The sequences are said to coalesce in this sequence. Before we introduce the theory underlying the coalescent process, let us examine a hypothetical gene tree illustrating it graphically. Figure 8.13 gives an example of a genealogy in which there are five sampled alleles in the present generation (bottom of tree) (after Hudson, 1990). As we go back in time, the number of ancestral alleles represented in the present

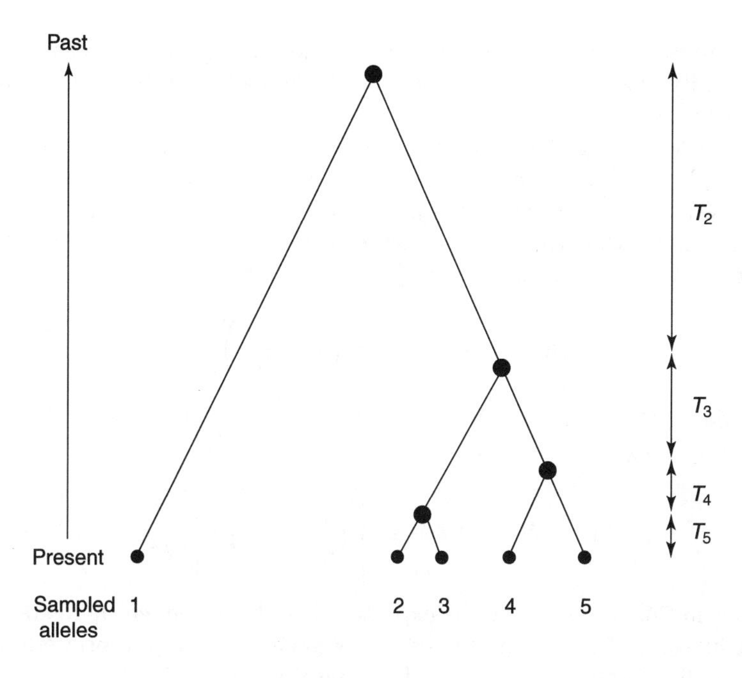

Figure 8.13. An example of a gene tree for five sampled alleles. The four large circles indicate coalescent events (after Hudson, 1990). T_i is the expected time in which there are i alleles, and these intervals are shown proportional to their expected time as given in expression 8.11c.

sample is reduced. Again, we are not yet considering mutation, and thus, ancestral alleles indicates identity by descent, as we discussed in Chapter 5. The first coalescence in this example is for alleles 2 and 3. Eventually, after four coalescent events, all of the alleles in the present generation are shown to have descended from one ancestral allele or MRCA. At first, this seems to contradict what we know from pedigree analysis where the number of ancestors for a given individual increases as we go back in time, and in general, for t generations, there are 2^t ancestors. However, in coalescence theory, we are tracing only the specific alleles that are transmitted between generations, and only one of the two alleles in a parent each generation is transmitted to a given offspring.

Now let us introduce the underlying theory used in the coalescence approach (Tajima, 1983; Hudson, 1990). The probability that two alleles descend from the same ancestral allele in the previous generation is $1/(2N)$; therefore, the probability that they came from different ancestral alleles in the previous generation is $1 - 1/(2N)$. The probability that three alleles descend from three ancestral alleles in the previous generation is the probability that the first two have different ancestry as above, $1 - 1/(2N)$, times the probability that the third allele has a different ancestry from the first two, $(2N - 2)/(2N) = 1 - 2/(2N)$. In general, the probability that n sampled alleles have n distinct ancestral alleles the previous generation is

$$\Pr(n) = \prod_{i=1}^{n-1} \left(1 - \frac{i}{2N}\right) \tag{8.11a}$$

This probability holds for each pair of subsequent generations so that the probability that n sampled alleles had n distinct ancestors t generations before is $[\Pr(n)]^t$.

Now let us consider the situation in which two alleles have the same ancestral allele—that is, they coalesce—$t + 1$ generations before the present. Because the probability that two alleles coalesce in any generation is $1/(2N)$ and assuming that they did not coalesce for the preceding t generations, the probability of coalescence is

$$\Pr(2)^t[1 - \Pr(2)] = \left(1 - \frac{1}{2N}\right)^t \frac{1}{2N}$$

Expanding this to n sampled alleles having $n - 1$ ancestral alleles $t + 1$ generations ago, we find that

$$\Pr(n)^t[1 - \Pr(n)] = \left[\prod_{i=1}^{n-1} (1 - \frac{i}{2N})\right]^t \frac{1}{2N} \tag{8.11b}$$

From this distribution, the expected time during which there are n distinct lineages, if time is measured in $2N$ generations (appropriate because genetic drift is the only factor influencing the alleles), is

$$E(T_n) = \frac{4N}{n(n-1)} \tag{8.11c}$$

Looking at Figure 8.13, we see that the expected time during which there are five, four, three, and two lineages are $2N/10$, $2N/6$, $2N/3$, and $2N$, respectively. In other words, the expected time increases as the number of lineages decreases. The expected time to coalescence of all of the n sampled alleles is

$$\begin{aligned} E(t) &= \sum_{i=2}^{n} E(T_i) \\ &= 4N\left(1 - \frac{1}{n}\right) \end{aligned} \tag{8.11d}$$

For example, the expected time to coalescence for five alleles is $3.2N$ generations.

The tree given in Figure 8.13 illustrates the expected times of coalescent events. What do randomly drawn coalescent trees look like? To generate trees, we can use random uniform numbers x drawn from the interval 0 to 1 and the continuous time approximation for coalescent time as

$$T_i = \frac{-2\ln(1-x)}{i(i-1)}$$

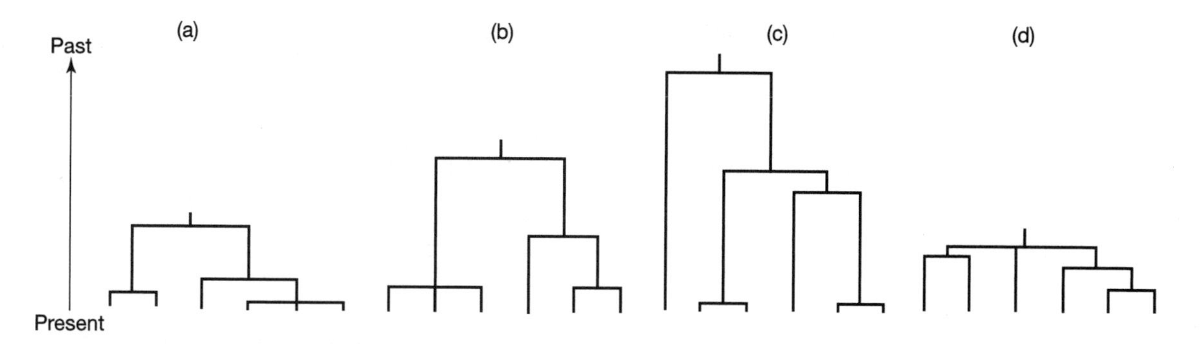

Figure 8.14. Four examples of coalescent trees for a sample size of $n = 6$ drawn on the same scale (Nordborg, 2001). Labels (1 to 6) indicating sample number should be randomly assigned to the tips of the tree.

(Hudson, 1990) where T_i is in terms of $2N$ generations. For example, if there is a sample of six sequences ($i = 6$) and the first random number $x = 0.22$, then $T_6 = 0.0166$. If the second $x = 0.57$, then $T_5 = 0.0844$, and so on. To illustrate the variation of neutral coalescent tress, Nordborg (2001) provided four examples for $n = 6$ (Figure 8.14). Obviously, there is great variation among these trees, both in their topology (the overall pattern of branching in the tree) and the lengths of the various branches. Because the lineages coalesce at a rate "n choose 2," coalescence events occur more rapidly when there are many lineages and more slowly as the number of lineages is reduced. In fact, the expected time during which there are only two branches is more than half the total time and the variability in T_2 among different trees accounts for much of the variability in total time to the MRCA (Nordborg, 2001).

The process by which sequences diverge is generally assumed to be **mutation**; that is, going down the gene tree over time, sequences accumulate differences by mutation. It is assumed that mutation occurs independently in different individuals and different generations. With these assumptions, mutations accumulate over time along lineages, and the sequences diverge from each other. Therefore, the greater the difference between sequences, the longer the expected time back to their common ancestor. Let us assume that T is the sum of the branch lengths of the gene tree, the total amount of time during which mutation could have occurred; then

$$T = \sum_{i=2}^{n} iT_i$$

Assuming that the mutation rate per generation is u and $\theta = 4Nu$ as before, then the total expected number of mutations, which is equal to the expected number of polymorphic sites, is

$$\begin{aligned} E(S) &= 2NuT \\ &= \frac{\theta}{2}\sum_{i=2}^{n} iT_i \\ &= \theta\sum_{i=1}^{n-1}\frac{1}{i} \end{aligned} \tag{8.11e}$$

For example, if $n = 5$, then $T = 25N/3$ generations. If we assume that we are measuring time in units of $2N$ generations, then $E(S) = \theta\ (25/12)$.

To illustrate how a simulation that includes mutation (under the IAM) can be carried out (see Hudson, 1990), say we have the tree given in Figure 8.15. Two mutations are generated using random numbers, and they are randomly placed on the branches of the tree as indicated. Assuming that the ancestral allele is A_1, mutation on the left-hand branch of the tree results in a new allele A_2 for sampled allele 1. Sampled alleles 2 and 3 are in lineages without mutation, so they remain A_1 alleles. Sampled alleles

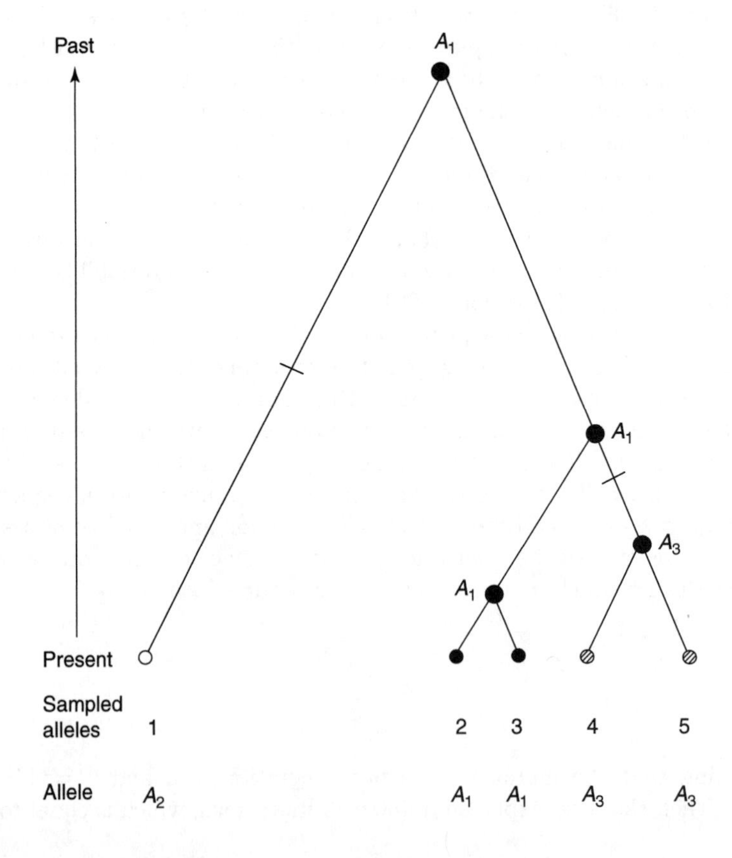

Figure 8.15. The same gene tree for five sampled alleles as given in Figure 8.13. The slashes indicate randomly generated mutations. The ancestral allele is A_1 and the descendant alleles after the first and second mutations are designated A_2 and A_3. The sample of five alleles contains two A_1 alleles, one A_2 allele, and two A_3 alleles.

4 and 5 are in a lineage that produced the mutation A_3. Overall, then for the five sampled alleles, the frequencies of alleles A_1, A_2, and A_3 are 0.4, 0.2, and 0.4, respectively. In this way, a sample of alleles reflects the equilibrium neutrality distribution—that is, the sample determined only by genetic drift and mutation.

Nordborg (2001) provided an instructive illustration of the effects of both genetic drift and mutation in a 10-generation example where $2N = 10$ (Figure 8.16). First, Figure 8.16a gives the complete genealogy illustrating the effect of genetic drift, as we gave in the example in Figure 8.12. Next, Figure 8.16b includes a mutation in generation 6 in the leftmost allele and follows its fate through generation 10. Also indicated are the initial variation in generation 0 (3 of the 10 alleles are different) and the loss of this variation over the next two generations. Because we do not need to consider the members of population that do not leave descendants, we can illustrate the genealogy by showing only the ancestors of the 10 alleles in generation 10 (Figure 8.16c). The number of ancestral alleles n from which the 10 alleles in the present generation are descended can be counted. Notice that 7 of the 10 alleles in the present generation are unchanged from the MRCA

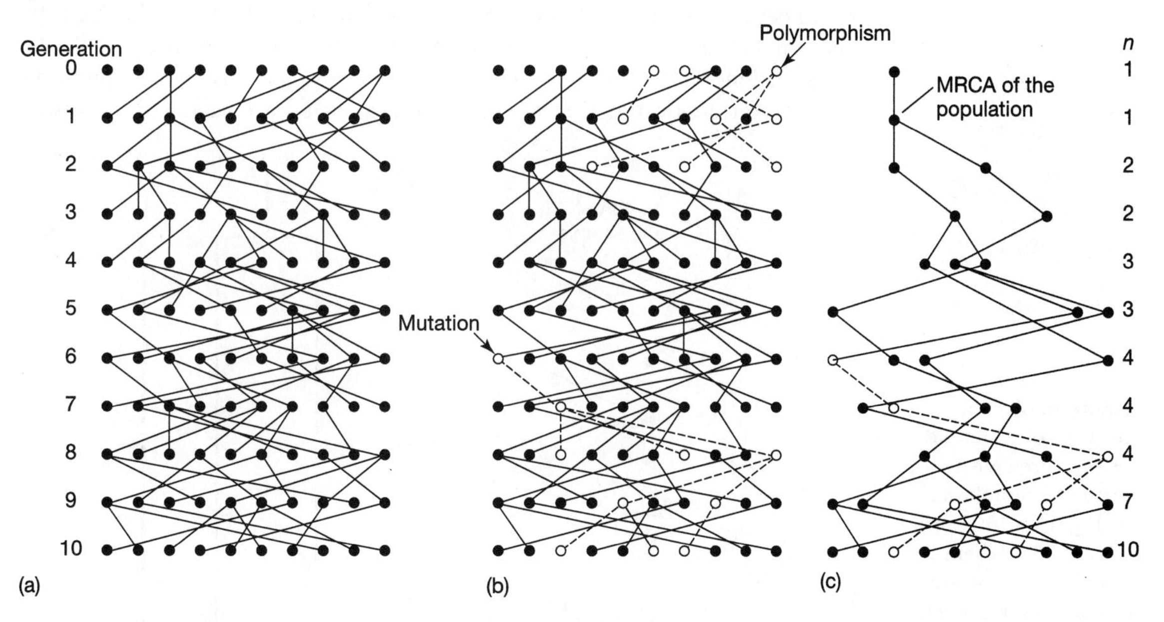

Figure 8.16. An illustration of the effects of both genetic drift and mutation in 10-generation example where $2N = 10$ (Nordborg, 2001) where it (a) gives the complete genealogy illustrating the effect of genetic drift, (b) includes a mutation in generation 6 and initial variation in generation 0, and (c) gives the genealogy showing only the ancestors of the 10 alleles in generation 10; n is the number of ancestral alleles from which the 10 alleles in the present generation are descended.

(in generation 1), whereas 3 descend from the mutation that occurred in generation 6.

We can also look at the genealogy of a sample of the three alleles from this population (Nordborg, 2001). Figure 8.17a gives the genealogy for the three leftmost alleles. One of the three alleles in this sample is descended from the mutation in generation 6. However, this tree is more complicated than necessary because the identity of all ancestors was retained throughout. If we ignore the generation-to-generation ancestry, then a genealogy as in Figure 8.17b can give coalescent events (this includes both branch lengths and the tree topology). The time to the first coalescence from three to two alleles T_3 is two generations, and the time to the second coalescence from two alleles to the MRCA T_2 is seven generations.

Refinements in the coalescence approach have been developed to include all of the different evolutionary factors, although some of algorithms to accomplish these effects are complicated. For example, various nonselective factors, such as variation in population size, gene flow, and self-fertilization, can be incorporated (Kaplan *et al.*, 1991; Slatkin, 2000; Nordborg, 2001). For example, if the variance in progeny number is greater than Poisson, then the effective population size is reduced, and coalescent times can be measured in scaled units that are lowered by the amount of change in the variance of reproduction (Nordborg, 2001). In addition, the effects of recombination, which spreads the ancestry of a mutation over different chromosomes, can be considered (Hudson,

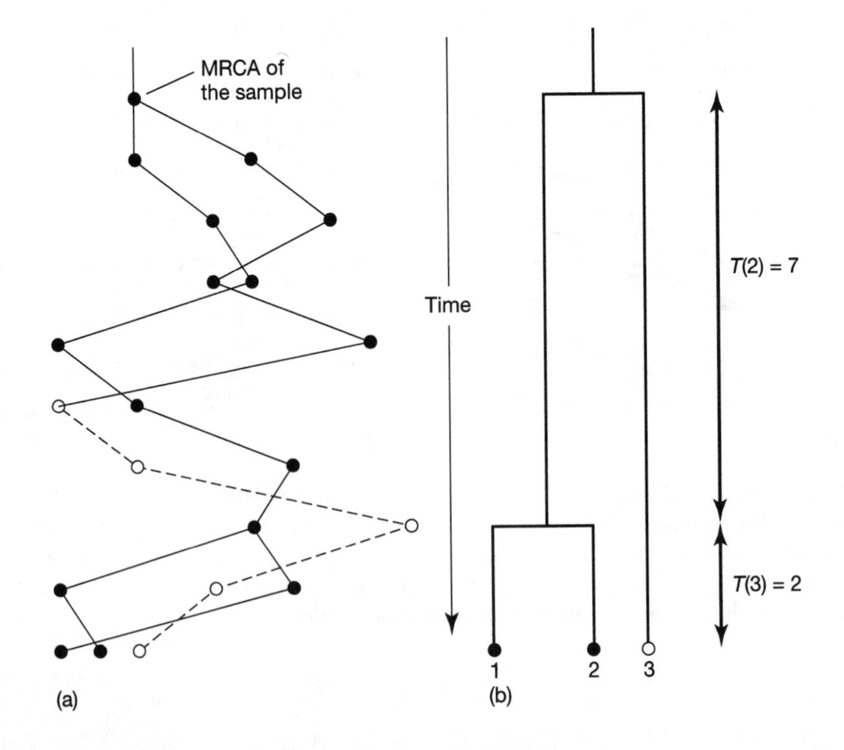

Figure 8.17. The genealogy of a sample of the three alleles from the population in Figure 8.16 (Nordborg, 2001) where it (a) gives the genealogy for the three leftmost alleles and (b) ignores the generation-to-generation ancestry, giving a genealogy with the coalescent events, including both branch lengths and the tree topology.

1983; Nordborg, 2001). A tree with recombination has both branches and coalescent events as one goes back in time. Finally, various types of selection, such as balancing selection, selective sweeps, and purifying selection can be incorporated (Takahata, 1990; Kaplan *et al.*, 1991; Nordborg, 2001). Nonetheless, combining all of these aspects into a practical application for nucleotide sequence data is easier said than done and can be theoretically difficult as well as computationally burdensome. For example, one of the best applications available is the GENETREE (http://taxonomy.zoology.gla.ac.uk/rod/genetree/genetree.html) program designed by R. Griffiths and colleagues, which applies coalescent theory and a maximum-likelihood (ML) approach to estimate the time to the MRCA and the N_e of a sample of sequences as well as the age of alleles at the locus. However, these estimates are limited by the assumption that there is no recombination or selection at the locus, given the almost infinite possibilities associated with these factors in a genealogy when utilizing a ML approach.

As a relatively straightforward example illustrating the potential versatility of the coalescent approach, let us examine the effect of changes in population size on the expected distribution of coalescent times. Remember that the number of coalescent events is proportional to the size of the population and that it increases as there is more genetic drift (lower population size). To illustrate the impact of historical demography, Garrigan *et al.* (2002) provided example genealogical trees for $n = 10$ when there is constant population size (as we have been considering until now), a decline in population size, and an increase in population size where the relative population size is indicated by the width of the thick lines in Figure 8.18.

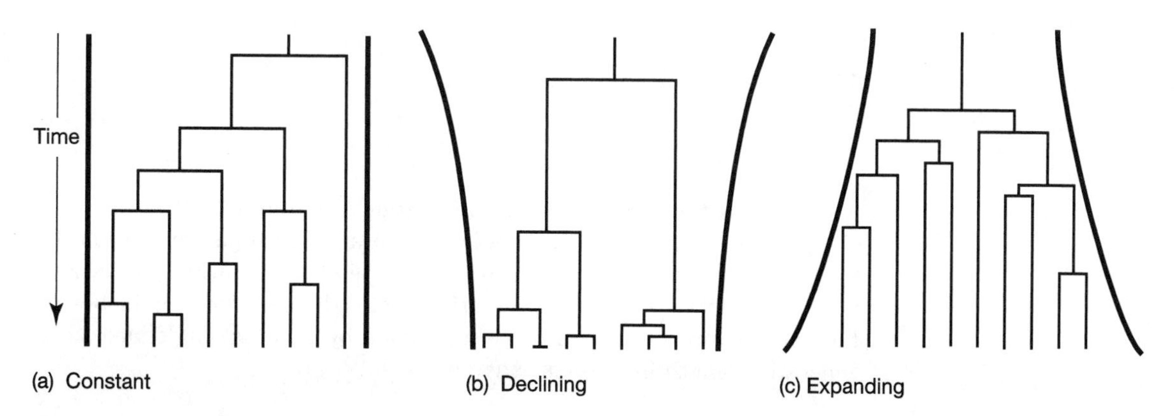

Figure 8.18. The theoretical distributions of coalescent times in genealogies with 10 contemporary samples under three scenarios of historical population change: (a) constant population size, (b) declining population size, and (c) increasing population size (Garrigan *et al.*, 2002). The distance between the thick lines indicates the relative population sizes at different times, and the genealogies reflect coalescent events in periods of relative small or large population sizes, respectively.

For the constant population size, the number of coalescent events is somewhat higher toward the tips of the branches because the expected times are shorter when n is larger (expression 8.11c). Compared with this, when the population is declining, the number of coalescent events in recent times is much greater, as shown at the bottom of Figure 8.18b. If N_e increases as we go back in time, then the coalescence times will become longer than those expected under constant population size. When the population is expanding, then there are fewer coalescent events in recent times relative to the constant population size expectation (Figure 8.18c). If N_e decreases going back in time, the coalescent times will become shorter than expected under constant population size.

b. Estimating Effective Population Size

Neutral theory and the coalescent approach can potentially be used to estimate the value of evolutionary parameters, such as effective population size and mutation rate. In the simplest form, at neutrality equilibrium, the level of diversity is a balance between mutation and genetic drift and is

$$\theta = 4N_e u$$

This expression can be solved for an estimate of the effective population size as

$$N_e = \frac{\theta}{4u}$$

For diversity data for mitochondrial sequences, then

$$\theta = 2N_{ef} u$$

and

$$N_{ef} = \frac{\theta}{2u}$$

where N_{ef} is the female effective population size (see p. 327).

In both of these estimates, u is the mutation rate per generation, but the mutation rate is estimated from the rate of substitution per year. To put this estimate on a per generation scale, the denominator in both of the above expressions need to be multiplied by T, which is defined as the generation length in years, so that

$$N_e = \frac{\theta}{4uT} \quad (8.12a)$$

and

$$N_{ef} = \frac{\theta}{2uT} \tag{8.12b}$$

For example, in 50 random, noncoding, nuclear DNA segments in humans, $\theta = 0.000882$ (Yu *et al.*, 2002). The divergence between humans and chimpanzees for these 50 segments is 0.01221 (Yu *et al.*, 2003) and, assuming that humans and chimpanzees diverged 6×10^6 years ago, the estimate of mutation rate per year is $u = 0.01221/[(2)(\times 10^6)] = 1.02 \times 10^{-9}$ (the 2 in the denominator is included because divergence is occurring in both lineages). Assuming that the generation length in humans is 20 years, then using expression 8.12*a*, $N_e = 0.000882/[(4)(1.02 \times 10^{-9})(20)] = 10,800$ (Yu *et al.*, 2003). Example 8.8 gives the estimates of the long-term female effective population sizes in three species of baleen whales using this approach and mtDNA data.

Example 8.8. The abundance of whales was assumed to be much greater before large-scale and intensive hunting began in the 19th century. Now most whale species worldwide are protected from hunting, and they have somewhat recovered in numbers from their middle-to-late-20th-century low numbers. However, the actual population level before this decline has only been estimated from whaling logbooks and other nonscientific reports, making an objective estimate of population size based on genetic variation important both to determine the health of contemporary populations and for population goals in recovery recommendations.

Roman and Palumbi (2003) estimated the level of diversity for three species of baleen whales: humpback, fin, and minke in the North Atlantic for part of the mtDNA control region (Table 8.15). Using an estimate of $u = 1.75 \times 10^{-8}$ per year that they obtained from whale species divergence data and estimated female generation lengths, the long-term N_{ef} using expression 8.12*b* for humpback, fin, and minke whales are approximately 34,300, 49,100, and 38,800, respectively. They suggested that the census number before hunting is sevenfold higher than the estimated N_{ef}: twofold higher to include males, twofold higher to account for an N_e/N ratio of 0.5, and 1.75-fold higher to account for the ratio of adults to juveniles. These estimates are 240,100 and 344,000 for humpback and fin whales, respectively, while the current census estimates for humpback whales are only 10,700 and for fin whales are 56,000. Only for minke whales are the estimated census number before hunting and the current census number of similar magnitude.

TABLE 8.15 The level of mtDNA diversity (θ), the estimated generation length (T), and the estimated female effective population sizes (N_{ef}) for three species of baleen whales (Roman and Palumbi, 2003). From these estimates, the estimated census number before hunting is extrapolated and is compared with current census estimates.

				Estimated census numbers	
Species	θ	T	N_{ef}	Before hunting ($N_e \times 7$)	Current
Humpback whale	0.0216	18	34,300	240,100	10,700
Fin whale	0.0430	25	49,100	344,000	56,000
Minke whale	0.0231	17	38,800	271,600	215,500

This estimate is different from the estimates given in Chapter 6 in several respects. First, this is an estimate of the equilibrium or average N_e over a long period and has been termed **evolutionary** or **long-term effective population size**. It is primarily determined by the cumulative impacts of mutation and genetic drift over many thousands of generations so that it may take an extremely long time to approach the equilibrium value given an extreme demographic event such as a bottleneck or founder effect. Second, although this estimate can give insight into the effective population size over evolutionary history, it may be unrelated to the present-day effective population size. For example, in an endangered species, the current effective population size may be small. However, the loss of variation from a much larger long-term effective population size only takes place over a number of generations so that an estimate of the long-term effective population size from a current population may reflect more the higher past equilibrium state and not the present endangered status. Thinking about it in another fashion, in current small populations, if they haven't been small too long, the previous long-term effective population size may be estimated to understand better the evolutionary context of the species.

The above approach estimates the long-term effective population size of a constant population size. As we discussed in Figure 8.18, changes in effective population size influence the distribution of coalescent events over genealogies. Kuhner *et al.* (1995, 1998) have developed a coalescent approach using sequence data that allows a joint estimate of the long term effective population size and whether the population is constant, growing, or declining in size (FLUCTUATE at http://evolution.genetics.washington.edu/lamarc/fluctuate.html). In Example 8.9, estimates of the long-term effective using both the equilibrium assumption and allowing for changes in size are given for three endangered fishes of the lower Colorado River.

Example 8.9. Of the four large fishes of the lower Colorado River, one is extirpated (Colorado squawfish or pike minnow), and three are endangered (humpback chub, bonytail chub, and razorback sucker) (Minckley *et al.*, 2003). Present estimates of the numbers of adult humpback chub, bonytail chub, and razorback sucker are approximately 3000, 100, and 4000, respectively. All four species were known historically to be in large numbers, and in fact, the extirpated squawfish (it is still present in low numbers in the upper Colorado River) were so abundant that early settlers used them for fertilizer. Despite such accounts, it is important to have genetically based estimates of their long-term effective population sizes to determine appropriate recovery goals for these endangered species.

Garrigan *et al.* (2002) examined mtDNA variation for these three species. For the humpback chub, in 18 samples, there were five different haplotypes observed with haplotypes AB and GB occurring in frequencies of 0.44 and 0.28 and with $\theta = 0.00195$ (Table 8.16). Figure 8.19a gives a maximum likelihood genealogy for these sequences (this is a representation of the coalescent times rather than a specific topology within the haplotypes). The rare haplotypes AH and LD were quite divergent from the other haplotypes. For the 16 bonytail chub, there were only three haplotypes, and they were nearly equally frequent with $\theta = 0.00179$. Figure 8.19b gives the coalescent genealogy for this sample and illustrates that the three haplotypes are somewhat divergent from each other. For these data, Tajima's $D = 1.909$ and is significant, potentially indicating the occurrence of a population bottleneck. For the razorback sucker, there were 13 different haplotypes. Although haplotype E occurred with a frequency of 0.59, $\theta = 0.01338$, the highest of the three species.

Brown *et al.* (1979) estimated that the mtDNA mutation rate per nucleotide per year for mammals was approximately 2×10^{-8}, and Kocher *et al.* (1989) found that the mutation rate in fishes was approximately five times lower than in mammals. Therefore, assuming that the mutation rate per year is 4×10^{-9}, a constant population size, $T = 10$ for the chubs and $T = 15$ for razorback suckers, and using expression 8.12*b*, the estimates of N_{ef} are approximately 24,400, 22,400, and 111,500 for humpback

TABLE 8.16 The estimated diversity for mtDNA genes in three endangered fishes of the lower Colorado River when a constant long-term effective population size is assumed and when growth or decline in the population is allowed (Garrigan *et al.*, 2002). For comparison, the estimates of long-term and of the current census numbers are given.

			Constant			Growth		Census	
	Gene	n	T	θ	N_{ef}	θ	N_{ef}	*Long term*	*Current*
Humpback chub	*ND2*	18	10	0.00195	24,400	0.00298	37,200 (constant)	488,000	3000
Bonytail chub	*ND2*	16	10	0.00179	22,400	0.00124	15,500 (declining)	448,000	100
Razorback sucker	*Cyt b*	49	15	0.01338	111,500	0.01881	156,800 (expanding)	2,230,000	4000

chub, bonytail chub, and razorback sucker, respectively (Table 8.16). If we allow population growth to vary and reflect the phylogenetic pattern of the mtDNA data using the approach of Kuhner *et al.* (1998), the estimate of growth is not significantly different from zero for humpback chub, is significantly declining for bonytail chub, and is significantly increasing for razorback suckers. These adjusted estimates of N_{ef} are somewhat higher for humpback chubs and razorback suckers and lower for bonytail chubs. If we assume that the long-term census number is approximately 20-fold higher than the constant population size N_{ef} estimate, 2-fold higher to include males and 10-fold higher to account for an N_e/N ratio of 0.1 (see p. 333), the long-term census estimates are over 400,000 for the chubs and over 2 million for the razorback sucker. From these estimates, it appears that the long-term census population size was at least several orders of magnitude larger than the present census numbers.

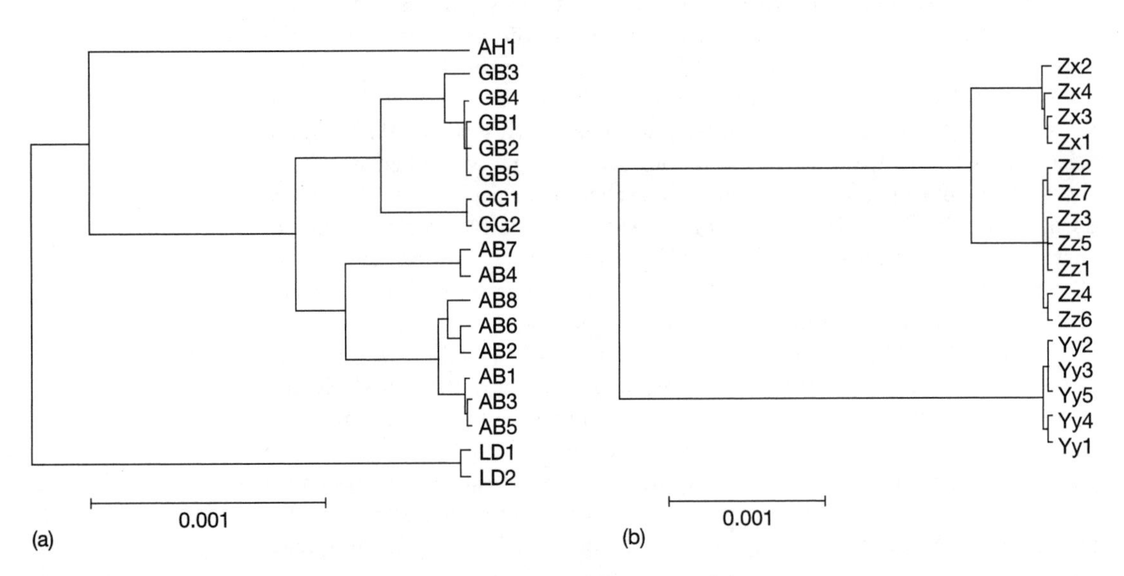

Figure 8.19. Coalescent genealogies for (a) humpback chub and (b) bonytail chub that maximize the likelihood of the mtDNA data. Branch lengths are scaled in terms of the number of substitutions per site, yet substitutions do not occur on each branch. The first two letters on the branches indicate the mtDNA sequence, and the subsequent number indicates the individual with that sequence. The distance between identical sequences represents the time, in generations, since a common ancestor. Courtesy of Garrigan, D., P. C. Marsh, and T. E. Dowling. 2002. Long term effective population size of three endangered Colorado river fish. *Anim. Cons.* 5:95–102. Reprinted with the permission of Cambridge University Press.

PROBLEMS

1. Calculate the mean fitness in a population in which there are 100 loci at equilibrium with selective disadvantages of $s_1 = s_2 = 0.05$, assuming multiplicative fitness over loci.
2. Assume that the heterozygosity in mice for X-linked microsatellite loci is 0.58 and that for autosomal microsatellite loci it is 0.65. Using the logic in Example 8.1, are these values consistent with those predicted from the infinite-allele model or from the stepwise mutation model?
3. Explain why, under the neutrality model, the rate of allele substitution is independent of the population size. If the rate of allele substitution is 10^{-5}, what is the expected time between new substitutions?
4. Assume that the amino acid sequence of the globin genes in two organisms differed at 30 of 100 sites and that the estimated time since their divergence was 200 million years. What are K_{aa}, its standard error, and k_{aa}?
5. Assume that the nucleotide sequence in two homologous genes from two organisms differed at 200 out of 500 sites and that the estimated time since their divergence was 5 million years. What are K, its standard error, and k?
6. Sum the data for the three codon positions in Table 8.3 and calculate the overall number of nucleotide substitutions. How does this differ from that calculated for each position individually?
7. Explain why expression 8.5*b* is true.
8. What is the probability that two alleles coalesced 10 generations before the present in a population of size $N = 50$? What is the expected time in a population of $N = 5$ that there are 10, 5, and 2 lineages?
9. Given the sequence of five codons CATAGGTCAATGTGT, calculate the number of potentially synonymous sites without looking at Table 8.12. Assuming that two sequences have the two codons CCTTTT and CATTCC, calculate the average number of observed synonymous differences.
10. In a sample of $2N = 100$, three alleles were observed with frequencies of 0.48, 0.35, and 0.17. What is the observed Hardy–Weinberg heterozygosity? Is this different than the homozygosity expected under the neutrality theory? (Solve expression 8.6*b* for $n = 3$.)
11. What are the expected values for the substitutions given in Table 8.9b and 8.9c? By comparing the observed and expected values, give explanations for these two sets of data.
12. Using the estimates of θ, T, and u given in Table 8.15, calculate the estimates of N_{ef} and N_e for the three whale species. These estimates are much larger than the estimated census numbers. How could this estimation procedure result in such an erroneous answer?
13. Two measures of codon usage are discussed in the text. Which of these do you prefer and why? Are there other characteristics that you think should be incorporated in a measure of codon usage?
14. Go through Figures 8.16 and 8.17 and explain the information and differences given in the different parts of the two figures.

15. On p. 416, it was stated about the neutral theory that "These simple but elegant mathematical predictions ... provide the basis for the most important developments in evolutionary biology in the past half century." Do you agree and why?

9

Gene Flow and Population Structure

> The most general conclusion is that evolution depends on a certain balance among its factors. There must be gene mutation, but an excessive rate gives an array of freaks, not evolution: there must be selection but too severe a process destroys the field of variability and thus the basis for further advance; prevalence of local inbreeding within a species has extremely important evolutionary consequences but too close inbreeding leads merely to extinction. A certain amount of crossbreeding is favorable but not too much. In this dependence on balance, the species is like a living organism. At all levels of organization, life depends on the maintenance of a certain balance among its factors.
>
> Sewall Wright (1932)

> The importance of gene flow in evolution is not fully agreed upon, although a variety of strong views are held. One view ... is that gene flow is common and that a small amount of gene flow among different parts of a species' range effectively unifies the species and affects significantly the genetic changes in each part of the range. The other view ... is that gene flow is uncommon and that natural selection acts more of less independently in each part of a species' range. While both of these views are probably correct for some species, there is no way to determine the importance of gene flow in natural populations because there is no direct way to estimate levels of gene flow. Indirect estimates of the level of gene flow obtained by measuring the movement of marked individuals are difficult to interpret. Levels of gene flow can be severely underestimated through the movement of unmarked individuals or unmarkable life stages or severely overestimated because individuals may move but not breed.
>
> Monty Slatkin (1981)

In the previous chapters, we have generally examined the effect of various evolutionary genetic factors in a single population. In most species, however, populations are often subdivided into smaller units because of geographical, ecological, or behavioral factors. For example, the populations of fish in pools, trees on mountains, and insects on host plants are subdivided because suitable habitats for these species are not continuous. Population subdivision can also result from behavioral factors, such as in troops of primates, among territorial birds, and in colonies of social insects.

When a population is subdivided, the amounts of genetic connectedness among the parts of the population can differ. This genetic connection depends primarily on the amount of genetically effective gene flow that takes place among the subpopulations or subgroups. When the amount of gene flow between groups is high, gene flow has the effect of homogenizing genetic variation over the groups. When gene flow is low, genetic drift, selection, and even mutation in the separate groups may lead to genetic differentiation.

Before we go further, it is important to realize that some types of migration, such as migration in birds that seasonally fly to the tropics, do not necessarily result in the exchange of genes between subpopulations. Also, movement or dispersal into another area for food, during which no mating or reproduction occurs, is not genetically effective gene flow. Even movement of reproductive adults may not result in any genetic contribution from them because of exclusion from breeding caused by behavioral or other factors. Because of the confusion surrounding these terms, Endler (1977) made an effort to distinguish among gene flow, migration, and dispersal. Following his recommendation, we use **gene flow** to indicate ***movement between groups that results in genetic exchange***.

Subdivision may take place on several different spatial or geographic scales. For example, within a watershed, there may be separated fish or plant groups that have a substantial amount of genetic exchange between them. On a larger scale, there may be genetic exchange between adjacent watersheds but in lesser amounts than between the groups within a watershed. On an even larger scale, there may be populations in quite separated watersheds that presumably have little direct exchange but may share some genetic history, depending on the amount of gene flow among the adjacent groups or occasional long-distance gene flow. This **hierarchical** representation is useful in describing the overall relationships of populations of an organism and documenting the spatial pattern of genetic variation. Recently, there has been increasing interest in landscape and geographic approaches to estimating historical and contemporary gene flow (Sork *et al.*, 1999; Epperson, 2003; Manel *et al.*, 2003). In such studies, often the assumption of discrete populations is relaxed. There is explicit accounting of landscape features and barriers, and sometimes geographical information system approaches are used to visualize spatial patterns. On p. 611, we discuss **phylogeography** (Avise, 2000), the joint use of phylogenetic techniques and geographic distributions, to understand spatial relationships and distributions of taxa.

In general, the subdivision of populations assumes that the various subpopulations are always present. Another view assumes that the population subdivisions may go extinct and then later may be subsequently recolonized from other subpopulations. For example, tide pools may provide a suitable habitat for particular species of fish; however, local extinction of these species within a tide pool might occur, and no fish would occupy the tide pool until recolonization from another tide pool occurred. ***When a population consists of a number of different population subdivisions in suitable habitat patches whose absence is determined by extinction and whose presence is determined by recolonization from the other subpopulations***, it is termed a **metapopulation** (Hanski, 1998). The dynamics of extinction and recolonization can make metapopulations different both ecologically and genetically from the traditional concept of a subdivided population. Sometimes any subdivided population is called a metapopulation (Hanski and Gilpin, 1997).

Gene flow is central to understanding evolutionary potential and mechanisms in several areas of applied population genetics. First, the potential for movement of genes from **genetically modified organisms** into related wild populations—that is, the gene flow of **transgenes** into natural populations can be estimated and addressed using population genetic models (Arnaud *et al.*, 2003; Ellstrand, 2003; Stewart *et al.*, 2003; Howard *et al.*, 2004). Second, invasion of nonnative plants and animals into new areas is one of the major challenges for maintaining native biota throughout the world. A number of factors appear to contribute to the **invasiveness** potential of nonnative species, including adaptive change and hybridization (gene flow) between nonnative taxa and between nonnative and native taxa (Ellstrand and Schierenbeck, 2000; Gaskin and Schaal, 2002). Finally, a number of endangered species are composed of only one, or a few, remaining populations with low fitness. It appears that gene flow from other populations of the same species can result in **genetic rescue** or **genetic restoration** of these populations by introducing new variation that allows the detrimental variation to be removed and adaptive change to be restored (Richards, 2000; Ebert *et al.*, 2002; Saccheri and Brakefield, 2002; Vilà *et al.*, 2003; see p. 512).

In this chapter, we first examine the effect of gene flow and population structure in an infinite population to determine the theoretical impact of gene flow acting by itself. Then we discuss some methods used to estimate gene flow and population structure. Next, we examine the joint effects of gene flow and genetic drift on the pattern and amount of genetic variation. Finally, we examine the joint effect of gene flow and selection to maintain genetic variation and form clines. For reviews of some of the recent developments in gene flow and population structure, see Hey and Machado (2003) and Charlesworth *et al.* (2003).

I. POPULATION STRUCTURE

A population may have substructure—differences in genetic variation among its constituent parts—for several different evolutionary reasons. For example, a population may have localized subpopulations in which there is genetic drift. Exchange of individuals may not have equal probabilities throughout a population, or selection may have different effects in different part of the population. In other words, all of the evolutionary factors that we have discussed can contribute to the structure of a population. To elucidate the effect of population structure on the pattern and amount of genetic variation, we begin by examining the effect of the simplest population structure, the continent–island model, and then consider a general model of population structure. Such general models may not precisely fit a particular biological example, but they give close approximations to many situations and allow an evaluation of the effect of limited gene flow.

a. The Continent–Island Model

There are a number of examples of effectively unidirectional gene flow, such as occurs from a continental to an island population. Such examples include species with populations on land islands and nearby large land masses (continents), aquatic species in ponds with a lake as the source of gene flow, and peripheral populations of any species that are constantly replenished by individuals from the main part of the species range.

To formulate a model for this situation, let us assume that an island population receives migrants from a large source (continental) population as shown in Figure 9.1a. Although reciprocal gene flow may occur, we

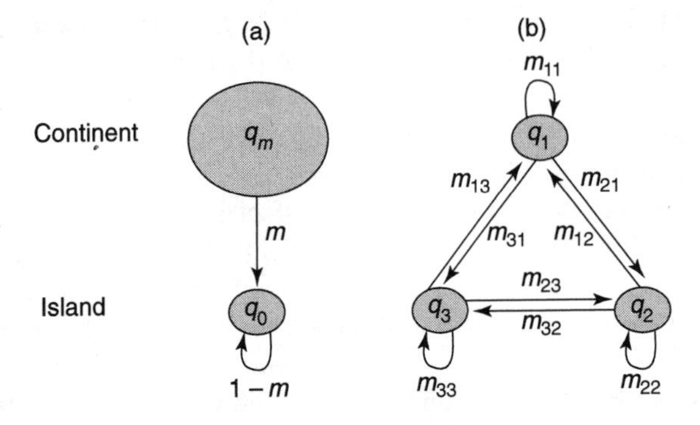

Figure 9.1. Illustration of (a) the continent–island model and (b) population structure with three subpopulations.

assume that it has a negligible effect on the allele frequency in the source population. We also assume that the island population is large enough that the effect of genetic drift is small relative to that of gene flow. Let the proportion of migrants moving into the island population each generation be m and the proportion of nonmigrants (residents) be $1-m$. If the frequency of A_2 in the migrants (the continent) is q_m and the frequency of A_2 on the island before gene flow is q_0, the allele frequency after gene flow is

$$\begin{aligned} q_1 &= (1-m)q_0 + mq_m \\ &= q_0 - m\,(q_0 - q_m) \end{aligned} \tag{9.1a}$$

The change in allele frequency after one generation of gene flow is then

$$\begin{aligned}\Delta q &= q_1 - q_0 \\ &= -m\,(q_0 - q_m) \qquad (9.1b)\end{aligned}$$

From this expression, it is obvious that there will be no change in allele frequency if $m = 0$ or if $q_0 = q_m$. Only the second alternative is of interest to us because a value of $m = 0$ indicates that there is no gene flow. Remember that both q_m and m are assumed to be constant over time and have values between zero and unity. If $q_0 < q_m$, the frequency of A_2 increases on the island, and if $q_0 > q_m$, the frequency decreases, indicating that there is a stable equilibrium frequency of A_2 at $q_m = q_0$. The effect of different allele frequencies in the migrants on Δq when $m = 0.1$ can be seen in Figure 9.2. The change in allele frequency increases linearly as the frequency moves away from the equilibrium value and reaches an absolute maximum at either zero or unity depending upon the value of q_m.

If gene flow continues, then in the second generation the allele frequency becomes

$$q_2 = (1 - m)\,q_1 + mq_m$$

When we substitute this in the expression for q_1 given above, the allele frequency becomes

$$q_2 = (1 - m)^2\,q_0 + \left[1 - (1 - m)^2\right]q_m$$

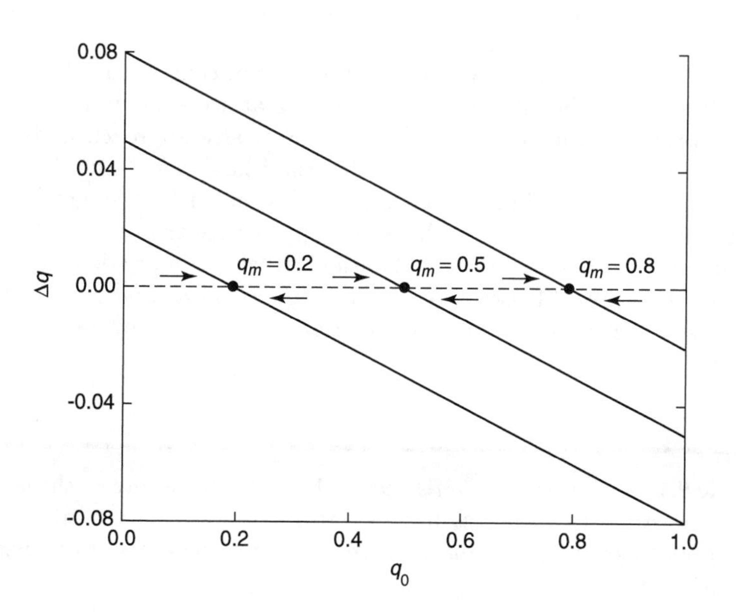

Figure 9.2. The change in allele frequency in the continent–island model for three different allele frequencies in the migrants when $m = 0.1$.

The general solution to this recursion equation relating the frequency of A_2 in generation t to that in the initial generation is

$$q_t = (1-m)^t q_0 + \left[1-(1-m)^t\right] q_m \qquad (9.1c)$$

As t increases, the first term in this expression approaches zero, and the second term approaches q_m so that q_t asymptotically approaches the equilibrium value of q_m.

To illustrate the rate of approach to the equilibrium, Figure 9.3 gives the allele frequency over time for two values of q_0, 0.1 and 0.9, when

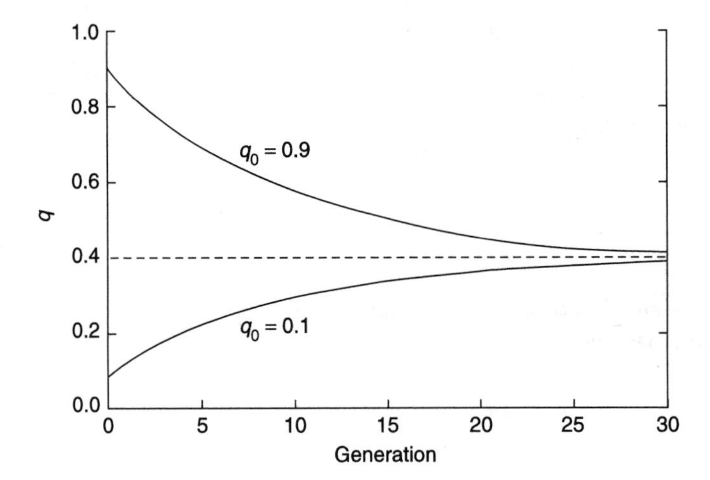

Figure 9.3. The allele frequency over time for two different initial allele frequencies for the continent–island model when $q_m = 0.4$ and $m = 0.1$.

$q_m = 0.4$ and $m = 0.1$. As we would expect from considering the Δq values in Figure 9.2, the allele frequency changes at a maximum rate initially and decreases as the equilibrium is asymptotically approached. Because gene flow is unidirectional in this model, the island population is eventually composed of individuals that have all descended from migrants. As a result, the allele frequency on the island approaches that of the continent as more and more individuals in the island population are descendants of migrants. Example 9.1 uses this approach to predict the proportion of red wolf ancestry in a reintroduced population with which coyotes are interbreeding.

Example 9.1. Red wolves (*Canis rufus*) historically occurred throughout southeastern North America, but by the 1960s, they were confined to a small population in Louisiana and Texas where there was hybridization with the much more abundant coyote (*C. latrans*). Wolves from this population were captured to start a captive population, and in 1987, this captive population was used to establish a wild population in eastern North Carolina (Phillips *et al.*, 2003). However, over the next decade, coyotes colonized this area, and in the late 1990s, it was estimated that approximately 15% of the litters in the newly established population were hybrid.

To generally predict the impact of this introgression by coyotes into the red wolf population on the extent of red wolf ancestry, we can use the continent–island model where the "island" red wolf population is receiving one-way gene flow from the "continent" coyote population. Let us assume that $q_0 = 1$—that is, initially that there is 100% red wolf ancestry in the red wolves—and $q_m = 0$—all of the coyote genes do not reflect red wolf ancestry. In hybrid litters, half the ancestry is from red wolves and half from coyotes; thus, let us assume that $m = 0.075$, or half the rate of observed hybrid litters. Assuming that the generation length in red wolves is approximately 5 years, then Figure 9.4 shows that the proportion of red wolf ancestry is expected to drop quickly from 100% to approximately 46% after 50 years.

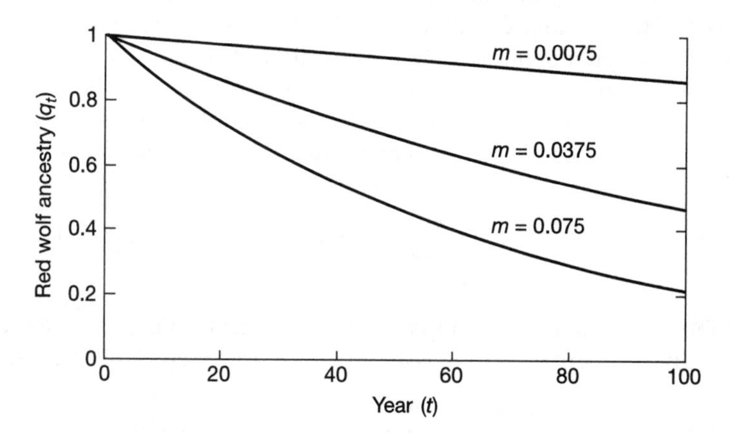

Figure 9.4. The predicted proportion of red wolf ancestry over time when there is gene flow from coyotes as might occur in a continent (coyote)–island (red wolf) model. The proportion of gene flow is assumed to be 0.075 without management intervention and either 0.0375 or 0.0075 if 50% or 90% of the hybrid litters are identified and eliminated.

However, if there are management actions to identify (Miller *et al.*, 2003) and eliminate hybrid litters, then the rate of gene flow can be significantly reduced. For example, if 50% or 90% of the hybrid litters are eliminated so that $m = 0.0375$ or 0.0075, then the red wolf ancestry remains approximately 68% and 93%, respectively, after 50 years. This heuristic treatment demonstrates that the red wolf ancestry in this re-established population is in great danger of being swamped by coyote introgression. In a detailed individual-based simulation that examined the specifics of mating for wolves, coyotes, and hybrids with different ancestry, the distribution of ancestry over individuals, and various management options, Fredrickson and Hedrick (2004) predicted that red wolf ancestry in this population will be lost without intensive management or strong positive-assortative mate choice within red wolves.

b. General Model

The continent–island model examines the allele frequency change only in the island population and assumes that only gene flow to the island is important. A more general model assumes that gene flow can occur among all parts of a substructured population. Assume that a population consists of k subpopulations and that the proportion of individuals migrating from subpopulation j to subpopulation i each generation is m_{ij}. As a result, there is a matrix of gene flow parameters, called the backward **migration matrix** by Bodmer and Cavalli-Sforza (1968), that describes the gene flow pattern among subpopulations (see Table 9.1). The proportion of nonmigrants (or residents) for subpopulation i is given by m_{ii}. Each row of this matrix sums to unity because it describes the proportion coming from every other possible subpopulation to that particular subpopulation or

$$\sum_{j=1}^{k} m_{ij} = 1.0$$

The columns of the matrix will generally not sum to unity (as they did in the transition matrix used to illustrate genetic drift on p. 310). A schematic

TABLE 9.1 The migration matrix indicating the proportion of gene flow from subpopulation j in generation t to subpopulation i in generation $t+1$.

Subpopulation in generation $t+1$	*Subpopulation in generation* t										*Total*
	1	2	3	·	·	·	j	·	·	·	k	
1	m_{11}	m_{12}	m_{13}	·	·	·	m_{1j}	·	·	·	m_{1k}	1
2	m_{21}	m_{22}	m_{23}	·	·	·	m_{2j}	·	·	·	m_{2k}	1
3	m_{31}	m_{32}	m_{33}	·	·	·	m_{3j}	·	·	·	m_{3k}	1
·	·	·	·				·				·	·
·	·	·	·				·				·	·
·	·	·	·				·				·	·
i	m_{i1}	m_{i2}	m_{i3}				m_{ij}				m_{ik}	1
·	·	·	·				·				·	·
·	·	·	·				·				·	·
·	·	·	·				·				·	·
k	m_{k1}	m_{k2}	m_{k3}				m_{kj}				m_{kk}	1

illustration of a population with three subpopulations and the resulting nine gene flow parameters, as in the upper left of Table 9.1, is given in Figure 9.1b.

Each of the subpopulations may have a different frequency of A_2, and thus, let us indicate the allele frequency in the jth subpopulation as q_j. Therefore, the frequency of A_2 in the ith subpopulation after gene flow is

$$q_i' = m_{i1}\,q_1 + m_{i2}\,q_2 + \ldots m_{ij}\,q_j + \ldots m_{ik}\,q_k$$
$$= \sum_{j=1}^{k} m_{ij}\,q_j \qquad (9.2a)$$

or the sum of the products of the proportion of migrants from the jth subpopulation into the ith subpopulation and the allele frequency in the migrants from the jth subpopulation.

We can symbolize the process of allele frequency change over all the subpopulations by using matrix notation (see p. 26). First, we can indicate the migration matrix as $\mathbf{M}$ and the vector of allele frequencies for the different subpopulations in generation t with $\mathbf{Q}_t$. Therefore,

$$\mathbf{Q}_{t+1} = \mathbf{M}\mathbf{Q}_t$$

and in general

$$\mathbf{Q}_t = \mathbf{M}^t\mathbf{Q}_0$$

The elements in the powered matrix $\mathbf{M}^t$, $m_{ij\cdot t}$, give the proportion of individuals in subpopulation i and generation t that have descended from subpopulation j in generation 0. From these equations, we can project the allele frequencies in any future generation, given constant m_{ij} values over time. Over a long period of time, the frequencies will converge and approach an asymptotic value that can be calculated from

$$q_i' = \sum_{j=1}^{k} m_{ij\cdot t}\,q_j \qquad (9.2b)$$

where $m_{ij\cdot t}$ is an element in the powered migration matrix with t large enough so that the elements have reached their asymptotic values. Example 9.2 presents a simple situation to illustrate this in which there is gene flow between three human populations in Sudan.

Example 9.2. Generally, to measure the amount of gene flow among subpopulations, one must identify individuals of different generations. As a result, many of the best-documented examples of population structure are in human populations. A simple example was provided by Roberts and Hiorns (1962) for three Sudanese populations. Table 9.2, which gives the proportion of gene flow among these groups, shows that a high proportion of these individuals do not migrate. For example, $m_{11} = 0.985$, or 98.5%, of the Nuer individuals do not migrate from one generation to the next. The estimated frequencies of the blood group allele M for the Nuer, Dinka, and Shilluk populations are 0.575, 0.567, and 0.505, respectively. If we use the migration matrix in Table 9.2, the allelic frequency projections over future generations result in the changes illustrated in Figure 9.5. Only a slow change in frequency occurs because the proportion of nonmigrants is high and there is little initial differentiation among the populations. In this case, the asymptotic frequency of allele M for the set of three populations is 0.546.

TABLE 9.2 The proportion of gene flow among three Sudanese populations (after Roberts and Hiorns, 1962).

	Source population		
Recipient population	*(1) Nuer*	*(2) Dinka*	*(3) Shilluk*
(1) Nuer	0.9850	0.0125	0.0025
(2) Dinka	0.0138	0.9775	0.0087
(3) Shilluk	0.0000	0.0098	0.9902

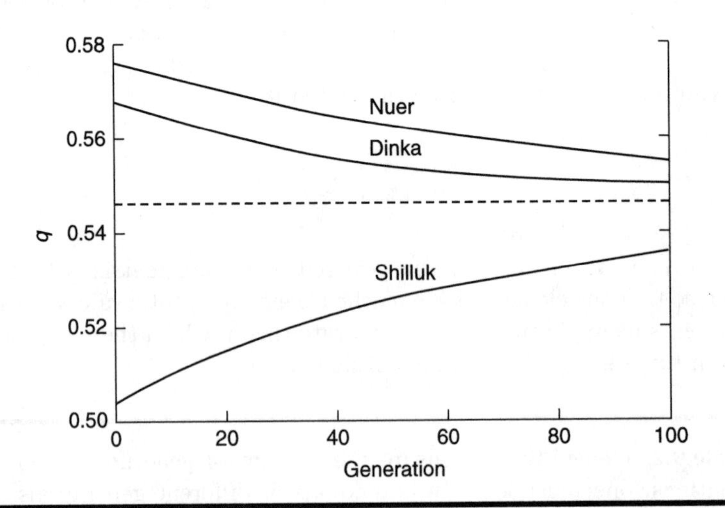

Figure 9.5. The change in allele frequency for the M allele in the Sudanese populations, using the migration matrix given in Table 9.2.

In Table 9.1 and Example 9.2, it was assumed that all individuals came from one of the known subpopulations. However, individuals may also migrate into the population from outside, thus resulting in exogamous matings—matings between members of the population and migrants from outside. The data below in Example 9.3 from a tropical tree species illustrate that a substantial proportion of the progeny in a population may have parents from outside of the known area. To allow for this, we can add an extra column to the migration matrix with elements m_{iE}, resulting in $k + 1$ columns. The allele frequency in subpopulation i using the migration matrix is then

$$q_i' = \sum_{j=1}^{E} m_{ij}\, q_j$$

where the last terms are $m_{iE}q_E$ and represent the product of the amount of gene flow into subpopulation i from outside and the allele frequency in these migrants. The asymptotic frequency in this case is determined by the allele frequency in the migrants as in the continent–island model. With a small amount of exogamy, the allele frequencies initially equalize among subpopulations and then gradually approach the allele frequency of the exogamous migrants. In other words, if there is gene flow from outside the connected group of subpopulations, they will eventually reflect the genetic constitution of these outside immigrants. This assumes that genetic drift and selection do not have significant effects on allele frequencies.

c. Wahlund's Principle

Sometimes population substructuring is not obvious, and as a result, a sample may consist of a group of heterogeneous subsamples from a population. For example, subpopulations may be separated by subtle physical or ecological barriers that limit movement between groups. ***When these subpopulations are lumped together and if there are differences in allele frequencies among these subsamples, there will be a deficiency of heterozygotes and an excess of homozygotes***, even if Hardy–Weinberg proportions exist within each subsample (Wahlund, 1928). As is shown below, this rather surprising relationship occurs because the square of the mean allele frequency over populations is less than the mean of the square of the allele frequencies over populations.

This effect, which is known as Wahlund's principle or the **Wahlund effect**, can be illustrated in the following way. Assume that p_i and q_i are the frequencies of the A_1 and A_2 alleles, respectively, in subpopulation i. The observed mean frequencies of the three genotypes, A_1A_1, A_1A_2, and A_2A_2, over k equal-sized subpopulations, assuming Hardy–Weinberg proportions within each subpopulation, are

$$\overline{P} = \frac{1}{k}\sum p_i^2$$
$$\overline{H} = \frac{1}{k}\sum 2p_iq_i$$
$$\overline{Q} = \frac{1}{k}\sum q_i^2$$

However, the expected proportions of the three genotypes may be different from these observed genotype proportions. The mean frequencies of the A_1 and A_2 alleles over all k subpopulations, assuming equal numbers in each subpopulation, are

$$\overline{p} = \frac{1}{k}\sum_{i=1}^{k} p_i$$

$$\overline{q} = \frac{1}{k}\sum_{i=1}^{k} q_i$$

Using these mean frequencies, we expect the proportions of the three genotypes A_1A_1, A_1A_2, and A_2A_2 to be $\overline{p}^2$, $2\overline{pq}$ and $\overline{q}^2$, respectively. Therefore, the difference between the observed and expected genotype frequency for A_2A_2 is

$$\begin{aligned}\overline{Q} - \overline{q}^2 &= \frac{1}{k}\sum q_i^2 - \overline{q}^2\\ &= \frac{1}{k}\sum q_i^2 - 2\overline{q}^2 + \overline{q}^2\\ &= \frac{1}{k}\left(\sum q_i^2 - 2\sum q_i\overline{q} + k\overline{q}^2\right)\\ &= \frac{1}{k}\sum (q_i - \overline{q})^2\end{aligned}$$

The last line is the definition of the variance in the frequency of A_2, V_q, over subpopulations. Therefore, the difference between the observed and expected frequency of A_2A_2 is equal to the variance of the allele frequency over subpopulations.

The observed genotype frequencies can then be written as

$$\begin{aligned}\overline{P} &= \overline{p}^2 + V_q\\ \overline{H} &= 2\overline{pq} - 2V_q\\ \overline{Q} &= \overline{q}^2 + V_q\end{aligned} \tag{9.3a}$$

As a result, the observed frequency of the homozygotes, as compared with that expected from a large homogeneous population, is increased by an amount V_q, and the observed frequency of the heterozygote is reduced by twice the variance. The same result would occur if populations were lumped over time with temporal variation in allele frequency.

To illustrate the Wahlund effect, let us assume that two subpopulations had different frequencies of allele A_2, $q_1 = 0.4$ and $q_2 = 0.8$. Assuming Hardy–Weinberg proportions within each subpopulation, the frequency of genotypes A_1A_1, A_1A_2, and A_2A_2 are 0.36, 0.48, and 0.16 in population 1 and 0.04, 0.32, and 0.64 in population 2. The observed genotype frequencies, $\overline{P}$, $\overline{H}$, and $\overline{Q}$, are then 0.2, 0.4, and 0.4, respectively; for example, $\overline{Q} = {}^1\!/_2\left(q_1^2 + q_2^2\right) = 0.4$. If there were Hardy–Weinberg proportions for the total population, because $\overline{p} = 0.4$ and $\overline{q} = 0.6$, then the expected frequencies of the genotypes would be 0.16, 0.48, and 0.36 for A_1A_1, A_1A_2, and A_2A_2. From expression 9.3a, then the variance is 0.04, and this deviation from Hardy–Weinberg proportions is due only to lumping samples that differ in allele frequencies.

From these calculations, it is apparent that a fairly large heterogeneity in allele frequencies must exist to result in a substantial departure from Hardy–Weinberg proportions due only to the Wahlund effect. For example, a difference in allele frequencies of 0.1 between two subpopulations results in an observed excess of each homozygote (V_q) of only 0.0025. The largest effect occurs when two different subpopulations are fixed for two different alleles and the observed excess of each homozygote (V_q) is 0.25 because there are no heterozygotes observed.

The effect of subdivision when there are more than two alleles at a locus is more complicated. Although the total frequency of all heterozygotes is reduced with multiple alleles, the frequency of particular heterozygotes may not be reduced. If there is a high correlation among the frequency of different alleles, then the frequency of some heterozygotes may actually be greater than expected, and the frequency of some homozygotes may be smaller (Li, 1969). For multiple alleles, the observed genotype frequencies are

$$\begin{aligned} \overline{P}_{ii} &= \overline{p}_i^2 + V_{p(i)} \\ \overline{P}_{ij} &= 2\overline{p}_i\overline{p}_j + 2COV_{p(i,j)} \end{aligned} \qquad (9.3b)$$

where $COV_{p(i,j)}$ is the covariance of the frequencies of alleles A_i and A_j and is

$$COV_{p(i,j)} = \frac{1}{k}\sum p_i p_j - \overline{p}_i\overline{p}_j$$

Sometimes it can be difficult to determine whether a deficiency of heterozygotes is the result of inbreeding or the Wahlund effect. However, the heterozygote frequency at all loci should be affected by inbreeding, whereas the heterozygote frequency at only those loci with allele frequency variation over subpopulations should be reduced by the Wahlund effect. Furthermore, when there are multiple alleles, all heterozygotes should be reduced in frequency from inbreeding, whereas some may be decreased and others unaffected or increased from the Wahlund effect.

d. Gametic and Zygotic Gene Flow

Gene flow can occur in different life stages and may have different effects depending on the age, sex, or other characteristics of the migrants. In particular, gene flow that occurs in the gametic stage, such as pollen or sperm, or in the zygotic stage, such as seeds or juveniles, can have quite different genetic consequences. For example, seeds that are stored in a cache by an animal and subsequently initiate subpopulations may deviate in frequency from subpopulation to subpopulation. As a result, there may be deviations from Hardy–Weinberg proportions due to the Wahlund effect—that is, there could be a general deficiency of heterozygotes and excess of homozygotes.

Gametic or pollen gene flow may be quite different in its effect on genotype proportions. Assume that the pollen pool is composed of migrant pollen with a frequency for A_2 of q_m and pollen from resident plants with a frequency for A_2 of q. In such a case, if a proportion m of the pollen is from migrants, then the overall frequency of A_2 in the pollen pool is

$$q_p = (1 - m)\,q + mq_m$$

Assuming that the frequency of A_2 in the female gametes is q, the frequencies of the three genotypes are then

$$\begin{aligned} P &= pp_p \\ H &= pq_p + p_pq \\ Q &= qq_p \end{aligned} \tag{9.4a}$$

These proportions are analogous to those given by expression 2.6a for different allele frequencies in female and male gametes. Obviously, pollen migration from an area with a different allele frequency may result in an increase in heterozygosity and a decrease in homozygosity. Therefore, seed and pollen gene flow may have opposite effects on genotype proportions.

It is useful to combine the two types of gene flow—pollen and seed (or gametic and zygotic)—into one overall measure of the amount of gene flow. Because the genetic content of pollen is haploid and that of seeds is diploid, then the overall gene flow is

$$m = \tfrac{1}{2}m_p + m_s \tag{9.4b}$$

where the amounts of pollen and seed gene flow are m_p and m_s, respectively. On p. 497, overall and maternal (mtDNA) F_{ST} values, measures of population subdivision, are used to estimate the ratio of pollen and seed gene flow.

II. ESTIMATION OF GENE FLOW AND POPULATION STRUCTURE

Estimating the amount of gene flow in most situations is rather difficult. For example, an estimate of movement of individual in insects or rodents into a population may be an overestimate of the gene flow because these individuals may leave no offspring or fewer offspring than those that have not moved. In other instances, it may be difficult to identify migrants that result in gene flow because they are present for only a short time but have still mated before they moved from the area. As a result to be genetically meaningful, measurements of the amount of dispersal should have associated with them an estimate of the probability that these dispersers will contribute to the population.

Direct estimates of the amount of gene flow can be obtained in organisms where different individuals can be identified. Direct observation of gene flow is possible in humans where birth records or other information is available or in other organisms where individual identification is possible or individual marks are used. Many approaches have been employed to mark individuals differentially, such as toe clipping in rodents, leg banding in birds, coded-wire tags in fish, and radio transponders in many different vertebrates. However, both movement of individuals and their incorporation into the breeding population are necessary for gene flow. Using highly variable genetic markers, it is now possible to identify parents genetically and thereby determine the spatial movement of gametes between generations without direct movement information of the parents (see p. 496). As we discuss on p. 647, individuals can be assigned to specific populations using genetic markers, thereby determining whether they are migrants or not.

In situations where the population has a continuous distribution, the variance of gene flow distance and the population density can be used to estimate the neighborhood size, as shown on p. 336. However, the distribution of gene flow distance may have various patterns, often with an excess of both nonmigrants and long-distance migrants. The overall description of such gene flow is complicated, and fitting theoretical distributions to observed data is often difficult (for an extensive discussion on human migration patterns, see Cavalli-Sforza and Bodmer, 1971; for a review of gene flow distributions in plants, see Cain *et al.*, 2000; Austerlitz *et al.*, 2004).

When the population is subdivided into discrete colonies or groups, then a migration matrix can be constructed. In human populations, generally, the birth places of the parents and their offspring are used to establish such a matrix. Because birth records in some populations are available for many generations, evaluation of the effect of gene flow patterns among villages or regions over long periods are possible. In most other populations, only short-term data at best are available, making estimates of variation in gene flow over time difficult.

Indirect measures of gene flow using genetic markers are useful to confirm behavioral or other observations or when these observations are inconclusive or impossible. The simplest indirect estimate is for hybrid populations where information exists on the parental populations. Most commonly, the number of migrants between groups is measured using techniques to evaluate population structure, such as F statistics or related measures. Slatkin (1985) suggested that the frequency of rare alleles over different groups could also be used to estimate gene flow. The use of these different measures has been evaluated by Slatkin and Barton (1989), Cockerham and Weir (1993), and Excoffier (2001). More recently, measures of gene flow based on coalescent methods are being used (Beerli and Felsenstein, 2001).

a. Hybrid Populations

Hybrid populations—***those that receive migrants from two or more other populations***—are often of interest because they are at a border between two species or they may be undergoing rapid genetic change. Such populations are generally the result of intermating or the mixture of two or more parental populations over an extended period of time. If there are only two parental populations, then the ***proportion of gene flow from the outside population*** (or the continent), sometimes called **admixture**, in one generation can be estimated by rearranging equation 9.1a so that

$$\hat{m} = \frac{q_0 - q_1}{q_0 - q_m}$$

If gene flow has occurred over a number of generations, the symbols in this expression can be changed so that

$$\hat{M} = \frac{q_A - q_H}{q_A - q_B} \tag{9.5a}$$

where $\hat{M}$ is the estimate of the total amount of gene flow of individuals from parental population B in the hybrid population H ($1 - \hat{M}$ is the proportion from parental population A) (Cavalli-Sforza and Bodmer, 1971). The allele frequencies in parental populations A and B and the hybrid population H are q_A, q_B, and q_H, respectively.

We can estimate the proportion of gene flow per generation over a number of generations by rearranging expression 9.1c so that

$$(1-\hat{m})^t = \frac{q_t - q_m}{q_0 - q_m}$$

This can be rewritten to reflect gene flow of populations A and B into population H as

$$(1-\hat{m})^t = \frac{q_H - q_B}{q_A - q_B}$$

and rearranged so that

$$\hat{m} = 1 - e^{\ln(1-\hat{M})/t} \tag{9.5b}$$

where $1 - \hat{M} = (q_H - q_B)/(q_A - q_B)$. The value of $\hat{m}$ here is the estimate of the proportion of gene flow per generation into the hybrid population from parental population B. Of course, this estimate of gene flow assumes that the effects of selection and genetic drift on allele frequencies are negligible relative to the amount of gene flow (see Long, 1991; Wang, 2003; and Choisy *et al.*, 2004, for approaches and developments in the methods used to estimate admixture contributions). Estimates of admixture are used in Example 9.3, which examines European gene flow into the African American population.

Example 9.3. The amount of gene flow (admixture) has been estimated in a number of human populations with ancestry from several populations. In particular, estimation of the amount of gene flow from Europeans in the African-American population was extensively examined using blood group loci (Reed, 1969; Adams and Ward, 1973). However, some of these loci may have been influenced by selection as well as gene flow, and the differences in allele frequency between ancestral European and African groups were often not large enough to yield good estimates. Parra *et al.* (1998) have found 10 of what they term population-specific alleles that show large differentiation between populations of African and European ancestry (see also Collins–Schramm *et al.*, 2002). These are all biallelic loci, and most of them are the result of the presence or absence of restriction site polymorphisms. Table 9.3 gives the frequencies of four of these alleles for Africans, Europeans, and six samples with predominantly African ancestry. The most extreme frequency difference is for the *FY-NULL* 1* marker, which was present in all of the Europeans examined and in none of the African individuals. The estimate of total gene flow (and the standard error) in the rightmost column is similar to that given in expression 9.5a but incorporates the effect of genetic drift and sampling variance (Long, 1991) and is combined over 10 loci.

TABLE 9.3 The allele frequencies of the four most extreme population-specific alleles in Africans, Europeans, and six U.S. populations of African descent (Parra *et al.*, 1998). In the rightmost column is the estimated European ancestral proportion for the populations of African descent, averaged over 10 population-specific alleles.

Population	*FY-NULL*1*	*OCA2*1*	*RB2300*1*	*GC-1F*	$\hat{M}^*$
African	0.000	0.098	0.920	0.824	
European	1.000	0.769	0.333	0.156	
African-American					
Maywood, IL	0.185	0.203	0.776	0.710	0.188 ± 0.014
New York	0.210	0.220	0.821	0.738	0.198 ± 0.021
Philadelphia	0.160	0.137	0.802	0.771	0.138 ± 0.019
Charleston, SC	0.112	0.208	0.888	0.765	0.116 ± 0.013
New Orleans	0.200	0.284	0.842	0.669	0.225 ± 0.016
Jamaica	0.065	0.091	0.870	0.790	0.068 ± 0.013

The lowest estimate of gene flow is for the Jamaican sample, followed by that from Charleston, South Carolina. Parra *et al.*, 2001, estimated the European ancestry in Gullah-speaking residents of South Carolina as even lower, $\hat{M} = 0.035$. There is significant variation in the estimated amount of European ancestry in the different northern United States samples (see Parra *et al.*, 1998, for a discussion of possible explanations). Parra *et al.* also estimated the gene flow that is either maternally or paternally specific by examining mtDNA and Y chromosome variants. For the nine United States samples for which there were data for both types of markers, estimated gene flow for mtDNA was 0.140 and for Y chromosomes was 0.248. Thus there appears to have been, historically, a sex-biased contribution of European ancestry, with more matings between European-American males and African-American females than vice versa.

Africans were brought in large numbers to the United States as slaves beginning at the end of the seventeenth century. Assuming that gene flow has occurred at a fairly continuous rate since that time and that the average generation length is about 25 years, then the elapsed number of generations (t) during which gene flow has occurred is approximately 12. With these assumptions, $\hat{m}$ from expression 9.5*b* would be 0.021 for New Orleans and 0.010 for Charleston. The hypothetical pattern of gene flow over time for *FY-NULL* 1* with these values would be as given in Figure 9.6. As expected, this model shows that gene flow has a greater effect on allele frequency in the early generations and that the frequency in the hybrid population reaches that of the European parental population B only asymptotically. Of course, these projections assume given rates of gene flow, and gene flow in the future may differ from these assumed values.

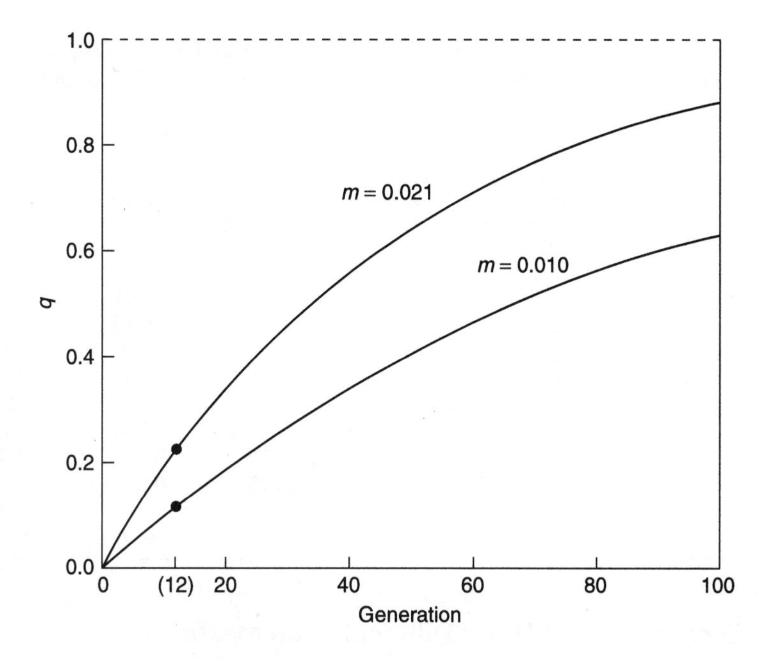

Figure 9.6. The hypothetical pattern of allele change over time for the African-American population, using two different gene flow estimates. The solid circles indicate present-day frequencies of the *FY-NULL*1* allele and the broken line is the asymptotic frequency.

When there are several ancestral sources for a population, then the allele frequency can be calculated using expression 9.2*a*. If a population has ancestral contributions from several populations, then

$$q_H = M_1 q_1 + M_2 q_2 + M_3 q_3 \cdots M_i q_i \tag{9.5c}$$

where M_i and q_i are the contributions and allele frequencies from the *i*th population.

For example, Cerda-Flores *et al.* (2002) estimated allele frequencies for the 13 core CODIS (Combined DNA Index System of the U.S. Federal Bureau of Investigation, see p. 639) microsatellite loci in a Mestizo (mixed race) population from northeastern Mexico. Previous studies have shown that in such populations there is ancestry from Europeans, Amerindians, and Africans. Table 9.4 gives the allele frequencies for two of these loci in the Mestizo population and from samples from Spain, Nigeria, and southwestern United States Amerindians. There is a general overlap in the presence of the alleles, but the representative ancestral populations differ in allele frequencies. Overall, Cerda-Flores *et al.* (2002) estimated that in this population 55%, 40%, and 5% of the ancestry are Spanish, Amerindian, and African, respectively. For example, if we examine allele 15 at locus D3S1358 (boldface in Table 9.4) using expression 9.5*c*, then the predicted $q_H = (0.55)(0.228) + (0.40)(0.653) + (0.05)(0.312) = 0.0402$ and the observed $q_H = 0.0.409$, only slightly different.

TABLE 9.4 The estimated allele frequencies for two microsatellite loci (D3S1358 and VWA) from a Mestizo (mixed race) population from northeastern Mexico and from representative ancestral populations from Spain, Amerindians, and Africa (Cerda-Flores *et al.*, 2002).

	D3S1358				*VWA*			
Allele	*Mestizo*	*Spain*	*Amerindian*	*Africa*	*Mestizo*	*Spain*	*Amerindian*	*Africa*
11	—	—	—	0.010				
12	0.004	—	—	—				
13	0.010	0.011	0.008	—	—	—	0.001	—
14	0.073	0.080	0.061	0.135	0.087	0.116	0.045	0.087
15	**0.409**	**0.228**	**0.653**	**0.312**	0.105	0.167	0.036	0.304
16	0.238	0.243	0.168	0.312	0.318	0.268	0.439	0.239
17	0.147	0.217	0.077	0.177	0.297	0.173	0.254	0.174
18	0.115	0.199	0.032	0.052	0.136	0.174	0.145	0.087
19	0.004	0.022	0.001	—	0.056	0.076	0.066	0.065
20					—	0.025	0.014	0.011
21					—	—	—	0.022
22					—	—	—	0.011

b. *F* Coefficients and Other Differentiation Measures

Several different approaches have been used to estimate the amount of differentiation in the subdivisions of a population. Most importantly, Wright (1951, 1965b) developed an approach to partition the genetic variation in a subdivided population that is commonly used and provides an obvious description of differentiation. This approach consists of three different F coefficients (these are correlation coefficients and are different from the F statistics used in the analysis of variance) used to allocate the genetic variability to the total population level (T), subpopulations (S), and individuals (I). These three coefficients, F_{ST}, F_{IT}, and F_{IS}, are interrelated so that

$$1 - F_{IT} = (1 - F_{ST})(1 - F_{IS})$$

$$F_{ST} = \frac{F_{IT} - F_{IS}}{1 - F_{IS}} \tag{9.6}$$

F_{ST} is a measure of the genetic differentiation over subpopulations and is always positive. F_{IS} and F_{IT} are measures of the deviation from Hardy–Weinberg proportions within subpopulations and in the total population, respectively, where positive values indicate a deficiency of heterozygotes and negative values indicate an excess of heterozygotes. There has been some controversy on the estimation of these values (Weir and Cockerham, 1984; Nei, 1986; Neigel, 1997), but the general properties of differentiation measures that we discuss here are valid for all these approaches. Excoffier (2001) discussed the different estimators, compared the different approaches, and pointed out their limitations.

Nei (1977, 1987) has shown how these coefficients can be expressed in terms of allele frequencies and observed and expected genotype frequencies. Let us consider the case where there are two alleles at a locus and there are k subpopulations. The frequency of genotype A_2A_2 in the ith subpopulation is

$$Q_i = q_i^2 + F_{ISi}\, q_i\, (1 - q_i)$$

where F_{ISi} is the fixation index in the ith subpopulation (see p. 95). We can rearrange this expression to read

$$F_{ISi} = \frac{Q_i - q_i^2}{q_i\,(1 - q_i)}$$

The fixation index over all subpopulations is then

$$F_{IS} = \frac{\overline{Q} - \overline{q^2}}{\overline{q} - \overline{q^2}} \qquad (9.7a)$$

where $\overline{Q} = \Sigma\, w_i Q_i$ is the weighted frequency of A_2A_2 in the total population, $\overline{q} = \Sigma\, w_i q_i$ is the weighted frequency of A_2 in the total population, and $\overline{q^2} = \Sigma\, w_i q_i^2$. Generally, equal weight is given to each subpopulation—that is, $w_i = 1/k$, unless information about different subpopulation sizes is known. The significance of F_{IS} can be calculated from a χ^2 test (expression 2.17b) as

$$\chi^2 = NF_{IS}^2 \qquad (9.7b)$$

where N is the number of individuals in the sample and there is one degree of freedom.

The frequency of A_2A_2 in the total population can be expressed as

$$\overline{Q} = (1 - F_{IT})\,\overline{q}^2 + F_{IT}\overline{q}$$

This expression can be rearranged so that

$$F_{IT} = \frac{\overline{Q} - \overline{q}^2}{\overline{q}\,(1 - \overline{q})} \qquad (9.7c)$$

(Note the differences in the means from expression 9.7a.) We can calculate the value of F_{ST} by substituting expressions 9.7a and 9.7c for F_{IS} and F_{IT} into expression 9.6 so that

$$F_{ST} = \frac{\overline{q^2} - \overline{q}^2}{\overline{q}\,(1 - \overline{q})}$$

This expression agrees with the usual one for F_{ST}, which is

$$F_{ST} = \frac{V(q)}{\bar{q}(1-\bar{q})} \tag{9.8a}$$

where $V(q)$ is the variance in the frequency of A_2. Note from expression 2.25*a* that a χ^2 test is

$$\chi^2 = 2NF_{ST} \tag{9.8b}$$

where $2N$ is the number of gametes in the sample and there is one degree of freedom (see Ryman and Jorde, 2001, for a discussion of statistical power when testing for genetic differentiation).

Two numerical examples are given in Table 9.5 to help clarify the meaning of the F coefficients. In the top example, the two subpopulations have

TABLE 9.5 Two hypothetical examples to illustrate the meaning of F coefficients.

Subpopulation	A_1A_1	A_1A_2	A_2A_2	q
1	0.25	0.5	0.25	0.5
2	0.35	0.3	0.35	0.5
	$F_{IS} = 0.2$	$F_{IT} = 0.2$	$F_{ST} = 0.0$	
1	0.25	0.5	0.25	0.5
2	0.49	0.42	0.09	0.3
	$F_{IS} = 0.0$	$F_{IT} = 0.0417$	$F_{ST} = 0.0417$	

the same allele frequencies so that there is no genetic differentiation among subpopulations, making $F_{ST} = 0$. However, both F_{IS} and F_{IT} are positive because of the deficiency of heterozygotes in subpopulation 2. In the bottom example, both subpopulations are in Hardy–Weinberg proportions so that $F_{IS} = 0$. However, because of the variation in allele frequencies between subpopulations, both F_{IT} and F_{ST} are positive.

Clark *et al.* (2003) estimated the amount of population differentiation among European Americans, African Americans, and Asians for 4833 SNPs (Figure 9.7). A large proportion of the SNPs indicated very low differentiation, and the mean F_{ST} value was 0.083. However, the distribution had a long tail with 10% of the SNPs having an F_{ST} value of > 0.18. The SNPs in the long tail indicate that there is strong differentiation among the groups for the genetic regions marked by these SNPs, potentially pointing out past selective events acting differentially in these populations.

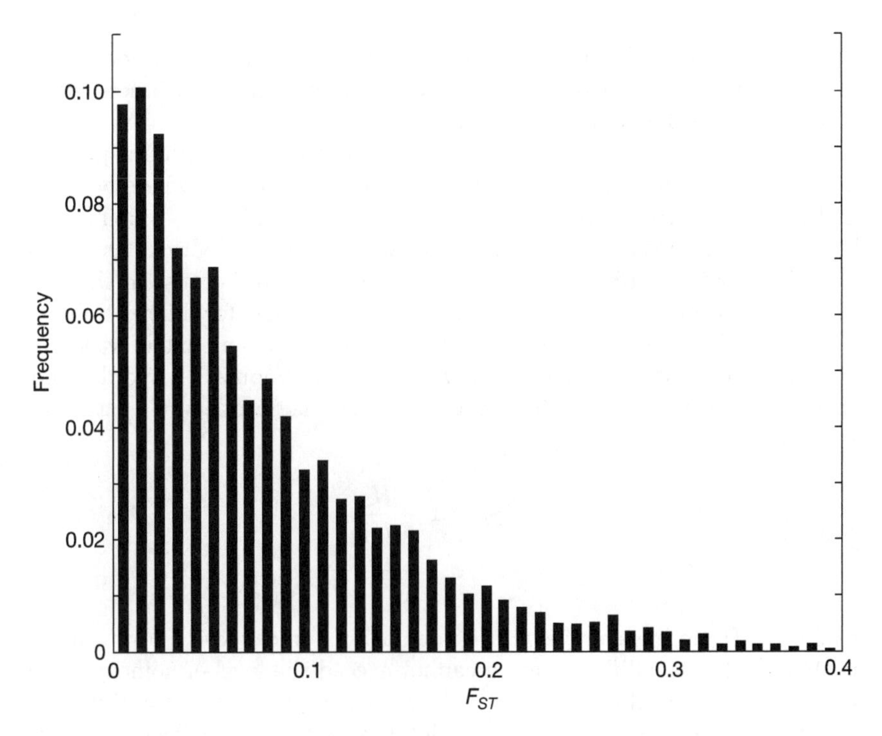

Figure 9.7. The observed amount of population differentiation measured by F_{ST} among human populations of European Americans, African Americans, and Asians for 4,833 SNPs (Clark *et al.*, 2003).

Nei (1977) also extended this analysis to multiple loci. For example, the average F coefficients can be calculated for multiple loci as

$$\overline{F}_{IS} = \frac{\overline{H}_S - \overline{H}_0}{\overline{H}_S}$$

$$\overline{F}_{IT} = \frac{\overline{H}_T - \overline{H}_0}{\overline{H}_T} \qquad (9.8c)$$

$$\overline{F}_{ST} = \frac{\overline{H}_T - \overline{H}_S}{\overline{H}_T}$$

where $\overline{H}_0$ is the average observed heterozygosity within a subpopulation over loci, $\overline{H}_S$ is the average expected heterozygosity within subpopulations over loci, and $\overline{H}_T$ is the average of the expected heterozygosity in the total population over loci.

As an estimate of F_{ST}, and assuming Hardy–Weinberg proportions, Nei (1973) defined the coefficient of gene differentiation as

$$G_{ST} = \frac{H_T - H_S}{H_T} \qquad (9.9a)$$

where H_S is the average subpopulation Hardy–Weinberg heterozygosity and $H_T = 1 - \sum \overline{p}_i^2$ for any number of alleles. Nei (1973, 1987) pointed out that although G_{ST} is a good measure of the relative differentiation among subpopulations, it is highly dependent on the amount of variation within subpopulations and in the total population.

The dependence of G_{ST} on the amount of genetic variation is particularly true for highly variable loci such as microsatellite loci where both H_S and H_T can approach unity. As a result, the G_{ST} can be very small even if the subpopulations have nonoverlapping sets of alleles (Hedrick, 1999b; see also Charlesworth, 1998). This seems counterintuitive because in the two-allele case, when the subpopulations are monomorphic for different alleles, $F_{ST} = 1$. However, G_{ST} measures the proportional amount of variation within subpopulations as compared with the total population and does not specify the identity of the alleles involved. The magnitude of G_{ST} can also be written as

$$\begin{aligned} G_{ST} &= 1 - \frac{H_S}{H_T} \\ &< 1 - H_S \end{aligned}$$

where $1 - H_S$ is the average within population homozygosity. From this, it is obvious that the differentiation cannot exceed the level of homozygosity, no matter what evolutionary factor is influencing the amount and pattern of variation. Obviously, when using highly polymorphic makers that make the level of homozygosity low, then the maximum G_{ST} must also be greatly reduced.

Jin and Chakraborty (1995) derived the predicted change of G_{ST} over subpopulations when they are all descended from a common ancestral population and completely isolated over time. Asymptotically, this value becomes

$$G_{ST} = \frac{(k-1)(1-H_S)}{k-1+H_S} \tag{9.9b}$$

where k is the number of subpopulations. At this limit, all populations have nonoverlapping sets of alleles and the genetic distance is maximized; for example, the standard genetic distance of Nei (1972) is ∞.

We can show how the maximum differentiation is related to mutation by assuming the equilibrium level of heterozygosity within populations for the infinite allele model as in expression 7.8c. Therefore, by substitution, expression 9.9b becomes

$$G_{ST} = \frac{k-1}{k(4N_eu+1)-1} \tag{9.9c}$$

To illustrate the maximum values of G_{ST} for different levels of mutation, Figure 9.8 plots G_{ST} for 2 and 10 populations. For high mutation rate loci, such as microsatellites, $4N_eu > 1$, and G_{ST} is relatively small. For low

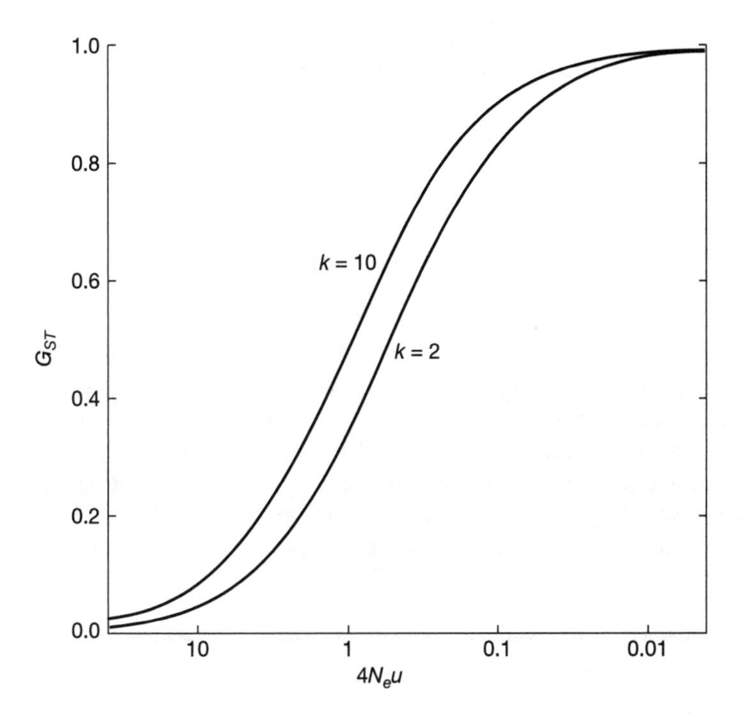

Figure 9.8. The asymptotic level of G_{ST} between completely isolated populations as a function of the mutation rate for two and ten populations.

mutation rate loci, such as allozymes or SNPs, $4N_eu < 0.1$, G_{ST} approaches unity. In other words, the size of G_{ST} for isolated populations is strongly influenced by the amount of variation (determined by mutation here) at a locus.

Because of the dependence of G_{ST} on the level of diversity, Nei (1973) proposed an absolute measure of gene differentiation called the minimum genetic distance, which is independent of gene diversity within populations and can be calculated as

$$\overline{D}_m = \frac{k}{k-1}(H_T - H_S) \tag{9.9d}$$

For example, Nei (1987, p. 192) suggested the high G_{ST} value (0.674) for allozyme loci in a study of kangaroo rats by Johnson and Selander (1971) was mainly the result of low gene diversity. In this case of nine populations, $H_T = 0.097$ and $H_S = 0.012$ so that $\overline{D}_m = 0.028$, a value similar to those Nei calculated for humans (0.019) and *Escherichia coli* (0.028).

In the foregoing discussion, we have assumed only one level of structure for the total population. Some populations have further levels of obvious structure. For example, subpopulations may be grouped into regions or divided into colonies (Weir and Cockerham, 1984; Nei, 1987). Given such a hierarchical array of groups, then the distribution of variation can be partitioned among these various levels. For example, if we assume that there is a logical regional level into which subpopulations can be placed, then we can calculate the additional measures

$$F_{SR} = \frac{H_R - H_S}{H_R} \tag{9.10a}$$

and

$$F_{RT} = \frac{H_T - H_R}{H_T} \tag{9.10b}$$

which partition the variation into the diversity among subpopulations within region and that among regions for the total population. With such hierarchical partitioning, it is possible to see at which level the largest amount of variation can be explained. For example, most of the variation may be among subpopulations in some species, whereas in other species, most of the variation may be among regional groups (see Example 9.4 for data for a microsatellite locus from three regions in the endangered Gila topminnow).

Example 9.4. The Gila topminnow was once the most common fish in the Gila River drainage of Arizona, but it is now present only in a few head-springs and remote locations because of water developments and the introduced western mosquito fish (Parker *et al.*, 1999). Hedrick *et al.*, (2001b) surveyed the amount of variation in all the remaining nine natural populations and found that there were five different groups. Table 9.6 gives the allele frequency for microsatellite locus *LL53* for the three regional groups in which there are two or more populations. The frequencies for the subpopulations within a region are very similar, and between regions they are much more different. We can quantify this pattern by calculating F_{SR} and F_{RT} from expressions 9.10*a* and 9.10*b*.

To do this, first calculate the Hardy–Weinberg heterozygosity within each of the seven subpopulations (given in the rightmost column). The mean of these seven values is $H_S = 0.263$. Next calculate the mean allele frequency within each regional group. Using these mean allele frequencies, the Hardy–Weinberg heterozygosity for each region can be calculated (also in the rightmost column). The weighted mean of these values is $H_R = [2(0.109) + 2(0.469) + 3(0.435)]/7 = 0.352$. Finally, the mean allele frequency is calculated (bottom row), and the Hardy–Weinberg heterozygosity of these frequencies is $H_T = 0.647$. With these values, the proportion of variation among subpopulations within regions is $F_{SR} = 0.253$, and the proportion of variation among regions is $F_{RT} = 0.456$. Therefore, nearly twice as much variation is partitioned among the regions as among the subpopulations within groups. Note that we have calculated these values for illustration purposes and without the appropriate corrections for sample size and numbers of groups (for references, see Neigel, 1997).

TABLE 9.6 The frequency of alleles at the LL53 microsatellite locus for the remaining natural populations of the Gila topminnow from three regional groups, Bylas Springs, Sonoita Creek Springs, and Sonoita Creek, and where — indicates allele is absent.

	Allele							
	138	142	144	146	148	150	164	*H*
Bylas Springs								
Bylas Spring I	—	—	—	1.000	—	—	—	0.000
Bylas Spring II	—	—	0.115	0.885	—	—	—	0.204
Mean	—	—	0.058	0.942	—	—	—	0.109
Sonoita Creek Springs								
Cottonwood Spring	—	0.278	0.722	—	—	—	—	0.401
Monkey Spring	—	0.988	—	—	—	0.012	—	0.024
Mean	—	0.633	0.361	—	—	0.006	—	0.469
Sonoita Creek								
Coalmine Canyon	—	0.725	0.250	—	0.025	—	—	0.411
Sonoita Creek	—	0.759	0.241	—	—	—	—	0.366
Red Rock Falls	0.025	0.700	—	—	—	—	0.275	0.434
Mean	0.008	0.728	0.164	—	0.008	—	0.092	0.435
Total mean	0.004	0.493	0.190	0.269	0.004	0.002	0.039	0.647

Extending these ideas to DNA sequence data, Lynch and Crease (1990) showed how nucleotide diversity information can be partitioned into within- and between-population components, Holsinger and Mason-Gamer (1996) showed how a hierarchical analysis can be applied to nucleotide diversity data, and Excoffier (2001) discussed AMOVA (analysis of molecular variance) and how it can be applied to populations with substructure (see AMOVA at http://www.bioss.sari.ac.uk/smart/unix/mamova/slides/frames.html and Arlequin at http://lgb.unige.ch/arlequin/software/2.000/doc/faq/faqlist.htm for software to carry out these calculations). Statistical tests for detecting DNA sequence differentiation of subpopulations have been given by Hudson *et al.* (1992) and Hudson (2000).

c. Sex Differences in Gene Flow

The gene flow pattern may differ substantially between females and males. In many organisms, the level of gene flow is often significantly higher in males, or in male gametes such as pollen, than in females. When there are different rates of gene flow of individuals of the two sexes, then the arithmetic average over the two sexes

$$\overline{m} = \tfrac{1}{2}(m_f + m_m)$$

where m_f and m_m are estimates of female and male gene flow, respectively, can generally be used.

Example 9.5 provides an account determining pollen gene flow, much of it from outside of the sampled area, using microsatellite loci in a tropical tree. In this study, there was no evidence of female gene flow. A study in chimpanzees that appeared to show high gene flow from males outside of the colony (Gagneux *et al.*, 1997) has now been shown (Vigilant *et al.*, 2001) to have a low rate of unknown paternity. The erroneous conclusions of the original study appear to have been the result of technical errors, leading to inaccurate paternity determination.

Example 9.5. Using highly variable genetic markers, it is now possible to determine the paternity of seeds in plant populations where there are relatively small numbers of potential fathers and thus to estimate the amount of pollen gene flow. Chase *et al.* (1996) used 9 microsatellite loci with 42 alleles to determine the pollen parents for 72 independent mating events in a canopy tree, *Pithecellobium elegans*, in Costa Rica. The pollen parents for 41 mating events were assigned unambiguously to one of 28 adult trees within the stand. For example, 6 pollen parents for seeds in tree 106 are indicated in Figure 9.9, one of which, tree 110, was over 300 meters away. Of

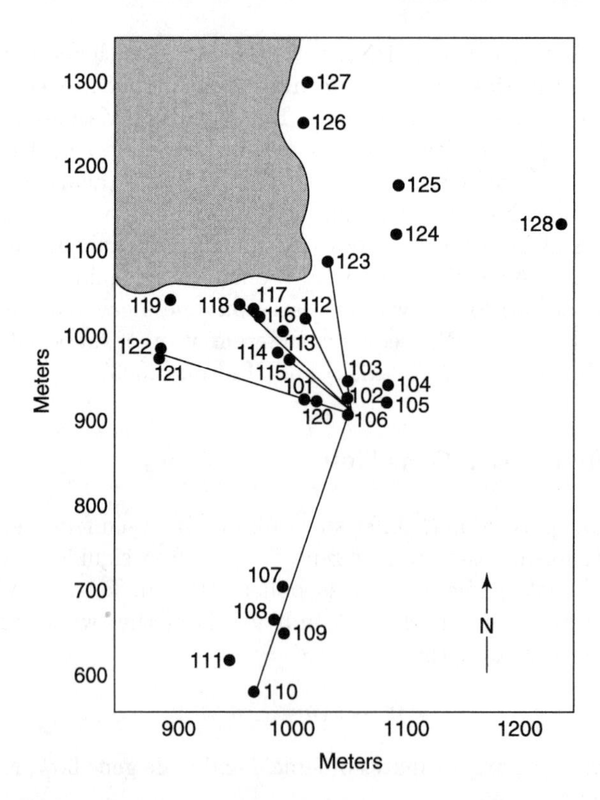

Figure 9.9. Map of the population of a large canopy tree, *Pithecellobium elegans*, in an area of Costa Rica, with individual adult plants indicated by numbers. Other unknown reproductive individuals exist in the partially degraded natural forest (shaded area) (Chase *et al*, 1996). The lines indicate known pollen flow events to tree 106 in the center of the map.

the remaining 31 matings, 9 matings appear to be the result of selfing (or possibly fertilization from an unknown identical tree outside the sampled area), 2 matings were unresolved and could have been from either of two trees within the sampled trees, and 20 matings were definitely not from the sampled trees. Because 20 of 72 genes from the male parents, and none from the female parents, are from outside, the average estimate of outside gene flow, excluding seed movement, is $\overline{m} = (0.00 + 0.28)/2 = 0.14$. The average distance for the known matings (excluding the probable selfing events) was 142 meters, much farther than the average distance of the nearest neighbor (27 m). The high average pollen gene flow distance is consistent with the known pollination biology of *P. elegans*, which is thought to be pollinated by hawkmoths with long foraging distances.

In plants, there may be gene flow from both gametes (nearly always pollen) and zygotes (seeds), and the effect of a given amount of gene flow is more effective from seeds because they are diploid and pollen is haploid. However, the amount of gene flow from pollen may often be much larger, which could overcome this twofold difference in genetic content. Ennos (1994) showed that the biparental or overall F_{ST} is related to that for maternally inherited markers, $F_{ST(m)}$, and paternally inherited markers, $F_{ST(p)}$, as

$$F_{ST} = \frac{F_{ST(m)}F_{ST(p)}}{F_{ST(m)} + F_{ST(p)} - 3F_{ST(m)}F_{ST(p)}} \tag{9.11a}$$

In conifers, mtDNA is maternally inherited, and cpDNA is paternally inherited. Latta and Mitton (1997) examined the variation in mtDNA, cpDNA, and a number of allozyme loci in populations of limber pine, *Pinus flexilus*, mainly in north central Colorado. The value of $F_{ST(m)} = 0.679$, high mainly because a southern sample had a high frequency of mtDNA type not found in the other populations, and the value of $F_{ST(p)} = 0.013$, low because many of the same cpDNA types were found in all populations. When these values are used, the predicted F_{ST} from expression 9.11*a* is 0.013; in other words, the high inferred level of pollen gene flow is predicted to result in a low overall F_{ST}. The median value for 10 allozyme loci was 0.016, not significantly different than that expected.

Using the level of overall F_{ST} and the maternal $F_{ST(m)}$, Ennos (1994) showed that (with a number of assumptions) the ratio of pollen gene flow (m_p) to seed gene flow (m_s) is

$$\frac{m_p}{m_s} = \frac{F_{ST(m)} - 2F_{ST} + F_{ST}F_{ST(m)}}{F_{ST}(1 - F_{ST(m)})} \tag{9.11b}$$

Table 9.7 gives the overall and maternal F_{ST} for eight different tree species and the estimates of the ratios of pollen to seed gene flow. In the oak and

TABLE 9.7 The overall (biparental) and maternal F_{ST} values for eight tree species, one oak (*Quercus*) and five pines (*Pinus*), and two mountain ash (*Sorbus*), and the estimated ratio of pollen to seed gene flow (Ennos, 1994; Latta and Mitton, 1997; Oddou Muratorio *et al.*, 2001; Bacles *et al.*, 2004).

Species	F_{ST}	$F_{ST(m)}$	m_p/m_s
Quercus petraea/Q. robar	0.037	0.88	196
Pinus contorta	0.061	0.66	28
P. radiata	0.13	0.83	31
P. attenuata	0.12	0.86	44
P. muricata	0.22	0.88	24
P. flexilis	0.013	0.68	159
Sorbus aucuparia	0.043	0.13	1.4
S. torminalis	0.11	0.34	2.2

pine species, the pollen gene flow is much higher, and in two species, the oak species complex and limber pine, it appears to be more than two orders of magnitude higher. On the other hand, in the two mountain ash species, pollen and seed gene flow rates are similar, apparently reflecting effective seed dispersal strategies by birds and mammals. This general theoretical approach has been extended by Hu and Ennos (1997) and Hamilton and Miller (2003).

Molecular data have suggested that there is female-biased dispersal in human populations (Bamshad *et al.*, 1998; Seielstad *et al.*, 1998) and male-biased dispersal in whales, great white sharks, and elephant seals (Lyrholm *et al.*, 1998; Pardini *et al.*, 2001; Fabiani *et al.*, 2003). In an examination of the effect of known different movements in the two sexes on the amount and pattern of genetic variation, Oota *et al.* (2001) examined human populations in Thailand (Example 9.6). The theoretical impact of sex differences in gene flow has been examined by Wang (1997), Berg *et al.* (1998), and Laporte and Charlesworth (2002).

Example 9.6. Sex differences in gene flow should result in differences in the amount and pattern of sex-specific markers within and between populations. For example, in humans, matrilocal and patrilocal societies are those in which women and men, respectively, stay in their birthplace and the other sex moves after marriage. As a result, in a matrilocal population, the genetic variation in women (and mtDNA variation) should be only from the local population, resulting in lower within-population and greater between-population variation. Genetic variation in men (and Y chromosome variation) should be a composite from other populations, resulting in higher within-population and lower between-population variation. For patrilocal populations, the opposite should be true with within-population mtDNA variation, a composite of other populations, and within-population Y chromosome variation from the local population.

To determine whether these predicted differences are actually observed, Oota *et al.* (2001) documented the amount of within- and between-population variation for mtDNA and Y chromosomes for three known matrilocal groups and three patrilocal groups in the hill tribes of northern Thailand. The haplotype diversity for the control region of mtDNA was higher in patrilocal groups than in the matrilocal groups, as predicted (Table 9.8). In addition, the diversity for nine microsatellite loci on the Y chromosome was higher in matrilocal groups than in the patrilocal groups, again as predicted. The genetic distance showed the opposite patterns (Table 9.8), as predicted. That is, the genetic distance was larger between matrilocal groups for mtDNA, and it was larger between patrilocal groups for Y chromosome markers.

TABLE 9.8 The mean diversity in mtDNA and Y chromosomes in matrilocal and patrilocal human populations from northern Thailand and the genetic distance in mtDNA and Y chromosomes between matrilocal and between patrilocal human populations (the larger values are given in boldface).

	mtDNA		*Y chromosome*	
	Matrilocal	*Patrilocal*	*Matrilocal*	*Patrilocal*
Diversity	0.860	**0.937**	**0.965**	0.863
Genetic distance	**0.290**	0.118	0.131	**0.451**

III. POPULATION STRUCTURE AND GENETIC DRIFT

The effect of gene flow is to keep the allele frequencies in different subpopulations similar. However, if the subpopulations are finite in size, then genetic drift may result in random differences among them, even with gene flow. The simplest model to examine the joint effects of gene flow and genetic drift assumes that migrants enter a number of equal-sized, finite populations in equal proportions from a source population. More realistically, subpopulations are distributed over space, and gene flow between them must depend to some extent on their distance from each other. For example, distance-dependent gene flow can be included in models where individuals are distributed in discrete groups, colonies, or villages; these are generally known as stepping-stone models. The general model of population structure that we discussed on p. 476 can be used to incorporate distant-dependent gene flow of varying amounts between subpopulations as in stepping-models models. Beerli and Felsenstein (2001) have provided a maximum likelihood estimate of a migration matrix (MIGRATE at http://evolution.genetics.washington.edu/lamarc.html).

a. The Continent–Island or Island Model

Let us examine the joint effects of gene flow and finite population size with a model that assumes that replicate island populations have N individuals each and receive a proportion of migrants m each generation from a continental population (Figure 9.10a). Wright (1940) called this the **island**

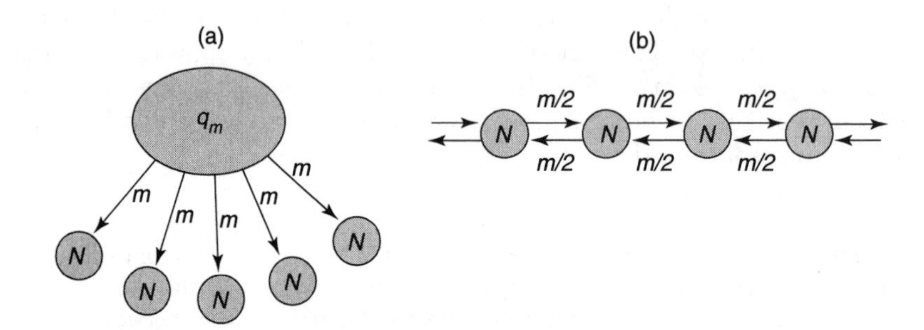

Figure 9.10. Representations of two different population structures with finite population size: (a) the continent–island or island model and (b) the one-dimensional stepping-stone model.

model because he assumed that there were ***many finite subpopulations (equivalent to the continent) that were the source of migrants as well as receive them***. When the amount of gene flow and the population size on the islands are both large, then the allele frequency on the islands will soon become similar to that on the continent—essentially the situation that we discussed earlier. However, if the population size on the islands is small and/or the rate of gene flow is low, then it is expected that genetic drift could result in chance changes in allele frequencies. As a result, the allele frequencies on the islands may differ significantly from each other and from the allele frequency in the migrants.

The effect of finite population size is to cause the allele frequencies in the subpopulations to drift apart, whereas gene flow serves to counteract this effect and keep their frequencies similar. We can evaluate these contrasting effects by examining the equilibrium value of the fixation index. Let us assume that there is a probability $1/(2N)$ that two alleles are identical by descent in the previous generation $t-1$ and a probability $1-1/(2N)$ that they are descended from different alleles in the previous generation (N is assumed to be the effective population size here). The expected homozygosity in generation t is then

$$f_t = \frac{1}{2N} + (1 - \frac{1}{2N})f_{t-1}$$

The probability of identity is modified by the probability that both alleles are not migrants, or $(1-m)^2$, so

$$f_t = \left[\frac{1}{2N} + \left(1 - \frac{1}{2N}\right) f_{t-1}\right] (1-m)^2 \qquad (9.12a)$$

If it is assumed that there is an equilibrium between gene flow bringing in new variation and finite population size reducing variation, then $f = f_t = f_{t-1}$. Furthermore, if we assume that f is equal to the equilibrium fixation index F_e or F_{ST}, then

$$F_{ST} = \frac{(1-m)^2}{2N - (2N-1)(1-m)^2} \tag{9.12b}$$

When $m = 0$, $F_{ST} = 1$ and when $m = 1$, $F_{ST} = 0$. If the terms with m^2 are ignored, then

$$F_{ST} = \frac{1-2m}{4Nm + 1 - 2m}$$

and when we ignore $2m$ in both the numerator and denominator, then

$$F_{ST} \approx \frac{1}{4Nm + 1} \tag{9.12c}$$

similar to the derivation given on p. 378 for mutation and genetic drift. When $m < 0.01$, expressions 9.12b and 9.12c give quite similar values (Waples, 1998). If we assume that there are k equivalent subpopulations, then an estimate of the differentiation among them (Slatkin, 1995) is

$$G_{ST} = \frac{1}{4Nm\left(\frac{k}{k-1}\right)^2 + 1} \tag{9.12d}$$

In general, it has been suggested that **one migrant per generation**, $Nm = 1$, is enough to prevent the effects of genetic drift among populations. If $Nm = 1$, then $F_{ST} = 0.2$ in equation 9.12c, a significant level of differentiation, even for very small sample sizes. This fairly substantial value and other considerations led Mills and Allendorf (1996) to recommend that $Nm = 1$ may be inadequate connectivity for natural populations, and they recommended higher levels of gene flow for management of endangered species. However, Wang (2004) has considered the theoretical assumptions underlying the one-migrant-per-generation recommendation and has found that they are generally robust when the effective number of migrants, $N_e m_e$, where the effective rate of gene flow m_e, which takes into account variance in migration, is substituted for m.

Expression 9.12c can be solved for an estimate of the number of migrants per generation as follows:

$$Nm = \frac{1 - F_{ST}}{4F_{ST}} \tag{9.12e}$$

This relationship has been widely used to estimate the number of migrants between populations. It is an approximation of a particular theoretical model at equilibrium and therefore should be used only as a general guideline to estimate the number of migrants (see the discussion in Waples, 1998; Gaggiotti *et al.*, 1999; Whitlock and McCauley, 1999; Neigel, 2002). In addition, when F_{ST} is small, then there can be bias in the estimation of Nm. For example, if $F_{ST} = 0.01$, then expression 9.12e gives an estimate of $Nm = 24.8$. Using expression 9.11b and assuming that $N = 50$, we get an estimate of $m = 0.29$ and $Nm = 14.6$, a value over 40% lower.

One concern about using estimates based on variation in allele frequency over groups is that they may be strongly influenced by the history of the populations and may not be at equilibrium. First, Wright (1943b) showed that if there is no gene flow between populations ($m = 0$ in expression 9.12a), then

$$F_{ST} = 1 - e^{-t/2N} \tag{9.13a}$$

This expression ranges from near 0 in the early generations and approaches unity when genetic drift over time has resulted in complete divergence between the populations. From this expression, the amount of F_{ST} is expected to increase at a nearly linear rate at low values and then asymptotically approach unity at high values.

We can include the effect of gene flow as well as genetic drift by iterating expression 9.12a to examine the rate of approach to equilibrium for different combinations of N and m. Figure 9.11 gives F_{ST} over time when the initial value of $F_{ST} = 0$ and $Nm = 1$. Obviously, when the effects of genetic drift and gene flow are large, the approach is fast, but when N is large and m is small, the approach can be quite slow. It is interesting to examine the difference in F_{ST} at a given point in time with and without gene flow. For example, with $N = 10,000$ using expression 9.13a after 1000 generations, $F_{ST} = 0.095$. With $m = 0.0001$ as in Figure 9.11, after 1000 generations $F_{ST} = 0.044$, only 46% as much.

Crow and Aoki (1984) showed that the time for G_{ST} to go half way to equilibrium is approximately

$$t_{0.5} \approx \frac{\ln(2)}{(2m + 1/2N)} \tag{9.13b}$$

showing explicitly that the rate of approach to equilibrium is faster as the values of m and N increase, that is, as the effects of gene flow and genetic drift increase. For example using this expression, when $N = 100$ and $m = 0.01$, $N = 1000$ and $m = 0.001$, and $N = 10,000$ and $m = 0.0001$, then it takes about 28,277 and 2773 generations, respectively, to go halfway to the equilibrium frequencies (see also Figure 9.11). As is apparent for $N =$ 1000 and $m = 0.001$ in Figure 9.11, the final approach to the equilibrium may be relatively slower than the time to go halfway to the equilibrium (Crow and Aoki, 1984).

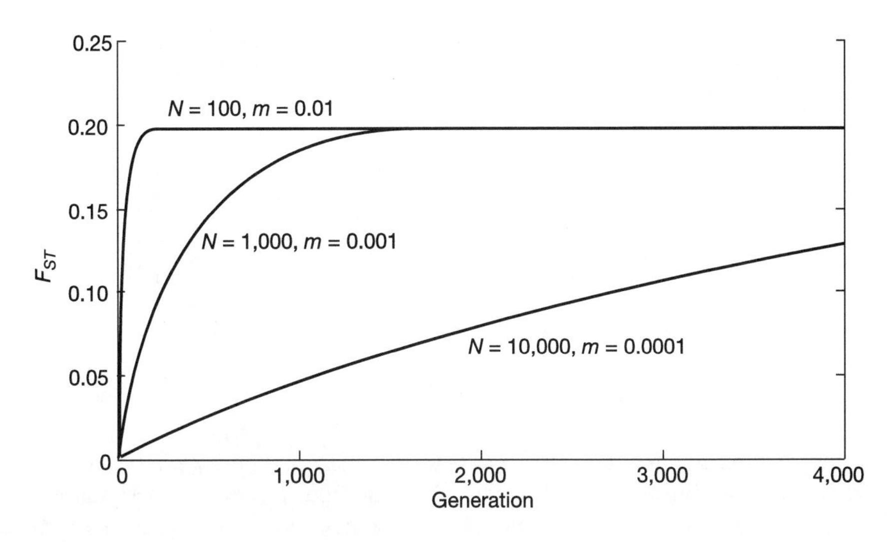

Figure 9.11. The amount of differentiation (F_{ST}) expected over generations for three different combinations of effective population size N and gene flow m.

It is also useful to point out the effect of subdivision on the effective population size. Wright (1943b) showed that for the island model

$$N_e = kN\left[1 + \frac{(k-1)^2}{4Nmk^2}\right]$$
$$= \frac{kN}{1 - F_{ST}} \tag{9.13c}$$

When the number of migrants, Nm, is large or F_{ST} is low, then $N_e \approx kN$, as would be expected for a random-mating population. However, when gene flow is low, $4Nm < 1$, then the effective population size may be larger than kN. For example, if $4Nm = 0.25$ ($F_{ST} = 0.8$), then $N_e \approx 5kN$ because the low level of gene flow allows each subpopulation to evolve independently.

Wright (1940) gave an explicit way of combining the effects of gene flow and genetic drift to predict the distribution of allele frequencies over islands. Assume that the frequency of A_2 in the migrants is constant and is equal to q_m. If we examine a large number of island populations, their average allele frequency will be q_m, but depending on the population size and the amount of gene flow, the distribution of allele frequency over islands will vary. For example, if N and m are large, then all island allele frequencies will closely approach q_m, and their distribution will have a large peak around q_m. The shape of the distribution of allele frequencies over islands is related to the size of $4Nmq_m$ and $4Nm(1 - q_m)$. If these values are both much greater than one, then the island frequencies will be very close to each other and to that of the continent. In fact, if $4Nm \gg 1$—that is, $m \gg 1/4N$—or there is much more than one migrant every four generations where Nm is the number of migrants, then there will be virtually no differentiation among the island populations (e.g., $Nm = 50$ in Figure 9.12). On the other

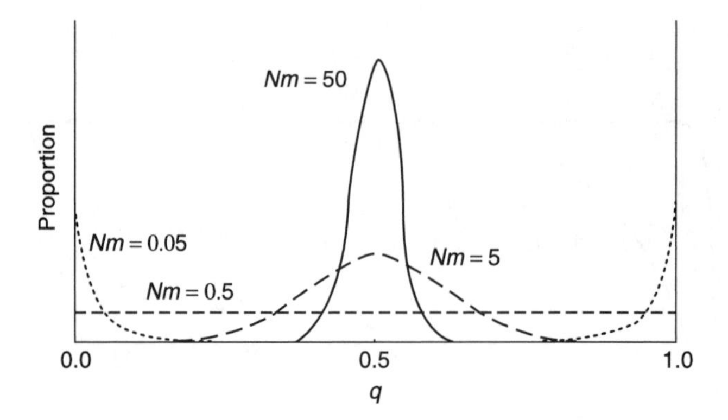

Figure 9.12. The equilibrium distribution of allele frequencies for the island model for different values of Nm where $q_m = 0.5$ (after Wright, 1969).

hand, if $4Nmq_m$ and $4Nm(1 - q_m)$ are less than one and q_m is nearly 0.5, then the distribution of alleles over island populations becomes U shaped, and most of the islands become fixed or nearly fixed for either A_1 or A_2 (e.g., $Nm = 0.05$ in Figure 9.12). These distributions are examples of stable gene-frequency distributions because it is assumed that the allele frequency in the migrants is constant and that there is a reintroduction of alleles by gene flow into populations that temporarily become fixed because of genetic drift.

b. The Stepping-Stone Model and Isolation by Distance

Let us assume that the populations have different positions in space or on a landscape. If ***the populations are arrayed in a one-dimensional spatial pattern and gene flow is restricted to adjacent populations***, then movement is approximated by a **linear stepping-stone model** (see Figure 9.10b). Natural populations that exist in a linear series of ponds in a watershed or tidal pools along a seashore may fit a linear stepping-stone model. Populations that exist on a habitat edge, for example, shallow water around a lake edge, may fit a circular stepping-stone model. Populations that inhabit oases in a desert or forest patches in farmland may fit a two-dimensional stepping-stone model (see Example 9.7 for data from the annual wild oats that approximate a two-dimensional stepping-stone model).

Example 9.7. Wild oats (*Avena fatua*) is genetically polymorphic for several morphological loci throughout central California (Jain and Marshall, 1967). This species occurs in many orchards, generally in small colonies within a meter of the base of the trees. Jain and Rai (1974) examined populations around prune trees in two orchards in two different years. Estimates of gene flow among the colonies surrounding each tree (the trees are about six meters apart) appear to be 1% or less. Furthermore, these populations

TABLE 9.9 The frequency of the recessive gray lemma color genotype in wild oats around trees in a prune orchard over two years (from Jain and Rai, 1974).

		Tree coordinate				
Tree coordinate	*Year*	1	2	3	4	5
1	1970	0.971	0.400	0.037	0.374	0.000
	1971	0.774	0.048	0.000	0.404	0.004
2	1970	0.053	0.360	0.068	0.021	0.000
	1971	0.091	0.562	0.174	0.000	0.079
3	1970	0.068	0.090	0.205	0.118	×
	1971	0.011	0.072	0.136	0.057	
4	1970	0.076	0.118	0.843	×	×
	1971	0.185	0.150	0.823		
5	1970	0.381	0.049	0.021	0.000	0.034
	1971	0.333	0.070	0.560	0.040	0.039

× = Tree missing

are quite small, the actual number of plants ranging between 17 and 191. Taking into account variation in size over the two years, self-fertilization, and the high variance in seed output among plants, Jain and Rai estimated the effective local population size to be between 14.6 and 40.3.

An example of their data for the locus determining lemma color is given for a sample matrix of colonies in Table 9.9. Note that there is a great deal of variation in allele frequency, even between adjacent colonies, but that there is a relatively high correlation over years ($r = 0.92$ for the whole orchard). Using the estimated effective population size, rate of gene flow, and self-fertilization, they found that the observed variation among colonies was consistent with the expected variation for a large population, strongly subdivided into numerous colonies.

Maruyama (1970) (see also Wang and Cabellero, 1999) showed that the effective population size in such a linearly subdivided population when there is high gene flow is

$$N_e \approx kN \tag{9.14}$$

the effective size of a randomly mating population.

In many populations, the individuals are distributed across the landscape in a pattern related to the distribution of suitable habitat. There may not be obvious substructure, but if there is distance-dependent gene flow, then the expected patterns of genetic variation may be generally similar to stepping-stone models. Wright (1943b) investigated the impact of gene flow when individuals were randomly distributed and there was **isolation by distance** between individuals. Slatkin (1991) and Rousset (1997) have suggested that the amount of genetic divergence as estimated by Nm

(expression 9.12e) or $F_{ST}/(1 - F_{ST})$, respectively, should change in a linear fashion with the inverse of geographic distance and the geographic distance, respectively, between pairs of populations along a linear habitat (these measures are scaled inverses of each other). There are a number genetic distance measures that have been used to examine the relationship with geographic distance (Paetkau *et al.*, 1997; Hardy *et al.*, 2003) and a variety of statistical approaches to examine genetic divergence as a function of geographic distance (Epperson, 2003). Example 9.8 discusses the relationship of geographic and genetic distance in populations of desert bighorn sheep.

Example 9.8. With the westward expansion of Europeans in North America, bighorn sheep (*Ovis canadensis*) were greatly reduced in both distribution and abundance. A number of factors, including hunting, habitat loss and modification, and diseases from domestic livestock, have been implicated in their decline. The remaining populations of desert bighorn sheep generally exist in isolated groups in mountain ranges surrounded by lower elevation, unsuitable habitat, not unlike that of a two-dimensional stepping-stone model. To examine the genetic divergence between many of the remaining populations in Arizona and southern California, Gutiérrez-Espeleta *et al.* (2000) determined genetic variation at 10 highly variable microsatellite loci.

Overall, they found extensive genetic variation in most of the populations, nearly as much as in a population from Alberta that had not undergone a great reduction in numbers. When the amount of genetic divergence, using the standard genetic distance measure of Nei (1972) (see p. 108), was compared with the geographic distance, there was a statistically significant linear relationship (Figure 9.13). Some of the populations compared were of the same putative subspecies, and some were of different subspecies. If the subspecies designation had an important biological basis, there should be a higher rate of genetic differentiation, for the same

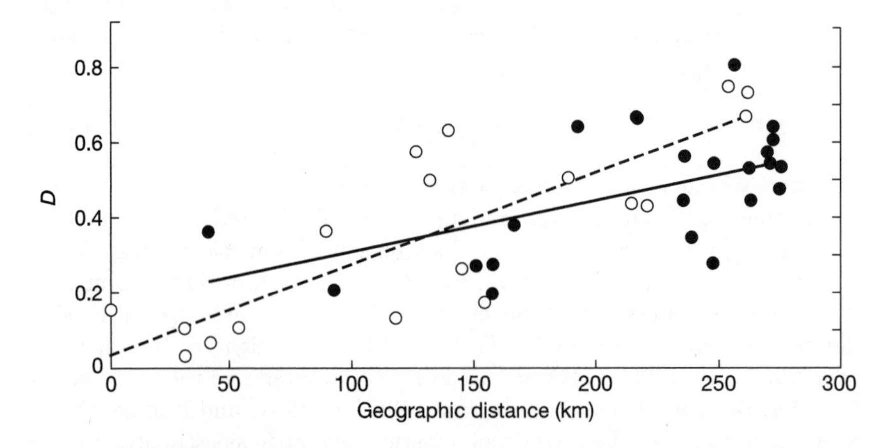

Figure 9.13. Genetic distance for microsatellite loci between populations of desert bighorn sheep plotted against geographic distance where the open circles (and broken line) are within putative subspecies and the closed circles (and solid line) are between putative subspecies (from Gutiérrez-Espeleta *et al.*, 2000).

geographic distance, when comparing locations between subspecies than for within subspecies. However, the slopes within and between subspecies are not significantly different, consistent with the explanation that genetic divergence between these groups is primarily determined by isolation by distance and not greatly influenced by subspecies designation.

c. Metapopulation

The term metapopulation is sometimes used to describe any population with spatial subdivision, in which case it would include the different population structures discussed above. However, in the traditional definition of a **metapopulation** given by Levins (1969), ***subpopulations exist in discrete habitat patches, and these subpopulations may turnover with extinctions and recolonizations from other patches***. At any time some proportion of the patches, determined by the rates of extinction and recolonization, may be unoccupied. In the traditional metapopulation model, when the rate of extinction exceeds the rate of recolonization, the total metapopulation is expected to go extinct.

The dynamics of extinction and recolonization can also greatly influence the effective population size, the amount of genetic variation in the metapopulation, and the distribution of genetic variation over the subpopulations (Slatkin, 1977; Hedrick and Gilpin, 1997; Whitlock and Barton, 1997; Nunney, 1999). Figure 9.14 gives a simulation example of the heterozygosity in a simple metapopulation of three patches each with 500 individuals (Hedrick and Gilpin, 1997). Here the heterozygosity is initially high in all three patches. The important sequence of events starts in generation 48, when the population in patch 2 goes extinct and is subsequently recolonized by a single fertilized female from patch 3 in generation 51, with a consequent reduction in heterozygosity. The next significant event occurs when empty patch 1 is recolonized from patch 2 with a founder population

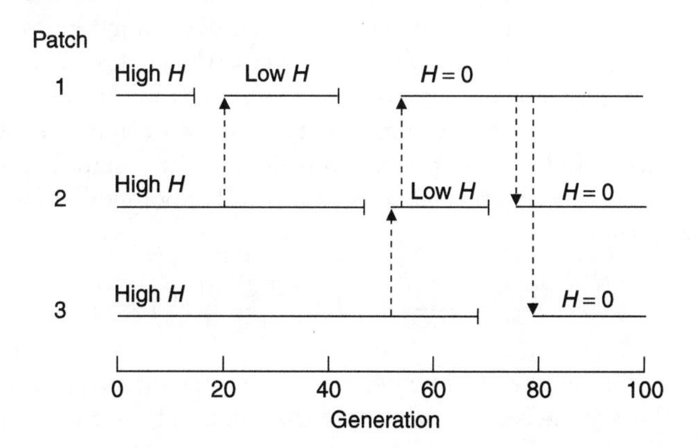

Figure 9.14. The level of heterozygosity (H) over time in a simulation of a population existing in three patches (Hedrick and Gilpin, 1997). The short vertical bars on the right-hand end of the horizontal lines indicate extinctions in a patch, and the arrows indicate recolonizations.

that has no genetic variation. Finally, when the population in patch 2 goes extinct in generation 71, the metapopulation has no variation, although there are still 500 individuals remaining in patch 1. These individuals subsequently colonize the other two patches, and the whole population at this point has no heterozygosity. All of these individuals can be traced back to individuals in patch 3 before generation 51.

A number of parameters can influence the rate of loss of genetic variation in a metapopulation model. For example, the source of the individuals that recolonize empty patches may be a group of individuals from a single patch (propagule pool mode of colonization) or a group from all of the occupied patches (migrant pool mode) (Slatkin, 1977). In addition, the number of founders in a recolonization event, the rates of extinction and recolonization, the size of a population within a patch, the number of patches, and the amount of gene flow between patches may influence the amount of genetic variation. For example, Hedrick and Gilpin (1997) showed that given recolonization from a singly fertilized female, an infinite population size within each patch, 20 patches, and no gene flow (except during recolonization), the expected effective metapopulation size is only approximately 150. This low effective population size is attributable to the low number of founders involved in each recolonization event. Because the present-day individuals may trace back to a few individuals in previous generations, the level of F_{ST} may be small. On the other hand, as shown in Example 9.9, the observed F_{ST} in recently colonized populations may be higher than that of older established populations. For further detailed theoretical discussion about genetic variation, N_e and F_{ST} in metapopulations, see Whitlock and Barton (1997), Nunney (1999), Wang and Caballero (1999), Pannell and Charlesworth (2000), Pannell (2003), Rousset (2003), and Wakeley (2004).

Example 9.9. The source of individuals that recolonize an empty patch greatly influences the impact of metapopulation dynamics on genetic variation. If the colonizers are drawn randomly from all the other occupied patches (migrant pool mode), then more genetic variation would be expected than if they were all drawn from a single colony (propagule pool mode). Whitlock and McCauley (1990) suggested that the type of colonizers could in principle be on any point on the continuum between these two modes. They defined ϕ as the probability that two colonizers came from the same patch, and if ϕ is 0, it is equivalent to the migrant pool mode, whereas if $\phi = 1$, it is equivalent to the propagule pool mode. Furthermore, they showed that

$$F'_{ST} = \frac{1}{2N_f} + \phi\left(1 - \frac{1}{2N_f}\right) F_{ST}$$

where N_f is the number of founders in a group recolonizing a patch, and F_{ST} and F'_{ST} are measured over established populations and recently colonized patches, respectively. If N_f, F_{ST}, and F'_{ST} can be estimated, then ϕ can be calculated by solving this expression.

Silene alba is a dioecious, short-lived perennial with high rates of extinction and colonization and whose distribution appears similar to that in metapopulation models (Antonovics *et al.*, 1994). McCauley *et al.* (1995) estimated F_{ST} in 11 established populations and F'_{ST} in 12 recently colonized populations for seven allozyme loci (Figure 9.15). They found that the vari-

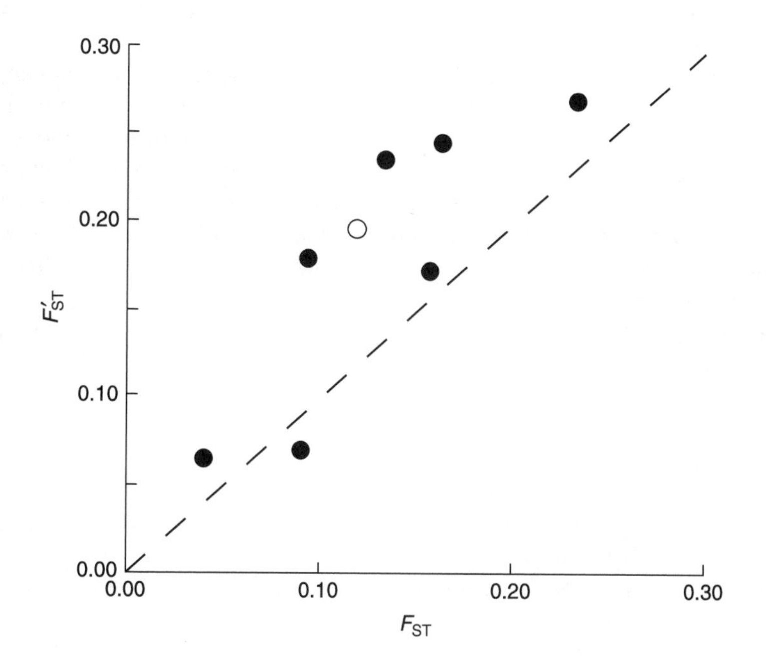

Figure 9.15. F_{ST} values estimated for 12 recently colonized (F'_{ST}) and 11 established (F_{ST}) populations of *Silene alba* from allele frequencies at seven polymorphic allozyme loci (solid circles), as well as an estimate combined across loci (open circle). Note that F'_{ST} is larger than F_{ST} at six of seven loci (after McCauley *et al.*, 1995).

ation over populations was higher for the 12 recently colonized populations for six of the seven loci and that, overall, $F_{ST} = 0.126$ and $F'_{ST} = 0.197$. The average estimate of the number of colonizers was 4.2, and from the above expression, $\phi = 0.73$. In other words, it appears that there is only limited mixing of individuals from different groups in the founders that established these new populations and that most of the founders probably came from one, or a few, source populations.

IV. GENE FLOW AND SELECTION

Gene flow and selection are frequently important in the same population, and this makes it necessary to consider their joint effects on genetic variation. Both gene flow and selection may be quite diverse in their effects on allele frequencies, and the two factors together may result in even more complex potential outcomes (Barton and Clark, 1990; Lenormand, 2002). In this section, we discuss only quite elementary combinations of gene flow and selection, but both the diversity of potential joint effects of these two factors and the general approach to exploring particular combinations of gene flow and selection should be readily apparent. The joint effects of gene flow and selection are particularly important when considering study of a hybridization between species (Harrison, 1993; Arnold, 1997). For a case study of a hybrid zone of red deer (elk) and sika deer that utilizes a number of different approaches, see Goodman *et al.* (1999).

In some cases, there may be a correlation between the tendency to migrate and the relative fitness of the migrants. For example, in marmots, voles, and other rodents, young males appear to migrate most frequently. In social animals, such individuals need a territory in which to breed and raise offspring. Therefore, a male without a suitable territory will have a very low probability of reproducing—and consequently, a very low fitness value. Among individuals of the same age and sex, a correlation may also exist between a particular genotype and a tendency to migrate. In the following discussion, however, we will assume that gene flow rates are not associated with particular genotypes.

a. The Continent–Island Model

Let us return to the continent–island model and investigate how selection affects the allele frequency in the island population. Assume that the change in frequency on the island is the joint effect of gene flow and selection. That is

$$\Delta q = \Delta q_s + \Delta q_m$$

where Δq_s and Δq_m are the changes in frequency caused by selection and gene flow, respectively. For algebraic simplicity, let us assume that the heterozygote is intermediate in fitness and the relative fitnesses of A_1A_1, A_1A_2, and A_2A_2 are 1, $1-s$, and $1-2s$, respectively. Then this expression becomes approximately

$$\begin{aligned}\Delta q &= -sq(1-q) - m(q-q_m)\\ &= sq^2 - (m+s)q + mq_m\end{aligned}$$

(Li, 1976). An equilibrium occurs when $\Delta q = 0$; there is no change in the frequency of A_2 on the island. The equilibrium frequency for the island population is found by solving this quadratic equation so that

$$q_e = \frac{1}{2s}\left\{(m+s) \pm \left[(m+s)^2 - 4msq_m\right]^{1/2}\right\} \tag{9.15a}$$

Selection will favor allele A_1 if s is positive or favor A_2 if s is negative.

There are three different general situations to consider, dependent on the relative magnitudes of gene flow and selection; $m \ll |s|$, $m = |s|$, and $m \gg |s|$. In other words, the rate of gene flow is much less, equal to, or much greater than the absolute amount of selection. Intuitively, as m increases with respect to $|s|$, then the effect of selection is overwhelmed by gene flow, and genetic differentiation does not occur in the island population. This is illustrated in an example from Li (1976) in Table 9.10 where $q_m = 0.4$ and gene flow and selection reflect the three situations above. If $m = |s|$, expression 9.15a simplifies to

$$q_e = (q_m)^{1/2} \qquad \text{or} \qquad q_e = 1 - (1 - q_m)^{1/2} \tag{9.15b}$$

when s is negative or positive, respectively. As shown in Table 9.10, if $m = |s|$ or $m \ll |s|$, selection results in a large amount of differentiation at equilibrium in the island population. When $m \gg |s|$, the frequency is much less affected by selection, and the island equilibrium frequency is very close to that of the migrants.

TABLE 9.10 The equilibrium allele frequencies for different levels of gene flow and selection where $q_m = 0.4$ (Li, 1976).

	$m \ll \lvert s\rvert$		$m = \lvert s\rvert$		$m \gg \lvert s\rvert$	
m	0.01	0.01	s	s	0.15	0.15
s	−0.15	0.15	$-m$	m	−0.01	0.01
q_e	0.961	0.026	0.632	0.225	0.416	0.384

Another important situation occurs when the allele frequency in the migrants is zero ($q_m = 0.0$). This may occur when a wingless variant or some other trait is advantageous on the island but disadvantageous (and, therefore, in very low frequency) in the mainland population or when there is introgression from a common taxum into a rarer one. Assume that selection on the island counterbalances the reduction in allele frequency due to gene flow so that the fitnesses of genotypes A_1A_1, A_1A_2, and A_2A_2 are 1, $1 + hs$, and $1 + s$, respectively. Therefore, the change in allele frequency caused by selection is

$$\Delta q_s = \frac{spq[q + h(1 - 2q)]}{1 + 2pqhs + q^2 s}$$

and the total change in allele frequency then becomes

$$\Delta q = -mq + \frac{spq[q + h(1 - 2q)]}{1 + 2pqhs + q^2 s}$$

Example 9.10 gives an application of this model to predicting the effect of introducing Texas pumas into the last population of Florida panthers.

Example 9.10. The Florida panther, *Puma concolor coryi*, is an endangered subspecies of puma or mountain lion that is isolated in southwestern Florida and is estimated to be represented by only about 50 to 70 wild individuals. The Florida panther phenotypically appears somewhat different from other mountain lions, having a flatter skull, a cowlick in the middle of the back, and often a distinctive 90-degree kink near the end of the tail (Belden, 1986). It also appears to have lowered fitness, with high levels of sperm abnormalities, a high proportion of undescended testicles (chryptochordism), and perhaps other congenital abnormalities (O'Brien *et al.*, 1990). As a result, a decision was made to introduce Texas pumas, the closest extant population source, into the Florida population to "genetically restore" the population to higher fitness. The recommendation was to introduce female Texas pumas (the males tend to wander much more than the females) in numbers such that the level of gene flow in the first generation would be 0.20, and 0.025 (about one animal) per generation thereafter (Seal, 1994).

Hedrick (1995c) theoretically evaluated the potential effect of the introduction by assuming that the Florida panthers were fixed either for an allele that reduced fitness or for an allele that gave them a fitness advantage and that the Texas pumas had the alternative alleles (there has been speculation that Florida panthers were specifically adapted to live in the swampy habitat in which the remaining animals are found). For example, Figure 9.16a gives the frequency of the Florida panther allele (A_2) when there is the suggested gene flow from Texas and the fitnesses of A_1A_1, A_1A_2, and A_2A_2 are 1, 1.2, and 1.2 (adaptive advantage in Florida); 1, 1, and 1 (effect of gene flow only); or 1, 1, and 0.5 (low fitness in Florida), respectively. For illustration, it is assumed that $h = 1$ for the adaptive allele, but h could be near zero and result in the allelic frequency dynamics similar to that in Figure 9.17. Obviously, the frequency of the low-fitness allele is greatly reduced over ten generations, whereas the adaptive allele remains at a high frequency even after ten generations. If we examine the expected relative fitnesses of these two loci (Fig. 9.16b), we find that for the detrimental locus the fitness is quickly increased, and that for the adaptive locus it is only slightly reduced. Overall, when these two loci are combined, it appears that the fitness will quickly increase and that with this level of gene flow, any adaptive alleles will not be swamped by the introduction of migrants from Texas.

The introduction was begun in 1995 with the movement of 8 wild-caught Texas females to Florida. Five of the eight introduced Texas females produced offspring with resident Florida panther males, and there are now a number F_1, F_2, and backcross offspring. It is now estimated that approximately 24.5% of the ancestry in the Florida panther population is

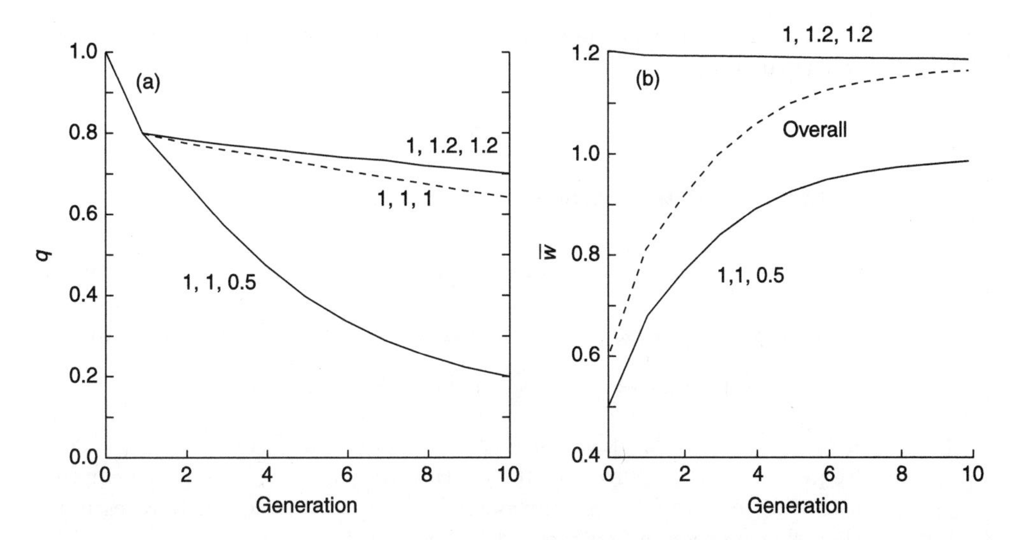

Figure 9.16. (a) The change in the frequency of A_2 (the "Florida panther" allele) over ten generations with gene flow of 0.2 in the first generation and 0.025 thereafter. Selection causes lowered fitness of A_2 (1, 1, 0.5), an adaptive advantage (1, 1.2, 1.2), or no selection (1, 1, 1). (b) The change in relative fitness over ten generations for the two different selective schemes, and for the two of them combined, over ten generations (after Hedrick, 1995c).

TABLE 9.11 The proportion of Florida panthers (sample size in parentheses) with a kinked tail, cowlick, or chryptorchidism with no Texas puma ancestry (first column) and those with Texas puma ancestry (F_1, F_2, backcross to Texas, and backcross to Florida) (from Roelke *et al.*, 1993; Land *et al.*, 2001).

		Texas ancestry				
	No Texas ancestry	F_1	F_2	*BC-Texas*	*BC-Florida*	*Total*
Kinked tail	0.88 (48)	0.00 (17)	0.00 (7)	0.00 (3)	0.20 (15)	0.07 (42)
Cowlick	0.93 (46)	0.20 (10)	0.00 (5)	0.00 (1)	0.60 (5)	0.24 (21)
Chryptorchidism	0.68 (22)	0.00 (2)	0.00 (2)	—	0.00 (1)	0.00 (5)

from the Texas animals (Maehr and Lacy, 2002). Of the animals with Texas ancestry, only 7% have a kinked tail (compared with 88% before the introduction), and the kinked-tail animals are backcrosses to Florida animals (Table 9.11). Similarly, but not as dramatic, only 24% have the cowlick (compared with 93% before). Only five males with Texas ancestry have been evaluated for chryptorchidism, and all have two descended testicles, a reduction from 68% chryptorchidism before to 0% now. In other words, the introduction of Texas pumas has already resulted in a substantial reduction of the frequency of the detrimental traits that had accumulated in the Florida panther.

When $h = 0.5$ (the heterozygote is intermediate in fitness), this reduces to the equation

$$\Delta q = -mq + \frac{spq}{2\,(1+sq)}$$

Setting $\Delta q = 0$, then there is a stable equilibrium when

$$q_e = \frac{s - 2m}{s\,(1+2m)} \tag{9.16a}$$

When $h \neq 0.5$ and Δq is set equal to zero, then we obtain the quadratic

$$s\,(1-2h)\,(m+1)\,q^2 + s\,(2hm - 1 + 3h)\,q + m - sh = 0$$

In general, there is only one feasible solution to this equation (an allele frequency between zero and unity). However, when h is zero or near zero, there may be two feasible solutions. If we let $h = 0.0$ (complete recessivity), then this expression becomes

$$s\,(m+1)\,q^2 - sq + m = 0$$

and the solutions to this equation are

$$q_e = \frac{s \pm \left[s^2 - 4s\,(m+1)\,m\right]^{1/2}}{2s\,(m+1)} \tag{9.16b}$$

The equilibrium with the smaller value is unstable, and the one with the higher value is stable. An example to illustrate the latter is given in Figure 9.17, where $s = 0.2$ and $m = 0.04$, and the Δq curves for gene flow and

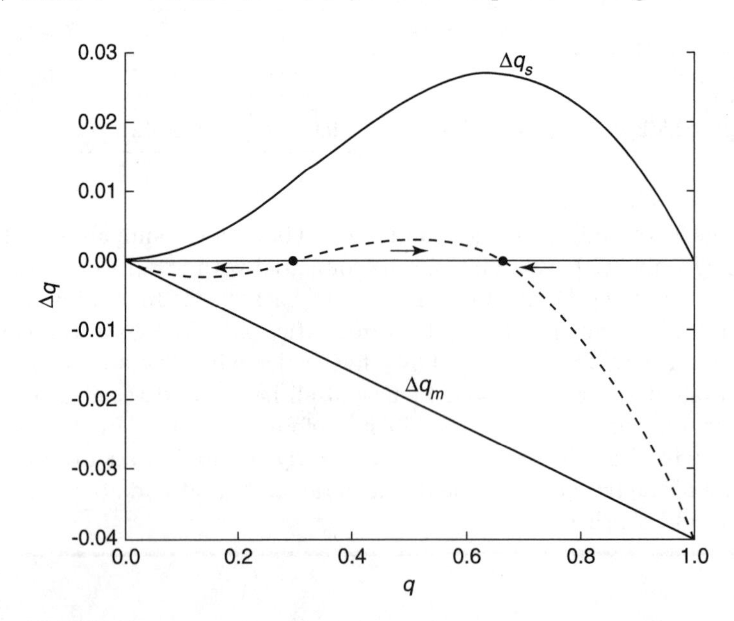

Figure 9.17. The change in allele frequency resulting from selection, Δq_s, and from gene flow, Δq_m, in the continent–island model where $s = 0.2$, $h = 0$, and $m = 0.04$. The broken line indicates the composite effect of selection and gene flow, and the circles indicate allele frequency equilibria.

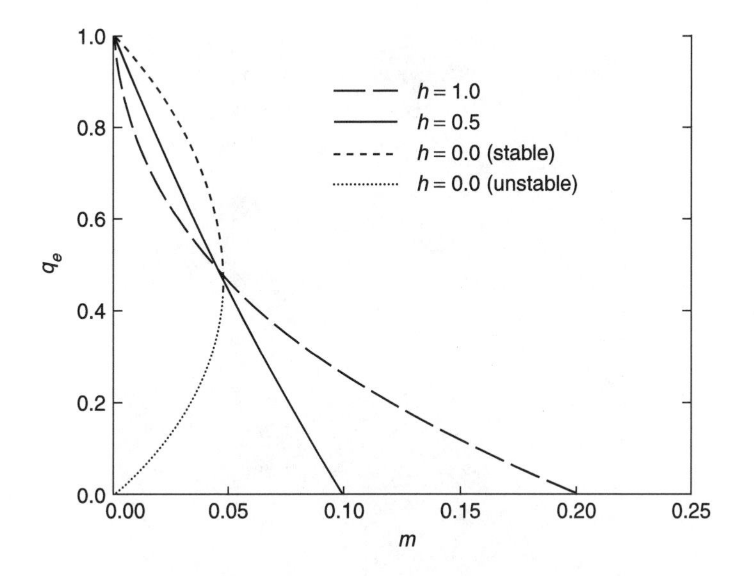

Figure 9.18. The equilibrium allele frequency when there is balance between selection and gene flow for three levels of dominance. When $h = 0.0$ there are two equilibria: a stable one, indicated by the short dashed line, and an unstable one, indicated by the dotted line.

selection are plotted separately and then together. Examining the Δq curve for gene flow and selection combined (the broken line), we find an unstable equilibrium at 0.28 and a stable equilibrium at 0.68.

In general, the equilibrium frequency on the island is a function of the amount of selective advantage and the level of dominance on the island and the amount of gene flow. If we assume that s is 0.2, then the equilibrium allele frequency on the island is given in Figure 9.18, assuming dominance ($h = 1.0$), additivity ($h = 0.5$), and recessivity ($h = 0.0$) of the variant. There is a difference in the equilibrium for the three levels of dominance, reflecting the different Δq values caused by selection. With high gene flow, however, the favorable variant completely disappears from the island for all levels of dominance. This **patch disappearance**, even in the presence of favorable selection, was first noticed by Haldane (1948). As pointed out above, when the favorable variant is recessive, there can be both stable and unstable equilibria. In the example in Figure 9.18, both these equilibria disappear when m is greater than 0.0475. For an application of these expressions, see the Example 9.11 about island populations of water snakes in Lake Eire.

Example 9.11. On several islands in Lake Erie, a high proportion of water snakes, *Nerodia sipedon*, exhibit polymorphism of color banding: There are both regular banded forms, as found on the mainland in the United States and Canada, and uniformly gray, unbanded forms (Fig. 9.19). Camin and Ehrlich (1958) suggested that this was the result of a balance between selection against the banded forms on flat limestone rocks of the islands because of predation by gulls, herons, and raptors and gene flow to the islands from the banded mainland populations.

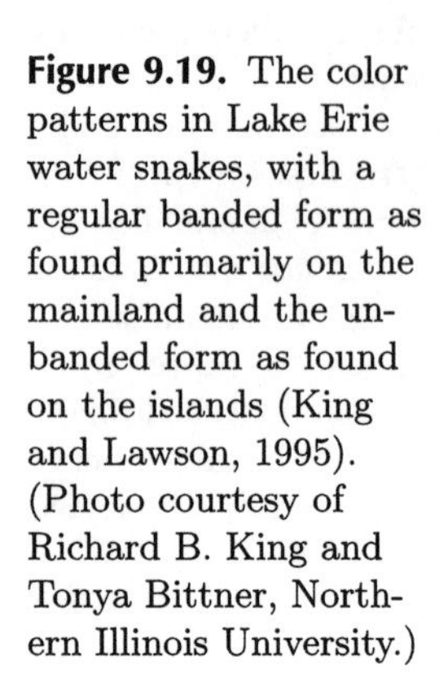

Figure 9.19. The color patterns in Lake Erie water snakes, with a regular banded form as found primarily on the mainland and the unbanded form as found on the islands (King and Lawson, 1995). (Photo courtesy of Richard B. King and Tonya Bittner, Northern Illinois University.)

More recent work suggests that the inheritance of banding may be determined primarily by variation at a single locus difference (King, 1993) and that the banded form is dominant over the unbanded. Estimates of present-day allele frequencies of the recessive unbanded allele on different islands range from 0.53 to 0.86 with a mean of 0.73 (King and Lawson, 1995). From capture–recapture studies, King and Lawson estimate that the selective advantage of unbanded individuals on the islands is between 0.11 and 0.28, and from variation in frequencies of allozyme alleles among the islands, they estimated $N_e m$ from the mainland to be about 12.8. Assuming that N_e was equal to an estimate of adult water snakes on the island, they obtained an approximate estimate of gene flow of $m = 0.01$.

Using these values of s and assuming that $m = 0.01$, expression 9.16b can be used to calculate the expected allele frequency of the unbanded allele on the islands. If $s = 0.11$, then there is an unstable equilibrium at 0.101 and a stable one at 0.890, and if $s = 0.28$, then there is an unstable equilibrium at 0.037 and a stable one at 0.953. For both levels of selection, the stable equilibrium frequencies are somewhat higher than the frequencies observed. There are a number of possible explanations for this difference. Perhaps s is lower or m higher than estimated. Or perhaps the populations have not yet come to an equilibrium state. Overall, this case study illustrates how the various evolutionary factors can be estimated and then put in a theoretical context to explain the observed patterns of genetic variation.

b. Variable Selection and Clines in a Substructured Population

A ***directional change in allele frequency over space or subpopulations***, or a **cline** in variation, is often thought to be the result of the combination of differential selection in various parts of the population in combination with population substructure. Such spatial variation in allele frequency may be temporally stable or changing with time, resulting in either a **stable cline** or a **transient cline**, respectively. Furthermore, there may be a large change in frequencies over a short distance forming a **stepped cline**, or the allele frequency change in the cline forms a gentle gradient or gradual cline (Endler, 1977; Barton and Hewitt, 1985).

To illustrate both stable and transient clines, let us again consider a population with k subpopulations. However, let us now assume that there is selection, for example, in preadult viability, that varies with subpopulation prior to gene flow. Therefore, the frequency of A_2 in the jth subpopulation after selection is

$$q_j' = \frac{1}{\overline{w}_j}\left(p_j q_j \, w_{12 \cdot j} + q_j^2 \, w_{22 \cdot j}\right) \tag{9.17a}$$

where $w_{11 \cdot j}$, $w_{12 \cdot j}$, and $w_{22 \cdot j}$ are the relative fitnesses of genotypes A_1A_1, A_1A_2, and A_2A_2, respectively, in subpopulation j, q_j is the frequency of allele A_2 in subpopulation j, and

$$\overline{w}_j = p_j^2 \, w_{11 \cdot j} + 2p_j q_j \, w_{12 \cdot j} + q_j^2 \, w_{22 \cdot j}$$

The allele frequency after gene flow in the ith subpopulation becomes

$$q_i'' = \sum_{j=1}^{k} m_{ij} \, q_j' \tag{9.17b}$$

where m_{ij} is the gene flow parameter as defined earlier. Depending on the relative fitness values and initial frequencies in the subpopulations, either transient or stable clines may result. Example 9.12 illustrates a transient cline with human data for the sickle-cell allele in Liberia and a stable cline for the thalassemia allele from Sardinia.

Example 9.12. Two interesting applications of the joint effects of gene flow and selection are to the distribution of the sickle-cell allele over a linear array of tribal populations in Liberia and to the distribution of the thalassemia allele in a linear series of towns in central Sardinia, Italy (see Figure 9.20). To simulate these situations, Livingstone (1969) used gene flow patterns in which most of the gene flow among a group of 40 populations was to adjacent populations (80%) and other nearby populations. The homozygotes with sickle-cell anemia or thalassemia were assumed to have

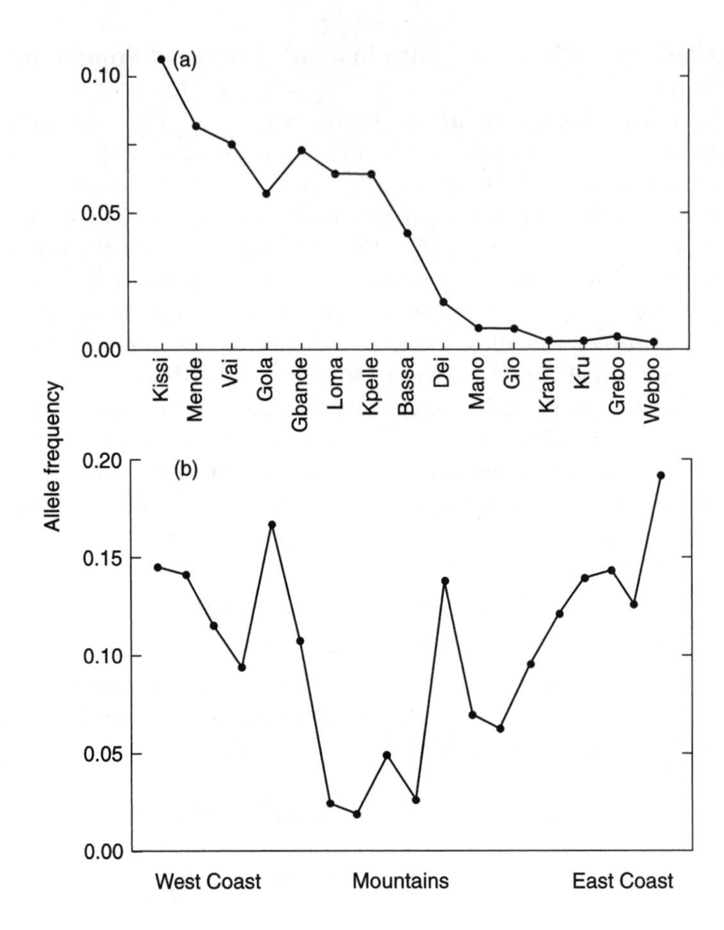

Figure 9.20. The observed allele frequencies (a) for the sickle-cell allele in some Liberian populations and (b) for the thalassemia allele in Sardinia (from Livingstone, 1969).

a relative fitness of zero, the other homozygotes a fitness of 1.0. The heterozygote fitness varied in different populations, mimicking the advantage for heterozygotes at both loci in malarial environments. In populations 1 through 30 for the sickle-cell case, the heterozygote fitness was 1.25, and in populations 31 through 40, it was 1.0. For the thalassemia simulation, the heterozygote fitness was varied from 1.25 in the terminal populations (1–8 and 33–40) and was decreased in several increments to 1.0 in the central populations (16–25).

The results of these simulations, as given in Figure 9.21, are consistent with the distributions observed in the African and Sardinian populations given in Figure 9.20. The simulation of the sickle-cell allele was initiated with the allele in low frequencies in populations 1–5. The movement of the cline through these populations as the allele increases in frequency describes what seems to be occurring in the Liberian tribes. For example, the curve after about 20 generations is similar to that observed in Liberia today. With this model, a stable, stepped cline should eventually develop near population 30. The simulation of the thalassemia cline was started

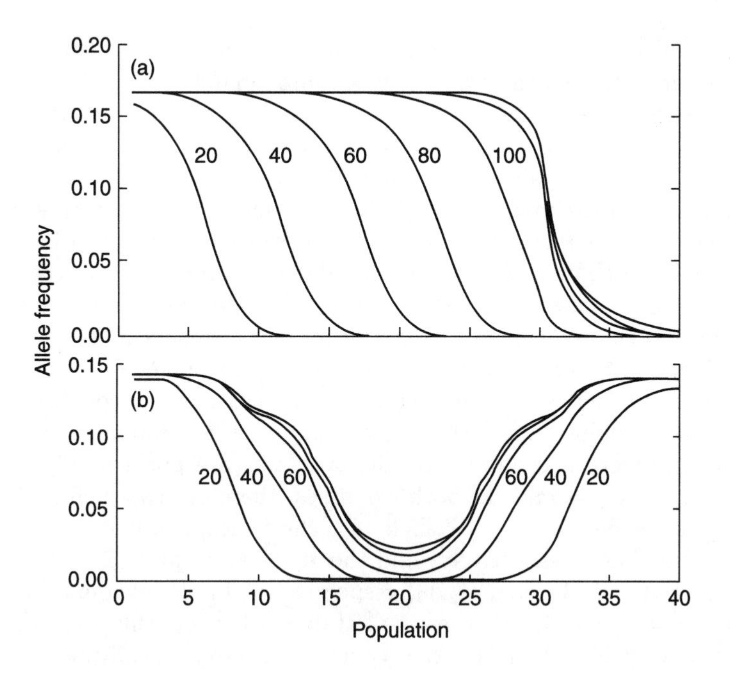

Figure 9.21. Simulated allele frequencies for (a) the sickle-cell allele and (b) the thalassemia allele (from Livingstone, 1969). The numbers of generations after initiation are indicated on the graphs every 20 generations.

nearer the observed values. Furthermore, because of the gradual change in fitnesses used in the thalassemia simulation, a much more gentle gradient in allelic frequencies is achieved.

These simulations provide a satisfactory explanation for the observed distribution of allelic frequencies, but different combinations of gene flow, selection, and other factors could also yield results consistent with the observations. Livingstone noted that small amounts of long-distance gene flow were quite important in determining the rate of movement and the shape of a transient or advancing cline, whereas the amount of gene flow among neighboring isolates was most important in determining the shape of stable clines. Livingstone (1989) used similar simulations that included mutation and multiple hemoglobin alleles with resistance to malaria in an effort to understand the worldwide distribution of these variants.

Both the rate of spread of a cline and the length of a stable cline in a continuously distributed population are functions of the amount of gene flow and selection. Fisher (1937) showed that rate of spread of a cline, given that $V^{1/2}$ is the standard deviation of the distance between parent and offspring at reproduction, is proportional to $s^{1/2}$, where s is the selective advantage of the favored allele. Furthermore, the length of the wave (the area where there is a large difference in allele frequencies) is proportional to $(V/s)^{1/2}$. Slatkin (1973) has shown that when there is a stable cline such that an allele has an advantage s in one area and a disadvantage $-s$ in another area, the width of the cline is approximately $(V/s)^{1/2}$.

The spatial selection model of Levene discussed on p. 210 assumes that an individual is as likely to go to any other niche (subpopulation) as return to the parental niche (subpopulation). In terms of the migration matrix, this means that $m_{ij} = 1/k$ for all values of i and j where k is the number of niches (subpopulations). The migration matrix can be used to incorporate both limited gene flow and different proportions of the various environments (Christiansen, 1974). With this extension, the conditions for a stable polymorphism are a function of the elements in the migration matrix and the proportions of the different environments, the c_i values. The general effect of limited gene flow is to broaden the conditions for a stable polymorphism more than with equal gene flow to all niches.

Two numerical examples to illustrate this are given in Figure 9.22, where $m = m_{12} = m_{21}$. In both examples, there are two niches with the fitness values favoring the A_1 allele in niche 1 (1.2, 1.0, and 0.8 for A_1A_1, A_1A_2, and A_2A_2, respectively) and the A_2 allele in niche 2 (0.8, 1.0, and 1.2 for A_1A_1, A_1A_2, and A_2A_2, respectively). In the example in Figure 9.22a, $c_1 = c_2 = 0.5$, and there is a stable polymorphism even when the m_{ij} values are 0.5 (there is harmonic mean heterozygote advantage). With high values of gene flow, the allele frequencies in the two niches become more similar and approach 0.5 as m approaches 0.5. When the gene flow is small, then there is still a global equilibrium allele frequency of 0.5, but the two niches have quite different allele frequencies. In Figure 9.22b, $c_1 = 0.375$ and $c_2 = 0.625$, and there is no stable polymorphism when the m_{ij} values are 0.5. Here the allele frequencies at low gene flow values are also quite

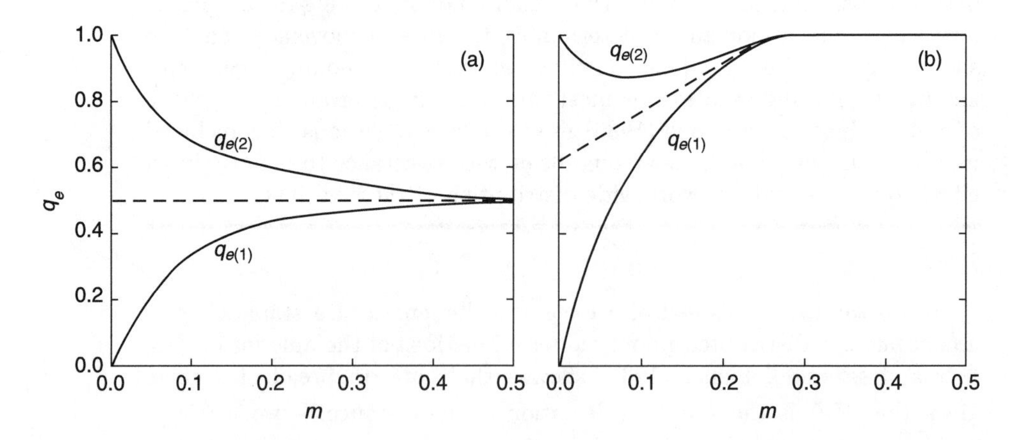

Figure 9.22. The global allele frequency equilibrium (broken line) and the equilibria within two niches with differential selection and limited gene flow (a) where the niches occur in equal proportions, $c_1 = c_2 = 0.5$, and (b) where niche 2 is more common, $c_2 = 0.625$ and $c_1 = 0.375$.

different in the two niches, and as gene flow increases, though the allele frequencies become closer, the global frequency goes to 1.0 because there is no stable polymorphism above a value of $m = 0.28$. In other words, limited gene flow between the niches gives a stable polymorphism that disappears as the amount of gene flow increases. Example 9.13 discusses a melanic polymorphism in pocket mice, in which there is strong selection for concealing coloration but also evidence for substantial gene flow from neutral markers.

Example 9.13. There are several melanic color forms of vertebrates for which the molecular basis has been determined (Theron *et al.*, 2001; Eizirik *et al.*, 2003; Mundy *et al.*, 2004). Nachman *et al.* (2003) have presented a detailed examination of the genetic variation in the rock pocket mouse (*Chaetodipus intermedius*), a light-colored mouse that generally lives on light-colored granite rocks but has melanic forms that live on black lava in several restricted sites in southwestern United States. Figure 9.23 shows both the normal recessive and dominant melanic forms from the Pinacate lava flow in Arizona on light colored rocks and dark lava. In this area, 16 of 18 mice caught on the dark lava were melanic, and 10 of 11 mice caught on light rock were light colored, strong evidence of adaptive Darwinian selection.

Investigation of molecular variation in the *Mc1r* gene, which is known to have variants that produce dark-colored house mice, was found to correlate nearly completely with the light and melanic phenotypes. Table 9.12

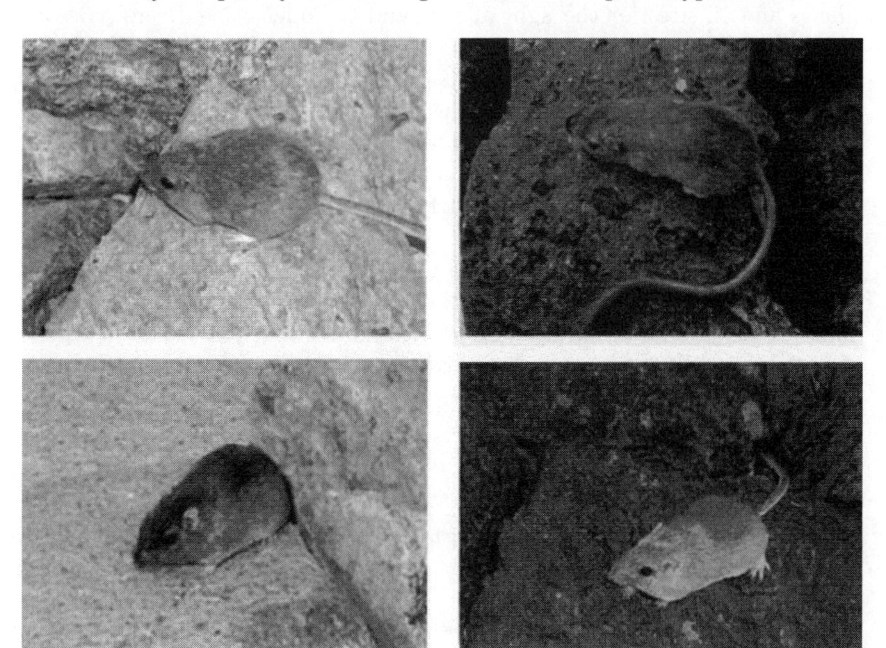

Figure 9.23. Light and melanic forms of *C. intermedius* on light rocks (left side) and melanic and light forms on dark lava rocks. (Courtesy of Nachman, M.W., H.E. Hoekstra, and S.L. D'Agostino. 2003. The genetic basis of adaptive melanism in pocket mice. *Proc. Natl. Acad. Sci. USA.* 100:5268–5273. ©2003 National Academy of Sciences, USA.)

TABLE 9.12 Polymorphism at the five nucleotide sites, each one resulting in an amino acid change, that differentiate melanic and light alleles at the *Mc1r* gene in the samples of *C. intermedius* from the Pinacate population (Nachman *et al.*, 2003). The number (N) of each genotype caught in the two different environments is given.

Phenotype	*Background*	N	*Genotype*	*Position* 52	325	478	633	699
Melanic	Dark	11	*DD*	T T	T T	T T	C C	C C
Melanic	Dark	5	*Dd*	T C	T C	T C	C T	C A
Melanic	Light	1	*Dd*	T C	T C	T C	C T	C A
Light	Dark	2	*dd*	C C	C C	C C	T T	A A
Light	Light	10	*dd*	C C	C C	C C	T T	A A

gives the genotypes at the five nucleotide positions that differentiated the melanic and light alleles. Approximately one-third of the melanic mice were heterozygous for the two different alleles. In addition, there was little variation among the melanic alleles compared to the amount of variation among light alleles—specifically, there was only 1 variable site among the melanic alleles and 13 sites for the light alleles, and the nucleotide diversity π was 0.0001 for the melanic alleles and 0.0021 for the light alleles. The lower variation among the melanic alleles is consistent with the expected pattern if selection has recently increased the frequency of the melanic allele (see p. 137). In a follow-up study, Hoekstra *et al.* (2004) examined coat color in a 35-km transect in which the center of the transect is approximately 10 km of black lava. They found a nearly complete association of coat color and rock color; in other words, there appears to be a stable cline, similar to but even more extreme than the cline for the thalassemia allele given in Figures 9.20 and 9.21. On the other hand, there was no association of mtDNA variation and background color, and estimates of the level of gene flow for this neutral marker were substantial.

PROBLEMS

1. Assume that for the continent–island model, $q_m = 0$, $m = 0.05$, and $q_0 = 0.8$. Draw a graph that shows how the allele frequency on the island changes over time.
2. Using the migration matrix in Table 9.2 and initial allele frequencies of 0.575, 0.567, and 0.505 in the Nuer, Dinka, and Shilluk, respectively, calculate the expected allele frequencies in these groups in the next three generations.

3. Assume that the allele frequencies of alleles A_1 and A_2 were 0.2 and 0.8 in one subpopulation and 0.4 and 0.6 in another subpopulation. Assuming that there are Hardy–Weinberg proportions within each subpopulation and that the subpopulations are of equal size, what would be the observed heterozygosity if these subpopulations were lumped? How does this compare to what would be expected if there were one random-mating population? Calculate the same values for the three heterozygotes when the subpopulations had frequencies of 0.2, 0.4, and 0.4 and 0.6, 0.2, and 0.2 for alleles A_1, A_2, and A_3. How does the observed heterozygosity compare to the expected heterozygosity for each heterozygote and for all heterozygotes combined?
4. Assuming that $q_A = 0.2$, $q_B = 0.6$, and $q_H = 0.3$, what is $\hat{M}$? Assuming that admixture has occurred for 10 generations, what is $\hat{M}$? What assumptions are important in such a calculation?
5. Calculate the F coefficients from the data in Table 9.5.
6. Assume that two populations both have five alleles in equal frequency but that they do not have any alleles in common. What is the G_{ST} value? Interpret your answer.
7. Calculate the F_{SR} and F_{RT} values for the data given in Table 9.6. Interpret your results. What further data, if any possible information were available, would you like to have to evaluate your interpretation?
8. Assume that estimates of $F_{ST(m)}$ and $F_{ST(p)}$ are 0.05 and 0.5. Calculate the expected F_{ST} and interpret your results. Now assume that estimates of $F_{ST(m)}$ and F_{ST} are 0.5 and 0.05. Calculate the expected ratio of pollen to seed gene flow and interpret your results. In mammals, if you had $F_{ST(m)}$ and $F_{ST(p)}$ based on mitochondrial variants and Y-chromosome variants, respectively, in a given generation, explain whether this would always be a good predictor for F_{ST}.
9. The data for wild oats given in Example 9.7 might be explained by a model that assumes variable selection under different trees. Design two experiments that might distinguish between this hypothesis and the nonselective explanation.
10. Compare the results from expressions 9.12*b* and 9.12*c* when $N = 500$ and $m = 0.1$ or $m = 0.001$ and interpret your results.
11. Whitlock and McCauley (1999) gave the illustration of two subpopulations in equal proportions with $Nm = 0.1$ and $Nm = 10$. What are the expected F_{ST} for the average Nm of these two values and the average F_{ST} for the two separate Nm values? Interpret your results.
12. Calculate the equilibrium values given in Table 9.10.
13. Do you think that the program to introduce Texas pumas into the Florida panther population is a good idea? Elaborate on your answer and suggest any modifications that you think would be appropriate.
14. Using the data given in Example 9.11, calculate the equilibria when $s = 0.11$ and 0.28. What do you think are the best explanations for the differences between these predicted values and those observed?
15. Do you think the theoretical predictions in Figure 9.21 are a good description of those observed in Figure 9.20? If you think they are, how would you support your answer? If you think they are not, how would you alter the theoretical model?

10

Linkage Disequilibrium and Recombination

> To every complex problem, there is a simple solution, and it is usually wrong.
>
> H. L. Mencken

> The fitness at a single locus ripped from its interactive context is about as relevant to real problems of evolutionary genetics as the study of the psychology of individuals isolated from their social context is to an understanding of man's sociopolitical evolution. In both cases context and interaction are not simply second-order effects to be superimposed on a primary monadic analysis. Context and interaction are of the essence.
>
> Richard Lewontin (1974)

> It is now generally understood that, as a consequence of selection, random genetic drift, co-ancestry, or gene flow, alleles at different loci may not be randomly associated with each other in a population. While this effect is generally regarded as a consequence of linkage, even genes on different chromosomes may be held temporarily or permanently out of random association by forces of selection, drift and nonrandom mating.
>
> Richard Lewontin (1988)

We have assumed in our earlier discussions that the transmission of alleles at a given locus was independent of the alleles at other loci (independent assortment) and that the fitnesses of the genotypes at a given locus were independent of the fitnesses of genotypes at other loci. These simplifications have been useful in many evolutionary contexts and in many cases may be valid. However, the documentation of extensive genetic variation on the molecular level, particularly of nucleotide sequence variation within genes and nearby regions, suggests that genetic variants may not be independent of each other. We generally talk about different loci in this chapter, but we also give some examples in which the same approaches can be used for different nucleotide sites within a gene. To understand the potential forces operating over different loci or different nucleotides, we need to be able to predict the maintenance and dynamics of genetic variation in a multiple-locus context that include the amount of recombination between loci. As more and more sequence data are generated, understanding the complexities of multigene (multisite) systems will become an important consideration in interpreting these data. Furthermore, detailed examination of nucleotide sequence data should shed light on the fine structure and organization of the genome and the significance of various evolutionary genetic factors that effect it.

With the mapping of the human genome, there has been great interest in finding the genetic variants that cause genetic diseases. One suggested approach is to determine the amount of statistical association between large numbers of SNPs in both individuals affected with a disease and a control group (Kruglyak, 1999). Differences in variants at the marker loci in a local genetic region between the two groups may provide a possible location for the disease-causing variant. The first two large surveys (Cargill *et al.*, 1999; Halushka *et al.*, 1999) developed the basic molecular techniques for this approach but also showed that the density of markers needs to be very high and that the potential disease variants may be population-specific (see discussion in Pritchard and Przeworski, 2001). On p. 658 we consider the approaches to identify disease-causing variants using multilocus techniques further.

When the genetic variation at two or more loci is considered simultaneously, the allele frequencies are insufficient to describe the genetic variation. Instead, multilocus gamete frequencies must be utilized because association of alleles within gametes may occur. Central to consideration of multilocus systems is a discussion of the evolutionary factors that influence gamete frequencies and **gametic disequilibrium** (generally called **linkage disequilibrium, LD**), the ***nonrandom association of alleles at different loci into gametes***. For nucleotide sequence data or closely linked loci, often the term **haplotype** is used, instead of gamete, to indicate the closely linked nucleotides or genes on a single copy of a chromosome. In human genetics, there is an effort to document the number of common haplotypes for chromosomal regions so that disease-causing variants can be identified (see p. 658).

As we will see, all of the evolutionary factors that we have discussed above—selection, inbreeding, genetic drift, gene flow, and mutation—can influence gamete frequencies and disequilibrium. In addition, the amount of recombination between loci is very important in affecting the extent of linkage disequilibrium. After we discuss the evolutionary forces influencing the amount of nonrandom association over loci, we examine some of the methods used for estimating linkage disequilibrium.

As an overall perspective, linkage disequilibrium (in random-mating populations) is generally observed only for very tightly linked loci or nucleotide sites. Generally for these loci, genetic drift and mutation appear to be the major factors generating disequilibrium, whereas recombination is either too low or has had insufficient time to break down these associations. The genetic distance over which disequilibrium is maintained appears to vary with genetic region, but in humans, it appears unlikely to extend beyond a few kilobases (Huttley *et al.*, 1999; Kruglyak, 1999), although there are other studies in which the region of disequilibrium is much greater (Pritchard and Przeworski, 2001).

If there is an interaction between the fitnesses at one locus and the fitnesses at other loci, one may need a fitness value for each multilocus genotype. In this discussion of multilocus selection, we examine **epistasis**, the ***interaction of fitness values at different loci***, and explore the implications of this phenomenon for linkage disequilibrium and the evolutionary process.

I. LINKAGE DISEQUILIBRIUM

Shortly after he derived what has become known as the Hardy–Weinberg principle, Weinberg (1909) noted that in a random-mating population, the alleles at two loci approach random association only asymptotically. Soon afterward, Jennings (1917) and Robbins (1918) gave the actual mode of approach to equilibrium frequencies for the two-locus model.

Because the rate of approach to random association is reduced by linkage, this nonrandom association was called **linkage disequilibrium** by Lewontin and Kojima (1960). However, this term is somewhat misleading because the loci involved may be unlinked (i.e., on different chromosomes or far apart on the same chromosome) and still be in linkage disequilibrium. The nonrandom association of alleles within gametes is a phenomenon that can be affected by linkage as well as other factors. In other words, the term **gamete disequilibrium** is a more accurate description of this phenomenon (a shortened form of the term gametic phase disequilibrium as used by Crow and Kimura, 1970). Also, linkage by itself is not sufficient to result in linkage disequilibrium. That is, alleles at linked loci are often not statistically associated so that they are not in linkage disequilibrium. However, the amount of linkage disequilibrium is generally a function of the rate of recombination. At the end of this section, we discuss some of the factors that influence the rate of recombination.

When there is more than one locus, there are more parameters that need to be considered than if it is assumed that there are multiple, independent single-locus systems. Assuming random mating and n different alleles at each of two loci, then there are $n - l$ independent allele frequencies for each locus or $2n - 2$ for two loci. On the other hand, to specify the frequencies of the multilocus gametes at these two loci, $n^2 - 1$ independent gamete frequencies must be known. In other words, for multilocus models, the allele frequencies at each locus are insufficient to describe the genetic variation; instead, multilocus gamete frequencies must be utilized because association of alleles within gametes may occur.

TABLE 10.1 The gametes (or haplotypes), the alleles, and their frequencies for a two-locus, two-allele model.

Gamete (haplotype)	*Frequency*	*Allele*	*Frequency*
A_1B_1	x_{11}	A_1	$p_1 = x_{11} + x_{12}$
A_1B_2	x_{12}	A_2	$p_2 = x_{21} + x_{22}$
A_2B_1	x_{21}	B_1	$q_1 = x_{11} + x_{21}$
A_2B_2	x_{22}	B_2	$q_2 = x_{12} + x_{22}$

a. The Linkage Disequilibrium Measure, *D*

Assume that a large random-mating population having discrete generations is segregating for two alleles at the A locus, say A_1 and A_2, and two alleles at the B locus, B_1 and B_2. The frequencies of the four possible gametes (or haplotypes), the x_{ij} values, and the four alleles are given in Table 10.1. Notice that p_1 and q_1 are used to indicate the frequency of alleles A_1 and B_1, respectively, and that $p_1 + p_2 = 1.0$, $q_1 + q_2 = 1.0$, and $\sum x_{ij} = 1.0$. If there is random association between alleles within gametes, then the frequency of each gamete is equal to the product of the frequencies of the alleles it contains—that is,

$$x_{11} = p_1q_1$$

$$x_{12} = p_1q_2$$

$$x_{21} = p_2q_1$$

$$x_{22} = p_2q_2$$

and a complete specification of gamete frequencies can be made using allele frequencies.

However, when there is nonrandom association of these alleles into gametes, we need to write the gamete frequencies as functions of these expected frequencies and a deviation (D) caused by the nonrandom association of alleles within gametes (see Table 10.2)

$$\begin{aligned} x_{11} &= p_1q_1 + D \\ x_{12} &= p_1q_2 - D \\ x_{21} &= p_2q_1 - D \\ x_{22} &= p_2q_2 + D \end{aligned} \tag{10.1}$$

TABLE 10.2 The frequencies of the four possible gametes when there are two alleles at each of two loci.

	A_1	A_2	*Total*
B_1	$x_{11} = p_1q_1 + D$	$x_{21} = p_2q_1 - D$	q_1
B_2	$x_{12} = p_1q_2 - D$	$x_{22} = p_2q_2 + D$	q_2
Total	p_1	p_2	1

From these equations, it is apparent that one can completely specify a two-locus, two-allele system in a random-mating population by knowing either three of the gamete frequencies, for example, x_{11}, x_{12}, and x_{21}, because $x_{22} = 1 - x_{11} - x_{12} - x_{21}$; or the frequency of an allele at both loci and the association between them, say p_1, q_1, and D. Both of these require three parameters, instead of the two, such as p_1 and q_1, that would be necessary if the alleles at the two loci were not associated.

The deviation, D, is the **linkage disequilibrium** parameter of Lewontin and Kojima (1960) and is a measure of the deviation from random association between alleles at different loci as defined by

$$D = x_{11} - p_1 q_1 \qquad (10.2a)$$

or the observed minus the expected frequency of gamete A_1B_1. By substituting values of p_1 and q_1 from Table 10.1, one may also write this expression as

$$D = x_{11}x_{22} - x_{12}x_{21} \qquad (10.2b)$$

With the expression written in this manner, one can see that D is the product of the frequencies of the coupling gametes minus the product of the frequencies of the repulsion gametes, where **coupling** and **repulsion** denote combinations of alleles with same subscript notation. The original use of the terms coupling and repulsion came from indication of the parental gamete types in a dihybrid cross.

How do the amount of linkage disequilibrium and the frequency of the two-locus gametes change over time? The approach used to derive these changes over one generation uses the information in Table 10.3, which gives the 10 possible genotypes, their frequencies, and the proportions of different

TABLE 10.3 The different genotypes when there are two alleles at each of two loci, their frequencies, and the expected proportion of the gametes produced in their progeny.

		Gametes of offspring			
Genotypes	*Frequencies*	A_1B_1	A_1B_2	A_2B_1	A_2B_2
A_1B_1/A_1B_1	x_{11}^2	x_{11}^2	—	—	—
A_1B_1/A_1B_2	$2x_{11}x_{12}$	$x_{11}x_{12}$	$x_{11}x_{12}$	—	—
A_1B_2/A_1B_2	x_{12}^2	—	x_{12}^2	—	—
A_1B_1/A_2B_1	$2x_{11}x_{21}$	$x_{11}x_{21}$	—	$x_{11}x_{21}$	—
A_1B_1/A_2B_2	$2x_{11}x_{22}$	$(1-c)x_{11}x_{22}$	$cx_{11}x_{22}$	$cx_{11}x_{22}$	$(1-c)x_{11}x_{22}$
A_1B_2/A_2B_1	$2x_{12}x_{21}$	$cx_{12}x_{21}$	$(1-c)x_{12}x_{21}$	$(1-c)x_{12}x_{21}$	$cx_{12}x_{21}$
A_1B_2/A_2B_2	$2x_{12}x_{22}$	—	$x_{12}x_{22}$	—	$x_{12}x_{22}$
A_2B_1/A_2B_1	x_{21}^2	—	—	x_{21}^2	—
A_2B_1/A_2B_2	$2x_{21}x_{22}$	—	—	$x_{21}x_{22}$	$x_{21}x_{22}$
A_2B_2/A_2B_2	x_{22}^2	—	—	—	x_{22}^2
	1	$x'_{11} = x_{11} - cD_0$	$x'_{12} = x_{12} + cD_0$	$x'_{21} = x_{21} + cD_0$	$x'_{22} = x_{22} - cD_0$

types of progeny that they generate. There are 10 genotypes because there are two types of double heterozygotes, the coupling (A_1B_1/A_2B_2) and the repulsion (A_1B_2/A_2B_1) genotypes.

To understand this derivation, note, for example, that A_1B_1/A_1B_1 genotypes produce only A_1B_1 gametes and that A_1B_1/A_1B_2 genotypes produce $\frac{1}{2}A_1B_1$ and $\frac{1}{2}A_1B_2$ gametes. For the double heterozygotes, recombination can produce gametes different from the parental gametes. For example, A_1B_2 and A_2B_1 gametes can be produced by recombination from A_1B_1/A_2B_2 individuals. The ***rate of recombination between loci A and B*** is symbolized by $\mathbf{c}$ and ranges from 0 when there is no recombination to 0.5 when there is independent assortment. (Sometimes the symbol r is used to indicate the rate of recombination, but because it can be confused with the commonly used measure of disequilibrium r^2 that we discuss in expression 10.4*b*, we use c to indicate the rate of recombination.) The maximum of 0.5 for c occurs because even with independent assortment between two loci, half the gametes produced are still the parental type. Because the frequency of A_1B_1/A_2B_2 individuals in the populations is $2x_{11}x_{22}$ and a proportion c of the gametes produced are recombinants, half of which are A_1B_2 and half A_2B_1, the gametes are produced in the proportions indicated in the fifth row of Table 10.3.

We can find the frequency of the gametes in the next generation by summing each of the columns. For example, the frequency of gamete A_1B_1 is

$$\begin{aligned} x'_{11} &= x_{11}^2 + x_{11}x_{12} + x_{11}x_{21} + (1-c)x_{11}x_{22} + cx_{12}x_{21} \\ &= x_{11}(x_{11} + x_{12} + x_{21} + x_{22}) - c(x_{11}x_{22} - x_{12}x_{21}) \\ &= x_{11} - cD_0 \end{aligned} \tag{10.3a}$$

where D_0 is the initial amount of linkage disequilibrium. The frequencies of the other gametes can be derived in a like manner and are given at the bottom of Table 10.3. The value of D after one generation is, therefore,

$$\begin{aligned} D_1 &= x'_{11}x'_{22} - x'_{12}x'_{21} \\ &= (x_{11} - cD_0)(x_{22} - cD_0) - (x_{12} + cD_0)(x_{21} + cD_0) \end{aligned}$$

which simplifies after some manipulation to

$$D_1 = (1-c)D_0 \tag{10.3b}$$

This simple recursive relationship then gives the general relationship

$$D_t = (1-c)^t D_0 \tag{10.3c}$$

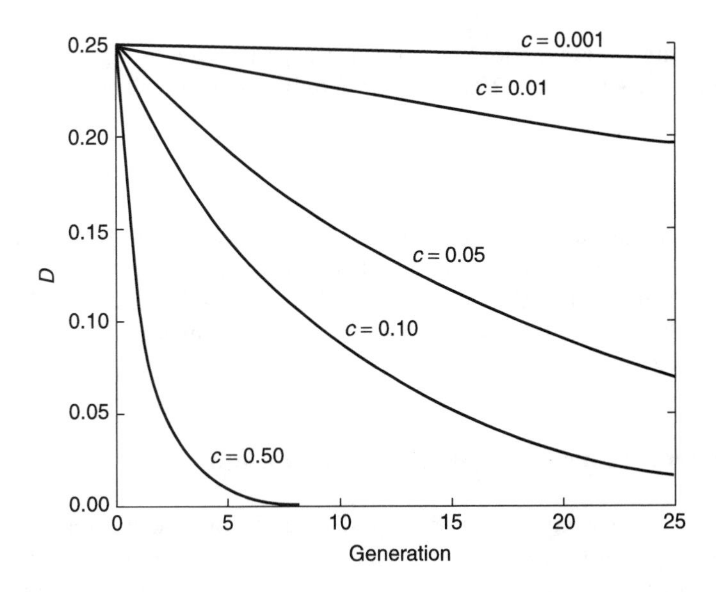

Figure 10.1. The decay of linkage disequilibrium D for different amounts of recombination (c).

where D_t is the value of D in the tth generation. This expression is approximately

$$D_t = e^{-ct} D_0 \tag{10.3d}$$

Changes in the gamete frequencies can take place only through recombination in double heterozygotes. With random proportions of genotypes, only $2(x_{11}x_{22} + x_{12}x_{21})$ of the gametes are potential recombinants. The other gametes are in either single or double homozygotes, where recombination does not produce different gametes.

D has a maximum value of 0.25 when there are only coupling gametes, $x_{11} = x_{22} = 0.5$ (and a minimum of -0.25 when there are only repulsion gametes, $x_{12} = x_{21} = 0.5$). Figure 10.1 is an example of the change of disequilibrium over time, assuming that $D_0 = 0.25$; it shows how the level of recombination affects the geometric decay of linkage disequilibrium. With no linkage or independent assortment ($c = 0.5$), most of disequilibrium is lost in five generations, whereas with tight linkage, a substantial proportion of the initial disequilibrium still remains after 25 generations. The disequilibrium does not go to zero immediately because new gametes can only be generated from double heterozygotes and many of the genotypes are double homozygotes or single heterozygotes. When the two loci are tightly linked—for example, $c = 0.001$—only approximately 2.5% of the disequilibrium has been lost after 25 generations. Molecular markers that are tightly linked may have very infrequent recombination, and thus, any disequilibrium present could remain for many generations.

To determine how long it takes for D_0 to decay to a given value D_t, we can solve expression 10.3c for the number of generations as

$$t = \frac{\ln (D_t/D_0)}{\ln (1-c)} \tag{10.3e}$$

where t is the number of generations until the disequilibrium is D_t. For example, if there is 10% recombination between the loci ($c = 0.1$), then the time until half of the initial disequilibrium has disappeared ($D_t/D_0 = 0.5$), the half-life, is 6.6 generations, whereas the time for 95% decay is 28.4 generations.

b. Other Measures of Linkage Disequilibrium

Although D as a measure of linkage disequilibrium has a number of good algebraic properties, it varies in size as a function of the frequencies of the constituent alleles. The maximum and minimum of 0.25 and −0.25, respectively, occur only when all the alleles have equal frequencies, and they are smaller when the allele frequencies are not 0.5. For example, if there is complete association of A_1 with B_1 but $x_{11} = 0.1$ and $x_{22} = 0.9$, then $D = 0.09$. Because of this and other reasons, other measures of linkage disequilibrium have been proposed for various contexts, although all appear to have some qualities that make then less than ideal (Hedrick, 1987c; Lewontin, 1988; Devlin and Risch, 1995).

For example, Lewontin (1964) suggested the parameter D', which is

$$D' = \frac{D}{D_{\max}} \tag{10.4a}$$

where $D_{\max}$ is the maximum D possible for a given set of allele frequencies at the two loci. $D_{\max}$ is equal either to the lesser of p_1q_2 or p_2q_1 if D is positive or to the lesser of p_1q_1 or p_2q_2 if D is negative. The advantage of this measure is that it has a range from −1.0 to 1.0, regardless of the allele frequencies at the two loci.

The square of the correlation coefficient

$$r^2 = \frac{D^2}{p_1p_2q_1q_2} \tag{10.4b}$$

was used by Hill and Robertson (1968) to measure linkage disequilibrium (the correlation coefficient, r, the square root of expression 10.4b with the same sign as D, has also been used). When the allele frequencies are the same at both loci, then r^2 has a range from 0.0 to 1.0, and r has a range from −1.0 to 1.0. When the allele frequencies are different at the two loci, then the ranges for both r^2 and r are somewhat smaller and a function of the allele frequencies. However, if we define $r' = r/r_{\max}$, where $r_{\max}$ is the maximum possible value of r for the given allele frequencies, then

$$r' = \frac{D/(p_1p_2q_1q_2)^{1/2}}{D_{\max}/(p_1p_2q_1q_2)^{1/2}} = D'$$

(Hedrick and Kumar, 2001). Therefore, if r^2 is standardized by the maximum r^2 value for the given allelic frequencies, then it is equal to $(D')^2$.

Because both D' and r are functions of D (the denominators of these measures do not change unless selection or other factors affect allele frequencies), they also decay exponentially in the same manner as D—that is, as a function of $(1 - c)^t$. The extent of measured linkage disequilibrium appears to vary with the estimated amount of recombination over different human chromosomes. For example, on chromosome 22, there are several regions of low LD and two regions of higher LD (Figure 10.2). Very similar patterns are seen using either D' or r^2 for four different samples (Dawson *et al.*, 2002). Consistent with the prediction of higher linkage disequilibrium when there is lower recombination, the two regions of high LD at 11–16 Mb and 21–27 Mb are in regions of exceptionally low recombination rates

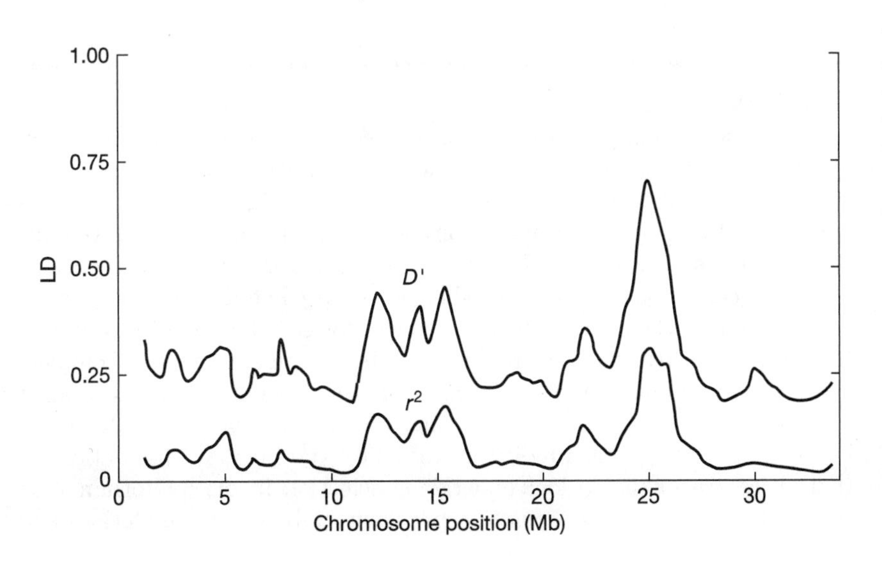

Figure 10.2. The extent of linkage disequilibrium between closely linked markers as a function of map position along human chromosome 22 as measured by D'(upper line) and r^2 (lower line) for the European ancestry CEPH sample. (Courtesy of *Nature*, 418: 2002, by Dawson, E., G.R. Abecasis, S. Bumpstead, *et al.* Reprinted with permission of Nature Publishing Group.)

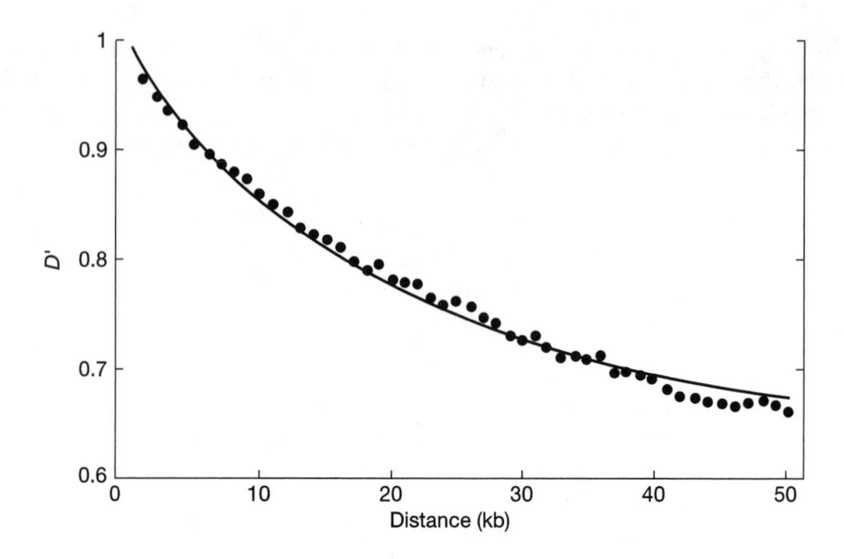

Figure 10.3. The decline of linkage disequilibrium as measured by D' between pairs of 24,056 SNPs on human chromosome 21 where the closed circles indicate the observed averages and the line shows the expected decline. (Courtesy of Innan, H., B. Padhukasahasram, and M. Nordborg. 2003. The pattern of polymorphism on human chromosome 21. *Genome Res.* 13:1158–1168.)

relative to the chromosome average. Figure 10.3 gives the observed D' for 24,046 SNPs at different distances from each other on 20 haplotypes from human chromosome 21 (Innan *et al.*, 2003). The average observed decline of D' with distance between pairs of SNPs from 0 to 50 kb is close to that expected. Example 10.1 shows the observed decay of D' over time for two allozyme loci in *D. melanogaster.*

Example 10.1. Clegg *et al.* (1978; 1980) carried out experiments in *D. melanogaster* designed to observe the decay of associations among alleles at marker loci within a chromosome. An example is given in Figure 10.4 for alleles at the two allozyme loci *Idh* and *Pgm*, genes 16.2 map units apart ($c = 0.162$) on the second chromosome. In this case, the observed disequilibrium as measured by D' was already near zero in all four replicates between generations 10 and 15, whereas the predicted theoretical decay, as indicated by the smooth line, was much slower. The essential results in this and the other examples of Clegg *et al.* was that there was complete and rapid randomization of alleles at the marker loci at a rate much faster than predicted. Clegg *et al.* suggest that the explanation for the difference between the theoretical prediction and the experimental observations is that there are unknown loci undergoing selection in disequilibrium with the marker loci in the *Drosophila* experiment. Subsequent simulations and

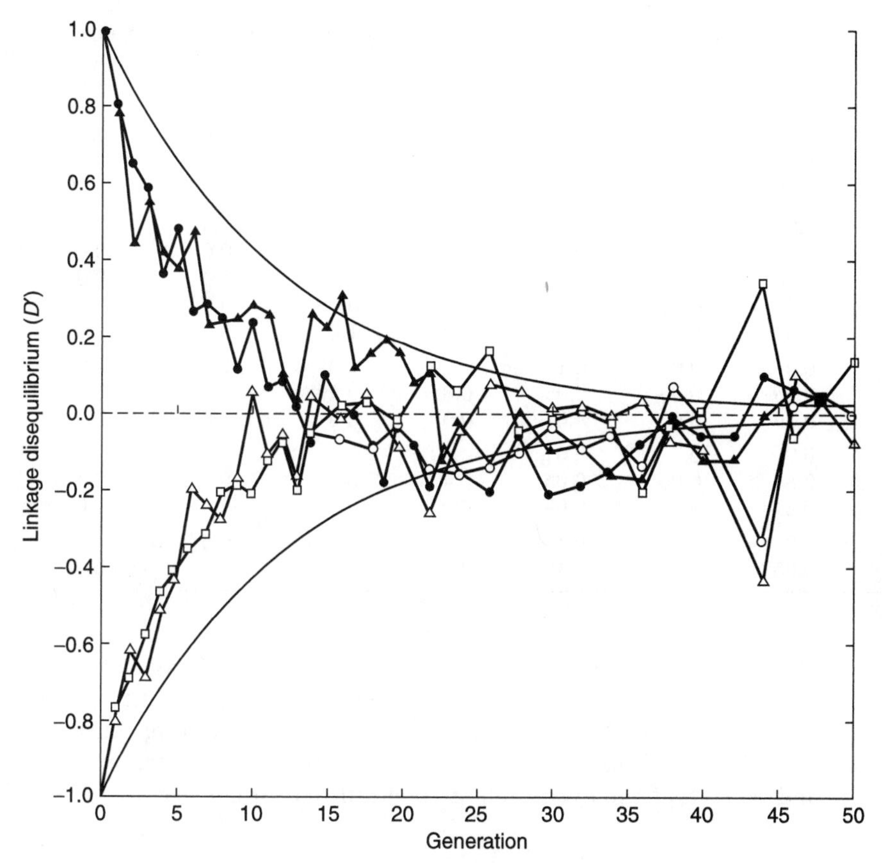

Figure 10.4. The observed decay of linkage disequilibrium (D') between two allozyme loci, and the theoretical decay given no selection is indicated by the smooth lines (after Clegg *et al.*, 1980). There are two replicates each for initial complete coupling and complete repulsion.

analytical work (Clegg, 1978; Asmussen and Clegg, 1982) demonstrated that the rate of decay of disequilibrium is generally accelerated by selection at associated loci.

Table 10.4 gives the values of these linkage disequilibrium measures for several combinations of gamete frequencies. The value of D is highly dependent on gamete frequencies, and when either p_1 or q_1 are much different from 0.5, its maximum value is greatly reduced. D', on the other hand, has the same range, regardless of the allele frequencies. The ranges of r^2 and r are also dependent on the allele frequencies when the allele frequencies at the two loci are different. For example, when $p_1 = 0.1$ and $q_1 = 0.5$ (bottom of Table 10.4), the maximum values of D and r^2 are only 0.05 and 0.111, respectively, whereas the maximum value of D' is 1.0.

TABLE 10.4 The values of various linkage disequilibrium measures for examples using different gamete frequencies.

Allele frequencies	*Gamete frequencies*				*Linkage disequilibrium measures*			
	x_{11}	x_{12}	x_{21}	x_{22}	D	r^2	r	D'
	0.5	0.0	0.0	0.5	0.25	1.0	1.0	1.0
	0.4	0.1	0.1	0.4	0.15	0.36	0.6	0.6
$p_1 = q_1 = 0.5$	0.25	0.25	0.25	0.25	0.0	0.0	0.0	0.0
	0.1	0.4	0.4	0.1	−0.15	0.36	−0.6	−0.6
	0.0	0.5	0.5	0.0	−0.25	1.0	−1.0	−1.0
	0.9	0.0	0.0	0.1	0.09	1.0	1.0	1.0
$p_1 = q_1 = 0.9$	0.85	0.05	0.05	0.05	0.04	0.198	0.445	0.444
	0.81	0.09	0.09	0.01	0.0	0.0	0.0	0.0
	0.0	0.9	0.1	0.0	−0.09	1.0	−1.0	−1.0
$p_1 = q_2 = 0.9$	0.05	0.85	0.05	0.05	−0.4	0.198	−0.445	−0.444
	0.09	0.81	0.01	0.09	0.0	0.0	0.0	0.0
$p_1 = 0.1$,	0.1	0.0	0.4	0.5	0.05	0.111	0.577	1.0
$q_1 = 0.5$	0.05	0.05	0.45	0.45	0.0	0.0	0.0	0.0

Asmussen *et al.* (1987) have examined the amount of association between a cytoplasmic gene (either mtDNA or cpDNA) and a nuclear gene (see also Basten and Asmussen, 1997). In the simplest situation, there are two cytoplasmic alleles or haplotypes, M_1 and M_2, and two nuclear alleles A_1 and A_2 so that the **cytonuclear disequilibrium** between them is

$$D_c = x_{11} - m_1 p_1$$

where x_{11} here is the frequency of the M_1A_1 gamete and m_1 is the frequency of M_1 mitochondrial haplotype. Because cytoplasmic and nuclear genes are unlinked, cytonuclear disequilibrium decays quickly at the same rate as unlinked nuclear genes so that $(1/2)^t$ of the initial disequilibrium is expected to be left after t generations. However, in recent hybrids, measurement of cytonuclear disequilibrium can be used to identify the types of matings and backcrosses because of the maternal inheritance of cytoplasmic genes (see Example 10.2 for an illustration from data from hybrid tree frogs).

Example 10.2. The association between alleles from mtDNA or cpDNA and nuclear genes is particularly useful in identifying the kinds of hybrid matings between groups, such as type 1 female × type 2 male or vice versa, because of different inheritance of the cytoplasmic (generally maternal) and nuclear (biparental) genes. For example, Lamb and Avise (1986) examined both mtDNA and allozyme variation in putative hybrids of two tree frog

TABLE 10.5 The genotype frequencies for mtDNA and the albumin locus for tree frogs of apparent hybrid origin in samples from Lamb and Avise (1986) between species *Hyla cineras* and *H. gratiosa*.

		Albumin			
		A_cA_c	A_cA_g	A_gA_g	*Total*
mtDNA	M_c	0.162	0.077	0.035	0.274
	M_g	0.141	0.380	0.204	0.725
	Total	0.303	0.457	0.239	

$$D_1 = 0.162 - (0.274)(0.303) = 0.079$$
$$D_2 = 0.077 - (0.457)(0.274) = -0.048$$
$$D_3 = 0.035 - (0.239)(0.274) = -0.030$$

species, *Hyla cineras* and *H. gratiosa*, that often breed in the same ponds. Generally, *H. cineras* males call from elevated perches on the shoreline of the ponds and *H. gratiosa* from the water surface, and females go to these ponds to mate. However, in disturbed habitats there are no perches, so *H. cineras* males call from ground level at the shore and appear to intercept and mate with both *H. cineras* and *H. gratiosa* females as they enter the ponds.

H. cineras and *H. gratiosa* differ diagnostically by several mtDNA restriction site polymorphisms and by alleles at several allozyme loci. For example, for mtDNA and the albumin locus, pure *H. cineras* have the genotype $M_cA_cA_c$ and pure *H. gratiosa* have $M_gA_gA_g$, where the subscripts c and g indicate alleles from the two species. Table 10.5 gives the frequencies for the six different genotypes in a sample of 142 individuals thought to be of hybrid origin (163 individuals that appear to be pure species types are omitted from the data presented in Asmussen *et al.*, 1987). Because there is genotypic association between mtDNA and albumin, three disequilibrium values, one for the association of each genotype with M_c, are calculated (Asmussen *et al.*, 1987). These values, which are the observed genotype frequencies minus the expected frequencies (bottom of Table 10.5), show that there is an excess of the pure species types, $M_cA_cA_c$ (D_1 is positive) and $M_gA_gA_g$ (D_3 is negative). In addition, there is an excess of the $M_gA_cA_g$ genotype (D_2 is negative), which is expected from crosses between *H. cineras* males and *H. gratiosa* females and confirms the behavioral observations.

c. Rate of Recombination, *c*

The **rate of recombination,** c, is generally ***estimated as the proportion of recombinant gametes produced from a parent with a known (or inferred) gamete constitution***. When two loci are close to each other, then the number of map units (number of centiMorgans, cM) between the loci is $100c$. However, when the loci are further apart, then multiple recombinations can occur between the loci, and there are fewer apparent nonrecombinants. Haldane (1919) proposed the following mapping function based on the Poisson distribution in which crossovers are assumed to occur randomly along the chromosome

$$c = \frac{1 - e^{-2m}}{2} \qquad (10.5a)$$

where m is the expected number of crossovers. Notice that when m is small c and m are similar ($m = 0.01$, then $c = 0.0099$) but that when m is large c asymptotically approaches 0.5 (for further discussion of mapping functions, see Zhao and Speed, 1996; Lynch and Walsh, 1998).

The amount of recombination can vary because of a number of factors (Robinson, 1996); in particular, the level of recombination may vary greatly with sex. In some organisms, such as *Drosophila* and *Bombyx* (bumblebees), there is no recombination in males; in Atlantic salmon, the rate of recombination appears to be nearly an order of magnitude higher in females than males (Moen *et al.*, 2004); and in humans, the level of recombination is approximately 60% higher in females than males for autosomal genes (see Table 10.6) (Kong *et al.*, 2002; Matise *et al.*, 2003). To accommodate this difference, c can be replaced with

$$\bar{c} = (c_f + c_m)/2 \qquad (10.5b)$$

where c_f and c_m are the recombination values in females and males, respectively. In organisms with recombination in only one sex, the approach to linkage equilibrium is half as fast as in organisms with recombination in both sexes. For X-linked genes and genes in haplo-diploid organisms, in which there is no recombination in males, the rate of decay is more complicated because the sexes may differ in gamete frequencies (Bennett and Oertel, 1965).

In addition to sex differences in recombination, the extent of recombination varies on different chromosomes (see Table 10.6 for human chromosomes) and between different regions of chromosomes. In fact, the terms **hot spots** and **cold spots** of recombination are used to indicate regions of higher- and lower-than-average recombination, respectively (Arnheim *et al.*, 2003; Kauppi *et al.*, 2004). Turning things around, the amount of linkage disequilibrium can be used to estimate the rate of recombination, particularly regions of locally high and low recombination (Stumpf and McVean, 2003; Ptak *et al.*, 2004). Example 10.3 discusses some of the factors influencing the rate of recombination and gives some data showing recombination variation across human chromosome 3 and large variation in recombination in sperm from different individuals.

TABLE 10.6 For each chromosome in the human genome, the physical length, the estimated genetic length in meioses from females, males, and the average over both sexes, and the ratio of the genetic to physical lengths (cM/Mb). The data are calculated from 1257 meiotic events from 146 Icelandic families using 5146 microsatellite markers (Kong *et al.*, 2002).

		Genetic length (cM)			
Chromosome	*Physical length (Mb)*	*Female*	*Male*	*Sex average*	*cM/Mb (sex average)*
1	282.6	345.4	195.1	270.3	0.96
2	252.5	325.4	189.6	257.5	1.02
3	224.5	275.6	160.7	218.2	0.97
4	205.4	259.1	146.5	202.8	0.99
5	199.2	260.2	151.2	205.7	1.03
6	190.9	241.6	137.6	189.6	0.99
7	168.5	230.3	128.4	179.3	1.06
8	158.1	209.9	107.9	158.9	1.01
9	150.2	198.2	117.2	157.7	1.05
10	145.6	218.1	133.9	176.0	1.21
11	153.0	195.5	109.4	152.4	1.00
12	153.4	206.6	135.5	171.1	1.12
13	100.4	155.9	101.3	128.6	1.28
14	87.1	142.4	94.6	119.5	1.36
15	87.2	155.0	102.6	128.8	1.48
16	106.4	149.6	108.1	128.9	1.21
17	89.4	161.5	108.6	135.0	1.51
18	89.4	142.57	98.6	120.6	1.35
19	69.4	126.8	92.6	109.7	1.58
20	59.4	122.0	74.7	98.4	1.66
21	30.0	76.4	47.3	61.9	2.06
22	31.2	82.8	49.0	65.9	2.11
X	156.8	179.9	—	179.0	1.14
Total or mean	3190.8	4460.0	2590.5	3614.7	1.13

Example 10.3. While the extent of recombination is known to vary as a result of a number of factors, the mechanism and the exact impact of most of these factors are not well known (Robinson, 1996; Arnheim *et al.*, 2003). As we discussed above, recombination often varies between the two sexes, usually with higher recombination in females than in males (see Table 10.6). In addition, selection for recombination can change the recombination rate (for a review, see Brooks, 1988), specific genes may influence recombination rates, and a number of environmental factors can influence the recombination level.

Recombination also varies across different chromosomal regions. For example, the telomeric and centromeric regions of the X chromosome in *D. melanogaster* have much lower recombination rates per megabase than the rest of the chromosome (Charlesworth, 1996). In addition, there appear to be hot spots and cold spots of recombination where the amount of recombination per megabase is either greatly elevated or reduced (Arnheim

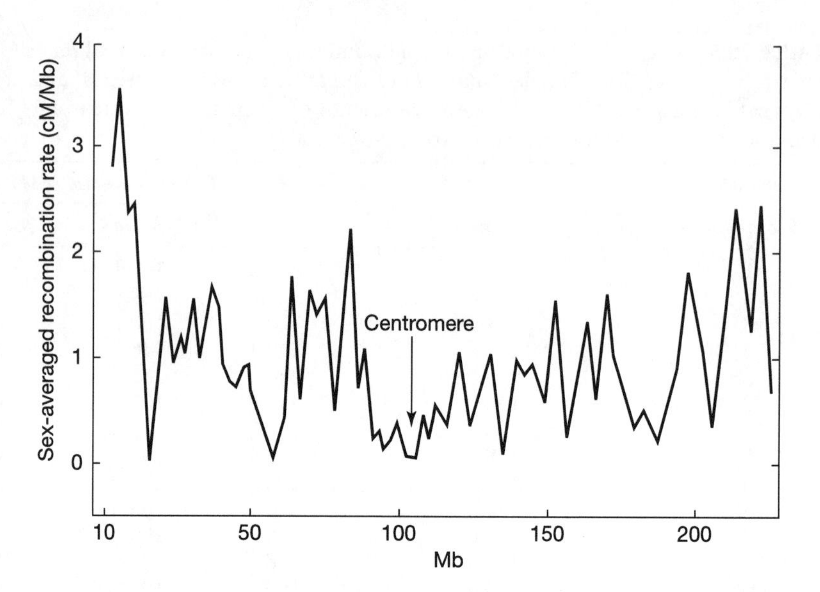

Figure 10.5. The sex-averaged rate of recombination (cM/Mb) for human chromosome 3 where the centromere is located just above 100 Mb. (Courtesy of *Nat. Genet.* 31: 2002, by Kong, A., D.F. Gudbjartsson, J. Sainz, *et al.* Reprinted with permission of Nature Publishing Group.)

et al., 2003). Figure 10.5 gives the estimated level of recombination (cM/Mb) for human chromosome 3 (Kong *et al.*, 2002). The estimated rate of recombination varies from greater than 3 cM/Mb at the telomere of the short arm (left side) to less than 0.1 cM/Mb at the centromere and the surrounding area. They found a number of peaks and valleys of recombination across the chromosome, a pattern they also found throughout the human genome.

In addition, there may be great variation in the recombination rate among different individuals (Yu *et al.*, 1996). Cullen *et al.* (2002) examined over 20,000 single sperm from 12 donors to determine the recombination rate in the 3.3-Mb region of the MHC in humans. They identified 325 recombinant chromosomes and estimated recombination rates for the different donors. The data in Table 10.7 show that there was an over sixfold difference in the estimated recombination rate for donors C (0.0071) and

TABLE 10.7 The number of recombinant sperm observed in six donors and the estimated recombination rate per Mb (the MHC region is assumed to 3.3 Mb) (from Cullen *et al.*, 2002).

Donor	*Number recombinants*	*Number sperm*	*c*	*cM/Mb*
A	51	2992	0.0170	0.516
B	22	1806	0.0122	0.369
C	6	842	0.0071	0.216
G	45	1605	0.0280	0.849
L	59	1350	0.0437	1.324
E	26	1733	0.0150	0.454
Total (12 donors)	325	20,031	0.0162	0.492

L (0.0437). Although recombination rates can generally be averaged over individuals, as they are over the two sexes, the finding of such high individual variation in recombination may have important implications, particularly when genes are tightly linked.

II. THE INFLUENCE OF GENETIC DRIFT, MUTATION, INBREEDING, AND GENE FLOW ON LINKAGE DISEQUILIBRIUM

When linkage disequilibrium has been observed in a population, it has often been attributed to some type of multilocus selection. This assumption may not be valid because a number of other factors can affect linkage disequilibrium. As illustrated in the preceding section, recombination influences the rate of approach to linkage equilibrium, although recombination cannot actually generate linkage disequilibrium. In addition, other factors besides multilocus selection, such as genetic drift, mutation, gene flow, and genetic hitchhiking, can actually generate linkage disequilibrium. In a fashion similar to linkage, inbreeding can also retard the rate of approach to linkage equilibrium. We first discuss these other factors (except for genetic hitchhiking, which we discuss later) and then return to a discussion of multilocus selection.

a. Genetic Drift

Multilocus models incorporating finite population size have shown that nonrandom associations may develop between alleles at different loci as a result of genetic drift. This effect has become much more important in sequence data in which the rate of recombination between loci or nucleotide sites is often very low (Pritchard and Przeworski, 2001). The easiest way to understand the effect of genetic drift on linkage disequilibrium is to envision the four gamete types for a two-locus, two-allele model as four alleles at one locus. As a result, genetic drift from generation to generation in a finite population alters the frequency of the gametes in a fashion similar to that of a single-locus, four-allele model. These stochastic fluctuations in a small population are almost certain to result in nonrandom associations between alleles at different loci. When recombination occurs between the two loci, the effect of genetic drift is less extreme because some gametes can be reconstituted from other pairs of gametes.

As with the expected allele frequency change at a single locus due to genetic drift, the expected value of disequilibrium due to two-locus genetic drift over many populations is zero. However, in any individual population, drift-generated linkage disequilibrium can be very great and either negative or positive in sign. As a result, the squared correlation coefficient, r^2, has been used to show the effect of finite population size on linkage disequilibrium. For an effective population size of N_e and a rate of recombination c between two loci, the expected value of r^2 is approximately

$$E(r^2) \approx \frac{1}{1 + 4N_e c} \tag{10.6a}$$

(Hill and Robertson, 1968; Ohta and Kimura, 1969). From this expression, it is obvious when $N_e c$ is large, $E(r^2)$ approaches 0.0, and when $N_e c$ is small, $E(r^2)$ approaches 1.0. For intermediate values of $N_e c$, Hill (1976) showed that expression 10.6a somewhat underestimates the true value of $E(r^2)$.

Nevertheless, this expression can be used to give a minimal estimate of the effect of population size on the average r^2 value. For example, if $N_e c$ is 1 (say $N_e = 1{,}000$, $c = 0.001$), then the expected value of r^2 is 0.20, a quite substantial amount of linkage disequilibrium. Even if the present population size is relatively large, a small founder population or a bottleneck in a previous generation could be responsible for considerable disequilibrium. If N_e was small at some time in the recent past, disequilibrium may still be present because of tight linkage, inbreeding, or other factors that retard the decay of linkage disequilibrium.

If $N_e c$ is large, then expression 10.6a becomes approximately

$$E(r^2) \approx \frac{1}{4N_e c} = \frac{1}{C} \tag{10.6b}$$

where $C = 4N_e c$ and is the population recombination rate, analogous to the population mutation rate $\theta = 4N_e u$ (Wall, 2000). The expected magnitude of linkage disequilibrium decreases as the size of C increases (Long and Langley, 1999; Pritchard and Przeworski, 2001), assuming that factors other than genetic drift do not greatly influence linkage disequilibrium. Because the effective population size is expected to be the same for all genes within a population, then the relative size of $C = 1/r^2$ for different regions of the genome should reflect differences in the rate of recombination, as in different regions of chromosome 22 in Figure 10.2. On the other hand, if c is known, then an estimate of the effective population size is $N_e = 1/4cr^2$ (see also Hayes *et al.*, 2003).

When the effect of the sample size N on disequilibrium is included, then

$$E(r^2) = \frac{(1-c)^2 + c^2}{2N_e c(2-c)} + \frac{1}{N} \tag{10.6c}$$

for populations segregating at both loci (Weir and Hill, 1980). This expression can be solved for an estimate of the effective population size as

$$N_e \approx \frac{(1-c)^2 + c^2 1}{2c(2-c)(r^2 - \frac{1}{N})} \tag{10.6d}$$

Assuming that the pair of loci is unlinked ($c = 0.5$), then expression 10.6c becomes

$$E(r^2) \approx \frac{1}{3N_e} + \frac{1}{N}$$

and expression 10.6d becomes

$$N_e \approx \frac{1}{3(r^2 - \frac{1}{N})} \tag{10.6e}$$

Although Hill (1981) states that such an estimate for a single locus is quite imprecise, Waples (1991a) showed that if a number of unlinked loci are available, for example, 10 or more, precision is greatly increased.

Slatkin (1994b) showed that under some conditions a population that is stable in size will have linkage disequilibrium between closely linked loci, whereas in a rapidly growing population, there will be much less equilibrium. He suggested that in a stable population, mutations are more likely to be on internal branches of a gene genealogy and represented by several descendants, whereas in a growing population, the gene genealogy is more star-like and the mutations are on terminal branches. Once the population begins to grow, genetic drift is less effective, and there are fewer coalescent events. However, the overall picture is more complicated because the predictions of Slatkin are only for new linkage disequilibrium; disequilibrium that was already present at the start of a population expansion will remain. Example 10.4 discusses data in different human populations that are consistent with these predictions.

Example 10.4. The extensive effort to map genomes has resulted in the mapping of a large number of variable loci (Dib *et al.*, 1996; Matise *et al.*, 2003). Examination of the amount of linkage disequilibrium between these markers can give an estimate of the background level of disequilibrium in a given population so that, for example, the level of disequilibrium between a marker locus and a disease phenotype can be adequately evaluated.

TABLE 10.8 The probability of significant linkage disequilibrium in samples from four populations for pairs of seven closely linked microsatellite loci (the complete locus name has a *DXS* prefix) on the human X chromosome (c is the recombination rate between a locus pair) (Laan and Paabo, 1997). Probabilities in boldface are significant at the 0.05 level without correction for multiple comparisons.

		P			
Locus pair	c	*Saami*	*Finns*	*Estonians*	*Swedish*
995–983	0.04	**0.01**	0.51	0.31	0.59
995–986	0.025	**0.00**	0.73	0.83	0.43
995–8092	0.024	0.10	0.12	0.73	0.42
995–8082	0.023	**0.00**	0.13	0.24	0.49
8037–983	0.02	0.30	0.68	0.10	0.92
1225–983	0.02	**0.00**	0.63	0.52	0.48
995–1225	0.02	**0.00**	0.15	0.59	0.56
995–8037	0.02	0.12	0.87	0.62	0.22
8082–983	0.017	**0.00**	0.56	0.73	0.08
8092–983	0.016	**0.00**	0.31	0.15	0.75
986–983	0.015	**0.00**	0.83	0.47	0.40
1225–986	0.005	**0.00**	0.39	0.69	0.45
8037–986	0.005	**0.00**	0.62	0.74	0.26
8037–8092	0.004	**0.00**	0.18	0.07	**0.03**
1225–8092	0.004	**0.00**	0.28	0.12	0.68
8037–8082	0.003	**0.01**	0.24	0.62	**0.03**
1225–8082	0.003	**0.00**	**0.00**	**0.00**	**0.00**
8082–986	0.002	**0.00**	0.09	0.14	0.62
8092–986	0.001	**0.00**	0.33	0.10	0.33
8082–8092	0.001	**0.00**	**0.04**	0.06	0.10
8037–1225	0.000	0.09	0.84	0.49	0.24

Laan and Paabo (1997) have examined the disequilibrium between tightly linked pairs of seven microsatellite loci on a 4-cM segment of the human X chromosome (Xq13) in samples of males from four groups: Saami (Laplanders) from Sweden ($N = 54$), Finns ($N = 80$), Estonians ($N = 45$), and Swedish ($N = 41$). Table 10.8 gives the probability of significant linkage disequilibrium using Fisher's exact test (see p. 587) for the 21 pairs of loci in the four samples. Eighteen of the 21 pairs are significant for the Saami sample, whereas only one to three comparisons are significant for the other three samples. In the Estonians and the Swedish, every man had a different X haplotype, and in the Finnish sample, 75 of the 80 men had different X haplotypes. On the other hand, there were only 32 X haplotypes in the sample of 54 Saami, but these are more different from each other than the haplotypes within the other samples. This difference does not appear to be the result of much lower diversity in the Saami, who had an overall diversity of 0.688 compared to 0.738, 0.734, and 0.717 for the Finns, Estonians, and Swedish, respectively.

Laan and Paabo (1997) showed with pairwise sequence differences from mtDNA that the Saami sample is consistent with that expected from a stable population, whereas the mtDNA data from the other three samples suggested that they had undergone a recent expansion (this interpretation is also consistent with historical and other data). They concluded, on the basis of the theory of Slatkin (1994b), that the higher disequilibrium found in the Saami compared to the other population, is the result of their relatively stable population. However, examining these findings, Freimer *et al.* (1997) point out that the most important factor influencing the level of background disequilibrium is probably the variation found in different genomic regions within and between populations. They also suggest caution in identifying the factors responsible for generating these striking results (see also Lonjou *et al.*, 1999).

b. Mutation

Mutation by itself probably generates only low amounts of linkage disequilibrium, but mutation, along with recombination and gene flow, is the source of new gamete types in a population. New gametes, produced by mutation, in turn can increase in frequency by genetic drift or selection, and this combination of factors may generate further linkage disequilibrium.

To understand the magnitude of linkage disequilibrium originated by mutation, assume that a population is segregating for two alleles, B_1 and B_2, at the B locus and that it is monomorphic for allele A_1 at the A locus. If there is a mutation from A_1 to A_2, then the gamete frequencies are as given in Table 10.9. The new mutant occurs on a chromosome containing either the B_1 or the B_2 allele, and thus, there are two possible arrays of gamete frequencies. If there is only one mutant allele assumed, then the initial frequency of A_2, as well as the gamete containing the mutant, is $p_2 = 1/(2N)$, where N is the population size.

TABLE 10.9 The gamete frequencies and linkage disequilibrium before and after a mutation at the A locus from A_1 to A_2.

	Gamete frequencies						
	A_1B_1	A_1B_2	A_2B_1	A_2B_2	D	r^2	D'
Before mutation	q_1	q_2	—	—	0.0	0.0	0.0
After mutation							
Mutant on B_1	$q_1 - p_2$	q_2	p_2	—	$-p_2q_2$	$\frac{p_2q_2}{p_1q_1}$	-1.0
Mutant on B_2	q_1	$q_2 - p_2$	—	p_2	p_2q_1	$\frac{p_2q_1}{p_1q_2}$	1.0

The linkage disequilibrium thus generated by mutation as measured by D is either $-p_2q_2$ or p_2q_1, depending on whether the mutation occurred on a chromosome with B_1 or B_2. These initial values are small because it is assumed that p_2 is small. Linkage disequilibrium as measured by r^2 is either p_2q_2/p_1q_1 or p_2q_1/p_1q_2, depending on whether the mutation occurred on a chromosome with B_1 or B_2 (Hedrick and Kumar, 2001). Again, these initial values are small because it is assumed that p_2 is small. However, if linkage disequilibrium is measured using D', which gives a proportion of the maximum disequilibrium possible with the given allele frequencies, then linkage disequilibrium is at the maximum possible and is either -1.0 or 1.0, depending on whether the mutant occurred on a B_1 or a B_2 chromosome, respectively. For example, if the mutation from A_1 to A_2 occurred on B_1 chromosome and $p_2 = 0.001$ and $q_1 = q_2 = 0.5$, then $D = -0.0005$, $r^2 = 0.001$, and $D' = -1.0$.

Such high standardized association may be used to identify associations of disease alleles, even if they are rare, with marker loci (de la Chapelle and Wright, 1998). In a specific case, high disequilibrium in African β-globin sickle-cell haplotypes has been used to suggest that there were at least three independent mutations to the sickle-cell allele (see Example 10.5).

Example 10.5. Perhaps the most often cited case of balancing selection is that of the human sickle-cell variant in areas with endemic malaria (Allison, 1964). However, the origin and spread of these mutants continue to be a focal point of research and discussion (Currat *et al.*, 2002). Examination in the early 1980s of the major haplotypes containing the sickle-cell mutant showed strong linkage disequilibrium between restriction sites (Antonarakis *et al.*, 1982; Pagnier *et al.*, 1984). In addition, there appears to be a very localized distribution of the three major sickle-cell haplotypes within Africa, such that 100% of the sickle-cell haplotypes in Benin are the Benin haplotype, 83% in the Central African Republic (CAR) are the CAR haplotype, and 81% in Senegal are the Senegal haplotype (Flint *et al.*, 1993). These findings led to the hypothesis that the sickle-cell mutant occurred independently at least three different times in Africa on different haplotypes in different areas (see discussion in Currat *et al.*, 2002).

Chebloune *et al.* (1988) examined a region 5′ to the β-globin gene and showed that the sequences on these haplotypes supported the earlier general findings. They found both differences in a segment of repeated sequences and ten variable nucleotide positions among the three haplotypes. Figure 10.6 shows the putative relationship between these haplotypes for eight nucleotides, including an ancestral sequence inferred from a chimpanzee sequence. Note that the three sickle-cell haplotypes differ from each other by at least four nucleotides, which suggests that the haplotypes containing the sickle cell are not recent derivations. Because malaria is thought to have been a major selective force only within the last few thousand years, these

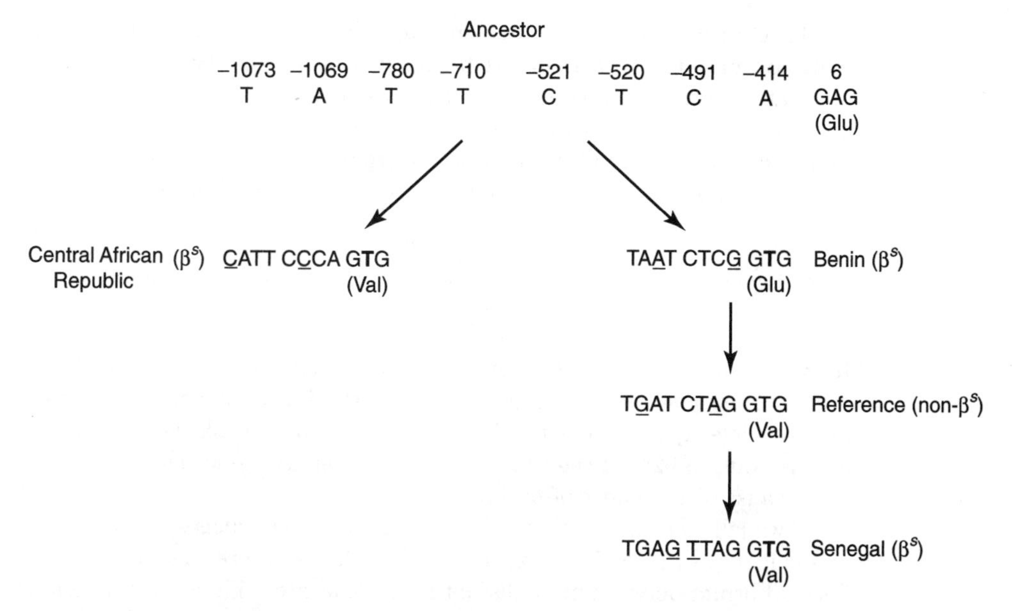

Figure 10.6. The sequence of eight variable nucleotides in the region 5′ to the β-globin gene and the putative relationship of the three sickle-cell haplotypes with the changed T in boldface (after Chebloune *et al.*, 1988). Also included is the codon that results in the sickle-cell mutant, replacement of glutamic by valine acid.

haplotypes, each containing the same sickle-cell mutant, appear to have evolved much earlier, an observation consistent with the tricentric origin of the sickle-cell mutant in Africa. Examining a well-defined ethnic sample in Sengal, Currat *et al.* (2002) estimated that the Senegal mutant arose approximately 45 to 70 generations ago or approximately 1125 to 1750 years ago, assuming 25 years per generation.

It is unlikely that recurrent mutation will cause a buildup of linkage disequilibrium between alleles at different genes because mutations are generally much less frequent events than recombination that would break up newly generated gametes. However, when the rate of recombination is low, as for tightly linked loci or nucleotide sites within loci, or nonexistent as for mtDNA or the nonrecombining part of the Y chromosome, linkage disequilibrium generated by mutation may be very important. It also is unlikely that subsequent mutants would always occur associated with the same allele. However, once a new gamete has been generated by mutation, it can increase (or be lost) by genetic drift. The initial stages of this process, until the other gamete is generated by recombination, would follow the same dynamics as a three-allele system with no selection. In addition, as discussed in the next two sections, gamete frequencies may be affected by multilocus selection or by genetic hitchhiking, potentially very important factors for new mutant alleles.

In addition, when the mutation rate is high, as for microsatellite loci, mutation may play a role in the breakdown of linkage disequilibrium. To illustrate, assume that B is a microsatellite locus and that mutation from normal allele A_1 to disease allele A_2 generates a new A_2B_1 gamete with frequency x_{21}. Assuming that mutation from B_1 to $\overline{B}_1$(not B_1) occurs at a rate u and assuming that there is no recombination, after one generation

$$x'_{21.1} = x_{21.0}(1 - u)$$

After t generations, the frequency of the A_2B_1 gamete is reduced to

$$x_{21.t} = x_{21.0}(1 - u)^t$$

If we assume that $u = 10^{-3}$ and $t = 1000$, then $x_{21.t}/x_{21.0} = 0.368$. In other words, just by mutation at locus B, the initial association between disease allele A_2 and microsatellite allele B_1 has been broken down. In fact, at this point, 63.2% of the time the mutant allele A_2 is not associated with B_1 because of the effect of mutation.

Awadalla *et al.* (1999) published a report that suggested that there was evidence of recombination in mtDNA because there was lower linkage disequilibrium between more distant nucleotide sites. There were a number of criticisms of this conclusion, but Hedrick and Kumar (2001) (see also p. 607) demonstrated that the pattern observed was consistent with mutation resulting in lower linkage disequilibrium between some sites and that invocation of recombination was not the most parsimonious explanation (these authors have since retracted their claim; McVean *et al.*, 2002; Piganeau and Eyre-Walker, 2004).

c. Inbreeding

If there is some degree of inbreeding, then the rate of decay of linkage disequilibrium is affected in a manner similar to that resulting from linkage. However, the similarity of linkage and inbreeding effects on the asymptotic rate of decay is somewhat misleading and not indicative of the whole decay process. The bases for the two effects are quite different in that for low levels of recombination, few recombinants are formed from the double heterozygotes. On the other hand, a high level of inbreeding, such as selfing, leads to a reduction of the proportion of double heterozygotes, the only genotypes from which new gametes are formed by recombination.

The inbreeding situation in which the rate of decay of linkage disequilibrium has been most thoroughly examined is partially selfing organisms. In this case, Weir *et al.* (1972) gave the asymptotic decay rate of linkage disequilibrium in terms of the proportion of selfing (S) and recombination (c) as

$$D_{t+1} = {}^1\!/_2 \left\{ \frac{1 + \lambda + S}{2} + \left[\left(\frac{1 + \lambda + S}{2} \right)^2 - 2S\lambda \right]^{1/2} \right\} D_t \qquad (10.7)$$

where $\lambda = 1 - 2c$ so that the proportions of selfing and recombination are on the same scale; that is, the range of the recombination effect is from 0.0 to 1.0, and high values of both λ and S indicate the greatest effect. To show their equivalency, if $S = 0$ (random mating), then expression 10.7 becomes

$$D_{t+1} = {}^1\!/\!_2(1 + \lambda)D_t$$

which is same as the recursion relationship in equation 10.3*b*. If the two genes assort independently of each other, then $c = 0.5$ ($\lambda = 0$) and

$$D_{t+1} = {}^1\!/\!_2(1 + S)\ D_t$$

If both partial selfing and linkage are present, then jointly they can reduce the rate of decay to nearly a standstill, as shown in Figure 10.7 for $c = 0.1$. The half-lives for decay of disequilibrium are extremely large when both linkage is tight and selfing is high so that, for example, when $S = 0.99$ and $c = 0.01$, it takes 3570 generations for the disequilibrium to be halved.

The effect of partial selfing on retarding the decay of linkage disequilibrium is also dependent on the initial genotype frequencies, which are not included in expression 10.7. For example, if $D = 0.25$, then all of the individuals in the population could be either A_1B_1/A_2B_2, or half could be A_1B_1/A_1B_1 and half A_2B_2/A_2B_2 (as was assumed in Figure 10.7); they could also be in binomial proportions with 0.25 A_1B_1/A_1B_1, 0.50 A_1B_1/A_2B_2, or 0.25 A_2B_2/A_2B_2. With random mating, all of these starting points reach Hardy–Weinberg proportions in one generation, but with a high proportion of selfing, it takes several generations to reach inbreeding equilibrium (as we discussed on p. 250) so that the decay of disequilibrium is described by expression 10.7.

In addition, for two loci, the inbreeding equilibrium genotype proportions are not simply the product of the single-locus proportions (Haldane,

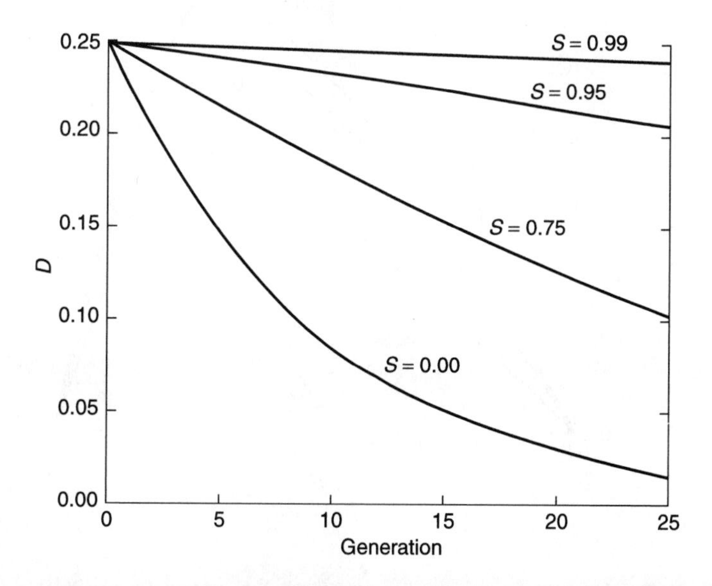

Figure 10.7. The decay of gamete disequilibrium for different proportions of selfing (S) when $c = 0.1$.

1949). In fact, inbreeding results in an excess of both double heterozygotes and double homozygotes and a deficiency of two-locus, single heterozygotes when compared with single-locus inbreeding equilibrium frequencies, an effect that has been termed **identity disequilibrium** (Weir and Cockerham, 1973). Bennett and Binet (1956) suggested that this effect is best measured by the extent of the difference (d) between the equilibrium frequency of the double heterozygotes and the product of the two single-locus equilibrium heterozygosities. Figure 10.8 gives the magnitude of this difference for different values of S and c and obviously demonstrates that the largest excess occurs for tight linkage and selfing rates from 0.6 to 0.8. For unlinked loci and low levels of inbreeding, the excess becomes small. However, when inbred and noninbred individuals are unknowingly lumped, such genotype associations are generated. Furthermore, when the inbred individuals have both lower heterozygosity and lower fitness than outbred individuals because of inbreeding depression, a positive association of heterozygosity and fitness can result (Ledig *et al.*, 1983; Hedrick, 1990b; Charlesworth, 1991).

Arabidopsis thaliana, an annual, colonizing, model plant species, appears to have at least 99% self-fertilization (Abbot and Gomes, 1989). Using the sequence information from the *A. thaliana* genome, Nordborg *et al.* (2002) have examined the extent of linkage disequilibrium between segregating sites at different map distances. For example, in a sample of 117 individuals from LaPorte County, Indiana, the extent of linkage disequilibrium (as measured by r^2) for 39 segregating sites is even high between unlinked loci (Figure 10.9). *A. thaliana* is thought to have been introduced to North America within the last 200 years; thus, part of the extensive disequilibrium may be from these recent founding events, but the very high

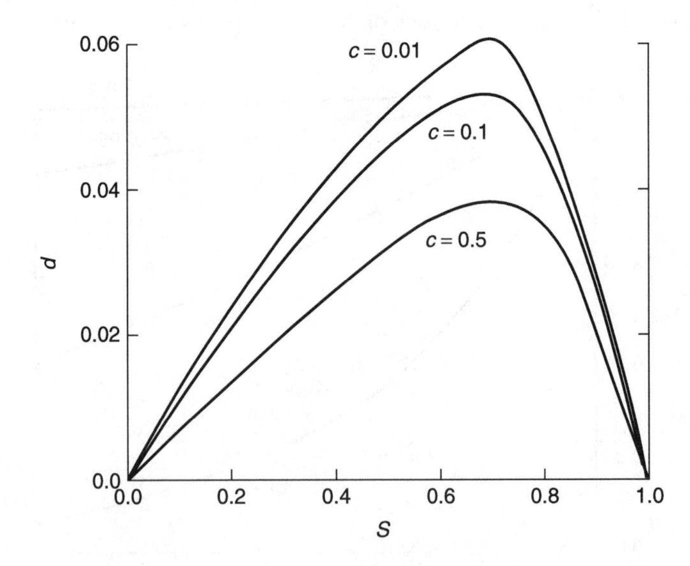

Figure 10.8. The difference (d) between the equilibrium frequency of double heterozygotes and the product of the single-locus equilibrium heterozygosities for several recombination levels and a range of selfing (Hedrick, 1990b).

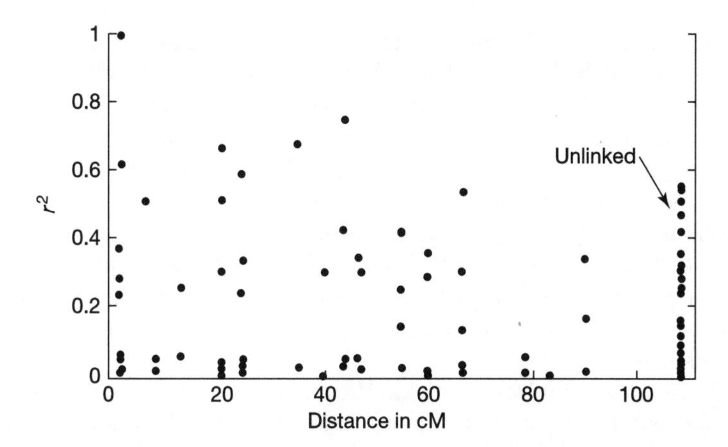

Figure 10.9. The amount of linkage disequilibrium, as measured by r^2, between segregating sites as a function of map distance (cM) in a natural population of *Arabidopsis thaliana* from Indiana. (Courtesy of *Nat. Genet.* 30: 2002, by Nordsborg, M., J.O. Borevitz, J. Bergelson, *et al.* Reprinted with permission of Nature Publishing Group.)

selfing would result in the retention of this high initial disequilibrium for many generations.

d. Gene Flow

Gene flow, or hybridization, between two populations can also cause linkage disequilibrium. An extreme example of the generation of linkage disequilibrium is in a population initiated from the cross of two populations fixed for different alleles: $A_1A_1B_1B_1 \times A_2A_2B_2B_2$. Here, the amount of disequilibrium in the progeny is at a maximum so that $D = 0.25$. A general expression that gives the amount of disequilibrium when two populations (x and y) are mixed in the proportions m_x and m_y ($m_y = 1 - m_x$) is

$$D = m_x m_y (p_{1\cdot x} - p_{1\cdot y})(q_{1\cdot x} - q_{1\cdot y}) \qquad (10.8a)$$

where $p_{1.x}$ and $p_{1.y}$ are the frequencies of the A_1 allele and $q_{1.x}$ and $q_{1.y}$ are the frequencies of the B_1 allele in populations x and y, respectively (Cavalli-Sforza and Bodmer, 1971). Obviously, for disequilibrium to be generated, the allele frequencies of both loci in the two populations must be different, and the greater the difference in allele frequencies and the more equal the contributions of the ancestral populations, the more potential disequilibrium.

Parra *et al.* (2001) examined a number of markers that had different allele frequencies in African and European samples and predicted the extent of linkage disequilibrium in African American populations. To illustrate the potential effect, allele 1 at the AT3 locus differed in frequency between African and European samples by 0.580, and allele Null*1 at locus FY differed by 0.999 (see also Lautenberger *et al.*, 2000). Using these and other loci, the estimated proportion of African ancestry (m_x) in the African American population from the Low Country, South Carolina, is 0.118 (see p. 484 for the approach used). Therefore, from expression 10.8a above, the expected initial disequilibrium from a combining these two populations is $D = (0.118)(0.882)(0.580)(0.999) = 0.060$. Parra *et al.* (2001) estimated that $D = 0.022$ in the current Low Country population between these markers, consistent with the generation of the initial disequilibrium by admixture and the map distance ($c = 0.22$) between the two loci on chromosome 1.

If there is disequilibrium within the ancestral populations, Prout (1973) and Nei and Li (1973) have shown that the total disequilibrium over k subpopulations is

$$D = \overline{D} + COV(p_1, q_1) \qquad (10.8b)$$

where

$$\overline{D} = \sum_{i=1}^{k} m_i D_i$$

is the average disequilibrium value from a population weighted by its contribution, m_i, over k subpopulations and $COV(p_1,q_1)$ is the covariance between the frequencies of alleles A_1 and B_1. For linkage disequilibrium to be generated by the mixing of k subpopulations, either the allele frequencies of the loci must co-vary or the subpopulations themselves must be in linkage disequilibrium. For two subpopulations, this expression becomes

$$D = m_x D_x + m_y D_y + m_x m_y (p_{1\cdot x} - p_{1\cdot y})(q_{1\cdot x} - q_{1\cdot y}) \qquad (10.8c)$$

If there is no disequilibrium in either subpopulation, ($D_x = D_y = 0.0$), then this expression reduces to expression 10.8a. Example 10.6 discusses linkage disequilibrium in the Lemba, a population from southern Africa that has both Bantu and Semitic ancestry.

Example 10.6. The Lemba are a southern African group who speak Bantu but claim that they have Jewish ancestry. In fact, molecular analysis indicates that 68% of the Lemba Y chromosomes appear to have Semitic origin; however, there is no evidence of female gene flow from mtDNA analysis. To determine the potential impact of admixture on linkage disequilibrium in the Lemba, Wilson and Goldstein (2000) examined 66 microsatellite loci spread over the X in samples of Lemba and from Bantu and Ashkenazi Jews to represent the ancestralpopulations.

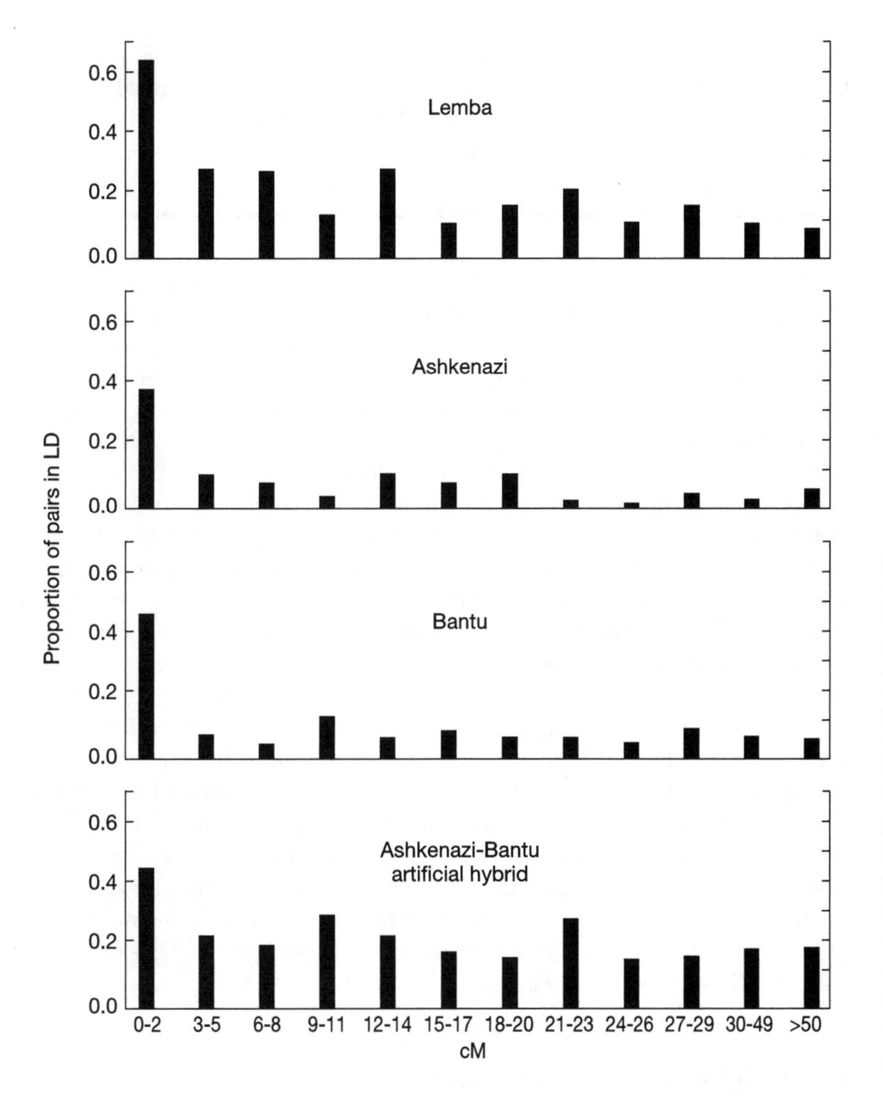

Figure 10.10. The proportion of pairs of 66 microsatellite loci at different genetic distances (cM) on the X chromosome that have significant linkage disequilibrium at the 0.05 level for samples of Lemba, Ashkenazi Jews, Bantu, and an artificial population composed of Bantu and Ashkenazi ancestry. (Wilson and Goldstein, 2000. Courtesy of *American Journal of Human Genetics.*)

In both the Bantu and Ashkenazi samples, there was only an excess of linkage disequilibrium when the loci were closely linked, that is, less than two map units apart (Figure 10.10). Overall, only 7.0% and 7.7% of the marker pairs had significant linkage disequilibrium at the 5% level in the Bantu and Ashkenazi samples. On the other hand, the Lemba had significant linkage disequilibrium for markers even 20 map units apart, and 13.8% of the marker pairs overall had significant linkage disequilibrium. To evaluate the potential effect that an admixed Bantu-Ashkenazi population would have on linkage disequilibrium, Wilson and Goldstein (2000) constructed an artificial population composed of the two ancestral samples. In this population, 20.5% of the marker pairs were in LD, and the pattern of LD was unrelated to map distance except for the tightly linked markers (Figure 10.10). In other words, the observed high LD in the Lemba appears generally consistent with fairly recent admixture between Bantu and Semitic ancestral populations.

The effect of population subdivision is to reduce the rate of decay; given that initially $D \neq 0$, then D approaches zero asymptotically. Because the subpopulations may have different allele frequencies, the frequency of heterozygotes (double heterozygotes too) is smaller than if there was one large population because of the Wahlund effect, and this condition reduces the chance of generating recombinant gametes. As shown by Nei and Li (1973) for two populations, the rate of decay is $1 - c$, the proportion of nonrecombination, or $(1 - 2m)^2$, where m is the gene flow rate to the other population, whichever is greater. In other words, if gene flow between populations is small, it can determine the rate of decay of disequilibrium. For example, if the two loci are unlinked ($c = 0.5$) and if $m < 0.141$, the decay of linkage disequilibrium is determined by population subdivision.

As applications of these concepts, Barton and Hewitt (1985) have suggested that disequilibrium can be used in examination of clines, Waples and Smouse (1990) have suggested that the observed amount of disequilibrium generated by admixture can be used to estimate the source of salmon stocks, and Chakraborty and Weiss (1988) have suggested that the amount of disequilibrium in a recently admixed population can be used to detect linkage between loci. In addition, it has been proposed that the linkage disequilibrium in human populations with ancestry from populations with different allele frequencies may be used to identify genes influencing complex genetic diseases (Stephens *et al.*, 1994; McKeigue, 1998; Pfaff *et al.*, 2001; Collins-Schramm *et al.*, 2002; Smith *et al.*, 2004).

III. MULTILOCUS SELECTION

In the 1960s and 1970s, multilocus selection was thought to be responsible for generating much of the linkage disequilibrium observed, and a great deal of theoretical research attempted to understand the impact of multilocus selection on genetic variation. In fact, theoretical investigations indicated that only under certain conditions is linkage disequilibrium generated and maintained by multilocus selection. Here, we concentrate on simple and what appear to be biologically meaningful selection models and their effect on linkage disequilibrium.

The knowledge obtained from genome projects may uncover the details of situations where multilocus selection is important. For example, the tightly linked members in a multigene family may be under selection that results in linkage disequilibrium, or polymorphic nucleotide sites within a coding region, and thereby very tightly linked, may be under multilocus selection. As we discuss below, the effect of multilocus selection on polymorphism is determined by the level of recombination c relative to the amount of interlocus interaction, or epistasis. If c is very small, for example, 0.0001, then conditions for multilocus equilibria are broader because the necessary amount of epistasis is less. On the other hand, although the levels of recombination and epistasis predict an equilibrium in an infinite population, their absolute levels may be so low that genetic drift may become a more important factor in most populations.

A category of genes that shows linkage disequilibrium are **supergenes**, a term introduced by Darlington and Mather (1949) to describe genes so closely linked that there is little crossing over between them. Generally, the genes composing a supergene are related in their adaptive role in the population. Examples include pin and thrum alleles that affect the flower structure of *Primula* in several ways (see p. 191) and the color and banding genes that affect the shell of *Cepaea* (see p. 206). Although these loci may have related fitness effects, it is generally thought (Ford, 1971) that the genes involved in a supergene were once not so closely linked and that selection, by either reducing recombination or capitalizing on a cytological aberration, brought the genes together.

Multigene families are also composed of several tightly linked genes and usually produce nearly identical gene products. Genes in a multigene family are thought to be closely linked because they have originated by tandem duplication, and over a period of time, they have diverged somewhat by mutation and genetic drift, or perhaps by differential selection, so that they are identified as similar but not as identical. Besides having a historical association, alleles at genes within a multigene family may still have factors that cause interaction such as similar regulatory control, physiological function, and substrate utilization. The MHC in vertebrates appears to both have the elements of a supergene (related but nonhomologous linked loci) and loci resulting from past duplications (Edwards and Hedrick, 1998; Beck and Trowsdale, 2000).

a. Fitness Epistasis

Before we can examine the effects of multilocus selection, we need a general means of envisioning multilocus fitness values. A comparison of the fitnesses of any two genotypes that differ at only a single locus provides a relative measure of the fitness of a single locus, given a constant genetic background. If this fitness value is the same for all genetic backgrounds, then the fitness of this locus is independent of the rest of the genome. Independence can be defined on a scale using either the differences or ratios of different fitnesses (see below), and if either relationship holds, one can say that there is no fitness interaction between the two loci.

If fitness comparisons are made on a ratio scale, then this implies that the fitness of a two-locus genotype can be expressed as the product of the values assigned to each single-locus genotype. Any fitness pattern that deviates from such a restriction is said to display fitness interaction or **epistasis on a multiplicative scale**. The multiplicative independence concept has intuitive appeal, particularly in discussions of selection resulting from differential viability. For example, consider an insect where 90% of the individuals with genotype A_1A_1 survive the larval stage and 80% of the individuals with genotype B_1B_1 survive the pupal stage. Then the overall probability of survival of an individual with the $A_1A_1B_1B_1$ genotype, relative to a genotype in which there was no death, is determined multiplicatively and is $(0.8)(0.9) = 0.72$.

If genotypes are compared using differences in fitness, then values can be assigned to each single-locus genotype, and the fitness of each two-locus genotype is the sum of the component single-locus values. Any deviation from this additive ideal is called **epistasis on an additive scale**. The concept of additive independence has its basis in quantitative genetics (Falconer and Mackay, 1996; Lynch and Walsh, 1998). In this case, one can think of selection on a trait such as differential fecundity so that egg or seed production is influenced by two loci, both of which result in more eggs or seeds produced in the lifetime of the individual. For example, genotype A_1A_1 may add two seeds, and B_1B_1 may add three seeds, relative to other genotypes, for a total of five more seeds for $A_1A_1B_1B_1$, as compared with other genotypes.

In order to determine the effects of epistatic selective forces on allele associations, we must have a measure of epistatic interaction (nonindependence of fitnesses). First, consider the haploid selection model where the two-locus relative fitnesses of the genotypes A_1B_1, A_1B_2, A_2B_1, and A_2B_2 are given by w_{11}, w_{12}, w_{21}, and w_{22}, respectively. The extent of **additive epistasis** can be measured by

$$E = w_{11} - w_{12} - w_{21} + w_{22} \tag{10.9a}$$

If fitnesses are additive, then $E = 0$, as in the first column of Table 10.9; otherwise $E \neq 0$. An analogous version for **multiplicative epistasis** is

$$\begin{aligned} E' &= \ln w_{11} - \ln w_{12} - \ln w_{21} + \ln w_{22} \\ &= \ln\left(\frac{w_{11}w_{22}}{w_{12}w_{21}}\right) \end{aligned} \tag{10.9b}$$

(Felsenstein, 1965). The second column of Table 10.10 gives an example where $E' = 0$; the values that satisfy $E = 0$ and $E' = 0$ are very similar.

TABLE 10.10 Fitness values for the haploid model with four different levels of epistasis.

Genotype		$E = 0.0$	$E' = 0.0$	$E' = 0.1$	$E' = -0.1$
A_1B_1	w_{11}	1.44	1.44	1.44	1.44
A_1B_2	w_{12}	1.22	1.2	1.142	1.262
A_2B_1	w_{21}	1.22	1.2	1.142	1.262
A_2B_2	w_{22}	1.0	1.0	1.0	1.0

For diploids, there are analogous measures of epistasis (Felsenstein, 1965; see also Wade *et al.*, 2001; see p. 563).

b. Directional Selection at Two Loci

In a majority of multilocus theoretical studies, the effects of linkage and selection on the existence and stability of polymorphic equilibria have received the most attention. However, directional selection, in which some alleles are increasing toward fixation, is also affected by linkage. One consequence of directional selection is that linkage disequilibrium is commonly generated during the substitution process, although it disappears when fixation of the favored alleles occurs.

To illustrate this phenomenon, let us first consider the two-locus, two-allele haploid model. Assume that the frequency of the haploid genotypes after meiosis is given as at the bottom of Table 10.3. If there is subsequent differential survival of genotypes, then the frequencies after selection become

$$\begin{aligned} x'_{11} &= w_{11}(x_{11} - cD)/\overline{w} \\ x'_{12} &= w_{12}(x_{12} + cD)/\overline{w} \\ x'_{21} &= w_{21}(x_{21} + cD)/\overline{w} \\ x'_{22} &= w_{22}(x_{22} - cD)/\overline{w} \end{aligned} \tag{10.10a}$$

where

$$\overline{w} = \sum_{i=1}^{2}\sum_{j=1}^{2} w_{ij}x_{ij}$$

Felsenstein (1965) has shown that directional two-locus selection will tend to generate linkage disequilibrium of the same sign as the multiplicative epistatic parameter in the haploid case. Thus, selection favoring alleles A_1 and B_1 such that $E' > 0$ generates positive linkage disequilibrium and selection with $E' < 0$ generates negative disequilibrium. This can be illustrated by comparing the change in linkage disequilibrium for the fitness arrays given in Table 10.10. The array in the second column assumes that there is a 0.2 selective advantage for both alleles A_1 and B_1 and that there is no epistasis on a multiplicative scale ($E' = 0.0$). The arrays in the third and fourth columns have the same fitness difference between the coupling genotypes but have epistasis so that $E' = 0.1$ and -0.1, respectively.

Assuming $D_0 = 0$, $p_1 = 0.1$, $q_1 = 0.1$, and $c = 0.01$, and the fitness values in Table 10.10, the linkage disequilibrium and allele frequencies over time are given in Figure 10.11. First, in Figure 10.11a, when $E' = 0.0$, there is no disequilibrium generated by selection. On the other hand, when $E' = 0.1$, D increases to a maximum of 0.073 in generation 14. The positive D generated when there is positive E' occurs because the repulsion A_1B_2 and A_2B_1 genotypes have a lower relative fitnesses (compared with $E' = 0$), which results in a deficiency in their frequencies and consequently a positive D. When $E' = -0.1$, D decreases to -0.076 in generation 15. Here the higher relative fitnesses of the A_1B_2 and A_2B_1 genotypes result in an excess in their frequencies and consequently a negative D. When $E = 0.0$, some negative disequilibrium is generated, as expected because $E' = -0.033$ with this fitness array.

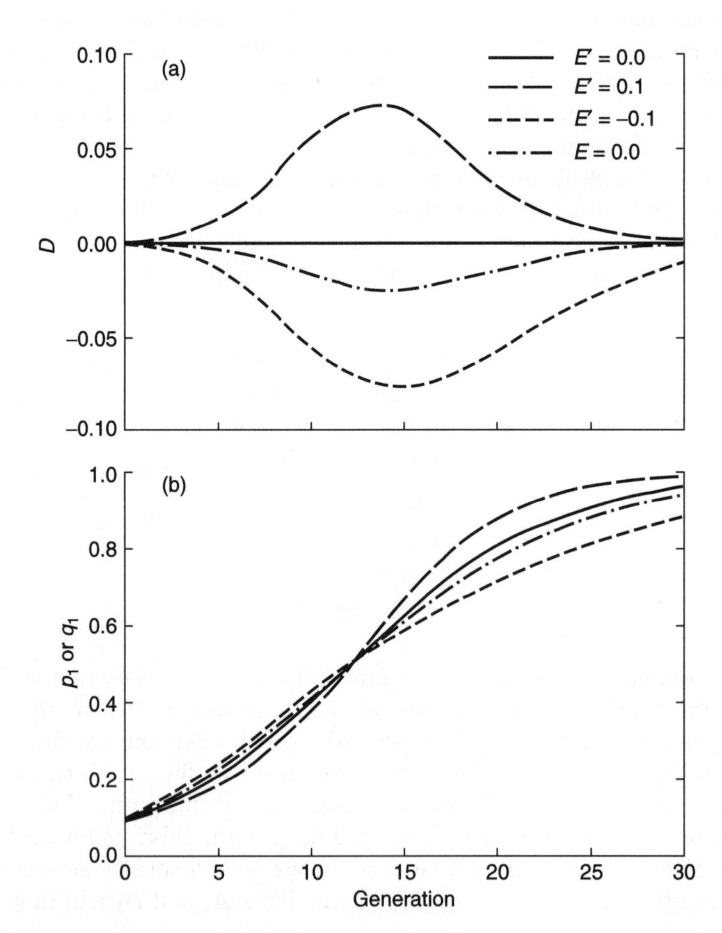

Figure 10.11. (a) The generation of linkage disequilibrium and (b) the changes in allele frequency for the four haploid fitness arrays given in Table 10.10, where E and E' indicate the amount of epistasis. In this example, initially $p_1 = q_1 = 0.1$, $D_0 = 0.0$, and $c = 0.01$.

The rate of allele frequency change is also affected, as shown in Figure 10.11b. The basis for the differences is somewhat complicated, but one can understand by realizing that at low frequencies, differential selection takes place primarily between the initial, low fitness A_2B_2 genotype and the repulsion genotypes, whereas at high frequencies, it occurs between the most favorable A_1B_1 genotype and the repulsion genotypes. Consider, for example, the situation where $E'= 0.1$. Here selection is initially slowed because A_1B_2 and A_2B_1 have only fitnesses of 1.142, compared with that of A_2B_2 of 1.0. However, when frequencies of A_1 and B_1 are high, selection occurs mainly between the repulsion genotypes and the A_1B_1 genotype, which has a fitness of 1.44. The difference in fitness here is larger, and thus, the allele frequency change is faster.

Felsenstein (1965) also showed that if the disequilibrium generated by epistatic selection is positive, tighter linkage accelerates the change in allele frequency, whereas if selection generates negative disequilibrium, tighter linkage decreases the rate of change in allele frequencies. We can perceive this concept intuitively because a large positive disequilibrium allows selection to act mainly on the gametes A_1B_1 and A_2B_2 and is more efficient if there is also positive epistasis. The more A_1B_2 and A_2B_1 gametes exist in the population, the greater is the "interference" to selection. Because tighter linkage reduces recombination, it should increase the efficiency of directional selection because it reduces this interference. However, if selection tends to make D negative ($E' < 0$), then looser linkage is needed to generate by recombination those A_1B_1 and A_2B_2 gametes lost. These general implications also are true for the analogous diploid models.

c. General Two-Locus Polymorphisms

We noted earlier that in the absence of selection and other evolutionary factors, linkage disequilibrium eventually disappears. For balanced polymorphisms where allele frequencies are kept intermediate and, therefore, heterozygosity is kept high, one intuitively might expect linkage disequilibrium to decrease rapidly because the alleles at various loci would be effectively randomized because of the influence of recombination. On the other hand, if epistasis exists, we might expect selection to favor some combinations of alleles at different loci over others, increasing their association in the population. Therefore, what we would like to know is the amount and type of selection required to counter the force of recombination. If such selection must be very strong, one may argue that linkage disequilibrium would be expected to be rare in nature or, if present, not necessarily caused by selection. If, however, the conditions necessary to resist randomization are suspected to be common, then, all else being equal, polymorphisms with large interlocus associations are also likely to be common in nature. To reach an understanding of the relative effects of epistatic selection and recombination, we introduce some conclusions drawn from the many theoretical investigations of two-locus, random-mating equilibrium models.

To examine these effects, we first use a general two-locus, two-allele fitness pattern for diploids that is given in Table 10.11a. The subscripts of genotype fitnesses are expressed in a way that reflects their component two-locus gametes. For example, $w_{11.21}$ indicates the fitness of the genotype composed of gametes A_1B_1 and A_2B_1. For simplicity, it is assumed that the fitnesses of the two double heterozygotes are equal—that is, $w_{11.22} = w_{12.21} = w_h$. Then, if we let

$$\overline{w}_{ij} = \sum_{k=1}^{2}\sum_{l=1}^{2} x_{kl} w_{ij.kl}$$

and

$$\overline{w} = \sum_{i=1}^{2}\sum_{j=1}^{2} x_{ij}\overline{w}_{ij}$$

the frequency of gamete A_1B_1 after selection and recombination becomes

$$\begin{aligned} x'_{11} &= [x_{11}^2 w_{11.11} + x_{11}x_{12}w_{11.12} + x_{11}x_{21}w_{11.21} + (1-c)x_{11}x_{22}w_{11.22} + \\ &\quad c x_{12}x_{21}w_{12.21}]/\overline{w} \\ &= [x_{11}(x_{11}w_{11} + x_{12}w_{12} + x_{21}w_{13} + x_{22}w_{14}) - cw_h(x_{11}x_{22} - x_{12}x_{21})]/\overline{w} \\ &= (x_{11}\overline{w}_{11} - cw_h D)/\overline{w} \end{aligned}$$

The frequencies of the other gametes after selection and recombination can similarly be derived so that

$$\begin{aligned} x'_{11} &= (x_{11}\overline{w}_{11} - cw_h D)/\overline{w} \\ x'_{12} &= (x_{12}\overline{w}_{12} + cw_h D)/\overline{w} \\ x'_{21} &= (x_{21}\overline{w}_{21} + cw_h D)/\overline{w} \\ x'_{22} &= (x_{22}\overline{w}_{22} - cw_h D)/\overline{w} \end{aligned} \qquad (10.10b)$$

TABLE 10.11 (a) The general viability fitness array and (b) the symmetrical viability array for the two-locus, two-allele diploid model.

(a)		A_1A_1	A_1A_2	A_2A_2
	B_1B_1	$w_{11.11}$	$w_{11.21}$	$w_{21.21}$
	B_1B_2	$w_{11.12}$	$w_{11.22} = w_{12.21}$	$w_{21.22}$
	B_2B_2	$w_{12.12}$	$w_{12.22}$	$w_{22.22}$

(b)		A_1A_1	A_1A_2	A_2A_2
	B_1B_1	a	b	a
	B_1B_2	b	1	b
	B_2B_2	a	b	a

When $x'_{ij} = x_{ij}$ for all gametes, that is, there is no change in gamete frequencies, the population is at an equilibrium.

The general results from these equations depend ultimately on single-locus, multiple-allele theory because a two-allele, two-locus case with no recombination ($c = 0$) is identical to a single-locus model with four alleles. Here we can assume that the results for tight linkage will not be too different from those with $c = 0$. However, unlike multiple-allele conditions, for two biallelic loci, generally two patterns of selection result in polymorphic equilibria: selection that results in an equilibrium with $D = 0$, regardless of the value of c, and selection that favors the increase of some two-locus gamete types and the decrease of other gametes, thereby resulting in linkage disequilibrium. Obviously, recombination is important only under the latter type of selection. In this case, recombination regenerates the less favored gametes, thus maintaining a polymorphism with all gamete types present. If c is increased beyond a critical level, then the polymorphism is lost or an equilibrium with $D = 0$ results.

Several other general statements can be made. First, for any two-locus, two-allele viability model, there exist at most two stable polymorphic equilibria for sufficiently tight linkage. Thus, in the heuristic terms of adaptive topographies (see Figure 10.12 and Wright, 1969), there should be no more than two adaptive and polymorphic peaks for a two-locus model, and in general, one equilibrium will have $D > 0$ and one $D < 0$. Second, in general, as recombination is increased, the equilibrium mean fitness is decreased. This generality has intuitive value because we know that selection favors

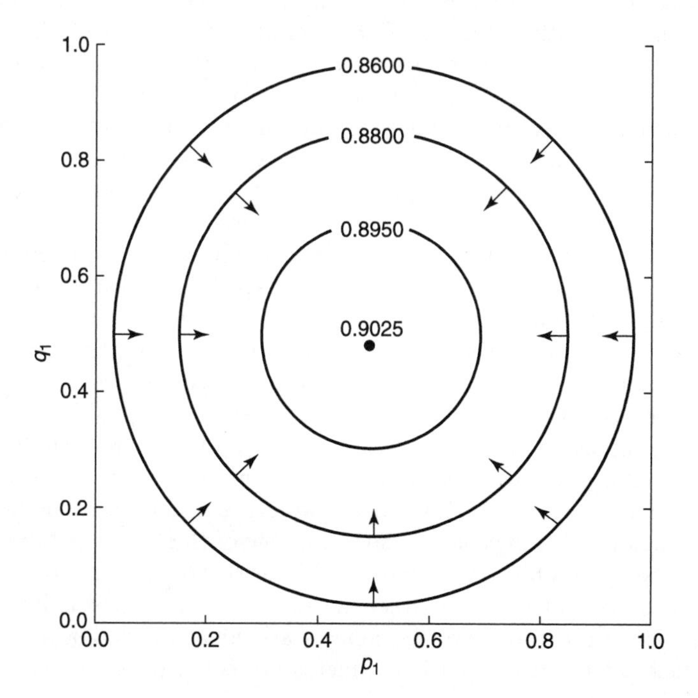

Figure 10.12. An adaptive topography where the relative fitnesses of the double homozygotes, single heterozygotes, and double heterozygotes are 0.81, 0.9, and 1.0, respectively. The numbers indicate the mean fitness on the contour lines, the arrows indicate the direction of allele frequency change, and the solid circle in the center indicates the highest fitness.

elimination of certain chromosomal types, whereas recombination counters this force. Thus, reducing recombination allows selection to increase the mean fitness of the population by increasing the frequency of particular gametes.

Some fitness models are of biological or general interest, and we now summarize some of the important results for them. First, with no epistasis on an additive scale, a single stable polymorphism with $D = 0$ is assured when heterozygote advantage prevails at each locus (Bodmer and Felsenstein, 1967). Under this model, the mean fitness, $\overline{w}$, always increases (or never decreases) during the course of selection, which implies that the final polymorphic equilibrium is a local fitness maximum. These results appear to conform to what we would expect of fitness independence; there is no interlocus association of alleles at equilibrium, and the final equilibrium is the state where the mean fitness is maximized.

On the other hand, with no epistasis on a multiplicative independence, and single-locus heterozygote advantage at both loci, two types of equilibria exist. For sufficiently loose linkage, a polymorphic equilibrium with $D = 0$ exists, and for sufficiently tight linkage, two more or less complementary polymorphic equilibria exist, one with $D > 0$ and one with $D < 0$. If the two complementary equilibria exist, then when the population starts with $D > 0$, it will eventually move to the $D > 0$ equilibrium, and when the population starts with $D < 0$, it will move to the $D < 0$ equilibrium. The fitness surface or adaptive topography (Wright, 1969) when $D = 0$ separates the regions of attraction for the two equilibria. However, if linkage is so tight that the $D \neq 0$ equilibria exist, the surface has little importance—for even if the population starts at $D = 0$, random fluctuations will certainly move it off the surface. Of course, if the recombination value is so large that the $D \neq 0$ equilibria do not exist, the population will move to the surface under the influence of recombination and remain there. On the surface, the mean population fitness increases over successive generations.

Figure 10.12 gives an example of an adaptive topography where the relative fitness of the double homozygotes, single heterozygotes, and double heterozygotes are 0.81, 0.9, and 1.0, respectively (no epistasis on a multiplicative scale and let us assume loose linkage). The mean fitness in this case is completely specified by the allele frequencies at the two loci and the relative fitnesses, given that $D_0 = 0$. The maximum fitness, or a peak, occurs when $p_1 = q_1 = 0.5$, and populations with frequencies that differ from these frequencies progress toward the peak as indicated by the arrows.

The simplest form of symmetric viability of the two-locus fitnesses, where genotypes opposite each other in the fitness matrix have the same relative viability (Table 10.11b), was investigated analytically by Lewontin and Kojima (1960). Because of the symmetry of the fitness pattern in this model, there are only symmetric equilibria—those with $p_1 = q_1 = 0.5$. Let us assume that the allele frequencies are all equal at equilibrium, the frequencies of the two coupling gametes are equal to each other, and the frequencies of the two repulsion gametes are equal to each other. In this case, the expressions in 10.10*b* can be simplified to one cubic equation. When recombination is above a certain level of additive epistasis—that is, $c > E/4$, where $E = a - 2b + 1$—an equilibrium where all gametes are equal in frequency is present. In other words

$$x_{ij \cdot e} = 1/4 \qquad D_e = 0 \tag{10.11a}$$

in this situation. However, if $c < E/4$, the critical recombination level, then there are two complementary equilibria possible such that the gamete frequencies and the linkage disequilibrium value are

$$\begin{aligned} x_{11 \cdot e} &= x_{22 \cdot e} = 1/4 \pm 1/4 \left(1 - \frac{4c}{a - 2b + 1}\right)^{1/2} \\ x_{12 \cdot e} &= x_{21 \cdot e} = 1/2 - x_{11 \cdot e} \\ D_e &= \pm 1/4 \left(1 - \frac{4c}{a - 2b + 1}\right)^{1/2} \end{aligned} \tag{10.11b}$$

To get an idea of the dynamics of gamete frequency change, let us examine a specific numerical example where $a = 0.9$ and $b = 0.93$. In this case, $E = 0.04$ so that c must be less than 0.01 for equilibria with $D \neq 0$ to occur. If it is assumed that $x_{11} = x_{22}$ and $x_{12} = x_{21}$ initially, then these pairs of gametes remain equal in frequency. The results of two different recombination values are given in Figure 10.13: $c = 0.0075$, which is below

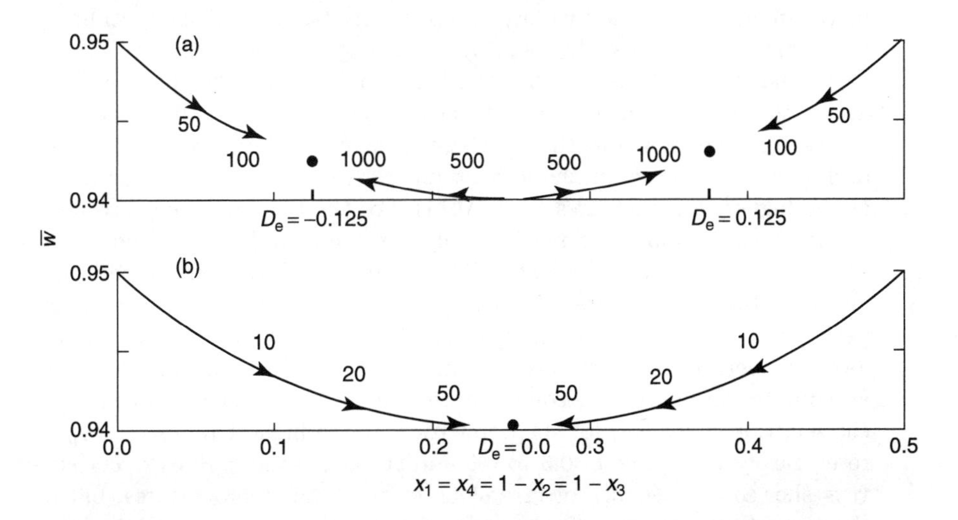

Figure 10.13. The mean fitness and the rate of change in gamete frequency for the fitness array in Table 10.11b when $a = 0.9$ and $b = 0.93$, (a) for $c = 0.0075$ and (b) for $c = 0.05$. The number of generations necessary for a given gamete frequency change are indicated, and the closed circles indicate equilibrium values.

the critical value, and $c = 0.05$, which is above the critical level. In Figure 10.13a ($c = 0.0075$), when populations are started with $x_{11} = x_{22} = 0.0$ or 0.24, they go to the $D < 0$ equilibrium ($D_e = -0.125$), whereas when populations are started at $x_{11} = x_{22} = 0.26$ or 0.5, they go to the $D > 0$ equilibrium ($D_e = 0.125$).

The time scales are quite different so that an approach to the equilibrium from $x_{11} = x_{22} = 0.0$ or 0.5 is much faster than from 0.24 or 0.26. The basis for this difference in rate is apparent from the mean fitnesses for different gamete frequencies. The linkage equilibrium values with $D \neq 0$ are a result of a balance in which selection is increasing the frequency of particular gametes and recombination is reducing them. When c is larger than the critical value, the approach to the $D = 0$ equilibrium is quite fast (Figure 10.13b). Example 10.7 gives data showing the dependence of the outcome of a long term *D. melanogaster* experiment on both recombination and the environment.

Example 10.7. The factors that may result in the balanced polymorphism at the *Adh* locus in *D. melanogaster* have been extensively examined (Van Delden, 1982). In addition, the α-*Gpdh* locus, which is 29.6 cM away from *Adh* on chromosome II, is also polymorphic in many populations. Even though these two loci are not closely linked, in a long-term greenhouse population, linkage disequilibrium was significant, and an inversion that resulted in no recombination between the loci was identified (Van Delden, 1984).

In an attempt to understand further the importance of the inversion on polymorphism of the two loci, samples were taken from the greenhouse population and used to establish three large cage populations kept at 20°C, 25°C, and 29.5°C. These populations were monitored for the polymorphism at the two allozyme loci for a number of generations, and there was substantial divergence among the three environments for both loci, with *Adh-S* declining in frequency at the lower temperatures and α-*Gpdh-S* declining at the highest temperatures (Fig. 10.14). For the two lower temperatures, the frequencies appear to reach a polymorphic equilibrium (although α-*Gpdh-S* nearly becomes fixed at 20°C), whereas at the higher temperature, although the two loci are still polymorphic at about 0.5, there still appears to be slow change. By generation 20 for the populations kept at the lower temperatures, the two loci are in linkage equilibrium, which suggests an equilibrium in which recombination had a greater effect than multilocus selection. However, for the population at the highest temperature, D' rose sharply to a value of 0.3 to 0.5 and stabilized there. Subsequent surveys showed no evidence of inversions at the lower temperatures, but at the highest temperature, the inversion frequency increased to 0.245. All of the inversion chromosomes carried both *Adh-S* and α-*Gpdh-F* alleles, and an additional 5% of the noninversion chromosomes also had these alleles.

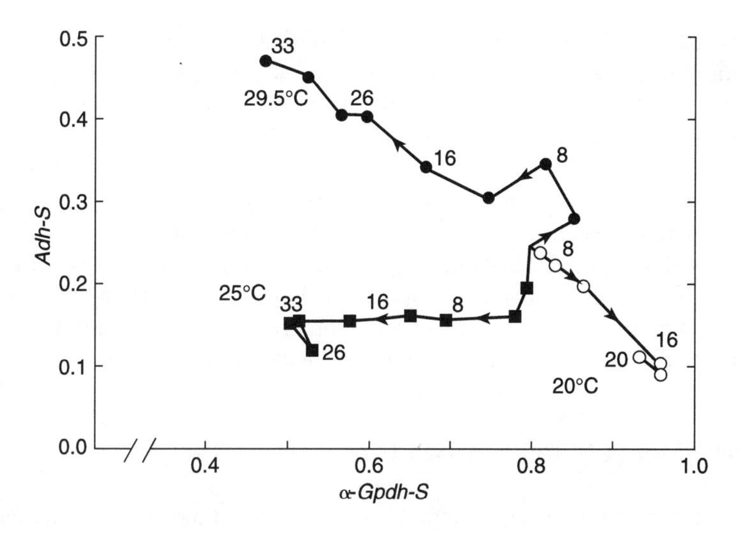

Figure 10.14. Changes in *Adh-S* and *α-Gpdh-S* allele frequencies in three populations kept at 20°C, 25°C, and 29.5°C, respectively. Numbers indicate generations (after Van Delden, 1984).

Although the target of selection that causes these changes is not clear, it is obvious that with lowered recombination caused by the presence of the inversion, an equilibrium with disequilibrium between the two allozyme loci is present.

From the foregoing theoretical considerations of selection and linkage disequilibrium, it would appear that alleles at many loci might show nonrandom associations. Finite population size, mutation, gene flow, selection, and genetic hitchhiking (see below) all can generate linkage disequilibrium, and tight linkage and high inbreeding can retard the rate of decay of linkage disequilibrium. However, if a nonrandom association is found and no historical information is available, it is difficult to attribute the linkage disequilibrium to a particular cause. For example, such different factors as a bottleneck or hybridization in the past or selection in the past or present could be the cause of disequilibrium. With tight linkage, as for nucleotide sequence data, or highly selfed populations, the relics of past events may remain for an extremely long period.

How can the cause of linkage disequilibrium be determined? This is an extremely difficult problem even if information about gamete frequencies over time and space is available. One approach to understand the cause of linkage disequilibrium in a particular instance is to eliminate various factors from consideration until only one factor remains. For example, Lewontin (1974) suggested that if significant linkage disequilibrium for a pair of loci is observed that is **consistent** among populations, then this can be attributed to selection. Consistency among populations does not necessarily mean that the disequilibrium values are the same sign in all populations because, as we have seen, multiple complementary equilibria can exist. Furthermore, as Lewontin indicated, genetic drift and gene flow (hybridization) must be excluded as causes. Ideally, to assign cause to linkage disequilibrium, one would like to know linkage disequilibrium over time and space; the history of the populations so that population size, gene flow, and nonrandom mating can be calculated; and the recombination distance between the loci (or sites) involved. Of course, the loci being examined may not be undergoing selection, and selection may be operating on loci associated with those being observed (genetic hitchhiking). If the disequilibrium between different pairs of loci or sites within a population is consistent, then this may be support for genetic drift because selection, genetic hitchhiking, mutation, and perhaps gene flow would have more locus-specific effects.

d. Selection at More than Two Loci

In situations where selection acts simultaneously on more than two loci, the possible results could be quite complicated. We might expect, for example, that if epistasis among loci is reinforced by more loci, then there could be extensive linkage disequilibrium with multilocus selection models. For example, Lewontin (1964) found that for multiplicative, heterotic five-locus models and tight linkage, there exist complementary gamete polymorphisms with interlocus associations. In addition, Lewontin uncovered two phenomena peculiar to multilocus models. First, there was a cumulative effect in which alleles at two loci far apart on a chromosome were held in disequilibrium because of associations with alleles at loci between them. Second, alleles at adjacent loci embedded in an associated complex of alleles possessed greater disequilibrium between them than would be predicted from two-locus theory alone.

Building on these studies, Franklin and Lewontin (1970) carried out a multilocus, finite population, computer simulation study. They observed a phenomenon similar to crystallization in which areas of high disequilibrium acted as seeds for the buildup of more associations. The regions near the loci in high disequilibrium experienced a greater degree of selection and thus have a higher rate of buildup of disequilibrium than those farther away. Eventually, the entire chromosomal segment crystallized, resulting in only several (complementary) chromosomal types in the population. Two examples of the gametes that resulted are given in Table 10.12 from simulations of a 36-locus model (Franklin and Lewontin, 1970). The two alleles at each locus are represented by 0 or 1 so that the two most common gametes in replicate 1 have different alleles at 33 of 36 loci. For this crystallization to occur, there must be large enough selection of the right type; the map distance between adjacent loci must be small enough to allow the intensified epistasis to counteract recombination, and nuclei of high disequilibrium must be generated by some mechanism. In a small population (as in the simulations of Franklin and Lewontin, relative to the number

TABLE 10.12 The two most common 36-locus gametes, where 0 and 1 represent different alleles at a given locus, and their frequencies in two different replicates (Franklin and Lewontin, 1970).

Replicate	*Gamete*	*Frequency*
1	011 010 110 011 000 110 101 011 011 110 001 101	0.441
	100 101 001 010 111 001 010 101 100 000 110 010	0.424
	Other gametes combined	0.135
2	001 100 100 011 101 011 111 100 100 111 000 111	0.440
	110 011 011 100 010 100 001 011 011 000 111 000	0.427
	Other gametes combined	0.133

of possible multilocus genotypes), genetic drift provided the seeds for the crystallization process.

Are there conditions in nature indicating that the crystallization phenomenon is common or even possible? Studies by Clegg *et al.* (1978, 1980), in which they carried out experiments in *D. melanogaster* designed to observe the decay of associations among alleles at marker loci within a chromosome, cast doubt upon this mega-association-of-chromosomes hypothesis. The essential results were that the rate of decay of disequilibrium was greater than that predicted from neutral theory, the opposite of that predicted by the Franklin and Lewontin model (see Figure 10.4). Subsequent simulations and analytical work (Clegg, 1978; Asmussen and Clegg, 1982) suggest that the explanation for the difference between this conclusion and that of Franklin and Lewontin stems from Clegg having fewer loci per map length and a lower intensity of selection per unit map length than did Franklin and Lewontin. One can argue that the similarity between simulations of Clegg and his experimental results provides more support for his choice of parameters. With more detailed information on tightly linked loci or nucleotide sites, situations having a high density of selectively maintained polymorphic loci may be identified.

IV. GENETIC BACKGROUND, GENETIC HITCHHIKING, AND RELATED TOPICS

There are a number of evolutionary phenomena impacted by selection at one or more loci. First, because alleles cannot be separated from their **genetic background**, association with alleles at other genes where selection is operating may result in an apparent fitness difference at genotypes at a neutral locus. Second, if there is a statistical association of alleles at a neutral locus with another locus undergoing selection, then the ***neutral allele may be carried along because of the selective advantage of the associated allele***—an effect known as **genetic hitchhiking**. This may be particularly true if this association is maintained for a period by tight linkage or nucleotide sites within loci, or by high selfing. Third, **selective sweeps**, in which ***a selectively advantageous allele increases in frequency and changes the frequency of variants in linkage disequilibrium***, have been suggested as an important factor reducing the amount of variation in nearby regions of chromosomes. Finally, recombination or **sexual reproduction** may be advantageous by producing new adaptive genotypes or by breaking down linkage disequilibrium between detrimental mutant alleles. Later we introduce these concepts and show how they are related to the multilocus approaches that we have considered.

a. Genetic Background

The complexities introduced by multilocus considerations may account for some phenomena encountered in evolutionary genetics. For example, what appears to be heterozygous advantage at a particular locus may in fact result from an association of alleles at a linked locus to the alleles at the observed (or marker) locus. Furthermore, unpredictable changes in allele frequencies may be the result of selection operating on alleles at an associated locus and not at the locus being observed. It is generally difficult to evaluate selection at a given locus, independent of the effects of associated alleles at other loci, but consideration of this genetic background effect is particularly important when the selective differences among genotypes at the locus under study are relatively small, as is the apparent situation for most molecular variants.

The most straightforward approach to illustrate the importance of association with alleles at other loci on fitness is to assume that alleles at the locus being examined, for example, locus A, are neutral with respect to each other. Let us assume that alleles at the A locus are associated with alleles at another locus, B, at which there is directional selection favoring allele B_1. The general array of fitnesses for this situation is given in Table 10.13. Let us examine two types of association: complete coupling in which there are only gametes A_1B_1 and A_2B_2 and complete repulsion in which there are only gametes A_1B_2 and A_2B_1. In both cases, there are only three genotypes possible in a population: for example, with complete repulsion A_1B_2/A_1B_2, A_1B_2/A_2B_1, and A_2B_1/A_2B_1. The genotypes and the fitnesses for the three A locus genotypes in these two situations are given at the bottom of Table 10.13. If we assume that there is complete coupling, then because of the association with selected locus B, it appears that there is directional selection favoring allele A_1. On the other hand, if there is complete repulsion, then it appears that there is selection favoring A_2. Therefore, depending on the type of association with locus B, it can appear that for neutral locus A, there is either selection favoring A_1 or A_2. Thomson (1977) gave a more complete discussion of such apparent selection (or pseudoselection).

TABLE 10.13 An example of apparent selection at a neutral locus (A) that varies in direction depending on the association with alleles at locus B.

	A_1A_1	A_1A_2	A_2A_2
B_1B_1	$1+s$	$1+s$	$1+s$
B_1B_2	1	1	1
B_2B_2	$1-s$	$1-s$	$1-s$
(a) **Complete coupling indicates selection favoring A_1**			
Genotype	A_1B_1/A_1B_1	A_1B_1/A_2B_2	A_2B_2/A_2B_2
Fitness	$1+s$	1	$1-s$
(b) **Complete repulsion indicates selection favoring A_2**			
Genotype	A_1B_2/A_1B_2	A_1B_2/A_2B_1	A_2B_1/A_2B_1
Fitness	$1-s$	1	$1+s$

To evaluate the effect of the genetic background on fitness estimates, it is useful to consider two alternative possibilities in which the genetic background is either random or constant with respect to the genotypes at the locus under consideration (the marker locus). If the genetic background is random with respect to the alleles at the marker locus, then associations between background alleles and alleles at the marker locus should not be important. When there is a constant genetic background, the selective effects from associated alleles would be the same for all genotypes at the marker locus. As a result, it seems more useful to measure fitness components or investigate genetic changes on an average or randomized background. Theoretically, if a number of independent alleles of each type were separately isolated from different sources, then the effect of association with background alleles would be eliminated. More precisely, the expected amount of association in the first generation, when n independent haplotypes are obtained from a parental population in which there is no association, declines as a function of $1/(n-1)$ (Ohta and Kimura, 1970).

In addition to changing allele frequency, selection occurring at other associated loci may result in what appears to be heterozygote advantage and a consequent polymorphic equilibrium. Such associations have been suggested as an explanation for the associations sometimes seen between heterozygosity and fitness (David, 1998). Ohta (1971) showed that **associative overdominance** might result in an ***apparent heterozygote advantage at a marker locus because of linkage disequilibrium with detrimental alleles and the neutral marker locus***. For example, assume that the A locus is a neutral marker locus and that there is disequilibrium with detrimental ($-$) and wild-type ($+$) alleles at two other linked loci so that there are only the two chromosomes, $A_1 + -$ and $A_2 - +$, in the population. Assuming that $-/-$ homozygotes have a selective disadvantage of s and complete dominance of the wild-type alleles, then the fitnesses of the three genotypes, $A_1 + -/A_1 + -$, $A_1 + -/A_2 - +$, and

$A_2 - +/A_2 - +$, are $1 - s$, 1, and $1 - s$, respectively. Although this results in the appearance of heterozygous advantage at the marker locus A, it occurs because of associated detrimental alleles. To evaluate the importance of intrinsic heterozygote advantage compared with the alternative explanations for heterozygosity–fitness association, associative overdominance, and identity disequilibrium, Savolainen and Hedrick (1995) examined populations of Scots pine where neither genetic drift nor inbreeding is thought important (see Example 10.8).

Example 10.8. In addition to intrinsic heterozygote advantage at the loci being studied, positive correlations between heterozygosity and fitness could be the result of identity disequilibrium or associative overdominance as discussed above. One approach to determining the importance of intrinsic heterozygote advantage is to examine the heterozygosity–fitness association in populations in which there is little genetic drift, so that associative overdominance would influence only a small part of the genome, and with little inbreeding so that identity disequilibrium is unlikely (Houle, 1989).

The population of adult Scots pine, *Pinus silvestris*, in Finland is very large (Muona and Harju, 1989), and there is extensive pollen gene flow such that over distances greater than 1000 km, there is no detectable differentiation for molecular markers (Karhu *et al.*, 1996). In addition, even though Scots pine is self-compatible, adult tree genotypes are in Hardy–Weinberg proportions and there is no genetic evidence of inbreds surviving to adults (Muona and Harju, 1989). As a result of this large, undifferentiated, noninbred population, it is unlikely that a heterozygosity–fitness association generated by associative overdominance or identity disequilibrium would be observed in Scots pine, because little genetic drift, population structure, or inbreeding is present.

To examine whether there was any heterozygosity–fitness association in Scots pine, Savolainen and Hedrick (1995) examined the association of heterozygosity at 12 allozyme and 6 quantitative traits related to fitness in three different populations. One of the populations was at treeline above the Arctic Circle in one of the most extreme and stressful environments in which Scots pine exists. The other two populations have been cloned, and quantitative measurements were taken on multiple ramets of the same clone so that an excellent measure of the intrinsic genotype value of the quantitative traits was possible. For the 156 tests of association, only 12 (7.7%) were significant at the 5% level. For these 12 significant values, in 6 the heterozygote had higher values and in the other 6 the homozygotes had higher values. A useful way to illustrate these results is given in Figure 10.15, in which the level of significance is given for the three populations and the 12 different loci. Obviously, there is no pattern for these results, such as a clustering near the 5%-significance vertical line for the different

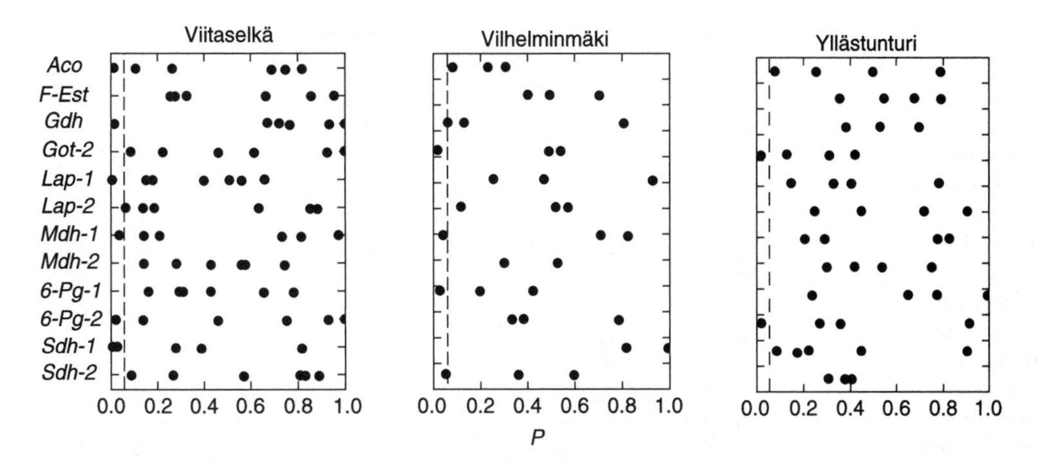

Figure 10.15. The probability of significance level for a difference between heterozygotes and homozygotes for 12 polymorphic loci in three different Scots pine populations. The individual points are the probability level for a given quantitative trait. The vertical broken lines indicate significance at the 5% level.

loci or populations. Overall, there was no evidence of intrinsic heterozygote advantage in these populations, even though there was high statistical power to detect such an association for many of the comparisons.

b. Genetic Hitchhiking

The impact of genetic hitchhiking or the change in frequency of neutral alleles because of an association (linkage disequilibrium) with a selected allele was first considered by Kojima and Schaffer (1967) and Maynard Smith and Haigh (1974). The magnitude of genetic hitchhiking is dependent on not only the extent of linkage or inbreeding but also on the amount and type of the initial linkage disequilibrium (Thomson, 1977; Kaplan *et al.*, 1989; Hedrick, 1980a). In other words, if there is no positive statistical association between the selected allele and the neutral allele, even if there is low recombination or high selfing, there can be no genetic hitchhiking.

As mentioned earlier, linkage and partial selfing have analogous effects in reducing the rate of decay of disequilibrium when there is no selection. Similarly, given initial disequilibrium, selection on one of two loci, and either tight linkage or high selfing, genetic hitchhiking can be a potent force in changing allele frequency and heterozygosity. For example, Figure 10.16 gives a comparison of linkage and partial selfing that results in a similar change in the frequency of neutral allele A_1; that is, A_1 reaches a final frequency of approximately 0.6 in both examples when there is selection at locus B. In this case, $c = 0.01$ for the linkage example, and

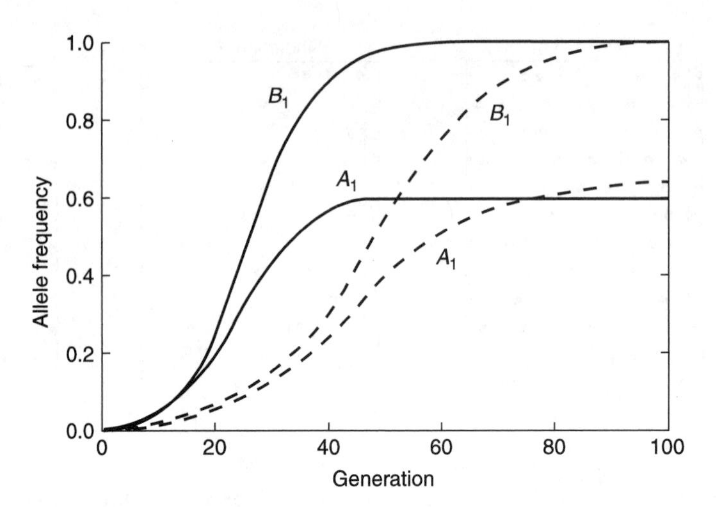

Figure 10.16. The change in allele frequency for a selectively favored allele (B_1) and a neutral allele (A_1) associated with it. The broken lines indicate the frequencies when A and B are linked ($c = 0.01$), and the solid lines indicate the changes when $S = 0.95$ and there is no linkage.

$S = 0.95$ for the partial-selfing example. Here the selection coefficient at the B locus is 0.2, there is additive gene action, and the initial gamete array has frequencies of 0.01 and 0.99 for gametes A_1B_1 and A_2B_2, respectively. Although the level of linkage and selfing chosen in this example results in similar total amounts of allele frequency change, partial selfing hitchhiking changes the allele frequency at the neutral locus much more quickly than linkage hitchhiking. This can be related back to the faster change of allele B_1 at the selected locus B for partial selfing (see p. 287 for a discussion). Genetic hitchhiking can also reduce the level of heterozygosity in a similar manner (see p. 576).

In the example in Figure 10.16, the rate of recombination c is much less that the level of selection s—that is, $c/s = 0.01/0.2 = 0.05$. In general, for hitchhiking to have a major impact, this ratio generally needs to be much less than unity (Barton, 2000). To visualize the dynamics and impact of different amounts of recombination on the extent of hitchhiking, let us assume that $s = 0.02$ at the B locus with additive gene action. Assuming that the initial gamete array has frequencies of 0.01 and 0.99 for gametes A_1B_1 and A_2B_2, respectively, Figure 10.17 gives the frequency of A_1 after 200 and 400 generations and the expected asymptotic frequency. Only when c is much less than s is there significant hitchhiking. For example, if $c = 0.002$, then the frequencies of A_1 after 200 generations, 400 generations, and asymptotically are 0.056, 0.205, and 0.422, respectively. Only when there is very tight linkage, for example, $c = 0.0002$, is hitchhiking change at the neutral allele nearly indistinguishable to the change at the selected allele.

To understand more precisely the extent of genetic hitchhiking, given an initial level of linkage disequilibrium, it is necessary to know the rate of frequency change for the selected allele, relative to the amount of

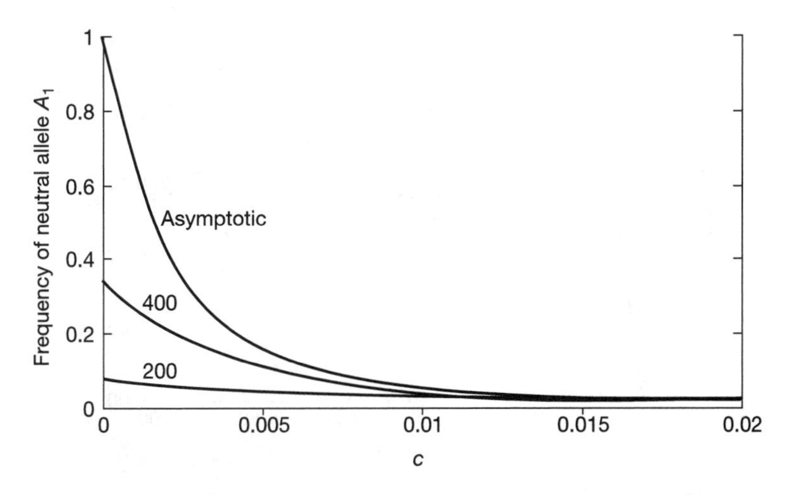

Figure 10.17. The frequency of neutral allele A_1 resulting from genetic hitchhiking with a selected allele B_1 that has a selective advantage of 0.02 and additive gene action. The level of recombination between the loci is c, and the frequency of A_1, which was initially 0.01, is given after 200 generations, 400 generations, and asymptotically.

recombination. For example, if a new favorable variant is recessive, then the initial rate of change may be slow at the initial low frequencies, allowing recombination to breakdown linkage disequilibrium. Therefore, even if there is a substantial selective advantage, then the expected amount of hitchhiking would be low because decay of the initial disequilibrium occurs before there is significant change in the frequency of the favorable variant.

In addition to influencing the allele frequency and heterozygosity at associated loci, genetic hitchhiking can result in changes in linkage disequilibrium when it influences the frequencies of two or more loci simultaneously (Thomson, 1977). The basis for this effect is the simultaneous hitchhiking of alleles at two loci that are in linkage disequilibrium with a third selected locus. Figure 10.18 illustrates the changes in allele frequencies and linkage disequilibrium for an example where three-locus gametes $A_1B_1C_1$, $A_2B_2C_1$, $A_1B_2C_2$, and $A_2B_1C_2$ all have initial gamete frequencies of 0.25 and the other four gametes have initial frequencies of 0 so that there is no initial linkage disequilibrium between any pair of loci but an initial three-locus disequilibrium (Hedrick 1980a). There is also a 20% selective advantage of genotype C_1C_1 over C_2C_2, with the heterozygote being intermediate.

The example in Figure 10.18 is for 99% self-fertilization, but tight linkage produces similar results. The advantageous C_1 allele increases quickly to a frequency near 1.0, whereas neutral alleles A_1 or B_1 both remain at their initial frequencies of 0.5. However, the disequilibrium between the two neutral loci, D_{AB}, rises rapidly from zero so that by generation 25 it is

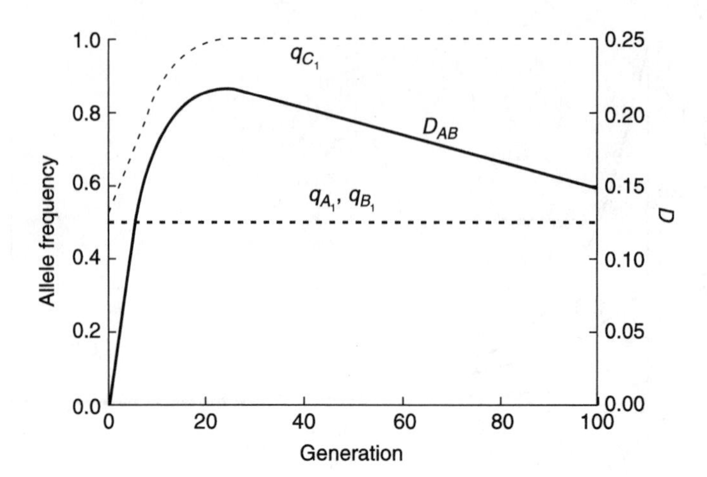

Figure 10.18. The change in allele frequencies (broken lines) and linkage disequilibrium (solid line) for selection only at the C locus and 99% partial selfing.

more than 85% of its maximum value ($D'_{AB} = 4D_{AB}$ in this case). This occurs because the favored C_1 is associated with the coupling gametes, A_1B_1 and A_2B_2, and as C_1 increases, it also increases the frequencies of these associated gametes. The disequilibrium between the neutral loci remains high for an extended period of time because of high selfing and eventually decays slowly at the rate expected for two neutral loci with high selfing. This example illustrates that genetic hitchhiking can generate linkage disequilibrium between neutral loci and that the increase in disequilibrium may occur with no change in allele frequency at the neutral loci—exactly the situation that has been suggested as one that would indicate multilocus selection. Example 10.9 discusses the changes in linkage disequilibrium observed for three linked allozyme in a highly selfing barley population.

Example 10.9. Allard and his co-workers (Allard *et al.*, 1972; Weir *et al.*, 1972, 1974) carried out detailed genetic studies in the highly selfed plant *Hordeum vulgare* (cultivated barley). They found very high associations between alleles at different allozyme loci in this species. Furthermore, in two populations, they observed that the amount of linkage disequilibrium actually increased over generations. More specifically, in one population the allele frequency changes at four esterase loci were rather small (an average of 0.086 during 26 generations) for the two most common alleles at each locus, but there were sizable changes in the values of linkage disequilibrium between pairs of these loci (Allard *et al.*, 1972). The pairwise linkage disequilibrium values for three linked esterase loci (E_A, E_B, E_C) for all the generations examined in this population, Composite Cross V (CCV), are plotted in Figure 10.19 (solid lines and open circles).

Results such as these have been used to argue for the existence of coadaptation in this highly self-fertilized species, but a somewhat different

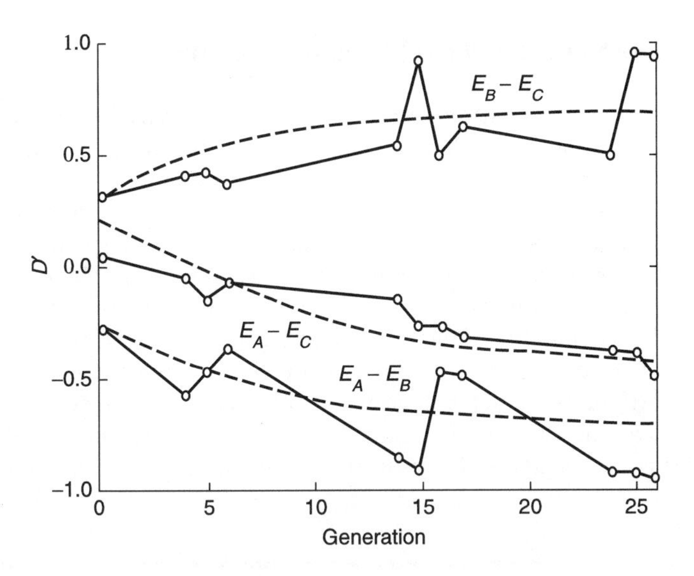

Figure 10.19. The observed changes in linkage disequilibrium for three esterase loci, E_A, E_B, and E_C (solid lines), and simulations to mimic these changes (broken lines) (after Hedrick and Holden, 1979).

hypothesis could explain these data. This explanation is based on hitchhiking primarily due to the mating system—that is, where selection occurs at another locus or loci in the genome and affects frequencies of alleles at the allozyme loci because of the high degree of self-fertilization.

Hedrick and Holden (1979) mimicked the observed changes in disequilibrium by assuming simple directional selection at a locus unlinked to the allozyme loci (broken lines in Figure 10.19). In these simulations, the reported recombination values were used (recombinations between E_A-E_B, E_A-E_C, and E_B-E_C were 0.0023, 0.0048, and 0.0061, respectively), and selfing was assumed to be the estimated value of 0.9943. Because the population was started from a series of crosses, a random distribution of gametes into zygotes was assumed in the initial generation. For the locus undergoing selection, there was a 20% selection difference between the homozygotes, and the heterozygote was assumed to be intermediate.

From the similarity of the observed values and the simulation, it appears that changes such as those observed in this population could be the result of hitchhiking of allozyme loci when there is directional selection at other loci in the genome. Because the experimental populations of barley represent the gene pools of a worldwide collection, it would be very unlikely that the original population was adapted to the Davis, California environment. Directional selection against one or more unfavorable alleles in the population would be an obvious and simple expectation. From these simulations, it appears that hitchhiking resulting from partial selfing is a realistic and simple alternative to explain the allele frequencies and linkage disequilibrium observed.

c. Selective Sweeps and Background Selection

The amount of selection operating on many loci identified by molecular techniques may be so small that these loci are essentially neutral in the context of selection occurring at other background loci.

In this case, potentially one of the most important effects of genetic hitchhiking is the reduction of heterozygosity of such molecular variation in areas of low recombination due to **selective sweeps**. Maynard Smith and Haigh (1974) first suggested that molecular polymorphism may be modified by hitchhiking of neutral alleles tightly linked to loci undergoing allele substitution. For closely linked neutral loci, such allele substitution could greatly decrease heterozygosity at other associated loci. As a simple example, assume that the polymorphic and neutral A_1 and A_2 alleles at the A locus have equal initial frequencies and that a new favorable mutant,

---

(Photo courtesy of Colin Atherton/University of Sussex. Used with permission.)

JOHN MAYNARD SMITH (1920-2004)

John Maynard Smith, originally inspired by the work and guidance of his mentor J. B. S. Haldane to study evolution, made important contributions to an incredible diversity of topics in theoretical biology (Nee, 2004; Szathmáry and Hammerstein, 2004). His early work focused on experimental research in inbreeding, aging, and fitness in *Drosophila* and his finding of a tradeoff between female fertility and longevity is now a fundamental aspect of life history evolution (Lewontin, 2004). Although he began his long career in theory in the 1950s, his training as an engineer and interest in natural history served to keep this research grounded in real world problems. In 1965, he became the founding dean of the School of Biological Sciences at the University of Sussex, where he remained for the rest of his career. He introduced game theory and the approach known as the **Evolutionary Stable Strategy** (**ESS**), particularly to understanding contests in animal behavior. He formulated the problem of the twofold fitness cost of sex and worked on various solutions to this quandary. He wrote a number of influential books, including *Mathematical Ideas in Biology* (Maynard Smith, 1968), *The Evolution of Sex* (Maynard Smith, 1978), *Evolutionary Genetics* (Maynard Smith, 1998), and *Animal Signals* (Maynard Smith and Harper, 2003). His research in recent years turned to various topics in molecular evolution, including the identification of recombination in bacteria and mtDNA. He will be remembered fondly by many for his insatiable interest in evolution, keen mind, and untiring social energy.

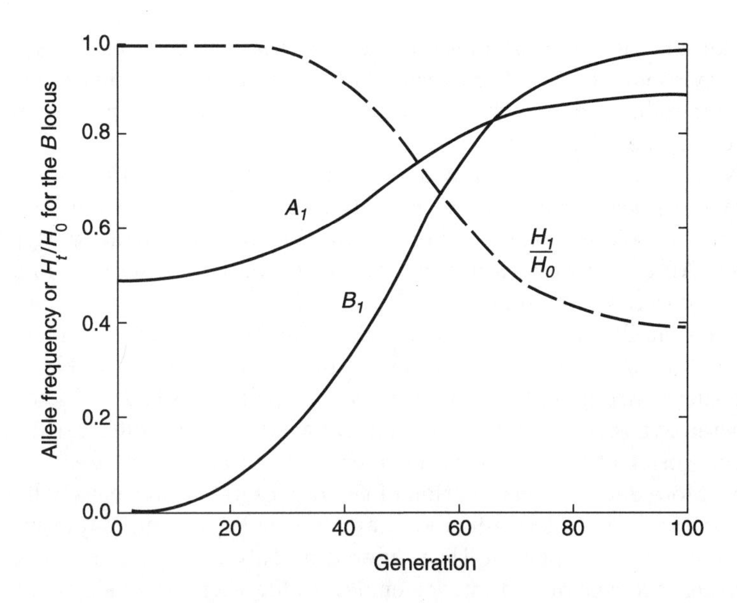

Figure 10.20. The allele frequency and the proportion of initial heterozygosity at the A locus over time when hitchhiking occurs as a result of selection at the B locus.

B_1, with a selective advantage of $s = 0.2$ is generated at the B locus on an A_1 haplotype with an initial frequency of 0.01. Because the two loci are tightly linked ($c = 0.01$), as the B_1 allele passes through the population, it reduces the heterozygosity at the A locus to approximately 40% of its level before the selective sweep (Figure 10.20).

If the recombination is much lower, for example, $c = 0.001$, then the heterozygosity at the A locus is reduced to nearly zero. To put this effect in a context for variation within, or near by, a single gene, Hudson *et al.* (1997) suggested that a selective sweep with $s = 0.01$ can reduce variation at sites up to 10 kb away from the selected locus (assuming recombination rates typical of *D. melanogaster*). Example 10.10 presents an example in which a recent selective sweep reduced the variation in a region around a locus that provides resistance to an antimalarial drug in the parasite that causes malaria.

Example 10.10. Some of the best examples of rapid adaptive change are the development of pathogen resistance to antibiotics and other medications. Because malaria affects over 500 million people worldwide each year and is a leading global cause of death, the use of antimalarial drugs is widespread, and consequently, the evolution of resistance to these drugs in the malarial parasites is also widespread. Pyrimethamine is an inexpensive antimalarial drug used in countries where there is resistance in the malarial parasite *Plasmodium falciparum* to the drug chloroquine. Pyrimethamine was introduced to the area along the Thailand-Myanmar border in the mid-1970s and resistance spread to fixation in approximately 6 years. Pyrimethamine resistance is the result of point mutations at the active site of the gene *dhfr* on chromosome 4 of *P. falciparum.* This gene codes for the enzyme dihydrofolate reductase, and the mutations alter the binding of pyrimethamine to the enzyme (Nair *et al.*, 2003).

Nair *et al.* (2003) examined variation at 33 dinucleotide microsatellite markers across chromosome 4, including 11 clustering around the *dhfr* gene in 61 isolates collected from a location on the Thailand-Myanmar border. All of the isolates had mutations at the *dhfr* gene that indicated that they were pyrimethamine resistant. There was very low variation for the microsatellite markers for approximately 12 kb (0.7 cM) immediately surrounding *dhfr* and reduced variation in a region of about 100 kb (6 cM) indicating a strong and recent selective sweep (Figure 10.21). In contrast, the heterozygosity beyond this region averaged 0.81, not different from the heterozygosity of 0.80 at 56 other microsatellite loci on different chromosomes. More detailed examination of the microsatellite loci closely flanking the resistant *dhfr* alleles showed that 80% had the same haplotype and 16% more had a haplotype that differed at only one of the six loci. This and other evidence provide strong evidence that there was a single origin of the resistant allele that caused the observed selective sweep. Resistance in

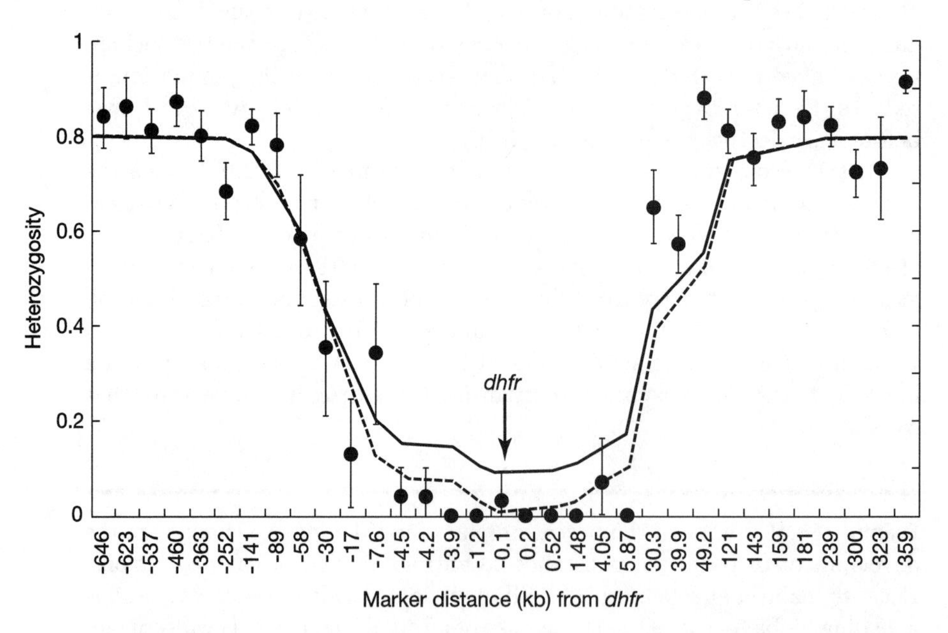

Figure 10.21. The observed (solid circles and one standard error bars) and expected (solid and broken lines for two different mutation rates) microsatellite heterozygosities around the *dhfr* locus in the malaria parasite *Plasmodium falciparum*, which provides resistance to the antimalarial drug pyrimethamine. (Courtesy of Nair, S., J.T. Williams, A. Brockman, *et al.* 2003. A selective sweep driven by pyrimethamine treatment in southeast Asian malaria parasites. *Molec. Biol. Evol.* 20: 1526–1536.)

P. falciparum to chloroquine, another antimalarial drug, is the result of a mutation at gene *pfcrt* on chromosome 7, but in this case, there appears to four independent origins of resistance that show a selective sweep pattern (Wootton *et al.*, 2002).

An alternative explanation for low variation in regions of low recombination is selection against detrimental mutants, closely linked to neutral variants—a phenomenon termed **background selection** by Charlesworth *et al.* (1993). In this case, the chromosome on which the new detrimental mutant occurred will eventually be lost by selection, and if there is no recombination between mutant and the neutral locus, the variation at the neutral locus on that chromosome will also be lost. The effect of each such event is small, but if the number of detrimental mutations is large and they are spread over the genome, they can have a significant cumulative effect. Hudson and Kaplan (1995) showed that this effect is approximately

$$\pi \approx \pi_0 e^{-u/c} \tag{10.12}$$

where π are π_0 are the nucleotide diversities with and without background selection, u is the mutation rate to detrimentals per kilobase in the region, c is the local recombination rate in the region. They compared this predicted effected to that observed for 15 genes on the third chromosome in *D. melanogaster* (Figure 10.22). The observed amount of diversity increases as the level of local recombination increases (solid circles). To determine generally their theoretical prediction approximated these observations, they examined situations where $u = 0.0002$ and $\pi_0 = 0.012$ (solid line) and where $u = 0.00004$ and $\pi_0 = 0.007$ (broken line). In general, these curves give a reasonable explanation of the observed relationship of diversity and recombination, and other models give even a better fit (Hudson and Kaplan, 1995).

Overall, both selective sweeps and background selection predict lower diversity in regions of low recombination as seen in a number of *Drosophila* examples (Berry *et al.*, 1991; Begun and Aquadro, 1992). However, because selective sweeps generally predict a change in the distribution of variants from neutrality with lower proportions of rare alleles while background selection does not (Braverman *et al.*, 1995), examining observed distributions appears to provide a way to distinguish between the two alternatives. Although both processes appear to have been important in given instances (Charlesworth, 1996; Begun and Whitley, 2000; Andolfatto, 2001; Innan and Stephan, 2003), the overall relevance of the two hypotheses is still not clear, particularly in genera other than *Drosophila*.

A useful perspective with which to interpret these empirical findings is the general one developed by Hill and Robertson (1966) when examining the limits to selection, given that there is tight linkage (Hey, 1999). Hill and

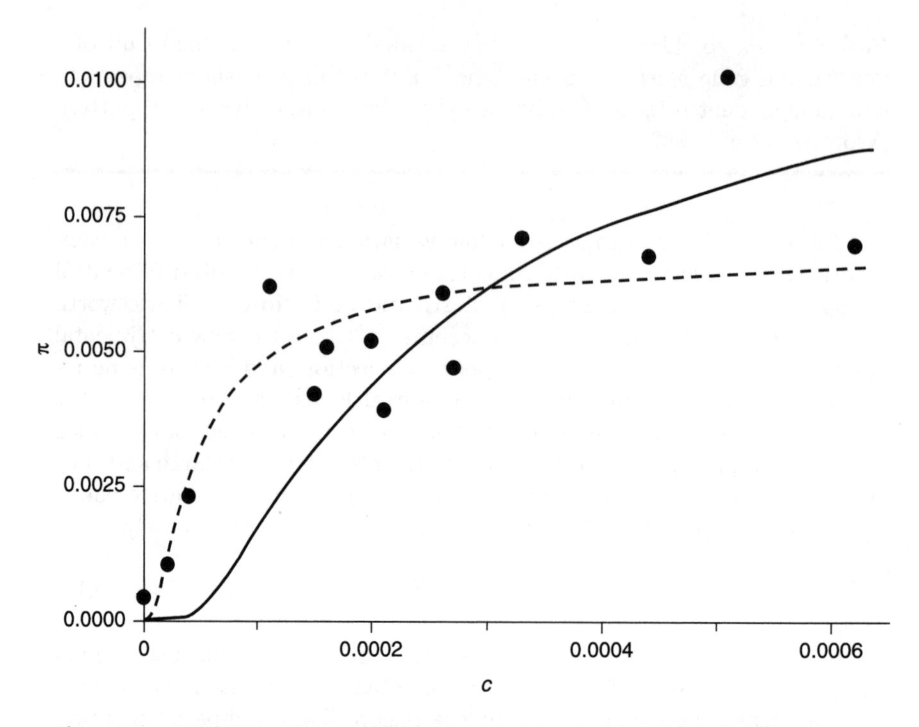

Figure 10.22. The observed nucleotide diversity (π) for different levels of recombination (c) for genes on the third chromosome of *D. melanogaster* (solid circles) and that predicted using expression 10.12 when $u = 0.0002$ and $\pi_0 = 0.012$ (solid line) and when $u = 0.00004$ and $\pi_0 = 0.007$ for background selection (broken line) (after Hudson and Kaplan, 1995).

Robertson showed than ***for two tightly linked loci (sites), directional selection at one locus influenced selection and the probability of fixation at the second locus***. They analogized this effect, sometimes called the **Hill–Robertson effect**, to an increase in genetic drift, that is, a decrease in effective population size.

d. Age of a Favorable Age

One way to estimate the age of a new favorable mutant is to assume that it occurred only once on a given haplotype and that it would remain associated with the alleles on that haplotype, except if recombination or mutation broke up this relationship. To illustrate, let us assume that the original allele is A_1, the favorable mutant is A_2, the mutant occurred on a haplotype with allele B_1 at a second locus, and the rest of the alleles at locus B are indicated as $\overline{B}_1$, or not B_1(see Table 10.14). After t generations, the

TABLE 10.14 The gamete frequencies before, after, and t generations after a favorable mutation at locus A from A_1 to A_2.

	Gamete frequencies			
	A_1B_1	$A_1\bar{B}_1$	A_2B_1	$A_2\bar{B}_1$
Before mutation	q_1	$q_{\bar{1}}$	—	—
After initial mutation A_1 to A_2	$q_1 - p_2$	$q_{\bar{1}}$	p_2	—
t generations after initial mutation	$x_{11.t}$	$x_{1\bar{1}.t}$	$x_{21.t}$	$x_{2\bar{1}.t}$

frequency of the A_2 allele is assumed to have increased, but most of the A_2 alleles are still assumed to be associated with B_1, that is, $x_{21.t} > x_{2\bar{1}.t}$.

The proportion of A_2 alleles that are on the ancestral haplotype after t generations is

$$\begin{aligned} P_t &= x_{21.t}/p_{2.t} \\ &\approx (1 - c' - u)^t \end{aligned} \tag{10.13a}$$

where $c' = x_{1\bar{1}}c$ and u is the mutation rate from B_1 to $\overline{B}_1$. Here c is the recombination rate between the two loci, and it is multiplied by the frequency of gametes that have a $\overline{B}_1$ allele; when there is a B_1 allele, recombination produces a A_1B_1 gamete that cannot be identified as a recombinant. This expression gives, after t generations, the proportion of haplotypes in which A_2 would still be associated with B_1 because neither recombination nor mutation has broken the association. Expression 10.13a can be solved for t, the number of generations since the mutation as

$$t = \frac{\ln(P_t)}{\ln(1 - c' - u)} \tag{10.13b}$$

Example 10.11 shows how this approach can be used to give an estimate of the age of allele *CCR5*-Δ32, which provides resistance to AIDS. For further discussion and other approaches to estimate the age of an allele, see Slatkin and Rannala (2000) and Slatkin (2002).

Example 10.11. The *CCR5* gene produces a chemokine receptor that serves as an entry port for the human immunodeficiency virus (HIV). A 32-base pair deletion mutation (Δ32) interrupts the coding region of the *CCR5* gene and results in nearly complete resistance to HIV infection in homozygous individuals and results in a delay of AIDS onset in heterozygotes compared with normal (++) individuals (Stephens *et al.*, 1998). Their survey of 4166 individuals in 38 human populations for the Δ32 allele showed a north-to-south cline with most northern populations having a frequency of greater than 10% while the variant is missing in east Asian, middle eastern, and Amerindian populations.

In order to estimate the age of the allele, Stephens *et al.* (1998) examined the variation at two microsatellite loci (*GAAT* and *AFMB*) closely linked to *CCR5* on chromosome 3. Table 10.15 gives the five Δ32 haplotypes and the seven + haplotypes. Because the haplotype Δ32-197-215 is by far the most frequent Δ32 haplotype at a frequency of 0.848, they assumed that it was the ancestral Δ32 haplotype and suggested that it was generated from the most common wild-type haplotype +-197-215 by mutation. The other Δ32 haplotypes can be generated by either recombination or mutation from the other haplotypes.

Stephens *et al.* (1998) estimated that $c = 0.0021$ between *CCR5* and *GAAT* and $c = 0.0072$ between *GAAT* and *AFMB*. Because approximately 0.637 of the + haplotypes are not 197–215 and approximately 0.479 are not 215, they estimated c' as $(0.637)(0.0021) + (0.479)(0.0072) = 0.00479$. Assuming that an upper limit to u is 0.001 and $P_t = 0.848$, then using expression 10.13*b*, $t = 28.3$. If the generation length is 25 years, then the mutant Δ32 would have originated approximately 708 years ago.

Because HIV has not existed long enough to result in enough selection to increase the Δ32 mutation, Stephens *et al.* (1998) suggested that it originally increased in frequency because it provided resistance to the bubonic plague. However, Galvani and Slatkin (2003) have shown that the extent of selection provided by smallpox over human history is more consistent with the amount necessary to increase the frequency of the Δ32 mutation. They also point out that geographic distribution of the Δ32 mutation is more consistent with the historical incidence of smallpox and that poxvirus enters leucocytes by using chemokine receptors, as does HIV. In addition, Mecsas *et al.* (2004) showed that CCR5 deficiency in mice does not protect against experimental infection by the plague bacteria.

TABLE 10.15 The frequencies of *CCR5* haplotypes observed in a sample of 46 haplotypes with the deletion Δ32 and 146 haplotypes with the + allele (Stephens *et al.*, 1998). The putative origin of the five Δ32 haplotypes is given.

CCR5	*GAAT*	*AFMB*	*Frequency*	*Type of haplotype*
Δ32	197	215	0.848	ancestral
Δ32	197	217	0.065	recombination
Δ32	193	215	0.043	recombination
Δ32	197	219	0.022	recombination or mutation
Δ32	197	213	0.022	mutation
+	197	215	0.363	
+	197	217	0.308	
+	193	215	0.137	
+	197	219	0.014	
+	193	217	0.144	
+	191	217	0.014	
+	191	215	0.021	

e. Advantages and Disadvantages of Recombination (Sex)

The impact of recombination on adaptation and the pattern and extent of genetic variation has long been an important evolutionary topic. For example, consideration of multilocus selection in a constant environment suggested that lower recombination generally leads to a higher mean fitness (Fisher, 1930; Lewontin, 1971). However, if the environment is changing or the fitness of the organism in the present environment could be increased, then it may be advantageous to have a greater variety of genotypes in the population as would result if there were substantial recombination. As an example, assume that gametes A_1B_1 and A_2B_2 are in high frequency because of multilocus selection in a given environment favoring these balanced gametes. Then assume that in a new environment, selection favors A_1B_2/A_1B_2 homozygotes. Because the A_1B_2 gametes (not the A_1 or B_2 alleles) are initially low in frequency, the rate of response would be reduced as compared with that of a population in linkage equilibrium.

In theory, ***sexual reproduction has a twofold disadvantage compared with asexual reproduction*** (Maynard Smith, 1978; Barton and Charlesworth, 1998). To illustrate the **disadvantage of sex**, assume that a population is stable in size so that each sexual female and each asexual female have two offspring. The two progeny from the sexual females are only half descended from their mother's genes (the other half from their father's), whereas the asexual females are entirely descended from their mother's genes—hence the twofold disadvantage to sex.

A number of hypotheses have been proposed to counter this disadvantage to sex, but the **advantage of recombination,** often referred to the **advantage of sex,** or sexual reproduction, is basic to the population genetic differences between sexual reproduction and asexual reproduction (Muller, 1932; Barton and Charlesworth, 1998). With asexual reproduction, pre-existing, well-adapted genotypes are maintained because there is no gametogenesis (segregation or recombination). With sexual reproduction, favorable gene combinations are broken up each generation, but because of recombination, there is the potential that new zygotes may be formed that have a higher fitness than any in the previous generation. As a result, selection in a sexual population may eventually produce a population with a phenotypic mean well above that of the range of phenotypes in the initial population, unlike an asexual population that can progress no further than the best genotype in the original population. Supporting this scenario, there is evidence for the advantage of sexual reproduction in adaptation in laboratory experiments (Rice and Chippendale, 2001b; Colegrave, 2002; Bachtrog, 2003).

Furthermore, in an asexual population, subsequent favorable mutations, in order to be incorporated into the population, must occur in the same lineage (Figure 10.23). In a sexual population, however, recombination provides a means for favorable mutants originating in different lineages to

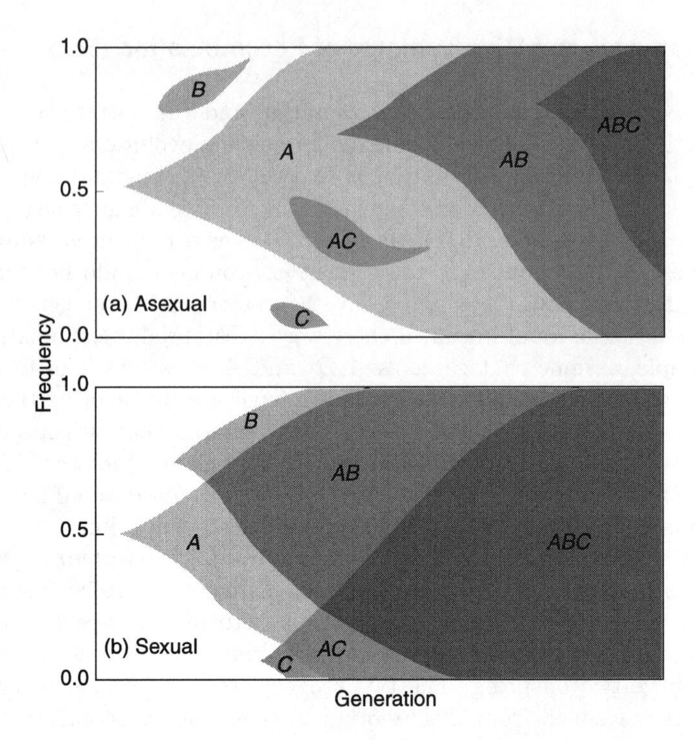

Figure 10.23. The frequency of different genotypes over time when there is (a) asexual or (b) sexual reproduction (Crow and Kimura, 1965).

be incorporated in the same individual. In other words, sex (through segregation and recombination) appears to be advantageous because it increases the rate of adaptation to changes in both the physical and biotic (predators, pathogens, etc.) environments. Because of the selective advantage that recombination affords in incorporating independent, favorable mutants, there may be selection for a higher recombination rate, particularly when linkage is tight or in periods of rapid evolutionary change (Otto and Barton, 1997).

Sexual reproduction may also be important in the elimination or reduction of detrimental variation from the population. Muller (1932) suggested that ***without sexual reproduction detrimental mutations can accumulate and the fitness will decline over time***, a phenomenon now known as **Muller's ratchet** (Kondrashov, 1988; Gillespie, 1998). To illustrate this theory, let us assume that at an initial time t, that the individuals in the population may have 0, 1, 2, 3, . . . detrimental mutations (Table 10.16). These mutations may be at completely different genes in different individuals, and we are assuming, for example, that a proportion of individuals, $y_{1.t}$, initially have one detrimental mutation at some gene. If it is assumed that the number in the 0 mutation class is small, then by chance, the 0 class can be lost, and at time t +1, the class with the lowest number of mutants left is the one in which individuals have one mutant. At this point, the ratchet, a wheel with teeth that mechanically prevents reversal, has made one turn up, and the fitness of the population has irretrievably

TABLE 10.16 A schematic representation of the operation of Muller's ratchet over time in an asexual population, where $y_{i \cdot t}$ is the proportion of individuals with i detrimental mutations at time t. Over time, the class with the fewest mutations is lost by chance and, because there is no recombination, cannot be restored. As a result, the ratchet turns only one way, and the fitness of the population declines as detrimental mutations accumulate.

	Time			
Number of mutants	t	$t+1$	$t+2$	$t+3$
0	$y_{0 \cdot t}$	0	0	0
1	$y_{1 \cdot t}$	$y_{1 \cdot t+1}$	0	0
2	$y_{2 \cdot t}$	$y_{2 \cdot t+1}$	$y_{2 \cdot t+2}$	0
3	$y_{3 \cdot t}$	$y_{3 \cdot t+1}$	$y_{3 \cdot t+2}$	$y_{3 \cdot t+3}$
⋮	⋮	⋮	⋮	⋮
i	$y_{i \cdot t}$	$y_{i \cdot t+1}$	$y_{i \cdot t+2}$	$y_{i \cdot t+4}$

declined. Now if the number of individuals with one mutant is small, then by time $t + 2$, this class may be lost and the ratchet has made two turns up. During this process, mutation to detrimentals is assumed to have increased the number of mutants so that the number of individuals in the lowest remaining class has declined.

According to this scenario, the fitness of the population will gradually decline, and the population will eventually go extinct because every individual has a large number of detrimental mutants. In theory, the time scale for this process may be very long (Charlesworth *et al.*, 1993; Gordo and Charlesworth, 2000), but in an experimental system, Rice (1994) observed significant degeneration of a nonrecombining chromosome over a relatively short time. However, if there is sexual reproduction, recombination can produce individuals with fewer detrimental mutants—individuals in classes that were previously lost—and the decline in fitness can be reversed. In contrast to the predictions for asexual species, it appears that bdelloid rotifers have existed for 40 million years without sexual reproduction, an observation supported by extreme molecular divergence of the two lineages within these diploid organisms (Welch and Meselson, 2000; Welch *et al.*, 2004).

V. ESTIMATION OF LINKAGE DISEQUILIBRIUM

When two or more polymorphic loci are examined in individuals in a population, then linkage disequilibrium can be estimated between these loci. When the genes are in the haploid state, such as for haploid organisms, mtDNA, cpDNA, Y chromosomes, X chromosomes in males (see Example 10.4), or males in haplo-diploid organisms, then the gamete frequencies and disequilibrium can be estimated directly. In organisms with few multiple heterozygotes, such as those with a high degree of selfing, multiple heterozygotes may be ignored with only a small effect on the estimate of linkage disequilibrium. In some organisms that can be bred in the laboratory or for which there are family groups, the phase of the gametes may be determined. Often, estimation of gamete frequencies cannot be carried out directly because the phase of the gametes in multiple heterozygotes cannot be determined, but it is still possible, as outlined below, to estimate the extent of linkage disequilibrium.

a. Direct Identification of Gametes

If it is possible to identify directly a random sample of n two-locus gametes, whether it be from a haploid organism, mtDNA, Y chromosomes, or other loci, then the numbers of the different gametes for two alleles at each locus are as in Table 10.17a. The maximum-likelihood estimates of the allele frequencies and the linkage disequilibrium are then

$$\begin{aligned} \hat{p}_1 &= \frac{n_{11} + n_{12}}{n} \\ \hat{q}_1 &= \frac{n_{12} + n_{22}}{n} \\ \hat{D} &= \frac{n_{11}n_{22} - n_{12}n_{21}}{n^2} \end{aligned} \tag{10.14a}$$

A test of $D = 0$ uses the likelihood ratio statistic and is

$$\begin{aligned} Q &= n\hat{r}^2 \\ &= \frac{n\hat{D}^2}{\hat{p}_1\hat{p}_2\hat{q}_1\hat{q}_2} \end{aligned} \tag{10.14b}$$

TABLE 10.17 (a) The numbers of different gametes when they can be directly identified. (b) The numbers of different phenotypes when both loci are codominant.

(a) **Gamete identification**

	A_1	A_2	*Total*
B_1	n_{11}	n_{21}	$n_{\cdot 1}$
B_2	n_{12}	n_{22}	$n_{\cdot 2}$
Total	$n_{1\cdot}$	$n_{2\cdot}$	n

(b) **Both loci codominant**

	A_1A_1	A_1A_2	A_2A_2	*Total*
B_1B_1	N_{11}	N_{21}	N_{31}	$N_{\cdot 1}$
B_1B_2	N_{12}	N_{22}	N_{32}	$N_{\cdot 2}$
B_2B_2	N_{13}	N_{23}	N_{33}	$N_{\cdot 3}$
Total	$N_{1\cdot}$	$N_{2\cdot}$	$N_{3\cdot}$	N

When $D = 0$ and N is large, Q is approximately distributed as a χ^2 variate with one degree of freedom (Hill, 1974).

Alternatively, the probability using Fisher's exact test that gamete frequencies as extreme as that observed can be calculated (Weir, 1996). Using the symbols in Table 10.17a, the probability of an array with n_{11}, n_{12}, n_{21}, and n_{22} of gametes A_1B_1, A_1B_2, A_2B_1, and A_2B_2, respectively, given the marginal numbers $n_{1.}$ and $n_{.1}$ is

$$\Pr\langle n_{11}, n_{12}, n_{21}, n_{22} \mid n_{1.}, n_{.1}\rangle = \frac{n_{1.}!n_{2.}!n_{.1}!n_{.2}!}{n_{11}!n_{12}!n_{21}!n_{22}!n!} \tag{10.14c}$$

where the vertical line in the left side of the equation symbolizes given. All possible samples with the same marginal totals are generated and their probabilities calculated. These are then ranked in ascending probability order, and the probability of the observed sample by chance is the cumulative probability up to and including the observed sample. Using the exact test, Lewontin (1995) showed that when the rarer allele at a site or locus is only present a few times that it is often impossible or very difficult to detect linkage disequilibrium. Although Fisher's exact test is now being widely used to examine disequilibrium between nucleotide sites, Zapata and Alvarez (1997) suggested that it may be conservative and sometimes result in an unnecessary loss of power. The proportion of significant associations they found using a modification of the exact test was similar to that using χ^2 test given above, a finding that led them to support the use of the simpler χ^2 test. Example 10.12 shows how the exact test is calculated for a small data set and then discusses disequilibrium in a large data set for the human lipoprotein lipase.

Example 10.12. As large numbers of sequences become available for many genes, one of the basic questions that arises concerns the amount and pattern of linkage disequilibrium among sites within these genes. One of the common approaches to quantifying disequilibrium is to calculate the exact test for all pairs of polymorphic sites. Weir (1996) provided a simple example of the exact test using restriction fragment data from the *Adh* locus in *D. melanogaster* from Langley *et al.* (1982). Table 10.18a gives the 17 gametes observed for two restriction sites, where $-$ and $+$ indicate the absence and presence of the restriction site. There are seven possible gametic arrays that satisfy the marginal totals (Table 10.18b). (These can be found by listing the seven numbers, 0 to 6, for the rarest haplotype and then filling in the other gamete numbers to satisfy the marginal numbers.) Using expression 10.14*c*, the probabilities of the arrays can be calculated and then ordered in terms of ascending probability. The array observed (in boldface) has the third lowest probability, and the cumulative probability including the observed array is significant at 0.042. For this array,

TABLE 10.18 (a) The numbers of gametes observed for the restriction sites *BamHI* and *XhoI* in the *Adh* locus in *D. melanogaster* for 17 chromosomes (Langley *et al.*, 1982). (b) The probability of the numbers of gametes given from expression 10.14*c* in ascending order and the cumulative probability (Weir, 1996).

(a)

		BamHI		
		−	+	*Total*
XhoI	−	0	6	6
	+	6	5	11
	Total	6	11	17

(b)

−−	−+	+−	++	*Probability*	*Cumulative probability*
6	0	0	11	0.000	0.000
5	1	1	10	0.005	0.005
0	**6**	**6**	**5**	**0.037**	**0.042**
4	2	2	9	0.067	0.109
1	5	5	6	0.224	0.333
3	3	3	8	0.267	0.600
2	4	4	7	0.400	1.000

$D = -0.125$ and using expression 10.14*b*, $Q = 5.06$, which is also significant at the 0.05 level.

Clark *et al.* (1998) examined the variation in 9.7 kb of the human lipoprotein lipase gene from 71 individuals from three populations: 24 African Americans from Mississippi, 24 Finns from North Karelia, and 23 nonhispanic whites from Minnesota. Of the 142 chromosomes examined, there were 88 different haplotypes, which fell into two major groups. These major groups were present in all three populations, and tests for homogeneity over populations were nonsignificant. Exact tests were calculated for all pairs of the 88 polymorphic nucleotide sites (Figure 10.24). Site pairs with a significant exact test ($P < 0.001$), without correction for multiple comparisons, are indicated by solid squares. There was a concentration of significant disequilibrium between pairs in the 3′ end of the gene starting in the intron before exon 7. However, because of rare nucleotides at one or both sites, a large number of polymorphic sites did not have the statistical power to detect a significant association (squares with a dot in center). A test for overall disequilibrium, based on the numbers of positive and negative disequilibrium observed (Lewontin, 1995), showed a significant excess of coupling disequilibria with association of rare alleles at a pair of sites. This extensive disequilibrium was found in spite of evidence suggesting past recombination.

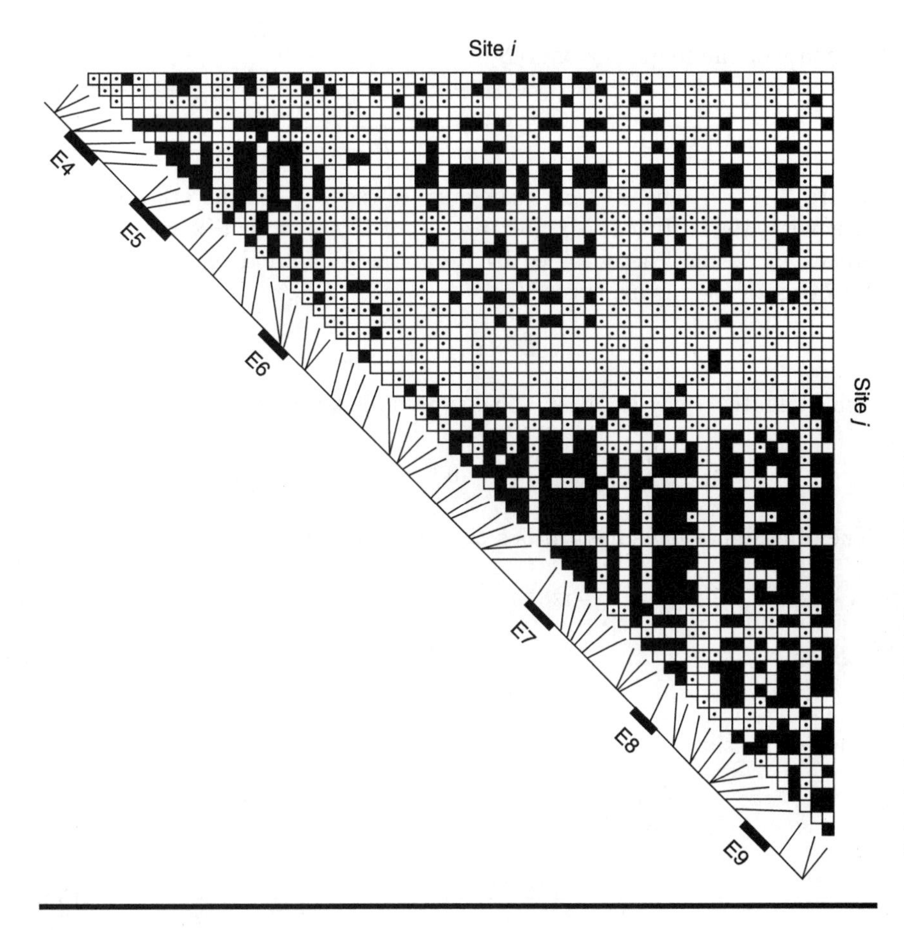

Figure 10.24. Significant linkage disequilibrium between the 88 polymorphic sites at the human lipoprotein lipase gene (solid squares). The diagonal line indicates the location of the site and the position of exons 4 to 9. The squares with a dot in the center are ones that lack the power to test for a significant association (Clark *et al.*, 1998).

b. Two Codominant Loci

If there are two codominant alleles at each of two polymorphic loci, then the gametes that make up each of the phenotypes can be identified except for the double heterozygote. For this phenotype, the two types of double heterozygotes, coupling or repulsion, are indistinguishable (unless a breeding test is carried out). As a result, only the nine observed phenotypic classes are given in Table 10.17b. For simplicity, we have used a somewhat different notation here with, for example, the first subscript 1, 2, or 3 indicating genotypes A_1A_1, A_1A_2, and A_2A_2 at the A locus. The number of individuals of each type is given so that, for example, N_{11} is the number of $A_1A_1B_1B_1$ individuals and N_{23} is the number of $A_1A_2B_2B_2$ individuals. To estimate disequilibrium where the phase of the double heterozygotes is unknown, we can use the approach referred to as the **expectation-maximization (EM) algorithm** to obtain the maximum-likelihood estimate of the frequency of A_1B_1

$$\hat{x}_{11} = \left(\frac{1}{2N}\right)\left(2N_{11} + N_{12} + N_{21} + \frac{N_{22}\hat{x}_{11}\hat{x}_{22}}{\hat{x}_{11}\hat{x}_{22} + \hat{x}_{12}\hat{x}_{21}}\right) \tag{10.15a}$$

where $\hat{x}_{12}$, $\hat{x}_{21}$, and $\hat{x}_{22}$ are analogous expressions for the other gametes (Bennett, 1965; Hill, 1974). Because there is no explicit solution to this equation, $\hat{D}$ must be obtained by iteration. A method suggested by Hill (1974) is to use the expression

$$\hat{D} = \hat{x}_{11} - \hat{p}_1\hat{q}_1 \tag{10.15b}$$

and evaluate $\hat{x}_{11}$ by iteration. This is possible because

$$\hat{p}_1 = \frac{N_{1.} + {}^1\!/_2 N_{2.}}{N} \qquad \text{and} \qquad \hat{q}_1 = \frac{N_{.1} + {}^1\!/_2 N_{.2}}{N}$$

The other following substitutions can be made

$$\begin{aligned}\hat{x}_{12} &= \hat{p}_1 - \hat{x}_{11}\\ \hat{x}_{21} &= \hat{q}_1 - \hat{x}_{11}\\ \hat{x}_{22} &= 1 - \hat{p}_1 - \hat{q}_1 + \hat{x}_{11}.\end{aligned}$$

Then expression 10.15a becomes

$$\hat{x}_{11} = \left(\frac{1}{2N}\right) [2N_{11} + N_{12} + N_{21} + \frac{N_{22}\hat{x}_{11}(1 - \hat{p}_1 - \hat{q}_1 + \hat{x}_{11})}{\hat{x}_{11}(1 - \hat{p}_1 - \hat{q}_1 - \hat{x}_{11}) + (\hat{p}_1 - \hat{x}_{11})(\hat{q}_1 - \hat{x}_{11})} \tag{10.15c}$$

The only unknown in this expression is $\hat{x}_{11}$, and we can calculate it by choosing a value of $\hat{x}_{11}$ for the right-hand side of the equation and then calculating the resulting value of $\hat{x}_{11}$. This calculated value then can be substituted into the right-hand side and the process repeated until the desired precision is obtained. A possible starting point for this iteration is

$$\hat{x}_{11} = \frac{2N_{11} + N_{12} + N_{21}}{2(N - N_{22})}$$

which is an estimate of the gamete frequency excluding the double heterozygotes. A note of caution is this: Weir and Cockerham (1979) showed that in some cases, depending on the starting point, this iteration procedure may converge to different values.

The χ^2 statistic with one degree of freedom is

$$\begin{aligned}Q &= N\hat{r}^2\\ &= \frac{N\hat{D}^2}{\hat{p}_1\hat{p}_2\hat{q}_1\hat{q}_2}\end{aligned} \tag{10.15d}$$

Example 10.13 gives an example of significant disequilibrium for two allozyme loci in the blue mussel using this approach.

Example 10.13. Mitton and Koehn (1973) examined simultaneously two allozyme loci, *Ap* and *Lap*, in a population of the blue mussel *Mytilus edulis*. These loci code for enzymes that have similar catalytic functions, but there was no information on their linkage relationship. The numbers that Mitton and Koehn found in the nine phenotypic classes are given in Table 10.19. Using the procedure described using expression 10.15*c*, the estimated frequency of gamete Ap^fLap^f ($\hat{x}_{11}$) is 0.302 and the estimated value of disequilibrium from expression 10.15*b* is −0.0246. In this case, the observed value of linkage disequilibrium is significant at the 5% level ($Q = 5.43$).

TABLE 10.19 The numbers of blue mussels in different phenotypic classes for two allozyme loci (Mitton and Koehn, 1973).

	Lap			
Ap	*ff*	*fm*	*mm*	*Total*
ff	61	96	49	206
fm	60	112	56	228
mm	49	34	17	100
Total	170	242	122	534
	$\hat{x}_{11} = 0.302$	$\hat{D} = -0.0246$	$Q = 5.43$	

---

c. Extension to Multiple Alleles

When there are more than two alleles per locus, it is difficult to express the degree of interlocus association in a single measure. The extension of D to multiple alleles at two loci for any pair of alleles becomes

$$\hat{D}_{ij} = \hat{x}_{ij} - \hat{p}_i\hat{q}_j \qquad (10.16a)$$

where $\hat{x}_{ij}$ is the estimate of the frequency of gamete A_iB_j in the population and $\hat{p}_i$ and $\hat{q}_i$ are the estimates of the frequencies of alleles A_i and B_j, respectively. For example, if there are three alleles at each locus, there are nine such values although they are not all independent of each other.

Several alternative approaches exist to condensing the information for multiple allele gametes (Hill, 1975; Hedrick, 1987c). A straightforward technique is the statistic

$$Q = n\sum_{i=1}^{k}\sum_{j=1}^{l}(\hat{D}_{ij}^2/\hat{p}_i\hat{q}_j) \qquad (10.16b)$$

where there are k and l alleles are at loci A and B, respectively, and n total gametes exist (Hill, 1975; Hedrick and Thomson, 1987). If all $D_{ij} = 0$, then Q is approximately chi-square distributed with $(k-1)(l-1)$ degrees of freedom. Using the standardized D' values given by expression 10.4*a* for each gamete, Hedrick (1987c) suggested that the measure

$$D' = \sum_{i=1}^{k}\sum_{j=1}^{l} p_i q_j |D'_{ij}| \tag{10.16c}$$

captured many of the important qualities necessary for an appropriate measure for multiple-allele loci (see also Zapata, 2000; Zapata *et al.*, 2001).

To estimate the amount of disequilibrium when there are multiple alleles and the phase of the double heterozygotes is unknown, the approach of Hill (1974) has been expanded and generalized (Excoffier and Slatkin, 1995; Long *et al.*, 1995; Slatkin and Excoffier, 1996). As shown theoretically for two alleles, Kalinowski and Hedrick (2001) also demonstrated for multiple allele microsatellite and MHC data in bighorn sheep that the EM algorithm may generate different estimates of the gamete frequencies, depending on the starting point.

When there are many alleles at a locus, then it difficult to calculate the exact probability, as shown above for two alleles, and the statistic in expression 10.16*b* may not behave well because of small expected frequencies. As a result, a simple Monte Carlo approach can be used to calculate the exact probability of significance (Hudson, 2001). Given that the observed frequencies for the ith allele at locus A is p_i and the jth allele at locus B is q_j and the n_{ij} two-locus gametes with these alleles are observed (or estimated), then the probability of observing this array is given by the multinomial

$$\Pr(n_{ij}|p_i, q_j) = \frac{n!}{\prod_{i,j} n_{ij}!}\prod_{i,j}(p_i q_j)^{n_{ij}} \tag{10.16d}$$

To calculate the exact probability of significance of this observed array, many random arrays of two-locus gametes can be generated with the constraint that the given number of alleles (allele frequencies) is retained at each locus and compared with the observed probability. To do this, first the probability for the observed array of gametes is generated using expression 10.16*d*. As an example, let us say that we observed in a sample of size n = 10 with three alleles at each locus $n_{11} = 1$, $n_{12} = 1$, $n_{13} = 1$, $n_{21} = 3$, $n_{22} = 2$, $n_{23} = 0$, $n_{31} = 1$, $n_{32} = 1$, and $n_{33} = 0$ (column 1 of Table 10.20a), and we get $\Pr(O) = 0.00082$. The exact probability is calculated by generating a large number of arrays with the same marginal numbers (allele frequencies), and the probability of observing each array from expression 10.16*d* is calculated. The proportion of random arrays that have a probability equal to, or more extreme, than that calculated from the observed array is the estimated exact probability for the observed array.

TABLE 10.20 (a) The observed array of gametes for two loci A and B, each with three alleles in a sample of size 10 (column 1), the ordered array for the alleles at each locus (column 2), and the first randomly permuted array of the alleles at the two loci (column 3). (b) The number of each type of gamete in the observed array (left column) and in the first randomly permuted array (right column) and the probabilities for each array.

	Observed	*Ordered*	*Randomly permuted*
(a) Gametes ($n = 10$)			
	A_2B_1	A_1 B_1	A_3 B_2
	A_1B_1	A_1 B_1	A_1 B_1
	A_3B_2	A_1 B_1	A_2 B_1
	A_2B_1	A_2 B_1	A_1 B_1
	A_3B_1	A_2 B_1	A_2 B_2
	A_1B_3	A_2 B_2	A_2 B_3
	A_2B_2	A_2 B_2	A_3 B_1
	A_2B_1	A_2 B_2	A_2 B_2
	A_1B_2	A_3 B_2	A_1 B_1
	A_2B_2	A_3 B_3	A_2 B_2

	Observed	*Randomly permuted*
(b) Number of gametes		
A_1B_1	1	3
A_1B_2	1	0
A_1B_3	1	0
A_2B_1	3	1
A_2B_2	2	3
A_2B_3	0	1
A_3B_1	1	1
A_3B_2	1	1
A_3B_3	0	0
	Pr(O) = 0.00082	Pr(E_1) = 0.00027

A straightforward approach to generate these random arrays is to first order the n alleles at the A locus so that there are $n_{1.}$ A_1 alleles ($p_1 = n_{1.}/n$), $n_{2.}$ A_2 alleles, $n_{3.}$ A_3 alleles, and so on (column 2 of Table 10.20b gives this ordering for both loci A and B). Now independently permute the order of the A locus alleles and the B locus alleles and the first set of randomly generated gametes is then the pairs of alleles in each row of column 3. In other words, $n_{11} = 3$, $n_{12} = 0$, $n_{13} = 0$, $n_{21} = 1$, $n_{22} = 3$, $n_{23} = 1$, $n_{31} = 1$, $n_{32} = 1$, and $n_{33} = 0$ with Pr (E_1) $= 0.00027$. The probability of this array is compared with the probability for the observed array and found to be more likely. If we generate 10,000 such randomly permuted arrays, we find that the exact probability is 0.669; that is, 66.9% of them have Pr values lower than that from the observed array, and we can say that the observed array is not statistically significant from what we would be expected to observe. Similarly, the genotypes at two loci can be randomly permuted (this does not necessitate estimation of gamete frequencies and ignores deviations from Hardy–Weinberg proportions), and the exact probability of the observed two-locus genotypic array can be estimated (Slatkin and Excoffier, 1996; Hudson, 2001). The same approach can be used to determine exact probabilities when there are more than two loci or more than two nucleotide sites.

PROBLEMS

1. Derive the expression for D given in equation 10.2b from that given in equation 10.2a. Derive the expression for the frequency of gamete A_1B_2 after one generation of random mating.
2. Given that $x_1 = 0.3$, $x_2 = 0.1$, $x_3 = 0.2$, and $x_4 = 0.4$, calculate D, D', and r.
3. Assuming that $c = 0.1$ and $S = 0.9$, what is D after three generations, given that it is initially 0.10?
4. If $c = 0.01$ and $N_e = 500$, what is the expected value of r^2?
5. Given that two populations are mixed so that $m_x = 0.2$ ($m_y = 0.8$) and the frequencies for an allele at one biallelic locus are 0.3 and 0.6 and for an allele at a second biallelic locus are 0.4 and 0.5 in the two populations, what is the expected value of D?
6. Do the data presented in Example 10.5 support the tricentric origin of the sickle-cell mutant? What further data would you like to have to examine this hypothesis?
7. Derive the general expression for the frequency of gamete A_1B_2 when there is selection (see expression 10.10b).
8. Assuming $a = 0.8$ and $b = 0.85$, write out the 3×3 matrix of fitnesses. What is the critical recombination level for these fitnesses? If $c = 0.01$, what are the frequencies of the gametes and the linkage disequilibria at the two complementary equilibria?
9. Given that the frequencies of gametes $++$, $+-$, $-+$, and $--$ are 0.15, 0.45, 0.35, and 0.05, calculate the frequencies of the 10 possible genotypes, assuming Hardy–Weinberg proportions. If $-/-$ genotypes have a selective disadvantage of 0.2, there is complete dominance of the wildtype alleles, and there is multiplicative fitnesses over loci, what are the average fitnesses for double homozygotes, single heterozygotes, and double heterozygotes?
10. Calculate the cumulative probabilities for the gametes in Table 10.18b.
11. Calculate the linkage disequilibrium for the data in Table 10.19, using the suggested starting frequency for the iteration. Calculate the value for expression 10.15d for these data.
12. Given that a strong linkage disequilibrium was observed between alleles at two loci in different populations, how would you design a research program to determine the cause of this association?
13. Think of two factors that may invalidate Muller's ratchet. How would you examine whether these factors are important in actual organisms?

14. Lewontin (1995) pointed out that even though there may be strong linkage disequilibrium between nucleotide sites, this association may not be statistically significant. As an example, assume that the gametes (and numbers) at two sites are AA (1), AT (6), TA (15), and TT (78). Calculate the cumulative probability of finding disequilibrium equal to or more extreme than that observed.
15. Discuss the differences between selective sweeps and background selection. Design an experiment that would provide data to show the relative importance of the two mechanisms on the amount of genetic variation.
16. Can the findings from two-locus models be directly used for two-nucleotide site examples with only a change in parameters (e.g., lower c)? Explain your answer.

11

Molecular Population Genetics and Evolution

> The steam of heredity makes phylogeny; in a sense it is phylogeny. Complete genetic analysis would provide the most priceless data for the mapping of this stream.
>
> George Gaylord Simpson (1945)

> Human individualization based on the genome exploits the fact that everyone except for identical twins is genetically distinguishable. Moreover, human genetic material is found in every nucleated cell in the body and can be recovered from samples as diverse as bone, blood stains, saliva residues, nasal secretions, and even fingerprints. DNA may be recovered from very old samples that have been well preserved, and DNA signatures may even be preserved over successive generations.
>
> Bruce Weir (2001)

> The future for understanding quantitative traits in terms of complex genetics rather than statistical descriptions is bright. The various genome projects are yielding very dense linkage maps for humans, model organisms, and species of agricultural importance that often show remarkable conservation of linkage groups across taxa. With the development of improved statistical methods for analysis of experimental crosses and pedigrees to detect segregating QTLs associated with molecular markers, and with the potential to resolve QTLs to the level of single genes, the description of the Mendelian genetic basis of quantitative variation is within reach.
>
> Douglas Falconer and Trudy Mackay (1996)

Molecular evolution and molecular population genetics are areas of important developments and extensive research and will be the focus of much evolutionary genetics research in the future. Throughout the book we have presented a number of new examples of findings from molecular genetics, and Chapter 8 specifically discussed the neutral theory and the coalescent approach used in analysis of much of molecular population genetics. In this chapter, we introduce three other areas of molecular evolution and molecular population genetics that are important evolutionary applications.

First, we discuss molecular phylogenetics and provide introductions to the UPGMA, neighbor-joining, and maximum parsimony phylogenetic approaches. In addition, we discuss the joint application of phylogenetic techniques and geographic source in the study of phylogeography. Next, we discuss paternity analysis and assignment, both for humans and for applications to natural populations of other organisms. Further, we discuss genetic determination of individual identity and its uses in humans for forensics and criminal identification and its application for identity of wild individuals to specific natural source populations. Finally, we provide a brief introduction to the linkage disequilibrium approaches used to identify specific genes that influence quantitative traits, or QTLs. We also discuss the use of haplotypes, and the HapMap, in human genetics and how they may be used to characterize the spatial pattern of genetic variation over the genome and potentially to identify genes involved in common diseases and other quantitative traits.

I. MOLECULAR PHYLOGENETICS

Analyzing DNA sequence data from an evolutionary perspective has resulted in the fast-growing field of molecular phylogenetics, which has as a main focus the reconstruction of evolutionary relationships for both genes and species. Here we provide an introduction to this rather technically sophisticated area and provide an understanding for some of its widely used concepts. The similarity and differences in DNA sequence among different organisms may be used to understand evolutionary relationships. Presumably, organisms that have similar DNA sequences are closely related, and those that have quite different sequences are only distantly related. Such predictions would be most accurate if the differences in DNA sequence among species were solely a function of the time since the divergence of the species; that is, if the evolution of these sequences is governed by the accumulation of changes by mutation and genetic drift as expected under the neutral theory.

The availability of DNA sequence data in many organisms provides a database to determine the phylogenetic relationships between species or other taxa that were not clear from other traits. It is generally assumed that DNA sequence data better reflect the true phylogenetic relationships between taxa than other data, such as morphology, because sequence data are less influenced by selective effects. Furthermore, differences between phylogenetic trees generated from molecular data and from other traits provide an opportunity to evaluate the effect of selective effects on other traits.

Although the detailed techniques of constructing phylogenetic trees from molecular data are beyond the scope of this book (for reviews, see Felsenstein, 1988; Nei, 1996; for advanced treatments, see Nei and Kumar, 2000; Felsenstein, 2004), we show the logic behind several of the commonly used approaches. A basic understanding of the methodology of tree building gives an appreciation of what information phylogenetic trees convey and how they may contribute to population genetics (Hillis *et al.*, 1994; Papadopoulos *et al.*, 1999). Phylogenetic trees consist of **nodes** and **branches** connecting the nodes (see examples below). External or terminal nodes indicate the taxa under consideration, whereas internal nodes connect related taxa. Often the lengths of branches are scaled to represent the amount of divergence between the taxa they connect.

The branching pattern of a phylogenetic tree is called its **topology**. Trees may be broadly categorized into **rooted trees**, those with an inferred common ancestor and therefore a given evolutionary direction, and **unrooted trees** that reflect no such assumptions. To provide a root for a tree, an **outgroup** is generally used—that is, a taxon that, from paleontological or other data, is assumed to have separated earlier than the taxa under study. With an appropriate outgroup, the correct temporal sequence of evolutionary divergence from the ancestral lineage is then apparent. Several large simulation studies to determine the ability of different approaches to recover trees with known topologies (Sourdis and Nei, 1988; Hillis, 1996) have demonstrated that when there is good statistical confidence in the topology of the chosen tree, then the major methods all work well and are reasonably comparable.

One of the important uses of genetic differences between genes, populations, or species is their organization in a biologically meaningful way. An arrangement of genes or taxa, which generally have a root or ancestral type at one side and the branches or evolutionarily derived types on the other, can illustrate putative ancestry or recently derived types. A major challenge is to find the correct topology of branching arrangements for a given array of groups, particularly when there are a number of groups. As the number of groups increases, the number of different rooted topologies increases very quickly; for 3 groups there are 3 different topologies, for 4 groups there are 15, for 5 groups there are 105, and for 10 groups there are over 34 million different possible rooted topologies (Felsenstein, 1978). Obviously, computers are needed to examine different trees when there are many groups, and an efficient approach to find the best fitting tree is important.

Before we proceed to a discussion of methods of phylogenetic tree construction, it is important to differentiate between **gene trees** and **species trees** (Pamilo and Nei, 1988; Nei and Kumar, 2000). A gene tree is constructed with DNA sequence data for a given gene, obtained from both variation within species for the gene and for homologous sequences in different species. A species (or population) tree represents the evolutionary history of a group of species. On a species tree, the splitting of one species into two species indicates the time of origin of the species. When a gene tree is reconstructed from a gene sampled from different species, this tree may not completely agree with the species tree. For example, if the genes are polymorphic within the species that diverged, then the times of divergence in the gene tree may be greater than the times of divergence of the species. In addition, the branching pattern of a gene tree may be different from that of the species tree.

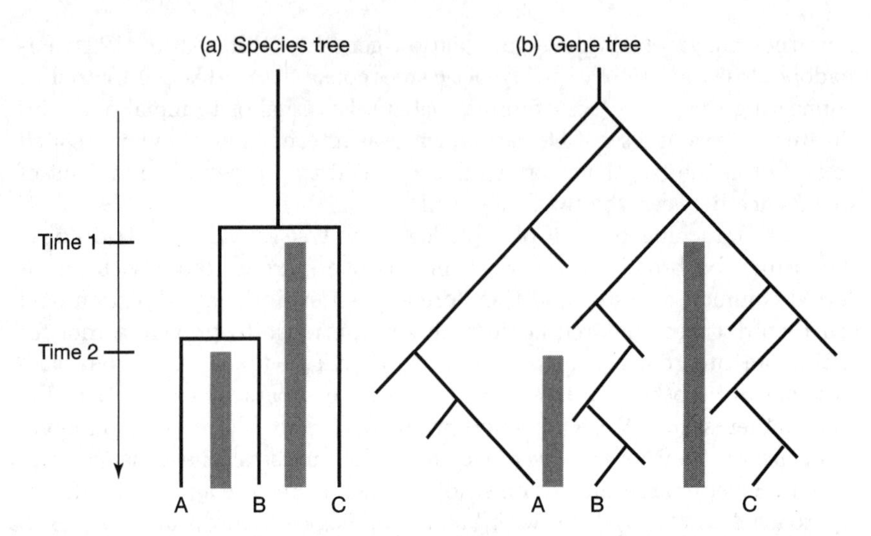

Figure 11.1. (a) A species tree that results from one barrier (shaded bar) splitting the ancestral species at time 1 and another barrier splitting part of the species at time 2. As a result, species A and B are most closely related. (b) A gene tree for a gene in the same species in which the ancestral species is polymorphic for several sequences. By chance, the surviving sequences in species B and C are more closely related than the surviving sequences in species A and B (after Kocher, 2003).

Let us illustrate this potential difference by examining a situation in which a species tree and a gene tree have different topologies For example, let us assume that at time 1 a physical barrier divides an ancestral species into two groups and that at time 2 a second barrier divides one of the groups again into two (Figure 11.1). As a result, there are three species A, B, and C, and a species tree has species A and B as the most closely related species because they split most recently, and species C is more distantly related to them (Figure 11.1a). On the other hand, let us assume that the ancestral species was polymorphic for several different sequences at a gene. By chance, present-day species B and C have sequences that are more closely related to each other, that is, diverged more recently, than do species A and B as shown in the gene tree in Figure 11.1b.

We introduce three different methods for construction of phylogenetic trees. The **unweighted pair group method (UPGMA)** was proposed originally by Sokal and Sneath (1963) for analyzing morphological data. UPGMA is an easy approach to understand, but it assumes a constant evolutionary rate over time for all lineages. The **neighbor-joining (NJ)** method, developed by Saitou and Nei (1987), is based on the minimum evolution principle. Both of these methods are called distance matrix methods and depend on a genetic distance measure between all of the pairs of taxa under examination. The third method utilizes **maximum parsimony**, an approach that finds the tree that requires the least number of mutational changes necessary to arrange the taxa. Another approach, which we do not discuss but is also widely used, is the **maximum likelihood method** (Felsenstein, 1981, 2004). In this method, the likelihood of observing a given group of sequences is maximized for each tree topology, and the topology with highest likelihood is chosen as the tree of choice. Related to likelihood methods, **Bayesian methods** that incorporate other (prior) information are also used to infer phylogenies (Huelsenbeck *et al.*, 2001; Felsenstein, 2004).

a. Unweighted Pair Group Method (UPGMA)

The UPGMA is an example of an approach that uses distance information (sometimes called phenetics), whereas maximum parsimony uses character states (sometimes called cladistics). In traditional taxonomy, proponents of these two approaches often have had virulent debates. For molecular data, in which the basic data are the presence of different nucleotides, some have suggested that approaches that use character-state methodology are more appropriate. However, because some nucleotide sites are not informative (see below), they are excluded from character-state approaches. Distance approaches, however, incorporate information from all of the variable sites (Saitou and Nei, 1987).

To illustrate the construction of a phylogenetic tree, let us assume that there are a number of **operational taxonomic units (OTUs)**, indicated by 1, 2, 3, . . ., that could be populations, species, or other biological groups. Assume also that the genetic distances between these groups have been measured using allele frequencies, nucleotide substitutions, restriction map information, or other data so that a matrix of pair-wise genetic distances is available where D_{ij} is the genetic distance between groups i and j. As an example, let us say that there are four OTUs and the genetic distances are as given in Table 11.1a. Now our goal is to organize these OTUs in a way that reflects biological relationships using these genetic distance values.

Using UPGMA to determine this organization, we begin by clustering the two groups that have the smallest genetic distance, combining them into a single new OTU. Then we calculate new estimates of genetic distance between this new unit and the other remaining units by taking the arithmetic average of their distances. For example, assume that there are four different OTUs and that populations 1 and 2 have the smallest genetic distance for all of the comparisons. These two can be combined into a new unit (12). The genetic distance between this unit and 3 or 4 is then

$$D_{(12)3} = \frac{1}{2}(D_{13} + D_{23})$$

and (11.1)

$$D_{(12)4} = \frac{1}{2}(D_{14} + D_{24})$$

TABLE 11.1 The genetic distance (a) between all pairs of four groups and (b) between three groups after groups 1 and 2 have been clustered.

(a)	*1*	*2*	*3*
2	D_{12}	—	—
3	D_{13}	D_{23}	—
4	D_{14}	D_{24}	D_{34}

(b)	*(12)*	*3*
3	$D_{(12)3}$	—
4	$D_{(12)4}$	D_{34}

There is now a new matrix of distance values using the three remaining OTUs, (12), 3, and 4; this matrix is given in Table 11.1b. Assume that D_{34} is the smallest genetic distance remaining so that a second cluster of 3 and 4 is formed. The last step in this simple example is to calculate the average distance between the two clusters, which is

$$D_{(12)(34)} = \frac{1}{2}(D_{(12)3} + D_{(12)4})$$

It is useful to give a numerical example of this procedure so that the branch lengths have a particular value and the phylogenetic tree can be drawn to scale. Table 11.2 gives mtDNA distances from an example provided by Nei and Kumar (2000) using original data from Brown *et al.* (1982). These data are from five primates, including humans, and were obtained to determine the relationship of these species. The distance values are based on the proportion of different sites in an 895-bp sequence, and the original data for the first 240 base pairs are given in Table 11.3. The proportion of different sites for each pair of species is then converted to a measure of the number of substitutions between the sequence using a formula similar to expression 8.4a but one that allows different rates of substitution for transitions and transversions, called the two-parameter model (Kimura, 1980) (see Figure 8.4 on p. 423). It is assumed that a single gene sequence represents each one of the different species.

First, note that D_{12}, the distance between humans and chimpanzees, is the smallest distance value of all the comparisons. Therefore, 1 and 2 are

TABLE 11.2 The estimated numbers of nucleotide substitutions for an 895 bp mtDNA sequence from five primate species (data from Brown, 1982, and estimation from Nei and Kumar, 2000).

	(1) Human	*(2) Chimpanzee*	*(3) Gorilla*	*(4) Orangutan*
(2) Chimpanzee	0.095	—	—	—
(3) Gorilla	0.113	0.118	—	—
(4) Orangutan	0.183	0.201	0.195	—
(5) Gibbon	0.212	0.225	0.225	0.222

```
Human       AAGCTTCACCGGCGCAGTCATTCTCATAATCGCCCACGGACTTACATCCTCATTACTATTCTGCCTAGCAAACTCAAACT    80
Chimpanzee  . . . . . . . . . . . . . . . . .A.T. .C. . . . . . . . . . . . . . . . . . . . . . . . . . . . . . . .T. . . . . . . . . . . . . . . . . . . . . .T.
Gorilla     . . . . . . . . . . . . . . . . . . .TG. . . .T. . . . .T. . . . . . . . . . . . . . . .A. . . . . . .T. . . . . . . . . . . . . . . . . . . . . . .
Orangutan   . . . . . . . . . . . . . . . . .AC. .CC. . . . .G. .T. . . . . T. . . . .C. . . . . . . . .CC. . . .G. . . . . . . . . . . . . . . . . . . . .
Gibbon      . . . . . .T. .A. .T. . .AC.G.C. . . . . . . . . . . . . . . . . . . . .A. .C. .T. .CC.G. . . . . . . . . . . .T. . . . . . . . . . . . .

Human       ACGAACGCACTCACAGTCGCATCATAATCCTCTCTCAAGGACTTCAAACTCTACCCACTAATAGCTTTTTGATGACTT    160
Chimpanzee  .T. . . . . . . .C. . . . . . . . . . . . . . . . . .T. . . . .C. . . . . . . . . . . . . . . . . . . . . . . . . . . . . . .C. . . . . . . . . . .C
Gorilla     . . . . . . . .A. .C. . . . . C. . . . . . . . . . . .T. . . . . . . . . . . . . . . .C. . . . .C. . . . . . . . . . . . . . . .CC. . . . . . . . . .
Orangutan   . . . . . . . .A. .C. . . . . C. . . . . . . . . . . . . . . . . . . . . . . . .C. . . . . . . . . . . . . . . . .C. . . . . . . .CC.C. . . . . . . . .
Gibbon      . . . . . . . .A. . . . . . . . C. . . . . . . . . . . . . . . .A. . . .G. . .G. .C. . .G.CT. . . . . . . . .G. . . . .C. .C. . . . . . . .C

Human       CTAGCAAGCCTCGCTAACCTCGCCTTACCCCCCACTATTAACCTACTGGGAGAACTCTCTGTGCTAGTAACCACGTTCTC    240
Chimpanzee  . . . . . . . . . . . . . . . . . . . . . . . . .C. . . . . . . .T. .C. . . . .T. .C. .A. .G. . . . . . . . .C. . . . . . . . . . . . .T.A. . . . .
Gorilla     . .G. . . . . . . . . . . .C. . . . . . . . . . . . . . . . . . . . . . .C. . . . . . . . . . . . .A. . . . . .G. . . . .C. .A. . . . . . . . . . . . .A. . . . .
Orangutan   . . . . . . . . . . . . . . .A. . . . . . . . .T. . .C. . . . .A. . . . .C. .C. . . . . .T. .A. . . . . . . . . . . . . .C. .A. . .A. .G. . .TA. . . . .
Gibbon      GC. . . . . . . . . . . . . . . . . . . . . . .C. . . . . . . . . . . . . . . . . . . . . .C. .A. .T. . . . . . . .TC. .A. . .A.GG. .T.C. . . . .
```

Table 11.3 The first 240 bp of the 895 bp used from mtDNA of five primate species to construct the phylogenetic trees in Figure 11.2 (data from Brown *et al.*, 1982)

TABLE 11.4 The number of nucleotide substitutions, D_{ij} values, between the five primate species in Table 11.2 after (a) species 1 and 2 are clustered, (b) after species 3 is clustered with cluster (12), and (c) after species 4 is clustered with cluster (123), using the UPGMA algorithm. Boldface indicates the lowest values.

(a)

	(12)	*3*	*4*
3	**0.115**	—	—
4	0.192	0.195	—
5	0.218	0.225	0.222

(b)

	(123)	*4*
4	**0.194**	—
5	0.222	0.222

(c)

	(1234)
5	0.222

clustered; the distances between this cluster and the other three taxa are calculated using expression 11.1, and the reduced matrix in Table 11.4a is obtained. Then, $D_{(12)3}$ is the smallest of the distance values in this matrix so that a new cluster 123 is formed. The distance between this cluster and taxa 4 and 5 is calculated and is given in Table 11.4b. Finally, because the distance between (123) and 4 is the smallest, these taxa form a new cluster. Table 11.4c gives the distance between cluster (1234) and 5.

The phylogenetic tree for these data is given in Figure 11.2a, where the numbers indicate branch lengths. The branch lengths are half of the genetic distance between two groups because evolutionary change in UPGMA is assumed to occur equally in all lineages. For example, the distance between humans and chimpanzees is 0.095, but the branch length from the time at which they split from a common ancestor is half (0.048) in each branch. Next, the distance from the human–chimpanzee cluster to gorillas is 0.115 (Table 11.3a). Taking half of this, 0.058, gives the time to this common ancestor, and the distance from here to the human–chimpanzee split is then 0.058 − 0.048 = 0.010. The same approach can be used to generate the other distances in the tree.

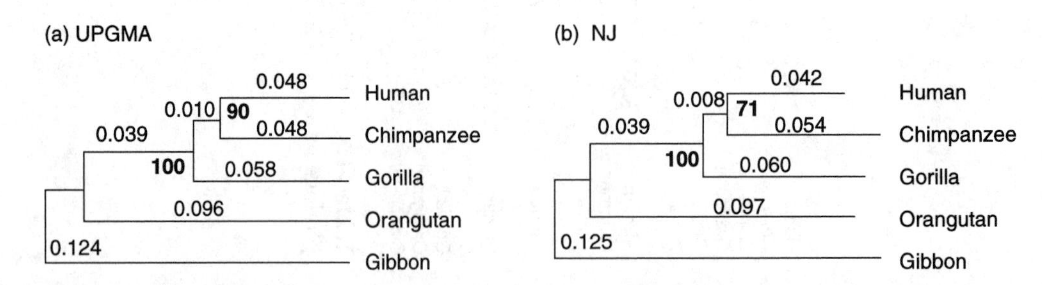

Figure 11.2. (a) UPGMA and (b) NJ trees for the mtDNA data given in Table 11.2. The numbers indicate branch lengths, and the boldface numbers indicate bootstrap confidence values for the appropriate nodes (Nei and Kumar, 2000).

b. Neighbor-Joining (NJ) Method

The neighbor-joining (NJ) method has become the distance approach of choice for many types of molecular data, partly because, unlike the UPGMA method, it can incorporate different rates of evolution in different lineages. Although NJ may sound similar to the approach used in UPGMA, and it does proceed in a stepwise fashion, what is minimized at each step is the sum of the branch lengths in the total tree. To start to find a NJ tree, we first assume that there is a star tree in which there is no clustering (Figure 11.3a). The first calculation is to determine the total length of this tree when all pairs of taxa are potentially nearest neighbors (Nei and Kumar, 2000). For example, when taxa 1 and 2 are nearest neighbors, the total length is

$$S_{12} = \frac{2T - R_1 - R_2}{2(N-2)} + \frac{D_{12}}{2} \tag{11.2a}$$

$$T = \sum_{ij} D_{ij} \quad \text{and} \quad R_1 = \sum_{i}^{N} D_{1i}$$

where N is the number of OTUs.

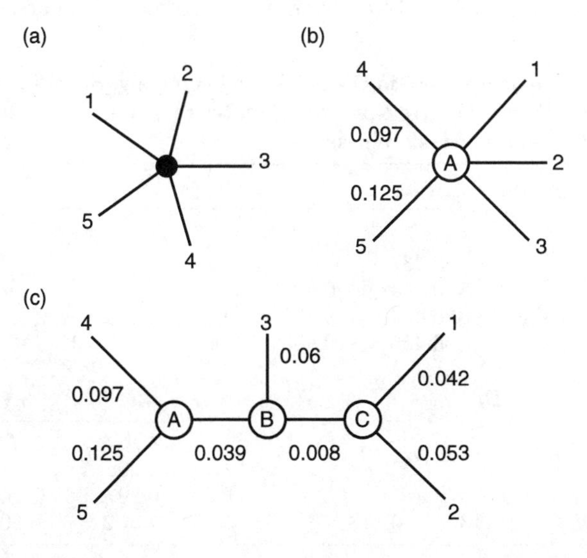

Figure 11.3. The sequential trees used in constructing the neighbor-joining tree, where (a) is the initial star tree, (b) is the tree after the first step, and (c) is the final tree.

All of the S_{ij} values are calculated, the smallest is determined, and a new node A that connects the closest taxa i and j is inserted. The branch lengths from this node A to taxa i and j are calculated as

$$b_{Ai} = \frac{1}{2(N-2)}[(N-2)D_{ij} + R_i - R_j]$$

and (11.2*b*)

$$b_{Aj} = \frac{1}{2(N-2)}[(N-2)D_{ij} + R_j - R_i]$$

Then the distances between the new node and the remaining k taxa are calculated as

$$D_{Ak} = \frac{1}{2}(D_{ik} + D_{jk} - D_{ij}) \quad (11.2c)$$

The whole sequence is repeated until there are only four taxa left and the lengths of the remaining branches can be calculated.

To illustrate this process, let us use the data from Table 11.2 again. First, using the expressions above, we can calculate $T = 1.789$, $R_1 = 0.603$, $R_2 = 0.639$, $R_3 = 0.651$ $R_4 = 0.801$, and $R_5 = 0.884$. Then by substitution of these values into expression 11.2*a* for different pairs of taxa, we get the array of S_{ij} values in Table 11.5a. The lowest value is that between taxa 4 and 5, and thus, we create a node A between these taxa (see Figure 11.3b). This initial joining is different from that in the UPGMA algorithm and is at first counterintuitive. However, it is understandable when we realize that the quantity being examined for a minimum is different; for UPGMA, it is the minimum distance between taxa, and for NJ, it is the minimum total branch length for the tree, given a pair of taxa are nearest neighbors.

TABLE 11.5 The arrays used in the neighbor-joining algorithm for the data given in Table 11.2 for all five taxa (a) and after taxa 4 and 5 are joined (b). Boldface indicates the lowest values.

(a)

	S_{ij}			
	1	*2*	*3*	*4*
2	0.437	—	—	—
3	0.444	0.440	—	—
4	0.454	0.457	0.452	—
5	0.455	0.455	0.453	**0.427**

(b)

	D_{ij}				S_{ij}		
	A	*1*	*2*		*A*	*1*	*2*
1	0.087	—	—	*1*	0.205	—	—
2	0.102	0.095	—	*2*	0.207	**0.202**	—
3	0.099	0.113	0.118	*3*	**0.202**	0.207	0.205

The branch lengths from expression 11.2*b* are $b_{A4} = 0.097$ and $b_{A5} = 0.125$. The new distances between node A and the other taxa can be calculated from expression 11.2*c* and are given on the left part of Table 11.5b. Now the process of calculating new S_{ij} values is repeated with these new distances (new T and R_i values based on these distances are calculated), resulting in the values on the right part of Table 11.5b. When there are four taxa remaining, there are two equally likely clusters using NJ, and in this case, they are $A3$ and 12. A new node B is created connecting A and 3 and the branch length from taxa B to 3, $b_{B3} = 0.060$. A new node C is created connecting 1 and 2 and the branch length from C to 1 is $b_{C1} = 0.042$ and from C to 2 is $b_{C2} = 0.053$. Finally, the distance from node B to node C is $D_{BC} = 0.008$. All of these values are incorporated into Figure 11.3c (which is not drawn to scale). Finally, the NJ tree can be drawn to scale, omitting the nodes as in Figure 11.2b. In this case, the topology is the same as the UPGMA tree, and the branch lengths are also very similar.

When there are more than a few taxa, then there are a very large number of tree topologies so that determining the statistical support various parts of a tree becomes critical. The most common approach is to determine the level of statistical confidence in a particular node of the tree by calculating **bootstrap** values (Felsenstein, 1985, 2004). Bootstrapping is an approach that resamples the original data to get significance levels. In this case, a sample is randomly obtained, a phylogenetic tree is drawn with this sample, and the tree is examined to see what clusters of OTUs it contains. Generally, this process is repeated many times, say 1000, and the percentage of times that a given cluster is present is indicated on the tree. For example, bootstrap values given in Figure 11.2 (from Nei and Kumar, 2000) indicate that for both trees, the cluster of humans, chimpanzees, and gorillas is highly supported (100% of the resampled trees had this cluster). The cluster of humans and chimpanzees is strongly supported (90%) for the UPGMA tree and not as much for the NJ tree (71%). Example 11.1 discusses the NJ tree for 45 complete human mtDNA sequences and illustrates that mutation is a significant factor influencing the phylogenetic history of the tree and linkage disequilibrium.

Example 11.1. Awadalla *et al.* (1999) published a surprising report that suggested that there was evidence of recombination in 45 complete human mtDNA sequences because there was lower linkage disequilibrium between more distant nucleotide sites. There were a number of criticisms of this conclusion (they have since retracted their conclusion; McVean *et al.*, 2002; Piganeau and Eyre-Walker, 2004), but examining the phylogenetic tree of the sequences illustrates that the observed pattern is consistent with mutation (Kumar *et al.*, 2000; Hedrick and Kumar, 2001).

Figure 11.4 (Kumar *et al.*, 2000) presents the NJ tree of the 45 mtDNA sequences (with 22 unique haplotypes) analyzed by Awadalla *et al.* (1999).

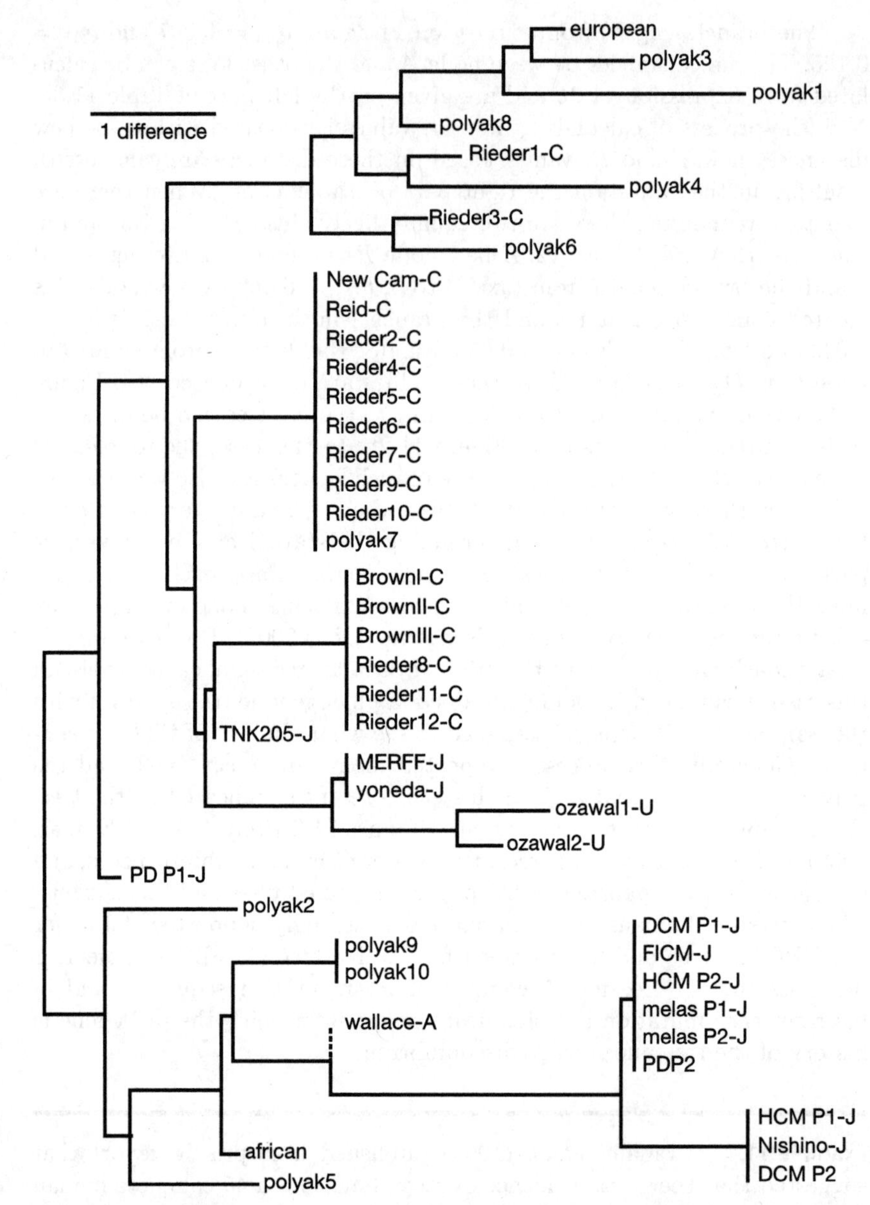

Figure 11.4. A NJ tree of 45 complete human mtDNA sequences based on the number of differences observed at the 14 sites analyzed (Kumar *et al.*, 2000). Branch lengths denote the actual number of differences per sequence. Identical sequences are joined by a vertical line, and the wallace-A sequence (connected by a dashed line) has missing data at 3 of the 14 sites.

Mapping the nucleotide substitutions on the phylogenetic tree reveals unique transitional changes at four nucleotide sites, parallel (independent) transitional changes at eight sites, and backward changes at two sites. For example, one of the pairs of nearby sites that shows high linkage disequilibrium (14,783 and 15,043) appears to be the result of unique transition mutations at two sites that are both present in the adjacent groups of six and three sequences at the bottom right of the tree. The rest of the haplotypes are of the ancestral type, resulting in 35 gametes of the ancestral coupling type (wallace-A sequence is not included) and 9 descendant gametes of the other coupling type, with consequently high linkage disequilibrium.

One of the pairs of distant sites that have low linkage disequilibrium (7028 and 14,783) appears to be the result of one of the same unique transition mutations at 14,783 and two parallel transition mutations at 7028 in the group of 10 sequences and the polyak 2 sequence. In this case, there are 24 sequences of the ancestral gamete, nine sequences of one repulsion gamete, 11 sequences of the other repulsion gamete, and none of other coupling gamete. This results in low linkage disequilibrium using the r^2 measure (Awadalla *et al.* 1999), but the presences of only three of the four possible gametes (and $|D'| = 1$) do not provide evidence of recombination. In other words, the pattern of linkage disequilibrium is consistent with the phylogenetic relationships of the sequences and suggests that the history of mutation, not distance-dependent recombination between pairs of sites, is the cause of the observed associations of nucleotides at pairs of sites in different haplotypes.

c. Maximum Parsimony

Maximum parsimony is a method that identifies the tree (or trees) that involves the smallest number of mutational changes necessary to explain the differences among the groups under investigation (Fitch, 1977). Such a goal is obviously appealing, and such trees often are very enlightening. However, as we show below in a simple example, there may be more than one tree with the same minimum number of mutations. In addition, finding the most parsimonious tree does not guarantee that it is the correct tree because if the rates of substitution vary along different lineages, parsimony may not give the correct tree topology.

To construct a maximum parsimony tree, we first have to identify **informative sites**, sites that support one phylogenetic tree over another. To illustrate the maximum parsimony method, let us use four taxa with the sequences given in Table 11.6. For four taxa, there are three possible

TABLE 11.6 Hypothetical sequences at seven nucleotide sites from four different taxa to illustrate maximum parsimony. Informative sites are indicated with *.

	Site						
Taxon	*1*	*2*	*3*	*4*	*5**	*6**	*7**
1	T	T	C	G	T	G	A
2	T	C	T	T	T	T	A
3	T	C	A	C	C	T	C
4	T	C	A	A	C	G	C

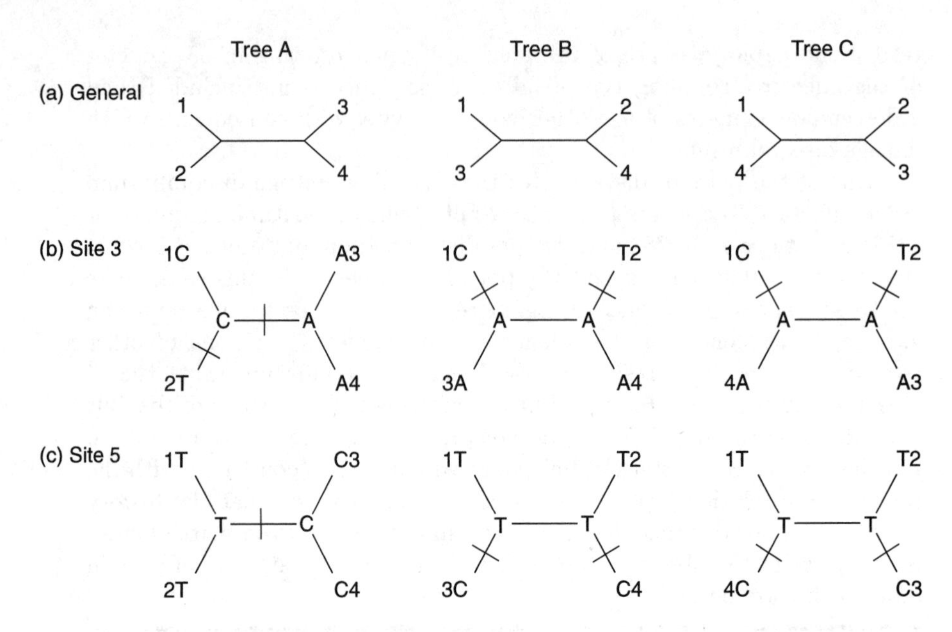

Figure 11.5. (a) The three possible, unrooted trees when there are four taxa. Examples of the least number of substitutions, indicated by slashes, for each tree given the nucleotide sequences in Table 11.6 for site 3 (b) and site 5 (c). The nucleotides at the terminal nodes are the ones given for each taxon in Table 11.6.

unrooted trees (Figure 11.5a), and our goal is to determine whether one of these trees is a more parsimonious explanation with the given data than are the others.

Let us go through these sites one by one to illustrate the possible types of data. Site 1 is not informative because all sequences are identical with T and thereby require no changes. Site 2 is also not informative because a single mutation in the branch leading to taxon 1 for any of the three trees will give this pattern. Site 3, although there are three different nucleotides present, is also not informative. Figure 11.5b shows that for all three possible trees, two changes are the minimum necessary to explain the relationships. (To work through this, begin with the nucleotides observed in each of the four taxa, and fill in the changes and the subsequent nucleotides at the nodes between them.) On the other hand, site 5 is an informative site (Figure 11.5c). In this case, tree A requires only one change between the two internal nodes of the tree, whereas trees B and C both require two changes. Site 6 is also informative, and for it, tree C requires only one change, whereas trees A and B require two. Finally, site 7 is like site 5 in that tree A requires only one change and trees B and C require 2. In general, for a site to be informative, it has to have at least two different nucleotides, and each of these nucleotides has to be present at least twice.

To determine the maximum parsimony tree, the informative sites are first identified, and then the minimum of changes necessary for each possible tree is calculated for each of these sites. The minimum number of changes over all of the informative sites is then summed for each of the trees, and the tree with smallest number of substitutions is the most parsimonious tree. In the example above for the informative sites, tree A required four substitutions, tree B required six, and tree C required five. Therefore, on the basis of these data, tree A is the most parsimonious. For a more complex tree based on parsimony principles, see Figure 1.8, which uses mtDNA data in pocket gophers.

d. Phylogeography

One of the first applications of molecular phylogenetic approaches was to determine simultaneously both the phylogenetic and spatial relationships among different mtDNA haplotypes or sequences. For example, Figure 1.8 (p. 37) presented the spatial distribution of pocket gopher mtDNA haplotypes organized in a phylogenetic network (Avise *et al.*, 1979), illustrating the close connection between genetic variants and geographic source. In the simplest framework, spatial distribution would be expected to mimic a temporal pattern; that is, sequences most distant spatially should also be most different in sequence just as sequences that diverged the longest time ago should also be most different in sequence. That is, populations distant in space, with very limited or no gene flow between them, should accumulate differences due to the effects of genetic drift and mutation, as well as selection. However, the connection between geographic distance and genetic patterns may be obscured by usually unknown nonselective events in the past, such as founder events and different gene flow patterns, as well as selective changes that may spread over populations with even minimal gene flow connection.

Avise (2000) termed the joint use of phylogenetic techniques and geographic distributions **phylogeography** and defined it as the "field of study concerned with the principles and processes governing the geographical distributions of genealogical lineages . . ." and suggested that "time and space are the jointly considered axes of phylogeography onto which (ideally) are mapped particular gene genealogies. . . ." Figure 11.6 gives a hypothetical gene genealogy illustrating restricted gene flow between two different regions separated by a dispersal barrier. In this example, the population split when the barrier arose, and then the divided populations separately evolved and dispersed within the two different regions.

Maternally inherited mtDNA data have been the workhorse for phylogeographic research because mtDNA does not recombine in most organisms and, as a result, shows a clearer phylogenetic record than many nuclear genes (chloroplast DNA and Y chromosomes theoretically are similarly

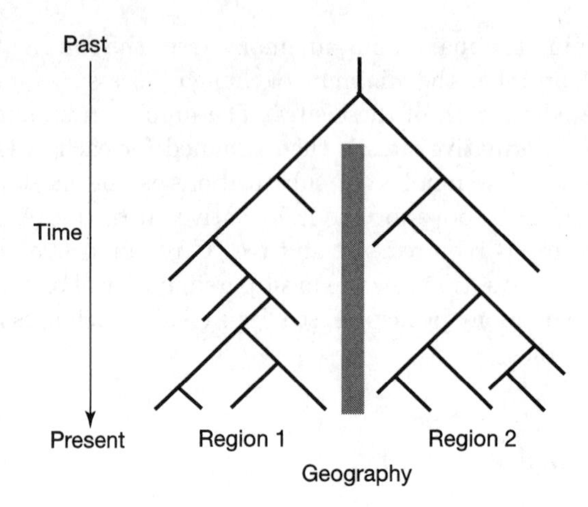

Figure 11.6. A gene tree for a species that has been split by a barrier (shaded bar) that restricts gene flow between regions 1 and 2. As a result, there is concordance between the phylogenetic tree and geographic region (after Avise, 2000).

useful). In addition, the effective population size for mtDNA (as well as for chloroplast DNA and Y chromosomes) is only approximately one-fourth that of nuclear genes (see p. 329) so that divergence occurs about four times faster than for nuclear genes. However, this faster rate of divergence and potentially differential gene flow for the two sexes may cause the signal for these uniparentally inherited genes to be different than the phylogenetic pattern for nuclear genes, which constitute a very large proportion of the genome. As a result, nuclear gene phylogeography over multiple independent loci may provide a more complete picture of the phylogenetic history of species and populations (Hare, 2001). For discussion of some of the theoretical and statistical aspects of phylogeography, and the examination of specific genetic and demographic processes, see Edwards and Beerli (2000), Knowles and Maddison (2002), and Knowles (2004).

Traditionally, it is assumed that the spatial differentiation of groups occurs by either the evolutionary scenarios of **dispersal** or **vicariance**. A dispersal interpretation of a present-day distribution suggests that it was the result of active or passive movement of the organism from an ancestral center of origin into new areas. A vicariance interpretation suggests that past geological or environmental events split the more-or-less continuous range of an organism. For example, the rise of a mountain range or the formation of a canyon could divide the distribution of terrestrial organisms, or the division of a watershed or body of water could split the range of aquatic organisms. Figure 11.7 illustrates how populations or species can be correlated with geographic variation as the result of dispersal or vicariance. In this example, there is complete concordance of the phylogeny of the taxa with the area under the vicariance model, whereas under the dispersal model, there is a strong but less concordant relationship of the phylogeny with geography.

As we discussed on p. 452 in the section on coalescence, specific lineages may be lost by chance due to genetic drift over time, a phenomenon known

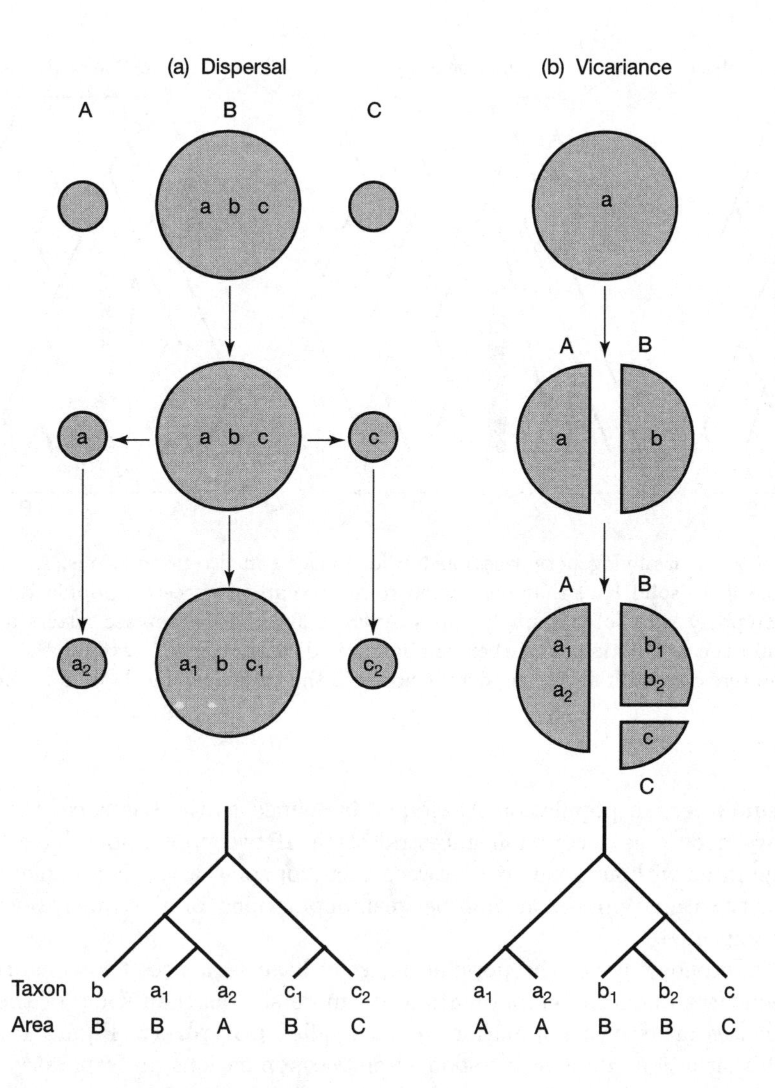

Figure 11.7. The phylogenetic relationships of spatially disjunct populations or species under (a) dispersal or (b) vicariance models (after Avise, 2000). The horizontal arrows in the dispersal model indicate movement into new areas from a center of origin. The open areas dividing up the circle in the vicariance model indicate barriers that have divided the range of the organism.

as **lineage sorting**. If we examine a nonrecombining part of the genome, such as mtDNA in most organisms, there are three general phylogenetic patterns for extant lineages in two populations that have descended from a common ancestral population. First, both of the two different populations may have lineages that are more closely related to all other lineages within the population than to any lineages in the other population, a state called **reciprocal monophyly**. This is illustrated in Figure 11.8a where, for example, lineages a1 and a2 in population A are closer to each other than they are to any sequence in population B (similarly, lineages b1 and b2 in population B are closer to each other than they are to any sequence in population A). All of the present-day sequences in population A are members of the same **clade**, that is, ***a group of phylogenetically clustered***

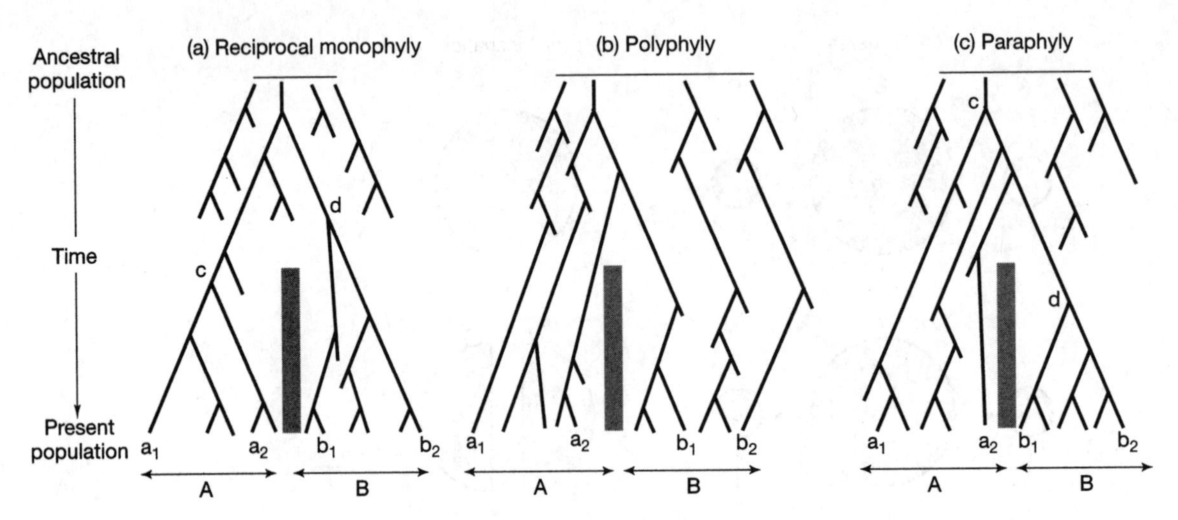

Figure 11.8. The three general types of phylogenetic relationship for lineages in two populations, A and B, separated by a barrier to gene flow (solid bar) that descended from a common ancestral population: (a) reciprocal monophyly, (b) polyphyly, and (c) paraphyly (after Avise, 2000). The lowercase letters a and b indicate the sequences within the populations and their similarity is discussed in the text for the three relationships. The lowercase letters c and d indicate important nodes in the trees and the clades descendant from these nodes.

sequences. In population A, clade c is defined as the sequences that all have node c as a common ancestral state. If two populations have been separated without gene flow between them for some time, then higher similarity within population than between populations for all sequences would be expected.

Second, both of the populations may have sequences for which some sequences in the other population are more similar than some of the sequences in its own population, a state called **polyphyly** (Figure 11.8b). This situation could result soon after two populations are separated and there has not been enough evolutionary time for the lineages to sort independently in the two populations (or if there is gene flow between the populations). For example, in Figure 11.8b, members of one of the lineages are present in both populations, and these are more similar to each other (for example, a2 in population A and b1 in population B) than some other sequence pairs within the same population, such as lineages a1 and a2 in population A. The third category is **paraphyly**, a state that can generally be thought of as the transitional stage between polyphyly and reciprocal monophyly. In this relationship, all of the lineages within one of the populations form a monophyletic group (the sequences in population B in Figure 11.8c) nested within the broader phylogenetic group, which contains all of the lineages in both present-day populations. That is, clade c encompasses all present-day sequences in both populations, and clade d, which encompasses all population B sequences, is a monophyletic group. Example 11.2 presents mtDNA data on disjunct populations in the tassel-eared squirrel that illustrates the potential impact of both vicariance and dispersal and the presence of reciprocal monophyly.

Example 11.2. The disjunct populations of the tassel-eared squirrel, *Sciurus aberti*, exist in the ponderosa pine forests in the higher elevations of the southwestern United States and northern Mexico, separated by lower elevation, unsuitable habitat. To examine the relative contributions of vicariance and dispersal hypotheses to the present-day distributions, Lamb *et al.* (1997) examined restriction site mtDNA variation from 22 populations range-wide. They found extensive variation between populations, and the 21 different observed haplotypes were distributed into distinct eastern (Mexico to Colorado and Utah) and western (Arizona and western New Mexico) reciprocally monophyletic clades (Figure 11.9).

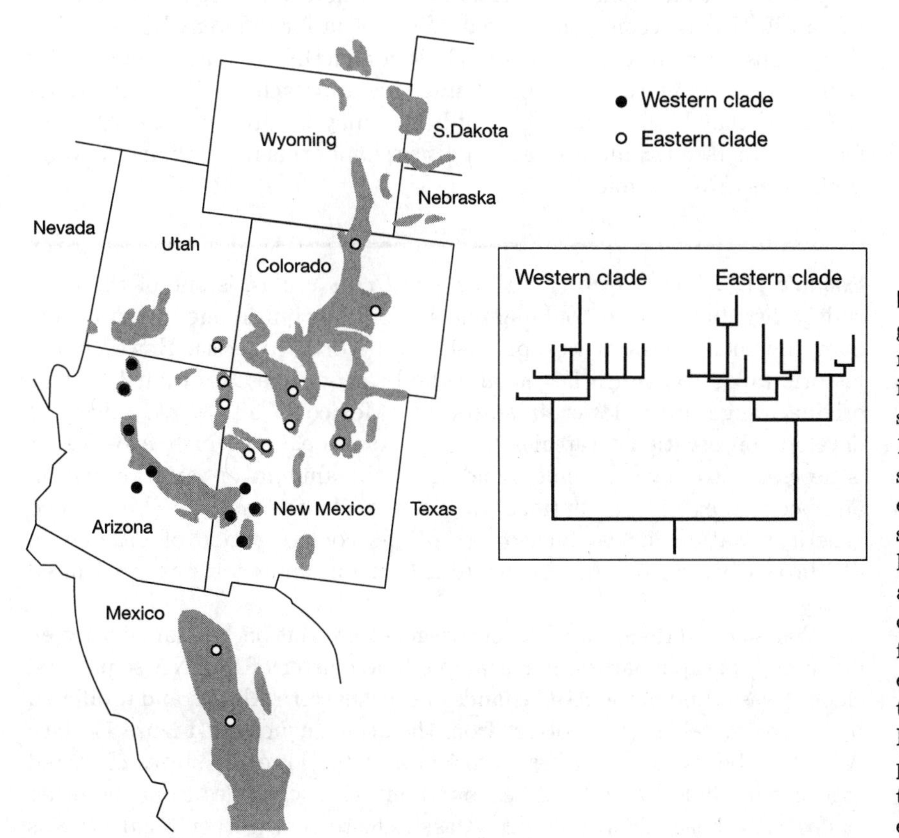

Figure 11.9. A phylogeographic pattern of mtDNA haplotypes in the tassel-eared squirrel (Lamb *et al.*, 1997; Avise, 2000). The shaded area shows the extent of present-day suitable ponderosa pine habitat, and the closed and open circles indicate sampling locations for the western and eastern clades, respectively. To the right is a NJ tree illustrating the phylogeographic pattern of the western and eastern clades.

One vicariance hypothesis suggested that the present distribution of ponderosa pines and tassel-eared squirrels reflects the remnant distribution of a vast late-Pleistocene (10,000 to 20,000 years ago) pine forest. However, information from packrat middens has shown that ponderosa pines have existed in only the northern regions of this area in the last 10,000 years. Therefore, it appears that northern dispersal has been important in determining the northern parts of the current distribution. On the other hand, the 1.8% estimated mtDNA divergence between the eastern and western clades suggests that much earlier vicariance events were responsible for the divergent evolution of these two lineages.

The distribution of genetic variation has been used to identify units of endangered species, whether they are populations, groups of populations, or species, that are important to protect from extinction. Waples (1991b) described the concept of the **evolutionarily significant unit (ESU)**, based on the ecological, historical, and genetic uniqueness of the group, as an entity that should be provided protection. Moritz (1994) subsequently suggested that an explicit phylogenetic definition would be useful to identify an ESU and recommended that the criteria for different ESUs is that the groups be reciprocally monophyletic for mtDNA sequences and differ significantly for nuclear variants. Using this approach, Waits *et al.* (1998) suggested that brown bears (grizzly bears) may be divided into three different ESUs (see Example 11.3 for a discussion of this research and findings from permafrost samples).

Example 11.3. The brown (grizzly) bear, *Ursus arctos*, is one of the most widely distributed terrestrial mammals, with a current range encompassing many northern areas of Europe, Asia, and North America. However, the historic range was much larger, and the bear occupies less than 1% of its original range in the lower 48 states and Mexico (Waits *et al.*, 1998). To develop conservation strategies for the brown bear in North America, it is necessary to have an understanding of the amount of variation within and between extant populations. As a result, Waits *et al.* (1998) examined genetic variation in 294 nucleotides of the control region of mtDNA in 317 brown bears from Alaska, western Canada, and northwestern United States.

Waits *et al.* (1998) found extensive genetic variation, primarily between different geographic areas. For example, one cluster of mtDNA sequences, clade I, was found in the ABC Islands of southeastern Alaska, and it differed by approximately 20 nucleotides from the other sequences (Figure 11.10a). All of the bears from southern Canada and the lower 48 states clustered together in clade IV and differed by about 17 nucleotides from the other clades. The bears from much of Alaska clustered together (clade II), as

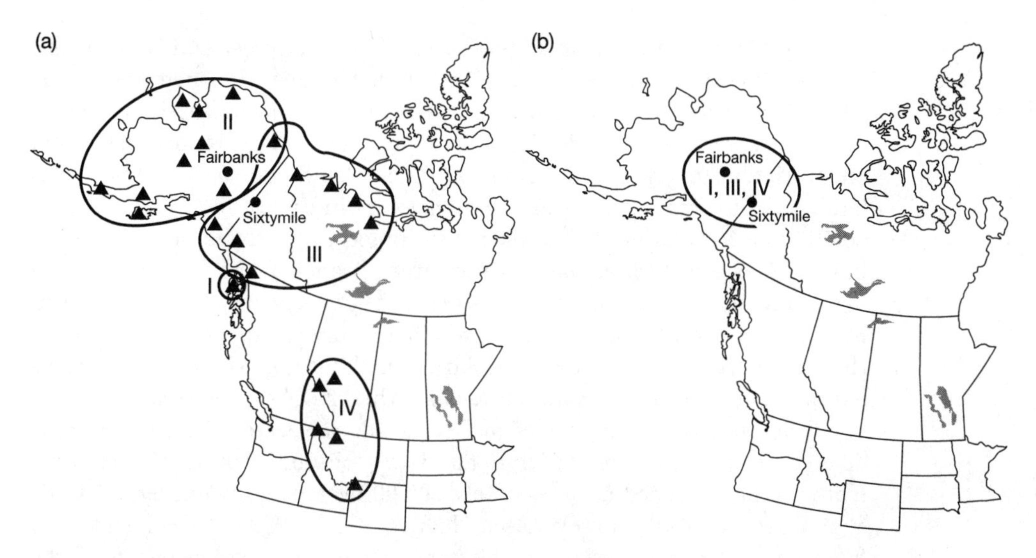

Figure 11.10. (a) The current (triangles indicate sampling locations) and (b) past (36,000 to 43,000 years ago) geographic distributions of mtDNA clades I, II, III, and IV in the brown bear (Waits *et al.*, 1998; Leonard *et al.*, 2000).

did the bears from eastern Alaska and northwestern Canada (clade III). As a result of the strong correspondence of mtDNA clades with geographic location, all except clades II and III were reciprocally monophyletic with each other, Waits *et al.* (1998) suggested that the three geographic groups from the ABC Islands (clade I), southern Canada and the lower 48 states (clade IV), and the other bears from Alaska and Canada (clades II and III) represent three different ESUs and should be conserved and managed separately.

Subsequently, Leonard *et al.* (2000) examined samples of brown bears preserved in the permafrost from eastern Alaska and northwestern Canada dating from 14,000 to 43,000 years ago. Surprisingly, they found mtDNA sequences from all of the ESUs in their permafrost samples taken from a relatively small area (Figure 11.10b). Specifically, they found samples of clades I, II–III, and IV, all in eastern Alaska or northwestern Canada dating from 36,000 to 43,000 years ago. The lack of congruence of contemporary and permafrost samples suggests that simple explanations for the current phylogenetic patterns, such as isolation for hundreds of thousands of years, are not warranted. In fact, Barnes *et al.* (2002) examined a larger sample of permafrost samples and suggested that major phylogeographic changes occurred 35,000 to 21,000 years ago and that there has been less change since then. On the other hand, although the mtDNA phylogeographic pattern observed by Waits *et al.* (1998) may have occurred over the past 20,000 to 30,000 years, this is certainly long enough for adaptive differences to have accumulated in the different groups.

The place of origin of species is often of great interest and controversy. For example, the place of origin of modern humans was suggested to be Africa because analysis of mtDNA variation indicated that modern humans recently arose "out of Africa" (Cann *et al.*, 1987). This report was based on restriction fragment data from 147 different individuals from five populations and is presented as parsimony network in Figure 11.11 (after Avise, 2000). The African haplotypes are distributed in both the most ancestral branch A of the tree as well as the larger branch B, therefore having a higher amount of variation than the other groups (Asians, Australians, Europeans, and New Guineans), all of which are only in branch B. In addition, the inferred root of the tree is in Africa, making the African populations paraphyletic to the other populations for the mtDNA gene tree.

When independent rates of mtDNA evolution were used to determine divergence, the common mtDNA ancestor, "the mtDNA Eve," was estimated to have existed approximately 200,000 years ago (Cann *et al.*, 1987; Vigilant *et al.* 1991; however, see Hedges *et al.*, 1992). As a result, this hypothesis of human origin that we descended from colonization of the whole world by a small group of Africans is generally referred to as the **recent African origin** hypothesis. Another hypothesis about the origin of humans is the **multiregional hypothesis**, which suggests that modern humans emerged gradually and simultaneously on different continents because of ongoing gene flow between them. In addition, there are suggestions that Africans settled other parts of the world more than once, that there were immigrations back to Africa from other parts of the world, and that there was population substructuring within continents. Takahata *et al.* (2001), Excoffier (2002), and Templeton (2002) provide discussion about these hypotheses and their genetic implications. Example 11.4 examines these human origin hypotheses using data from a nuclear gene.

The place of origin of domesticated livestock and their progenitor species or subspecies has been the subject of extensive investigation using molecular markers (Bruford *et al.*, 2003). It appears that most livestock were domesticated from one or more related ancestral species or subspecies in either southwest Asia (cattle, pigs, sheep, and goats), east Asia (buffalo, pigs, and yak), or the Andean Mountains (alpaca and llama). On the other hand, domestic horses were apparently domesticated multiple times in more northerly regions (Vilá *et al.*, 2001), and similarly, dogs were apparently domesticated multiple times from wolves (Vilá *et al.*, 1997). Understanding the origin of domesticated animals may be important for protection of endangered breeds of livestock and the remaining wild progenitors of domesticated livestock, as well as identification of genetic variation in livestock or their progenitors that may be of agricultural or economic significance.

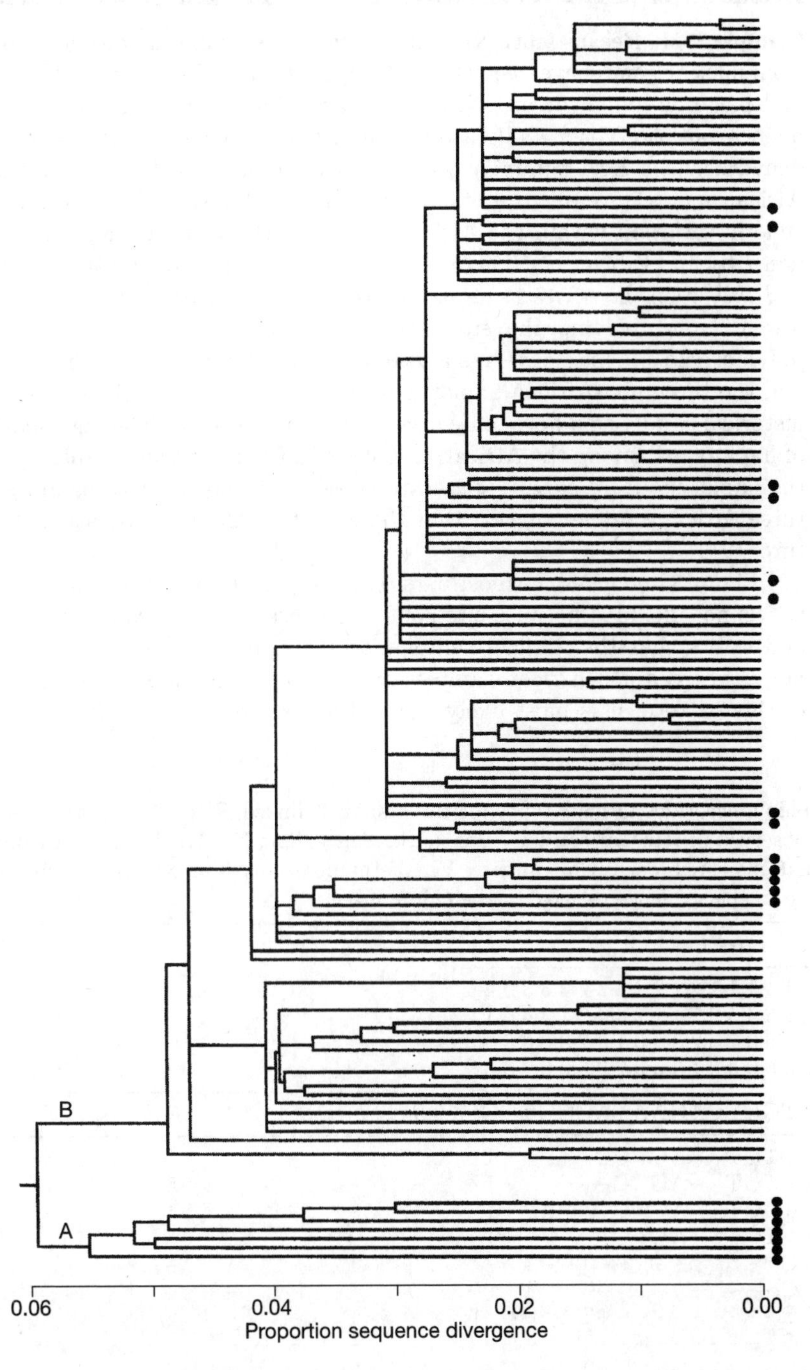

Figure 11.11. Parsimony network of the 147 human mtDNA haplotypes from Cann *et al.* (1987) (after Avise, 2000). The black circles indicate haplotypes from Africans, which are in both the ancestral branch A and spread throughout descendant branch B. Haplotypes for individuals from other continents are spread throughout branch B.

Example 11.4. Because mtDNA reflects the ancestry of only one gene, data from nuclear genes have been thought important to substantiate differences and patterns found for mtDNA. Harris and Hey (1999) examined the sequence variation for a 4200-base region of the X-linked *PDHA1* (pyruvate dehydrogenase $E1\alpha$ subunit) gene in a sample of 16 African and 19 non-African males. Overall, they found 25 polymorphic sites, but the level of genetic variation within the African sample was much higher than in the non-African (Table 11.7). There were 23 polymorphic sites and the estimate of θ was 6.9 in the African sample, whereas in the non-African sample there were only two polymorphic sites and the estimate of θ was only 0.57. One polymorphic site (site 544) was a fixed difference between the eight African and the three non-African haplotypes, which suggests that there has been historical subdivision between Africans and non-Africans. The high sharing of haplotypes among the African samples and the presence of haplotype C, the haplotype most closely related to those in the non-African samples, in three African samples suggest that there is little African population structure.

Figure 11.12 gives the most parsimonious gene tree for these haplotypes (B1 is not included because site 3306 is not used in the tree construction) and shows that the African samples occur on both sides of the deepest node, whereas non-African samples are restricted to one side. As a result of this pattern, it is most likely that Africa is the source of the ancestral

TABLE 11.7 The polymorphic nucleotides in haplotypes for the human X-linked *PDHA1* gene and the number of each haplotype observed in four African (Ba, Bantu; Se, Senegalese; Kh, Khoisan; Py, Pygmy) and four non-African populations (Fr, French; Ch, Chinese; Vi, Vietnamese; Mo, Mongolian) (Harris and Hey, 1999). * indicates a gap in the sequence relative to the chimpanzee.

	Base Position								
	1111111222222233333 44								
	1456702334591112449134612								
	5947903053732366187800800								
Haplotype	7442152069361506795967889	*African*				*Non-African*			
Chimpanzee	CCGGTTATGCCGAGAATACGGCGCC	*Ba*	*Se*	*Kh*	*Py*	*Fr*	*Ch*	*Vi*	*Mo*
A	--ACCC--TGT--AC-CC-----T-	–	–	–	–	–	2	1	1
B	--ACCC--TGT--AC-C------T-	–	–	–	–	5	5	4	–
B1	--ACCC--TGT--AC-C---A--T-	–	–	–	–	1	–	–	–
C	---CCC--TGT--AC-C------T-	1	2	–	2	–	–	–	–
D	-A-----C--*-T-----T--T---	–	1	–	1	–	–	–	–
E	TA-----C----------T--T---	1	–	–	–	–	–	–	–
F	-A----CC----------TA-----	–	1	–	–	–	–	–	–
G	-A-----C-------G--T---C-T	1	–	–	–	–	–	–	–
H	-A----CC--*----G--T---C--	–	2	–	–	–	–	–	–
I	-A-----C--*A------T-A-C--	1	–	2	–	–	–	–	–
J	-A-----C--*-------T------	–	–	1	–	–	–	–	–

sequence. By using a chimpanzee sequence, Harris and Hey estimated the time of the divergence of the human *PDHA1* tree to be about 1.9 million years before present (myr BP). Although these new data are at odds with mtDNA data in some respects, the incorporation of mtDNA data, *PDHA1* data, and other nuclear data into a comprehensive view of human origins now seems possible.

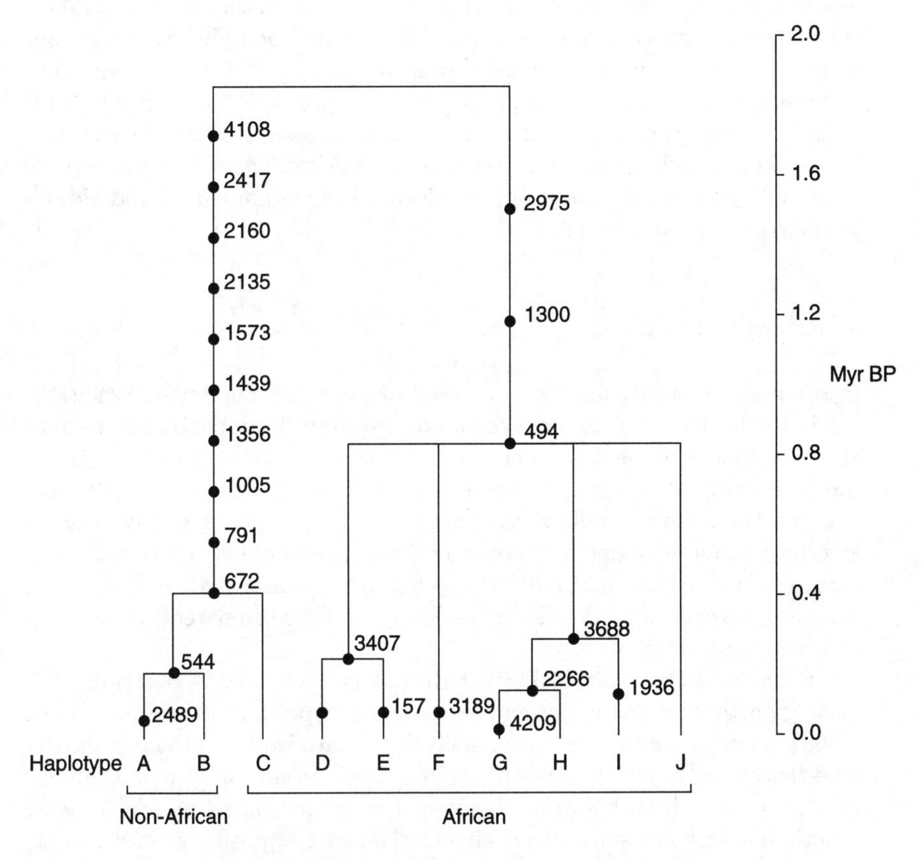

Figure 11.12. The maximum-parsimony tree for the *PDHA1* gene for the sites given in Table 11.7 where the sites having individual mutations are indicated on the appropriate branches. The resulting haplotypes are indicated at ends of the terminal branches (after Harris and Hey, 1999).

II. PATERNITY ANALYSIS AND INDIVIDUAL IDENTITY

Genetics has long been used to determine differences between individuals and relationships between them. Some important applications have been to differentiate between identical and fraternal twins (Vogel and Motulsky,

1997), between different inbred crop varieties (Gorman and Kiang, 1977), between breeds of livestock (Blott *et al.*, 1999; Werner *et al.*, 2004), within and between dog breeds (Denise *et al.*, 2003; Parker *et al.*, 2004), and for identification of pathogenic strains of bacteria and other organisms (Urwin and Maiden, 2003). When a number of blood group markers were discovered in humans, investigation of the paternity of putative fathers became feasible. With the discovery of many highly variable markers in many different organisms, genetics has been used for a number of additional situations to identify individual differences and relationships. The highest visibility of these uses is in forensics, to determine the probable innocence or guilt of individuals accused of crimes. For example, lawyer Barry Scheck and Innocence Project in New York have used DNA evidence to help free more than 100 prisoners. In addition, we discuss the assignment of individuals to their population of origin.

a. Paternity Exclusion

In the past, paternity analysis was used primarily in humans in situations where the father of a child is unknown or disputed. Variable blood group or allozyme loci or *HLA* types were used to exclude paternity or to estimate the probability of paternity of particular individuals. Recently, highly variable markers, such as microsatellites or RFLP markers at highly variable loci, have made it possible to give more precise estimates and to carry out paternity analysis in many other species. As we discussed in Chapter 5, genetic data can be used to infer the level and type of relatedness in many different organisms.

Behavioral observations have traditionally been used to determine the mating patterns in many nonhuman vertebrate species. For example, Lack (1968) stated, based on extensive behavioral data in birds, that "well over nine-tenths (93%) of all passerine subfamilies are normally monogamous. . . . Polyandry is unknown." However, determination of paternity using genetic markers has shown that almost the exact opposite is true; that is, "true genetic monogamy occurs in only 14% of surveyed passerine species, and that genetic polyandry occurs regularly in the 86% of species . . ." (Griffith *et al.*, 2002).

In particular, **extra-pair paternity**, that is, fertilizations resulting from copulations outside of the recognized pair bond, are often much higher than normal behavioral observations would predict in birds. For example, a large proportion of socially monogamous bird species are actually sexually promiscuous, based on genetic data. The most extreme such example is the socially monogamous reed bunting in which extra-pair offspring are 55% of all offspring and are in 86% of all broods (Dixon *et al.*, 1994). Obviously, paternity analysis using genetic markers has completely revolutionized the thinking about mating systems in birds and has also provided paternity data for many other species in which it was previously unavailable (for example, Bishop *et al.*, 2004; Krützen *et al.*, 2004). Example 11.5 provides some data on multiple paternity from a green sea turtle population.

Example 11.5. Understanding the importance of various factors that influence female reproductive strategies is of great interest in evolutionary ecology. For example, some potential advantages of **multiple paternity** for a single female are that fertilization is assured and that there is greater genetic variation in her offspring. On the other hands, there may be costs for multiple paternity, such as increased risk of mortality from disease, predation, or interrupted foraging. To examine whether multiple mating was beneficial or not in the green sea turtle, Lee and Hays (2004) examined measures of fitness in clutches fathered by either single or multiple males.

To determine whether the eggs in green turtle nests on Ascension Island were from a single or multiple fathers, Lee and Hays (2004) screened five polymorphic microsatellite loci. The maternal female genotype was determined directly from the individual laying the clutch. Paternal alleles were then inferred from the genotypes of the offspring in a clutch. For seven of the clutches examined, there were only one or two different paternal alleles observed for all five microsatellite loci, consistent with a single father for the clutch. On the other hand, for 10 other clutches (Table 11.8), more than two paternal alleles were observed at a number of loci. For example, in nest TP48, three or four different paternal alleles were identified at four of the five microsatellite loci. As a result, at least two different fathers must have fertilized eggs within this clutch. For four of the clutches, five or six different paternal alleles were observed, indicating that at least three males fertilized eggs within these clutches.

TABLE 11.8 For 10 green sea turtle nests with multiple paternity, the clutch sample size (N), the number of paternal alleles at each of five microsatellite loci, and the inferred number of fathers (Lee and Hays, 2004).

		Locus					
Nest	*N*	*CM58*	*CM3*	*CC7*	*CC117*	*CM84*	*Inferred number of fathers*
TP48	23	3	2	3	4	4	2
TP51	31	4	3	4	4	5	3
TP53	59	4	2	4	3	5	3
TT1	56	2	2	4	3	5	3
TT4	12	4	3	3	3	3	2
TT6	19	3	2	4	4	1	2
TT9	51	5	3	5	6	3	3
TT10	46	5	3	5	6	5	3
TT11	16	3	2	4	3	4	2
TT13	16	3	3	3	3	4	2

Lee and Hays (2004) then measured clutch size, proportion of eggs fertilized, and proportion of eggs hatching and surviving in the clutches with single and multiple paternity. They found, in contrast to expectations, no correlation of these fitness measures and multiple paternity, suggesting that multiple maternity was not significantly beneficial to these female green turtles. They suggested that multiple paternity may be largely the result of "male coercion, where females have given in to harassment as a means of reducing their overall costs."

To illustrate the general genetic basis of paternity analysis, let us begin with a locus that has only two alleles in a population. When only one gene is used, it is difficult to exclude an individual as a parent because many individuals who might be potential parents have alleles that are indistinguishable. An example of mother–child genotype combinations and the excluded paternal genotypes is given in Table 11.9 for a biallelic, codominant locus, such as the *MN* blood group or an allozyme locus. Given the genotype of the mother and the offspring, one may exclude certain genotypes as the father, an approach called **paternity exclusion**. For example, given that the genotypes of both mother and offspring are A_1A_1, then the father could be either A_1A_1 or A_1A_2, and only the paternal genotype A_2A_2 can be excluded. At most, only one genotype can be excluded with a biallelic locus, and when both the mother and child are heterozygotes, no paternal genotypes can be excluded.

Let us illustrate how the probability of exclusion can be derived for the simplest case, a two-allele codominant system. Table 11.9 gives the probability of exclusion for each mother–offspring combination. For example, when both the mother and child are A_1A_1, an event that has a frequency of p_1^3, when mating is random, only A_2A_2 genotypes, which have a frequency of p_2^2, can be excluded. The joint probability of this mother–offspring combination and the paternal genotype is therefore $p_1^3p_2^2$. The probabilities for

TABLE 11.9 The possible genotypes and probabilities (in parentheses) at a biallelic, codominant locus for various mother–offspring combinations and the excluded paternity types and probabilities of exclusion.

Mother	*Offspring*	*Excluded paternity types*	*Probability of exclusion*
$A_1A_1(p_1^2)$	$A_1A_1(p_1)$	$A_2A_2(p_2^2)$	$p_1^3p_2^2$
	$A_1A_2(p_2)$	$A_1A_1(p_1^2)$	$p_1^4p_2$
$A_1A_2(2p_1p_2)$	$A_1A_1(p_1/2)$	$A_2A_2(p_2^2)$	$p_1^2p_2^3$
	$A_1A_2(\frac{1}{2})$	—	—
	$A_2A_2(p_2/2)$	$A_1A_1(p_1^2)$	$p_1^3p_2^2$
$A_2A_2(p_2^2)$	$A_1A_2(p_1)$	$A_2A_2(p_2^2)$	$p_1p_2^4$
	$A_2A_2(p_2)$	$A_1A_1(p_1^2)$	$p_1^2p_2^3$

(Photo courtesy of University of Pittsburgh.)

CHING CHUN LI (1912–2003)

C.C. Li was born in northeastern China, educated in the United States, and returned to China in 1941 to organize a modern genetics research program (Chakravarti, 2004). During this period, he wrote the first of his population genetics books (Li, 1948), a work that first made the population genetics theories of Fisher, Wright, and Haldane accessible to a larger audience. However, in 1949, the most defining event in C.C. Li's life occurred, that is, his confrontation with Chinese followers of the Russian antigeneticist Trofim Lysenko. He was forced to resign his university position because of his defense of genetics (Li, 1949) and fled on foot with his family to Hong Kong. H.J. Muller intervened on his behalf and Li obtained a position at the University of Pittsburgh where he spent the rest of his scientific career in what became the Department of Biostatistics in the School of Public Health. When Sewall Wright was asked by an editor at the University of Chicago Press who could write a book on population genetics, he suggested C.C. Li (Speiss, 1983). As a result, he wrote the population genetics book (Li, 1955) that dominated the field for nearly 25 years. This book formed the basis of population genetics theory for geneticists who could not understand many of the mathematical details of the original publications. Li made contributions to nearly every aspect of theoretical population genetics, including inbreeding, adaptive selection, gene flow, paternity analysis, and recessive diseases (Speiss, 1983), although nearly all were focused on human population genetics problems. His terrible experiences with the followers of Lysenko led him to comment on a current challenge from nonscientists, "I think the first requirement is the depoliticization of the classroom. We shall teach science. Creation 'science' is not science" (Li, 1999).

all possible mother–offspring combinations can be calculated, and the sum of these is the total probability of exclusion for this locus. If we sum these probabilities, then the overall probability for the kth locus is

$$\begin{aligned}\mathrm{Pr}_k &= p_1^3p_2^2 + p_1^4p_2 + p_1^2p_2^3 + p_1^3p_2^2 + p_1p_2^4 + p_1^2p_2^3 \\ &= p_1p_2[(p_1+p_2)^3 - (p_1^2p_2 + p_1p_2^2)] \\ &= p_1p_2(1 - p_1p_2) \qquad (11.3a)\end{aligned}$$

The maximum probability of exclusion occurs when $p_1 = p_2 = 0.5$, in which case $\mathrm{Pr}_k = 0.1875$. When the alleles are substantially different in frequency, then the Pr_k values are much smaller, and there is little probability of exclusion because the common allele is shared among many individuals.

When a number of genetic loci are used jointly, the overall exclusion probability can be substantially higher. The probability of excluding a falsely accused male for m independent genes is

$$\mathrm{Pr} = 1 - \prod_{k=1}^{m} (1 - \mathrm{Pr}_k) \tag{11.3b}$$

(Chakraborty *et al.*, 1974). If there were 10 loci with two equally frequent alleles, then the expected probability of exclusion would be 0.875. However, if there were 100 such loci, as for single nucleotide polymorphisms, then the probability of exclusion would become $1 - (9.6 \times 10^{-10})$, or very close to 1.

With a relatively low number of useful biallelic loci, the probability of exclusion is not very high. Besides increasing the number of loci, another resolution for this problem is to use highly variable loci, initially such as *HLA* haplotypes but more recently RFLPs from hypervariable loci or microsatellite loci (Chakraborty *et al.*, 1999). The paternal genotypes that can be excluded for different possible mother–offspring combinations for a locus with multiple alleles are given in Table 11.10. For example, the probability of an A_iA_i mother is p_i^2 and the probability of her having an A_iA_i child from a random man in the population is p_i. This excludes all men that do not have an A_i allele, and their frequency is $(1 - p_i)^2$. Therefore, the combined probability of exclusion for this trio, given in the first line of Table 11.10, is the product of these probabilities or $p_i^3(1 - p_i)^2$.

TABLE 11.10 Paternity exclusion genotypes and probabilities for different mother–offspring combinations at a locus with multiple alleles (after Weir, 1996). A_i, A_j, and A_k indicate different alleles in the mother–offspring combinations, and A_x and A_y indicate alleles in excluded paternity types. To the right of the mother, offspring, and excluded paternity genotypes are given their probabilities. For the offspring, the probability is that given for the mother's genotype.

Mother	*Offspring*	*Excluded paternity types*			*Probability of exclusion*
$A_iA_i(p_i^2)$	$A_iA_i(p_i)$	A_xA_y	$x, y, \neq i$	$(1 - p_i)^2$	$p_i^3(1 - p_i)^2$
	$A_iA_j(p_j)$	A_xA_y	$x, y, \neq j$	$(1 - p_j)^2$	$p_i^2p_j(1 - p_j)^2$
A_iA_j $(2p_ip_j)$	$A_iA_i(p_i/2)$	A_xA_y	$x, y, \neq i$	$(1 - p_i)^2$	$p_i^2p_j(1 - p_i)^2$
	$A_jA_j(p_j/2)$	A_xA_y	$x, y, \neq j$	$(1 - p_j)^2$	$p_ip_j^2(1 - p_j)^2$
	$A_iA_j(p_i + p_j)/2$	A_xA_y	$x, y, \neq i, j$	$(1 - p_i - p_j)^2$	$p_ip_j(p_i + p_j)(1 - p_i - p_j)^2$
	$A_iA_k(p_k/2)$	A_xA_y	$x, y, \neq k$	$(1 - p_k)^2$	$p_ip_jp_k(1 - p_k)^2$
	$A_jA_k(p_k/2)$	A_xA_y	$x, y, \neq k$	$(1 - p_k)^2$	$p_ip_jp_k(1 - p_k)^2$

Adding the probabilities from the seven different trios in Table 11.10 for n alleles gives the probability of exclusion of

$$\Pr_k = \sum_{i=1}^{n} p_i(1-p_i)^2 - \frac{1}{2}\sum_{i=1}^{n}\sum_{i \neq j}^{n} p_i^2 p_j^2 (4 - 3p_i - 3p_j) \qquad (11.4a)$$

(Weir, 1996). The maximum probability of exclusion occurs when the frequencies of the alleles are equal; that is, they are equal to $1/n$, and then

$$\Pr_k = 1 - \frac{2n^3 + n^2 - 5n + 3}{n^4} \qquad (11.4b)$$

When there are 20 alleles with equal frequency, the probability of exclusion is 0.898, slightly higher for this one locus than if there were 10 loci with two equally frequency alleles.

As an illustration of paternity exclusion from an actual example, Table 11.11 gives the genotypes of a mother brown (grizzly) bear, her three cubs, and the alleged father of two of the cubs from northwestern Alaska at eight microsatellite loci (Craighead *et al.*, 1995). The genotypes of both offspring 1 and 2 are compatible with the mother–father combination; below we show how the likelihood that this male is the father of these cubs can be calculated. On the other hand, alleles at five loci in cub 3 (boldfaced)

TABLE 11.11 The genotypes at eight microsatellite loci for a mother brown (grizzly) bear, her three offspring, and the alleged father of her offspring (Craighead *et al.*, 1995). The alleles in cub 3 that were not found in the mother–alleged father combination are boldfaced.

		Offspring (cubs)			
Locus	*Mother*	*1*	*2*	*3*	*Alleged father*
A	184, 192	184, 194	184, 192	184, 194	184, 194
B	156, 160	152, 160	152, 160	160, **164**	152, 152
C	105, 113	105, 111	105, 105	111, 113	105, 111
D	172, 177	172, 172	172, 177	172, **178**	172, 177
L	155, 159	155, 157	159, 161	155, **155**	157, 161
M	208, 208	208, 212	208, 208	208, 212	208, 212
P	153, 153	153, 159	153, 159	153, **157**	159, 161
X	135, 137	135, 137	133, 135	137, **141**	133, 137

were not found in the father of cubs 1 and 2. As a result, he can be excluded as the father of cub 3, making this litter an apparent instance of multiple paternity. Although 35 adult males were genotyped in the population, the male that fathered cub 3 was not found. Highly variable loci, such as *HLA* genes in humans, have also been used to determine other putative relationships, for example, grandpaternity in missing Argentinean children (Example 11.6).

Example 11.6. As we have noted, highly variable *HLA* haplotypes were extensively used in human paternity determination before the development of other highly variable markers. When necessary, the same approach can be extended to other relatives. A major impetus for this extension was the disappearance of at least 9000 Argentinians between 1975 and 1983. Among the missing were a number of children who had been abducted by the military and police or who had been born to kidnapped women in captivity. In many cases, the parents of these children were among the missing Argentinians, most of whom are now presumed to have been killed. It appears that many of the Argentinian children lived with military couples who claimed to be their biological parents but may have actually obtained them after they were kidnapped.

If all four grandparents are alive, and if the child has a haplotype that none of them has, grandpaternity can be excluded. But if the haplotypes in the child are also present in the grandparents, then the probability of grandpaternity can be calculated. For example, if the shared haplotypes are rare in the population, then the probability of grandpaternity may be quite high. Figure 11.13 gives a pedigree of a young Argentinian kidnap victim,

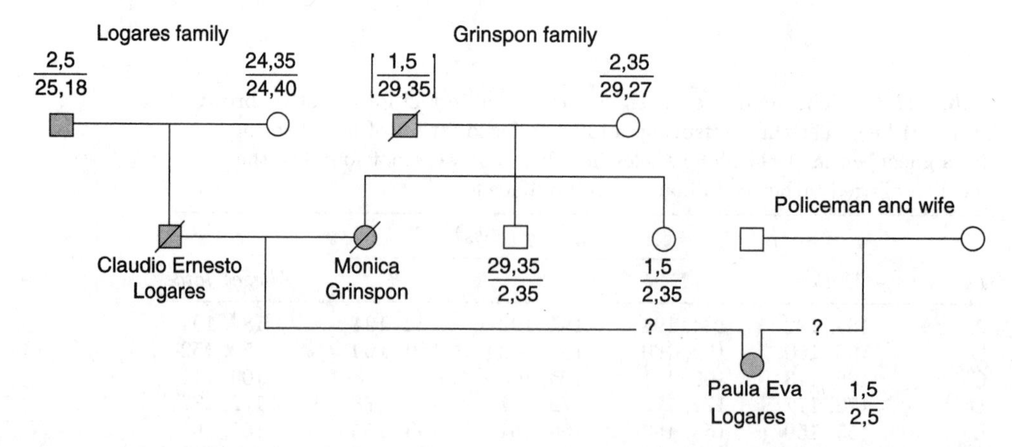

Figure 11.13. A pedigree (after Diamond, 1987) showing how the grandpaternity was determined, using *HLA* haplotypes, for Paula Eva Logares (slashes indicate deceased individuals and shading indicates ancestry of haplotypes in Paula Eve Logaros). The policeman and his wife refused to be tested. For simplicity, haplotype A2-B5 is indicated as 2,5 and so on.

Paula Eva Logares, who was raised by a policeman and his wife (Diamond, 1987). In this case, the putative parents were both dead or missing, and the child's *HLA* haplotypes were found in the putative grandparents. One of her *HLA* haplotypes, A2-B5, was present in her paternal grandfather. The genotype of her maternal grandfather (reconstructed from his other children) contains the other haplotype, A1-B5. On the basis of these data and a grandpaternity probability of over 0.999, it is highly likely that Paula Eva Logares is the grandchild of the Logares and Grinspon families.

Markers for the Y chromosome can also be used to determine paternity in one generation, and because of their patrilineal inheritance, they can be used to trace male-to-male inheritance over generations. Similarly, mtDNA haplotypes can be used for maternity determination, and because of their matrilineal inheritance, they can be used to trace female-to-female inheritance over generations (Budowle *et al.*, 2003). Because the Y chromosome has no recombination, multilocus haplotypes can be used to infer ancestry a number of generations in the past. For example, the relative homogeneity of their Y chromosome haplotypes has been used to suggest that Jewish priests (Cohanim) appear to be descended from a single lineage approximately 2000 to 3000 years ago (Thomas *et al.*, 1998). In addition, Y chromosome haplotypes have been used to investigate the putative paternity of President Thomas Jefferson for the children of one of his slaves (see Example 11.7).

Example 11.7. One of the most celebrated paternity determinations in recent years was the use of molecular markers to evaluate whether Thomas Jefferson, the third president of the United States, was the probable father of the children of one of his slaves, Sally Hemmings (Foster *et al.*, 1998). Thomas Jefferson was suspected of being the father of at least two of her children, her first son Thomas (Woodson), born in 1790, and her last son Eston (Hemmings), born in 1808. If Thomas Jefferson was their father, then male descendants of these sons, who are the result of unbroken male-to-male lineages, should have the same Y chromosome as Jefferson, barring mutation.

Because Thomas Jefferson had no surviving direct male descendants, descendants of his paternal uncle, Field Jefferson, were used to establish the Y-chromosome haplotype of Thomas Jefferson. The first row of Table 11.12 gives, for five descendants of Field Jefferson, the alleles at the biallelic marker, the five microsatellite loci, and the minisatellite locus that were informative (six other biallelic markers and six other microsatellite loci were examined). Actually, one descendant differed from the others by one repeat at one microsatellite locus, probably the result of a mutation.

TABLE 11.12 The Y haplotypes at the informative loci inferred from male descendants or descendants of male relatives for President Thomas Jefferson, the last and first sons of his slave Sally Hemmings, and the sons of Jefferson's sister (Foster *et al.*, 1998).

	Type of marker						
	Biallelic	*Microsatellite*					*Minisatellite*
		1	2	3	4	5	
President Thomas Jefferson	0	15	4	9	10	15	A_1
Eston (last son of Sally Hemmings)	0	15	4	9	10	15	A_1
Thomas (first son of Sally Hemmings)	1	14	5	10	13	13	A_2
Samuel and Peter Carr (sons of Jefferson's sister)	1	14	5	10	13	13	A_3

In the second and third rows are the Y haplotypes of the descendants of the two sons of Sally Hemmings who are suspected to have been fathered by Jefferson. The Y haplotype of the descendant of Eston appears identical to that from Jefferson. The probability of this occurring by chance is very low, because the Jefferson-type Y chromosome has been found only in this lineage and not been found in 670 other European men sampled (Foster *et al.*, 1998).

The only likely alternative to President Jefferson as the father would be another relative with the Jefferson-type Y chromosome. Abbey (1999) and Marshall (1999) suggest that the younger brother of President Jefferson, Randolph (or any of Randolph's five sons), who would have the Jefferson-type Y chromosome, could have been the father of Eston. On the other hand, the Y genotype of the male descendants of the first son of Sally Hemmings, Thomas (Woodson), differs from the Jefferson Y chromosome at seven markers, demonstrating that Thomas Jefferson does not appear to be the father of her first son. It has also been alleged that one of the sons of Jefferson's sister, Samuel or Peter Carr, is the father of Sally's first son. On the bottom line of Table 11.12 is the Y-chromosome type of relatives having the Carr Y chromosome; it matches that of Thomas (Woodson) except for the minisatellite locus.

b. Paternity Assignment

In a number of situations in natural populations, such as in many nonhuman vertebrates, paternity exclusion except for one putative father may not be possible. When multiple males are not excluded, then a statistically based method to assign paternity is needed. The likelihood approach has been shown to be an efficient way to evaluate human relationships as well as paternity assignment in polygynous natural populations.

The **likelihood** approach determines the likelihood of alternative hypothesis given the observed data (Edwards, 1972). In general, the likelihood L of a hypothesis given the data D is indicated as $L(H|D)$ where the vertical line is read as given (the symbols and approach here follows that in Marshall *et al.*, 1998). The likelihood of one hypothesis, for example, hypothesis 1 (H_1), relative to hypothesis 2 (H_2), is called the likelihood ratio and is

$$L(H_1, H_2|D) = \frac{\Pr(D|H_1)}{\Pr(D|H_2)} \tag{11.5a}$$

where $\Pr(D|H_i)$ is the probability of obtaining the data D given hypothesis i (H_i). For paternity analysis, the data D are the genotypes of the offspring, mother, and putative (or alleged) father at a particular locus. H_1 is that the alleged male is the true father, and this is compared with H_2, which is that the alleged father is an unrelated male drawn at random from the population.

Let G_O, G_M, and G_{AF} be the genotypes of the offspring, the mother, and the alleged father, respectively. The likelihood that the mother and the alleged father are the parents of the offspring is

$$L(H_1|G_O, G_M, G_{AF}) = T(G_O|G_M, G_{AF}) \Pr(G_M) \Pr(G_{AF}) \tag{11.5b}$$

where the Mendelian segregation or transmission probability is $T(G_O|G_M, G_{AF})$, and $\Pr(G_M)$ and $\Pr(G_{AF})$ are population frequencies of the mother's and the alleged father's genotypes, respectively. The likelihood that the mother is the parent of the offspring and that the father is a random individual from the population is

$$L(H_2|G_O, G_M, G_{AF}) = T(G_O|G_M) \Pr(G_M) \Pr(G_{AF}) \tag{11.5c}$$

The ratio of these two likelihoods is

$$\begin{aligned} L(H_1, H_2|G_O, G_M, G_{AF}) &= \frac{T(G_O|G_M, G_{AF}) \Pr(G_M) \Pr(G_{AF})}{T(G_O|G_M) \Pr(G_M) \Pr(G_{AF})} \\ &= \frac{T(G_O|G_M, G_{AF})}{T(G_O|G_M)} \end{aligned} \tag{11.5d}$$

This ratio L is the likelihood of paternity for a given male relative to a random male from the population and is called the **paternity index** in human paternity testing (Pena and Chakraborty, 1994).

The transmission probabilities and the likelihood ratios for all of the compatible genotype combinations for an autosomal, codominant locus are given in Table 11.13. For example, in the fourth row, the offspring, mother, and alleged father have genotypes of A_1A_1, A_1A_x, and A_1A_x, respectively. $T(G_O|G_M, G_{AF})$ in this case is 1/4 because A_1 needs to be transmitted

TABLE 11.13 Transmission probabilities and likelihood ratios for all compatible offspring–mother–alleged father trios (after Marshall *et al.*, 1998). A_x indicates any allele other than A_1 and A_y indicates any allele other than A_1 or A_2.

Offspring (G_O)	*Mother* (G_M)	*Alleged father* (G_{AF})	$T(G_O \mid G_M, G_{AF})$	$T(G_O \mid G_M)$	$L(H_1, H_2)$
A_1A_1	A_1A_1	A_1A_1	1	p_1	$1/p_1$
A_1A_1	A_1A_1	A_1A_x	1/2	p_1	$1/2p_1$
A_1A_1	A_1A_x	A_1A_1	1/2	$p_1/2$	$1/p_1$
A_1A_1	A_1A_x	A_1A_x	1/4	$p_1/2$	$1/2p_1$
A_1A_2	A_2A_2	A_1A_1	1	p_1	$1/p_1$
A_1A_2	A_2A_2	A_1A_x	1/2	p_1	$1/2p_1$
A_1A_2	A_2A_y	A_1A_1	1/2	$p_1/2$	$1/p_1$
A_1A_2	A_2A_y	A_1A_x	1/4	$p_1/2$	$1/2p_1$
A_1A_2	A_1A_2	A_1A_1	1/2	$(p_1 + p_2)/2$	$1/(p_1 + p_2)$
A_1A_2	A_1A_2	A_1A_y	1/4	$(p_1 + p_2)/2$	$1/[2(p_1 + p_2)]$
A_1A_2	A_1A_2	A_1A_2	1/2	$(p_1 + p_2)/2$	$1/(p_1 + p_2)$

from both heterozygous parents and $T(G_O|G_M)$ is $p_1/2$ because A_1 needs to be transmitted both from the heterozygous mother and from a random father from the population that has an A_1 frequency of p_1.

Fung (2003) provided an example of a paternity case that illustrates the use of this approach. Table 11.14 gives the genotypes for the offspring, mother, and alleged father trio for two different microsatellite loci. These trios are equivalent to those in the eighth and ninth rows of Table 11.13 for loci TH01 and D7S820, respectively. Using the estimated frequency of allele TH01–8 of $p_8 = 0.04$, then $L(H_1, H_2) = 12.5$ and using the estimated frequencies of allele D7S820–7 ($p_7 = 0.01$) and of allele D7S820–9 ($p_9 = 0.06$), then $L(H_1, H_2) = 14.3$ for locus D7S820. The overall paternity index is the product of the values for each locus, or $(12.5)(14.3) = 178.8$.

Approaches using **Bayesian inference**, which allows the use of prior information about the parameters of interest, are being widely applied in genetics and molecular evolution (Huelsenbeck *et al.*, 2001; Beaumont and Rannala, 2004). The **Bayes' theorem** has long been used in paternity identification because it allows the use of prior information about the alleged father, unrelated to the genetic data, and this information is then modified based on the observed genetic data. Let us assume that the prior probability of paternity for an alleged man is $\Pr(C)$ and that the genetic data about paternity is indicated by D. Using the Bayes' theorem, then

$$\frac{\Pr(C|D)}{\Pr(\overline{C}|D)} = \left[\frac{\Pr(D|C)}{\Pr(D|\overline{C})}\right]\left[\frac{\Pr(C)}{\Pr(\overline{C})}\right] \tag{11.6a}$$

TABLE 11.14 A disputed paternity case with data from two microsatellite loci (Fung, 2003).

Locus	*Offspring*	*Mother*	*Alleged father*	$L(H_1, H_2)$
TH01	7, 8	7, 9	8, 10	$1/2p_8 = 1/[(2)(0.04)] = 12.5$
D7S820	7, 9	7, 9	9, 9	$1/(p_7 + p_9) = 1/(0.01 + 0.06) = 14.3$

(Weir, 1996) where the overbar indicates "not." The left-hand side of this equation is the ratio of posterior probabilities, that is, the probability that the man is the father, given the data (and the prior probability information) divided by probability that the man is not the father, given the data. The right-hand side is the product of likelihood ratio L or paternity index (as we gave in equation 11.5d) in the first set of brackets and the prior probability ratio in the second set of brackets. This expression can be rearranged to give the posterior probability that the man is the father, given the data, as

$$\Pr(C|D) = \frac{L\Pr(C)}{L\Pr(C) + 1 - \Pr(C)} \qquad (11.6b)$$

(Weir, 1996). Often, unless there is good evidence pointing to the man as the father, then the prior probability $\Pr(C)$ is assumed to $1/N$ where there are N possible fathers, or examined for a range of values. If the alleged father is as likely to be the father as a random man drawn from the population, then $L = 1$ and $\Pr(C|D) = \Pr(C)$; that is, the posterior probability is equal to prior probability, as expected. Using the data above from Fung (2003) where $L = 178.8$ and assuming $\Pr(C) = 0.5$, then the posterior probability of paternity $\Pr(C|D) = 0.994$.

Figure 11.14 is a summary of the posterior probabilities of paternity using haplotypes for *HLA-A* and *HLA-B* in 590 cases of disputed pater-

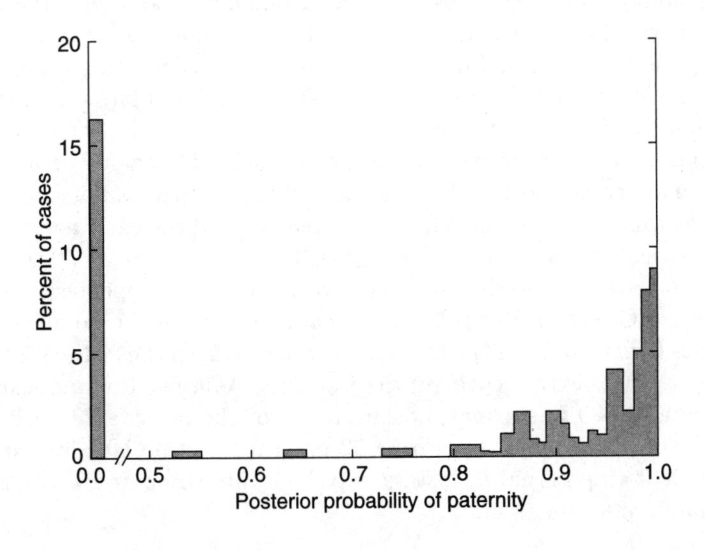

Figure 11.14. The probability of paternity calculated for 590 cases of disputed paternity using the *HLA* system (after Terasaki, 1978).

nity. The usefulness of a highly variable system is illustrated in the highly bimodal distribution, with one group having a probability of 0.0 (excluded as parents) and another group having a probability between 0.85 and 1.0. In all of the 1000 cases he examined, Terasaki (1978) found that 25% were exclusions (0.0 probability of paternity), and 64% had a probability of paternity of greater than 0.9. With more detailed identification of *HLA* alleles using DNA sequences and more *HLA* loci included in the haplotypes, from this one highly variable system alone, nearly all cases now being examined either are exclusions or have a very high posterior probability of paternity.

Paternity exclusion needs to be carefully applied, particularly when there are multiple possible parents that may not be excluded. Using strict paternity exclusion criteria (a single mismatch excludes a putative parent) may result in false exclusions because of mutation, genotyping errors, or null alleles (Marshall *et al.*, 1998; Jones and Arden, 2003). These problems may become an even greater concern when more loci are used because the probability of errors or mutations also increases. Example 11.8 presents a case in which there was paternity exclusion at one locus, apparently because of a mutation, but the estimate of the paternity index, excluding this locus, showed that there is very high likelihood of the alleged father being the true father.

Example 11.8. Thangaraj *et al.* (2004) analyzed a case of disputed paternity using 15 microsatellite loci and found a mismatch at only one of the loci (D21S11); that is, an allele at this locus in the child was not found in either the mother or the alleged father (Table 11.15). Except for the mismatch, the paternity index was substantial at a number of loci, and the overall value for the 14 loci, excluding the mismatched one, was 2.4×10^{10}. To investigate this situation further, they then examined six microsatellite loci on the Y chromosome in the male child and the alleged father and found a complete match at all six loci.

Finally, to characterize the mismatch allele, Thangaraj *et al.* (2004) cloned and sequenced all of the alleles of the mother, child, and alleged father at this locus. The 29 allele in the mother and the child had the same repeat structure, $(TCTA)_6(TCTG)_5(TCTA)_{10}$. The two 29 alleles in the father appeared to be the same but with a different repeat structure of $(TCTA)_4(TCTG)_6(TCTA)_{11}$. The repeat structure of the 30 allele in the child was $(TCTA)_4(TCTG)_6(TCTA)_{12}$, differing from the father's 29 alleles only by a single extra repeat in the last unit. As a result, it appears that the new allele 30 was a mutation from one of the father's 29 alleles. The rate of mutation at this locus from 52 paternity testing labs was reported to be 0.0024 in paternal meiosis, a very high rate compared with 12 other microsatellite loci examined.

TABLE 11.15 The genotypes of the mother, child, and alleged father for 15 microsatellite loci along with the paternity index for each locus (Thangaraj *et al.*, 2004). Allele 30 for locus 3 (indicated by an asterisk) in the child was not found in either the mother or the alleged father.

	Genotype			
Locus	*Mother*	*Child*	*Alleged father*	*Paternity index*
1	12/15	12/13	13/14	9.4
2	12/19	13/19	13/19	3.2
3	29/32	29/30*	29/29	0.0
4	8/9	7/9	7/10	5.0
5	16/17	15/16	15/17	1.4
6	24/26	23/24	23/25	2.9
7	8/11	8/8	8/11	1.1
8	12/15	12/14	11/14	3.4
9	18/21	18/18	19/19	15.2
10	9/12	10/12	9/10	2.1
11	12/12	11/12	11/12	2.3
12	12/13	10/13	10/10	18.9
13	9/13	9/13	9/13	8.5
14	11/13	13/13	12/13	7.8
15	10/13	13/14	14/14	111.1

In situations where the genotype of the mother is unknown, the likelihood ratio is

$$L(H_1, H_2|G_O, G_{AF}) = \frac{T(G_O|G_{AF})\,\Pr(G_{AF})}{\Pr(G_O)\,\Pr(G_{AF})} = \frac{T(G_O|G_{AF})}{\Pr(G_O)} \quad (11.7)$$

The denominator in this case is the frequency of the offspring genotype in the population. The transmission probabilities and likelihood values for all the compatible alleged father–offspring pairs are given in Table 11.16. Of course, if there are no known candidates for the father and one wants to determine which female is the mother among several candidates, the same approach can be used by substituting the alleged mother for the alleged father in the above equation and Table 11.16.

As we stated above, when multiple loci are used, the likelihood ratios may be multiplied together to obtain an overall paternity index. Or the natural logarithm of the combined likelihood ratios can be taken to obtain the likelihood of odds (LOD) score (Edwards, 1972; Meagher, 1986; this is different than the LOD score used in genetic mapping, which uses the $\log_{10}$). Therefore, a LOD score of zero indicates that the alleged father is

TABLE 11.16 Transmission probabilities and likelihood ratios for all compatible offspring–alleged father pairs (after Marshall *et al.*, 1998). A_x indicates any allele other than A_1, and A_y indicates any allele other than A_1 or A_2.

Offspring (G_O)	*Alleged father* (G_{AF})	$T(G_O \vert G_{AF})$	$P(G_O)$	$L(H_1, H_2)$
A_1A_1	A_1A_1	p_1	p_1	$1/p_1$
A_1A_1	A_1A_x	$p_1/2$	p_1	$1/2p_1$
A_1A_2	A_1A_1	p_2	$2p_1p_2$	$1/2p_1$
A_1A_2	A_1A_y	$p_2/2$	$2p_1p_2$	$1/4p_1$
A_1A_2	A_1A_2	$(p_1+p_2)/2$	$2p_1p_2$	$(p_1+p_2)/4p_1p_2$

likely to be the father of the offspring as a random male from the population, and a positive LOD score indicates that the alleged male is more likely to be the father than a random male.

Marshall *et al.* (1998) suggested a statistical approach to discriminate between nonexcluded males in which Δ is the difference in LOD scores between the most likely male and the next most likely male or

$$\Delta = LOD_1 - LOD_2 \tag{11.8}$$

where LOD_1 and LOD_2 are the scores for the most likely male and the next most likely male, respectively. They suggested that LOD_2 have a minimum score of 0 so that Δ is not falsely high.

Marshall *et al.* (1998) developed a program called CERVUS (http://helios.bto.ed.ac.uk/evolgen/cervus/cervus.html) to calculate Δ values and their statistical significance (see Jones and Ardren, 2003, for a list of available software to determine paternity in different situations). Slate *et al.* (2000) retrospectively evaluated the statistical confidence values of CERVUS by comparing their previous results in 364 red deer using 12 loci to that obtained from 84 microsatellite loci. They found that the actual confidence of CERVUS-assigned paternities from the 84 loci was not significantly different from that predicted earlier.

c. Individual Identity

Molecular information from highly variable markers is being widely used to identify individuals (or evidence) for forensic purposes (Evett and Weir, 1998; Butler, 2001). Applications of DNA technology to criminal cases have resulted in the identification of criminals out of a large pool of potential suspects and have also resulted in the exoneration of previously convicted individuals. With more genetic loci available, the focus in recent years has changed from excluding specific people as being the source of evidence from blood, bone, saliva, semen, nasal secretions, and fingerprints to making probability statements about genetic profiles if these people were not the source of the evidence (Weir, 2001).

The most straightforward forensic situation is when DNA is obtained from a sample at the crime scene; let us call it genotype G_C because it is from the crime scene, and it is thought that the sample is from the perpetrator of the crime. Subsequently, DNA is obtained from a suspect in the crime; let us call it genotype G_S, usually from blood or saliva, and it is found not to match the crime scene sample, that is, $G_C \neq G_S$. This lack of a match of genotypes is similar to paternity exclusion in that any incongruencies between the crime sample and the suspect's sample can lead to exoneration. Of course, this conclusion assumes no errors in determining the genotypes in the sample and from the suspect and that the crime scene sample came from the perpetrator.

Another common forensic situation is where the DNA obtained from a sample at the crime scene, again thought to be from the perpetrator of the crime, and DNA from a suspect in the crime is found to match, that is, $G_C = G_S$. How much evidence is this that the suspect was the source of the crime scene sample? A simple answer is that the probability of the suspect having the same DNA genotype as found at the crime scene is equal to the population probability of that genotype, or a random individual from the population. In other words, if the matching genotype of the sample and the suspect are rare, then the evidence against the suspect is strong because it is unlikely that a random individual would have the same genotype. On the other hand, if the matching genotype is common, then the evidence against the suspect is weak.

Before we briefly discuss how forensic probabilities are calculated, let us first show how the frequency of a particular multilocus genotype can be calculated. Let us initially assume Hardy–Weinberg proportions so that the frequency of a genotype at locus k is

$$\begin{aligned} P_k &= p_i^2 \text{ for homozygote } A_iA_i \\ P_k &= 2p_ip_j \text{ for heterozygote } A_iA_j \end{aligned} \tag{11.9a}$$

where p_i and p_j are the population frequencies of the alleles in the suspect. For example, if $p_i = 0.08$ and $p_j = 0.19$, then the probability that the suspect's genotype A_iA_j would be randomly drawn from the population is only $2p_ip_j = 0.030$. Then, over all loci examined and assuming no linkage disequilibrium, the probability of the multilocus genotype observed in the suspect is the product over all k loci or

$$P = \prod_{k}^{m} P_k \tag{11.9b}$$

and is known as the **match probability**. In other words, P is the match probability that a random individual from a population would have the genotype observed in the suspect. Example 11.9, about the murder trial of O.J. Simpson, discusses some aspects of using genetic evidence in trials.

Example 11.9. One of the most notorious cases in which DNA evidence played a pivotal role was the trial of football player–actor O.J. Simpson, who was accused of murdering his estranged wife, Nicole Simpson, and her friend, Ronald Goldman. DNA from these three individuals was compared with DNA samples from bloodstains obtained from various locations associated with the crime. DNA information from 45 bloodstain samples was presented at the trial, and all 400 alleles from the different loci among the 45 samples appear to have come from either O.J. Simpson or one of the two victims.

The only controversy was for item 29, a bloodstain from the steering wheel of O.J. Simpson's Bronco (Weir, 1995). Table 11.17 gives the alleles found for some of loci examined for this item, and it is apparent that more than one individual contributed to this item. From data for all three loci, it appears that both O.J. Simpson and at least one of the victims contributed to this item and that none of the principals could be excluded as contributors. However, the defense argued that the absence of allele 1.3 at the *DQα* locus on item 29 was evidence that the 4 allele must have been contributed by someone other than the three principles because it did not contain the complete genotype for Ronald Goldman. Testimony from the lab carrying out this analysis noted that the 4 allele was significantly weaker in intensity that the other alleles and suggested that there was much less DNA from the contributor with the 4 allele. They suggested that the 1.3 allele might not have been observed because of low amount of DNA (from Ronald Goldman) in this sample.

Calculations similar to that in expression 11.9*b* for multiple genetic markers resulted in prosecutor Marcia Clark stating that "the chance that a random person would have the profile found on the rear gate at Bundy would be 1 in 57 billion" (Weir, 1995). Given that there are only five billion people on earth, presentation of such a low probability can be considered the prosecutor's effort to suggest that it is inconceivable that someone else left this DNA evidence at the crime scene, or one can realize that the probability estimated from expression 11.9*b* refers to all possible genotypes that

TABLE 11.17 The alleles found on item 29, a bloodstain on the steering wheel of O.J. Simpson's Bronco and the genotypes for O.J. Simpson and the murder victims, Nicole Simpson and Ronald Goldman (Weir, 1995).

Locus	*Item 29*	*O.J. Simpson*	*Nicole Simpson*	*Ronald Goldman*
HBGG	A, B, C	B, C	A, B	A, A
Gc	A, B, C	B, C	A, C	A, A
DQα	1.1, 1.2, 4	1.1, 1.2	1.1, 1.1	1.3, 4

may be formed randomly from the constituent alleles, whereas the current world population is itself only a sample from this collection. In fact, Bruce Weir (1995), an expert witness for the prosecution, stated that "the need for presenting probability numbers in cases where one identifiable profile is present appears to me to be superfluous." Although many considered the DNA evidence to be overwhelmingly against him, Simpson was acquitted of the charges in October, 1995.

Obviously, if a number of highly variable loci are used, than the value of P quickly becomes very small. This is the logic behind the 13 highly polymorphic microsatellite loci used as **CODIS** (Combined DNA Index System) by the U.S. Federal Bureau of Investigation (Butler, 2001). Table 11.18 gives the number of alleles and the expected heterozygosity for these loci, and they have a high average of 8.8 alleles per locus and a heterozygosity of 0.783 in Caucasians. All of these loci are on different chromosomes except CSF1PO and D5S818, which are far apart on chromosome 5. As shown by Chakraborty *et al.* (1999), this battery of 13 loci has very high power for forensic and identification purposes.

Because of the genome and sequencing projects in other organisms, similar numbers of microsatellite loci are available or can be generated for strain and individual identification in virtually any organism. An application using data from this approach confirmed that Dolly, the sheep, was

TABLE 11.18 The 13 CODIS microsatellite loci and the number of alleles and expected heterozygosity in a sample of approximately 200 United States Caucasian individuals from the Federal Bureau of Investigation (Budowle *et al.*, 1999). Also given are the estimates of differentiation for the 13 loci among 11 Caucasian populations across Europe using F_{ST} (Budowle and Chakraborty, 2001) and for nine loci among the five major human groups—African, Caucasian, Asian, Native American, and Oceanic—using G_{ST} (Sun *et al.*, 2003).

Locus	*Number of alleles*	H_E	F_{ST} *(Europeans)*	G_{ST} *(five major groups)*
D3S1358	7	0.797	0.0049	0.027
VWA	8	0.813	0.0022	0.043
FGA	12	0.862	0.0029	0.019
D8S1179	10	0.799	0.0042	0.019
D21S11	11	0.856	0.0017	0.025
D18S51	13	0.878	0.0021	0.016
D5S818	7	0.683	0.0025	0.037
D13S317	7	0.773	0.0001	0.062
D7S820	9	0.808	0.0027	0.022
CSFiPO	10	0.736	−0.0004	—
TPOX	6	0.623	0.0011	—
TH01	7	0.785	0.0074	—
D16S539	8	0.769	0.0040	—
Mean	8.8	0.783	0.0028	0.030

produced clonally from mammary tissue from her mother as was reported (Example 11.10).

Sometimes when the suspect is confronted with the evidence that their genotype matches that found in a sample at the crime scene, they confess

---

Example 11.10. The first cloned adult mammal, Dolly the sheep, was produced by somatic cell nuclear transfer from mammary tissue from a six-year-old Finn Dorset ewe. Because some questions were raised as to whether Dolly was derived from embryonic or fetal cells, Ashworth *et al.* (1998) examined nine polymorphic microsatellite loci in Dolly and the population of Finn Dorset sheep at the Hannah Research Institute in Scotland. Table 11.19 gives Dolly's genotypes and the frequency of these genotypes, assuming Hardy–Weinberg proportions, in a sample of 44 Finn Dorset sheep. Because Dolly's genotype at some of these loci is expected to be quite rare, the overall probability that her genotype (and that of her mother) came from another local Finn Dorset sheep (the product of the genotype frequencies for the individual loci) is only 5.13×10^{-11}. From this examination, it appears very unlikely that Dolly was derived from another Finn Dorset sheep because of experimental error. Further, the probability that Dolly was derived from a fetal cell from her mother (and an unknown ram), in which case half of her alleles would be like Dolly, is approximately 10^{-6}. In addition, using independent minisatellite loci to determine DNA fingerprints, Signer *et al.* (1998) calculated the probability that Dolly came from another local Finn Dorset sheep as 6×10^{-10} and the probability of her coming from a fetal cell as less than 10^{-6}. Overall, these results strongly support the assertion that Dolly was derived from her mother's mammary cell.

TABLE 11.19 The genotypes of the sheep, Dolly, for nine microsatellite loci, and the frequency of these genotypes in Finn Dorset sheep (Ashworth *et al.*, 1998).

Locus	*Number of alleles in population*	*Dolly's genotype*	*Genotype frequency in population*
TGLA53	7	151/151	0.302
SPS115	4	248/248	0.048
TGLA126	7	118/126	0.082
TGKA122	9	190/190	0.005
ETH3	4	104/106	0.206
ETH225	4	148/150	0.025
FCB11	6	124/126	0.036
MAF209	4	109/121	0.194
FCB128	5	208/208	0.240

---

to the crime. On the other hand, when the suspect claims that the match between them and the crime scene sample is a coincidence, then it is useful to compare the probabilities under the two different explanations, that is, when the sample at the crime scene came from the suspect, event S, or when the sample at the crime scene came from someone else besides the suspect, event $\overline{S}$ (not S). One approach is to compare these conditional probabilities in the likelihood ratio

$$L = \frac{\Pr(G_C = G_S|S)}{\Pr(G_C = G_S|\overline{S})} \tag{11.10a}$$

Of course, if the suspect is the perpetrator of the crime, then the probability that the genotype at the crime scene and that of the suspect would match is $\Pr(G_C = G_S|S) = 1$. If it is also assumed that the genotypes are independent, then

$$\begin{aligned} L &= \frac{1}{\Pr(G_C = G_S|\overline{S})} \\ &= \frac{1}{\Pr(G_C)} \end{aligned} \tag{11.10b}$$

If the probability of the genotype observed at the crime scene is small, then the likelihood that the suspect left the DNA evidence at the crime scene is high. A similar approach has been used to identify victims from disasters, such as the attack on the World Trade Center (Example 11.11).

Example 11.11. DNA profiles have been used to identify of victims from airplane crashes (Goodwin *et al.*, 1999) and in the attack on the World Trade Center (WTC) on September 11, 2001 (Brenner and Weir, 2003). In the WTC disaster, 2792 people are known to have died, and approximately 15,000 body parts were recovered. Identification can be formulated in a standard forensic framework (Evett and Weir, 1998) with hypotheses as

H_1: the WTC sample is from victim X.
H_0: the WTC sample is not from victim X.

Based on the DNA evidence E the likelihood ratio becomes

$$L = \frac{\Pr(E|H_1)}{\Pr(E|H_0)}$$

The investigation adopted the approach that there are two ways in which a DNA match can occur. A body part sample is identified either if it is sufficiently similar to a known biological sample from the victim—these samples are mostly recovered from toothbrushes (78%) but also from hair—or if the sample is sufficiently similar (using different standards) to a DNA profile from a relative.

One of the complications is that more than one body part sample could be from the same victim. In this case the hypotheses are

H_1: two WTC samples are from the same victim.
H_0: two WTC samples are from different victims.

For example, Table 11.20 gives the genotypes obtained from a toothbrush sample from a victim and two body part sample from the WTC. The first sample is identical to the known biological sample, whereas the second sample differs from the toothbrush sample at two genes. Both of these appear to be the result of **allele dropout** or loss of amplification of a second allele in a heterozygote because of a degraded sample. In this case, it was concluded that sample 2 was also from this victim.

TABLE 11.20 Genotypes for the 13 CODIS microsatellite loci from a toothbrush of a World Trade Center victim and two body part samples recovered from the disaster site (Brenner and Weir, 2003). The differences between the known genotype of the victim and that in sample 2 are boldfaced.

		World Trade Center samples	
Locus	*Victim*	*1*	*2*
D3S1358	14, 18	14, 18	14, 18
VWA	16, 17	16, 17	16, 17
FGA	22, 24	22, 24	22, 24
D8S1179	13, 13	13, 13	13, 13
D21S11	28, 29	28, 29	28, 29
D18S51	12, 13	12, 13	**12, 12**
D5S818	11, 12	11, 12	11, 12
D13S317	8, 14	8, 14	8, 14
D7S820	11, 12	11, 12	11, 12
CSFiPO	11, 12	11, 12	11, 12
TPOX	8, 8	8, 8	8, 8
TH01	7, 8	7, 8	**8, 8**
D16S539	11, 13	11, 13	—

The use of these general approaches in forensics raised a number of concerns (Lewontin and Hartl, 1991) and resulted in the convening of two National Research Council panels (National Research Council, 1992, 1996) to examine these issues. First, expression 11.9*a* assumes that there are

Hardy–Weinberg proportions at each locus, and that expression 11.9*b* assumes that there is no association between loci (no linkage disequilibrium, see Chapter 10). However, these assumptions are unlikely to influence the general calculations greatly because there appears to be close to Hardy-Weinberg proportions (Budowle *et al.*, 2001) and linkage equilibrium (between unlinked loci) in most human populations (Sun *et al.*, 2003).

Another concern is whether the DNA is compared with a proper subpopulation group. In the case of the data for Dolly presented in Example 11.10, that group was a sample of sheep of the same breed in the same flock. There may be important differences in allele frequencies in different ethnic subpopulations, and these differences may influence the calculations. For example, matches may be higher when a person is compared with their own ethnic group than if they are compared with the whole population. Others suggested that the low amount of population substructure generally observed would have a relatively minor effect on the calculations. The first report (National Research Council, 1992) reached a compromise between these views using a modification of expression 11.9*a* called the **ceiling principle**. Under this procedure, each allele frequency was equal to the larger of either 0.10 or the upper 95% of the highest frequency of the allele observed in surveys of several racial groups.

Not all population geneticists thought this compromise appropriate (Weir, 1993; Devlin *et al.*, 1993). As a result, the second panel (National Research Council, 1996) suggested that another modification of the product formula be used in calculating the probability of a match. When there is population structure, the expected frequency of the genotypes becomes

$$P' = \prod [p_i^2 + p_i(1 - p_i)F_{ST}] \prod [2p_ip_j(1 - F_{ST})] \qquad (11.11a)$$

where the first term on the right side of the equation is used for the m homozygotes and the second for the n heterozygotes. F_{ST} here indicates the amount of substructure among the major racial groups. The panel pointed out that if F_{ST} is greater than zero, that for homozygotes $2p_i > p_i^2 + p_i(1 - p_i)F_{ST}$ and for heterozygotes $2p_ip_j > 2p_ip_j(1 - F_{ST})$. Therefore, they recommended that probabilities be based on

$$P'' = \prod 2p_i \prod 2p_ip_j \qquad (11.11b)$$

which is always larger than that give by P'. In addition to recommending this general approach for calculating probabilities, the panel also stated that all probability values should be accompanied by the appropriate 95% confidence intervals and gave other formulas to use when there was extensive inbreeding or population substructure.

Specifically, the National Research Council (1996) recommended that a conservative value of population differentiation, $F_{ST} = 0.01$, be used, although they suggested that $F_{ST} = 0.03$ might be used until more data were obtained for isolated populations such as Native Americans (see also Curran *et al.*, 2003: Sun *et al.*, 2003). Budowle and Chakraborty (2001) estimated the amount of differentiation for 11 European populations for the 13 CODIS loci and found that F_{ST} was less than 0.01 for all loci and that the mean value was 0.0028 (Table 11.18). In other words, the observed level of heterogeneity appears to be quite low and probably of little consequence for these calculations in Caucasian populations, consistent with theoretical predictions of Li and Chakravarti (1994). Low levels were also observed for nine loci among African populations (0.0018), slightly higher values in Asian populations (0.0048), and significantly higher ones among Native American (0.041) and among Oceanic (0.027) populations (Sun *et al.*, 2003). Finally, if variation among these five major human groups are examined for these nine loci (Table 11.18), then the level of differentiation is 0.030, the level suggested by the National Research Council (1996).

An estimator of individual identification that has been used in natural animal populations is the **probability of identity, P_{ID}**, which for locus k is

$$P_{ID.k} = \sum_i p_i^4 + \sum_i \sum_{i<j} (2p_i p_j)^2 \qquad (11.12a)$$

This is the probability that two individuals (or samples) drawn at random from the same population will have the same genotype (Paetkau and Strobeck, 1994). The first term is the probability of identity for homozygotes, and the second term is the probability of identity of heterozygotes. The probability of identity is the square of the match probability for each genotype, summed over all possible genotypes. Assuming independence over loci, then the estimate over all m loci is

$$P_{ID} = \prod_k^m P_{ID.k} \qquad (11.12b)$$

Individual identity has been used to estimate population size, monitor individual movements, identify predatory animals that have attacked livestock or humans, and identify individuals in wildlife forensics (Waits *et al.*, 2001). In particular, genetic forensic techniques are now widely used to investigate trade in parts or products from endangered species. As an example, genetic analysis of meat of an individual protected blue/fin whale hybrid harpooned near Iceland in 1989 was found in the Japanese whale market (Cipriano and Palumbi, 1999).

Natural populations of many plants and animals are more likely to contain related individuals or to have substantial substructure than human populations. For comparison to the above situation where the individuals are not assume to be related, Waits *et al.* (2001) calculated the probability of identity between full siblings at locus k as

$$P_{ID.sib.k} = \tfrac{1}{4}(1 - \sum_i p_i^4) + \tfrac{1}{2}[\sum_i p_i^2 + (\sum_i p_i^2)^2] \qquad (11.12c)$$

and

$$P_{ID.sib} = \prod_k^m P_{ID.sib.k} \qquad (11.12d)$$

Waits *et al.* (2001) then compared the observed probability of identity for populations of bears and wolves with that expected from random individuals and from full sibs (Figure 11.15). The observed P_{ID} was estimated by calculating the proportion of all possible pairs of individuals that have identical genotypes within a populations. The P_{ID} values were calculated by sequentially adding one locus at a time in descending order of heterozygosity. The observed P_{ID} values are much higher than that expected between unrelated individuals but less than expected between full sibs. Waits *et al.* (2001) suggested that the tendency of brown bear cubs to remain with their mothers for several years and the family pack structure of wolves result in the presence of close relatives in both data sets and the high observed P_{ID} levels.

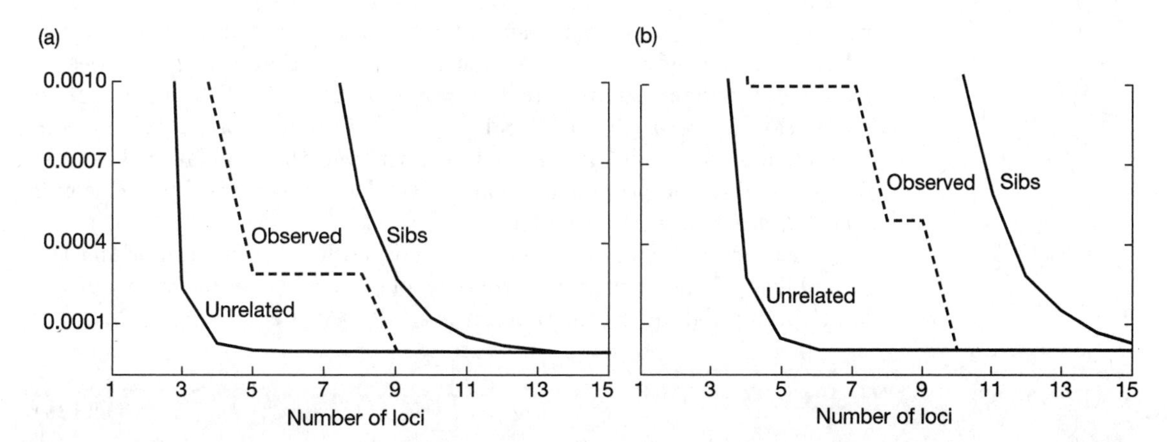

Figure 11.15. The unrelated, sib, and observed probabilities of identity for microsatellite loci in (a) brown bears and (b) wolves (after Waits *et al.*, 2001). Loci on the horizontal axis are in order of decreasing heterozygosity.

An increasing number of studies use **noninvasive** genetic sampling of hair or scat samples to identify individual wild animals that are either difficult or impossible to capture. Such samples may have low amounts or low-quality DNA that can result in genotyping errors or contamination (Taberlet *et al.*, 1999). For example, errors in genotyping may result in artifactual genotypes and a consequent overestimate of population size (Waits and Leberg, 2000; Creel *et al.*, 2003). Miller *et al.* (2002) suggested a maximum-likelihood approach to minimize such errors, and Paetkau (2003) recommended that many problems can be overcome with careful laboratory protocols and technique. Eggert *et al.* (2003) found that their population size estimate of forest elephants using microsatellite loci was close to the most common indirect method, dung counts.

d. Population Assignment

As we discussed in Chapter 9, there are a number of indirect genetic approaches that have been used to measure interpopulation dispersal. In addition, using multilocus genotypes at highly variable loci can be used to assign individuals to various groups (Davies *et al.*, 1999). The simplest approach for population assignment (Paetkau *et al.*, 1995) uses the estimated allele frequencies in different populations to calculate a measure similar to the match probability discussed above as

$$\begin{aligned} P_{k.l} &= p_{i.l}^2 \text{ for homozygote } A_iA_i \text{ in population } l \\ P_{k.l} &= 2p_{i.l}p_{j.l} \text{ for heterozygote } A_iA_j \text{ in population } l \end{aligned} \qquad (11.13a)$$

where $p_{i.l}$ and $p_{j.l}$ are the frequencies of the alleles at locus k in population l found in the individual. For example, assume that the frequencies of two alleles in population 1 are $p_{i.1} = 0.08$ and $p_{j.1} = 0.19$, and those in population 2 are $p_{i.2} = 0.28$ and $p_{j.2} = 0.36$. The probability that the individual A_iA_j would be randomly drawn from the population 1 is only 0.030, whereas the probability that it would be drawn from population 2 is 0.202, nearly seven times higher.

Then, over all loci examined, assuming no linkage disequilibrium, the probability of the multilocus genotype observed in the individual being found in population l is the product over all m loci or

$$P_l = \prod_k^m P_{k.l} \qquad (11.13b)$$

More precisely, Waser and Strobeck (1998) suggested that the test individual be removed from the estimate of allele frequency for the population it was sampled from and that the $\log_{10}P_l$ be used as the measure. In addition, they recommended that if an allele in the test individual is not present in one of the populations that $1/(2N)$ (N is the sample size for a given population) be added to this allele frequency in all populations.

To illustrate how this approach works, Waser and Strobeck (1998) simulated two populations connected by a low probability of gene flow (one disperser every other generation, $Nm = 0.5$) and calculated the log likelihood values for individuals found in the two populations. Figure 11.16a gives the scatterplot for the individuals found in the two different populations for a given generation. Virtually all of the individuals had a higher log likelihood for the population in which they were found. Only four individuals were misassigned. Two individuals found in population A had slightly higher values in population B, and one individual found in population B had a slightly higher value in population A (these individuals were probably descendants of previous dispersers). One individual found in population B had a much higher value in population A (it was probably an immigrant into population B). Carmichael *et al.* (2001) used this approach in wolves of the Canadian northwest and found evidence of movement of two wolves from Banks Island to a mainland site (Figure 11.16b). Paetkau *et al.* (2004) examined the statistical power and accuracy to detect recent migrants using genetic assignment methods, and Miller *et al.* (2003) developed assignment tests to estimate the extent of red wolf and coyote ancestry in the reintroduced red wolf population. An interesting application of assignment tests is in a case of a fishing competition fraud (Example 11.12).

Example 11.12. Competitions to catch the largest fish in a given lake occur in many countries. However, when a competitor presented a 5.5-kg Atlantic salmon to the judges for a fishing contest in Lake Saimaa, Finland, its size and preparation arose suspicions. To determine whether the suspect fish was legitimately caught, the judges had a genetic analysis carried out, comparing the genotype of the fish to that expected for a salmon from Lake Saimaa and several other reference populations.

Using seven microsatellite loci, Primmer *et al.* (2000) calculated the log likelihood that the suspect fish came from the landlocked Lake Saimaa. For comparison, they also calculated the probability that random fish taken from Lake Saimaa or from three different anadromous (migrate to the sea) salmon populations came from Lake Saimaa (Figure 11.17a). Obviously, the log likelihood that the suspect fish came from Lake Saimaa was very low (-22.3), and the probability that it came from the other populations was also very low. Primmer *et al.* (2000) concluded that the suspect fish probably came from the Baltic or Norwegian Seas, and as a result, the fisherman

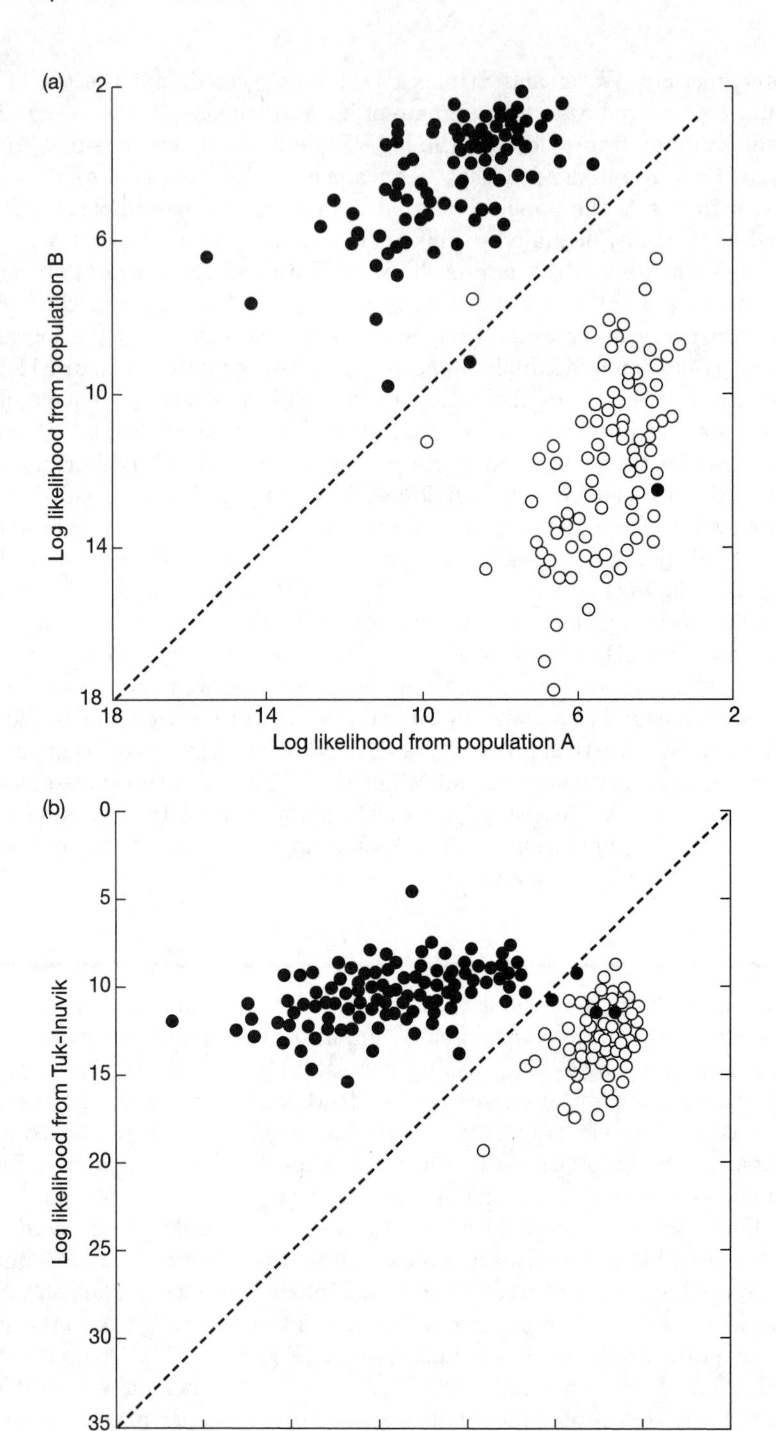

Figure 11.16. (a) The probability (given as negative $\log_{10}$ likelihood) that individuals found in population A (open circles) came from populations A or B and probability that individuals found in population B (closed circles) came from populations A or B (after Waser and Strobeck, 1998). The four misassigned individuals had either slightly larger log likelihood values in the nonnatal population or had values similar to those observed for individuals born in the other population. (b) The probability that wolves found on Banks Island (open circles) or on the mainland site Tuk-Inuvik (closed circles) came from the two populations (Carmichael *et al.* 2001). Two Tuk-Inuvik wolves appear to be migrants from Banks Island (closed circles below the diagonal line).

was consequently disqualified from the contest. When confronted with the evidence, he confessed to having purchased the salmon at a local fish shop.

As a general perspective as to how different this fish was from what would be expected, we can examine Figure 11.17b (Cornuet *et al.*, 1999).

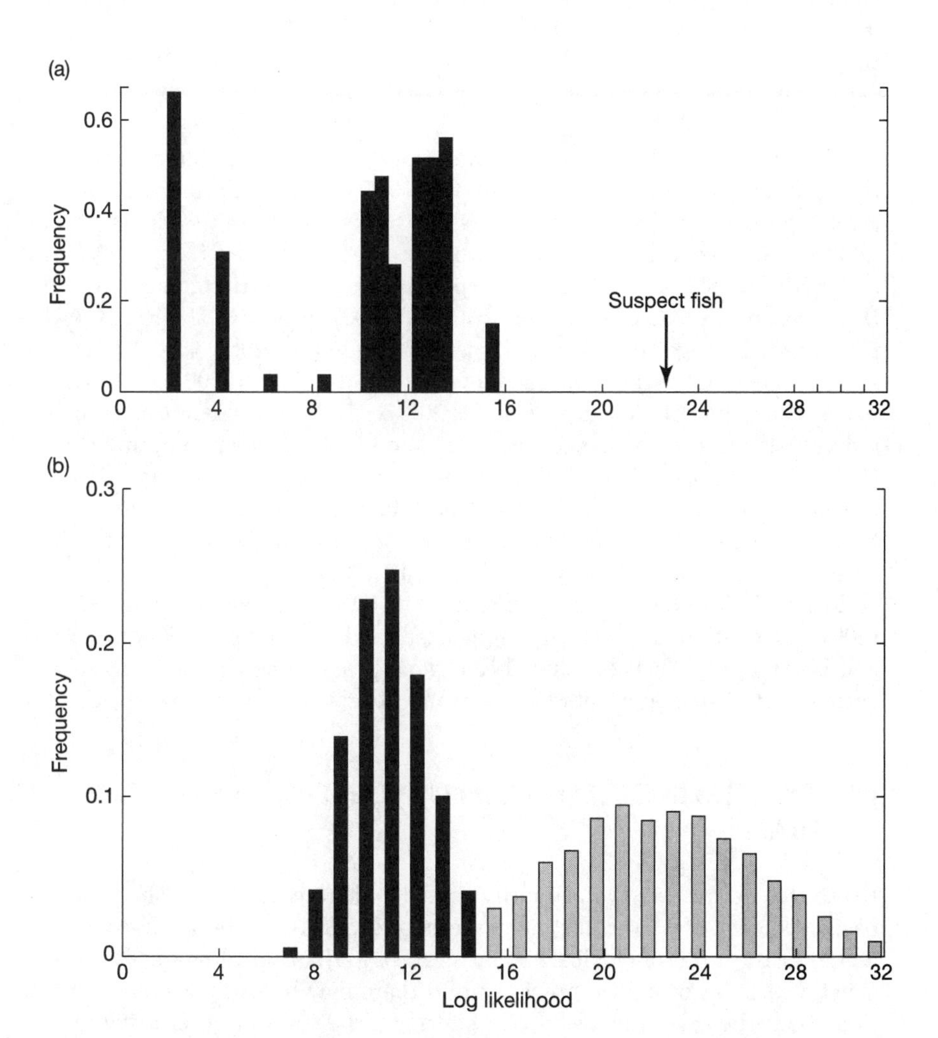

Figure 11.17. (a) The probability (given as $-\log_{10}$ likelihood) that the suspect fish came from Lake Saimaa (arrow) (after Primmer *et al.*, 2000). Also given is the probability that a random fish from Lake Saimaa came from Lake Saimaa (two leftmost histogram bars) and the probability that a random fish from three other reference populations came from Lake Saimaa (other histogram bars). (b) The probability that a random individual came from the population from which it was taken (histogram with dark bars to left) and the probability that a random individual from a population that has been isolated from the ancestral population for 200 generations is from the ancestral population (after Cornuet *et al.*, 1999).

Here the histogram on the left gives the log likelihood values for individuals drawn at random from a population. The histogram to the right gives the log likelihood for individuals drawn from a population that has been isolated for 200 generations with $N_e = 1000$ and $u = 5 \times 10^{-4}$. The mean value for this group is similar to that for the suspect fish, suggesting that it comes from a source that has been isolated from Lake Saimaa for some time.

Cornuet *et al.* (1999) found that the partial Baysian method based on the approach of Rannala and Mountain (1997) was slightly more accurate than the frequency-based approach we have discussed above. Manel *et al.* (2002) evaluated the approach as implemented by Cornuet *et al.* (1999) in GENECLASS (http://www.montpellier.inra.fr/CBGP/softwares/) and the Baysian approach of Pritchard *et al.* (2000) in STRUCTURE (www.pritch.bsd.uchicago.edu) using large microsatellite data sets from 10 species. Generally, the approach of Pritchard *et al.* (2000) performed better than that of Cornuet *et al.* (1999), with the median proportion of individuals correctly assigned being 0.61 and 0.36 for the two approaches. However, the approach of Pritchard *et al.* requires that the true population of origin was sampled. Manel *et al.* (2002) concluded that individuals from two different populations with $F_{ST} > 0.15$ can be assigned to the correct population with very high certainty (99.9%), given that there are 10 loci with $H > 0.6$ and a sample of 30 to 50 individuals. Eldridge *et al.* (2001) found that all three methods identified dispersing rock wallabies, and Berry *et al.* (2004) concluded that genetic assignments were consistent with known dispersal events from capture–recapture data in grand skinks.

III. IDENTIFYING GENES INFLUENCING QUANTITATIVE TRAITS

Up to this point, we have primarily considered traits for which the genetic basis is known. However, for many traits of evolutionary importance, such as life history, morphological, behavioral, or physiological characters, the genetic basis is not known precisely, and there may be little or no information about the genes involved, their effects on the trait, or other attributes. In addition, the underlying genetic cause of many complex human diseases, such as diabetes, heart disease, and cancer, and the genetic basis of the many economically important traits in plants and animals used in agricultural are generally not known. Such complex genetic traits are generally known as **quantitative traits** because either their phenotypes are distributed on a continuous scale or their underlying genetic determination is assumed to be (see p. 53 for an introduction).

In the past, it was generally very difficult to determine the number of genes and their effects on quantitative traits. However, with the great effort being put into projects to sequence the genome of various organisms, many genetic markers have become available. These genetic markers then have been used to identify the location of genes influencing quantitative traits, called **quantitative trait loci (QTLs)**. Sometimes these QTLs appear to map genetically to the location of **candidate genes**—particular genes recognized as likely to be related to the trait by either their function or structure. A number of QTLs with major effects on evolutionary important traits have been identified, and it seems highly likely that many more will be characterized in the near future (see the discussion of QTLs that effect life span in *D. melanogaster*, muscle mass and backfat in pigs, and terrestrial or aquatic forms in salamanders on p. 55). For example, genes involved with speciation in *Drosophila* have been identified (Presgraves *et al.*, 2003; Wu and Ting, 2003) and QTLs responsible for recent adaptation in threespine sticklebacks located (Colosimo *et al.*, 2004; Cresko *et al.*, 2004). As the introductory quote for this chapter by Falconer and Mackay (1996) stated, there is great optimism that many genes that influence quantitative traits, such as those important in agricultural and human traits and diseases, can be identified.

Quantitative genetics has a rich history of empirical, experimental and theoretical research, primarily based on a statistical description of quantitative genetic traits (Falconer and Mackay, 1996; Lynch and Walsh, 1998). Although the exact genes determining quantitative traits were not generally known, a statistical connection between the phenotype and its genetic basis was assumed. A large number of unknown genes of small, additive effects have often been assumed in these treatments, but extensions of these approaches can accommodate more biologically realistic assumptions. In any case, these techniques will still provide invaluable heuristic tools and a general framework to understand and predict change and stasis for quantitative traits.

Let us begin the discussion of genes influencing quantitative traits by introducing a statistical approach that has been used to estimate the number of genes affecting a trait. When directional selection is practiced in a population for a number of generations, the amount of response can be envisioned as the cumulative response for all of the loci affecting the trait combined. Using a number of assumptions—such as that all loci have equal effects on the trait and that no epistasis and no linkage exist—we can obtain an estimate of the number of loci from a directional selection experiment (Lynch and Walsh, 1998). This is the approach used by Dudley and Lambert (2004) to estimate the minimum number of genes affecting oil and protein content in corn, 56 and 123, respectively, discussed on p. 54. However, these approaches do not include the effect of new mutations arising during the course of the experiment, an effect that may greatly increase the selection response (Hill, 1982).

The number of genes can also be estimated from the means and variances of parental lines, F_1, F_2, and backcrosses for two populations that have diverged by either artificial or natural selection (Lande, 1981; Cockerham, 1986; Lynch and Walsh, 1998). In the simplest approach, the minimum estimate of the number of effective genes is

$$n_e = \frac{(\overline{P}_1 - \overline{P}_2)^2}{8V} \tag{11.14}$$

where $\overline{P}_1$ and $\overline{P}_2$ are the phenotypic means of the two parental populations and V is the genetic variance of the F_2 generation between the two populations. If it is assumed that there are n unlinked, additive genes all with the same effect that differ between the two populations, then $\overline{P}_1 - \overline{P}_2 = 2n$ and $V = n/2$, and by substitution, then $n_e = n$. The basis of this relationship is that as the difference between the two populations is split between more and more genes, then the F_2 genetic variance is expected to be smaller and smaller. Generally, if genes are linked, have unequal effects, are not additive, or are epistatic, then the gene number estimate is lower so that such estimates are generally assumed to be minimum estimates.

Lande (1981) examined a number of data sets in different organisms and concluded that at least 5 to 10 genes are responsible for differences between populations. For example, he analyzed data from two varieties of tomatoes that greatly differed in fruit weight. The scaled log difference between the two varieties was 1.826 grams, and an estimate of the F_2 genetic variance is the F_2 phenotypic variance (0.0570) minus the F_1 phenotypic variance (0.0144), or 0.0426. This calculation assumes that the parental tomato varieties were homozygous but different so that the F_1 is also genetically invariant (but heterozygous at some genes). Therefore, any F_1 phenotypic variance is environmental, and if this is subtracted from the F_2 phenotypic variance, only the genetic variance should remain. In this case, $n_e = (1.826)^2/[(8)(0.0426)] = 9.8$, or 10 or more genetic factors appear responsible for the observed differences.

a. Identifying QTLs by Crosses

Although the above approach is useful to give a general idea about how many genes would be expected to influence a trait, to determine the number and location of QTLs, a genetic mapping approach is used. The traditional statistical techniques used for mapping the location of genes are relatively straightforward but become complex when there are many markers and QTLs (for details of these different approaches, see Lynch and Walsh, 1998). Most commonly, two different lines that are different in their phenotypic value for a trait can be crossed, and segregation for a marker locus (or loci) and the phenotypic values in F_2 or backcross progeny can be used to identify the location of QTLs. When lines that differ in both marker loci and QTLs are crossed, linkage disequilibrium is generated between the loci; it is then used to detect the presence and location of QTLs.

To illustrate the simplest approach, called **single-marker analysis**, let us assume that the parents were fixed for different alleles at both the

TABLE 11.21 The frequency of genotypes in the F_2 generation and the means for different marker classes when gene A contributes to the trait of interest having genotypic values of a, d, and $-a$ for genotypes A_1A_1, A_1A_2, and A_2A_2, gene M is a linked marker locus, and c is the amount of recombination between them.

Genotype	*Frequency*	*Genotypic value*
M_1A_1/M_1A_1	$(1-c)^2/4$	a
M_1A_1/M_1A_2	$c(1-c)/2$	d
M_1A_2/M_1A_2	$c^2/4$	$-a$
M_1A_1/M_2A_1	$c(1-c)/2$	a
M_1A_1/M_2A_2	$(1-c)^2/2$	d
M_2A_1/M_1A_2	$c^2/2$	d
M_2A_1/M_2A_2	$c(1-c)/2$	$-a$
M_2A_1/M_2A_1	$c^2/4$	a
M_2A_1/M_2A_2	$c(1-c)/2$	d
M_2A_2/M_2A_2	$(1-c)^2/4$	$-a$

marker locus M and locus A that influences the quantitative trait of interest so that their genotypes are M_1A_1/M_1A_1 and M_2A_2/M_2A_2, complete linkage disequilibrium. The F_1 genotype is then M_1A_1/M_2A_2, and the expected genotypes in the F_2 progeny are given in Table 11.21, where c is the amount of recombination between the loci. It is easiest to use the parameterization that assumes that genotypes A_1A_1, A_1A_2, and A_2A_2 have values of a, d, and $-a$ where a is the additive effect of allele A_1 and d is the measure of dominance (difference from complete additivity) (Falconer and Mackay, 1996). We can then calculate the expected phenotypic values for the three marker genotypes.

Using the sum of the products of the genotypic values and the frequencies for the three lines in Table 11.21 that have genotype M_1M_1, the contribution of genotype M_1M_1 to the F_2 is

$$\begin{aligned} M_1M_1 &= a(1-c)^2/4 + dc(1-c)/2 - ac^2/4 \\ &= \tfrac{1}{4}[a(1-2c) + 2dc(1-c)] \end{aligned}$$

Because $1/4$, $1/2$, and $1/4$ of the genotypes in the F_2 are M_1M_1, M_1M_2, and M_2M_2, the contribution for each genotype must be divided by these proportions to give

$$\begin{aligned} M_1M_1 &= a(1-2c) + 2dc(1-c) \\ M_1M_2 &= d[(1-c)^2 + c^2] \qquad (11.15a) \\ M_2M_2 &= -a(1-2c) + 2dc(1-c) \end{aligned}$$

An estimate of the additive effect can then be obtained from

$$\tfrac{1}{2}(M_1M_1 - M_2M_2) = a(1-2c) \qquad (11.15b)$$

For example, if $c = 0$, which means no recombination between the marker locus and the QTL (the marker locus ***is*** the QTL), then this estimate is exactly a. However, if the recombination level between the marker locus and the QTL is c, then the effect is underestimated by an amount $(1 - 2c)$ because recombination, in producing the F_2 progeny, would have reduced the association between the alleles at the two loci. In a similar manner, an estimate of d can be obtained from

$$M_1M_2 - \tfrac{1}{2}(M_1M_1 - M_2M_2) = d(1-c)^2 \qquad (11.15c)$$

In this case, the estimate of d is underestimated by $(1 - c)^2$ because of recombination. Example 11.13 discusses the first example of an association between a marker locus and quantitative trait for seed size in a bean.

Example 11.13. The first example of an association between a marker locus and a quantitative trait (Sax, 1923) was in the bean *Phaseolus vulgaris*, for a pigment gene and seed size (from Falconer and Mackay, 1996). The two parental lines used in the cross greatly differed in seed size, the line fixed for pigment marker allele M_1 having a seed weight of 48 centigrams (cg) and the line fixed for M_2 having a seed weight of only 21 cg. The mean seed weights for the three genotypes in the F_2 were

Genotype	M_1M_1	M_1M_2	M_2M_2
Seed weight	30.7	28.3	26.4

To estimate the additive effect, from expression 11.15*b* we get $a(1 - 2c) = 0.5(30.7 - 26.4) = 2.15$ cg. The difference in seed weight between the two F_2 homozygotes, 4.3 cg, explains about 16% of the 27-cg difference between the seed weights in the two parental lines. It is possible that this large effect was due to the pigment gene used as the marker; that is, the marker gene had a pleiotropic effect on seed weight. If the effect was due to another locus linked to a pigment gene, then the effect may actually have been somewhat larger if recombination, in producing the F_2, reduced the association. For example, if the recombination level value between the QTL and the marker was $c = 0.1$, then $a = 2.69$. Using expression 11.15*c*, $d(1 - c)^2 = -0.25$ cg. As a result, it appears that there is very little dominance for the gene affecting seed weight.

If a single marker is used, then the estimates of the additive and dominance effects are confounded with the amount of recombination. On the other hand, if multiple markers are used and their map positions are known, then each pair of adjacent markers can be examined for an association with a QTL. The approach, called **interval mapping** (or flanker–marker analysis), allows independent estimation of the position and the effect of the QTL (Lynch and Walsh, 1998).

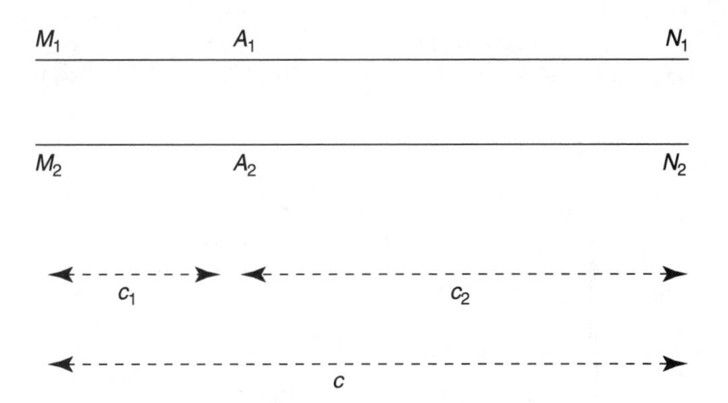

Figure 11.18. The rates of recombination between two marker loci, genes M and N, and a QTL, gene A.

To illustrate, Figure 11.18 gives the position of two marker loci, M and N, with a known level of recombination c between them, that are on either side of the QTL, again indicated as gene A. The unknown recombination between genes M and A is c_1, and the unknown recombination between genes A and N is c_2. Because $c_1 < c_2$ in this example, if F_2 progeny are observed, then A_1 should be more associated with M_1 than with N_1 (also, A_2 should be more associated with M_2 than with N_2). If A_1A_1 results in a higher value of the trait than A_2A_2 and the heterozygote is intermediate, then this would result in a higher mean for M_1M_1 homozygotes than for N_1N_1 homozygotes because of the tighter linkage of genes M and A than for A and N. In this way, both c_1 ($c_2 = c - c_1$ and c is known) and the effect of gene A on the trait can be estimated (Lynch and Walsh, 1998). Generally, more sophisticated approaches are employed that use more than two markers simultaneously (Lynch and Walsh, 1998).

To present graphically the statistical significance of an association with particular map locations, a likelihood map for QTL positions along the mapped chromosomes is often given. This approach was first given by Lander and Botstein (1989) who used likelihood of odds (LOD) scores, values that are related to the likelihood-ratio statistic. The $\log_{10}$ LOD score at a particular location is related to the probability that a QTL is located at a particular position. Generally, a LOD score needs to be greater than 2 for significance, but often LOD scores of 3 or higher are necessary for significance because of multiple tests. The higher the LOD score at a location, the more likely that it is the position of a QTL.

For example, it has long been known that there is substantial genetic variation influencing height in humans resulting in very high (around 0.8) estimates of heritability (see p. 55). With the large number of genetic markers available in the human genome, genome-wide scans have been conducted to determine the location of QTLs affecting height. Figure 11.19 gives the LOD scores for height over the human genome from a segregation analysis of 200 Dutch families using 344 evenly spaced autosomal markers

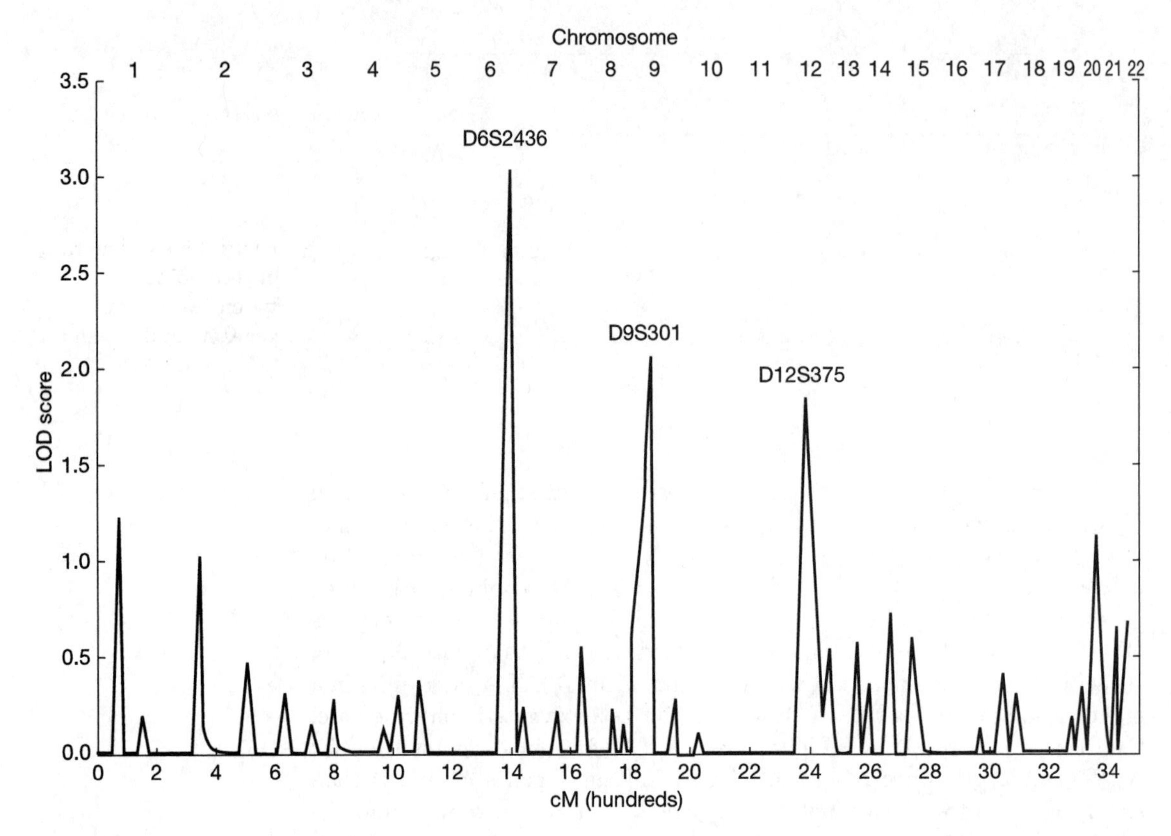

Figure 11.19. The results of a genome-wide scan for genes influencing height in humans as indicated by the LOD score over the 22 different autosomal chromosomes (across top) and over the cumulative genetic map (across bottom) (Xu *et al.*, 2002). The three peaks are identified by the microsatellite markers with highest LOD scores.

(Xu *et al.*, 2002). Three regions, on chromosomes 6, 9, and 12, gave statistically significant LOD scores. All three of these regions were also identified as influencing height in an independent study from four other European populations (Hirschhorn *et al.*, 2001).

With the availability of many markers in humans (and some other organisms), there have been a large number of studies attempting to locate genes for many complex behavioral traits. For example, studies have purported to locate genes for manic depression, schizophrenia, and homosexuality, although these findings have not been supported by later studies (see Example 11.14 for a discussion of the data related to a purported homosexuality gene). When there are many markers, then there are a large number of statistical tests carried out, and the probability of some significant results is high even when the significance level is set low. In other words, the probability of a false positive (type I) error, detecting an association when it really is only a chance effect, is high when many tests are conducted. Generally, a correction for multiple tests, such as the Bonferroni correction, can be used to reduce the α level used for significance. However, because many markers may be linked, a permutation test may be more appropriate (Churchill and Doerge, 1994). As a result, the LOD score necessary for statistical significance may be high if there are many tests.

Example 11.14. Hamer *et al.* (1993) presented data, derived from 40 families with gay brothers, that suggested there was a gene on the X chromosome for homosexuality, a finding that was greeted with a great deal of press coverage and editorial opinion. Rice *et al.* (1999) examined the same X chromosome region, Xq28, for 52 different gay male sibling pairs. Unlike the earlier study, there was no increased sharing of four highly variable microsatellite markers, marking a 12.5-cM segment, in the gay brothers over the 50% sharing expected. When LOD scores were calculated for the region, Rice *et al.* found that they were all negative and none were statistically significant (Figure 11.20). In fact, these LOD scores are so low that they can strongly exclude the presence of such a gene in this region in their sample. Because in theory there would be strong selection against a homosexuality gene, these negative results are not surprising. It is not clear whether the significance found by Hamer *et al.* resulted from type I error or occurred because they selected their study group in some way that biased their results, but it appears that this is another case of "mistaken identity" of a QTL for a complex behavioral trait in humans.

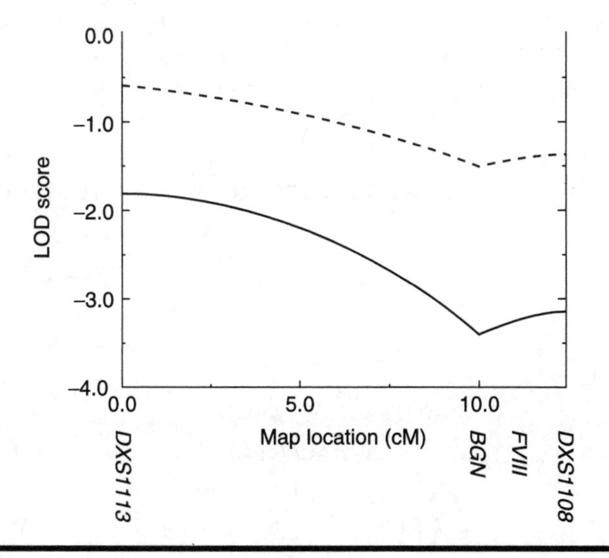

Figure 11.20. The LOD scores over the 12.5-cM region of the X chromosome examined by Rice *et al.* (after Rice *et al.*, 1999). The two lines indicate different values obtained using different ratios for homosexual orientation in the brothers of gay index subjects, as compared to the population frequency.

This brief introduction to locating QTLs using crosses illustrates that linkage disequilibrium between the QTL and a marker is an essential component. In addition, many of the approaches used in population genetics to quantify genetic variation and to predict the effects of evolutionary factors such as selection, inbreeding, genetic drift, gene flow, and mutation should be applicable to these genes influencing important evolutionary traits once they are identified. In other words, the population genetic principles that we have discussed throughout the book will provide the framework for understanding the evolution of QTLs and thereby quantitative traits.

b. Haplotype Mapping

When the Human Genome Project was nearing completion, some of the researchers involved were persuaded that to utilize this sequence information efficiently the construction of a haplotype map of the human genome was necessary and the International HapMap Project was conceived (http://www.hapmap.org/abouthapmap.html). The goal of this project is "to develop a haplotype map of the human genome, the HapMap, which will describe the common patterns of human DNA sequence variation." A main focus of the HapMap is to facilitate the identification of genes that cause common human diseases, such as diabetes, cancer, stroke, heart disease, depression, and asthma, and to discover the DNA sequence variants that contribute to common disease risk.

One of the findings of the human genome project is that about 1 base in 300 is variable or that about 10 million bases (SNPs) are variable in the world's population (The International HapMap Consortium, 2003). Most of these SNP variants have arisen by a rare mutation (the rate is approximately 10^{-8} per site per generation) on a particular chromosome and thereby are initially in linkage disequilibrium with other variants on that chromosome or haplotype. To illustrate, the top of Figure 11.21 gives three

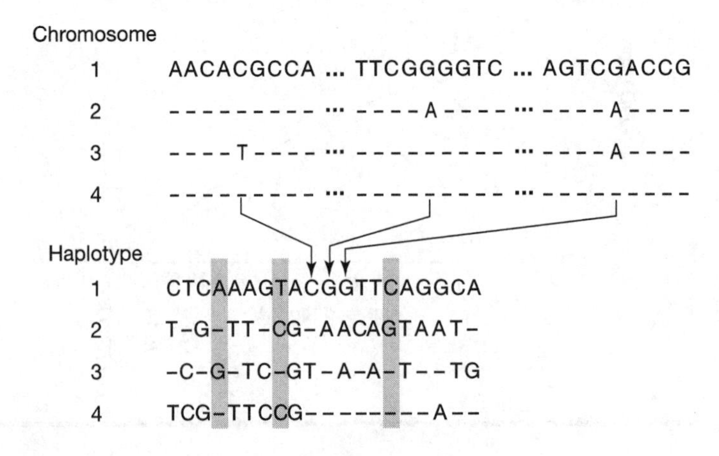

Figure 11.21. An example of variation at three SNPs (dashes indicate identity to the base in chromosome 1) for four different sampled chromosomes. Below are the 20 SNPs that vary over a 6000 base sequence and the three tag SNPs indicated by a shaded box that identify the sequences of the four haplotypes (after The International HapMap Consortium, 2003).

short stretches of a 6-kb sequence, each with a SNP in the center for four chromosomes from different individuals (The International HapMap Consortium, 2003). Over the 6-kb sequence, there are 20 SNPs, which form four different haplotypes as shown in the bottom of Figure 11.21. Also indicated in the shaded boxes are three **tagSNPs** that can be used to differentiate the four different haplotypes. That is, haplotypes 1, 2, 3, and 4 have bases ATC, ACG, GTC, and ACC, respectively, at the three tagSNPs and can be used instead of the 20 nucleotide sequences for all of the SNPs to distinguish between the haplotypes (Carlson *et al.*, 2004, evaluate an approach to determine the maximally informative set of tagSNPs).

Estimates are that much of the information about genetic variation across the human genome could be provided by 200,000 to 1,000,000 tagSNPs instead of examining all 10-million SNPs (Gabriel *et al.*, 2002; The International HapMap Consortium, 2003) (however, more SNPs may be necessary; Couzin, 2004). In other words, because of the linkage disequilibrium between closely linked SNPs, nearly the same information can be obtained by genotyping only the tag SNPs, rather than genotyping all of the SNPs in a region. Table 11.22 gives the most common haplotypes for three genes and their frequencies in large samples of Europeans (Johnson *et al.*, 2001). The sum of the frequencies of the haplotypes for genes *CFLAR*, *CASP10*, and *GAD2* are 0.99, 0.98, and 0.94, respectively, indicating that these few haplotypes constitute a very high proportion of all of the haplotypes at these genes. These common haplotypes can be ascertained by only 2, 4, and 2 tag SNPs for genes *CFLAR*, *CASP10*, and *GAD2*, respectively.

Gabriel *et al.* (2002) provided a definition of a **haplotype block** based on linkage disequilibrium. They suggested that a block is a region in which there is little or no evidence of recombination and that if the upper 95% confidence bound of $|D'| < 0.9$ between pairs of sites that there is evidence of historical recombination. They then defined a haplotype block as a region in which $< 5\%$ of the comparisons among SNPs show strong evidence of recombination.

Often these haplotype blocks that appear are separated by regions across which there is low linkage disequilibrium. For example, in one of the first major studies to document this pattern, Daly *et al.* (2001) used 103

TABLE 11.22 The most common European haplotypes in three genes and their frequencies (Johnson *et al.*, 2001). The tagSNPs are indicated by an asterisk.

CFLAR	*Frequency*	*CASP10*	*Frequency*	*GAD2*	*Frequency*
* *		* * * *		* *	
C A A T G G	0.46	C C A A C G A T G G G	0.44	A C C A G C G A C T C G T	0.45
- - G - T A	0.44	T - G - - - - A - C A	0.39	- - - - - - - - T - - - -	0.28
T - G - T A	0.09	T - G - - A - A - C A	0.07	- T A G A - - - T - - - A	0.13
		T - G - - - G A - C A	0.06	G - - - - - - - G A G T C -	0.08
		- T G G T - - - A - A	0.02		
Total	0.99		0.98		0.94

TABLE 11.23 The sequence and frequency of the six leftmost haplotype blocks on human chromosome 5 described by Daly *et al.* (2001). These blocks can be arranged in four ancestral long-range haplotypes (indicated as 1 to 4 on left), and a dash indicates a base identical to that in long-range haplotype 1.

	Block (length)					
	1 *(84 kb)*	*2* *(3 kb)*	*3* *(14 kb)*	*4* *(30 kb)*	*5* *(25 kb)*	*6* *(11 kb)*
Sequence						
1	GGACAACC	TTACG	CGGAACGA	CGCGCCCGGAT	CCAGC	CCGAT
2			GACTGTCG	TTGC- - - - - C-	-A-C-	
3	AATTCGTG	CCCAA	- - C- - - - -	-TGCTATAACG	G-GCT	-T- -C
4				-TGC- - -AACC	- - -C-	ATAC-
Frequency						
1	0.76	0.77	0.36	0.37	0.35	0.41
2			0.26	0.14	0.09	
3	0.18	0.19	0.28	0.19	0.13	0.38
4				0.21	0.35	0.18
Total	0.94	0.96	0.90	0.91	0.92	0.97

SNPs to determine the haplotype structure across 500 kb of chromosome 5. Table 11.23 gives the sequence for the SNPs in the 6 of the 11 blocks that Daly *et al.* (2001) reported and the frequencies of the common haplotypes. The numbers to the left indicate the four ancestral long-range haplotypes that they name. Some of these blocks are nearly 100 kb in length, and only a few haplotypes account for nearly all of the chromosomes sampled within a block. For example, in blocks 1 and 2, the two most common haplotypes constitute 94% and 96% of the total. In addition, these two haplotypes differ at all 13 sites for these blocks.

The blocks appear to be separated by regions of recombination, suggesting that there are local hotspots of recombination between the blocks (Daly *et al.*, 2001; Jeffreys *et al.*, 2001; McVean *et al.*, 2004). Although there is detectable recombination between the blocks, there is still some linkage disequilibrium over blocks so that the haplotypes within blocks can be assigned to one of four ancestral long-range haplotypes. Daly *et al.* (2001) stated that 38% of the chromosomes examined could be categorized as one of these four long-range haplotypes across the entire 500-kb region.

The stated goal of the HapMap project is "to develop a research tool that will help investigators discover the genetic factors that contribute to susceptibility to disease, to protection against illness and to drug response." Given the haplotype–block model of the genome and information about the most frequent haplotypes in a block, then it should be possible to scan across the region to determine an association of the disease phenotype and tag SNPs. This **association mapping** should be able to identify specific haplotypes with a high risk for the disease (Risch, 2000).

TABLE 11.24 The 13 SNPs for gene *GAD2* from Johnson *et al.* (2001) indicating the four most common haplotypes and their frequencies. Asterisks indicate variants at SNP 4 on haplotype 3 and at SNP 9 of haplotypes 2 and 3 discussed in the text as putative factors in causing a common disease.

	SNP													
Haplotype	*1*	*2*	*3*	*4*	*5*	*6*	*7*	*8*	*9*	*10*	*11*	*12*	*13*	*Frequency*
1	A	C	C	A	G	C	G	A	C	T	C	G	T	0.45
2	-	-	-	-	-	-	-	-	T*	-	-	-	-	0.28
3	-	T	A	G*	A	-	-	-	T*	-	-	-	A	0.13
4	G	-	-	-	-	-	-	G	A	G	T	C	-	0.08

The haplotype–block model suggests that disease alleles within a haplotype block would be in linkage disequilibrium with the known SNPs, or possibly some SNPs themselves may be causal for a disease. To understand some of the implications of this model, as a hypothetical example, let us look again at gene *GAD2* from Johnson *et al.* (2001) with the SNPs and haplotypes numbered (Table 11.24). First, if we assume that the G at SNP 4 in haplotype 3 is causal for the disease, then association studies should show that haplotype 3 has a higher disease risk because it is that only common haplotype with this base at this position. However, given the association of haplotype 3 with the disease, five bases at SNPs 2, 3, 4, 5, and 13 are only found in this haplotype, and because they appear to be in complete linkage disequilibrium, the causative SNP cannot be specifically identified without more information. Second, let us assume that haplotype 2 is associated with the disease phenotype. In this case, only the T at SNP 9 is different among these haplotypes and would therefore be considered the causative variant. However, T at SNP 9 is also present in haplotype 3. If haplotype 3 is not also associated with the disease, then one would assume that another unknown SNP may be the causal variant for this disease.

The haplotype mapping approach generally assumes that the risk of common genetic diseases is strongly influenced by common variants that increase susceptibility to the diseases, or the **common disease–common variant** (**CDCV**) hypothesis (Lohmueller *et al.*, 2003). Under this hypothesis, characterization of the common haplotypes in different haplotype blocks would be sufficient for initial identification of these common disease alleles. However, there are some concerns about this hypothesis and its generality. For example, common variants are generally expected to be older, and as a result, there will have been more time for them to recombine (Weiss and Clark, 2002). Therefore, the linkage disequilibrium signal in the surrounding haplotype needed to detect common disease variants may not be very strong. In addition, if there are any detrimental fitness effects related to the disease, then disease variants would be expected to be in low frequency. Indeed, Pritchard (2001) demonstrated, using a quantitative genetics disease model, that the much of genetic variance is expected to be the result of a number of disease susceptibility variants that are each low in frequency.

Overall, the intense effort in the HapMap Project to understand the detailed structure of the genome has already provided a detailed view of local linkage disequilibrium for some human genome regions. To explain these observed patterns, extensive theory has incorporated evolutionary genetic factors such as mutation, recombination, genetic drift, population structure, and gene flow. For example, there are studies that suggest that genetic drift or population history could be responsible (Stumpf and Goldstein, 2003; Zhang *et al.*, 2003a; Zhang *et al.*, 2003b; Anderson and Slatkin, 2004) and/or that spatial variation in recombination is important (Anderson and Slatkin, 2003; Wall and Pritchard, 2003). In all of these considerations, it is important to keep in mind that the rate of mutation per site and the rate of recombination, even in recombinational hotspots, are quite low. Therefore, stochastic effects play a large role in any particular observed pattern of local linkage disequilibrium and probably explain much of the variation and unpredictability of particular haplotype patterns.

PROBLEMS

1. Explain when a species tree and a gene tree may be different. How often do you think this may occur?
2. Three different approaches are given to construct phylogenetic trees in this chapter. What criteria would you use to evaluate which approach is the best one?
3. Construct the tree given in Figure 11.2a (without looking at it), using the UPGMA method and the data in Table 11.2.
4. Assume that you have the first three taxa in Table 11.6 and the fourth taxa is replaced with one that has the sequence ATATTGC. What are the informative sites? What is the most parsimonious tree?
5. What is reciprocal monophyly? Draw a phylogenetic tree that illustrates reciprocal monophyly, and draw one that does not.
6. Does it surprise you that contemporary and permafrost data in brown bears are so different? Explain you answer and discuss how contemporary data should be evaluated given this finding.
7. Assume that there are two alleles with frequencies 0.82 and 0.18. What is the probability of paternity exclusion? Why is it so low? If there are 20 such identical loci, what is the probability of exclusion?
8. Explain why the likelihoods in Table 11.14 are calculated as they are. Why is the value for locus TH01 higher than for locus D7S820?
9. In the O. J. Simpson trial, the only controversy about the DNA evidence was for item 29 (see Table 11.17). What do you think is the explanation, given the present evidence? What further analysis would you recommend to resolve this controversy?

10. Using expressions 11.12*a* and 11.12*c*, and assuming that $p_1 = p_2 = 0.5$, what is the probability of identity for two individuals when they are drawn at random from a population or are full sibs?

11. From the data in Table 11.19, calculate the probability of the multilocus genotype for Dolly. Can you think of any other explanation, besides that she was produced clonally, why Dolly would have the same genotype at these loci as her mother?

12. The F_2 progeny from a cross between inbred lines that differed in size gave values for the marker genotypes M_1M_1, M_1M_2, and M_2M_2 of 22.3, 20.1, and 15.3 cm, respectively. What are the estimates of a and d, assuming that there is no recombination between the marker locus and the locus influencing the trait? Later it was found that a locus that potentially influences size is 5 map units from the marker locus. What are the estimates of a and d, assuming that this is the QTL affecting size?

13. Example 11.14 presents recent data suggesting that, contrary to an earlier study, there is not a gene for homosexuality on the X chromosome. What further work could be carried out to determine which of these conclusions is correct?

14. Why do you think that the definition used to determine the extent of a haplotype block is appropriate? How would you improve it?

15. Give a procedure (algorithm) to determine the tagSNPs in a group of haplotypes. Do you identify the same tagSNPs in Table 11.22 with your approach?

Bibliography

Online Sites for Data Analysis (see Table 2.1)

Arlequin, http://lgb.unige.ch/arlequin/ (495)
DNAsp, http://www.ub.es/dnasp/ (429)
Genepop, http://wbiomed.curtin.edu.au/genepop/
LAMARC, http://evolution.genetics.washington.edu/lamarc.html
MEGA, http://evolgen.biol.metro-u.ac.jp/MEGA (443)
PAML, http://abacus.gene ucl.ac.uk/software/paml.html (443, 448)
PAUP, http://paup.csit.fsu.edu/
PHYLIP, http://evolution.genetics.washington.edu/phylip.html
PowerMarker, http://152.14.14.57/

Online Sites for Population Genetics Models and Prediction (see Table 3.1)

Populus (D. Alstad), http://www.cbs.umn.edu/populus/
PopG (J. Felsenstein), ftp://evolution.gs.washington.edu/pub/popgen/popg.html
AlleleA1 (J. Herron), http://faculty.washington.edu/~herronjc/SoftwareFolder/software.html
EvoTutor (A. Lemmon), http://www.evotutor.org/Software.html
PopGen (J. Aspi), http://cc.oulu.fi/~jaspi/popgen/popgen.htm
WinPop (P. Nuin), http://evol.biology.mcmaster.ca/paulo

Online Sites for Estimation of Genetic Parameters

AMOVA (analysis of molecular variance)
(http://www.bioss.sari.ac.uk/smart/unix/mamova/slides.frames.html) (485)

CERVUS (paternity) (http://helios.bto.ed.ac.uk/evolgen/cervus/cervus.html) (636)
FLUCTUATE (long-term effective population size), http://evolution.genetics.washington.edu/lamarc/fluctuate.html (M. Kuhner and J. Felsenstein) (464)
FSTAT (Program to Estimate and Test Gene Diversities and Fixation Indices, Version 2.9.3), http://www.unil.ch/izea/softwares/fstat.html (J. Goudet)
GENECLASS (population assignment), http://www.montpellier.inra.fr/CBGP/softwares/ (J.-M. Cornuet) (650)
GENTREE (time to MRCA), http://taxonomy.zoology.gla.ac.uk/rod/genetree/genetree.html, (R. Griffiths and colleagues) (461)
PM2000 (inbreeding), http://www.vortex9.org/pm2000.html (R. Lacy) (267)
MIGRATE (migration matrix), http://evolution.genetics.washington.edu/lamarc.html (P. Beerli and J. Felsenstein) (499)
MLTR (multilocus outcrossing), http:/www.genetics.forestry.ubc.ca/ritland/programs (K. Ritland) (254)
RELATEDNESS (coefficient of relatedness), http:/www.gsofnet.us/GSoft.html (D. Queller) (274)
STRUCTURE (population assignment), www.pritch.bsd.uchicago.edu (J. Pritchard) (651)
Watterson homozygosity test for neutrality, http://allele5.biol.berkeley.edu under Software (G. Thomson) (430)

Other Online Sites

Evolution Directory, EvolDir, http://life.biology.mcmaster.ca/~brian/evoldir.html (61)
Online Mendelian Inheritance in Man, OMIM, http://ncbi.nlm.nih.gov/OMIM (362, 391)

Printed References

Abbey, D. M. 1999. The Thomas Jefferson paternity case. *Nature* 397:32. (630)
Abbot, R. J., and M. F. Gomes. 1989. Population genetic structure and outcrossing rate of *Arabidopsis thaliana* (L.) Heynh. *Heredity* 62:411–418. (550)
Adams, J., and R. H. Ward. 1973. Admixture studies and the detection of selection. *Science* 180:1137–1143. (485)
Agrawal, A., and C. M. Lively. 2002. Infection genetics: gene-for-gene versus matching alleles models and all points in between. *Evol. Ecol. Res.* 4:79–90. (230)
Agresti, A. 1996. *An Introduction to Categorical Data Analysis.* New York: John Wiley and Sons. (155)
Akashi, H. 1995. Inferring weak selection from patterns of polymorphism and divergence at "silent" sites in *Drosophila* DNA. *Genetics* 139:1067–1076. (418)
Akashi, H. 1999. Inferring the fitness effects of DNA mutations from polymorphism and divergence data: statistical power to detect directional selection under stationarity and free recombination. *Genetics* 151:221–238. (418, 428, 451)
Allard, R. W., and J. Adams. 1969. The role of intergenotypic interactions in plant breeding. *Proc. XII Int. Congr. Genet.* 3:349–370. (224)
Allard, R. W., A. L. Kahler, and B. S. Weir. 1972. The effect of selection on esterase allozymes in a barley population. *Genetics* 72:489–503. (574)
Allendorf, F. W. 1986. Genetic drift and the loss of alleles versus heterozygosity. *Zool. Biol.* 5:181–190. (347)

Allison, A. C. 1956. The sickle-cell and haemoglobin C genes in some African populations. *Ann. Hum. Genet.* 21:67–89. (156, 158, 161)

Allison, A. C., 1964. Polymorphism and natural selection in human populations. *Cold Spring Harbor Symp. Quant. Biol.* 29:137–149. (546)

Anderson, E. C., and M. Slatkin. 2004. Population-genetic basis of haplotype blocks in the 5q31 region. *Am. J. Hum. Genet.* 74:40–49. (662)

Anderson, E. C., E. G. Williamson, and E. A. Thompson. 2000. Monte Carlo evaluation of the likelihood for N_e from temporally spaced samples. *Genetics* 156:2100–2118. (340)

Anderson, W. W. 1969. Polymorphism resulting from the mating advantage of rare male genotypes. *Proc. Natl. Acad. Sci. USA* 64:190–197. (193)

Anderson, W. W., J. Arnold, D. G. Baldwin, *et al.* 1991. Four decades of inversion polymorphism in *Drosophila pseudoobscura. Proc. Natl. Acad. Sci. USA* 88:10367–10371. (28)

Anderson, W. W., and C. E. King. 1970. Age-specific selection. *Proc. Natl. Acad. Sci. USA* 66:780–786. (205)

Andersson, M. 1994. *Sexual Selection.* Princeton, NJ: Princeton University Press. (189)

Andolfatto, P. 2001. Adaptive hitchhiking effects on genome variability. *Curr. Opin. Genet. Dev.* 11:635–641. (579)

Antonarakis, S. E., C. D. Boehm, P. J. C. Giardia, and H. H. Kazazian. 1982. Nonrandom association of polymorphic restriction sites in the β globin gene cluster. *Proc. Natl. Acad. Sci. USA* 79:137–141. (546)

Antonovics, J., P. Thrall, A. Jarosz, and D. Stratton. 1994. Ecological genetics of metapopulations: the *Silene–Ustilago* plant–pathogen system (pp. 146–170). In L. Real (ed.). *Ecological Genetics.* Princeton, NJ: Princeton University Press. (509)

Aquadro, C. F., and B. D. Greenberg. 1983. Human mitochondrial DNA variation and evolution: analysis of nucleotide sequences from seven individuals. *Genetics* 103:287-312. (434)

Ardlie, K. G., and L. M. Silver. 1998. Low frequency of t haplotypes in natural populations of house mice (*Mus musculus domesticus*). *Genetics* 52:1185–1196. (199)

Arkush, K. D., A. R. Giese, H. L. Mendonca, *et al.* 2002. Resistance to three pathogens in the endangered winter-run chinook salmon (*Oncorhynchus tshawytscha*): effects of inbreeding and major histocompatibility complex genotypes. *Can. J. Fish. Aquat. Sci.* 59:966–975. (153–154, 285)

Arnaud, J.-F., F. Viard, M. Delescluse, and J. Cuguen. 2003. Evidence for gene flow via seed dispersal from crop to wild relatives in *Beta vulgaris* (Chenopodiaceae): consequences for the release of genetically modified crop species with weedy lineages. *Proc. R. Soc. Lond. B.* 270:1565–1571. (471)

Arnheim, N., P. Calabrese, and M. Nordborg. 2003. Hot and cold spots of recombination in the human genome: the reason we should find them and how this can be achieved. *Am. J. Hum. Genet.* 73:5–16. (538–539)

Arnold, M. 1997. *Natural Hybridization and Evolution.* Oxford: Oxford University Press. (510)

Ashworth, D., M. Bishop, K. Campbell, *et al.* 1998. DNA microsatellite analysis of Dolly. *Nature* 394:329. (640)

Asmussen, M. A., J. Arnold, and J. C. Avise. 1987. Definition and properties of disequilibrium statistics for associations between nuclear and cytoplasmic genotypes. *Genetics* 115:755–768. (536–537)

Asmussen, M. A., R. A. Cartwright, and H. G. Spencer. 2004. Frequency-dependent selection with dominance: a window onto the behavior of the mean fitness. *Genetics* 167:499-512. (223)

Asmussen, M. A., and M. T. Clegg. 1982. Rates of decay of linkage disequilibrium under two-locus models of selection. *J. Math. Biol.* 14:37–70. (535, 567)

Asmussen, M. A., L. U. Gilliland, and R. B. Meagher. 1998. Detection of deleterious genotypes in multigenerational studies. II. Theoretical and experimental dynamics with selfing and selection. *Genetics* 149:727–737. (288)

Austerlitz, F., C. W. Dick, C. Dutech, *et al.* 2004. Using genetic markers to estimate the pollen dispersal curve. *Mol. Ecol.* 13:937–954. (483)

Avent, N. D., and M. E. Reid. 2000. The Rh blood group system: a review. *Blood* 95:375–387. (180)

Avise, J. C. 2000. *Phylogeography: The History and Formation of Species.* Cambridge, MA: Harvard University Press. (470, 611–615, 619)

Avise, J. C. 2004. *Molecular Markers, Natural History, and Evolution*, 2nd ed. Sunderland, MA: Sinauer Associates. (31, 36)

Avise, J. C., C. Giblin-Davidson, J. Laerm, *et al.* 1979. Mitochondrial DNA clones and matriarchal phylogeny within and among geographic populations of the pocket gopher, *Geomys pinetis. Proc. Natl. Acad. Sci. USA* 76:6694–6698. (36, 611)

Awadalla, P., A. Eyre-Walker, and J. Maynard-Smith. 1999. Linkage disequilibrium and recombination in hominid mitochrondrial DNA. *Science* 286:2524–2525. (548, 607–609)

Ayala, F. J. 1972. Frequency-dependent mating advantage in *Drosophila. Behav. Genet.* 2:85–91. (193)

Ayala, F. J. 1977. "Nothing in biology makes sense except in the light of evolution." Theodosius Dobzhansky: 1900–1975. *J. Hered.* 68:3–10. (28)

Bachtrog, D. 2003. Adaptation shapes patterns of genome evolution of sexual and asexual chromosome in *Drosophila. Nat. Genet.* 34:215–219. (583)

Bacles, C. F. E., A. J. Lowe, and R. A. Ennos. 2004. Genetic effects of chronic habitat fragmentation on tree species: the case of *Sorbus aucuparia* in a deforested Scottish landscape. *Mol. Ecol.* 13:573–584. (498)

Baglione, V., D. Canestrari, J. M. Marcos, and J. Ekman. 2003. Kin selection in cooperative alliances of carrion crows. *Science* 300:1947–1949. (277)

Ballou, J. 1983. Calculating inbreeding coefficients from pedigrees (pp. 509–520). In C. M. Schonewald-Cox, S. M. Chambers, B. MacBryde, and L. Thomas (eds.). *Genetics and Conservation.* Menlo Park, CA: Benjamin\Cummings. (267)

Ballou, J. D. 1997. Ancestral inbreeding only minimally affects inbreeding depression in mammalian populations. *J. Hered.* 88:169–178. (289)

Ballou, J. D., M. Gilpin, and T. J. Foose (eds.). 1995. *Population Management for Survival and Recovery.* New York: Columbia University Press. (269, 303)

Ballou, J. D., and R. C. Lacy. 1995. Identifying genetically important individuals for management of genetic variation in pedigreed populations (pp. 76–111). In J. D. Ballou, M. Gilpin, and T. J. Foose (eds.). *Population Management for Survival and Recovery.* New York: Columbia University Press. (269)

Bamshad, M. J., W. S. Watkins, M. E. Dixon, *et al.* 1998. Female gene flow stratifies Hindu castes. *Nature* 395:651–652. (498)

Bamshad, M., and S. P. Wooding. 2003. Signatures of natural selection in the human genome. *Nat. Rev. Genet.* 4:99–111. (429)

Baquero, F., and J. Blazquez. 1997. Evolution of antibiotic resistance. *Trends Ecol. Evol.* 12:482–487. (116)

Barbadilla, A., L. M. King, and R. C. Lewontin. 1996. What does electrophoretic variation tell us about protein variation? *Mol. Biol. Evol.* 13:427–432. (32)

Barnes, I., P. Matheus, B. Shapiro, *et al.* 2002. Dynamics of Pleistocene population extinctions in Beringian brown bears. *Science* 295:2267–2270. (617)

Barrett, S. C. H. 1990. The evolution and adaptive significance of heterostyly. *Trends Ecol. Evol.* 5:144–148. (191)

Barrett, S. C. H. 2002. The evolution of plant sexual diversity. *Nat. Rev. Genet.* 3:274–284. (200)

Barrett, S. C. H., and D. Charlesworth. 1991. Effects of a change in the level of inbreeding on the genetic load. *Nature* 352:522–524. (286, 289)

Barton, N. H. 2000. Genetic hitchhiking. *Phil. Trans. R. Soc. Lond B.* 355:1553–1562. (572)

Barton, N. H., and B. Charlesworth. 1998. Why sex and recombination? *Science* 281:1986–1990. (583)

Barton, N. H., and A. Clark. 1990. Population structure and process in evolution (pp. 115–173). In K. Wohrmann and E. Jain (eds.). *Population Biology: Ecological and Evolutionary Viewpoints.* New York: Springer-Verlag. (510)

Barton, N. H., and G. M. Hewitt. 1985. Analysis of hybrid zones. *Annu. Rev. Ecol. Syst.* 16:113–148. (517, 554)

Barton, N. H., and S. Rouhani. 1991. The probability of fixation of a new karyotype in a continuous population. *Evolution* 45:499–517. (390)

Basten, C. J., and M. A. Asmussen. 1997. The exact test for cytonuclear disequilibria. *Genetics* 146:1165–1171. (536)

Bataillon, T. 2000. Estimation of spontaneous genome-wide mutation rate parameters: whither beneficial mutations. *Heredity* 84:497–501. (400)

Bateman, A. J. 1952. Self-incompatibility systems in angiosperms. I. Theory. *Heredity* 6:285–310. (204)

Battaglia, B. 1958. Balanced polymorphism in *Tisbe reticulata*, a marine copepod. *Evolution* 12:358–364. (205)

Beattie, A. J., and D. C. Culver. 1979. Neighborhood size in *Viola. Evolution* 33:1226–1229. (336)

Beaumont, M. A. 2003. Estimation of population growth or decline in genetically monitored populations. *Genetics* 164:1139–1160. (340)

Beaumont, M. A., and B. Rannala. 2004. The Bayesian revolution in genetics. *Nat. Genet.* 5:251–261. (632)

Beck, S., and J. Trowsdale. 2000. The human major histocompatibility complex: lessons from the DNA sequence. *Annu. Rev. Genom. Hum. Genet.* 1:117–137. (40, 555)

Beerli, P., and J. Felsenstein. 2001. Maximum likelihood estimation of a migration matrix and effective population sized in n subpopulations by using a coalescent approach. *Proc. Natl. Acad. Sci. USA* 98:4563–4568. (484, 499)

Begon, M., M. Mortimer, and D. J. Thompson. 1996. *Population Ecology: A Unified Study of Animal and Plants*, 3rd ed. Cambridge, MA: Oxford University Press. (318)

Begun, D. J., and C. F. Aquadro. 1992. Levels of naturally occurring DNA polymorphism correlate with recombination rates in *D. melanogaster. Nature* 356:519–520. (579)

Begun, D. J., A. J. Betancourt, C. H. Langley, and W. Stephan. 1999. Is the fast/slow allozyme variation at the *Adh* locus of *Drosophila melanogaster* an ancient balanced polymorphism? *Mol. Biol. Evol.* 16:1816–1819. (438)

Begun, D. J., and P. Whitley. 2000. Reduced X-linked nucleotide polymorphism in *Drosophila simulans. Proc. Natl. Acad. Sci. USA* 97:5960–5965. (579)

Belden, R. C. 1986. Florida panther recovery plan implementation—1983 progress report (pp. 159–172). In S. D. Miller and D. D. Everett (eds.). *Cats of the World: Biology, Conservation, and Management.* Kingville, TX: Caesare Leberg Wildlife Research Institute. (512)

Bengtsson, B. O., and W. F. Bodmer. 1976. On the increase of chromosome mutations under random mating. *Theor. Popul. Biol.* 9:260–281. (390)

Bennett, J. H. 1965. Estimation of the frequencies of linked gene pairs in random mating populations. *Am. J. Hum. Genet.* 17:51–53. (589)

Bennett, J. H. (ed.). 1971–1974. *Collected Papers of R. A. Fisher*, vols. 1–5. Adelaide, Australia: University of Adelaide Press. (115)

Bennett, J. H., and F. E. Binet. 1956. Association between Mendelian factors with mixed selfing and random mating. *Heredity* 10:51–56. (550)

Bennett, J. H., and C. R. Oertel. 1965. The approach to random association of genotypes with random mating. *J. Theor. Biol.* 9:67–76. (538)

Bennett, R. L., A. G. Motulsky, A. Bittles, *et al.* 2002. Genetic counseling and screening of consanguineous couples and their offspring: recommendation of the National Society of Genetic Counselors. *J. Genet. Counsel.* 11:97–119. (281)

Berg, L. M., M. Lascoux, and P. Pamilo. 1998. The infinite island model with sex-differentiated gene flow. *Heredity* 81:63–68. (498)

Bergelson, J., M. Kreitman, E. A. Stahl, and D. Tian. 2001. Evolutionary dynamics of plant *R*-genes. *Science* 292:2281–2285. (229)

Bernstein, F. 1930. Über die Erblichkeit der Blutgruppen. *Z. Abst. u. Vererb.* 54:400–426. (88)

Berry, A. J., J. Ajioka, and M. Kreitman. 1991. Lack of polymorphism on the *Drosophila* fourth chromosome resulting from selection. *Genetics* 129:1111–1117. (579)

Berry, O., M. D. Tocher, and S. D. Sarre. 2004. Can assignment tests measure dispersal? *Mol. Ecol.* 13:551–561. (650)

Berry, R. J. 1990. Industrial melanism and peppered moths (*Biston betularia* (L.)). *Biol. J. Linnean Soc.* 9:301–322. (138)

Berthier, P., M. A. Beaumont, J. M. Cornuet, and G. Luikart. 2002. Likelihood-based estimation of the effective population size using temporal changes in allele frequencies: a genealogical approach. *Genetics* 160:741–751. (340)

Betancourt, A. J., and D. C. Presgraves. 2002. Linkage limits the power of natural selection in *Drosophila. Proc. Nat. Acad. Sci. USA* 99:13616–13620. (452)

Biémont, C., C. Nardon, G. Deceliere, *et al.* 2003. Worldwide distribution of transposable element copy number in natural populations of *Drosophila simulans. Evolution* 57:159–167. (358)

Bijlsma, R., J. Bundgaard, and A. C. Boerema. 2000. Does inbreeding affect the extinction risk of small populations? Predictions from *Drosophila. J. Evol. Biol.* 13:502–514. (280)

Bijlsma, R., J. Bundgaard, and W. F. Van Putten. 1999. Environmental dependence of inbreeding depression and purging in *Drosophila melanogaster. J. Evol. Biol.* 12:1125–1137. (282)

Bishop, J. A. 1973. An experimental study of the cline of industrial melanism in *Biston betularia* (L.) (Lepidoptera) between urban Liverpool and rural North Wales. *J. Anim. Ecol.* 41:209–243. (44)

Bishop, J. G., A. M. Dean, and T. Mitchell-Olds. 2000. Rapid evolution in plant chitinases: Molecular targets of selection in plant-pathogen coevolution. *Proc. Natl. Acad. Sci. USA* 97:5322–5327. (449)

Bishop, J. M., J. U. M. Jarvis, A. C. Spinks, *et al.* 2004. Molecular insight into patterns of colony composition and paternity in the common mole-rat *Cryptomys hottentotus hottentotus. Mol. Ecol.* 13:1217–1229. (623)

Black, F. L., and P. W. Hedrick. 1997. Strong balancing selection at HLA loci: evidence from segregation in South Amerindian families. *Proc. Natl. Acad. Sci. USA* 94:12452–12456. (154–155, 182)

Black, F. L., and F. M. Salzano. 1981. Evidence for heterosis in the HLA system. *Am. J. Hum. Genet.* 33:894–899. (154)

Blaxter, M. 2003. Two worms are better than one. *Nature* 426:395–396. (36)

Blott, S. C., J. L. Williams, and C. S. Haley. 1999. Discriminating among cattle breeds using genetic markers. *Heredity* 82:613–619. (622)

Blouin, M. S. 2003. DNA-based methods for pedigree reconstruction and kinship analysis in natural populations. *Trends Ecol. Evol.* 18:503–511. (271–272, 274)

Blumenfeld, O. O., and C. H. Huang. 1997. Molecular genetics of glycophorin *MNS* variants. *Transfus. Clin. Biol.* 4:357–365. (69)

Bodmer, W. F. 1965. Differential fertility in population genetics. *Genetics* 51:411–424. (175)

Bodmer, W. F., and L. L. Cavalli-Sforza. 1968. A migration matrix model for the study of random genetic drift. *Genetics* 59:565–592. (476)

Bodmer, W. F., and L. L. Cavalli-Sforza. 1976. *Genetics, Evolution, and Man.* San Francisco: W. H. Freeman. (243)

Bodmer, W. F., and J. Felsenstein. 1967. Linkage and selection: theoretical analysis of the deterministic two-locus random mating model. *Genetics* 57:237–265. (562)

Bonnell, M. L., and R. K. Selander. 1974. Elephant seals: genetic variation and near extinction. *Science* 134:908–909. (303, 348)

Box, J. F. 1978. *R. A. Fisher: The Life of a Scientist.* New York: John Wiley. (115)

Boyd, L., and F. A. Houpt. 1994. *Przewalski's Horse: The History and Biology of an Endangered Species.* Albany, NY: State University New York Press. (303, 446)

Brakefield, P. M. 1987. Industrial melanism: do we have the answers? *Trends Ecol. Evol.* 2:117–122. (138)

Braverman, J. M., R. R. Hudson, N. L. Kaplan, *et al.* 1995. The hitchhiking effect on the site frequency spectrum of DNA polymorphism. *Genetics* 140:783–796. (436, 579)

Brenner, C. H., and B. S. Weir. 2003. Issues and strategies in the DNA identification of World Trade Center victims. *Theor. Popul. Biol.* 63:173–178. (641–642)

Briskie, J. V., and M. Mackintosh. 2004. Hatching failure increases with severity of population bottlenecks in birds. *Proc. Nat. Acad. Sci. USA* 101:558–561. (280)

Bromham, L., and D. Penny. 2003. The modern molecular clock. *Nat. Rev. Genet.* 4:216–1224. (418)

Brookfield. J. F. Y. 1996. A simple new method for estimating null allele frequency from heterozygote deficiency. *Mol. Ecol.* 5:453–455. (89)

Brooks, L. D. 1988. The evolution of recombination rates (pp. 87–105). In R. Michod and B. Levins (eds.). *The Evolution of Sex: An Examination of Current Ideas.* Sunderland, MA: Sinauer. (539)

Brown, A. H. D. 1970. The estimation of Wright's fixation index from genotypic frequencies. *Genetica* 41:399–406. (151)

Brown, A. H. D. 1979. Enzyme polymorphism in plant populations. *Theor. Popul. Biol.* 15:1–42. (254)

Brown, W. M., M. George, and A. C. Wilson. 1979. Rapid evolution of animal mitochondrial DNA. *Proc. Natl. Acad. Sci. USA* 76:1967–1971. (465)

Brown, W. M., E. M. Prager, A. Wang, and A. C. Wilson. 1982. Mitochondrial DNA sequences of primates: tempo and mode of evolution. *J. Mol. Evol.* 18:225–239. (602–603)

Bruck, D. 1957. Male segregation ratio advantage as a factor in maintaining lethal alleles in wild populations of house mice. *Proc. Natl. Acad. Sci. USA* 43:152–158. (198)

Bruford, M. W., D. G. Bradley, and G. Luikart. 2003. DNA markers reveal the complexity of livestock domestication. *Nat. Rev. Genet.* 4:900–910. (618)

Brussard, P. F. 1975. Geographic variation in North American colonies of *Cepea nemoralis. Evolution* 29:402–410. (207)

Budowle, B., M. W. Allard, M. R. Wilson, and R. Chakraborty. 2003. Forensics and mitochrondrial DNA: application, debates, and foundations. *Annu. Rev. Genomics Hum. Genet.* 4:119–141. (629)

Budowle, B., and R. Chakraborty. 2001. Population variation at the CODIS core short tandem repeat loci in Europeans. *Legal Med.* 3:29–33. (639, 644)

Budowle, B., T. R. Moretti, A. L. Baumstrark, *et al.* 1999. Population data on the thirteen CODIS core short tandem repeat loci in African Americans, U. S. Caucasians, Hispanics, Bahamians, Jamaicans, and Trinidadians. *J. Forensic Sci.* 44:1277–1286. (639)

Budowle, B., B. Shea, S. Niezgoda, and R. Chakraborty. 2001. CODIS STR loci data from 41 sample populations. *J. Forensic. Sci.* 46:453–489. (643)

Bunce, M., T. H. Worthy, T. Ford, *et al.* 2003. Extreme reversed sexual size dimorphism in the extinct New Zealand moa *Dinornis. Nature* 425:172–175. (42)

Bundgaard, J., and F. B. Christiansen. 1972. Dynamics of polymorphisms. 1. Selection components of *Drosophila melanogaster. Genetics* 71:439–460. (174)

Buonagurio, D. A., S. Nakada, J. D. Parvin, *et al.* 1986. Evolution of human influenza A viruses over 50 years: rapid uniform rate of change in NS genes. *Science* 232:980–982. (425)

Burch, C. L., and L. Chao. 1999. Evolution in small steps and rugged landscapes in the RNA virus $\phi 6$. *Genetics* 151:921–927. (388)

Buri, P. 1956. Gene frequency in small populations of mutant *Drosophila. Evolution* 10:367–402. (306–308, 316–317)

Burt, A. 1995. The evolution of fitness. *Evolution* 49:1–8. (133)

Butler, J. M. 2001. *Forensic DNA Typing.* New York: Academic Press. (636, 639)

Byers, D. L., and D. M. Waller. 1999. Do plant populations purge their genetic load: effects of population size and mating history on inbreeding depression. *Annu. Rev. Ecol. Syst.* 30:479–513. (289)

Caballero, A. 1994. Developments in the prediction of effective population size. *Heredity* 73:657–679. (334–335)

Cain, A. J., L. M. Cook, and J. D. Currey. 1990. Population and morph frequency in a long-term study of *Cepea nemoralis. Proc. R. Soc. Lond. B.* 240:231–250. (207)

Cain, A. J., and J. D. Currey. 1963. Area effects in *Cepaea. Phil. Trans. R. Soc. Lond. B.* 246:1–181. (207)

Cain, A. J., and P. M. Sheppard. 1954. Natural selection in *Cepaea. Genetics* 39:89–116. (206)

Cain, M. L., B. G. Milligan, and A. E. Strand. 2000. Long-distance seed dispersal in plant populations. *Am. J. Bot.* 87:1217–1227. (483)

Camin, J. H., and P. R. Ehrlich. 1958. Natural selection in water snakes (*Natrix sipedon* L.) on islands in Lake Erie. *Evolution* 12:504–511. (515)

Cann, R. L., M. Stoneking, and A. C. Wilson. 1987. Mitochondrial DNA and human evolution. *Nature* 325:31–36. (618–619)

Cargill, M., D. Altshuler, J. Ireland, *et al.* 1999. Characterization of single-nucleotide polymorphism in coding regions of human genes. *Nat. Genet.* 22:231–238. (526)

Carlini, D. B., and W. Stephan. 2003. In vivo introduction of unpreferred synonymous codons into the *Drosophila Adh* gene results in reduced levels of ADH protein. *Genetics* 163:239–243. (451)

Carlson, C. S., M. A. Eberle, M. J. Rieder, *et al.* 2004. Selecting a maximally informative set of single-nucleotide polymorphisms for association analyses using linkage disequilibrium. *Amer. J. Hum. Genet.* 74:106–120. (659)

Carmichael, L. E., J. A. Nagy, N. C. Larter, and C. Strobeck. 2001. Prey specialization may influence patterns of gene flow in wolves of the Canadian Northwest. *Molec. Ecol.* 10:2787–2798. (647–648)

Carr, D. E., and Dudash. 2003. Recent approaches into the genetic basis of inbreeding depression in plants. *Phil. Trans. R. Soc. Lond. B.* 358:1071–1084. (282)

Carrington, M., G. W. Nelson, M. P. Martin, *et al.* 1999. *HLA* and HIV-1: heterozygote advantage and *B*35-Cw*04* disadvantage. *Science* 238:1748–1752. (232)

Cavalli-Sforza, L. L., and W. F. Bodmer. 1971. *The Genetics of Human Populations.* San Francisco: W. H. Freeman. (89, 103, 109, 156, 264, 483–484, 551)

Cavalli-Sforza, L. L., P. Menozzi, and A. Piazza. 1994. *The History and Geography of Human Genes.* Princeton, NJ: Princeton University Press. (109, 182)

Cerda-Flores, R., B. Budowle, L. Jin, *et al.* 2002. Maximum likelihood estimates of admixture in northeastern Mexico using 13 short tandem repeat loci. *Am. J. Hum. Biol.* 14:429–439. (487–488)

Chakraborty, R., and M. Nei. 1977. Bottleneck effects on average heterozygosity and genetic distance with the stepwise mutation model. *Evolution* 31:347–356. (343)

Chakraborty, R., M. Shaw, and W. J. Schull. 1974. Exclusion probability: the current state of the art. *Am. J. Hum. Genet.* 26:477–488. (626)

Chakraborty, R., and K. M. Weiss. 1988. Admixture as a tool for finding linked genes and detecting that difference from allelic association between loci. *Proc. Natl. Acad. Sci. USA* 85:9119–9123. (554)

Chakraborty, R. D. N. Stivers, B. Su, *et al.* 1999. The utility of short tandem repeat loci beyond human identification: implications for development of new DNA typing systems. *Electrophoresis* 20:1682–1696. (626, 639)

Chakraborty, R., Y. Zhong, L. Jin, and B. Budowle. 1994. Nondetectability of restriction fragments and independence of DNA fragment sizes within and between loci in RFLP-typing of DNA. *Am. J. Hum. Genet.* 55:391–401. (89)

Chakravarti, A. 2004. Ching Chun Li (1912–2003): a personal remembrance of a hero of genetics. *Am. J. Hum. Genet.* 74:789–792. (625)

Charlesworth, B. 1971. Selection in density-regulated populations. *Ecology* 52:469–474. (205)

Charlesworth, B. 1994. *Evolution in Age-structured Populations*, 2nd ed. London: Cambridge University Press. (205)

Charlesworth, B. 1996. Background selection and patterns of genetic diversity in *Drosophila melanogaster. Genet. Res.* 8:131–149. (539, 579)

Charlesworth, B. 1998. Measures of divergence between populations and the effect of forces that reduce variability. *Mol. Biol. Evol.* 15:538–543. (492)

Charlesworth, B., and D. Charlesworth. 1987. Inbreeding depression and its evolutionary consequences. *Annu. Rev. Ecol. Syst.* 18:237–268. (200, 282)

Charlesworth, B., D. Charlesworth, and N. H. Barton. 2003. The effects of genetic and geographic structure on neutral variation. *Annu. Rev. Ecol. Evol. Syst.* 34:99–125. (471)

Charlesworth, D. 1991. The apparent selection on neutral marker loci in partially inbreeding populations. *Genet. Res.* 57:159–175. (550)

Charlesworth, D. 2003. Effects of inbreeding on the genetic diversity of populations. *Phil. Trans. Roy. Soc. Lond. B.* 358:1051–1070. (334–335)

Charlesworth, D., and B. Charlesworth. 1999. The genetic basis of inbreeding depression. *Genet. Res.* 74:329–340. (283)

Charlesworth, D., B. K. Mable, M. H. Schierup, *et al.* 2003. Diversity and linkage of genes in the self-incompatibility gene family in *Arabidopsis lyrata. Genetics* 164:1519–1535. (204)

Charlesworth, D., M. T. Morgan, and B. Charlesworth. 1993. Mutation accumulation in finite populations *J. Hered.* 84:321–325. (579, 585)

Chase, M. R., C. Moller, R. Kessell, and K. S. Bawa. 1996. Distant gene flow in tropical trees. *Nature* 383:398–399. (496)

Chebloune, Y., J. Pagnier, G. Trabuchet, *et al.* 1988. Structural analysis of the 5' flanking region of the β-globin gene in African sickle cell anemia patients: further evidence for three origins of the sickle cell mutation in Africa. *Proc. Natl. Acad. Sci. USA* 85:4431–4435. (546–547)

Chen, S. L., W. Lee, A. K. Hottes, *et al.* 2004. Codon usage between genomes is constrained by genome-wide mutational processes. *Proc. Nat. Acad. Sci. USA* 101:3480–3485. (452)

Christiansen, F. B. 1974. Sufficient conditions for protected polymorphism in a subdivided population. *Am. Nat.* 108:157–166. (520)

Christiansen, F. B., and O. Frydenberg. 1973. Selection component analysis of natural polymorphisms using population samples including mother: offspring combinations. *Theor. Popul. Biol.* 4:425–445. (148)

Christiansen, F. B., O. Frydenberg, and V. Simonsen. 1973. Genetics of *Zoarces* populations. IV. Selection component analysis of an esterase polymorphism using population samples including mother: offspring combinations. *Hereditas* 73:291–304. (148)

Choisy, M., P. Franck, and J.-M. Cornuet. 2004. Estimating admixture proportions with microsatellites: comparison of methods based on simulated data. *Mol. Ecol.* 13:955–968. (485)

Churchill, G. A., and T. W. Doerge. 1994. Empirical threshold values for quantitative trait mapping. *Genetics* 138:963-971. (657)

Cipriano, F., and S. R. Palumbi. 1999. Genetic tracking of a protected whale. *Nature* 397:307–308. (644)

Clark, A. G., M. Aguade, T. Prout, *et al.* 1995. Variation in sperm displacement and its association with accessory gland protein loci in *Drosophila melanogaster*. *Genetics* 139:189–201. (188)

Clark, A. G., R. Nielsen, J. Signorovitch, *et al.* 2003. Linkage disequilibrium and inference of ancestral recombination in 538 single-nucleotide polymorphism clusters across the human genome. *Amer. J. Hum. Genet.* 73:285-300. (490–491)

Clark, A. G., K. M. Weiss, C.A. Nickerson, *et al.* 1998. Haplotype structure and population genetic inferences from nucleotide sequence variation in human lipoprotein lipase. *Am. J. Hum. Genet.* 63:595–612. (588–589)

Clark, R. W. 1968. *JBS: The Life and Work of J. B. S. Haldane.* New York: Coward-McCann. (216)

Clarke, B. C. 1960. Divergent effects of natural selection on two closely related polymorphic snails. *Heredity* 14:423–443. (206)

Clarke, B. C., and D. R. S. Kirby. 1966. Maintenance of histocompatibility polymorphisms. *Nature* 211:999–1000. (182)

Clarke, B. C., L. Partridge, and A. Robertson. 1988. *Frequency-Dependent Selection.* New York: Cambridge University Press. (221)

Clarke, C. A., and P. M. Sheppard. 1966. A local survey of the distribution of industrial melanic forms in the moth *Biston betularia* and estimates of the selective values of these in an industrial environment. *Proc. R. Soc. Lond. B.* 165:424–439. (153)

Clegg, M. T. 1978. Dynamics of correlated genetic systems. II. simulation studies of chromosomal segments under selection. *Theor. Popul. Biol.* 13:1–23. (535, 567)

Clegg, M. T. 1980. Measuring plant mating systems. *Bioscience* 30:814–818. (255)

Clegg, M. T., and M. L. Durbin. 2000. Flower color variation: a model for the experimental study of evolution. *Proc. Natl. Acad. Sci. USA* 97:7016–7023. (291)

Clegg, M. T., J. F. Kidwell, and C. R. Horch. 1980. Dynamics of correlated genetic systems. V. Rates of decay of linkage disequilibrium in experimental populations of *Drosophila melanogaster*. *Genetics* 94:217–224. (534, 567)

Clegg, M. T., J. F. Kidwell, and M. G. Kidwell. 1978. Dynamics of correlated genetic systems. III. Behavior of chromosomal segments under lethal selection. *Genetica* 48:95–106. (534, 567)

Clegg, M. T., J. F. Kidwell, M. G. Kidwell, and N. J. Daniel. 1976. Dynamics of correlated genetic systems. 1. Selection in the region of the *glued* locus of *Drosophila melanogaster*. *Genetics* 83:793–810. (125)

Cleghorn, T. E. 1960. MNSs gene frequencies in English blood donors. *Nature* 187:701. (69)

Cockerham, C. C. 1986. Modifications in estimating the number of genes for a quantitative character. *Genetics* 114:659–664. (651)

Cockerham, C. C., and P. M. Burrows. 1971. Populations of interacting autogenous components. *Am. Nat.* 105:13–29. (224)

Cockerham, C. C., P. M. Burrows, S. S. Young, and T. Prout. 1972. Frequency dependent selection in randomly mating populations. *Am. Nat.* 106:493–515. (223)

Cockerham, C. C., and B. S. Weir. 1993. Estimation of gene flow from F-statistics. *Evolution* 47:855–863. (484)

Colegrave, N. 2002. Sex releases the speed limit on evolution. *Nature* 420:664–666. (583)

Collevatti, R. G., D. Grattapaglia, and J. D. Hay. 2001a. High resolution microsatellite based analysis of the mating system allows the detection of significant biparental inbreeding in *Caryocar brasiliense*, an endangered tropical tree species. *Heredity* 86:60–67. (258–259)

Collevatti, R. G., D. Grattapaglia, and J. D. Hay. 2001b. Population genetic structure of the endangered tropical tree species *Caryocar brasiliense*, based on variability at microsatellite loci. *Mol. Ecol.* 10:349–356. (259)

Collins-Schramm, H. E., C. M. Phillips, D. J. Opertio, *et al.* 2002. Ethnic-difference markers for use in mapping by admixture linkage disequilibrium. *Am. J. Hum. Genet.* 70:737–750. (485, 554)

Colosimo, P. F., C. L. Peichel, K. Nereng, *et al.* 2004. The genetic architecture of parallel armor plate reduction in threespine sticklebacks. *PLoS Biol.* 2: 635–641. (651)

Connor, J. L., and M. J. Bellucci. 1979. Natural selection resisting inbreeding depression in captive wild house mice (*Mus musculus*). *Evolution* 33:929–940. (281)

Cook, L. M. 1998. A two-stage model for *Cepaea* polymorphism. *Phil. Trans. Soc. Lond. B*.353:1577–1593. (206–207)

Cook, L. M. 2000. Changing views on melanic moths. *Biol. J. Linn. Soc.* 69:431–441. (138)

Cook, L. M., R. L. H. Dennis, and G. S. Mani. 1999. Melanic morph frequency in the peppered moth in the Manchester area. *Proc. Roy. Soc. Lond. B.* 266:293–297. (138–139)

Cook, L. M., and C. W. A. Pettitt. 1998. Morph frequencies in the snail *Cepaea nemoralis*: changes with time and their interpretation. *Biol. J. Linnean Soc.* 64:137–150. (207)

Cooke, B. D., and F. Fenner. 2002. Rabbit haemorrhagic disease and the biological control of wild rabbits, *Oryctolagus cuniculus*, in Australia and New Zealand. *Wildlife Res.* 29:689–706. (228)

Cornuet, J. M., and G. Luikart. 1996. Description and evaluation of two tests for detecting recent bottlenecks. *Genetics* 144:2001–2014. (348, 431)

Cornuet, J.-M., S. Piry, G. Luikart, *et al.* 1999. New methods employing multilocus genotypes to select of exclude populations as origins of individuals. *Genetics* 153:1989–2000. (649–650)

Couzin, J. 2003. Sequencers examine priorities. *Science* 301:1176–1177. (35)

Couzin, J. 2004. Consensus emerges on HapMap strategy. *Science* 304:671–672. (659)

Cowie, R. H., and J. S. Jones. 1998. Gene frequency changes in *Cepaea* snails on the Marlborough Downs over 25 years. *Biol. J. Linnean Soc.* 65:233–255. (207)

Cox, E. C., and T. C. Gibson, 1974. Selection for high mutation rates in chemostats. *Genetics* 77:169–184. (403–404)

Coyne, J. A. 2002. Evolution under pressure. *Nature* 418:19–20. (138)

Craighead, L., D. Paetkau, H. V. Reynolds, *et al.* 1995. Microsatellite analysis of paternity and reproduction in Arctic grizzly bears. *J. Hered.* 86:255–261. (627)

Creel, S., G. Spong, J. L. Sands, *et al.* 2003. Population size estimation in Yellowstone wolves with error-prone noninvasive microsatellite genotypes. *Mol. Ecol.* 12:2003–2009. (646)

Cresko, W. A., A. Amores, C. Wilson, *et al.* 2004. Parallel genetic basis for repeated evolution of armor loss in Alaskan threespine stickleback populations. *Proc. Natl. Acad. Sci. USA* 101:6050–6055. (651)

Crnokrak, P., and S. C. H. Barrett. Perspective: purging the genetic load: a review of the experimental evidence. *Evolution* 56:2347–2358. (289)

Crow, J. F. 1954. Breeding structure of populations. II. Effective population number (pp. 543–556). In O. Kempthorne, T. Bancroft, J. Gowen, and J. Lush (eds.). *Statistics and Mathematics in Biology.* Ames: Iowa State University Press. (325)

Crow, J. F. 1986. *Basic Concepts in Population, Quantitative, and Evolutionary Genetics.* New York: W. H. Freeman. (143, 260)

Crow, J. F. 1988. The ultraselfish gene. *Genetics* 118:389–391. (197)

Crow, J. F. 1990. R. A. Fisher: a centennial view. *Genetics* 124:207–211. (115)

Crow, J. F. 1992. Centennial: J. B. S. Haldane, 1892–1964. *Genetics* 130:1–6. (216)

Crow, J. F. 1995. Motoo Kimura (1924–1994). *Genetics* 140:1–5. (413)

Crow, J. F. 1997. Birth defects, jimson weeds and bell curves. *Genetics* 147:1–6. (19)

Crow, J. F. 2000. The origins, patterns and implication of human spontaneous mutation. *Nat. Rev. Genet.* 1:40–47. (400, 402)

Crow, J. F. 2002. Perspective: here's to Fisher, additive genetic variance, and the fundamental theorem of natural selection. *Evolution* 56:1313–1316. (133)

Crow, J. F., and K. Aoki. 1984. Group selection for a polygenic behavioral trait: estimating the degree of population subdivision. *Proc. Natl. Acad. Sci. USA* 81:6073–6077. (502)

Crow, J. F., and C. Denniston. 1988. Inbreeding and variance effective population numbers. *Evolution* 42:482–495. (319, 325)

Crow, J. F., and M. Kimura. 1965. Evolution in sexual and asexual populations. *Am. Nat.* 99:439–450. (584)

Crow, J. F., and M. Kimura. 1970. *An Introduction to Population Genetics Theory.* New York: Harper & Row. (61, 260–261, 299, 308, 329, 376, 413, 527)

Crow, J. F., and N. E. Morton. 1955. Measurement of gene frequency drift in small populations. *Evolution* 9:202–214. (326)

Crow, J. E, and Temin, R. G. 1964. Evidence for the partial dominance of recessive lethal genes in natural populations of *Drosophila. Am. Nat.* 98:21–23. (395–396)

Crozier, R. H., and P. Pamilo. 1996. *Evolution of Social Insect Colonies: Sex Allocation and Kin Selection.* Oxford: Oxford University Press. (279)

Culver, M., P. W. Hedrick, K. Murphy, *et al.* 2004. Estimation of bottleneck size in Florida panthers. (unpublished manuscript). (327–328)

Cullen, M., S. P. Perfetto, W. Klitz, *et al.* 2002. High-resolution patterns of meiotic recombination across the human major histocompatibility complex. *Amer. J. Hum. Genet.* 71:759-776. (540)

Cummings, C. J., and H. Y. Zoghbi. 2000. Fourteen and counting: unraveling trinucelotide repeat diseases. *Hum. Mol. Genet.* 9:909–916. (393)

Cunningham, E. P., J. J. Dooley, R. K. Splan, and D. G. Bradley. 2001. Microsatellite diversity, pedigree relatedness and the contributions of founder lineages to thoroughbred horses. *Anim. Genet.* 32:360–364. (242)

Curran, J. M., J. S. Buckleton, and C. M. Triggs. 2003. What is the magnitude of the subpopulation effect? *Forensic Sci. Int.* 135:1–8. (644)

Currat, M., G. Trabuchet, D. Rees, *et al.* 2002. Molecular analysis of the β-globin gene cluster in the Niokholo Mandenka population reveals a recent origin of the β^S Senegal mutation. *Am. J. Hum. Genet.* 70:207–223. (546–547)

Curtsinger, J. W., P. W. Service, and T. Prout. 1994. Antagonistic pleiotropy, reversal of dominance, and genetic polymorphism. *Am. Nat.* 144:210–228. (184–186)

Daborn, P. J., J. L. Yen, M. R. Bogwitz, *et al.* 2002. A Single P450 allele associated with insecticide resistance in *Drosophila. Science* 297:2253–2256. (118)

Daly, M. J., J. D. Rioux, S. F. Schaffner, *et al.* 2001. High-resolution haplotype structure in the human genome. *Nat. Genet.* 29:229–232. (659–660)

Daniels, G. 2002. *Human Blood Groups*, 2nd ed. Malden, MA: Blackwell. (69, 88, 180–181)

Daniels, S. B., K. R. Peterson, L. D. Strausbaugh, *et al.* 1990. Evidence for horizontal transmission of the P transposable element between *Drosophila* species. *Genetics* 124:339–355. (360)

Darlington, C. D., and K. Mather. 1949. *The Elements of Genetics.* London: Allen and Unwin. (555)

Darwin, C. 1859. *On the Origin of Species by Means of Natural Selection.* London: Murray. (114, 407, 412)

Darwin, C. 1871. *The Descent of Man and Selection in Relation to Sex.* New York: Appleton. (188)

David, P. 1998. Heterozygosity: fitness correlation: new perspectives on old problems. *Heredity* 80:531–537. (411, 569)

Davies, N., F. X. Villablanca, and G. K. Roderick. 1999. Determining the source of individuals: multilocus genotyping in nonequilibrium population genetics. *Trends Ecol. Evol.* 14:17–21. (646)

Davison, A., and B. Clark. 2000. History or current selection? A molecular analysis of 'area effects' in the land snail *Cepaea nemoralis. Proc. R. Soc. Lond. B.* 2267:1399–1405. (207)

Dawson, E., G. R. Abecasis, S. Bumpstead, *et al.* 2002. A first-generation linkage disequilibrium map of human chromosome 22. *Nature* 418:544–548. (533)

Dayoff, M. O., R. M. Schwartz, and B. C. Orcutt. 1978. A model of evolutionary change in proteins (pp. 345–352). In M. O. Dayoff (ed.). *Atlas of Protein Sequence and Structure.* Washington, DC: National Biomedical Research Foundation. (421)

de la Chapelle, A., and F. A. Wright. 1998. Linkage disequilibrium mapping in isolated populations: the example of Finland revisited. *Proc. Natl. Acad. Sci. USA* 95:12416–12423. (300, 546)

De Luca, M., N. V. Roshina, G. L. Gieger-Thornberry, *et al.* 2003. Dopa decarboxylase (*Ddc*) affects variation in *Drosophila* longevity. *Nat. Genet.* 34:429–433. (55–56)

de Meaux, J., and T. Mitchell-Olds. 2003. Evolution of plant resistance at the molecular level: ecological context of species interactions. *Heredity* 91:345–352. (229)

Deiniger, P. L., J. V. Moran, M. A. Batzer, and H. H. Kazazian. 2003. Mobile element and mammalian genome evolution. *Curr. Opin. Genet. Dev.* 13:651–658. (359)

Deng, H.-W., and M. Lynch. 1997. Inbreeding depression and inferred deleterious-mutation parameters in *Daphnia. Genetics* 147:147–155. (400)

Denise, S., E. Johnston, J. Halverson, *et al.* 2003. Power of exclusion for parentage verification and probability of match for identity in American kennel club breeds using 17 canine microsatellite markers. *Anim. Genet.* 35:14–17. (622)

Denniston, C. 1978. Small population size and genetic diversity (pp. 281–289). In S. A. Temple (ed.). *Endangered Birds: Management Techniques for Preserving Endangered Species.* Madison, WI: University of Wisconsin Press. (347)

Denver, D. R., K. Morris, M. Lynch, *et al.* 2000. High direct estimate of the mutation rate in the mitochondrial genome of *Caenorhabditis elegans*. *Science* 289:2342–2344. (423)

Derome, N., K. Métayer, C. Montchamp-Moreau, and M. Veuille. 2004. Signature of selective sweep associated with the evolution of *sex-ratio* drive in *Drosophila simulans*. *Genetics* 166:1357–1366. (197)

de Visser, J. A. G. M., C. W. Zeyl, P. J. Gerrish, *et al.* 1999. Diminishing returns from mutation supply rate in asexual populations. *Science* 283:404–406. (405)

Devlin, B., and N. Risch. 1995. A comparison of linkage disequilibrium measures for fine-scale mapping. *Genomics* 29:311-322. (532)

Devlin, B., N. Risch, and K. Roeder. 1993. Statistical evaluation of DNA fingerprinting: a critique of the NRC's report. *Science* 259:748–749, 837. (643)

Diamond, J. M. 1987. Abducted orphans identified by grandpaternity testing. *Nature* 327:552–553. (628–629)

Dib, C., S. Faure, C. Fizames, *et al.* 1996. A comprehensive map of the human genome based on 5264 microsatellites. *Nature* 380:152–154. (12, 101, 414, 543)

Dickerson, R. C. 1971. The structure of cytochrome C and the rates of molecular evolution. *J. Mol. Evol.* 1:26–45. (420–421)

Dieffenbach, E. W., and G. S. Dveksler (eds.). 2003. *Pcr Primer: A Laboratory Manual*, 2nd ed. New York: Cold Spring Harbor Laboratory. (36)

Di Rienzo, A., A. C. Peterson, J. C. Garza, *et al.* 1994. Mutational processes of simple-sequence repear loci in human populations. *Proc. Natl. Acad. Sci. USA* 91:3166–3170. (383)

Dixon, A., D. Ross, S. L. C. O'Malley, and T. Burke. 1994. Paternal investment inversely related to degree of extra-pair paternity in the reed bunting. *Nature* 371:698–700. (622)

Dobzhansky, T. 1941. *Genetics and the Origin of Species*. New York: Columbia University Press. (28)

Dobzhansky, T. 1955. A review of some fundamental concepts and problems of population genetics. *Cold Spring Harbor Symp. Quant. Biol.* 20:1–15. (28, 147, 173, 408)

Dobzhansky, T. 1970. *Genetics of the Evolutionary Process*. New York: Columbia University Press. (48, 147)

Dobzhansky, T. 1973. Nothing in biology makes sense except in the light of evolution. *Am. Biol. Teacher* 35:125–129. (28)

Dod, B., C. Litel, P. Makoundou, *et al.* 2003. Identification and characterization of *t* haplotypes in wild mice populations using molecular markers. *Genet. Res.* 81:103–114. (199)

Doebley, J., A. Stec, and C. Gustus. 1995. *Teosinte branched1* and the origin of maize: Evidence for epistasis and the evolution of dominance. *Genetics* 141:333–346. (56)

Dowrick, V. P. J. 1956. Heterostyly and homostyly in *Primula obconica*. *Heredity* 10:219–236. (191)

Drake, G. J. C., L. J. Kennedy, H. K. Auty, *et al.* 2004. The use of reference strand-mediated conformational analysis for the study of cheetah (*Acinonyx jubatus*) feline leucocyte antigen class II *DRB* polymorphisms. *Mol. Ecol.* 13:221–229. (382)

Drake, J. W., B. Charlesworth, D. Charlesworth, and J. F. Crow. 1998. Rates of spontaneous mutations. *Genetics* 148:1667–1686. (358, 398, 400, 402–403, 405, 426)

Driscoll, C. A., M. Menotti-Raymond, G. Nelson, *et al.* 2002. Genomic microsatellites as evolutionary chronometers: a test in wild cats. *Genome Res.* 12:414–423. (381–382)

Dronamraju, K. R. (ed.). 1990. *Selected Genetic Papers of J. B. S. Haldane*. New York: Garland Publishers. (216)

Duarte, L. C., C. Bouteiller, P. Fontanillas, *et al.* 2003. Inbreeding in the greater white-toothed shrew, *Crocidura russula*. *Evolution* 57:638–645. (276–277)

Dudley, J. W., and R. J. Lambert. 2004. 100 Generations of selection for oil and protein in corn. *Plant Breeding Rev.* 24(part 1):79–100. (54, 651)

Dubrova, Y. E., G. Grant, A. A. Chumak, *et al.* 2002. Elevated minisatellite mutation rate in the post-Chernobyl families from Ukraine. *Am. J. Hum Genet.* 71:801-809. (392–393)

Dubrova, Y. E., V. N. Nesterov, N. G. Krouchinsky, *et al.* 1996. Human minisatellite mutation rate after the Chernobyl accident. *Nature* 380:683–686. (392–393)

Durand, D., K. Ardlie, L. Buttel, *et al.* 1997. Impact of migration and fitness on the stability of lethal *t*-haplotype polymorphism in *Mus musculus*: a computer study. *Genetics* 145:1093–1108. (199)

Duvernell, D. D., and W. F. Eanes. 2000. Contrasting molecular population genetics of four hexokinases in *Drosophila melanogaster*, *D. simulans* and *D. yakuba*. *Genetics* 156:1191-1201. (410)

Dybdahl, M. F., and A. Storfer. 2003. Parasite local adaptation: red queen versus suicide king. *Trends Ecol. Evol.* 18:523–530. (230)

Eanes, W. F. 1999. Analysis of selection on enzyme polymorphisms. *Annu. Rev. Ecol. Syst.* 30:301–326. (410)

Eanes, W. F., M. Kirchner, and J. Yoon. 1993. Evidence for adaptive evolution of the *G6PD* gene in the *Drosophila melanogaster* and *D. simulans* lineages. *Proc. Natl. Acad. Sci. USA* 90:7475–7479. (440–441)

Ebert, D. 1998. Experimental evolution of parasites. *Science* 282:1432–1435. (228)

Ebert, D., C. Haag, M. Kirkpatrick, *et al.* 2002. A selective advantage to immigrant genes in a *Daphnia* metapopulation. *Science* 295:485–488. (471)

Eckl, K. M., H. P. Steven, G. G. Lestringant, *et al.* 2003. Mal de Meleda (MDM) caused by mutation in the gene for SLURP-1 in patients from Germany, Turkey, Palestine, and the United Arab Emirates. *Hum. Genet.* 112:50–56. (245)

Edwards, A. W. F. 1972. *Likelihood.* Cambridge: Cambridge University Press. (631, 635)

Edwards, A. W. F. 2002. The fundamental theorem of natural selection. *Theor. Popul. Biol.* 61:335–337. (133)

Edwards, S. V., and P. Beerli. 2000. Perspective: gene divergence, population divergence, and the variance in coalescence time in phylogeography studies. *Evolution* 54:1839–1854. (612)

Edwards, S. V., and P. W. Hedrick. 1998. Evolution and ecology of MHC molecules: from genomics to sexual selection. *Trends Ecol. Evol.* 13:305–311. (40, 153, 555)

Efron, B., and R. J. Tibshirani. 1993. *An Introduction to the Bootstrap.* New York: Chapman & Hall. (21)

Eggert, L. S., J. A. Egert, and D. S. Woodruff. 2003. Estimating population sizes for elusive animals: the forest elephants of Kakum National Park, Ghana. *Mol. Ecol.* 12:1389–1402. (646)

Egid, K., and J. L. Brown. 1989. The major histocompatibility complex and female mating preferences in mice. *Anim. Behav.* 38:548–550. (195)

Ehrman, L. 1967. Further studies on genotype frequency and mating success in *Drosophila*. *Am. Nat.* 101:415–424. (192)

Eizirik, E., N. Yukhi, W. E. Johnson, *et al.* 2003. Molecular genetics and evolution of melanism in the cat family. *Curr. Biol.* 13:448–453. (521)

Eldridge, M. D. B., J. E. Kinnear, and M. L. Onus. 2001. Source population of dispersing rock-wallabies (*Petrogale lateralis*) identified by assignment tests on multilocus genotypes data. *Mol. Ecol.* 10:2867–2876. (650)

Elena, S. F., V. S. Cooper, and R. E. Lenski. 1996. Punctuated evolution caused by selection of rare beneficial mutations. *Science* 272:1802–1804. (361)

Ellegren, H. 1999. Inbreeding and relatedness in Scandinavian grey wolves *Canis lupus*. *Hereditas* 130:239–244. (242)

Ellegren, H. 2004. Microsatellites: simple sequences with complex evolution. *Nat. Rev. Genet.* 5:435-445. (12, 380)

Ellegren, H., and A. K. Fridolfsson. 1997. Male-driven evolution of DNA sequences in birds. *Nat. Genet.* 17:182–184. (402)

Ellner, S., and N. G. Hairston. 1994. Role of overlapping generations in maintaining genetic variation in a fluctuating environment. *Am. Nat.* 14:403–417. (217)

Ellstrand, N. 2003. *Dangerous Liaisons: When Cultivated Plants Mate with Their Wild Relatives.* Baltimore, MD: Johns Hopkins University Press. (471)

Ellstrand, N. C., and K. A. Schierenbeck. 2000. Hybridization as a stimulus for the evolution of invasiveness in plants? *Proc. Natl. Acad. Sci. USA* 97:7043–7050. (471)

Emerson, S. 1939. A preliminary survey of the *Oenothera organensis* population. *Genetics* 24:524–537. (202–203)

Endler, J. A. 1977. *Geographic Variation, Speciation, and Clines.* Princeton, NJ: Princeton University Press. (470, 517)

Endler, J. A., 1980. Natural selection on color patterns in *Poecilia reticulata. Evolution* 34:76–91. (189)

Ennos, R. A. 1994. Estimating the relative rates of pollen and seed migration among plant populations. *Heredity* 72:250–259. (497–498)

Epling, C., and T. Dobzhansky. 1942. Genetics of natural populations. IV. Microgeographic races in *Linanthus parryae. Genetics* 27:317–332. (217)

Epling, C., H. Lewis, and F. M. Ball. 1960. The breeding group and seed storage: a study in population dynamics. *Evolution* 14:238–255. (217)

Epperson, B. K. 1990. Spatial autocorrelation of genotypes under directional selection. *Genetics* 124:757–771. (292)

Epperson, B. K. 2003. *Geographical Genetics.* Princeton, NJ: Princeton University Press. (470, 506)

Epperson, B. K., and M. T. Clegg. 1986. Spatial autocorrelation analysis of flower color polymorphisms within substructured populations of morning glory (*Ipomoea purpurea*). *Amer. Nat.* 128:1302–1311. (291)

Estes, S., and M. Lynch. 2003. Rapid fitness recovery in mutationally degraded lines of *Caenorhabditis elegans. Evolution* 57:1022–1030. (388–390)

Estes, S., P. C. Phillips, E. R. Denver, *et al.* 2004. Mutation accumulation in populations of varying size: the distribution of mutational effects for fitness correlates in *Caenorhabditis elegans. Genetics* 166:1269–1279. (400)

Estroup, A., P. Jarne, and J.-M. Cornuet. 2002. Homoplasy and mutation model at microsatellite loci and their consequences for population genetics analysis. *Mol. Ecol.* 11:1591–1604. (383)

Evett, I. W., and B. S. Weir. 1998. *Interpreting DNA Evidence.* Sunderland, MA: Sinauer Associates. (636, 641)

Ewens, W. J. 1972. The sampling theory of selectively neutral alleles. *Theor. Popul. Biol.* 3:87–112. (429)

Ewens, W. J. 1979. *Mathematical Population Genetics.* Berlin: Springer-Verlag. (430)

Excoffier, L. 2001. Analysis of population subdivision (pp. 271–307). In D. J. Balding, M. Bishop, and C. Cannings (eds.). *Handbook of Statistical Genetics.* New York: John Wiley. (484, 488, 495)

Excoffier, L. 2002. Human demographic history: refining the recent African origin model. *Curr. Opin. Genet. Dev.* 12:675–682. (618)

Excoffier, L., and M. Slatkin, 1995. Maximum-likelihood estimation of molecular haplotype frequencies in a diploid population. *Mol. Biol. Evol.* 12:921–927. (592)

Fabiani, A., A. R. Hoelzel, F. Galimberti, and M. M. C. Muelbert. 2003. Long-range paternal gene flow in the southern elephant seal. *Science* 299:676. (298)

Falconer, D. S. 1949. The estimation of mutation rates from incompletely tested gametes, and the detection of mutations in mammals. *J. Genet.* 49:226–234. (395)

Falconer, D. S., and T. Mackay. 1996. *Introduction to Quantitative Genetics*, 4th ed. Harlow, UK: Addison-Wesley Longman. (53, 130, 556, 597, 651, 653–654)

Fay, J. C., and C.-I. Wu. 2003. Sequence divergence, functional constraint, and selection in protein evolution. *Annu. Rev. Genomics Hum. Genet.* 4:213–235. (429, 436)

Fay, J. C., G. J. Wyckoff, and C.-I. Wu. 2001. Positive and negative selection on the human genome. *Genetics* 158:1227–1234. (442)

Fay, J. C., G. J. Wyckoff, and C.-I. Wu. 2002. Testing the neutral theory of molecular evolution with genomic data from *Drosophila*. *Nature* 415:1024–1026. (442)

Felsenstein, J. 1965. The effect of linkage on directional selection. *Genetics* 52:349–363. (556–557, 559)

Felsenstein, J. 1971. Inbreeding and variance effective numbers in populations with overlapping generations. *Genetics* 68:581–597. (335)

Felsenstein, J. 1978. The number of evolutionary trees. *Syst. Zool.* 27:27–33. (599)

Felsenstein, J. 1981. Evolutionary trees from DNA sequences: a maximum likelihood approach. *J. Mol. Evol.* 17: 368–376. (601)

Felsenstein, J. 1985. Confidence limits on phylogenies: an approach using the bootstrap. *Evolution* 39:783–791. (607)

Felsenstein, J. 1988. Phylogenies from molecular sequences: inference and reliability. *Annu. Rev. Genet.* 22:521–565. (598)

Felsenstein, J. 1995. PHYLIP: Phylogeny Inference Package, version 3.572. Seattle, WA: University of Washington. (423)

Felsenstein, J. 2001. *Theoretical Evolutionary Genetics*. Seattle, WA: Department of Genetics, University of Washington. (79)

Felsenstein, J. 2004. *Inferring Phylogenies*. Sunderland, MA: Sinauer. (598, 601, 607)

Fenner, F., and B. Fantini. 1999. *Biological Control of Vertebrate Pests: The History of Myxomatosis: An Experiment in Evolution*. Wallingford, UK: CABI Publ. (228)

Ferré, J., and J. Van Rie. 2002. Biochemistry and genetics of insect resistance to *Bacillus thuringiensis*. *Annu. Rev. Entomol.* 47:501–533. (116)

Fischer, J., B. Bouadjar, R. Heilig, *et al.* 2001. Mutation in the gene encoding SLURP-1 in Mal de Meleda. *Hum. Mol. Genet.* 10:875–880. (245)

Fisher, R. A. 1930, 1958. *The Genetic Theory of Natural Selection*. New York: Dover. (115, 133, 189, 278, 367, 385, 453, 583)

Fisher, R. A. 1937. The wave of advance of an advantageous gene. *Ann. Eugen.* 7:355–369. (519)

Fisher, R. A. 1941. Average excess and average effect of a gene substitution. *Ann. Eugen.* 11:53–63. (280)

Fisher, R. A. 1953. Population genetics. *Proc. R. Soc. B.* 141:510–523. (61)

Fisher, R. A. 1958. *Statistical Methods for Research Workers*. New York: Hafner. (79)

Fitch, W. M. 1977. On the problem of discovering the most parsimonious tree. *Am. Nat.* 111:223–257. (609)

Flint, J., R. M. Harding, J. B. Clegg, and A. J. Boyce. 1993. Why are some genetic diseases common? Distinguishing selection from other processes by molecular analysis of globin gene variant. *Hum. Genet.* 91:91–117. (546)

Flor, H. 1956. The complementary genic systems in flax and flax rust. *Adv. Genet.* 8:29:54. (229–230)

Flores, C., and W. Engels. 1999. Microsatellite instability in *Drosophila spellchecker1* (Muts homolog) mutants. *Proc. Natl. Acad. Sci. USA* 96:2964–2969. (403)

Ford, E. B. 1940. Polymorphism and taxonomy (pp. 493–513). In L. Huxley (ed.). *The New Systematics.* Oxford, UK: Clarendon Press. (28, 102)

Ford, E. B. 1971. *Ecological Genetics.* London: Chapman & Hall. (43, 191, 555)

Forster, P. 2003. To err is human. *Annu. Hum Genet.* 67:2–4. (9)

Forster, L., P. Forster, S. Lutz-Bonengel, *et al.* 2002. Natural radioactivity and human mitochondrial DNA mutations. *Proc. Natl. Acad. Sci. USA* 99:13950–13954. (400)

Foster, E. A., M. A. Jobling, P. G. Taylor, *et al.* 1998. Jefferson fathered slave's last child. *Nature* 396:27–28. (629–630)

Foster, G. G., M. J. Whitten, T. Prout, and R. Gill. 1972. Chromosome rearrangement for the control of insect pests. *Science* 176:875–880. (144–145)

Frank, S. A. 1994. Recognition and polymorphism in host-parasite genetics. *Phil. Trans. R. Soc. Lond. B.* 346:283–293. (230)

Frank, S. A., and M. Slatkin. 1992. Fisher's fundamental theorem of natural selection. *Trends Ecol. Evol.* 7:92–95. (133)

Frankham, R. 1995. Effective population size/adult population size ratios in wildlife: a review. *Genet. Res.* 66:95–107. (333, 342)

Franklin, I. R., and R. C. Lewontin. 1970. Is the gene the unit of selection? *Genetics* 65:707–734. (410, 566–567)

Fredrickson, R., and P. Hedrick. 2002. Body size in endangered Mexican wolves: effects of inbreeding and cross-lineage matings. *Anim. Cons.* 5:39–43. (285)

Fredrickson, R., and P. Hedrick. 2004. Dynamics of hybridization and introgression in red wolves and coyotes. (unpublished manuscript). (476)

Freimer, N. B., S. K. Service, and M. Slatkin. 1997. Expanding on population studies. *Nat. Genet.* 17:371–373. (545)

Frelinger, J. A. 1972. The maintenance of transferrin polymorphism in pigeons. *Proc. Natl. Acad. Sci. USA* 69:326–329. (139, 232)

Frelinger, J. A., and J. F. Crow. 1973. Transferrin polymorphism and Hardy–Weinberg ratios. *Am. Nat.* 107:314–317. (151, 176, 233)

Fry, J. D. 2001. Rapid mutational declines of viability in *Drosophila. Genet. Res.* 77:53–60. (399)

Fry, J. D. 2004. On the rate and linearity of viability declines in *Drosophila* mutation-accumulation experiments: genomic mutation rates and synergistic epistasis revisited. *Genetics* 166:797–806. (399)

Fry, J. D., P. D. Keightly, S. L. Heinsohn, and S. V. Nuzhdin. 1999. New estimates of the rates and effects of mildly deleterious mutation in *Drosophila melanogaster. Proc. Natl. Acad. Sci. USA.* 96:574–579. (398–399)

Fu, Y.-X. 1997. Statistical tests of neutrality of mutations against population growth, hitchhiking and background selection. *Genetics* 147:915–925. (436)

Fu, Y.-X., and H. Huai. 2003. Estimating mutation rate: how to count mutations? *Genetics* 164:797–805. (391)

Fu, Y.-X., and W.-H. Li. 1993. Maximum likelihood estimation of population parameters. *Genetics* 134:1261–1270. (436)

Fu, Y.-X., and W.-H. Li. 1999. Coalescing into the 21st century: an overview and prospects of coalescent theory. *Theor. Popul. Biol.* 56:1–10. (453)

Fung, W. K. 2003. User-friendly programs for easy calculations in paternity testing and kinship determinations. *Forensic Sci. Intern.* 136:22–34. (632–633)

Gabriel, S. B., S. F. Schafner, H. Nguyen, *et al.* 2002. The structure of haplotype blocks in the human genome. *Science* 296:2225–2229. (659)

Gaggiotti, O. E., O. Lange, K. Rassmann, and C. Gliddon. 1999. A comparison of two indirect methods for estimating average levels of gene flow using microsatellite data. *Mol. Ecol.* 8:1513–1520. (502)

Gagneux, P., D. S. Woodruff, and C. Boesch. 1997. Furtive mating in female chimpanzees. *Nature* 387:358–359. (496)

Gaines, M. S., L. R. McClenaghan, and R. K. Rose. 1978. Temporal patterns of allozymic variation in fluctuating populations of *Microtus ochrogaster*. *Evolution* 32:723–739. (33–34)

Galvani, A. P., and M. Slatkin. 2003. Evaluating plague and smallpox as historical selective pressures for the *CCR5*–Δ32 HIV-resistance allele. *Proc. Natl. Acad. Sci. USA* 100:15276–15279. (582)

Garcia-Dorado, A., A. Cabellero, and J. F. Crow. 2003. On the persistence and pervasiveness of a new mutation. *Evolution* 57:2644–2646. (388)

Garrigan, D., and P. W. Hedrick. 2003. Perspective: detecting adaptive molecular evolution, lessons from the MHC. *Evolution* 57:1707–1722. (41, 434, 440, 442)

Garrigan, D., P. C. Marsh, and T. E. Dowling. 2002. Long-term effective population size of three endangered Colorado River fishes. *Anim. Cons.* 5:95–102. (387, 461, 465)

Gaskin, J. F., and B. A. Schaal. 2002. Hybrid *Tamarix* widespread in U.S. invasion and undetected in native Asian range. *Proc. Natl. Acad. Sci. USA* 99:11256–11259. (471)

Gerber, A. S., R. Loggins, S. Kumar, and T. E. Dowling. 2001. Does nonneutral evolution shape observed patterns of DNA variation in animal mitochondrial genomes? *Annu. Rev. Genet.* 35:539–566. (429)

Giannelli, F., P. M. Green, K. A. High, *et al.* 1997. Haemophilia B: database of point mutations and short additions and deletions, 7th ed. *Nucleic Acids Res.* 25:133–135. (391)

Gigord, L. D. B., M. R. Macnair, and A. Smithson. 2001. Negative frequency-dependent selction maintains a dramatic flower color polymorphism in the rewardless orchid *Dactylorhiza sambucina* (L.) Soò. *Proc. Natl. Acad. Sci USA.* 98:6253-6255. (224–225)

Gillespie, J. H. 1978. A general model to account for enzyme variation in natural populations. V. The SAS-CFF model. *Theor. Popul. Biol.* 14:1–45. (410)

Gillespie, J. H. 1989. Lineage effects and the index of dispersion of molecular evolution. *Mol. Biol. Evol.* 6:636–647. (426)

Gillespie, J. H. 1991. *The Causes of Molecular Evolution.* New York: Oxford University Press. (410)

Gillespie, J. H. 1998. *Population Genetics: A Concise Guide.* Baltimore, MD: Johns Hopkins University Press. (584)

Giraud, A., I. Matic, O. Tenaillon, *et al.* 2001. Costs and benefits of high mutation rates: adaptive evolution of bacteria in the mouse gut. *Science* 291:2606–2608. (405)

Glass, B. (ed.). 1980. *The Roving Naturalist.* Philadelphia: American Philosophical Society. (28)

Glaubitz, J. C., O. E. Rhodes, and J. A. DeWoody. 2003. Prospects for inferring pairwise relationships with single nucleotide polymorphisms. *Mol. Ecol.* 12:1039–1047. (274, 276)

Glemin, S. 2003. How are deleterious mutations purged? Drift versus nonrandom mating. *Evolution* 57:2678–2687. (289)

Goldman, N., and Z. Yang. 1994. A codon-based model on nucleotide substitution for protein-coding DNA sequences. *Mol. Biol. Evol.* 11:725–736. (448)

Goldstein, D. B., and D. D. Pollock. 1997. Launching microsatellites: a review of mutation processes and methods of phylogenetic inference. *J. Hered.* 88:335–342. (108)

Good, A. G., G. A. Meister, H. W. Brock, *et al.* 1989. Rapid spread of transposable P elements in experimental populations of *Drosophila melanogaster*. *Genetics* 122: 387–396. (360)

Goodman, S. J. 1997. R_{ST} Calc: a collection of computer programs for calculating estimates of genetic differentiation from microsatellite data and determining their significance. *Mol. Ecol.* 6:881–885. (108)

Goodman, S. J., N. H. Barton, G. Swanson, *et al.* 1999. Introgression through rare hybridization: a genetic study of a hybrid zone between red and sika deer (genus *Cervus*) in Argyll, Scotland. *Genetics* 152:355–371. (510)

Goodwin, W., A. Linacre, and P. Vanezis. 1999. The use of mitochondrial DNA and short tandem repeat typing in the identification of air crash victims. *Electrophoresis* 20:1707–1711. (641)

Gordo, I., and B. Charlesworth. 2000. The degeneration of asexual haploid populations and the speed of Muller's ratchet. *Genetics* 154:1379–1387. (585)

Gorman, M. B., and Y. T. Kiang. 1977. Variety-specific electrophoretic variants of four soybean enzymes. *Crop Sci.* 17:963–965. (622)

Gould, F., A. Anderson, A. Jones, *et al.* 1997. Initial frequency of alleles for resistance to *Bacillus thuringiensis* toxins in field populations of *Heliothis virescens*. *Proc. Natl. Acad. Sci. USA* 94:3519–3523. (116)

Grant, B. S. 2002. Sour grapes of wrath. *Science* 297:940–941. (138)

Grant, B. S., D. F. Owen, and C. A. Clarke. 1996. Parallel rise and fall of melanic peppered moths in America and Britain. *J. Hered.* 87:351–357. (138–139)

Grant, B. S., and L. L. Wiseman, 2002. Recent history of melanism in American peppered moths. *J. Hered.* 93:86–90. (138)

Grantham, R. C., C. Gautier, M. Gouy, *et al.* 1980. Codon catalog usage and the genome hypothesis. *Nucleic Acids Res.* 8:r49–r62. (452)

Greaves, J. H., R. Redfern, P. B. Ayres, and J. E. Gill. 1977. Warfarin resistance: a balanced polymorphism in the Norway rat. *Genet. Res.* 30:257–263. (118, 139)

Gribel, R., and J. D. Hay. 1993. Pollination ecology of *Caryocar brasiliense* (Caryocaraceae) in Central Brazil cerrado vegetation. *J. Trop. Ecol.* 9:199–211. (259)

Griffith, S. C., I. P. F. Owens, and K. A. Thuman. 2002. Extra pair paternity in birds: a review of interspecific variation and adaptive function. *Mol. Ecol.* 11:2195–2212. (622)

Guo, S. W., and E. A. Thompson. 1992. Performing the exact test of Hardy–Weinberg proportion for multiple alleles. *Biometrics* 48:361–372. (93–94)

Gutiérrez-Espleta, G. A., S. T. Kalinowski, W. M. Boyce, and P. W. Hedrick. 2000. Genetic variation and population structure in desert bighorn sheep: implications for conservation. *Cons. Genet.* 1:3–15. (506)

Haldane, J. B. S. 1919. The combination of linkage values and the calculation of distance between the loci of linked factors. *J. Genet.* 8:299–309. (538)

Haldane, J. B. S. 1937. The effect of variation on fitness. *Am. Nat.* 71:337–349. (375)

Haldane, J. B. S. 1940. The conflict between selection and mutation of harmful recessive genes. *Ann. Eugen.* 10:417–421. (371)

Haldane, J. B. S. 1948. The theory of a cline. *J. Genet.* 48:277–284. (515)

Haldane, J. B. S. 1949. The association of characters as a result of inbreeding and linkage. *Ann. Eugen.* 15:15–23. (550)

Haldane, J. B. S. 1956. The estimation of viabilities. *J. Genet.* 54:294–296. (152)

Haldane, J. B. S. 1964. A defense of beanbag genetics. *Perspect. Biol. Med.* 7:343–359. (216)

Haldane, J. B. S., and S. D. Jayakar. 1963. Polymorphism due to selection of varying direction. *J. Genet.* 58:237–242. (215, 218, 234)

Halushka, M. K., J.-B. Fan, K. Bentley, *et al.* 1999. Patterns of single-nucleotide polymorphisms in candidate genes for blood-pressure homeostasis. *Nat. Genet.* 22:239–247. (526)

Hamer, D. H., S. Hu, V. L. Magnuson, *et al.* 1993. A linkage between DNA markers on the X chromosome and male sexual orientation. *Science* 261:321–327. (657)

Hamilton, M. B., and J. R. Miller. 2003. Comparing relative rates of pollen and seed gene flow in the island model using nuclear and organelle measures of populations structure *Genetics* 144:1933–1940. (498)

Hamilton, W. D. 1964. The genetical evolution of social behavior. *J. Theor. Biol.* 7:1–16, 17–52. (273, 278–279)

Hamilton, W. D. 1972. Altruism and related phenomena, mainly in the social insects. *Annu. Rev. Ecol. Syst.* 3:193–232. (273)

Hamilton, W. D. 1993. Inbreeding in Egypt and in this book: a childish perspective (pp. 429–450). In N. W. Thornhill (ed.). *The Natural History of Inbreeding and Outbreeding.* Chicago: University of Chicago Press. (237)

Hamilton, W. D., and M. Zuk. 1982. Heritable true fitness and bright birds: a role for parasites? *Science* 218:384–387. (189)

Hammer, M. F., F. Blackmer, D. Garrigan, *et al.* 2003. Human population structure and its effects on sampling Y chromosome sequence variation. *Genetics* 164:1495–1509. (434–435)

Hanski, I. 1998. Metapopulation dynamics. *Nature* 396:41–49. (471)

Hanski, I., and M. Gilpin (eds.). 1997. *Metapopulation Dynamics: Ecology, Genetics, and Evolution.* New York: Academic Press. (471)

Hardy, G. H. 1908. Mendelian proportions in a mixed population. *Science* 28:49–50. (63)

Hardy, O. J., N. Charbonnel, H. Freville, and M. Heuertz. 2003. Microsatellite allele sizes: a simple test to assess their significance on genetic differentiation. *Genetics* 163:1467–1482. (506)

Hare, M. P. 2001. Prospects for nuclear gene phylogeography. *Trends Ecol. Evol.* 16:700–706. (612)

Harper, J. L. 1977. *Population Biology of Plants.* New York: Academic Press. (224)

Harris, D. J. 2003. Can you bank on GenBank? *Trends Ecol. Evol.* 18:317–319. (9)

Harris, E. E., and J. Hey. 1999. X chromosome evidence for ancient human histories. *Proc. Natl. Acad. Sci. USA* 96:3320–3324. (620–621)

Harris, H. 1966. Enzyme polymorphism in man. *Proc. R. Soc. Lond. B.* 164:298–310. (30–31, 407)

Harris, H., and D. A. Hopkinson. 1972. Average heterozygosity in man. *Ann. Hum. Genet.* 36:9–20. (34–35, 100)

Harrison, R. G. 1993. *Hybrid Zones and the Evolutionary Process.* Oxford, UK: Oxford University Press. (510)

Hartl, D. L., E. N. Moriyama, and S. Sawyer. 1994. Selection intensity for codon bias. *Genetics* 138:227–234. (451)

Hayes, B. J., P. M. Visscher, H. C. McPartlan, and M. E. Goddard. 2003. Novel multilocus measure of linkage disequilibrium to estimate past effective population size. *Genome Res.* 13:635–643. (542)

Hayman, B. I. 1953. Mixed selfing and random mating when homozygotes are at a disadvantage. *Heredity* 7:185–192. (290)

Hebert, P. D. N. 1974. Enzyme variability in natural populations of *Daphnia magna.* III. Genotypic frequencies in intermittent populations. *Genetics* 77:335–341. (71–72)

Hedgecock, D., V. Chow, and R. S. Waples. 1992. Effective population numbers of shellfish broodstock estimated from temporal variance in allelic frequencies. *Aquaculture* 88:215–232. (326)

Hedges, S. B., and S. Kumar. 2003. Genomic clocks and evolutionary times cales. *Trends Genet.* 19:200–206. (418)

Hedges, S. B., S. Kumar, K. Tamura, and M. Stoneking. 1992. Human origins and analysis of mitochondrial DNA sequences. *Science* 255:737–739. (618)

Hedrick, P. W. 1972. Maintenance of genetic variation with a frequency-dependent selection model as compared to the overdominant model. *Genetics* 72:771–775. (354)

Hedrick, P. W. 1976. Simulation of X-linked selection in *Drosophila*. *Genetics* 83:551–571. (166–167)

Hedrick, P. W. 1980a. Hitchhiking: a comparison of linkage and partial selfing. *Genetics* 94:791–808. (571, 573)

Hedrick, P. W. 1980b. The establishment of chromosomal variants. *Evolution* 35:322–332. (390)

Hedrick, P. W. 1986a. Genetic polymorphism in heterogeneous environments: a decade later. *Annu. Rev. Ecol. Syst.* 17:535–566. (206, 210)

Hedrick, P. W. 1986b. Average inbreeding or equilibrium inbreeding? *Am. J. Hum. Genet.* 38:965–970. (264)

Hedrick, P. W. 1987a. Population genetics of intragametophytic selfing. *Evolution* 41:137–144. (246)

Hedrick, P. W. 1987b. Estimation of the rate of partial inbreeding. *Heredity* 58:161–166. (259)

Hedrick, P. W. 1987c. Gametic disequilibrium measures: proceed with caution. *Genetics* 117:331–341. (532, 591–592)

Hedrick, P. W. 1990a. Genotypic-specific habitat selection: a new model and its application. *Heredity* 65:145–149. (213–214)

Hedrick, P. W. 1990b. Mating systems and evolutionary genetics (pp. 83–114). In K. Wohrmann and E. Jain (eds.). *Population Biology: Ecological and Evolutionary Viewpoints.* New York: Springer-Verlag. (550)

Hedrick, P. W. 1992. Female choice and variation in the major histocompatibility complex. *Genetics* 132:575–581. (195)

Hedrick, P. W. 1993. Sex-dependent habitat selection and genetic polymorphism. *Am. Nat.* 141:491–500. (177, 179)

Hedrick, P. W. 1994. Purging inbreeding depression and the probability of extinction: full-sib mating. *Heredity* 73:363–372. (289, 387)

Hedrick, P. W. 1995a. Genetic polymorphism in a temporally varying environment: effects of delayed germination or diapause. *Heredity* 75:164–170. (217)

Hedrick, P. W. 1995b. Elephant seals and the estimation of a population bottleneck. *J. Hered.* 86:232–235. (348)

Hedrick, P. W. 1995c. Gene flow and genetic restoration: the Florida panther as a case study. *Cons. Biol.* 9:996–1007. (512–513)

Hedrick, P. W. 1996. Bottleneck(s) or metapopulation in cheetahs. *Cons. Biol.* 10:897–899. (382)

Hedrick, P. W. 1997. Neutrality or selection. *Nature* 387:138. (183–184)

Hedrick, P. W. 1998. Maintenance of genetic polymorphism: spatial selection and self-fertilization. *Am. Nat.* 152:145–150. (291)

Hedrick, P. W. 1999a. Antagonistic pleiotropy and genetic polymorphism: a perspective. *Heredity* 82:126–133. (184–186)

Hedrick, P. W. 1999b. Perspective: highly variable loci and their interpretation in evolution and conservation. *Evolution* 53:313–318. (343, 492)

Hedrick, P. W. 2001. Invasion of transgenes in salmon or other genetically modified organisms into natural populations. *Can. J. Fish. Aquat. Sci.* 58:841–844. (186)

Hedrick, P. W. 2002a. Pathogen resistance and genetic variation at MHC loci. *Evolution* 56:1902–1908. (232, 234)

Hedrick, P. W. 2002b. Lethals in finite populations. *Evolution* 56:654–657. (282, 286, 386–387)

Hedrick, P. W. 2003a. A heterozygote advantage. *Science* 302:57. (149)

Hedrick, P. W. 2003b. Hopi Indians, "cultural" selection, and albinism. *Am. J. Phys. Anthrop.* 121:151–156. (187)

Hedrick, P. W. 2004. Estimation of relative fitnesses from relative risk data and the predicted future of hemoglobin alleles *S* and *C*. *J. Evol. Biol.* 17:221–224. (155, 158, 163)

Hedrick, P. W., and F. L. Black. 1997. HLA and mate selection: no evidence in south Amerindians. *Am. J. Hum. Genet.* 61:505–511. (196)

Hedrick, P. W., and C. C. Cockerham. 1986. Partial inbreeding: equilibrium heterozygosity and the heterozygosity paradox. *Evolution* 40:856–861. (262–264)

Hedrick, P. W., and M. E. Gilpin. 1997. Genetic effective size of a metapopulation (pp. 166–182). In I. Hanski and M. Gilpin (eds.). *Metapopulation Dynamics: Ecology, Genetics, and Evolution.* New York: Academic Press. (507–508)

Hedrick, P. W., M. E. Ginevan, and E. P. Ewing. 1976. Genetic polymorphism in heterogeneous environments. *Annu. Rev. Ecol. Syst.* 7:1–32. (206, 210, 213)

Hedrick, P. W., G. A. Gutiérrez-Espeleta, and R. N. Lee. 2001a. Founder effect in an island population of bighorn sheep. *Mol. Ecol.* 10:851–857. (344)

Hedrick, P. W., D. Hedgecock, and S. Hamelberg. 1995. Effective population size in winter-run chinook salmon. *Cons. Biol.* 9:615–624. (337)

Hedrick, P. W., and L. Holden. 1979. Hitchhiking: an alternative to coadaptation for the barley and slender wild oat examples. *Heredity* 43:79–86. (575)

Hedrick, P. W., and S. T. Kalinowski. 2000. Inbreeding depression in conservation biology. *Ann. Rev. Ecol. Syst.* 31:139–162. (280, 285)

Hedrick, P. W., and S. Kumar. 2001. Mutation and linkage disequilibrium in human mtDNA. *Eur. J. Hum. Genet.* 9:969–972. (533, 546, 548, 607)

Hedrick, P. W., and V. Loeschcke. 1996. MHC and mate selection in humans? *Trends Ecol. Evol.* 11:24. (196)

Hedrick, P. W., P. S. Miller, E. Geffen, and R. Wayne. 1997. Genetic evaluation of the three captive Mexican wolf lineages. *Zool. Biol.* 16:47–69. (267, 303)

Hedrick, P. W., and E. Murray. 1983. Selection and measures of fitness (pp. 61–104). In M. Ashburner, H. L. Carson, and J. N. Thompson (eds.). *Genetics and Biology of Drosophila*, vol. 3. New York: Academic Press. (147)

Hedrick, P. W., and J. D. Parker. 1997. Evolutionary genetics and genetic variation of haplodiploids and X-linked genes. *Annu. Rev. Ecol. Syst.* 28:55–83. (168, 271, 414–415)

Hedrick, P. W., and K. M. Parker. 1998a. MHC variation in the endangered Gila topminnow. *Evolution* 52:194–199. (37, 89–90)

Hedrick, P. W., and K. M. Parker. 1998b. Corrigendum. *Evolution* 52:932. (90)

Hedrick, P.W., K. M. Parker, and R. Lee. 2001b. Genetic variation in the endangered Gila and Yaqui topminnows: microsatellite and MHC variation. *Mol. Ecol.* 10:1399–1412. (494)

Hedrick, P. W., K. M. Parker, E. L. Miller, and P. S. Miller. 1999b. Major histocompatibility complex variation in the endangered Przewalski's horse. *Genetics* 152:1701–1710. (446)

Hedrick, P. W., V. K. Rashbrook, and D. Hedgecock. 2000. Effective population size of winter-run Chinook salmon based on microsatellite analysis of returning spawners. *Can. J. Fish. Aquat. Sci.* 57:2368–2373. (337)

Hedrick, P. W., O. Savolainen, and K. Kärkkäinen. 1999a. Factors influencing the extent of inbreeding depression: an example from Scots pine. *Heredity* 82:441–450. (374–375, 387)

Hedrick, P. W., and G. Thomson. 1983. Evidence for balancing selection at HLA. *Genetics* 104:449–456. (432)

Hedrick, P. W., and G. Thomson. 1987. A two-locus neutrality test: application to humans, *E. coli*, and lodgepole pine. *Genetics* 112:135–156. (591)

Hedrick, P. W., and G. Thomson. 1988. Maternal–fetal interactions and the maintenance of HLA polymorphism. *Genetics* 119:205–212. (182)

Hedrick, P. W., T. S. Whittam, and P. Parham. 1991. Heterozygosity at individual amino acid sites: extremely high levels for *HLA-A* and *-B* genes. *Proc. Natl. Acad. Sci. USA* 88:5897–5901. (41)

Hendry, A. P., and M. T. Kinnison. 1999. Perspective: the pace of modern life: measuring rates of contemporary microevolution. *Evolution* 53:1637–1653. (116)

Herlihy, C. R., and C. G. Eckert. 2002. Genetic cost of reproductive assurance in a self-fertilizing plant. *Nature* 416:320–323. (280)

Hey, J. 1999. The neutralist, the fly and the selectionist. *Trends Ecol. Evol.* 14:35–38. (579)

Hey, J., and C. A. Machado. 2003. The study of structured populations: new hope for a difficult and divided science. *Nat. Rev. Genet.* 4:535–543. (453, 471)

Hill, W. G. 1974. Estimation of linkage disequilibrium in randomly mating populations. *Heredity* 33:229–239. (587, 589–590, 592)

Hill, W. G. 1975. Tests for association of gene frequencies at several loci in random mating diploid populations. *Biometrics* 31:881–888. (591)

Hill, W. G. 1976. Non-random association of neutral linked genes in a finite population (pp. 339–376). In S. Karlin and E. Nevo (eds.). *Population Genetics and Ecology.* New York: Academic Press. (542)

Hill, W. G. 1981. Estimation of effective population size from data on linkage disequilibrium. *Genet. Res.* 38:209–216. (543)

Hill, W. G. 1982. Predictions of response to artificial from new mutations. *Genet. Res.* 40:255–278. (651)

Hill, W. G. 1995. Sewall Wright's "systems of mating." *Genetics* 143:1499–1506. (304)

Hill, W. G., and A. Robertson. 1966. The effect of linkage on limits to artificial selection. *Genet. Res.* 8:269–294. (452, 579)

Hill, W. G., and A. Robertson. 1968. Linkage disequilibrium in finite populations. *Theor. Appl. Genet.* 38:226–231. (532, 542)

Hillis, D. M. 1996. Inferring complex phylogenies. *Nature* 383:130–131. (599)

Hillis, D. M., J. P. Hulsenbeck, and C. W. Cunningham. 1994. Application and accuracy of molecular phylogenies. *Science* 264:671–677. (599)

Hillis, D. M., D. Moritz, and B. K. Mable (eds.). 1996. *Molecular Systematics*, 2nd ed. Sunderland, MA: Sinauer. (31, 36)

Hirschhorn, J. N., C. M. Lindgren, M.J. Daly, *et al.* 2001. Genomewide linkage analysis of stature in multiple populations reveals several regions with evidence of linkage to adult height. *Am. J. Hum. Genet.* 69:106–116. (656)

Hoekstra, H. E., K. E. Drumm, and M. W. Nachman. 2004. Ecological genetics of adaptive color polymorphism in pocket mice: geographic variation in selected and neutral genes. *Evolution* 58:1329–1341. (522)

Hoelzel, A. R. (ed.). 1998. *Molecular Genetic Analysis of Populations.* Oxford, UK: Oxford University Press. (31, 36)

Hoelzel, A. R., R. C. Fleischer, C. Campagna, *et al.* 2002. Impact of a population bottleneck on symmetry and genetic diversity in the northern elephant seal. *J. Evol. Biol.* 15:567–575. (303, 348–349)

Hoffman, A. A., and S. O'Donnell. 1990. Heritable variation in resource use in *Drosophila* in the field (pp. 177–193). In J. S. F. Barker, W. T. Starmer, and R. McIntyre (eds.). *Ecological and Evolutionary Genetics of Drosophila.* New York: Plenum. (214)

Hofreiter, M., D. Serre, H. N. Poinar, *et al.* 2001. Ancient DNA. *Nat. Rev. Genet.* 2:353–359. (42)

Holland, B., and W. R. Rice. 1998. Perspective: chase-away sexual selection: antagonistic selection versus resistance. *Evolution* 52:1–7. (189)

Holsinger, K. E., and R. J. Mason-Gamer. 1996. Hierarchical analysis of nucleotide diversity in geographically structured populations. *Genetics* 142:629–639. (495)

Hori, M. 1993. Frequency-dependent natural selection in the handedness of scale-eating cichlid fish. *Science* 260:216–219. (221–223)

Houle, D. 1989. Allozyme-associated heterosis in *Drosophila melanogaster*. *Genetics* 123:789–801. (570)

Howard, R. D., J. A. DeWoody, and W. M. Muir. 2004. Transgenic male mating advantage provides opportunity for Trojan gene effect in a fish. *Proc. Natl. Acad. Sci. USA* 101:2934–2938. (471)

Hu, X.-S., and R. A. Ennos. 1997. On estimation of the ratio of pollen to seed flow among plant populations. *Heredity* 79:541–552. (498)

Huai, H., and R. C. Woodruff. 1998. Clusters of new identical mutants and the fate of underdominant mutations. *Genetica* 102/103:489–505. (390)

Huang, Q.-Y., F.-H. Xu, H. Shen, *et al.* 2002. Mutation patterns at dinucleotide microsatellite loci in humans. *Am. J. Hum. Genet.* 70:625–634. (397, 400–401)

Huang, S. L., M. Singh, and K. Kojima. 1971. A study of frequency-dependent selection observed in the *esterase-6* locus of *Drosophila melanogaster* using a conditioned media method. *Genetics* 68:97–104. (223)

Hudson, R. R. 1983. Properties of a neutral allele model with intragenic recombination. *Theor. Popul. Biol.* 23:183–201. (461)

Hudson, R. R. 1990. Gene genealogies and the coalescent process. *Oxford Surveys Evol. Biol.* 7:1–44. (453–455, 457–458)

Hudson, R. R. 2000. A new statistic for detecting genetic differentiation. *Genetics* 155:2011–2014. (495)

Hudson, R. R. 2001. Linkage disequilibrium and recombination (pp. 309–324). In D. J. Balding, M. Bishop, and C. Cannings (eds.). *Handbook of Statistical Genetics.* New York: John Wiley and Sons. (592, 594)

Hudson, R. R. 2002. Generating samples under a Wright-Fisher neutral model of genetic variation. *Bioinformatics* 18:337–338. (431)

Hudson, R. R., D. D. Boos, and N. L. Kaplan. 1992. A statistical test for detecting geographic subdivision. *Mol. Biol. Evol.* 9:138–151. (495)

Hudson, R. R., and N. L. Kaplan. 1995. Deleterious background selection with recombination. *Genetics* 141:1605–1617. (579–580)

Hudson, R. R., M. Kreitman, and M. Aguade. 1987. A test of neutral molecular evolution based on nucleotide data. *Genetics* 116:153–159. (436)

Hudson, R. R., A. G. Saez, and F. J. Ayala. 1997. DNA variation at the *Sod* locus of *Drosophila melanogaster*: an unfolding story of natural selection. *Proc. Natl. Acad. Sci. USA*. 94:7725–7729. (577)

Huelsenbeck, J. P., F. Ronquist, R. Nielsen, and J. P. Bollback. 2001. Bayesian inference of phylogeny and its impact on evolutionary biology. *Science* 294:2310–2314. (601, 632)

Hughes, A. L., and M. K. Hughes. 1995. Natural selection on the peptide-binding regions of major histocompatibility complex molecules. *Immunogenetics* 42:233–243. (196)

Hughes, A. L., and M. Nei. 1988. Pattern of nucleotide substitution at major histocompatibility complex class I loci reveals overdominant selection. *Nature* 335:167–170. (447)

Hughes, D. 2003. Exploiting genomics, genetics and chemistry to combat antibiotic resistance. *Nat. Rev. Genet.* 4:422–441. (116)

Hughes, K. A. 1995. The inbreeding decline and average dominance of genes affecting male life-history characters in *Drosophila melanogaster*. *Genet. Res.* 65:41–52. (400)

Hughes, K. A., L. Du, F. H. Rodd, and D. N. Reznick. 1999. Familiarity leads to female mate preference for novel males in the guppy, *Poecilia reticulata*. *Anim. Behav.* 58:907–916. (192)

Hurst, H. L., C. E. Payne, C. M. Nevison, *et al.* 2001. Individual recognition in mice mediated by major urinary proteins. *Nature* 414:631–634. (196)

Husband, B., and D. Schemske. 1996. Evolution of the magnitude and timing of inbreeding depression in plants. *Evolution* 50:54–70. (281–282, 286)

Huttley, G. A., M. W. Smith, M. Carrington, and S. J. O'Brien. 1999. A scan for linkage disequilibrium across the human genome. *Genetics* 152:1711–1722. (527)

Huynen, L., C. D. Millar, R. P. Scofield, and D. M. Lambert. 2003. Nuclear DNA sequences detect species limits in ancient moa. *Nature* 425:175–178. (42)

Ikemura, T. 1982. Correlation between the abundance of yeast transfer RNAs and the occurrence of the respective codons in protein genes: difference in synonymous codon choice patterns of yeast and Escherichia coli with reference to the abundance of isoaccepting transfer RNAs. *J. Mol. Biol.* 158:573–597. (451)

Ikemura, T. 1985. Codon usage and tRNA content in unicellular and multicellular organisms. *Mol. Biol. Evol.* 2:13–34. (450)

Imaizumi, Y., M. Nei, and T. Furusho. 1970. Variability and heritability of human fertility. *Ann. Hum. Genet.* 33:251–259. (325–326)

Imhof, M., and C. Schlotterer. 2001. Fitness effects of advantageous mutations in evolving *Escherichia coli* populations. *Proc. Natl. Acad. Sci. USA* 98:1113–1117. (361)

Ingvarsson, P. K. 2004. Population subdivision and the Hudson-Kreitman-Aguade test: testing for deviations from the neutral model in organelle genomes. *Genet. Res.* 83:31–39. (437)

Innan, H., B. Padhukasahasram, and M. Nordborg. 2003. The pattern of polymorphism on human chromosome 21. *Genome Res.* 13:1158–1168. (534)

Innan, H., and W. Stephan. 2003. Distinguishing the hitchhiking and background selection models. *Genetics* 165:2307–2312. (579)

Ioerger, T. R., A. G. Clark, and T.-H. Kao. 1990. Polymorphism at the self-incompatibility locus in Solanaceae predates speciation. *Proc. Natl. Acad. Sci. USA* 87:9732–9735. (204)

Jacob, S., M. K. McClintock, B. Zelano, and C. Ober. 2002. Paternally inherited *HLA* alleles are associated with women's choice of male odor. *Nat. Genet.* 30:175–179. (196)

Jaenike, J. 1985. Genetic and environmental determinants of food preference in *Drosophila tripuncata*. *Evolution* 39:362–369. (214)

Jaenike, J. 2001. Sex chromosome meiotic drive. *Annu. Rev. Ecol. Syst.* 32:25–49. (197)

Jain, S. K. 1976. Patterns of survival and microevolution in plant populations (pp. 49–90). In S. Karlin and E. Nevo (eds.). *Population Genetics and Ecology.* New York: Academic Press. (44, 45)

Jain, S. K., and D. R. Marshall. 1967. Population studies in predominantly self-pollinating species. X. Variation in natural populations of *Avena fatua* and *A. barbata*. *Am. Nat.* 101:19–33. (254–255, 504)

Jain, S. K., and K. N. Rai. 1974. Population biology of *Avena.* IV. Polymorphism in small populations of *Avena fatua*. *Theor. Appl. Genet.* 44:7–11. (504–505)

James, J. W. 1965. Simultaneous selection for dominant and recessive mutants. *Heredity* 20:142–144. (159)

Jeffreys, A. J., L. Jauppi, and R. Neumann. 2001. Intensely punctuate meiotic recombination in the class II region of the major histocompatibility complex. *Nat. Genet.* 29:217–222. (660)

Jeffreys, A. J., and C. A. May. 2004. Intense and highly localized gene conversion activity in human meiotic crossover hot spots. *Nat. Genet.* 36:151–156. (362)

Jennings, H. S. 1917. The numerical results of diverse systems of breeding, with respect to two pairs of characters, linked or independent, with special relation to the effects of linkage. *Genetics* 2:97–154. (527)

Jin, L., and R. Chakraborty. 1995. Population structure, stepwise mutation, heterozygote deficiency and their implications in DNA forensics. *Heredity* 74:274–285. (492)

Johnson, G. C. L., L. Esposito, B. J. Barratt, *et al.* 2001. Haplotype tagging for the identification of common disease genes. *Nat. Genet.* 29:233–237. (659, 661)

Johnson, W. E., and R. K. Selander. 1971. Protein variation and systematics in kangaroo rats (genus *Dipodomys*). *Syst. Zool.* 20:377–405. (493)

Johnston, M. O., and D. J. Schoen. 1995. Mutation rates and dominance levels of genes affecting total fitness in two angiosperm species. *Science* 276:226–229. (400)

Jones, A. G., and W. R. Arden. 2003. Methods of parentage analysis in natural populations. *Mol. Ecol.* 12:2511–2523. (634, 636)

Jones, A. G., G. Rosenqvst, A. Berglund, and J. C. Avise. 1999. Clustered microsatellite mutations in the pipefish *Syngnathus typhle*. *Genetics* 152:1057–106. (391)

Jones, J. S., B. H. Leith, and P. Rawlings. 1977. Polymorphism in *Cepaea*: a problem with too many solutions. *Annu. Rev. Ecol. Syst.* 8:109–143. (206)

Joron, M., and P. M. Brakefield. 2003. Captivity masks inbreeding effects on male mating success in butterflies. *Nature* 424:191–194. (280, 282)

Joron, M., and J. L. B. Mallet. 1998. Diversity in mimicry: paradox or paradigm? *Trends Ecol. Evol.* 13:461–466. (229)

Jukes, T. H., and C. R. Cantor. 1969. Evolution of protein molecules (pp. 21–132). In H. N. Munro (ed.). *Mammalian Protein Metabolism.* New York: Academic Press. (421–422)

Kärkkäinen, K., V. Koski, and O. Savolainen. 1996. Geographical variation in the inbreeding depression of Scots pine. *Evolution* 50:111–119. (279, 374)

Kärkkäinen, K., H. Kuittinen, R. Van Treuen, *et al.* 1999. Genetic basis of inbreeding depression in *Arabis petraea*. *Evolution* 53:1354–1365. (282)

Kalinowski, S. T. 2002. How many alleles per locus should be used to estimate genetic distances? *Heredity* 88:62–65. (109)

Kalinowski, S. T., and P. W. Hedrick. 1999. Detecting inbreeding depression is difficult in captive endangered species. *Anim. Cons.* 2:131–136. (285)

Kalinowski, S. T., and P. W. Hedrick. 2001. Estimation of gametic disequilibrium for loci with multiple alleles: basic approach and an application using data from bighorn sheep. *Heredity* 87:698–708. (592)

Kalinowski, S. T., P. W. Hedrick, and P. S. Miller. 1999. No evidence for inbreeding depression in Mexican and red wolves. *Cons. Biol.* 13:1371–1377. (284–285)

Kalinowski, S. T., P. W. Hedrick, and P. S. Miller. 2000. Reduction of inbreeding depression in the Speke's gazelle captive breeding program: a search for causes. *Cons. Biol.* 14:1374–1384. (289)

Kalinowski, S. T. and R. S. Waples. 2002. Relationship of effective to census size in fluctuation populations. *Cons. Biol.* 16:129–136. (299, 333)

Kaplan, N. L., R. R. Hudson, and C. H. Langley. 1989. The "hitchhiking effect" revisited. *Genetics* 123:887–899. (571)

Kaplan, N., R. R. Hudson, and M. Lizuka. 1991. The coalescent process in models with selection, recombination, and geographic subdivision. *Genet. Res.* 57:83–91. (460–461)

Karju, A., P. Hurme, M. Karjalinen, *et al.* 1996. Do molecular markers reflect patterns of differentiation in adaptive traits? *Theor. Appl. Genet.* 93:215–221. (570)

Kauppi, L., A. J. Jeffreys, and S. Keeney. 2004. Where the crossovers are: recombination distributions in mammals. *Nat. Rev. Genet.* 5:413–424. (538)

Kazazian, H. H. 2004. Mobile elements: drivers of genome evolution. *Science* 303:1626–1632. (359)

Keller, L., and K. G. Ross. 1999. Major gene effects on phenotypes and fitness: the relative roles of *Pgm-3* and *GP-9* in introduced populations of the fire ant *Solenopsis invicta*. *J. Evol. Biol.* 12:672–680. (168)

Keller, L. F., and P. Arcese. 1998. No evidence for inbreeding avoidance in a natural population of song sparrows (*Melospiza melodia*). *Am. Nat.* 152:380–392. (270)

Keller, L. F., and D. M. Waller. 2002. Inbreeding effects in wild populations. *Trends Ecol. Evol.* 17:230–241. (279)

Keightley, P. D., and A. Caballero. 1997. Genomic mutation rates for lifetime reproductive output and lifespan in *Caenorhabditis elegans*. *Proc. Natl. Acad. Sci. USA* 94:3823–3827. (400)

Keith, T. P., L. D. Brooks, R. C. Lewontin, *et al.* 1985. Nearly identical distributions of *xanthine dehydrogenase* in two populations of *Drosophila pseudoobscura*. *Mol. Biol. Evol.* 2:206–216. (430)

Kelly, J. K., and J. H. Willis. 2002. A manipulative experiment to estimate biparental inbreeding in monkeyflowers. *Int. J. Plant Sci.* 163:575–579. (259)

Kerr, B., M. A. Riley, M. W. Feldman, and B. J. M. Bohannan. 2002. Local dispersal promotes biodiversity in a real-life game of rock-paper-scissors. *Nature* 418:171–174. (227)

Kettlewell, B. 1973. *The Evolution of Melanism.* London: Oxford University Press. (44, 138, 153)

Kidwell, M. G. 1983. Evolution of hybrid dysgenesis determinants in *Drosophila melanogaster*. *Proc. Natl. Acad. Sci. USA* 80:1655–1659. (359)

Kidwell, J. F., M. T. Clegg, F. M. Stewart, and T. Prout. 1977. Regions of stable equilibria for models of differential selection in the two sexes under random mating. *Genetics* 85:171–183. (179)

Kidwell, M. G., and D. R. Lisch. 2001. Perspective: transposable elements, parasitic DNA, and genome evolution. *Evolution* 55:1–24. (358–359, 361)

Kidwell, M. G., K. Kimura, and D. M. Black. 1988. Evolution of hybrid dysgenesis potential following P element contamination in *Drosophila melanogaster*. *Genetics* 119:815–828. (360)

Kile, B. T., K. E. Hentges, A. T. Clark, *et al.* 2003. Functional genetic analysis of mouse chromosome 11. *Nature* 425:81–86. (51–52)

Kim, Y. 2004. Effect of strong directional selection on weakly selected mutation at linked sites: implication for synonymous codon usage. *Mol. Biol. Evol.* 21:286–294. (452)

Kimura, M. 1954. Process leading to quasi-fixation of genes in natural populations due to random fluctuation of selection intensities. *Genetics* 39:280–295. (413)

Kimura, M. 1962. On the probability of fixation of mutant genes in a population. *Genetics* 47:713–719. (349)

Kimura, M. 1967. On the evolutionary adjustment of spontaneous mutation rates. *Genet. Res.* 9:23–34. (405)

Kimura, M. 1968. Evolutionary rate at the molecular level. *Nature* 217:624–626. (407, 411, 416, 418)

Kimura, M. 1969. The rate of molecular evolution considered from the standpoint of population genetics. *Proc. Natl. Acad. Sci. USA* 63:1181–1188. (419, 433)

Kimura, M. 1980. A simple method for estimating evolutionary rates of base substitutions through comparative studies of nucleotide sequences. *J. Mol. Evol.* 16:111–120. (355, 423, 602)

Kimura, M. 1983. *The Neutral Theory of Molecular Evolution.* Cambridge, UK: Cambridge University Press. (1, 388, 408, 417, 420, 436)

Kimura, M. 1994. *Population Genetics, Molecular Evolution, and Neutral Theory: Selected Papers.* Chicago: University of Chicago Press. (413)

Kimura, M., and J. F. Crow. 1963. The measurement of effective population numbers. *Evolution* 17:279–288. (320, 325)

Kimura, M., and J. F. Crow. 1964. The number of alleles that can be maintained in a finite population. *Genetics* 49:725–738. (378, 412)

Kimura, M., and T. Ohta. 1969. The average number of generations until extinction of an individual mutant gene in a finite population. *Genetics* 63:701–709. (388)

Kimura, M., and T. Ohta. 1971. *Theoretical Aspects of Population Genetics.* Princeton, NJ: Princeton University Press. (290, 303, 315, 349, 377)

Kimura, M., and T. Ohta. 1972. On the stochastic model for estimation of mutational distance between homologous proteins. *J. Mol. Evol.* 2:87–90. (423)

King, J. L. 1967. Continuously distributed factors affecting fitness. *Genetics* 55:483–492. (409)

King, J. L., and T. H. Jukes. 1969. Non-Darwinian evolution. *Science* 164:788–798. (407, 418, 449)

King, R. A., R. K. Willaert, R. M. Schmidt, *et al.* 2003. *MC1R* mutations modify the classic phenotype of oculocutanteous albinism type 2 (OCA2). *Am. J. Hum. Genet.* 73:638–645. (391)

King, R. B. 1993. Color pattern variation in Lake Erie water snakes: inheritance. *Can. J. Zool.* 71:1985–1990. (516)

King, R. B., and R. Lawson. 1995. Color-pattern variation in Lake Erie water snakes: the role of gene flow. *Evolution* 49:885–896. (516)

Kingman, J. F. C. 1982. The coalescent. *Stochastic Proc. Applic.* 13:235–248. (453)

Kirkness, E. F., V. Bafna, A. L. Happern, *et al.* 2003. The dog genome: survey sequencing and comparative analysis. *Science* 301:1898–1903. (35)

Kirkpatrick, M., and N. H. Barton. 1997. The strength of indirect selection on female mating preferences. *Proc. Natl. Acad. Sci. USA* 94:1282–1286. (189)

Kirkpatrick, M., and P. Jarne. 2000. The effects of a bottleneck on inbreeding depression and the genetic load. *Am. Nat.* 155:154–167. (286)

Kirkup, B. C., and M. A. Riley. 2004. Antibiotic-mediated antagonism leads to a bacterial game of rock-paper-scissors *in vivo*. *Nature* 428:412–414. (227)

Klekowski, E. J. 1979. The genetics and reproductive biology of ferns (pp. 133–170). In A. F. Dyer (ed.). *The Experimental Biology of Ferns.* London: Academic Press. (246)

Klekowski, E. J., and P. J. Godfrey. 1989. Aging and mutation in plants. *Nature* 340:389–391. (396)

Knoppien, P. 1985. Rare male mating advantage: a review. *Biol. Rev.* 60:81–117. (192)

Knowles, L. L. 2004. The burgeoning field of statistical phylogeography. *J. Evol. Biol.* 17:1–10. (612)

Knowles, L. L., and W. P. Maddison. 2002. Statistical phylogeography. *Mol. Ecol.* 11:2623–2635. (612)

Kocher, T. D. 2003. Fractious phylogenies. *Nature* 423:489–409. (600)

Kocher, T. D., W. K. Thomas, A. Meyer, *et al.* 1989. Dynamics of mitochondrial DNA evolution in animals: Amplification and sequencing with conserved primers. *Proc. Natl. Acad. Sci. USA* 86:6196–6200. (465)

Kodaira, M., C. Satoh, K. Hiyama, and K. Toyama. 1995. Lack of effects of atomic bomb radiation on genetic instability of tandem-repetitive elements in human germ cells. *Am. J. Hum. Genet.* 57:1275–1283. (393)

Kohn, M. H., H.-J. Pelz, and R. K. Wayne. 2000. Natural selection mapping of the warfarin-resistance gene. *Proc. Natl. Acad. Sci. USA* 97:7911–7915. (118)

Kohn, M. H., H.-J. Pelz, and R. K. Wayne. 2003. Locus-specific genetic differentiation at *Rw* among warfarin-resistant rat (*Rattus norvegicus*) populations. *Genetics* 164:1055–1070. (118)

Kojima, K. 1971. Is there a constant fitness for a given genotype? No! *Evolution* 25:281–285. (410)

Kojima, K., and H. E. Schaffer. 1967. Survival processes of linked mutant genes. *Evolution* 21:518–531. (571)

Kondrashov, A. 1988. Deleterious mutation and the evolution of sexual reproduction. *Nature* 339:300–301. (584)

Kong, A., D. F. Gudbjartsson, J. Sainz, *et al.*, 2002. A high-resolution recombination map of the human genome. *Nat. Genet.* 31:241–247. (538–540)

Korber, B., M. Muldoon, J. Theiler, *et al.* 2000. Timing the ancestor of the HIV-1 pandemic strains. *Science* 288:1789–1796. (425–426)

Korves, T., and J. Bergelson. 2004. A novel cost of *R* gene resistance in the presence of disease. *Am. Nat.* 163:489–504. (229)

Krebs, C. 2002. *Ecology: The Experimental Analysis of Distribution and Abundance*, 5th ed. New York: Prentice Hall. (318, 326)

Kreitman, M. 1983. Nucleotide polymorphism at the *alcohol dehydrogenase* locus of *Drosophila melanogaster. Nature* 304:411–417. (31, 38, 105, 438)

Kreitman, M. 2000. Methods to detect selection in populations with applications to the human. *Annu. Rev. Genomics Hum. Genet.* 1:539–559. (429)

Kreitman, M., and H. Akashi. 1995. Molecular evidence for natural selection. *Annu. Rev. Ecol. Syst.* 26:403–422. (407)

Kreitman, M., and R. R. Hudson. 1991. Inferring the evolutionary histories of the *Adh* and *Adh-dup* loci in *Drosophila melanogaster* from patterns of polymorphism and divergence. *Genetics* 127:565–582. (437–439)

Krieger, M.J. B., and K. G. Ross. 2002. Identification of a major gene regulating complex social behavior. *Science* 295:328–332. (168)

Krüger, O., J. Lindström, and W. Amos. 2001. Maladaptive mate choice maintained by heterozygote advantage. *Evolution* 55:1207–1214. (293–294)

Kruglyak, L. 1999. Prospects for whole-genome linkage disequilibrium mapping of common disease genes. *Nat. Genet.* 22:139–144. (526–527)

Krützen, M., L. M. Barre, R. C. Conner, *et al.* 2004. "O father: where are thou?" Paternity assessment in an open fission-fusion society of wild bottlenose dolphins (*Thursiops* sp.) in Shark Bay, Western Australia. *Mol. Ecol.* 13:1975–1990. (623)

Kuhn, T. 1962. *The Structure of Scientific Revolutions.* Chicago: University of Chicago Press. (1)

Kuhner, M. K., J. Yamato, and J. Felsenstein. 1995. Estimating effective population size and mutation rate from sequence data using Metropolis–Hastings sampling. *Genetics* 140:1421–1430. (464)

Kuhner, M. K., J. Yamato, and J. Felsenstein. 1998. Maximum likelihood estimation of population growth rates based on the coalescent. *Genetics* 149:419–434. (464, 466)

Kumar, S., and S. B. Hedges. 1998. A molecular timescale for vertebrate evolution. *Nature* 392:917–920. (420)

Kumar, S., P. W. Hedrick, T. Dowling, and M. Stoneking. 2000. Questioning evidence for recombination in human mitochondrial DNA. *Science* 288:1931a. (607–608)

Kumar, S., and S. Subramanian. 2002. Mutation rates in mammalian genomes. *Proc. Natl. Acad. Sci. USA.* 99:803–808. (418)

Laan, M., and S. Paabo. 1997. Demographic history and linkage disequilibrium in human populations. *Nat. Genet.* 17:435–438. (544–545)

Lack, D. 1968. *Ecological Adaptations for Breeding in Birds.* London: Methuen Ltd. (622)

Lacy, R. C., and J. D. Ballou. 2001. *Population Management 2000, version 1.175.* Brookfield, IL: Chicago Zoological Society. (267, 269)

Laikre, L., P. E. Jorde, and N. Ryman. 1998. Temporal change of mitochondrial DNA haplotype frequencies and female effective size in a brown trout (*Salmo trutta*) population. *Evolution* 52:910–915. (339–340)

Laikre, L., and N. Ryman. 1991. Inbreeding depression in a captive wolf (*Canis lupus*) population. *Cons. Biol.* 5:33–40. (242)

Laikre, L., N. Ryman, and E. A. Thompson. 1993. Hereditary blindness in a captive wolf (*Canis lupus*) population: frequency reduction of a deleterious allele in relation to gene conservation. *Cons. Biol.* 7:592–601. (353)

Lamb, T., and J. C. Avise. 1986. Directional introgression of mitochondrial DNA in a hybrid population of tree frogs: the influence of mating behavior. *Proc. Natl. Acad. Sci. USA* 83:2526–2530. (536–537)

Lamb, T., T. R. Jones, and P. J. Wettstein. 1997. Evolutionary genetics and phylogeography of tassel-eared squirrels (*Sciurus aberti*). *J. Mammal.* 78:117–133. (615)

Land, D., M. Cunningham, M. Lotz, and D. Shindle. 2001. Florida panther genetic restoration and management: annual report 2000–2001. Tallahassee: Florida Fish and Wildlife Conservation Commission. (513)

Lande, R. 1979. Effective deme sizes during long–term evolution estimated from rates of chromosomal rearrangement. *Evolution* 33:234–251. (390)

Lande, R. 1981. The minimum number of genes contributing to quantitative variation between and within populations. *Genetics* 99:541–553. (651–652)

Lande, R. 1994. Risk of population extinction from fixation of new deleterious mutation. *Evolution* 48:1460–1469. (388)

Lande, R. 1995. Mutation and conservation. *Cons. Biol.* 9:782–791. (388)

Lande, R. 1998. Risk of population extinction from fixation of deleterious and reverse mutations. *Genetica* 102/103:21–27. (388)

Lande, R., and S. J. Arnold. 1985. Evolution of mating preferences and sexual dimorphism. *J. Theoret. Biol.* 117:651–664. (189)

Lande, R., and G. F. Barrowclough. 1987. Effective population size, genetic variation, and their use in population management (pp. 87–124). In M. Soulé (ed.). *Viable Populations for Conservation.* Cambridge, UK: Cambridge University Press. (327, 331)

Lande, R., and D. W. Schemske. 1985. The evolution of self-fertilization and inbreeding depression in plants. I. Genetic models. *Evolution* 39:24–40. (280, 372)

Lande, R., D. W. Schemske, and S. T. Schultz. 1994. High inbreeding depression, selective interference among loci, and the threshold selfing rate for purging recessive lethal mutations. *Evolution* 48:965–978. (280)

Lander, E. S., and D. Botstein. 1989. Mapping Mendelian factors underlying quantitative traits using RFLP linkage maps. *Genetics* 121:185–199. (655)

Landsteiner, K., and A. S. Weiner. 1940. An agglutinable factor in human blood recognized by immune sera for rhesus blood. *Proc. Soc. Exp. Biol. N.Y.* 43:223. (28)

Langley, C. H., E. Montgomery, and W. F. Quattlebaum. 1982. Restriction map variation in the *Adh* region of *Drosophila. Proc. Natl. Acad. Sci. USA* 79:5631–5635. (587–588)

Laporte, V., and B. Charlesworth. 2002. Effective population size and population subdivision in demographically structured populations. *Genetics* 162:501–519. (498)

Latta, R. G., and J. B. Mitton. 1997. A comparison of population differentiation across four classes of gene marker in limber pine (*Pinus flexilis* James). *Genetics* 146:1153–1163. (497–498)

Launey, S. and D. Hedgecock. 2001. High genetic load in the Pacific oyster *Crassostrea gigas. Genetics* 159:255–265. (47)

Lautenberger, J. A., J. C. Stephens, S. J. O'Brien, and M. W. Smith. 2000. Significant admixture linkage disequilibrium across 30 cM around the *FY* locus in African Americans. *Am. J. Hum. Genet.* 66:969–978. (552)

Lawrence, M. J. 2000. Population genetics of the homomorphic self-incompatibility polymorphisms in flowering plants. *Ann. Bot.* 85:221–226. (204)

Leberg, P. L. 2002. Estimating allelic richness: effects of sample size and bottlenecks. *Mol. Ecol.* 11:2445–2449. (104)

Ledig, F. T., R. P. Guries, and B. A. Bonefield. 1983. The relation of growth to heterozygosity in pitch pine. *Evolution* 37:1227–1238. (550)

Lee, P. L. M., and G. C. Hays. 2004. Polyandry in a marine turtle: females make the best of a bad job. *Proc. Natl. Acad. Sci. USA* 101:6530–6535. (623–624)

Leeflang, E. P., S. Tavare, P. Marjoram, *et al.* 1999. Analysis of germline mutation spectra at the Huntington's disease locus supports a mitotic mutation mechanism. *Hum. Mol. Genet.* 8:173–183. (393–394)

Leigh, E. 1970. Natural selection and mutability. *Am. Nat.* 104:301–305. (405)

Leips, J., and T. F. C. Mackay. 2002. The complex genetic architecture of *Drosophila* life span. *Exper. Aging Res.* 28:361–390. (55, 57)

Lenormand, T. 2002. Gene flow and the limits to natural selection. *Trends Ecol. Evol.* 17:183–189. (510)

Leonard, J. A., R. K. Wayne, and A. Cooper. 2000. Population genetics of Ice Age brown bears. *Proc. Natl. Acad. Sci. USA* 97:1651–1654. (617)

Lerner, I. M. 1954. *Genetic Homeostasis.* Edinburgh: Oliver and Boyd. (411)

Levene, H. 1949. On a matching problem arising in genetics. *Ann. Math. Stat.* 20:91–94. (92)

Levene, H. 1953. Genetic equilibrium when more than one ecological niche is available. *Am. Nat.* 87:331–333. (210, 212)

Levin, B. R., M. L. Petras, and D. I. Rasmussen. 1969. The effect of migration on the maintenance of a lethal polymorphism in the house mouse. *Am. Nat.* 103:647–661. (199)

Levin, D. A., K. Ritter, and N. C. Ellstrand. 1979. Protein polymorphism in the narrow endemic *Oenothera organensis. Evolution* 33:534–542. (202)

Levine, L. (ed.). 1995. *Genetics of Natural Populations: The Continuing Importance of Theodosius Dobzhansky.* New York: Columbia University Press. (28)

Levins, R. 1969. Some demographic and genetic consequences of environmental heterogeneity for biological control. *Bull. Entomol. Soc. Am.* 15:237–240. (507)

Lewontin, R. C. 1963. The role of linkage in natural selection. *Proc. XI Int. Congress Genet.* pp. 517–525. (1)

Lewontin, R. C. 1964. The interaction of selection and linkage. 1. General considerations; heterotic models. *Genetics* 49:49–67. (532, 566)

Lewontin, R. C. 1968. The effect of differential viability on the population dynamics of *t* alleles in the house mouse. *Evolution* 22:262–273. (199)

Lewontin, R. C. 1971. The effect of genetic linkage on the mean fitness of a population. *Proc. Natl. Acad. Sci. USA* 68:984–986. (583)

Lewontin, R. C. 1974. *The Genetic Basis of Evolutionary Change.* New York: Columbia University Press. (5, 29, 46, 53, 114, 136, 147, 208, 525, 565)

Lewontin, R. C. 1978. Adaptation. *Sci. Am.* 239:3–13. (114)

Lewontin, R. C. 1988. On measures of gametic disequilibrium. *Genetics* 120:849–852. (525, 532)

Lewontin, R. C. 1991. Electrophoresis in the development of evolutionary genetics: milestone or millstone? *Genetics* 128:657–662. (410)

Lewontin, R. C. 1995. The detection of linkage disequilibrium in molecular sequence data. *Genetics* 140:377–388. (587–588)

Lewontin, R. C. 2004. In memory of John Maynard Smith (1920–2004). *Science* 304:979. (576)

Lewontin, R. C., and C. C. Cockerham. 1959. The goodness-of-fit test for detecting selection in random mating populations. *Evolution* 13:561–564. (150–151)

Lewontin, R. C., and L. C. Dunn. 1960. The evolutionary dynamics of a polymorphism in the house mouse. *Genetics* 45:705–722. (199)

Lewontin, R. C., L. R. Ginsburg, and S. N. Tuljapurkar. 1978. Heterosis as an explanation for large amounts of genic polymorphism. *Genetics* 88:149–170. (158, 409)

Lewontin, R. C., and D. L. Hartl. 1991. Population genetics in forensic DNA typing. *Science* 254:1745–1750. (642)

Lewontin, R. C., and J. L. Hubby. 1966. A molecular approach to the study of genic heterozygosity in natural populations. II. Amount of variation and degree of heterozygosity in natural populations of *Drosophila pseudoobscura. Genetics* 54:595–609. (30–32, 407–408, 410)

Lewontin, R. C., and K. Kojima. 1960. The evolutionary dynamics of complex polymorphisms. *Evolution* 14:450–472. (527, 529, 562)

Lewontin, R. C., J. A. Moore, W. B. Provine, and B. Wallace. 1981. *Dobzhansky's Genetics of Natural Populations I–XLIII.* New York: Columbia University Press. (28)

Li, C. C. 1948. *An Introduction to Population Genetics.* Peking: National Peking University Press. (625)

Li, C. C. 1949. Genetics dies in China. *J. Hered.* 41:90. (625)

Li, C. C. 1955. *Population Genetics.* Chicago: University of Chicago Press. (140, 629)

Li, C. C. 1969. Population subdivision with respect to multiple alleles. *Ann. Hum. Genet.* 33:23–29. (481)

Li, C. C. 1970. Table of variance of ABO gene frequency estimates. *Ann. Hum. Genet.* 34:189–194. (88–89)

Li, C. C. 1976. *First Course in Population Genetics.* Pacific Grove, CA: Boxwood Press. (79, 81, 190, 248, 260–261, 271, 510–511)

Li, C. C. 1999. Remarks on receiving the ASHG award: science and science education. *Am. J. Hum. Genet.* 64:16–17. (629)

Li, C. C., and A. Chakravarti. 1994. DNA profile similarity in a subdivided population. *Hum. Hered.* 44:100–109. (644)

Li, C. C., D. E. Weeks, and A. Chakravarti. 1993. Similarity of DNA fingerprints due to chance and relatedness. *Hum. Hered.* 43:45–52. (274)

Li, G., and D. Hedgecock. 1998. Genetic heterogeneity, detected by PCR-SSCP, among samples of larval Pacific oysters (*Crassostrea gigas*) supports the hypothesis of large variance in reproductive success. *Can. J. Fish. Aquat. Sci.* 55:1025–1033. (326)

Li, W.-H. 1997. *Molecular Evolution.* Sunderland, MA: Sinauer. (361, 400, 427–428, 446)

Li, W.-H., and D. Graur. 1991. *Fundamentals of Molecular Evolution.* Sunderland, MA: Sinauer. (450)

Li, W.-H., C.-I. Wu, and C.-C. Luo. 1985. A new method for estimating synonymous and nonsynonymous rates of nucleotide substitution considering the relative likelihood of nucleotide and codon changes. *Mol. Biol. Evol.* 2:150–174. (443, 447)

Li, W.-H., S. Yi, and K. Makova. 2002. Male-driven evolution. *Curr. Opin. Genet. Dev.* 12:650–656. (402)

Lindsley, D. L., and G. G. Zimm. 1992. *The Genome of Drosophila melanogaster*. San Diego, CA: Academic Press. (46, 362)

Liu, R., A. M. Ferrenberg, L. U. Gilliland, *et al.* 2003. Detection of deleterious genotypes in multigenerational studies. III. Estimation of selection components in highly selfing populations. *Genet. Res.* 82:41–53. (288)

Livingstone, F. B. 1969. Gene frequency clines of the β hemoglobin locus in various human populations and their simulation by models involving differential selection. *Hum. Biol.* 41:223–236. (518–519)

Livingstone, F. B. 1989. Simulation of the diffusion of the β-globin variants in the old world. *Hum. Biol.* 61:297–309. (519)

Lohmueller, K. E., C. L. Pearce, M. Pike, *et al.* 2003. Meta-analysis of genetic association studies supports a contribution of common variant to susceptibility to common disease. *Nat. Genet.* 33:177–182. (661)

Long, A. D., and C. H. Langley. 1999. The power of association studies to detect the contribution of candidate gene loci to variation in complex traits. *Genome Res.* 9:720–731. (542)

Long, J. C. 1991. The genetic structure of admixed populations. *Genetics* 127:417–428. (485)

Long, J. C., R. C. Williams, and M. Urbanek. 1995. An EM algorithm and testing strategy for multiple-locus haplotypes. *Am. J. Hum. Genet.* 56:799–810. (592)

Lonjou, C., A. Collins, and N. E. Morton. 1999. Allelic association between marker loci. *Proc. Natl. Acad. Sci. USA* 96:1621–1626. (545)

Louis, E. J., and E. R. Dempster. 1987. An exact test for Hardy–Weinberg and multiple alleles. *Biometrics* 43:805–811. (93–94)

Luikart, G., and J.-M. Cornuet. 1998. Empirical evaluation of a test for identifying recently bottlenecked populations from allele frequency data. *Cons. Biol.* 12:228–237. (348)

Luikart, G., and J.-M. Cornuet. 1999. Estimating the effective number of breeders from heterozygote excess in progeny. *Genetics* 151:1211–1216. (341)

Lynch, M., J. Blanchard, D. Houle, *et al.* 1999. Perspective: spontaneous deleterious mutation. *Evolution* 53:645–663. (400)

Lynch, M., J. Conery, and R. Burger. 1995. Mutation meltdowns in sexual populations. *Evolution* 49:1067–1088. (388)

Lynch, M., and J. S. Conery. 2003. The origins of genome complexity. *Science* 302:1401–1404. (35, 361)

Lynch, M., and T. J. Crease. 1990. The analysis of population survey data on DNA sequence variation. *Mol. Biol. Evol.* 7:377–394. (495)

Lynch, M., and W. Gabriel. 1990. Mutation load and the survival of small populations. *Evolution* 44:1725–1737. (388)

Lynch, M., and K. Ritland. 1999. Estimation of pairwise relatedness with molecular markers. *Genetics* 152:1753–1766. (274)

Lynch, M., and B. Walsh. 1998. *Genetics and Analysis of Quantitative Traits*. Sunderland, MA: Sinauer. (53, 116, 130, 279, 281, 286, 538, 556, 651–652, 654–655)

Lyon, M. F. 2003. Transmission ratio distortion in mice. *Annu. Rev. Genet.* 37:393–408. (197, 199)

Lyrholm, T., O. Leimar, B. Johanneson, and U. Gyllensten. 1998. Sex-based dispersal in sperm whales: contrasting mitochondrial and nuclear genetic structure of global populations. *Proc. R. Soc. Lond. B.* 266:347–354. (498)

Mable, B. K., M. H. Schierup, and D. Charlesworth. 2003. Estimating the number, frequency, and dominance of *S*-alleles in a natural population of *Arabidopsis lyrata* (Brassicaceae) with sporophytic control of self-incompatibility. *Heredity* 90:422–431. (203–204)

MacCluer, J. W., J. L. VandeBerg, B. Read, and O. A. Ryder. 1986. Pedigree analysis by computer simulation. *Zool. Biol.* 5:147–160. (269)

Maehr, D. S., and R. C. Lacy. 2002. Avoiding the lurking pitfalls in Florida panther recovery. *Wildlife Soc. Bull.* 30:971–978. (513)

Majerus, M. E. N. 1998. *Melanism: Evolution in Action.* Oxford, UK: Oxford University Press. (138)

Mandel, S. P. H. 1970. The equivalence of different sets of stability conditions for multiple allelic systems. *Biometrics* 26:840–845. (160)

Manel, S., P. Berthier, and G. Luikart. 2002. Detecting wildlife poaching: identifying the origin of individuals with Bayesian assignment tests and multilocus genotypes. *Cons. Biol.* 16:650–659. (650)

Manel, S., M. K. Schwartz, G. Luikart, and P. Taberlet. 2003. Landscape genetic: combining landscape ecology and population genetics. *Trends Ecol. Evol.* 18:189–197. (470)

Mani, G. S., and M. E. N. Majerus. 1993. Peppered moth revisited: analysis of recent decreases in melanic frequency and predictions for the future. *Biol. J. Linnean Soc.* 48:157–165. (138)

Markow, T., P. W. Hedrick, K. Zuerlein, *et al.* 1993. HLA polymorphism in the Havasupai: evidence for balancing selection. *Am. J. Hum. Genet.* 53:943–952. (432)

Marshall, E. 1999. Which Jefferson was the father? *Science* 283:153–154. (630)

Marshall, H. D., and K. Ritland. 2002. Genetic diversity and differentiation of Kermode bear populations. *Mol. Ecol.* 11:685–697. (83)

Marshall, T. C., J. Slate, L.E. B. Kruuk, and J. M. Pemberton. 1998. Statistical confidence for likelihood-based paternity inference in natural populations. *Mol. Ecol.* 7:639–655. (632, 634, 636)

Maruyama, T. 1970. On the rate of decrease of heterozygosity in circular stepping stone models of populations. *Theor. Popul. Biol.* 1:101–119. (505)

Maside, S., A. W. Lee, and B. Charlesworth. 2004. Selection on codon usage in *Drosophila americana. Curr. Biol.* 14:150–154. (452)

Maside, X., C. Bartolome, and B. Charlesworth. 2002. *S*-element insertions are associated with the evolution of the *Hsp70* genes in *Drosophila melanogaster. Curr. Biol.* 12:1686–1691. (359)

Masters, B. S., B. G. Hicks, L. S. Johnson, and L. A. Erb. 2003. Genotype and extra-pair paternity in the house wren: a rare-male effect? *Proc. R. Soc. Lond. B.* 270:1393–1397. (192)

Mateu, E., F. Calafell, M. D. Ramon, *et al.* 2002. Can a place of origin of the main cystic fibrosis mutation be identified? *Am. J. Hum. Genet.* 70:257–264. (391)

Matise, T. C., R. Sachidanandam, A. G. Clark, *et al.* 2003. A 3.9-centimorgan-resolution human single-nucleotide polymorphism linkage map and screening set. *Am. J. Hum. Genet.* 73:271–284. (538, 543)

Maynard Smith, J. 1964. Group selection and kin selection. *Nature* 201:1145–1147. (278)

Maynard Smith, J. 1968. *Mathematical Ideas in Biology.* London: Cambridge University Press. (576)

Maynard Smith, J. 1978. *The Evolution of Sex.* London: Cambridge University Press. (576, 583)

Maynard Smith, J. 1998. *Evolutionary Genetics*, 2nd ed. Oxford: Oxford University Press. (576)

Maynard Smith, J., and J. Haigh. 1974. The hitch-hiking effect of a favorable gene. *Genet. Res.* 23:23–35. (571, 576)

Maynard Smith, J, and D. Harper. 2003. *Animal Signals.* Oxford: Oxford University Press. (576)

Mayr, E. 1963. *Animal Species and Evolution.* Cambridge, MA: Harvard University Press. (28, 216)

McCauley, D. E., J. Raveill, and J. Antonovics. 1995. Local founding events as determinants of genetic structure in a plant metapopulation. *Heredity* 75:630–636. (509)

McCauley, D. E., D. P. Whitter, and L. M. Reilly. 1985. Inbreeding and the rate of self-fertilization in a grape fern, *Botrychium dissectum. Am. J. Bot.* 72:1978–1981. (246)

McCune, A. R., R. C. Fuller, A. A. Aquilina, *et al.* 2002. A low genomic number of recessive lethals in natural populations of bluefin killifish and zebrafish. *Science* 296:2398–2401. (46–48)

McDonald, J., and M. Kreitman. 1991. Adaptive protein evolution at the *Adh* locus in *Drosophila. Nature* 351:652–654. (440–441)

McDonald, J. H. 1996. Detecting non-neutral heterogeneity across a region of DNA sequence in the ratio of polymorphism to divergence. *Molec. Biol. Evol.* 13:253–260. (438)

McDonald, J. H. 1998. Improved tests for heterogeneity across a region of DNA sequence in the ratio of polymorphism to divergence. *Molec. Biol. Evol.* 15:377–384. (438)

McHaffie, H. S., C. J. Legg, and R. A. Ennos. 2001. A single gene with pleiotropic effects accounts for the Scottish endemic taxon *Athyrium distentifolium* var. *flexile. New Phytol.* 152:491–500. (246–247)

McKeigue, P. M. 1998. Mapping genes that underlie ethnic differences in disease risk: methods for detecting linkage in admixed populations, by conditioning on parental admixture. *Am. J. Hum. Genet.* 63:241–251. (554)

McKenzie, J. A. 1996. *Ecological and Evolutionary Aspects of Insecticide Resistance.* New York: Academic Press. (116)

McKusick, V. A. 1978. *Medical Genetic Studies of the Amish.* Baltimore, MD: Johns Hopkins University Press. (301)

McKusick, V. A. 1998. *Mendelian Inheritance in Man*, 12th ed. Baltimore, MD: Johns Hopkins University Press. (362, 391)

McPherson, M. J., S. G. Möller, R. Beynon, and C. Howe. 2000. *PCR.* New York: Springer Verlag. (36)

McVean, G., P. Awadalla, and P. Fearnhead. 2002. A coalescent-based method for detecting and estimating recombination from gene sequences. *Genetics* 160:1231–1241. (548, 607)

McVean, G. A. T., and B. Charlesworth. 1999. A population genetic model for the evolution of synonymous codon usage: patterns and predictions. *Genet. Res.* 47:145–158. (451)

McVean, G. A. T., S. R. Myers, S. Hunt, *et al.* 2004. The fine-scale structure of recombination rate variation in the human genome. *Science* 304:581–584. (660)

Mead, S., M. P. H. Stumpf, J. Whitfield, *et al.* 2003. Balancing selection at the prion protein gene consistent with prehistoric kurulike epidemics. *Science* 300:640–643. (139, 149)

Meagher, T. R. 1986. Analysis of paternity within a natural population of *Chamaelirium luteum.* I. Identification of most-likely male parent. *Am. Nat.* 128:199–215. (635)

Mecsas, J., G. Franklin, W. A. Kuziel, *et al.* 2004. CCR5 mutation and plague protection. *Nature* 427:606. (582)

Menotti-Raymond, M., and S. J. O'Brien. 1993. Dating the genetic bottleneck of the African cheetah. *Proc. Natl. Acad. Sci. USA* 90:3172–3176. (382)

Merrill, C., L. Bayraktaroglu, A. Kusano, and B. Ganetsky. 1999. Truncated RanGAP encoded by the *Segregation Distorter* locus of *Drosophila. Science* 283:1742–1745. (197)

Mikko, S., and L. Andersson. 1995. Extensive MHC class II *DRB3* diversity in African and European cattle. *Immunogenetics* 42:408–413. (446)

Milinkovitch, M. C., D. Monteyne, J. P. Gibbs, *et al.* 2004. Genetic analysis of a successful repatriation programme: giant Galápagos tortoises. *Proc. Roy. Soc. Lond. B* 271:341–345. (303)

Milkman, R. D. 1967. Heterosis is a major cause of heterozygosity in nature. *Genetics* 55:493–495. (409)

Miller, C. R., J. Adams, and L. P. Waits. 2003. Pedigree based assignment tests for reversing coyote (*Canis latrans*) introgression into the wild red wolf (*Canis rufus*) population. *Mol. Ecol.* 12:3287–3301. (475, 647)

Miller, C. R., P. Joyce, and L. P. Waits. 2002. Assessing allelic dropout and genotype reliability using maximum likelihood. *Genetics* 160:357–366. (646)

Miller, P. S., and P. W. Hedrick. 1993. Inbreeding and fitness in captive populations: lessons from *Drosophila. Zool. Biol.* 12:333–351. (282)

Milligan, B. G. 2003. Maximum-likelihood estimation of relatedness. *Genetics* 163:1153–1167. (274)

Mills, L. S., and F. W. Allendorf. 1996. The one-migrant-per-generation rule in conservation and management. *Cons. Biol.* 10:1509–1518. (501)

Minckley, W. L., P. C. Marsh, J. E. Deacon, *et al.* 2003. A conservation plan for native fishes of the lower Colorado River. *Bioscience* 53:219–234. (465)

Mitton, J. B. 1998. *Selection in Natural Populations.* Oxford, UK: Oxford University Press. (410)

Mitton, J. B., and R. K. Koehn. 1973. Population genetics of marine pelecypods. III. Epistasis between functionally related isoenzymes of *Mytilus edulis. Genetics* 73:478–496. (591)

Modiano, D., G. Luoni, B. S. Sirima, *et al.* 2001. Haemoglobin C protects against clinical *Plasmodium falciparum* malaria. *Nature* 414:305–308. (156–157, 162–163)

Moen, T., B. Hoyheim, H. Munck, and L. Gomez-Raya. 2004. A linkage map of Atlantic salmon (*Salmo salar*) reveals an uncommonly large difference in recombination rate between the sexes. *Anim. Genet.* 35:81–92. (538)

Moloney, D., S. F. Slaney, M. Oldridge, *et al.* 1996. Exclusive paternal origin of new mutations in Apert syndrome. *Nat. Genet.* 13:48–53. (402)

Montoya, B., and G. Gates. 1975. Bighorn capture and transplant in Mexico. *Desert Bighorn Council Trans.* 19:28–32. (344)

Moran, P. A. P. 1962. *The Statistical Process of Evolutionary Theory.* Oxford, UK: The Clarendon Press. (68)

Moritz, C. 1994. Defining "Evolutionarily Significant units" for conservation. *Trends Ecol. Evol.* 9:373–375. (616)

Morton, N. E. 1971. Population genetics and disease control. *Soc. Biol.* 18:243–251. (371, 373)

Morton, N. E., J. F. Crow, and H. J. Muller. 1956. An estimate of the mutational damage in man from data on consanguineous marriages. *Proc. Natl. Acad. Sci. USA* 42:855–863. (283–284, 388)

Mourant, A. E., A. C. Kopec, and K. Domantiewska-Sobczak. 1976. *The Distribution of Human Blood Groups and Other Polymorphisms*, 2nd ed. New York: Oxford University Press. (182)

Mousseau, T. A., and D. A. Roff. 1987. Natural selection and the heritability of fitness components. *Heredity* 59:181–197. (55, 56)

Mukai, T. 1964. The genetic structure of natural populations of *Drosophila melanogaster*. I. spontaneous mutation rate of polygenes controlling viability. *Genetics* 50:1-19. (398–399)

Mukai, T. 1977. Genetic variance for viability and linkage disequilibrium in natural populations of *Drosophila melanogaster* (pp. 97–112). In F. B. Christiansen and T. M. Fenchel (eds.). *Measuring Selection in Natural Populations*. New York: Springer-Verlag. (411)

Mukai, T., S. I. Chigusa, L. E. Mettler, and J. F. Crow. 1972. Mutation rate and dominance of genes affecting viability in *Drosophila melanogaster*. *Genetics* 72:335–355. (361, 398)

Mukai, T., and S. Nagano. 1983. The genetic structure of natural populations of *Drosophila melanogaster*. XVI. Excess of additive genetic variance of viability. *Genetics* 105:115–134. (51, 52)

Muller, H. J. 1932. Some genetic aspects of sex. *Am. Nat.* 66:118–138. (583–584)

Muller, H. J. 1950. Our load of mutations. *Am. J. Hum. Genet.* 2:111–176. (143)

Mundy, N. I., N. S. Badcock, T. Hart, *et al.* 2004. Conserved genetic basis of a quantitative plumage trait involved in mate choice. *Science* 303:1870–1873. (521)

Mungall, A. J., S. A. Palmer, S. K. Sims, *et al.* 2003. The DNA sequence and analysis of human chromosome 6. *Nature* 425:805–811. (13, 41)

Muona, O., and A. Harju. 1989. Effective population sizes, genetic variability, and mating system in natural stands and seed orchards of *Pinus sylvestris*. *Silvae Genetica* 38:221–228. (570)

Murphy, K. 1998. The ecology of the cougar (*Puma concolor*) in the northern Yellowstone ecosystem: interactions with prey, bears, and humans (PhD Dissertation). Moscow, ID: University of Idaho. (328)

Nachman, M. W., V. L. Bauer, S. L. Crowell, and C. F. Aquadro. 1998. DNA variability and recombination rates at X-linked loci in humans. *Genetics* 150:1133–1141. (413–414)

Nachman, M. W., W. M. Brown, M. Stoneking, and C. F. Aquadro. 1996. Nonneutral mitochondrial DNA variation in humans and chimpanzees. *Genetics* 142:953–963. (440–441)

Nachman, M. W., and S. L. Crowell. 2000. Contrasting evolutionary histories of two introns of the Duchenne muscular dystrophy gene, *Dmd*, in humans. *Genetics* 155:1855–1864. (438)

Nachman, M. W., H. E. Hoekstra, and S. L. D'Agostino. 2003. The genetic basis of adaptive melanism in pocket mice. *Proc. Natl. Acad. Sci. USA* 100:5268–5273. (521–522)

Nair, S., J. T. Williams, A. Brockman, *et al.* 2003. A selective sweep driven by pyrimethamine treatment in southeast Asian malaria parasites. *Mol. Biol. Evol.* 20:1526–1536. (578)

Nasrallah, J. B. 2002. Recognition and rejection of self in plant reproduction. *Science* 296:305–308. (204)

National Research Council. 1992. *DNA Technology in Forensic Science*. Washington, DC: National Academies Press. (642–643)

National Research Council. 1996. *DNA Forensic Science: An Update*. Washington, DC: National Academies Press. (642–644)

Navarro, S., and N. H. Barton. 2003. Accumulation postzygotic isolation genes in parapatry: a new twist on chromosomal speciation. *Evolution* 57:447–459. (390)

Nee, S. 2004. Professor John Maynard Smith 1920–2004. *Trends. Evol. Ecol.* 19:345–346. (576)

Nei, M. 1968. The frequency distribution of lethal chromosomes in finite populations. *Proc. Natl. Acad. Sci. USA* 60:517–524. (386)

Nei, M. 1969. Heterozygous effects and frequency changes of lethal genes in populations. *Genetics* 63:669–680. (386)

Nei, M. 1972. Genetic distance between populations. *Am. Nat.* 106:283–292. (108, 492, 506)

Nei, M. 1973. Analysis of gene diversity in subdivided populations. *Proc. Natl. Acad. Sci. USA* 70:3321–3323. (491, 493)

Nei, M. 1977. *F*-statistics and analysis of gene diversity in subdivided populations. *Ann. Hum. Genet.* 41:225–233. (489, 491)

Nei, M. 1986. Definition and estimation of fixation indices. *Evolution* 40:643–645. (488)

Nei, M. 1987. *Molecular Evolutionary Genetics.* New York: Columbia University Press. (97, 107, 109, 343, 345, 361, 408, 424, 428, 489, 491, 493)

Nei, M. 1995. Motoo Kimura (1924–1994). *Mol. Biol. Evol.* 12:719–722. (413)

Nei, M. 1996. Phylogenetic analysis in molecular evolutionary genetics. *Annu. Rev. Genet.* 30:371–403. (598)

Nei, M., and T. Gojobori. 1986. Simple methods for estimating the numbers of synonymous and nonsynonymous nucleotide substitutions. *Mol. Biol. Evol.* 3:418–426. (443)

Nei, M., and D. Graur. 1984. Extent of protein polymorphism and the neutral mutation theory. *Evol. Biol.* 17:73–118. (413)

Nei, M., and Y. Imaizumi. 1966. Genetic structure of human populations. II. Differentiation of blood group gene frequencies among isolated populations. *Heredity* 21:183–190. (335)

Nei, M., and S. Kumar. 2000. *Molecular Evolution and Phylogenetics.* New York: Oxford University Press. (104–105, 400, 421, 423–424, 443, 445, 598–599, 602, 604–605, 607)

Nei, M., and W.- H. Li. 1973. Linkage disequilibrium in subdivided populations. *Genetics* 75:213–219. (552, 554)

Nei, M., T. Maruyama, and R. Chakraborty. 1975. The bottleneck effect and genetic variability in populations. *Evolution* 29:1–10. (383–384)

Nei, M., and A. K. Roychoudhury. 1974. Sampling variances of heterozygosity and genetic distance. *Genetics* 76:379–390. (97–98)

Nei, M., and A. K. Roychoudhury. 1993. Evolutionary relationships of human populations on a global scale. *Mol. Biol. Evol.* 10:927–943. (109)

Nei, M., and F. Tajima. 1981. Genetic drift and estimation of effective population size. *Genetics* 98:625–640. (339)

Neigel, J. E. 1997. A comparison of alternative strategies for estimating gene flow from genetic markers. *Annu. Rev. Ecol. Syst.* 28:105–128. (488, 502)

Neigel, J. E. 2002. Is F_{ST} obsolete? *Cons. Genet.* 3:167–173. (494)

Neuhauser, C. 2001. Mathematical models in population genetics (pp. 153–177). In D. J. Balding, M. Bishop, and C. Cannings (eds.). *Handbook of Statistical Genetics.* New York: John Wiley and Sons. (366)

Nielsen, R. 2001. Statistical tests of selective neutrality in the age of genomics. *Heredity* 86:641–647. (429)

Nielsen, R., D. R. Tarpy, and H. K. Reeve. 2003. Estimating effective paternity number in social insects and the effective number of alleles in a population. *Mol. Ecol.* 12:3157–3164. (104)

Nielsen, R, and D. M. Weinreich. 1999. The age of nonsynonymous and synonymous mutations in mtDNA and implications for the mildly deleterious theory. *Genetics* 153:497–506. (441)

Nordborg, M. 2001. Coalescent theory (pp. 179– 212). In D. J. Balding, M. Bishop, and C. Cannings (eds.). *Handbook of Statistical Genetics.* New York: John Wiley and Sons. (453, 457, 459–461)

Nordborg, M., J. O. Borevitz, J. Bergelson, *et al.* 2002. The extent of linkage disequilibrium in *Arabidopsis thaliana. Nat. Genet.* 30:190–193. (551)

Norio, R. 2003. Finnish disease heritage. I. Characteristics, causes, background. *Hum. Genet.* 12:441–456. (300)

Nunney, L. 1993. The influence of mating system and overlapping generations on effective population size. *Evolution* 47:1329–1341. (326)

Nunney, L. 1996. The influence of variation in female fecundity on effective population size. *Biol. J. Linnean Soc.* 59:411–425. (326)

Nunney, L. 1999. The effective size of a hierarchically structured population. *Evolution* 53:1–10. (507–508)

Nunney, L., and E. M. Baker. 1993. The role of deme size, reproductive patterns, and dispersal in the dynamics of *t*-lethal haplotypes. *Evolution* 47:1342–1359. (199)

Nunney, L., and D. R. Elam. 1994. Estimating the effective population size of conserved populations. *Cons. Biol.* 8:175–184. (335)

O'Brien, S. J. (ed.). 1993. *Genetic Maps*, 6th ed. Cold Spring Harbor, NY: Cold Spring Harbor Press. (8)

O'Brien, S. J., M. E. Roelke, J. Howard, *et al.* 1990. Genetic introgression within the Florida panther (*Felis concolor coryi*). *Natl. Geograph. Res.* 6:485–494. (212)

Oddou-Muratorio, S., R. J. Petit, B. Le Guerroue, *et al.* 2001. Pollen- versus seed-mediated gene flow in a scattered forest tree species. *Evolution* 55:1123–1135. (498)

Ogasawara, K., R. Yabe, M. Uchikawa, *et al.* 2001. Recombination and gene conversion-like events may contribute to *ABO* gene diversity causing various phenotypes. *Immunogenetics* 53:190–199. (88)

Ohta, T. 1971. Associative overdominance caused by linked detrimental mutations. *Genet. Res.* 18:277–286. (569)

Ohta, T. 1973. Slightly deleterious mutant substitutions in evolution. *Nature* 246:96–98. (352, 418)

Ohta, T. 1980. *Evolution and Variation of Multigene Families.* Berlin: Springer Verlag. (362)

Ohta, T. 1995. Synonymous and nonsynonymous substitutions in mammalian genes and the nearly neutral theory. *J. Mol. Evol.* 40:56–63. (426)

Ohta, T., and M. Kimura. 1969. Linkage disequilibrium due to random genetic drift. *Genet. Res.* 13:47–55. (542)

Ohta, T., and M. Kimura. 1970. Development of associative overdominance through linkage disequilibrium in finite populations. *Genet. Res.* 16:165–177. (569)

Ohta, T., and M. Kimura. 1971. On the constancy of the evolutionary rate of cistrons. *J. Mol. Evol.* 1:18–25. (426)

Ohta, T., and M. Kimura. 1973. A model of mutation appropriate to estimate the number of electrophoretically detectable alleles in a finite population. *Genet. Res.* 22:201–204. (380–381)

Okuda, H., H. Suganuma, T. Kamesaki, *et al.* 2000. The analysis of nucleotide substitutions, gaps, and recombination events between *RHD* and *RHCE* genes through complete sequencing. *Biochem. Biophy. Res. Commun.* 274:670–683. (180)

Onishi, O. 1977. Spontaneous and ethyl methanesulfonate-induced mutations controlling viability in *Drosophila melanogaster*. II. Homozygous effects of polygenic mutations. *Genetics* 87:547–556. (398)

Oota, H., W. Settheetham-Ishida, D. Tivavech, *et al.* 2001. Human mtDNA and Y-chromosome variation is correlated with matrilocal versus patrilocal residence. *Nat. Genet.* 29:20–21. (498–499)

Orr, H. A. 2003. The distribution of fitness effects among beneficial mutations. *Genetics* 163:1519–1526. (361)

Ostertag, E. M., J. L. Goodier, Y Zhang, and H. H. Kazazian. 2003. SVA elements are non-autonomous retrotransposons that cause disease in humans. *Am. J. Hum Genet.* 73:1444–1451. (358)

Ota, T., and M. Nei. 1994a. Divergent evolution and evolution by the birth-and-death process in the immunoglobin VH gene family. *Mol. Biol. Evol.* 11:469–482. (362)

Ota, T., and M. Nei. 1994b. Variances and covariances of the number of synonymous and nonsynonymous substitutions per site. *Mol. Biol. Evol.* 11:613–619. (445)

Otto, S. P., and N. H. Barton. 1997. The evolution of recombination: removing the limits to natural selection. *Genetics* 147:879–906. (584)

Otto, S. P., and M. C. Whitlock. 1997. The probability of fixation in populations of changing size. *Genetics* 146:723–733. (385)

Owen, D. F. 1966. Polymorphism in pleistocene land snails. *Science* 152:71–72. (43)

Paetkau, D. 2003. An empirical exploration of data quality in DNA-based population inventories. *Mol. Ecol.* 12:1375–1387. (646)

Paetkau, D., W. Calvert, I. Stirling, and C. Strobeck. 1995. Microsatellite analysis of population structure in Canadian polar bears. *Mol. Ecol.* 4:347–354. (646)

Paetkau, D., R. Slade, M. Burdens, and A. Estoup. 2004. Genetic assignment methods for the direct, real-time estimation of migration rate: a simulation-based exploration of accuracy and power. *Mol. Ecol.* 13:55–65. (647)

Paetkau, D., and C. Strobeck. 1994. Microsatellite analysis of genetic variation in black bear populations. *Mol. Ecol.* 3:489–495. (644)

Paetkau, D., L. P. Waits, P. L. Clarkson, *et al.* 1997. An empirical evaluation of genetic distance statistics using microsatellite data from bear (Ursidae) populations. *Genetics* 147:1943–1957. (101–102, 108, 345, 506)

Pagnier, J., J. G. Mears, O. Dunda-Belkhodja, *et al.* 1984. Evidence for the multicentric origin of the sickle cell hemoglobin gene in Africa. *Proc. Natl. Acad. Sci. USA* 81:1771–1773. (546)

Paland, S., and B. Schmid. 2003. Population size and the nature of genetic load in *Gentianella germanica*. *Evolution* 57:2242–2251. (286)

Palmer, M. S., A. J. Dryden, J. T. Hughes, and J. Collinge. 1991. Homozygous prion protein genotype predisposes to sporadic Creutzfeldt–Jakob disease. *Nature* 352:340–342. (149)

Palumbi, S. R. 2001. Humans as the world's greatest evolutionary force. *Science* 293:1786–1790. (116)

Pamilo, P. 1979. Genic variation at sex-linked loci: quantification of regular selection models. *Hereditas* 91:129–133. (168)

Pamilo, P., and M. Nei. 1988. Relationships between gene trees and species trees. *Mol. Biol. Evol.* 5:568–583. (599)

Pannell, J. R. 2003. Coalescence in a metapopulation with recurrent local extinction and recolonization. *Evolution* 57:949–961. (508)

Pannell, J. R., and B. Charlesworth. 2000. Effects of metapopulation processes on measures of genetic diversity. *Phil. Trans. R. Soc. Lond.* B 355:1851–1864. (508)

Papadopoulos, D., D. Schneider, J. Meier-Eiss, *et al.* 1999. Genomic evolution during a 10,000-generation experiment with bacteria. *Proc. Natl. Acad. Sci. USA* 96:3807–3812. (599)

Pardini, A. T., C. S. Jones, L. R. Noble, *et al.* 2001. Sex-biased dispersal of great white sharks. *Nature* 412:139–140. (498)

Parham, P., K. L. Arnett, E. J. Adams, *et al.* 1997. Episodic evolution and turnover of *HLA-B* in the indigenous human populations of the Americas. *Tissue Antigens* 50:219–232. (432)

Parham, P., and T. Ohta. 1996. Population biology of antigen presentation by MHC class I molecules. *Science* 272:67–74. (154)

Parker, H. G., L. V. Kim, N. B. Sutter, *et al.* 2004. Genetic structure of the purebred domestic dog. *Science* 304:1160–1164. (622)

Parker, K. M, R. J. Sheffer, and P. W. Hedrick. 1999. Molecular variation and evolutionarily significant units in the endangered Gila topminnow. *Cons. Biol.* 13:108–116. (90, 494)

Parra, E. J., R. A. Kittles, G. Argyropoulos, *et al.* 2001. Ancestral proportions and admixture dynamics in geographically defined African Americans living in South Carolina. *Am. J. Phys. Anthrop.* 114:18–29. (486, 551–552)

Parra, E. J., A. Marcini, J. Akey, *et al.* 1998. Estimating African American admixture proportions by use of population-specific alleles. *Am. J. Hum. Genet.* 63:1839–1851. (485–486)

Partridge, L. 1988. The rare-male effect: what is its evolutionary significance? *Phil. Trans. R. Soc. Lond. B* 319:525–539. (192)

Pena, S. D. J., and R. Chakraborty. 1994. Paternity testing in the DNA era. *Trends Genet.* 10:204–209. (631)

Penn, D. J. 2002. The scent of genetic compatibility: sexual selection and the major histocompatibility complex. *Ethology* 108:1–21. (196)

Penrose, L. S. 1949. The meaning of fitness in human populations. *Ann. Eugen.* 14:301–304. (175)

Petit, R. J., A. El Mousadik, and O. Pons. 1998. Identifying populations for conservation on the basis of genetic markers. *Cons. Biol.* 12:844–855. (103)

Pfaff, C. L., E. J. Parra, C. Bonilla, *et al.* 2001. Population structure in admixed populations: effect of admixture dynamics on the pattern of linkage disequilibrium. *Am. J. Hum. Genet.* 68:198–207. (554)

Phillips, M. K., V. G. Henry, and B. T. Kelley. 2003. Restoration of the red wolf (pp. 272–288). In L. D. Mech and L. Boitani (eds.). *The Ecology, Behavior, and Conservation of the Wolf.* Chicago: University of Chicago Press. (475)

Pielak, J., and N. H. Barton. 1997. The spread of an advantageous alleles across a barrier: the effects of random drift and selection against heterozygotes. *Genetics* 145:493–504. (390)

Piganeau, G., and A. Eyre-Walker. 2003. Estimation the distribution of fitness effects from DNA sequence data: implication for the molecular clock. *Proc. Natl. Acad. Sci. USA* 100:10335–10340. (441)

Piganeau, G., and A. Eyre-Walker. 2004. A reanalysis of the indirect evidence for recombination in human mitochondrial DNA. *Heredity* 92:282–288. (548, 607)

Platt, J. R. 1964. Strong inference. *Science* 146:347–353. (4)

Powell, J. R. 1997. *Progress and Prospects in Evolutionary Biology: The Drosophila Model.* New York: Oxford University Press. (47)

Powell, J. R., and E. N. Moriyama. 1997. Evolution of codon usage bias in *Drosophila. Proc. Natl. Acad. Sci. USA* 94:7776–7790. (451)

Presgraves, D. C., L. Balagopalan, S. M. Abmayr, and H. A. Orr. 2003. Adaptive evolution drives divergence of a hybrid inviability gene between two species of *Drosophila*. *Nature* 423:715–719. (651)

Primmer, C. R., M. T. Koskinen, and J. Piironen. 2000. The one that did not get away: individual assignment using microsatellite data detects a case of fishing competition fraud. *Proc. R. Soc. Lond. B.* 267:1699–1704. (647, 649)

Pritchard, J. K. 2001. Are rare variants responsible for susceptibility to complex diseases? *Am. J. Hum. Genet.* 69:124–137. (661)

Pritchard, J. K., and M. Przeworski, 2001. Linkage disequilibrium in humans: models and data. *Am. J. Hum. Genet.* 69:1–14. (526–527, 541–542)

Pritchard, J. K., M. Stephens, and P. Donnelly. 2000. Inference of population structure using multilocus genotype data. *Genetics* 155:945–959. (650)

Prout, T. 1965. The estimation of fitness from genotypic frequencies. *Evolution* 19:546–551. (148)

Prout, T. 1967. Selective forces in *Papilio glaucus*. *Science* 156:534. (208)

Prout, T. 1968. Sufficient conditions for multiple niche polymorphisms. *Am. Nat.* 102:493–496. (213)

Prout, T. 1969. The estimation of fitnesses from population data. *Genetics* 63:949–967. (177)

Prout, T. 1971. The relation between fitness components and population prediction in *Drosophila*. 11. Population prediction. *Genetics* 68:151–167. (142)

Prout, T. 1973. Appendix to J. B. Mitton and R. C. Koehn. Population genetics of marine pelecypods. III. Epistasis between functionally related isoenzymes in *Mytilus edulis*. *Genetics* 73:487–496. (552)

Provine, W. B. 1971. *The Origins of Theoretical Population Genetics*. Chicago: University of Chicago Press. (1)

Provine, W. B. 1986. *Sewall Wright and Evolutionary Biology*. Chicago: University of Chicago Press. (304)

Ptak, S. E., and M. Przeworski. 2002. Evidence for population growth in humans is confounded by fine-scale population structure. *Trends Genet.* 18:559–563. (436)

Ptak, S. E., K. Voelpel, and M. Przeworski. 2004. Insights into recombination from patterns of linkage disequilibrium in humans. *Genetics* 167:387–397. (538)

Pudovkin, A. I., D. V. Zaykin, and D. Hedgecock. 1996. On the potential for estimating the effective number of breeders from heterozygote excess in progeny. *Genetics* 144:383–387. (341, 347)

Purser, A. F. 1966. Increase in heterozygote frequency with differential fertility. *Heredity* 21:322–327. (74)

Queller, D. C., and K. F. Goodnight. 1989. Estimating relatedness using genetic markers. *Evolution* 43:258–275. (274–275)

Queller, D. C., R. Zacchim, R. Cervo, *et al.* 2000. Unrelated helpers in a social insect. *Nature* 405:784–787. (279)

Qvarnstrom, A., and E. Forsgren. 1998. Should females prefer dominant males? *Trends Ecol. Evol.* 13:498–501. (189)

Rainey, P. B., and M. Travisano. 1998. Adaptive radiation in a heterogeneous environment. *Nature* 394:69–72. (209)

Ralls, K., J. D. Ballou, and A. Templeton. 1988. Estimates of lethal equivalents and the cost of inbreeding in mammals. *Cons. Biol.* 2:185–193. (284)

Ralls, K., J. D. Ballou, B. A. Rideout, and R. Frankham. 2000. Genetic management of chondrodystrophy in California condors. *Anim. Cons.* 3:145–153. (353)

Ralls, K., K. Baugh, and J. Ballou. 1979. Inbreeding and juvenile mortality in small populations of ungulates. *Science* 206:1101–1103. (284)

Ralls, K., P. H. Harvey, and A. M. Lyles. 1986. Inbreeding in natural populations of birds and mammals (pp. 35–56). In M. Soulé (ed.). *Conservation Biology.* Sunderland, MA: Sinauer. (269)

Rambaut, A., D. Posada, K. A. Crandall, and E. C. Holmes. 2004. The causes and consequences of HIV evolution. *Nat. Rev. Genet.* 5:52–61. (425)

Rannala, B., and J. Mountain. 1997. Detecting immigration by using multilocus genotypes. *Proc. Natl. Acad. Sci. USA* 94:9197–9201. (650)

Ranson, H., C. Claudianos, F. Ortelli, *et al.* 2002. Evolution of supergene families associated with insecticide resistance. *Science* 298:179–181. (117)

Reed, T. E. 1969. Caucasian genes in American Negroes. *Science* 165:762–768. (485)

Remington, D. L., and D. M. O'Malley. 2000. Whole-genome characterization of embryonic stage inbreeding depression in a selfed loblolly pine family. *Genetics* 155:337–348. (47)

Rice, G., C. Anderson, N. Risch, and G. Ebers. 1999. Male homosexuality: Absence of linkage to microsatellite markers at Xq28. *Science* 284:665–667. (657)

Rice, W. R. 1989. Analyzing table of statistical tests. *Evolution* 43:223–225. (25)

Rice, W. R. 1992. Sexually antagonistic genes: experimental evidence. *Science* 256:1436–1439. (189)

Rice, W. R. 1994. Degeneration of a nonrecombining chromosome. *Science* 263:230–232. (585)

Rice, W. R., and A. K. Chippindale. 2001a. Intersexual ontogenetic conflict. *J. Evol. Biol.* 14:685–693. (189–190)

Rice, W. R., and A. K. Chippindale. 2001b. Sexual recombination and the power of natural selection. *Science* 294:555–559. (583)

Richards, C. M. 2000. Inbreeding depression and genetic rescue in a plant metapopulation. *Am. Nat.* 155:383–394. (471)

Richman, A. D., and J. R. Kohn. 1999. Self-incompatibility alleles from *Physalis*: implications for historical inference from balanced genetic polymorphisms. *Proc. Natl. Acad. Sci. USA* 96:168–172. (204)

Risch, N. 2000. Searching for genetic determinants in the new millennium. *Nature* 405:847–856. (660)

Risch, N., E. W. Reich, M. M. Wishnick, and J. G. McCarthy. 1987. Spontaneous mutation and parental age in humans. *Am. J. Hum. Genet.* 41:218–248. (402)

Risch, N., H. Tang, H. Katzenstein, and J. Ekstein. 2003. Geographic distribution of disease mutation in the Ashkenazi Jewish population supports genetic drift over selection. *Am. J. Hum. Genet.* 72:812–822. (391)

Ritland, K. 1984. The effective proportion of self-fertilization with consanguineous matings in inbred populations. *Genetics* 106:139–152. (257)

Ritland, K. 1996. Estimators for pairwise relatedness and individual inbreeding coefficients. *Genet. Res.* 67:175–185. (274)

Ritland, K. 2002. Extensions of models for the estimation of mating systems using n independent loci. *Heredity* 88:221–228. (254, 258)

Ritland, K., and F. R. Ganders. 1985. Variation in the mating system of *Bidens menziesii* (Asteraceae) in relation to population structure. *Heredity* 55:235–244. (257–258)

Ritland, K., C. Newton, and H. D. Marshall. 2001. Inheritance and population structure of the white-phased "Kermode" black bear. *Curr. Biol.* 11:1468–1472. (82–84)

Robbins, R. B. 1918. Some applications of mathematics to breeding problems. II. *Genetics* 3:73–92. (527)

Roberts, D. F., and R. W. Hiorns. 1962. The dynamics of social intermixture. *Am. J. Hum. Genet.* 14:261–277. (478)

Robertson, A. 1962. Selection for heterozygotes in small populations. *Genetics* 47:1291–1300. (353–354)

Robertson, A. 1965. The interpretation of genotypic ratios in domestic animal populations. *Anim. Prod.* 7:319–324. (74)

Robinson, W. P. 1996. The extent, mechanism, and consequences of genetic variation for recombination rate. *Am. J. Hum. Genet.* 59:1175–1183. (538–539)

Roelke, M. E., J. S. Martenson, and S. J. O'Brien. 1993. The consequences of demographic reduction and genetic depletion in the endangered Florida panther. *Curr. Biol.* 3:340–350. (513)

Roff, D. A. 2002. Inbreeding depression: tests of the overdominance and partial dominance hypotheses. *Evolution* 56:768–775. (283)

Roman, J., and S. R. Palumbi. 2003. Whales before whaling in the north Atlantic. *Science* 301:508–510. (463–464)

Rose, M. R. 1982. Antagonistic pleiotropy, dominance and genetic variation. *Heredity* 48:63–78. (184–185)

Rose, M. R. 1991. *The Evolutionary Biology of Aging.* Oxford: Oxford University Press. (184)

Rost, S., A. Fregin, V. Ivaskeviclus, *et al.* 2004. Mutations in *VKORC1* cause warfarin resistance and multiple coagulation factor deficiency type 2. *Nature* 427:537–541. (118)

Roughgarden, J. 1971. Density-dependent natural selection. *Ecology* 52:453–468. (205)

Rousset, F. 1997. Genetic differentiation and estimation of gene flow from *F*-statistics under isolation by distance. *Genetics* 145:1219–1228. (505)

Rousset, F. 2003. Effective size in simple metapopulation models. *Heredity* 91:107–111. (508)

Rousset, F., and M. Raymond. 1995. Testing heterozygote excess and deficiency. *Genetics* 140:1413–1419. (93)

Rousset, F., and M. Raymond. 1997. Statistical analyses of population genetic data: new tools, old concepts. *Trends Ecol. Evol.* 12:313–317. (93)

Rowley, I., E. Russell, and M. Brooker. 1993. Inbreeding in birds (pp. 304–328). In N. W. Thornhill (ed.). *The Natural History of Inbreeding and Outbreeding.* Chicago: University of Chicago Press. (269)

Russell, L. B., and W. L. Russell. 1996. Spontaneous mutations recovered as mosaics in the mouse specific-locus test. *Proc. Natl. Acad. Sci. USA* 93:13072–13077. (364, 390, 395)

Russell, C. 2002. *Spirit Bear: Encounters with the White Bear of the Western Rainforest.* Toronto, CA: Key Porter Books. (83)

Ruwende, C., S. C. Khoo, R. W. Snow, *et al.* 1995. Natural selection of hemizygotes for G6PD deficiency in Africa by resistance to severe malaria. *Nature* 376:246–249. (168)

Ryan, M. J. 1998. Sexual selection, receiver biases, and the evolution of sex differences. *Science* 281:1999–2003. (189)

Ryman, N., and P. E. Jorde. 2001. Statistical power when testing for genetic differentiation. *Mol. Ecol.* 100:2361–2373. (490)

Saccheri, I. J., and P. M. Brakefield. 2002. Rapid spread of immigrant genomes into inbred populations. *Proc. R. Soc. Lond. B.* 269:1073–1078. (471)

Salamon, H., W. Klitz, S. Easteal, *et al.* 1999. Evolution of HLA class II molecules: allelic and amino acid site variability across populations. *Genetics* 152:393–400. (433)

Saitou, N., and M. Nei. 1987. The neighbor-joining method: a method for reconstructing phylogenetic trees. *Mol. Biol. Evol.* 4:406–425. (600–601)

Salceda, V. M., and W. W. Anderson. 1988. Rare male mating advantage in a natural population of *Drosophila pseudoobscura*. *Proc. Natl. Acad. Sci. USA* 85:9870–9874. (192)

Sanger, F., G. M. Air, B. G. Barrell, *et al.* 1977. Nucleotide sequence of bacteriophage FX174 DNA. *Nature* 265:687–695. (37)

Sanjuan, R., A. Moya, and S. F. Elena. 2004. The distribution of fitness effects caused by single-nucleotide substitutions in an RNA virus. *Proc. Natl. Acad. Sci. USA*. 101:8396–8401. (361)

Sargent T. D., C. D. Millar, and D. M. Lambert. 1998. The "classical" explanation of industrial melanism. *Evol. Biol.* 30:299–322. (138)

Sarich, V. M., and A. C. Wilson. 1973. Generation time and genomic evolution in primates. *Science* 179:1144–1147. (426)

Satoh, C., and M. Kodaira. 1996. Effects of radiation on children. *Nature* 383:226. (393)

Saunders, M. A., M. F. Hammer, and M. W. Nachman. 2002. Nucleotide variability at *G6pd* and the signature of malarial selection in humans. *Genetics* 162:1849–1861. (168)

Savolainen, O., and P. W. Hedrick. 1995. Heterozygosity and fitness: no association in Scots pine. *Genetics* 140:755–766. (570)

Sawyer, S. A., and D. L. Hartl. 1992. Population genetics of polymorphism and divergence. *Genetics* 132:1161–1176. (441)

Sax, K. 1923. The association of size differences with seed-coat pattern and pigmentation in *Phaseolus vulgaris*. *Genetics* 8:522–560. (654)

Schaeffer, S. W., M. P. Goetting-Minesky, M. Kovacevic, *et al.* 2003. Evolutionary genomics of inversions in *Drosophila pseudoobscura*: evidence for epistasis. *Proc. Natl. Acad. Sci. USA* 100:8319–8324. (28)

Schemske, D. W., and P. Bierzychudek. 2001. Perspective: evolution of flower color in the desert annual *Linanthus parryae*: Wright revisited. *Evolution* 55:1269–1282. (217–218)

Schierup, M. H. 1998. The number of self-incompatibility alleles in a finite, subdivided population. *Genetics* 149:1153–1162. (202)

Schierup, M. H., X. Vekemans, and F. B. Christiansen. 1997. Evolutionary dynamics of sporophytic self-incompatibility alleles in plants. *Genetics* 147:835–846. (204)

Schlager, G., and M. M. Dickie. 1971. Natural mutation rates in the house mouse: estimates for five specific loci and dominant mutations. *Mut. Res.* 11:89–96. (363, 390, 392)

Schlenke, T. A., and D. J. Begun. 2004. Strong selective sweep associated with a transposon insertion in *Drosophila simulans*. *Proc. Nat. Acad. Sci. USA* 101:1626–1631. (359)

Schlötterer, C., R. Ritter, B. Harr, and G. Brem. 1998. High mutation rate of a long microsatellite allele in *Drosophila melanogaster* provides evidence for allele-specific mutation rates. *Mol. Biol. Evol.* 15:1269–1274. (397)

Schug, M. D., C. M. Hutter, K. A. Wetterstrand, *et al.* 1998. The mutation rates of di-, tri-, and tetranucleotide repeats in *Drosophila melanogaster*. *Mol Biol. Evol.* 15:1751–1760. (397)

Schull, W. J., and J. V. Neel. 1965. *The Effects of Inbreeding on Japanese Children.* New York: Harper & Row. (281)

Seal, U. S. 1994. A plan for genetic restoration and management of the Florida panther (*Felis concolor coryi*). Report to the U. S. Fish and Wildlife Service. Conservation Breeding Specialist Group. Apple Valley, MN: SSC/IUCN. (303)

Seal, U. S., E. T. Thorne, S. H. Anderson, *et al.* (eds.). 1989. *Conservation Biology of the Black-Footed Ferret.* New Haven, CT: Yale University Press. (212)

Searle, A. G. 1949. Gene frequencies in London's cats. *J. Genet.* 49:214–220. (78)

Seielstad, M. T., E. Minch, and L. L. Cavalli-Sforza. 1998. Genetic evidence for a higher female migration rate in humans. *Nat. Genet.* 20:278–280. (498)

Selander, R. K. 1966. Sexual dimorphism and differential niche utilization in birds. *Condor* 68:113–151. (177)

Selander, R. K. 1976. Genetic variation in natural populations (pp. 21–45). In F. Ayala (ed.). *Molecular Evolution.* Sunderland, MA: Sinauer. (32)

Selander, R. K., and D. W. Kaufman. 1975. Genetic structure of populations of the brown snail (*Helix aspera*). 1. Microgeographic variation. *Evolution* 29:385–401. (33)

Sharp, P. M., E. Cowe, D. G. Higgins, *et al.* 1988. Codon usage patterns in *Escherichia coli, Bacillus subtilis, Saccharomyces cerevisiae, Shizosaccharomyces pombe, Drosophila melanogaster* and *Homo sapiens*: a review of the considerable within-species diversity. *Nucleic Acids Res.* 16:8207–8211. (450, 452)

Sharp, P. M., T. M. F. Tuohy, and K. R. Mosurski. 1986. Codon usage in yeast: cluster analysis clearly differentiates highly and lowly expressed genes. *Nucleic Acids Res.* 14:5125–5143. (449)

Shaw, F. H., C. J . Geyer, and R. G. Shaw. 2002. A comprehensive model of mutations affecting fitness and inferences for *Arabidopsis thaliana. Evolution* 56:453–463. (400)

Shiang, R., L. M. Thompson, Y. Z. Zhu *et al.* 1994. Mutations in the transmembrane domain of FGFR3 cause the most common genetic form of dwarfism, achondroplasia. *Cell* 78:335–342. (402)

Shuster, S. M., and M. J. Wade. 2003. *Mating Systems and Strategies.* Princeton, NJ: Princeton University Press. (188)

Signer, E. N., Y. E. Dubrova, A. J. Jeffreys, *et al.* 1998. DNA fingerprinting Dolly. *Nature* 394:329–330. (640)

Simmons, M. J., and J. F. Crow. 1977. Mutations affecting fitness in *Drosophila* populations. *Ann. Rev. Genet.* 11:49-78. (361, 387)

Simonsen, K. I., G. A. Churchill, and C. F. Aquadro. 1995. Properties of statistical tests of neutrality for DNA polymorphism data. *Genetics* 141:413–428. (436)

Simpson, G. G. 1945. The principles of classification and a classification of mammals. *Bull. Am. Mus. Nat. Hist.* 85:1–350. (597)

Sinervo, B. 2001. Runaway social games, genetic cycles, driven by alternative male and female strategies, and the origin of morphs. *Genetica* 112–113:417–434. (87)

Sinervo, B., and C. M. Lively. 1996. The rock-paper-scissors game and the evolution of alternative male strategies. *Nature* 380:240–243. (86, 227)

Sittman, K., H. A. Abplanalp, and R. A. Fraser. 1966. Inbreeding depression in Japanese quail. *Genetics* 54:371–379. (280)

Slade, R. W., C. Moritz, A. R. Hoelzel, and H. R. Burton. 1998. Molecular population genetics of the southern elephant seal *Mirounga leonina. Genetics* 149:1945–1957. (322)

Slate, J., T. Marshall, and J. Pemberton. 2000. A retrospective assessment of the accuracy of the paternity inference program CERVUS. *Mol. Ecol.* 9:801–808. (636)

Slatkin, M. 1973. Gene flow and selection in a cline. *Genetics* 75:735–756. (519)

Slatkin, M. 1977. Gene flow and genetic drift in a species subject to frequent local extinctions. *Theor. Popul. Biol.* 12:253–262. (507–508)

Slatkin, M. 1981. Estimating levels of gene flow in natural populations. *Genetics* 99:323–335. (469)

Slatkin, M. 1985. Rare alleles as indicators of gene flow. *Evolution* 39:53–65. (484)

Slatkin, M. 1991. Inbreeding coefficients and coalescence times. *Genet. Res.* 58:167–185. (505)

Slatkin, M. 1994a. An exact test for neutrality based on the Ewens sampling distribution. *Genet. Res.* 64:71–74. (430)

Slatkin, M. 1994b. Linkage disequilibrium in growing and stable populations. *Genetics* 137:331–336. (543, 545)

Slatkin, M. 1995. A measure of population subdivision based on microsatellite allele frequencies. *Genetics* 139:457–462. (501)

Slatkin, M. 1996. A correction to the exact test based on the Ewens sampling distribution. *Genet. Res.* 68:259–260. (430)

Slatkin, M. 2000. A coalescent view of population structure (pp. 418–429). In R. S. Singh and C. B. Krimbas (eds.). *Evolutionary Genetics: From Molecules to Morphology.* Cambridge, UK: Cambridge University Press. (460)

Slatkin, M. 2002. The age of alleles (pp. 233–260). In M. Slatkin and M. Veuille (eds.). *Modern Developments in Theoretical Population Genetics: The Legacy of Gustave Malecot.* Oxford: Oxford University Press. (581)

Slatkin, M., and N. H. Barton. 1989. A comparison of three methods for estimating average levels of gene flow. *Evolution* 43:1358–1368. (484)

Slatkin, M., and L. Excoffier. 1996. Testing for linkage disequilibrium in genotypic data using the Expectation-Maximization algorithm. *Heredity* 76:377–383. (592, 594)

Slatkin, M., and B. Rannala. 2000. Estimating allele age. *Annu. Rev. Genome Hum. Genet.* 1:225–249. (581)

Slatkin, M., and M. Veuille. 2002. *Modern Developments in Theoretical Population Genetics: The Legacy of Gustave Malecot.* Oxford: Oxford University Press. (453)

Smith, M. W., N. Patterson, J. A. Lautenberger, *et al.* 2004. A high-density admixture map for disease gene discovery in African Americans. *Am. J. Hum. Genet.* 74:1001–1013. (554)

Smith, N. G. C. 2003. Are radical and conservative substitution rates useful statistics in molecular evolution? *J. Mol. Evol.* 57:467–478. (449)

Smith, N. G. C., and A. Eyre-Walker. 2002. Adaptive protein evolution in *Drosophila. Nature* 415:1022–1024. (442)

Smith, P., M. Berdoy, R. H. Smith, and D. W. MacDonald. 1993. A new aspect of warfarin resistance in wild rats: benefits in the absence of poison. *Funct. Ecol.* 7:190–194. (118)

Smith, T. B. 1993. Disruptive selection and the genetic basis of bill size polymorphism in the African finch *Pyrenestes. Nature* 363:619–620. (214)

Snow, A. A., D. Pilson, L. H. Rieseberg, *et al.* 2003. A Bt transgene reduces herbivory and enhances fecundity in wild sunflowers. *Ecol. Applic.* 13:279–286. (116)

Snyder, M., and M. Gerstein. 2003. Defining genes in the genomics era. *Science* 300:258–260. (6)

Sokal, R. R., and F. J. Rohlf. 1995. *Biometry,* 3rd ed. San Francisco: W. H. Freeman. (4, 17, 24)

Sokal, R. R., and P. H. A. Sneath. 1963. *Principles of Numerical Taxonomy.* San Francisco: Freeman Publ. (600)

Soodyall, H., T. Jenkins, A. Mukherjee, *et al.* 1997. The founding mitochondrial DNA lineages of Tristan da Cunha islanders. *Am. J. Phys. Anthrop.* 104:157–166. (301–302)

Soodyall, H., A. Nebel., B. Morat, and T. Jenkins. 2003. Genealogy and genes: tracing the founding fathers of Tristan da Cunha. *Eur. J. Hum. Genet.* 11:705–709. (302)

Sork, V. L., J. Nason, D. R. Campbell, and J. F. Fernandez. 1999. Landscape approaches to historical and contemporary gene flow in plants. *Trends Ecol. Evol.* 14:219–224. (470)

Soulé, M. 1986. (ed.). *Conservation Biology.* Sunderland, MA: Sinauer. (280)

Sourdis, J., and M. Nei. 1988. Relative efficiencies of the maximum parsimony and distance-matrix methods in obtaining the correct phylogenetic tree. *Mol. Biol. Evol.* 5:298–311. (599)

Spencer, H. G., and R. W. Marks. 1988. The maintenance of single-locus polymorphism. I. Numerical studies of a viability selection model. *Genetics* 120:605–613. (409)

Spencer, H. G., and R. W. Marks. 1991. The maintenance of single-locus polymorphism. IV. Models with mutation from existing alleles. *Genetics* 130:211–221. (409)

Spencer, W. P. 1957. Genetic studies on *Drosophila mulleri*. 1. Genetic analysis of a population. *Univ. Texas Publ.* 5721:186–205. (46)

Spiess, E. B. 1983. Ching Cun Li, courageous scholar of population genetics, human genetic, and biostatics: a living history essay. *Am. J. Med. Genet.* 16:603–630. (625)

Stahl, E. A., G. Dwyer, R. Mauricio, *et al.* 1999. Dynamics of disease resistance polymorphism at the *Rpm1* locus of *Arabidopsis*. *Nature* 400:667–671. (229, 291)

Stearns, S. C. 1992. *The Evolution of Life Histories.* Oxford: Oxford University Press. (184)

Stein, L. D., Z. Bao, D. Blasiar, *et al.* 2003. The genome sequence of *Caenorhabditis briggsae*: a platform for comparative genomics. *PLoS Biol.* 1:166–192. (36)

Stephens, J. C., D. Briscoe, and S. J. O'Brien. 1994. Mapping by admixture linkage disequilibrium in human populations: limits and guidelines. *Am. J. Hum. Genet.* 55:809–824. (554)

Stephens, J. C., D. E. Reich, E. B. Goldstein, *et al.* 1998. Dating the origin of the *CCR5-Δ32* AIDS-resistance allele by the coalescence of haplotypes. *Am. J. Hum. Genet.* 62:1507–1515. (581–582)

Stern, C. 1973. *Human Genetics.* San Francisco: W. H. Freeman. (392)

Stewart, C. N., M. D. Halfhill, S. I. Warwick. 2003. Transgene introgression from genetically modified crops to their wild relatives. *Nat. Rev. Genet.* 4:806–817. (471)

Storey, J. D., and R. Tibshirani. 2003. Statistical significance for genomewide studies. *Proc. Natl. Acad. Sci. USA* 100:9440–9445. (25)

Strand, M., T. A. Prolla, R. M. Liskay, and T. Petes. 1993. Destabilization of tracts of simple repetitive DNA in yeast by mutations affecting DNA mismatch repair. *Nature* 365:274–276. (403)

Stumpf, M. P. H., and D. B. Goldstein. 2003. Demography, recombination hotspot intensity, and the block structure of linkage disequilibrium. *Curr. Biol.* 13:1–8. (662)

Stumpf, M. P. H., and G. A. T. McVean. 2003. Estimating recombination rates from population-genetic data. *Nat. Rev. Genet.* 4:959–968. (538)

Subramaniam, B., and M. D. Rausher. 2000. Balancing selection on a floral polymorphism. *Evolution* 54:691–695. (291–292)

Sun, G., S. T. McGarvey, R. Bayoumi, *et al.* 2003. Global genetic variation at nine short tandem repeat loci and implications on forensic genetics. *Eur. J. Hum. Genet.* 11:39–49. (639, 643–644)

Sved, J. A., T. E. Reed, and W. F. Bodmer. 1967. The number of balanced polymorphisms that can be maintained in a natural population. *Genetics* 55:469–481. (409)

Swanson, W. J., Z. Yang, M. F. Wolfner, and C. F. Aquadro. 2000. Positive Darwinian selection drives the evolution of several female reproduction proteins in mammals. *Proc. Natl. Acad. Sci. USA* 98:2509–2514. (449)

Sweigart, A., K. Karoly, A. Jones, and J. H. Willis. 1999. The distribution of individual inbreeding coefficients and pairwise relatedness in a population of *Mimulus guttatus*. *Heredity* 83:625–632. (259)

Szathmáry, E., and P. Hammerstien. 2004. John Maynard Smith (1920–2004). *Nature* 429:258–259. (576)

Tabashnik, B. E., Y.-B. Liu, N. Finson, *et al.* 1997. One gene in diamondback moth confers resistance to four *Bacillus thuringiensis* toxins. *Proc. Natl. Acad. Sci. USA* 94:1640–1644. (116)

Taberlet, P., L. Waits, and G. Luikart. 1999. Noninvasive genetic sampling: look before you leap. *Trends Ecol. Evol.* 14:323–327. (646)

Tajima, F. 1983. Evolutionary relationship of DNA sequences in finite populations. *Genetics* 105:437–460. (455)

Tajima, F. 1989. Statistical method for testing the neutral mutation hypothesis by DNA polymorphism. *Genetics* 123:585–595. (433–434, 436)

Tajima, F. 1993. Simple methods for testing the molecular evolutionary clock hypotheses. *Genetics* 135:599–607. (426)

Takahata, N. 1990. A simple genealogical structure of strongly balanced allelic lines and trans-species polymorphism. *Proc. Natl. Acad. Sci. USA* 87:2419–2423. (461)

Takahata, N. 1991. Statistical models of the overdispersed molecular clock. *Theor. Popul. Biol.* 39:329–344. (426)

Takahata, N., S. H. Lee, and Y. Satta. 2001. Testing multiregionality of modern human origin. *Mol. Biol. Evol.* 18:172–183. (618)

Takezaki, N., and M. Nei. 1996. Genetic distances and reconstruction of phylogenetic trees from microsatellite DNA. *Genetics* 144:389–399. (108)

Tanaka, M. M., C. T. Bergstrom, and B. R. Levin. 2003. The evolution of mutator genes in bacterial populations: the roles of environmental change and timing. *Genetics* 164:843–854. (405)

Taylor, D. R., and P. K. Ingvarsson. 2003. Common features of segregation distortion in plants and animals. *Genetica* 117:27–35. (197)

Taylor, M., and R. Feyereisen. 1996. Molecular biology and evolution of resistance to toxicants. *Mol. Biol. Evol.* 13:719–734. (117)

Templeton, A. R. 2002. Out of Africa again and again. *Nature* 416:45–51. (618)

Templeton, A. R., and B. Read. 1983. The elimination of inbreeding depression in a captive herd of Speke's gazelle (pp. 241–261). In C. M. Schonewald, S. M. Chambers, B. MacBryde, *et al.* (eds.). *Genetics and Conservation.* Menlo Park, CA: Benjamin/Cummings. (289)

Terasaki, P. I. 1978. Resolution by HLA testing of 1,000 paternity cases not excluded by ABO testing. *J. Family Law* 16:543–557. (633–634)

Thangaraj, K., A. G. Reddy, and L. Singh. 2004. Mutation in the STR locus D21S11 of father causing allele mismatch in the child. *J. Forensic Sci.* 49:99–103. (634–635)

The International HapMap Consortium. 2003. The International HapMap Project. *Nature* 426:789–796. (658–659)

The International SNP Map Working Group. 2001. A map of human genome sequence variation containing 1.42 million single nucleotide polymorphisms. *Nature* 409:928–933. (39)

Theron, E., K. Hawkins, E. Bermingham, *et al.* 2001. The molecular basis of an avian plumage polymorphism in the wild: a melanocortin-1-receptor point mutation is perfectly associated with the melanic plumage morph of the banaquit, *Coereba flaveola*. *Curr. Biol.* 11:550–557. (521)

The U.S.-Venezuela Collaborative Research Project and N. S. Wexler. 2004. Venezuelan kindreds reveal that genetic and environmental factors modulate Huntington's disease age of onset. *Proc. Nat. Acad. Sci. USA* 101:3498–3503. (393)

Thijssen, H. W. 1995. Warfarin-based rodenticides: mode of action and mechanism of resistance. *Pestic. Sci.* 43:73–78. (118)

Thomas, M. G., K. Skorecki, H. Ben-Ami, *et al.* 1998. Origins of Old Testament priests. *Nature* 394:138–140. (629)

Thomas, M. L., J. H. Harger, D. K. Wagener, *et al.* 1985. HLA sharing and spontaneous abortion in humans. *Am. J. Obstet. Gynecol.* 151:1053–1058. (182)

Thompson, E. A. 1975. The estimate of pairwise relationships. *Ann. Hum. Genet.* 38:41–48. (272)

Thompson, J. N. 1994. *The Coevolutionary Process.* Chicago: University of Chicago Press. (228)

Thompson, J. N. 1998. Rapid evolution as an ecological process. *Trends Ecol. Evol.* 13:329–332. (116)

Thompson, J. N., and J. J. Burdon. 1992. Gene-for-gene coevolution between plants and parasites. *Nature* 360:121–125. (229)

Thomson, G. 1977. The effect of a selected locus on linked neutral loci. *Genetics* 85:753–788. (568, 571, 573)

Thurz, M. R., H. C. Thomas, B. M. Greenwood, and A. V. S. Hill. 1997. Heterozygote advantage for HLA class-II type in hepatitis B virus infection. *Nat. Genet.* 17:11–12. (232)

Tian, D., M. B. Traw, J. Q. Chen, *et al.* 2003. Fitness costs of R-gene-mediated resistance in *Arabidopsis* thaliana. Nature 423:74–77. (229)

Tiemann-Boege, I., W. Navidi, R. Grewal, *et al.* 2002. The observed human sperm mutation frequency cannot explain the achondroplasia paternal age effect. *Proc. Natl. Acad. Sci. USA* 99:14952–14957. (402)

Tishkoff, S. A., R. Varkonyl, N. Cahinhinan, *et al.* 2001. Haplotype diversity and linkage disequilibrium at human *G6PD*: recent origin of alleles that confer malarial resistance. *Science* 293:455–462. (168)

Tishkoff, S. A., and B. C. Verrelli. 2003. Patterns of human genetic diversity: implications for human evolutionary history and disease. *Ann. Rev. Genom. Hum. Genet.* 4:293–340. (436)

Tompkins, R. 1978. Genetic control of axolotl metamorphosis. *Am. Zool.* 18:313–319. (57)

True, J. R. 2003. Insect melanism: the molecules matter. *Trends Ecol. Evol.* 18:640–647. (138)

Turelli, M. 1981. Temporally varying selection on multiple alleles: a diffusion analysis. *J. Math. Biol.* 13:115–129. (235)

Turelli, M., D. W. Schemske, and P. Bierzychudek. 2001. Stable two-allele polymorphisms maintained by fluctuation fitnesses and seed banks: protecting the blues in *Linanthus parryae. Evolution* 55:1283–1298. (217–218)

Turner, B. J., J. F. Elder, T. F. Laughlin, *et al.* 1992. Extreme clonal diversity and divergence in populations of a selfing hermaphroditic fish. *Proc. Natl. Acad. Sci. USA* 89:10643–10647. (269)

Turner, T. F., L. A. Salter, and J. R. Gold. 2001. Temporal-method estimates of N_e from highly polymorphic loci. *Cons. Genet.* 2:297–308. (340)

Turner, T. F., J. P. Wares, and J. R. Gold. 2002. Genetic effective size is three orders of magnitude smaller than adult census size in an abundant, estuarine-dependent marine fish (*Sciaenops ocellatus*). *Genetics* 162:1329–1339. (340)

Urwin, R., and M. C. J. Maiden. 2003. Multi-locus sequence typing: a tool for global epidemiology. *Trends Microbiol.* 11:479–487. (622)

Uyenoyama, M. K. 2000. Evolutionary dynamics of self-incompatibility alleles in Brassica. *Genetics* 156:351–359. (204)

Uyenoyama, M. K. 2003. Genealogy-dependent variation in viability among self-incompatibility genotypes. *Theor. Popul. Biol.* 63:281–293. (204)

Uyenoyama, M. K., K. E. Holsinger, and D. M. Waller. 1993. Ecological and genetic factors directing the evolution of self-fertilization (pp. 327–381). In D. Futumya and J. Antonovics (eds.). *Oxford Surveys in Evolutionary Biology*, vol. 9. Oxford, UK: Oxford University Press. (280)

Vacher, C., D. Bourguet, F. Rousset, *et al.* 2003. Modeling the spatial configuration of refuges for a sustainable control of pests: a case study of Bt cotton. *J. Evol. Biol.* 16:378–387. (116)

Van Delden, W. 1982. The alcohol dehydrogenase polymorphism in *Drosophila melanogaster*: selection at an enzyme locus. *Evol. Biol.* 15:187–222. (564)

Van Delden, W. 1984. The *alcohol dehydrogenase* polymorphism in *Drosophila melanogaster*, facts and problems (pp. 127–142). In K. Wohrmann and V. Loeschcke (eds.). *Population Biology and Evolution.* Berlin: Springer-Verlag. (565)

Van Laere, A.-S., M. Nguyen, M. Braunschweig, *et al.* 2003. A regulatory mutation in *IGF2* causes a major QTL effect on muscle growth in the pig. *Nature* 425:832–836. (56)

van Noordwijk, A. J., and W. Scharloo. 1981. Inbreeding in an island population of the great tit. *Evolution* 35:674–688. (268–269)

Van Valen, L. 1973. A new evolutionary law. *Evol. Theory.* 1:1–30. (189)

Vazquez, J. F., T. Perez, J. Albornoz, and A. Dominguez. 2000. Estimation of microsatellite mutation rates in *Drosophila melanogaster. Genet. Res.* 76:323–326. (397)

Vekemans, X., M. H. Scheirup, and F. B. Christiansen. 1998. Mate availability and fecundity selection in multiallelic self-incompatibility systems in plants. *Evolution* 52:19–29. (202)

Verrelli, B. C., and W. F. Eanes. 2001. Clinal variation for amino acid polymorphisms at the *Pgm* locus in *Drosophila melanogaster. Genetics* 157:1649–1663. (410)

Verrelli, B. C., J. H. McDonald, G. Argyropoulos, *et al.* 2002. Evidence for balancing selection from nucleotide sequence analyses of human *G6PD. Am. J. Hum. Genet.* 71:1112–1128. (168, 440)

Veuille, M., and L. M. King. 1995. Molecular basis of polymorphism at the esterase 5B locus in *Drosophila pseudoobscura. Genetics* 88:255–262. (32)

Veuille, M., and M. Slatkin. 2002. Introduction: the Malecot lineage in population genetics (pp. 1–6). In M. Slatkin and M. Veuille (eds.). *Modern Developments in Theoretical Population Genetics: The Legacy of Gustave Malecot.* Oxford: Oxford University Press. (4)

Viard, F., C. Doums, and P. Jarne. 1997a. Selfing, sexual polymorphism and microsatellites in the hermaphroditic freshwater snail *Bulinus truncatus. Proc. R. Soc. Lond. B.* 264:39–44. (254)

Viard, F., F. Justy, and P. Jarne. 1997b. Population dynamics inferred from temporal variation at microsatellite loci in the selfing snail *Bulinus truncatus. Genetics* 146:973–982. (256–257)

Vigilant, L., M. Hofreiter, H. Siedel, and C. Boesch. 2001. Paternity and relatedness in wild chimpanzee communities. *Proc. Natl. Acad. Sci. USA* 98:12890–12895. (496)

Vigilant, L., M. Stoneking, H. Harpending, *et al.* 1991. African populations and the evolution of human mitochondrial DNA. *Science* 253:1503–1507. (618)

Vilá, C., P. Savolainen, J. E. Maldonado, *et al.* 1997. Multiple and ancient origins of the domestic dog. *Science* 276:1687–1689. (618)

Vilá, C., J. A. Leonard, A. Götherström, *et al.* 2001. Widespread origins of domestic horse lineages. *Science* 291:474–477. (618)

Vilá, C., A.-K. Sundqvist, O. Flagstand, *et al.* 2003. Rescue of a severely bottlenecked wolf (*Canis lupus*) population by a single immigrant. *Proc. R. Soc. Lond. B.* 270:91–97. (471)

Voelker, R. A., H. E. Schaffer, and T. Mukai. 1980. Spontaneous allozyme variations in *Drosophila melanogaster*: rate of occurrence and nature of the mutants. *Genetics* 94:961–968. (396)

Vogel, F., and A. G. Motulsky. 1997. *Human Genetics*, 2nd ed. New York: Springer-Verlag. (46, 244–245, 293, 395, 400, 621)

Vogl, C., A. Karhu, G. Moran, and O. Savolainen. 2002. High resolution analysis of mating systems: inbreeding in natural populations of *Pinus radiata*. *J. Evol. Biol.* 15:433–439. (255–256)

Voss, S. R., and H. B. Shaffer. 1997. Adaptive evolution via a major gene effect: paedomorphosis in the Mexican axolotl. *Proc. Natl. Acad. Sci. USA* 94:14185–14189. (57–58)

Vucetich, J. A., T. A. Waite, and L. Nunney. 1997. Fluctuating population size and the ratio of effective to census population size. *Evolution* 51:2017–2021. (333)

Wade, M. J. 2000. Opposing levels of selection can cause neutrality: mating patterns and maternal–fetal interactions. *Evolution* 54:290–292. (184)

Wade, M. J., R. G. Winther, A. F. Agrawal, and C. J. Goodnight. 2001. Alternative definitions of epistasis: dependence and interaction. *Trends Ecol. Evol.* 16:498–504. (557)

Wahlund, S. 1928. Zusammensetzung von Populationen und Korrelation-serscheinungen von Standpunkt der Verebungslehre aus betrachtet. *Hereditas* 11:65–106. (479)

Waits, J. L., and P. L. Leberg. 2000. Biases associated with population estimation using molecular tagging. *Anim. Cons.* 3:191–199. (646)

Waits, L. P., G. Luikart, and P. Taberlet. 2001. Estimating the probability of identity among genotypes in natural populations: cautions and guidelines. *Mol. Ecol.* 10:249–256. (644–645)

Waits, L. P., S. L. Talbot, R. H. Ward, and G. F. Shields. 1998. Mitochondrial DNA phylogeography of the North American brown bear and implications for conservation. *Cons. Biol.* 12:408–417. (616–617)

Wakeley, J. 2004. Metapopulation models for historical inference. *Mol. Ecol.* 13:865–875. (508)

Wall, J. D. 2000. A comparison of estimators of the population recombination rate. *Mol. Biol. Evol.* 17:156–163. (542)

Wall, J. D., and J. K. Pritchard. 2003. Assessing the performance of the haplotype block model of linkage disequilibrium. *Am. J. Hum. Genet.* 73:502–515. (662)

Wallace, B. 1970. *Genetic Load.* Englewood Cliffs, NJ: Prentice-Hall. (143)

Wang, J. 1997. Effective size and F-statistics of subdivided populations. II. Dioecious species. *Genetics* 146:1453–1463. (498)

Wang, J. 2001. A pseudo-likelihood method for estimating effective population size from temporally spaced samples. *Genet. Res.* 78:243–257. (340)

Wang, J. 2002. An estimator for pairwise relatedness using molecular markers. *Genetics* 160:1203–1215. (274)

Wang, J. 2003. Maximum-likelihood estimation of admixture proportions from genetic data. *Genetics* 164:747–765. (485)

Wang, J. 2004. Application of the one-migrant-per-generation rule to conservation and management. *Cons. Biol.* 18:332–343. (501)

Wang, J., and A. Caballero. 1999. Developments in predicting the effective size of subdivided populations. *Heredity* 82:212–226. (505, 508)

Wang, J., W. G. Hill, D. Charlesworth, and B. Charlesworth. 1999. Dynamics of inbreeding depression due to deleterious mutation in small populations: mutation parameters and the inbreeding rate. *Genet. Res.* 74:165–178. (289)

Wang, R.-L., A. Stec, J. Hey, *et al.* 1999. The limits of selection during maize domestication. *Nature* 398:236–239. (439)

Waples, R. S. 1989. A generalized method for estimating population size from temporal changes in allele frequency. *Genetics* 121:379–391. (338, 340–341)

Waples, R. S. 1991a. Genetic methods for estimating the effective size of cetacean populations (pp. 279–300). In A. R. Hoezel (ed.). *Genetic Ecology of Whales and Dolphins.* Cambrigde, UK: International Whaling Commission (Special Issue No. 13). (543)

Waples, R. S. 1991b. Pacific salmon, *Oncorhynchus* spp., and the definition of "species" under the Endangered Species Act. *Mar. Fish. Rev.* 53:11–22. (616)

Waples, R. S. 1998. Separating the wheat from the chaff: patterns of genetic differentiation in high gene flow species. *J. Hered.* 89:438–450. (501–502)

Waples, R. S. 2002. Definition and estimation of effective population size in the conservation of endangered species (pp. 147–168). In S. R. Beissinger and D. R. McCullough (eds.). *Population Viability Analysis.* Chicago: University of Chicago Press. (333)

Waples, R. S., and P. E. Smouse. 1990. Gametic disequilibrium analysis as a means of identifying mixtures of salmon populations. *Am. Fish. Soc. Symp.* 7:439–458. (554)

Waser, P. M., and C. Strobeck. 1998. Genetic signatures of interpopulation dispersal. *Trends Ecol. Evol.* 13:43–44. (647–648)

Watt, W. B. 1994. Allozymes in evolutionary genetics: self-imposed burden or extraordinary tool? *Genetics* 136:11–16. (410)

Watterson, G. A. 1978. An analysis of multi-allelic data. *Genetics* 88:171–179. (429–430, 432)

Wayne, M. L., and K. L. Simonsen. 1998. Statistical tests of neutrality in the age of weak selection. *Trends Ecol. Evol.* 13:236–240. (428)

Weber, D. S., B. S. Stewart, J. C. Garza, and N. Lehman. 2000. An empirical genetic assessment of the severity of the northern elephant seal population bottleneck. *Curr. Biol.* 10:1287–1290. (303)

Wedekind, C., T. Seebeck, F. Bettens, and A. Paepke. 1995. MHC-dependent mate preferences in human. *Proc. R. Soc. Lond. B.* 260:245–249. (196)

Weill, R., G. Lutfalla, K. Mogensen, *et al.* 2003. Insecticide resistance in mosquito vectors. *Nature* 423:136–137. (117)

Weinberg, W. 1908. Über den Nachweis der Vererbung beim Menschen. *Jahresh. Verein f. Vaterl. Naturk. in Wüttemberg* 64:368–382. (63)

Weinberg, W. 1909. Über Vererbungsgesetze beim Menschen. *Z. Abst. V. Vererb.* 1:277–330. (527)

Weir, B. S. 1993. Forensic population genetics and the National Research Council (NRC). *Am. J. Hum. Genet.* 52:437–440. (643)

Weir, B. S. 1995. DNA statistics in the Simpson matter. *Nat. Genet.* 11:365–368. (25, 638–639)

Weir, B. S. 1996. *Genetic Data Analysis II.* Sunderland, MA: Sinauer. (79, 93, 587–588, 627, 633)

Weir, B. S. 2001. Forensics (pp. 721–739). In D. J. Balding, M. Bishop, and C. Cannings (eds.). *Handbook of Statistical Genetics*, New York: John Wiley and Sons. (597, 637)

Weir, B. S., R. W. Allard, and A. L. Kahler. 1972. Analysis of complex allozyme polymorphisms in a barley population. *Genetics* 72:505–523. (548, 574)

Weir, B. S., R. W. Allard, and A. L. Kahler. 1974. Further analysis of complex allozyme polymorphisms in a barley population. *Genetics* 78:911–919. (574)

Weir, B. S., and C. C. Cockerham. 1973. Mixed selfing and random mating at two loci. *Genet. Res.* 21:247–261. (550)

Weir, B. S., and C. C. Cockerham. 1979. Estimation of linkage disequilibrium in randomly mating populations. *Heredity* 42:105–111. (589)

Weir, B. S., and C. C. Cockerham. 1984. Estimating *F*-statistics for the analysis of population structure. *Evolution* 38:1358–1370. (488, 493)

Weir, B. S., and W. G. Hill. 1980. Effect of mating structure on variation in linkage disequilibrium. *Genetics* 95:477–488. (543)

Weiss, K. M. and A. G. Clark. 2002. Linkage disequilibrium and the mapping of complex human traits. *Trends Genet.* 18:19–24. (661)

Welch, D. M., and M. Meselson. 2000. Evidence for the evolution of bdelloid rotifers without sexual reproduction or genetic exchange. *Science* 288:1211–1215. (585)

Welch, D. M., M. P. Cummings, D. M. Hillis, and M. Meselson. 2004. Divergent gene copies in the asexual class Bdelloidea (Rotifera) separated before the bdelloid radiation or within bdelloid families. *Proc. Natl. Acad. Sci. USA* 101:1622–1625. (585)

Werner, F. A. O., G. Durstewitz, F. A. Habermann *et al.* 2004. Detection and characterization of SNPs useful for identity control and parentage testing in major European dairy breeds. *Anim. Genet.* 35:44–49. (622)

West Eberhard, M. J. 1975. The evolution of social behavior by kin selection. *Q. Rev. Biol.* 50:1–33. (279)

White, M. J. D. 1978. *Modes of Speciation.* San Francisco: W. H. Freeman. (390)

Whitlock, M. C., and N. Barton. 1997. The effective size of a subdivided population. *Genetics* 146:427–441. (507–508, 523)

Whitlock, M. C., and D. E. McCauley. 1990. Some population genetic consequences of colony formation and extinction: genetic correlation within founding groups. *Evolution* 44:1717–1724. (508)

Whitlock, M. C., and D. E. McCauley. 1999. Indirect measures of gene flow and migration: $F_{ST} \neq 1/(4Nm + 1)$. *Heredity* 82:117–125. (502, 523)

Whitten, M. J. 1979. The use of genetically selected strains for pest replacement or suppression (pp. 31–40). In M. A. Hoy and J. J. McKelvey (eds.). *Genetics in Relation to Insect Management.* New York: Rockefeller Foundation. (144)

WHO. 1998. Demographic data for health situation assessment and projections. Geneva: WHO HST/HSP. (157)

Widemo, F., and S. A. Saether. 1999. Beauty is in the eye of the beholder: causes and consequences of variation in mating preferences. *Trends Ecol. Evol.* 14:26–31. (189)

Wilkinson, G. S., and C. L. Fry. 2001. Meiotic drive alters sperm competitive ability in stalk-eyed flies. *Proc. R. Soc. Lond. B.* 268:2559–2564. (197)

Wilkinson, G. S., D. C. Presgraves, and L. Crymes. 1998. Male eye span in stalk-eyed flies indicates genetic quality by meiotic drive suppression. *Nature* 391:276–279. (188)

Wilson, J. F., and D. B. Goldstein. 2000. Consistent long-range linkage disequilibrium generated by admixture in a Bantu-Semitic hybrid population. *Am. J. Hum. Genet.* 67:926–935. (552–553)

Woodruff, R. C., H. Huai, and J. N. Thompson. 1996. Clusters of identical new mutations in the evolutionary landscape. *Genetica* 98:149–160. (390–391)

Woolf, C. M., and F. C. Dukepoo. 1969. Hopi Indians, inbreeding and albinism. *Science* 164:30–37. (85, 186)

Wootton, J. C., X. Feng, M. T. Ferdig, *et al.* 2002. Genetic diversity and chloroquine selective sweeps in *Plasmodium falciparum. Nature* 418:320–323. (579)

Workman, P. L. 1969. The analysis of simple genetic polymorphism. *Hum. Biol.* 41:97–114. (97, 150)

Workman, P. L., and S. F. Jain. 1966. Zygotic selection under mixed random mating and self-fertilization: theory and problems of estimation. *Genetics* 54:159–171. (290)

Workman, P. L., and J. D. Niswander. 1970. Population studies on southwestern Indian tribes. II. Local genetic differentiation in the Papago. *Am. J. Hum. Genet.* 22:24–29. (107)

Wray, G. A., M. W. Hahn, E. Abouheif, *et al.* 2003. The evolution of transcriptional regulation in eukaryotes. *Mol. Biol. Evol.* 20:1377—1419. (438)

Wright, F. 1990. The "effective number of codons" used in a gene. *Gene* 87:23–29. (451)

Wright, S. 1917. Color inheritance in mammals. *J. Hered.* 8:224–235. (304)

Wright, S. 1922. Coefficients of inbreeding and relationship. *Am. Nat.* 56:330–338. (271)

Wright, S. 1931. Evolution in Mendelian populations. *Genetics* 16:97–159. (240, 318, 323, 338, 352, 453)

Wright, S. 1932. The roles of mutation, inbreeding, crossbreeding, and selection in evolution. *Proc. 6th Int. Congress Genet.* 1:356–365. (469)

Wright, S. 1937. The distributions of gene frequencies in populations. *Proc. Natl. Acad. Sci. USA.* 23:307–320. (386)

Wright, S. 1940. Breeding structure of populations in relation to speciation. *Am. Nat.* 74:232–248. (500, 503)

Wright, S. 1943a. An analysis of local variability of flower color in *Linanthus parryae.* *Genetics* 28:139–156. (217)

Wright, S. 1943b. Isolation by distance. *Genetics* 28:114–138. (336, 502–503, 505)

Wright, S. 1951. The genetical structure of populations. *Ann. Eugen.* 15:323–354. (488)

Wright, S. 1955. Classification of the factors of evolution. *Cold Spring Harbor Symp. Quant. Biol.* 20:16–24. (114)

Wright, S. 1965a. The distribution of self-incompatibility alleles in populations. *Evolution* 18:609–619. (202)

Wright, S. 1965b. The interpretation of population structure by F-statistics with special regard to systems of mating. *Evolution* 19:395–420. (485)

Wright, S. 1968, 1969, 1977, 1978. *Evolution and Genetics of Populations*, vols. 1–4. Chicago: University of Chicago Press. (221, 303–304, 504, 562)

Wu, C.-I., and C. T. Ting. 2003. Genes and speciation. *Nat. Rev. Genet.* 5:114–122. (651)

Xu, J., E. R. Bleecher, J. Jongepier, *et al.* 2002. Major recessive gene(s) with considerable residual polygenic effect regulating adult height: confirmation of genomewide scan results for chromosomes 6, 9, and 12. *Am. J. Hum. Genet.* 71:646–650. (656)

Xu, X., M. Peng, Z. Fang, and X. Xu. 2000. The direction of microsatellite mutations is dependent upon allele length. *Nat. Genet.* 24:396–399. (401)

Yang, Z., and J. P. Bielawski. 2000. Statistical methods for detecting molecular adaptation. *Trends Ecol. Evol.* 15:496–503. (443, 448)

Yang, Z., R. Nielsen, N. Goldman, and A.-M. K. Petersen. 2000. Codon-substitution models for variable selection pressure at amino acid sites. *Genetics* 155:431–449. (449)

Yi, S., D. L. Ellsworth, and W.-H. Li. 2002. Slow molecular clocks in Old World monkeys, apes, and humans. *Mol. Biol. Evol.* 19:2191–2191. (427–428)

Yi, Z., N. Garrison, O. Cohen-Barak, *et al.* 2003. A 122.5-kilobase deletion of the P gene underlies the high prevalence of oculocutaneous albinism type 2 in the Navajo population. *Am. J. Hum. Genet.* 72:62–72. (391)

Yip, S. P. 2002. Sequence variation at the human *ABO* locus. *Ann. Hum. Genet.* 66:1–27. (88)

Yokoyama, S., and L. E. Hetherington. 1982. The expected number of self-incompatibility alleles in finite plant populations. *Heredity* 48:299–303. (202)

Yoon, S.-R., L. Dubeau, M. de Young, *et al.* 2003. Huntington disease expansion mutations in humans can occur before meiosis is completed. *Proc. Natl. Acad. Sci. USA* 100:8834–8838. (394)

Yu, J., L. Lazzeroni, J. Qin, *et al.* 1996. Individual variation in recombination among human males. *Am. J. Hum. Genet.* 59:1186–1192. (540)

Yu, N., F. C. Chen, S. Ota, *et al.* 2002. Larger genetic differences within Africans than between Africans and Eurasians. *Genetics* 161:269–274. (463)

Yu, N., M. I. Jensen-Seaman, L. Chemnick, *et al.* 2003. Low nucleotide diversity in chimpanzees and bonobos. *Genetics* 164:1511–1518. (463)

Yuhki, N., and S. J. O'Brien. 1990. DNA variation of the mammalian major histocompatibility complex reflects genomic diversity and population history. *Proc. Natl. Acad. Sci. USA* 87:835–840. (382)

Zangenberg, G., M.-M Huang, N. Arnheim, and H. Ehrlich. 1995. New *HLA-DPB1* alleles generated by interallelic gene conversion detected by analysis of sperm. *Nat. Genet.* 10:407–414. (362)

Zapata, C. 2000. The D' measure of overall gametic disequilibrium between pairs of multiallelic loci. *Evolution* 54:1809–1812. (592)

Zapata, C., and G. Alvarez. 1997. On Fisher's exact test for detecting gametic disequilibrium between DNA polymorphisms. *Ann. Hum. Genet.* 61:71–77. (587)

Zapata, C., C. Carollo, and S. Rodriguez. 2001. Sampling variance and distribution of the D' measure of overall gametic disequilibrium between multiallelic loci. *Ann. Hum. Genet.* 65:395–406. (592)

Zar, J. H. 1999. *Biostatistical Analysis*, 4th ed. Upper Saddle River, NJ: Prentice-Hall. (4, 17, 24)

Zehl, C., and J. A. G. M. de Visser. 2001. Estimates of the rate and distribution of fitness effects of spontaneous mutation in *Saccharomyces cerevisiae*. *Genetics* 157:53–61. (400)

Zehl, C., M. Mizesko, and J. A. G. M. de Visser. 2001. Mutational meltdown in laboratory yeast populations. *Evolution* 55:909–917. (389)

Zeng, L.-W., J. M. Comeron, B. Chen, and M. Kreitman. 1998. The molecular clock revisited: the rate of synonymous vs. replacement change in *Drosophila*. *Genetica* 102/103:369–382. (426)

Zhang, J. 2000. Rates of conservative and radical nonsynonymous nucleotide substitutions in mammalian nuclear genes. *J. Mol. Evol.* 50:56–68. (449)

Zhang, J. 2003. Evolution by gene duplication: an update. *Trends Ecol. Evol.* 18:292–298. (361)

Zhang, J., W. L. Rowe, A. G. Clark, and K. H. Buetow. 2003a. Genomewide distribution of high-frequency, completely mismatching SNP haplotype pairs observed to be common across human populations. *Am. J. Hum. Genet.* 73:1073–1081. (662)

Zhang, K., J. M. Akey, N. Wang, *et al.* 2003b. Randomly distributed crossovers may generate block-like patterns of linkage disequilibrium: an act of genetic drift. *Hum. Genet.* 113:51–59. (662)

Zhao, H., and T. P. Speed. 1996. On genetic mapping functions. *Genetics* 142:1369–1377. (538)

Zuckerlandl, E., and L. Pauling. 1965. Evolutionary divergence and convergence in proteins (pp. 97–166). In V. Bryson and H. J. Vogel (eds.). *Evolving Genes and Proteins.* New York: Academic Press. (419)

Answers to Numerical Problems

CHAPTER 1

1. 0.0, 0.667, 0.667
2. 1.5, 202.5
3. 28.7, 1.70, 1.30, 4.53, 0.16
4. 143.2, 104.7, 59.3
5. 0.111, 0.374
6. 0.606, 0.076
7. 0.25, 0.5, 0.25; 0.375, 0.25, 0.375
11. 0.012, 0.006
14. 0.333

CHAPTER 2

1. 0.726; 8.8, 46.5, 61.7
2. $p_1 = 0.103$, $p_2 = 0.203$, $p_3 = 0.694$, $\chi^2 = 2.95$, consistent with Hardy-Weinberg proportions
3. Generation 1—0.12, 0.56, 0.32; generation 2—0.16, 0.48, 0.36
4. q_m—0.0, 0.3, 0.15, 0.225, 0.1875; q_f —0.3, 0.15, 0.225, 0.1875, 0.20625; 7 generations
5. 17.0, 17.4
6. 0.087
7. 0.02, 0.039, 0.0015
8. 0.024, 0.098, 0.073, 0.512, 0.098, 0.024, 0.171; 0.700
9. 0.228, 0.004, 0.768
10. 0.441, 1536
11. Site 1—$\chi^2 = 3.38$, site 2—$\chi^2 = 0.14$, overall—$\chi^2 = 3.52$, none are significant; 0.033, 0.000, 0.016
12. 0.533, 0.266
14. 0.032
15. 0.0052

CHAPTER 3

3. 0.663; 0.0, 0.663, 0.663, 1.0
5. 2.0, 1.0, 1.6; 0.625; 0.0, 0.0, 1.0, 1.0
6. q_f = 0.20, 0.194, 0.175, 0.163; q_m = 0.20, 0.167, 0.161, 0.145
10. 0.143; approximately 0.54 using expression 3.6*c*
13. 0.9; yes
14. 95, 491, 64.2
17. $v_{11} = 0.792$ and $v_{22} = 0.891$

CHAPTER 4

2. $q_m = 0.043$, $q_f = 0.048$, $0.1111 > s_m > 0.0909$
5. 0.4, 0.35, 0.325
9. 0.4 to 0.5
10. $q_S = 0.267$, $q_F = 0.256$

CHAPTER 5

1. 1.995
2. 0.3, 0.206, 0.075
3. 0.5, 0.3, 0.15, 0.16, 0.08; 0.5, 0.275, 0.174, 0.128, 0.108
4. 0.143, 0.043
6. Complete selfing—0.5, 0.25, 0.125, 0.0625, 0.0312, 0.0156; sib mating—0.5, 0.5, 0.375, 0.312, 0.25, 0.203; double first-cousin mating—0.5, 0.5, 0.5, 0.438, 0.406, 0.375
7. 0.027, 0.973
8. 0.0781
9. 0.0, 0.0, 0.1875, 0.125
10. 0.0045, 0.105
11. 0.4, 1.02
12. 0.291, 0.506, 0.203, 0.456; 0.481, 0.253, 0.266, 0.392

CHAPTER 6

1. 60, 599
2. 0.276, 0.144, 0.159, 0.144, 0.276, $H_1 = 0.1875$; 0.332, 0.107, 0.120, 0.107, 0.332, $H_2 = 0.1406$
3. 37.9, 379, 3790

4. 0.992, 0.999

5. 0.112, 0.479

6. 3.33, 2.14

7. 12.1

8. 50, 5, 36.4

9. 0.110, 0.210, 0.396

10.

1	0.301	0.056	0.003	0
0	0.421	0.238	0.042	0
0	0.221	0.374	0.200	0
0	0.052	0.262	0.421	0
0	0.004	0.069	0.332	1

Generation 1—0.272, 0.140, 0.159, 0.147, 0.281; generation 2—0.324, 0.103, 0.120, 0.111, 0.342

CHAPTER 7

2. 0.909; 41, 240

3. 0.25, 0.531

4. 0.00158

6. 0.00415, 0.005; 0.00225, 0.0025; 0.000954, 0.001

7. 0.0016, 0.00009, 0.0063

8. 1.25×10^{-5}, 1.25×10^{-5}; 2.5×10^{-6}, 2.5×10^{-6}; 2×10^{-6}, 4×10^{-7}

9. 8000, 16.6; 400, 10.6

10. 0.091, 0.5, 0.909; 0.087, 0.423, 0.782

12. 0.02, 0.1, 0.2

14. 0.00005, 0.000005

15. 2×10^{-5}

CHAPTER 8

1. 0.080

2. Infinite-allele model, $\theta_X/\theta_A = 0.74$; stepwise-mutation model, $\theta_X/\theta_A = 0.65$

3. 10^5 generations

4. 0.357, 0.065, 8.9×10^{-10}

5. 0.571, 0.0469, 5.7×10^{-8}

6. 0.501

8. 0.022; 0.222, 1, 10

9. 2.5, 1.0

10. 0.618, θg 0.43, f= 0.699

11.

3.2	5.8	7.9	4.1
20.8	38.2	27.1	13.9

CHAPTER 9

2. Nuer — 0.575, 0.574, 0.574; Dinka—0.567, 0.566, 0.566; Shilluk—0.506, 0.506, 0.507

3. $\overline{H} = 0.4$, $2\overline{pq} = 0.42$; $\overline{P}_{12} = 0.2$, $2\overline{p}_1\overline{p}_2 = 0.24$, $\overline{P}_{13} = 0.2$, $2\overline{p}_1\overline{p}_3 = 0.24$, $\overline{P}_{23} = 0.2$, $2\overline{p}_2\overline{p}_3 = 0.18$, $\overline{H} = 0.6$ versus 0.66

4. $\hat{M} = 0.25$, $\hat{m} = 0.0284$

6. 0.111

8. 0.053, 17

10. 0.0042, 0.0050; 0.333, 0.333

11. 0.047, 0.369

CHAPTER 10

2. 0.1, 0.5, 0.408

3. 0.0953

4. 0.048

5. 0.0048

8.

0.8	0.85	0.8
0.85	1.0	0.85
0.8	0.85	0.8

0.025; $x_{1\cdot e} = x_{4\cdot e} = 0.444$,
$x_{2\cdot e} = x_{3\cdot e} = 0.056$, $D_e = 0.194$; $x_{1\cdot e} = x_{4\cdot e} = 0.056$, $x_{2\cdot e} = x_{3\cdot e} = 0.444$, $D_e = -0.194$

9. 0.0225, 0.135, 0.2025, 0.105, 0.015, 0.315, 0.045, 0.1225, 0.035, 0.0025; 0.927, 0.95, 1.0

14. $0.283 + 0.406 = 0.689$

CHAPTER 11

4. 2, 6, and 7; tree C in Figure 11.5

7. 0.126, 0.932

10. 0.375, 0.594

12. 3.5, 1.3; 3.89, 1.44

Subject Index

Entries in boldface indicate definition or introduction of term